Advanced Engineering Physics

Advanced Engineering Physics

Harish Parthasarathy

Asstt. Professor,
Netaji Subhas Institute of Technology (NSIT)
New Delhi

ANSHAN LTD

Ane Books India

First Published in 2007 by

ANSHAN LTD
6 Newlands Road
Tunbridge Wells
Kent.
TN4 9AT. UK
Tel: +44 (0) 1892 557767
Fax: +44 (0) 1892 530358
e-mail: info@anshan.co.uk
Website: www.anshan.co.uk

Published in arrangement with
Ane Books India

4821, Parwana Bhawan 1st Floor
24 Ansari Road, Darya Ganj, New Delhi -110 002, India
Tel: 91 (011) 2327 6843-44, 2324 6385
Fax: 91 (011) 2327 6863
e-mail: anebooks@vsnl.com
Website: www.anebooks.com

ISBN-10: 1 905740 03 4
ISBN-13: 978 1 905740 03 1

British Library Cataloguing in Publication Data
A catalogue record for this book is available from the British Library

This edition not for sale in India, Pakistan, Nepal, Sri Lanka and Bangladesh

Printed at Gopsons Paper Ltd, Noida (U.P.), India

To my parents,

Sheela and Govind

The Author is grateful to
Prof. R. Senani for his constant
encouragement and support and providing
the lab facilities. He is also grateful to
Vipin Vats for helping with the
typesetting work

PREFACE

This book is intended to serve as an advanced book for the course Engineering Physics as taught in engineering colleges all over the country and abroad at the undergraduate level as well as a reference book for applied mathematicians, theoretical physicists and electrical and mechanical engineers intending to apply signal processing techniques to problems of physics. The chapterwise description of the book is as follows:

Chapter I is about classical mechanics. The kinematical quantities are introduced and the description of their dynamical behaviour under the action of forces in accordance with the Newtonian laws of motion is formulated. The notion of work as defined by a line integral of the force field along a path is discussed. The Newtonian laws of motion are applied to a system of particles moving under their mutual interaction as well as under the influence of external forces is introduced. It is then explained how motion of a system of particles in a potential field leads to energy conservation and how in in the absence of external forces, the momentum of the system of particles is conserved. Motion under constraints is then introduced leading to the famous D'Alembert's equations of motion which can be taken as the starting point for Lagrange's variational principle in mechanics. While discussing constrained motion, we also consider the problem of a particle moving on a curved surface. Lagrange's variational principle involving the derivation of Newtonian laws by minimizing the action functional is then introduced and how this leads to the Hamilton equations of motion which is a system of coupled first order ordinary differential equations for the position and momentum variables. The Hamilton equations of motion can also be taken as a version of nonlinear state variable equations in system theory. Poisson bracket relations are then introduced explaining the fundamental bracket relations between the canonical coordinates and momenta. These relations prove to be the starting point for the formulation of the Heisenberg equations of motion in quantum mechanics as explained in Dirac's book. The motion of spinning top is then introduced as a prototype example of rigid body motion. The description of rigid body motion involves the used of the moment of inertia tensor and and the angular velocity tensor. These are discussed in detail. The description of a spinning top also involves the introduction of the Euler angles for parameterizing rotations and these ideas are introduced. Then we come to the last section, namely waves where vibrating string, vibrating membrane and interference of light is introduced. This section also contains some ideas about lasers, polarization of light and holography. A variety of study projects is introduced at the end. The projects deal with numerical integration of the Newtonian equations of motion, the Hamilton-Jacobi formulation of classical mechanics, and Lie group ideas in mechanics.

The second chapter is about fluid dynamics. We start with the equation of continuity which is about fluid mass conservation. Then, the Navier-Stokes equation for the fluid is derived. This is a field theoretic version of the Newtonian equations of motion taking into account not only the external forces like gravity but also internal pressure forces. The Navier-Stokes equation along with the equation of continuity provide a complete description of the fluid motion. The modification of the fluid equations of motion in the presence of viscous terms is then considered. Velocity potential for irrotational flows is also discussed and how the velocity potential can be used to describe all two dimensional flows. Use of the theory of functions of a complex variable to construct the velocity potential is then introduced. Bernoulli's equation which is the energy conservation equation for a fluid taking into account velocity potential and pressure is then discussed. This equation has important practical applications especially in aircraft lift problems. The diffusion equation for the vorticity is then derived. If thermal energy of a fluid is taken into account, then one can arrive at an energy conservation equation for the fluid. This is derived. The derivation of this equation combines ideas of fluid dynamics and thermodynamics. Examples of fluid flow problems are then considered like flow past a cylinder, flow past a sphere and gravity waves in an ocean. The study projects deal with boundary layers and magnetohydrodynamics which is the motion of a charged fluid in the presence of electric and magnetic field. This is particularly important in the analysis of electromagnetic wave propagation in the ionosphere. Magnetodydrodynamics involves a combination of fluid dynamics and the Maxwell field equations.

The third chapter is about electromagnetism. We start with the fundamental equations of electrostatics, namely Coulomb's law and Gauss' law, and then discuss some practical electrostatic field computation problems which includes a section on capacitance. Formulation of the equations of electrostatics in terms of Poisson's equation with a boundary condition is then considered and uniqueness of the solution proved. How the equations of electrostatics can be expressed in a more elegant form for dielectric materials using the polarization and displacement vector fields is then discussed. We then introduce the fundamental equations of magnetostatics, namely Ampere's law combined with the no monopole condition and using these two arrive at the Poisson's equation for the magnetic vector potential. The solution to these equations leads to the Biot-Savart law and this is discussed. Some practical applications of the Biot-Savart law to the magnetic field computation is then discussed. The field produced by a little bar magnet is considered and for dealing with properties of magnetic materials, we introduce the magnetization vector. Faraday's law of induction involving the generation of electric fields from time varying magnetic fields is considered and some applications of it are discussed. Next we discuss how Ampere's law violates the charge conservation principle for nonsteady currents and charge distributions and how it can be corrected by adding a displacement current correction term. The addition of this term leads to the complete set of Maxwell equations and this set predicts the existence of electromagnetic waves traveling at the speed of light. This is discussed in detail. The complete solution to the Maxwell equations in terms of retarded potentials is then discussed. This discussion is based on the Feynman lectures on physics, volume II. The study projects range over ideas like numerical integration of the wave equation, summarizing the basic equations of electrostatics, numerical computation of the electrostatic field from the charge distribution, iterative solution to the electric field when the medium has spatially varying permittivity, writing down the Maxwell equations in a curvilinear system of coordinates, summarizing all properties about electromagnetic potentials, transmission line theory involving writing down partial differential equations for the current and voltage along a transmission line, deriving the Snell's laws of reflection and refraction for an electromagnetic wave incident on the boundary of two dielectric media, formulation of the Maxwell field equations in the language of differential forms, propagation of electromagnetic waves inside a conductor, describing scattering of electromagnetic waves in a medium having spatio-temporally varying permittivity, applying finite element techniques to waveguide problems, computation of the power flow in an electromagnetic wave and many other problems.

The fourth chapter is about the special and general theories of relativity. The chapter begins by explaining how the Newtonian formulation of the laws of mechanics violates some of the basic principles of relativity. Michelson's experiment is then described explaning how it predicts the constancy of the velocity of light. The basic postulates of relativity by Einstein leading to the Lorentz transformation law replacing the Gailiean transformation law are described. The most general form of the Lorentz transformation as a product of spatial rotations and reflections and Lorentz boosts is written down. An important effect called the Thomas precession arising as one of the predictions of the Lorentz transformation equations is explained. This effect states that a particle moving along a curved trajectory appears to have its frame of reference rotated. This rotation is a purely relativistic effect. The phenomena of length contraction and time dilation starting from the Lorentz transformation equations is then explained. The modification of the Galilean law of velocity addition due to relativistic effects is also discussed. This modification guarantees that the velocity of no particle can exceed that of the speed of light. How relativity leads to the famous equation of Einstein $E = mc^2$ is then explained. This is shown to be a consequence of the variation of the mass of a particle with its velocity. The invariance of the Maxwell equations under Lorentz transformations is discussed in the light that Maxwell's equations are correct from a relativistic standpoint unlike the Newtonian equations of motion. The remaining part of the chapter discusses Einstein's celebrated general theory of relativity. We start this section by introducing the principle of equivalence which leads to the fact that the presence of a gravitational field modifies the metric of space time from Minkowskian flat space-time to Riemannian curved space-time. The motion of particles in a curved space time specified by a Riemannian metric is then derived. The basic elements of Riemannian geometry like parallel displacement, covariant derivatives, Riemann Christoffel curvature tensor, Ricci tensor and general tensor transformation laws are then introduced. The physical reason for developing tensor calculus in detail here is that the laws of physics according to the general principle of relativity must be invariant not merely under uniform motion but under any arbitrary transformation of the space-time coordinates. We then introduce the field equations of Einstein using the properties of the Ricci tensor and the energy momentum tensor of matter. These equations are second order highly nonlinear partial differential equations for the metric tensor and they prove to be a correction to the Newtonian inverse square law of gravitation. The Schwarzchild solution to the Einstein field equations for a spherical distribution of matter is then discussed and how this leads to the existence of blackholes from which light and particles cannot escape is described. The study projects develop Riemannian geometry in greater detail and also discuss the formulation of the equations of motion of particles in special relativity. Some projects also discuss general relativistic modifications of the Maxwell equations of electrodynamics which describe the interaction between gravity and electricity.

The fifth chapter is about quantum mechanics. Just as position and momentum are the fundamental kinematical quantities in classical mechanics, the wave function is the fundamental kinematical quantity in the quantum theory. The idea of a wave function for a system of particles and its probabilistic interpretation due to Max Born has been explained at the beginning. To describe more general quantum states, we need the concept of a density matrix or a density operator and this has been discussed illustrating how it can be used to describe a mixed state quantum mechanical system. Observables in quantum mechanics are Hermitian operators. This has been introduced stating how one can obtain the probability distribution for the values assumed by it in a given quantum state. The formula for the average value of an observable in a mixed quantum state has been derived. The fundamental Heisenberg uncertainty relation for observables that do not commute has been derived. This shows that two observables which do not commute cannot be simultaneously measured. The De-Broglie relation relating the momentum of a quantum particle to the wavelength of the matter wave associated with the quantum particle has been derived. The fundamental Planck relation for the frequency of a photon released when the quantum system makes a transition from a higher energy state to a lower one has been explained. Then we come to Bohr's model of the atom. Based on semiclassical arguments, Bohr was able to arrive at the correct energy spectrum of the Hydrogen atom. This argument is based on quantization of the angular momentum of the electron. Using the De-Broglie and Planck equations, one can arrive at the Schrodinger equation for a free particle and also for a particle moving in a potential field. In this equation, the kinetic energy appears as a second order partial differential operator in the spatial variables and the total energy appears as a first order temporal derivative. This equation was first derived by Schrodinger and is the starting point for modern quantum theory. This has been explained. How the Schrodinger equation can be used to obtain the energy levels of a free particle in a box, a harmonic oscillator and the Hydrogen atom has been explained. The solution to these problems are based on eigenvalue theory of second order differential equations. An alternate formulation to the Schrodinger wave equation is the Heisenberg equations of wave mechanics. These equations have been derived and their equivalence to the Schrodinger formalism has been explained. If the energy of a potential barrier exceeds the energy of a particle, then classical mechanics prevents the particle from crossing over the barrier. However quantum mechanics does not prevent such a cross over. This has been explained in the section on tunnelling. When one combines relativity with quantum mechanics, then one arrives at the Dirac relativistic wave equation for the electron. This has been derived. The wave function acquires four components. This has been derived. Spin of the electron arises as a relativistic effect in this equation. The study projects deal with numerical simulation of quantum systems, many electron and many nucleon problems in quantum mecbanics, angular momenta and their addition in the quantum theory, qubits, quantum computation and realization of quantum gates using physical systems, finite state quantum systems, Quantum motion of a charged particle in an electromagnetic field, the role of group theory in quantum mechanics, perturbation theory and simulation of quantum systems using the TMS DSP processor.

The sixth chapter is about circuit theory. The chapter begins with the description of the current-voltage relations for a resistor, capacitor and and inductor which form the basic elements of any linear circuit. These relations are developed starting from the field theoretic concepts such as flux, charge, electric field and current density. We next introduce the reader to the Kirchchoff voltage and current relations. These relations are respectively circuit theory versions of the equations of charge conservation and the fact that the line integral of the electric field around a closed loop is zero. The highlight of the chapter is the graph theoretic analysis of circuits. This technique is particularly suited to programming a computer for performing the analysis of a circuit. The graph theoretic formulation is based on expressing the Kirchchoff current relations using cutset matrices and the Kirchchoff voltage relations using fundamental loop matrices. Then follows the discussion on mutual inductance which involves the linking of magnetic field between two closely spaced coils. Different kinds of controlled sources are then introduced. These involve a circuit element along with a voltage or a current source when the voltage or the current generated by the source is proportional to the voltage or current across the element. Circuits with sinusoidal excitation are then discussed. The idea of representing the response of the circuit to a sinusoid in terms of the circuit transfer function is discussed here. Then comes the use of Laplace transforms in circuit analysis. The idea is first to represent the circuit dynamics by means of a system of coupled first order differential equations, ie, as a state variable system. Laplace transforming this system leads to a transfer function relationship between the input transform and the output transform. The initial conditions of the circuit also come into picture here. The solved problems of this chapter include computation of charge from charge density, computation of current from current density and vice versa, deriving current from charge, the dynamics of a time dependent capacitance, computing the magnetic energy stored in an inductor, two port networks involving controlled sources, writing down the circuit equations in the Laplace transform domain, computation of initial conditions in a circuit, transistor as a non-linear two port network. Some study projects are also included here. One of them is about performing a perturbation theoretic analysis of a diode resistor capacitor circuit. Another study project deals with the analysis of equipment in a basic electronics laboratory.

The seventh chapter is a small one on optics. The chapter begins with how Fourier analysis can be used to represent light of all frequencies. We then discuss the cylindrical wave equation with application to propagation of light inside an optical fibre. Geometrical optics is then considered where the Eikonal equation is derived for the phase of the light wave. The phenomenon of total internal reflection of light is explained. One of the study projects deals with writing down the Maxwell equations in the cylindrical coordinate system. This has applications to optical fibres, the other project deals with optical fibres whose permittivity is a function of the radial distance from the fibre axis.

The eighth chapter is also a small one on solid state physics. The quantum mechanical wave equation in a periodic Lattice is formulated. The idea of reciprocal lattice vectors is given. This has applications to representing periodic functions inside a crystal using Fourier series. Diffraction theory is then elucidated. This gives a relationship between the wavelength of an incident wave and that scattered by the crystal lattice in terms of the reciprocal lattice vectors. Applications of group theory to crystal structure is discussed. Crystals having permutational, rotational and reflectional symmetry are discussed in detail. The idea of a group representation is also mentioned. Vibrational waves propagating inside a crystal lattice are discussed. The Einstein-Debye theory of specific heat of a solid is also explained. The idea is to regard the solid as a collection of harmonic oscillators and to derive the specific heat from the partition function of the solid. Using the Fermi-Dirac statistics, we next derive an expression for the specific heat of a metal consisting of free electrons. The chapter includes just one study project, namely deriving the structure of some of the important matrix groups.

The ninth chapter is about basic digital signal processing. The chapter is an introduction to signal and system theory and has been included because of the importance of the methods developed by electrical engineers in the physical sciences. The chapter begins with signal classification where aspects such as discrete and continuous time, periodic and aperiodic signals, signal energy and power are introduced. Then the general theory of linear systems is developed. Examples of linear systems from electromagnetic field theory are given. The notion of time invariance is then introduced. Time invariant systems described by differential equations with constant coefficients is considered, the convolution sum and integral representation of linear time invariant systems is discussed, methods for solving linear differential equations with constant coefficients is considered, the sampling theorem for analog systems is also discussed. This theorem plays an important role in processing analog signals using digital techniques. Filters based on frequency selectivity are introduced next where the characterization of filters using the Fourier transform is discussed. The Fast Fourier transform for discrete time signals as an important digital signal processing tool is introduced next. Then we talk about random signals in discrete time and quantization noise in a digital filter. The latter plays an important role in analyzing the errors that propagate through a digital filter that uses delay elements and multipliers implemented using digital computers that have finite memory. Filter design is the next issue addressed. Then we talk about basic adaptive filter theory. This topic deals broadly with identifying the parameters of systems that vary slowly with time based on the most recent measurements. The subject of adaptive filters was initiated by Bernhard Widrow.

The tenth chapter is about probability theory. The basic tools required for probability are introduced here. These are probability spaces, distribution and density functions, moments of a random variable, characteristic function, independence of random variables and notion of random processes with examples. These are dealt with here. This chapter plays a fundamental role in applications involving problems where the signal waveforms fluctuate wildly with time and in problems where measurement errors prevent exact identifiability of system parameters. The most important application of random process theory is perhaps to the identification of a system by exciting all its degrees of freedom with random inputs and matching the moments of the output computed from the system dynamics with the estimated moments. The material discussed in this chapter will explain how this program can be carried out.

The eleventh and final chapter is one of the most important applications of probability theory to physics, namely statistical mechanics. This subject deals with systems that can occupy one of several states in accord with a probability distrbibution. The computation of thermodynamical quantities such as pressure, energy, entropy and enthalpy of a gas using the ideas of statistical mechanics is discussed here. All computations in statistical mechanics revolve around the partition function and the maximum entropy principle. These are discussed in this chapter. Some of the study projects of this chapter deal with the Boltzmann kinetic transport equation. This is the fundamental equation that governs the dynamics of a gas of ions.

Acknowledgements: The author would like to thank Mr.Vipin Vats for helping with the typsetting of the manuscript. He is also grateful to Professor Raj Senani for providing the facilities and congenial environment in the Department.

Contents

5   Quantum mechanics   275

Advanced Engineering Physics

Harish Parthasarathy

Chapter 1

Mechanics

1.1 Displacement, velocity, acceleration

The starting point for the solution of any problem of mechanics is to define the displacement of a particle. The displacement can be linear or angular. For example, to describe the motion of a particle on the ground, we erect a cartesian xy frame and relative to this frame, define the x and y components of the displacement. If $\hat{x}$ and $\hat{y}$ are unit vectors directed along the x and y axis, then the position of the particle is specified by the vector $\mathbf{r} = x\hat{x} + y\hat{y}$. The motion of the particle is specified then by specifying x and y as functions of time, i.e., $\mathbf{r}(t) = x(t)\hat{x} + y(t)\hat{y}$. If motion takes place just along one dimension, then we need specify simply a single coordinate, i.e. x as a function of time. If motion takes place in three dimensions, then we need three cartesian coordinates and we write down the displacement vector as $\mathbf{r}(t) = x(t)\hat{x} + y(t)\hat{y} + z(t)\hat{z}$. While formulating the equations in the general theory of relativity, we do not restrict our dimension to four (which is the dimension of space-time), we prefer to work in an n dimensional space. If the space is Euclidean, we can erect cartesian axes along each of the coordinate axes and specify the position at time t as an n dimensional vector: $\mathbf{r}(t) = \sum_{i=1}^{n} x_i(t)\mathbf{e}_i$. In this equation, $\mathbf{e}_i$ is the vector along the ith coordinate axis and $x_i(t)$ is the coordinate along this axis. We can also view $x_i(t)$ as the projection of the displacement $\mathbf{r}(t)$ along the ith coordinate axis. Sometimes a different coordinate system proves useful. The other different coordinate systems usually employed are (1) the cylindrical system (2) the spherical system. Any coordinate system can be defined completely by specifying the different coordinates as functions of the cartesian coordinates. Thus, if (q_1, q_2, q_3) is a curvilinear system of coordinates, we specify the $q_i's$ as functions of x, y, z, i.e., $q_1(x, y, z), q_2(x, y, z), q_3(x, y, z)$. Then the problem arises as to what are the unit vectors along the three curvilinear axes: If we change q_1 slightly keeping q_2 and q_3 constant, the resulting direction in which the position vector changes defines the unit vector along the q_1 axis. Likewise for q_2 and q_3. This means that we first solve the algebraic equations $q_1 = q_1(x, y, z), q_2 = q_2(x, y, z), q_3 = q_3(x, y, z)$ to obtain x, y, z as functions of

q_1, q_2, q_3 , i.e. $x(q_1, q_2, q_3), y(q_1, q_2, q_3), z(q_1, q_2, q_3)$ and then form the position vector

$$\mathbf{r}(q_1, q_2, q_3) = x(q_1, q_2, q_3)\hat{x} + y(q_1, q_2, q_3)\hat{y} + z(q_1, q_2, q_3)\hat{z}$$

Then, the unit vector along q_i is given by

$$\mathbf{e}_{q_i} = \frac{\frac{\partial \mathbf{r}}{\partial q_i}}{|\frac{\partial \mathbf{r}}{\partial q_i}|}, i = 1, 2, 3$$

Note that in except for the cartesian system, the $\mathbf{e}'_{q_i} s$ are functions of (q_1, q_2, q_3), or equivalently, functions of the position. They vary from point to point in space. That is why we say that the triplet $(\mathbf{e}_{q_1}, \mathbf{e}_{q_2}, \mathbf{e}_{q_3})$ form a local coordinate system of axes at each point in space. To express the position vector of a point completely in the curvilinear system, we must use only the variables $q'_i s$ and the unit vectors $\mathbf{e}_{q_i}$. This can be achieved by considering the equations

$$|\frac{\partial \mathbf{r}}{\partial q_i}|\mathbf{e}_{q_i} = \frac{\partial x}{\partial q_i}\hat{x} + \frac{\partial y}{\partial q_i}\hat{y} + \frac{\partial \mathbf{r}}{\partial q_i}\hat{z}, i = 1, 2, 3$$

as three algebraic equations in the variables $\hat{x}, \hat{y}, \hat{z}$ and solving this system of algebraic equations to express $\hat{x}, \hat{y}, \hat{z}$ as linear combinations of $\mathbf{e}_{q_i}, i = 1, 2, 3$. The solution can be achieved via Cramer's determinants rule. We write the solution as

$$\hat{x} = \alpha_{11}(q_1, q_2, q_3)\mathbf{e}_{q_1} + \alpha_{12}(q_1, q_2, q_3)\mathbf{e}_{q_2} + \alpha_{13}(q_1, q_2, q_3)\mathbf{e}_{q_3}$$

$$\hat{y} = \alpha_{21}(q_1, q_2, q_3)\mathbf{e}_{q_1} + \alpha_{22}(q_1, q_2, q_3)\mathbf{e}_{q_2} + \alpha_{23}(q_1, q_2, q_3)\mathbf{e}_{q_3}$$

$$\hat{z} = \alpha_{31}(q_1, q_2, q_3)\mathbf{e}_{q_1} + \alpha_{32}(q_1, q_2, q_3)\mathbf{e}_{q_2} + \alpha_{33}(q_1, q_2, q_3)\mathbf{e}_{q_3}$$

Then, the position vector can be expressed in the curvilinear system as

$$\mathbf{r}(q_1, q_2, q_3) = x(\alpha_{11}\mathbf{e}_{q_1} + \alpha_{12}\mathbf{e}_{q_2} + \alpha_{13}\mathbf{e}_{q_3}) + y(\alpha_{21}\mathbf{e}_{q_1} + \alpha_{22}\mathbf{e}_{q_2} + \alpha)_{23}\mathbf{e}_{q_3}) + z(\alpha_{31}\mathbf{e}_{q_1} + \alpha_{32}\mathbf{e}_{q_2} + \alpha_{33}\mathbf{e}_{q_3})$$

$$= (x\alpha_{11} + y\alpha_{21} + z\alpha_{31})\mathbf{e}_{q_1} + (x\alpha_{12} + y\alpha_{22} + z\alpha_{32})\mathbf{e}_{q_2}(x\alpha_{13} + y\alpha_{23} + z\alpha_{33})\mathbf{e}_{q_3}$$

For the cylindrical coordinate system, we have the following relationships between the cartesian coordinates (x, y, z) and the cylindrical coordinates (ρ, ϕ, z) of a point. ρ is the perpendicular distance of the point in question from the z axis. Thus by Pythagoras' theorem, $\rho = \sqrt{x^2 + y^2}$. ϕ is the angle made by the perpendicular drawn from the point to the z axis with the xz plane. Thus, $\phi = tan^{-1}(y/x)$. These relationships can be inverted to get $x = \rho.cos\phi, y = \rho.sin\phi, z = z$. The position vector of the point is thus given by

$$\mathbf{r}(\rho, \phi, z) = \rho.cos\phi\hat{x} + \rho.sin\phi\hat{y} + z\hat{z}$$

and the unit vectors along the ρ, ϕ, z directions at the given point are obtained as

$$\hat{\rho} = \frac{\frac{\partial \mathbf{r}}{\partial \rho}}{|\frac{\partial \mathbf{r}}{\partial \rho}|} = \hat{x}.cos\phi + \hat{y}.sin\phi$$

$$\hat{\phi} = \frac{\frac{\partial \mathbf{r}}{\partial \phi}}{\left|\frac{\partial \mathbf{r}}{\partial \phi}\right|} = -\hat{x}.sin\phi + \hat{y}.cos\phi$$

$$\hat{z} = \hat{z}$$

Note that while differentiating the vector valued function $\mathbf{r}$ of ρ, ϕ, z, we've kept the unit vectors $\hat{x}, \hat{y}$ and $\hat{z}$ as constant. Given two vectors expressed in terms of the cartesian system as $\mathbf{A} = A_x\hat{x} + A_y\hat{y} + A_z\hat{z}$ and $\mathbf{B} = B_x\hat{x} + B_y\hat{y} + B_z\hat{z}$, we can draw rays representing these vectors from the origin. The distance between the endpoints of these vectors is given by

$$d(\mathbf{A}, \mathbf{B}) = \sqrt{(A_x - B_x)^2 + (A_y - B_y)^2 + (A_z - B_z)^2}$$

while the distance of the endpoints of $\mathbf{A}$ and $\mathbf{B}$ from the origin are respectively given by

$$d(\mathbf{A}, \mathbf{0}) = \sqrt{A_x^2 + A_y^2 + A_z^2}$$

$$d(\mathbf{B}, \mathbf{0}) = \sqrt{B_x^2 + B_y^2 + B_z^2}$$

By a well known result in trigonometry, the angle between the two vectors is given by θ where

$$cos(\theta) = \frac{d(\mathbf{A}, \mathbf{0})^2 + d(\mathbf{B}, \mathbf{0})^2 - d(\mathbf{A}, \mathbf{B})^2}{2d(\mathbf{A}, \mathbf{0})d(\mathbf{B}, \mathbf{0})}$$

Substituting for the distances and simplifying the result leads to the expression

$$A_xB_x + A_yB_y + A_zB_z = |\mathbf{A}||\mathbf{B}|cos(\theta)$$

where $|\mathbf{A}| = d(\mathbf{A}, \mathbf{0})$ and $|\mathbf{B}| = d(\mathbf{B}, \mathbf{0})$ are respectively the lengths of the vectors $\mathbf{A}$ and $\mathbf{B}$. The quantity $A_xB_x + A_yB_y + A_zB_z$ is called the scalar or dot product of the vectors $\mathbf{A}$ and $\mathbf{B}$ and is denoted by $\mathbf{A}.\mathbf{B}$. One very important property of the scalar product is that it is invariant under rotations. A rotation is a 3×3 real matrix $\mathbf{R}$ satisfying the property $\mathbf{R}^T\mathbf{R} = \mathbf{R}\mathbf{R}^T = \mathbf{I}$. If vectors are represented by column vectors, i.e. the vector $\mathbf{A}$ is represented by the column vector $[A_x, A_y, A_z]^T$, then applying the rotation $\mathbf{R}$ to this vector changes it to $\mathbf{R}\mathbf{A}$. The scalar product between vectors $\mathbf{A}$ and $\mathbf{B}$ is $\mathbf{A}^T\mathbf{B} = \mathbf{B}^T\mathbf{A}$, when the vectors are represented in column form. Then, the scalar product between the rotated vectors $\mathbf{R}\mathbf{A}$ and $\mathbf{R}\mathbf{B}$ is given by

$$(\mathbf{R}\mathbf{A})^T(\mathbf{R}\mathbf{B}) = \mathbf{A}^T\mathbf{R}^T\mathbf{R}\mathbf{B} = \mathbf{A}^T\mathbf{I}\mathbf{B} = \mathbf{A}^T\mathbf{B}$$

proving that the scalar product is invariant under rotations. Taking $\mathbf{B} = \mathbf{A}$ shows in particular that the length of the vector $\sqrt{\mathbf{A}^T\mathbf{A}}$ is also invariant under rotations. Thus, the angle between vectors is invariant under rotations. Two vectors are orthogonal if the angle between them is $\pi/2$ radians. This happens when and only when their scalar product equals zero. It is clear that at each point, the unit vectors $\hat{\rho}, \hat{\phi}, \hat{z}$ of the cylindrical coordinate system are orthogonal. Specifically,

$$\hat{\rho}.\hat{\phi} = (\hat{x}.cos\phi + \hat{y}.sin\phi).(-\hat{x}.sin\phi + \hat{y}.cos\phi) = cos\phi(-sin\phi) + sin\phi(cos\phi) = 0$$

$$\hat{\rho}.\hat{z} = (\hat{x}.cos\phi + \hat{y}.sin\phi).\hat{z} = cos\phi.(0) + sin\phi.(0) + 0.1 = 0$$

$$\hat{\phi}.\hat{z} = (-\hat{x}sin\phi + \hat{y}cos\phi).\hat{z} = -sin\phi.(0) + cos\phi.(0) + 0.1 = 0$$

The unit vectors $\hat{x}, \hat{y}, \hat{z}$ of the cartesian system by their very definition are orthogonal. The position vector of a point in the cylindrical system is given by

$$\mathbf{r} = \rho\hat{\rho} + z\hat{z}$$

This can be seen either through a diagram showing how to arrive at the point $\mathbf{r}$ via a displacement ρ in the direction $\hat{\rho}$ followed by a displacement z in the direction $\hat{z}$. This can also be seen by starting with the cartesian expression

$$\mathbf{r} = x\hat{x} + y\hat{y} + z\hat{z} = \rho.cos\phi\hat{x} + \rho.sin\phi\hat{y} + z\hat{z}$$

$$= \rho(\hat{x}.cos\phi + \hat{y}.sin\phi) + z\hat{z} = \rho\hat{\rho} + z\hat{z}$$

Note that whereas the unit vectors $\hat{x}, \hat{y}, \hat{z}$ of the cartesian system are constant everywhere, the unit vectors $\hat{\rho}, \hat{\phi}$ of the cylindrical system vary from point to point. At each point of space the three vectors $\hat{\rho}, \hat{\phi}, \hat{z}$ form a right handed orthogonal system of axes. The coordinates of a point in the spherical polar system are r, θ, ϕ. These are related to the cartesian coordinates by the transformations

$$x = r.cos\phi.sin\theta, y = r.sin\phi.sin\theta, z = r.cos\theta$$

with the inverse relations

$$r = \sqrt{x^2 + y^2 + z^2}, \theta = cos^{-1}(z/\sqrt{x^2 + y^2 + z^2}), \phi = tan^{-1}(y/x)$$

The unit vectors along the three coordinate directions are

$$\hat{r} = \frac{\partial\mathbf{r}}{\partial r}/|\frac{\partial\mathbf{r}}{\partial r}| = \hat{x}.cos\phi, sin\theta + \hat{y}.sin\phi.sin\theta + \hat{z}.cos\theta$$

$$\hat{\theta} = \frac{\partial\mathbf{r}}{\partial\theta}/|\frac{\partial\mathbf{r}}{\partial\theta}| = \hat{x}.cos\phi.cos\theta + \hat{y}.sin\phi.cos\theta - \hat{z}.sin\theta$$

$$\hat{\phi} = \frac{\partial\mathbf{r}}{\partial\phi}/|\frac{\partial\mathbf{r}}{\partial\phi}| = -\hat{x}.sin\phi + \hat{y}.cos\phi$$

These are obtained by differentiating the vector

$$\mathbf{r}(r, \theta, \phi) = r.cos\phi.sin\theta\hat{x} + r.sin\phi.sin\theta\hat{y} + r.cos\theta\hat{z}$$

with respect to the three parameters r, θ, ϕ. Note that during differentiation, the unit vectors $\hat{x}, \hat{y}, \hat{z}$ are kept constant. Once again, the unit vectors $\hat{r}, \hat{\theta}, \hat{\phi}$ vary with position but at each point, the three are mutually orthogonal and form a right handed system of axes. The position vector of a point in the spherical system is $\mathbf{r} = r\hat{r}$. This follows from the above expression for $\hat{r}$, or equivalently from a picture which shows that the position $\mathbf{r}$ can be arrived at by moving radially a distance r from the origin. By radially here we mean in the direction $\hat{r}$. Suppose we want to describe the motion of a particle as time evolves. We can then erect a set of three mutually orthogonal coordinate axes at a point and define the cartesian components of the point at each time instant with reference to these axes. Thus, the x, y, z coordinates of the position of the particle become functions of time and the position at time t is given by

$$\mathbf{r}(t) = x(t)\hat{x} + y(t)\hat{y} + z(t)\hat{z}$$

The rate of change of position is the velocity vector:

$$\mathbf{v}(t) = \frac{d\mathbf{r}(t)}{dt} = \frac{dx(t)}{dt}\hat{x} + \frac{dy(t)}{dt}\hat{y} + \frac{dz(t)}{dt}\hat{z}$$

Note that

$$\mathbf{v}(t)dt = dx(t)\hat{x} + dy(t)\hat{y} + dz(t)\hat{z}$$

equals the displacement of the particle in an infinitesimal duration of time dt. If the velocity vector is given as a function of time, then we can determine the displacement vector of the particle in the duration $[t_1, t_2]$ by integration:

$$\mathbf{r}(t_2) - \mathbf{r}(t_1) = \int_{t_1}^{t_2} \mathbf{v}(t)dt = \int_{t_1}^{t_2} dx(t)\hat{x} + dy(t)\hat{y} + dz(t)\hat{z}$$

The velocity vector can also be computed in any arbitrary curvilinear system of coordinates. Let the coordinates be q_1, q_2, q_3 so that the position is specified as a function of these coordinates $\mathbf{r}(q_1, q_2, q_3)$. The equation of motion is specified by giving q_1, q_2, q_3 as functions of t. Then the velocity is given by

$$\mathbf{v}(t) = \frac{d\mathbf{r}}{dt} = \sum_{i=1}^{3} \frac{\partial \mathbf{r}}{\partial q_i} \frac{dq_i}{dt}$$

The unit vectors along the three coordinates are

$$\mathbf{e}_i = \frac{\partial \mathbf{r}}{\partial q_i} \Big/ \left| \frac{\partial \mathbf{r}}{\partial q_i} \right|$$

and the Lame's coefficients are

$$H_i = \left| \frac{\partial \mathbf{r}}{\partial q_i} \right|, i = 1, 2, 3$$

In terms of these, we have

$$\mathbf{v}(t) = \sum_{i=1}^{3} H_i \frac{dq_i}{dt} \mathbf{e}_i$$

Specializing this formula to the cylindrical and spherical systems gives

$$\mathbf{v}(t) = \frac{d\rho}{dt}\hat{\rho} + \rho\frac{d\phi}{dt}\hat{\phi} + \frac{dz}{dt}\hat{z}$$

$$\mathbf{v}(t) = \frac{dr}{dt}\hat{r} + r\frac{d\theta}{dt}\hat{\theta} + r\sin\theta\frac{d\phi}{dt}\hat{\phi}$$

These are of course equivalent to the formulas for the line element in the two systems

$$d\mathbf{r} = dr\hat{r} + rd\theta\hat{\theta} + r\sin\theta d\phi\hat{\phi}$$

It is useful at this stage to have expressions for the rate of change of a vector in the cylindrical and spherical coordinates. In cylindrical coordinates, the vector as a function of time is specified as

$$\mathbf{F}(t) = F_\rho(t)\hat{\rho} + F_\phi(t)\hat{\phi} + F_z(t)\hat{z}$$

Note that $\hat{\rho}$ and $\hat{\phi}$ are also functions of time. The total differential of $\mathbf{F}$ is given by

$$d\mathbf{F}(t) = dF_\rho\hat{\rho} + F_\rho\hat{\phi}d\phi + dF_\phi\hat{\phi} - F_\phi\hat{\rho}d\phi + dF_z\hat{z}$$

where we've used the identities $\frac{\partial\hat{\rho}}{\partial\phi} = \hat{\phi}$ and $\frac{\partial\hat{\phi}}{\partial\phi} = -\hat{\rho}$. From this, we obtain the following expression for the rate of change of the vector $\mathbf{F}(t)$:

$$\frac{d\mathbf{F}(t)}{dt} = \left(\frac{dF_\rho}{dt} - F_\phi\frac{d\phi}{dt}\right)\hat{\rho} + \left(F_\rho\frac{d\phi}{dt} + \frac{dF_\phi}{dt}\right)\hat{\phi} + \frac{dF_z}{dt}\hat{z}$$

In spherical polar coordinates, we have

$$\mathbf{F}(t) = F_r\hat{r} + F_\theta\hat{\theta} + F_\phi\hat{\phi}$$

Then using

$$\hat{\theta} = \hat{x}.cos\phi.cos\theta + \hat{y}.sin\phi.cos\theta - \hat{z}.sin\theta$$

$$\hat{r} = \hat{x}.cos\phi.sin\theta + \hat{y}.sin\phi.sin\theta + \hat{z}.cos\theta$$

$$\hat{\phi} = -\hat{x}.sin\phi + \hat{y}.cos\phi$$

we get

$$d\hat{r} = \hat{\theta}.d\theta + sin\theta\hat{\phi}.d\phi$$

$$d\hat{\theta} = -\hat{r}d\theta + \hat{\phi}.sin\theta d\phi$$

$$d\hat{\phi} = (-\hat{\theta}.cos\theta - \hat{r}.sin\theta)d\phi$$

we can evaluate the differential of the vector by plugging in the expression

$$d\mathbf{F} = dF_r\hat{r} + dF_\theta\hat{\theta} + dF_\phi\hat{\phi} + F_r d\hat{r} + F_\theta d\hat{\theta} + F_\phi d\hat{\phi}$$

We leave the simplification of the final expression to the student. The acceleration of a particle in motion is defined as the rate of change of velocity or rate of change of rate of change of position:

$$\mathbf{a}(t) = \frac{d^2\mathbf{r}(t)}{dt^2} = \frac{d^2x}{dt^2}\hat{x} + \frac{d^2y}{dt^2}\hat{y} + \frac{d^2z}{dt^2}\hat{z}$$

$$= \frac{d\mathbf{v}}{dt} = \frac{dv_x}{dt}\hat{x} + \frac{dv_y}{dt}\hat{y} + \frac{dv_z}{dt}\hat{z}$$

We now give the expression for the acceleration of a particle in the cylindrical and spherical polar coordinate systems. These are based upon the formula for differentiating a vector in the two systems. In cylindrical coordinates,

$$\mathbf{v} = \frac{d\rho}{dt}\hat{\rho} + \rho\frac{d\phi}{dt}\hat{\phi} + \frac{dz}{dt}\hat{z}$$

$$\mathbf{a} = \frac{d\mathbf{v}}{dt} = \frac{d^2\rho}{dt^2}\hat{\rho} + \frac{d\rho}{dt}\frac{d\hat{\rho}}{dt}$$

$$+ \frac{d}{dt}(\rho\frac{d\phi}{dt})\hat{\phi} + \rho\frac{d\phi}{dt}\frac{d\hat{\phi}}{dt} + \frac{d^2z}{dt^2}\hat{z}$$

Now observe that

$$\frac{d\hat{\rho}}{dt} = \frac{\partial\hat{\rho}}{\partial\phi}\frac{d\phi}{dt} = \hat{\phi}\frac{d\phi}{dt}$$

$$\frac{d\hat{\phi}}{dt} = \frac{\partial\hat{\phi}}{\partial\phi}\frac{d\phi}{dt} = -\hat{\rho}\frac{d\phi}{dt}$$

Thus,

$$\mathbf{a} = (\rho'' - \rho\phi'^2)\hat{\rho} + ((\rho\phi')' + \rho'\phi')\hat{\phi} + z''\hat{z}$$

$$= (\rho'' - \rho\phi'^2)\hat{\rho} + (\rho\phi'' + 2\rho'\phi')\hat{\phi} + z''\hat{z}$$

In a similar manner acceleration in the spherical polar coordinate system can be derived:

$$\mathbf{v} = r'\hat{r} + r\theta'\hat{\theta} + r.sin(\theta)\phi'\hat{\phi}$$

$$\mathbf{a} = \mathbf{v}'' = r''\hat{r} + r'(\hat{r})' + (r\theta')'\hat{\theta} + r\theta'(\hat{\theta})' + (r.sin(\theta)\phi')'\hat{\phi} + (r.sin(\theta)\phi')(\hat{\phi})'$$

$$(\hat{r})' = \frac{d\hat{r}}{dt} = \hat{\theta}\theta' + sin(\theta)\hat{\phi}\phi'$$

$$(\hat{\theta})' = -\hat{r}\theta' + \hat{\phi}.sin\theta\phi'$$

$$\hat{\phi}' = (-\hat{\theta}.cos\theta - \hat{r}.sin\theta)\phi'$$

We leave the job of expressing the acceleration in final form to the student.

1.2   Newton's laws of motion

There are three laws. The first states that every body continues to remain in its state of rest or uniform rectilinear motion unless it be compelled to change its state by an external force. The second law states that the total force on a body is proportional to the mass of the body and its acceleration or equivalently, the acceleration experienced by a body is proportional to the external force experienced by the body and inversely proportional to the mass of the body. In effect, the first and the third law can be derived from the second which forms the foundation of the entire mechanics of Newton, i.e., of non-relativistic mechanics. A spring that obeys Hooke's law can be used to verify Newton's second law. For such a spring, if x denotes the displacement from equilibrium, the net force experienced by a mass attached to it is proportional to the displacement and acts in a direction opposite to the displacement. When released at this point, one can measure the acceleration and verify Newton's second law. In fact, Newton's second law can also provide us with the definition of mass. Suppose we take a mass M. Attach it to a spring which we call as standard and stretch it by one unit of length. We consider the force exerted by the spring on the mass as one unit of force. We then measure the acceleration of the body and let it be a units. We then define the mass M to have the value of $1/a$ units. We now take another mass m and attach it to the same spring and stretch the spring by one unit and measure its acceleration a'. Then, $m = 1/a'$. Thus, $m/M = a/a'$. We notice that this ratio in fact actually corresponds to our intuitive picture of mass. If m consists of two identical copies of M clubbed together, then we will find that $a/a' = 2$. If there is a force field permeating all of space, i.e. at each point $\mathbf{r}$ of space, a body when placed at that point, experiences a force $\mathbf{F}(\mathbf{r})$, then the equation of motion of the body acquires the form $m\frac{d^2\mathbf{r}}{dt^2} = \mathbf{F}(\mathbf{r})$, when m is the mass of the body. If this force field varies with time, i.e. $\mathbf{F}(t,\mathbf{r})$, then the equation of motion can be cast as

$$m\frac{d^2\mathbf{r}(t)}{dt^2} = \mathbf{F}(t,\mathbf{r}(t))$$

The solution to this system of three coupled nonlinear ordinary differential equation yields the trajectory of the particle. The force field in some instances can also depend upon the velocity, i.e. we write $\mathbf{F}(t,\mathbf{r},\mathbf{v})$. Then the equation of motion has the form

$$\frac{d\mathbf{r}}{dt} = \mathbf{v}, m\frac{d\mathbf{v}}{dt} = \mathbf{F}(t,\mathbf{r},\mathbf{v})$$

or equivalently as

$$m\frac{d^2\mathbf{r}}{dt^2} = \mathbf{F}(t,\mathbf{r},\frac{d\mathbf{r}}{dt})$$

An example of a force field which depends upon position and velocity is that of spring with damping friction. The force field as a function of displacement x and velocity v is $F(x,v) = -kx - \gamma v$ and the equation of motion has the form $m\frac{d^2x}{dt^2} + \gamma\frac{dx}{dt} + kx = 0$. If motion takes place in three dimensions, then we have one such equation for each cartesian component of the displacement. In vector notation, $\mathbf{F} = -k\mathbf{r} - \gamma\mathbf{v}$ and the equation of motion is

$$m\frac{d^2\mathbf{r}}{dt^2} + \gamma\frac{d\mathbf{r}}{dt} + k\mathbf{r} = \mathbf{0}$$

1.3   Work, potential and motion of a particle in a potential field

Suppose $\mathbf{F}(\mathbf{r})$ is a force field. The work done by this force in displacing a particle along a trajectory $s \to \gamma(s)$ from $\gamma(0)$ to $\gamma(1)$ is defined as the line integral of $\mathbf{F}$ along this trajectory, i.e., $\int_0^1 \mathbf{F}(\gamma(s)).\gamma'(s)ds$. The usual notation for this work is to consider the starting point $\gamma(0) = \mathbf{r}_1$ and the final point $\gamma(1) = \mathbf{r}_2$ and express the work as $W(1,2|\gamma) = \int_\gamma \mathbf{F}.d\mathbf{r}$. Suppose that this integral is independent of the path chosen, i.e., it depends only upon the two endpoints. Then, it easily follows that the line integral of $\mathbf{F}$ around any closed curve is zero and hence by Stokes' theorem, $\nabla \times \mathbf{F} = \mathbf{0}$. This means that $\mathbf{F} = -\nabla V$ where V is a scalar field. Then, we can write $W(1,2|\gamma) = V(\mathbf{r}_1) - V(\mathbf{r}_2)$, i.e., the work done by the force along any curve joining points $\mathbf{r}_1$ and $\mathbf{r}_2$ equals the difference between V evaluated at these two points. In particular, if V vanishes at ∞, (this will happen provided that $\mathbf{F}$ decays faster than $1/r$ as $r \to \infty$), then the work done by the force in transporting a particle from $\mathbf{r}$ to ∞ along any path equals $V(\mathbf{r})$. Hence, $V(\mathbf{r})$ can be called the potential of the force at the point $\mathbf{r}$. The potential is defined only for conservative forces, i.e., for forces whose curl vanishes everywhere. The equation of motion of a particle moving in a conservative force field $\mathbf{F} = -\nabla V$ is given by

$$m\frac{d^2\mathbf{r}}{dt^3} = -\nabla V(\mathbf{r})$$

or in terms of components,

$$m\frac{d^2x}{dt^2} = -\frac{\partial V(x,y,z)}{\partial x}$$
$$m\frac{d^2y}{dt^2} = -\frac{\partial V(x,y,z)}{\partial y}$$
$$m\frac{d^2z}{dt^2} = -\frac{\partial V(x,y,z)}{\partial z}$$

One characteristic feature of such a motion is that the total energy defined as the sum of the kinetic and potential energies:

$$E = \frac{m}{2}|\frac{d\mathbf{r}}{dt}|^2 + V(\mathbf{r})$$

is conserved along the trajectory of the particle. We shall see the proof of this in a later section.

1.4   Motion of a system of particles

In order to describe the motion of a system of particles having masses $m_1, ..., m_N$, we need two things. One the external force on each particle and two the force exerted by each particle on the other. Let $\mathbf{F}_{ext,i}$ denote the external force on the ith particle. This will in general be a function of the coordinates of the ith particle but it may also in some instances depend upon the coordinates of the other particles. We ignore the argument here. Let $\mathbf{F}_{ij}$ denote the force of the ith particle upon the jth particle. Then the sum total of the internal and external forces on the ith particle is given by

$$\mathbf{F}_i = \mathbf{F}_{ext,i} + \sum_{j=1, j \neq i}^{n} \mathbf{F}_{ji}$$

and the equation of motion of the system is given by

$$m_i \frac{d^2 \mathbf{r}_i}{dt^2} = \mathbf{F}_i, i = 1, 2, ..., n$$

The internal force $\mathbf{F}_{ij}$ in almost all cases is just a function of the positions of the ith and j^{th} particles. Of course if forces of electromagnetic origin are also considered, then the external force will also depend upon the velocity of the particles and the internal force will depend upon the velocities of both the particles. This is because the force of a magnetic field upon a moving charge depends upon the velocity of the particle and the magnetic field produced by a moving charge depends upon the velocity of the charge. These issues are discussed in the chapter on electromagnetic fields.

1.5   Conservation of momentum, energy and angular momentum

Consider a system of particles moving in external forces $\mathbf{F}_{ext,i}$ and internal forces $\mathbf{F}_{ij}$ with the equal action-reaction assumption, i.e. $\mathbf{F}_{ij} = -\mathbf{F}_{ji}$. In other words, the force exerted by the ith particle upon the jth equals the negative of the force exerted by the jth particle upon the ith. We shall also assume in addition that the force exerted by the ith particle upon the jth is directed along the line joining the two particles. This assumption fails to hold good in the case of electromagnetic forces owing to the $\mathbf{v} \times \mathbf{B}$ factor. The equations of motion are

$$m_i \frac{d^2 \mathbf{r}_i}{dt^2} = \mathbf{F}_{ext,i} + \sum_{j=1, j \neq i}^{n} \mathbf{F}_{ji}$$

Adding up all these equations gives us

$$\frac{d^2}{dt^2} \sum_{i=1}^{n} m_i \mathbf{r}_i = \sum_{i=1}^{n} m_i \frac{d^2 \mathbf{r}_i}{dt^2} = \mathbf{F}_{ext}$$

where $\mathbf{F}_{ext} = \sum_{i=1}^{n} \mathbf{F}_{ext,i}$ is the net external force on the system of particles. Note that $\sum_{i \neq j} \mathbf{F}_{ij} = \mathbf{0}$ since each pair (i,j) has a corresponding pair (j,i) which gives terms that cancel. The quantity $\mathbf{R}_{cm} = \sum m_i \mathbf{r}_i / \sum m_i$ is called the centre of mass of the system. $M = \sum m_i$ is the total mass of the system and we can write the above equation as

$$M \frac{d^2 \mathbf{R}_{cm}}{dt^2} = \mathbf{F}_{ext}$$

Equivalently, letting $\mathbf{v}_i = \frac{d\mathbf{r}_i}{dt}$, the velocity of the ith mass, we have

$$M \frac{d\mathbf{R}_{cm}}{dt} = \sum m_i \mathbf{v}_i = \sum \mathbf{p}_i = \mathbf{P}$$

where $\mathbf{p}_i$ is the momentum of the ith mass and $\mathbf{P}$ is the momentum of whole system. The momentum of a particle relative to a frame of reference equals its mass times it velocity relative to the frame. This is a definition. Newton's law for a particle can also be stated as the rate of change of the particle's momentum equals the net external force on it. The above equation shows that this also holds for a system of particles, i.e.,

$$\frac{d\mathbf{P}}{dt} = \mathbf{F}_{ext}$$

In particular, if there is no net external force on the system, i.e., $\mathbf{F}_{ext} = \mathbf{0}$, then $\mathbf{P}$ is a constant of the motion, i.e., it does not vary with time. This is the law of conservation of linear momentum. To obtain the law of conservation of angular momentum, we need first of all to introduce the notions of torque and angular momentum. If a system of particles having masses $m_1, ..., m_n$ has positions at a certain instant of time $\mathbf{r}_1, ..., \mathbf{r}_n$ respectively and these particles are moving with velocities $\mathbf{v}_1, ..., \mathbf{v}_n$ respectively, then the angular momentum of this system about the origin is defined as

$$\mathbf{L} = \sum_{i=1}^{n} m_i \mathbf{r}_i \times \mathbf{v}_i = \sum_{i=1}^{n} \mathbf{r}_i \times \mathbf{p}_i$$

where $\mathbf{p}_i$ is the momentum of the ith particle. The angular momentum about a point $\mathbf{a}$ moving with velocity $\mathbf{v}_a$ is defined by replacing $\mathbf{r}_i$ and $\mathbf{v}_i$ with the relative position and velocity of the ith particle with respect to $\mathbf{a}$, i.e.,

$$\mathbf{L}_a = \sum_{i=1}^{n} m_i (\mathbf{r}_i - \mathbf{a}) \times (\mathbf{v}_i - \mathbf{v}_a)$$

The rate of change of angular momentum can be computed:

$$\frac{d\mathbf{L}}{dt} = \sum m_i \mathbf{r}_i \times \frac{d\mathbf{v}_i}{dt} = \sum \mathbf{r}_i \times \mathbf{F}_i$$

since $\frac{d\mathbf{r}_i}{dt} \times \mathbf{v}_i = \mathbf{v}_i \times \mathbf{v}_i = \mathbf{0}$ and $m_i \frac{d\mathbf{v}_i}{dt} = \mathbf{F}_i$ equals the force acting on the ith particle according to Newton's second law of motion. The quantity $\tau = sum_{i=1}^{n} \mathbf{r}_i \times \mathbf{F}_i$ is called the torque of the forces on the system of particles. Note that the force $\mathbf{F}_i$ on the ith particle equals the sum of the external force and the internal forces on the ith particle, i.e., $\mathbf{F}_i = \mathbf{F}_{ext,i} + \sum_{j=1, j \neq i}^{n} \mathbf{F}_{ji}$. Thus,

$$\tau = \sum_{i=1}^{n} \mathbf{r}_i \times \mathbf{F}_{ext,i} + \sum_{j \neq i} \mathbf{r}_i \times \mathbf{F}_{ji}$$

The second term is the torque due to the internal forces. We shall show that this torque is zero assuming that the force between two particles is directed along the line joining them and that the action-reaction principle is satisfied:

$$\sum_{j \neq i} \mathbf{r}_i \times \mathbf{F}_{ji} = \sum_{i \neq j} \mathbf{r}_j \times \mathbf{F}_{ij}$$

$$= \frac{1}{2} \sum_{i \neq j} \mathbf{r}_i \times \mathbf{F}_{ji} + \mathbf{r}_j \times \mathbf{F}_{ij}$$

$$= \frac{1}{2} \sum_{i \neq j} \mathbf{r}_i \times \mathbf{F}_{ji} - \mathbf{r}_j \times \mathbf{F}_{ji}$$

$$= \frac{1}{2} \sum_{i \neq j} (\mathbf{r}_i - \mathbf{r}_j) \times \mathbf{F}_{ji} = 0$$

since $\mathbf{F}_{ji}$ is along $\mathbf{r}_i - \mathbf{r}_j$. This means that the rate of change of angular momentum for a system of particles equals the net torque on the system due to external forces. This constitutes a fundamental theorem of mechanics. Note that it is a consequence of Newton's second law of motion combined with two assumptions about the internal forces, namely the equal action-reaction principle and the assumption that the force between two particles is directed along the line joining the two particles. In particular, if the net torque is zero, then total angular momentum is conserved. The equation of energy conservation states that for motion in a potential field, the net energy of the system of particles defined as the sum of the kinetic and potential energies is conserved. For a single particle, we say that motion takes place in a potential field if the force field in which it moves is obtained as the gradient of a scalar function. The equation of motion then has the form

$$m \frac{d^2 \mathbf{r}}{dt^2} = -\nabla V(\mathbf{r})$$

The total energy of the particle is defined as

$$E(t) = \frac{m}{2} | \frac{d\mathbf{r}(t)}{dt} |^2 + V(\mathbf{r}(t))$$

Its rate of change is

$$\frac{dE}{dt} = m(\frac{d\mathbf{r}}{dt}, \frac{d^2\mathbf{r}}{dt^2}) + (\nabla V(\mathbf{r}), \frac{d\mathbf{r}}{dt})$$

$$= (\frac{d\mathbf{r}}{dt}, m\frac{d^2\mathbf{r}}{dt^2} + \nabla V(\mathbf{r})) = 0$$

Here, the notation $(\mathbf{a}, \mathbf{b})$ stands for the scalar or dot product of the two vectors $\mathbf{a}, \mathbf{b}$. The fact that for motion in a potential, the total energy is conserved is called the equation of energy conservation. This holds good only if the potential does not explicitly depend upon time. When the potential has explicit time dependence, i.e. $\mathbf{F}(\mathbf{r}, t) = -\nabla V(\mathbf{r}, t)$, we get by an easy computation

$$\frac{dE}{dt} = \frac{\partial V}{\partial t}$$

The equation of energy conservation can be generalized to a system of particles moving in a potential field. By this we mean that the force on the ith particle has the form

$$\mathbf{F}_i(\mathbf{r}_1, ..., \mathbf{r}_n) = \nabla_i V(\mathbf{r}_1, ..., \mathbf{r}_n)$$

where $\mathbf{r}_i$ is the position of the ith particle and ∇_i denotes the gradient with respect to the coordinates of the ith particle $\mathbf{r}_i$. Specifically, writing $\mathbf{r}_i = x_i\hat{x} + y_i\hat{y} + z_i\hat{z}$, we have

$$\nabla_i V = \frac{\partial V}{\partial x_i}\hat{x} + \frac{\partial V}{\partial y_i}\hat{y} + \frac{\partial V}{\partial z_i}\hat{z}$$

and the equations of motion are

$$m_i\frac{d^2\mathbf{r}_i}{dt^2} = -\nabla_i V(\mathbf{r}_1, ..., \mathbf{r}_n)$$

Here, $\mathbf{r}_i's$ are functions of time. The energy of the system is defined as

$$E = \sum_{i=1}^{n} \frac{m_i}{2}|\frac{d\mathbf{r}_i}{dt}|^2 + V(\mathbf{r}_1, ..., \mathbf{r}_n)$$

Its time rate of change is given by

$$\frac{dE}{dt} = \sum m_i(\frac{d\mathbf{r}_i}{dt}, \frac{d^2\mathbf{r}_i}{dt^2}) + \sum_{i=1}^{n}(\frac{d\mathbf{r}_i}{dt}, \nabla_i V)$$

$$= \sum m_i(\frac{d\mathbf{r}_i}{dt}, \frac{d^2\mathbf{r}_i}{dt^2} + \nabla_i V) = 0$$

proving energy conservation. One of the important examples where angular momentum is conserved is when motion takes place in a central potential. If $V(r)$ denotes the potential, then the equation of motion of the particle is given by

$$m\frac{d^2\mathbf{r}}{dt^2} = -V'(r)\hat{r}$$

and we easily deduce that

$$\mathbf{r} \times \frac{d^2\mathbf{r}}{dt^2} = \mathbf{0}$$

or equivalently, $\mathbf{L} = \mathbf{r} \times \frac{d\mathbf{r}}{dt}$ is a constant. For this problem the energy conservation equation assumes the form

$$\frac{m}{2}|\frac{d\mathbf{r}}{dt}|^2 + V(r) = E$$

and when motion takes place in a plane, then the angular momentum and energy conservation equations assume the form

$$r^2\frac{d\phi}{dt} = \beta$$

$$\frac{m}{2}((\frac{dr}{dt})^2 + r^2(\frac{d\phi}{dt})^2) = E$$

where β is a constant equal to angular momentum per unit mass and E is a constant equal to the energy of the particle.

1.6 Examples of the solution of Newton's equations of motion

Example 1: The first example deals with motion in one dimension in a potential $V(x)$. The equation of motion reads

$$m\frac{d^2x}{dt^2} + V'(x) = 0$$

and a first integral of the motion is

$$\frac{m}{2}(\frac{dx}{dt})^2 + V(x) = E$$

This can be expressed in the form of an integral equation for the trajectory

$$t + C = \pm \int_0^x \frac{dx}{\sqrt{2(E - V(x))/m}}$$

where C is a constant.

Example 2: A special case of example 1 is motion of a mass connected to a spring obeying Hooke's law. The equation of motion is

$$m\frac{d^2x}{dt^2} + kx = 0$$

We write $k = m\omega^2$ so that

$$\frac{d^2x}{dt^2} + \omega^2 x = 0$$

which can be solved easily by setting $x(t) = exp(st)$. The two values of s resulting from substituting this into the differential equation are $s = \pm i\omega$. Thus, the general solution to the motion is

$$x(t) = A exp(i\omega t) + B exp(-i\omega t)$$

Since the solution must be real, we must have $B = \bar{A}$, i.e., B must equal the complex conjugate of A. Taking $A = C exp(i\phi)/2, B = C exp(-i\phi)/2$, with C and ϕ real gives us the general real solution

$$x(t) = C.cos(\omega t + \phi)$$

The corresponding velocity is

$$v(t) = \frac{dx}{dt} = -C\omega.sin(\omega t + \phi)$$

The equation of energy conservation reads

$$mv^2/2 + m\omega^2 x^2/2 = E$$

We find on substituting the solution that

$$E = mC^2\omega^2/2$$

or

$$C = \frac{1}{\omega}\sqrt{2E/m}$$

    Example 3: This example deals with motion of a particle in a uniform force field in one dimension. This is once again a special case of example 1 obtained by choosing $V(x) = -Fx$. Letting $a = F/m$, we find the equation of motion is

$$\frac{d^2x}{dt^2} = a$$

This motion is equivalent to uniformly accelerated motion. Integrating once gives

$$\frac{dx}{dt} = v(t) = at + C$$

If $v(t_0) = v_0$, then $v_0 = at_0 + C$ and hence

$$v(t) = v_0 + a(t - t_0)$$

another integration gives

$$x(t) = v_0 t + a(t - t_0)^2/2 + D$$

If $x(t_0) = x_0$, then

$$x_0 = v_0 t_0 + D$$

so that

$$x(t) = x_0 + v_0(t - t_0) + a(t - t_0)^2/2$$

This is the equation of the trajectory. One can choose the origin of time and position so that $x_0 = 0, t_0 = 0$. Then, we get the standard formulas

$$v(t) = v_0 + at, x(t) = v_0 t + at^2/2$$

By eliminating t between these two equations gives us a relationship between velocity and distance:

$$x = v_0(v - v_0)/a + (v - v_0)^2/2a$$

or

$$ax = v_0 v - v_0^2 + (v^2 + v_0^2 - 2vv_0)/2 = (v^2 - v_0^2)/2$$

or

$$v^2 - v_0^2 = 2ax$$

These formulas relating time, distance covered and velocity are usually dealt with in basic physics textbooks in school.

Example 4: Motion of a projectile in the uniform gravitational field of the earth. Let the x axis be horizontal and the y axis be vertical. The projectile is thrown with an initial velocity u at an angle θ to the horizontal from the origin. After time t, let $(x(t), y(t))$ denote the position of the projectile. The equations of motion are

$$\frac{d^2 x}{dt^2} = 0 \qquad \frac{d^2 y}{dt^2} = -g$$

Integration with appropriate initial conditions $x(0) = y(0) = 0, x'(0) = u.cos\theta, y'(0) = u.sin\theta$ gives

$$x(t) = u.cos\theta t, y(t) = u.sin(\theta)t - gt^2/2$$

Eliminating t between these two equations gives the equation of the trajectory

$$y = x.tan(\theta) - gx^2/2u^2 cos^2(\theta)$$

which is the equation of a parabola. The maximum height reached can be obtained by setting the derivative of $y(t)$ equal to zero. This gives

$$u.sin\theta - gt = 0$$

which means that the time at which the maximum height is reached is given by

$$t = u.sin(\theta)/g$$

The maximum height itself is

$$y_{max} = u.sin(\theta)u.sin(\theta)/g - gu^2 sin^2(\theta)/2g^2 = u^2 sin^2(\theta)/2g$$

The time at which the projectile hits the ground is obtained by setting $y(t) = 0$. This equation has two solutions, one is $t = 0$ which corresponds to the time of projection. The other is

$$t = 2u.sin(\theta)/g$$

and the horizontal distance covered is obtained by plugging in this expression for the time into the expression for $x(t)$:

$$x_{max} = u.cos(\theta).2u.sin(\theta)/g = u^2 sin(2\theta)/g$$

Example 5: Circular motion. Consider a particle moving in a circular orbit of radius R with uniform angular velocity. The position of the particle can be expressed as $\mathbf{r}(t) = R.cos(\omega t)\hat{x} + R.sin(\omega)\hat{y}$. The acceleration is obtained by taking the double time derivative of this:

$$\mathbf{a}(t) = \frac{d^2\mathbf{r}(t)}{dt^2} = -R\omega^2(\hat{x}.cos(\omega t) + \hat{y}.sin(\omega t))$$

It is clear from this expression that the acceleration is directed toward the centre. If m be the mass of the particle, then the force required to sustain the motion is given by $f = mR\omega^2$ and the direction of this force is towards the centre. For example if we tie a stone to a thread and rotate it in a circular orbit, then this centripetal force is provided by the tension in the string. Another example is a cyclist cycling on a circular velodrome. The floor of the velodrome makes an angle α to the horizontal. If the cyclist moves with an angular velocity ω on this velodrome, then there are two forces at play. One gravitational force and two the normal reaction of the slanting floor on the cyclist. Denoting the mass of the cyclist plus cycle by M and the normal reaction by N, we see that the centripetal force is $N.sin(\alpha)$. This must equal $MR\omega^2$. The Vertical force on the cyclist is $-Mg + N.cos\alpha$. This must equal zero. We thus obtain two equations $N.sin(\alpha) = MR\omega^2$, $N.cos(\alpha) = Mg$. Taking the ratio gives $tan(\alpha) = R\omega^2/g$. This determines the angular velocity required to be generated by the cyclist to remain in balance on the velodrome. Another example of circular motion is the same as discussed earlier, i.e., a stone tied to a thread but the thread makes an angle α with the vertical. Let L be the length of the thread and ω the angular velocity of rotation. Then, the radius of the orbit is $R = L.sin\alpha$. If T denotes the tension in the thread, then the centripetal force equals $T.sin\alpha$. This must equal $MR\omega^2$, i.e.. $T.sin\alpha = M.Lsin\alpha\omega^2$, or $T = ML.\omega^2$. The vertical component of the tension is balanced by the weight of the particle, i.e., $T.\cos\alpha = Mg$. Eliminating T between these two equations gives $\sec\alpha = L\omega^2/g$ and this equation determines the angular velocity required to maintain the particle in a circular orbit. As a final example of circular motion, we consider the motion of a satellite around the earth moving in a circular orbit of radius r. The centripetal force is provided by the gravitational attraction of the earth. The relevant equation is therefore $mr\omega^2 = GMm/r^2$ where m is the mass of the satellite and M the mass of the earth. This gives $\omega = \sqrt{GM/r^3}$ and for the time period, we obtain the expression $T = 2\pi/\omega = \sqrt{4\pi^2 r^3/GM}$ or equivalently, $T^2/r^3 = 4\pi^2/GM$, which is Kepler's third law of motion for circular orbits.

Example 6: Radial motion of a body in the inverse square attractive law. Consider a rocket moving radially away from the earth's surface upwards. If r is the distance from the earth's centre, then the equation of motion reads $\frac{d^2r}{dt^2} = -GM/r^2$ where M is the earth's mass. Multiplying by $\frac{dr}{dt}$ and integrating gives the energy conservation equation

$$\frac{1}{2}(dr/dt)^2 - GM/r = E$$

where $E = \frac{1}{2}v_0^2 - GM/R$ with R as the earth's radius and v_0 as the initial velocity of the rocket on the earth's surface. The escape velocity can be determined. This is the minimum initial velocity that would guarantee that the rocket escapes away to ∞. For this to happen, clearly $E = 0$ since the minimum requires $(dr/dt)_{r=\infty} = 0$ and $GM/r = 0$ when $r = \infty$. Thus, the escape velocity is given by $v_0 = \sqrt{2GM/R}$.

Example 7: Two body central force problem. Consider a body of mass m and a body of mass M moving in the mutual gravitational field. Let $\mathbf{r}_1$ denote the position of m and $\mathbf{r}_2$ the position of M after time t. The equations of motion are

$$\frac{d^2\mathbf{r}_1}{dt^2} = -\frac{GM(\mathbf{r}_1 - \mathbf{r}_2)}{|\mathbf{r}_1 - \mathbf{r}_2|^3}$$

$$\frac{d^2\mathbf{r}_2}{dt^2} = -\frac{Gm(\mathbf{r}_2 - \mathbf{r}_1)}{|\mathbf{r}_2 - \mathbf{r}_1|^3}$$

From these two we easily see that

$$m\frac{d^2\mathbf{r}_1}{dt^2} + M\frac{d^2\mathbf{r}_2}{dt^2} = \mathbf{0}$$

which implies

$$\mathbf{R}(t) = \mathbf{R}_0 + \mathbf{V}_0 t$$

where $\mathbf{R}(t) = (m\mathbf{r}_1 + M\mathbf{R}_2)/(m + M)$ is the position of the centre of mass. Here, $\mathbf{R}_0$ is the initial position of the center of mass and $\mathbf{V}_0$ is its initial velocity. Subtracting the two equations gives

$$\frac{d^2}{dt^2}(\mathbf{r}_1 - \mathbf{r}_2) = -G(M + m)(\mathbf{r}_1 - \mathbf{r}_2)/|\mathbf{r}_1 - \mathbf{r}_2|^3$$

We let $M_0 = M + m$, the total mass of the system. Then, the relative motion dynamics has the same law associated with motion about a fixed centre of mass M_0. Letting $\mathbf{r} = \mathbf{r}_1 - \mathbf{r}_2$ gives us

$$\frac{d^2\mathbf{r}}{dt^2} = -GM_0\mathbf{r}/r^3 = GM_0\nabla(1/r)$$

Taking the scalar product on both sides with $\frac{d\mathbf{r}}{dt}$ and integrating gives us the equation of energy conservation

$$\frac{1}{2}|\frac{d\mathbf{r}}{dt}|^2 - GM_0/r = E$$

or if we assume motion to take place in a plane, then we can introduce plane polar coordinates r, ϕ leading to the equation

$$\frac{1}{2}(r'^2 + r^2\phi'^2) - GM_0/r = E$$

where prime denotes derivative with respect to time. The equation of motion also leads to conservation of angular momentum as can be seen from

$$\frac{d}{dt}(\mathbf{r} \times \frac{d\mathbf{r}}{dt}) = \mathbf{r} \times \frac{d^2\mathbf{r}}{dt^2} = -GM_0\mathbf{r} \times \mathbf{r}/r^3 = \mathbf{0}$$

or

$$\mathbf{r} \times \frac{d\mathbf{r}}{dt} = \mathbf{L}$$

where $\mathbf{L}$ is a constant. This equation implies $\mathbf{L}.\mathbf{r} = 0$ which says that relative motion takes place in a plane whose normal is along the direction $\mathbf{L}$. This justifies our using of plane polar coordinates to describe the relative motion. For plane motion, we have

$$\mathbf{r} = r\hat{r}, \frac{d\mathbf{r}}{dt} = \frac{d}{dt}(r\hat{r}) = r'\hat{r} + r\phi'\hat{\phi}$$

so that

$$\mathbf{r} \times \frac{d\mathbf{r}}{dt} = r^2\phi'\hat{z}$$

and hence $r^2\phi' = \beta$ a constant where $\mathbf{L} = \beta\hat{z}$, i.e., the direction of $\mathbf{L}$ is along the $\hat{z}$ axis. From the energy and angular momentum conservation equations, we get

$$\frac{1}{2}((dr/d\phi)^2 + r^2) - GM_0r^3/\beta^2 = Er^4/\beta^2$$

which is the differential equation for the orbit. Setting $r = 1/u$ gives

$$(du/d\phi)^2 + u^2 - 2GM_0u/\beta^2 = 2E/\beta^2$$

Differentiating this equation once again gives

$$\frac{d^2u}{d\phi^2} + u - GM_0/\beta^2 = 0$$

whose solution is

$$u(\phi) = GM_0/\beta^2 + A.cos(\phi - \phi_0)$$

This is the equation of an ellipse in polar coordinates.

1.7   Motion under constraints

Suppose that a system of N particles has k degrees of freedom where $k \leq N$. An example where this occurs is that of a rigid body. A spinning top for example has only five degrees of freedom, two describe the location of the pivot on the plane and three specify the angle of rotation of the top. The three Euler angles may for example be used to describe the rotational degrees of freedom. The top itself however consist of an infinite number of particles. Another example is that of a classical model for a molecule consisting of two atoms connected to each other by a rigid rod. The motion of this molecule in three dimensional space can be described by means of three coordinates for the location of the centre of gravity of the molecule, two coordinates for the orientation of the molecule in space and one for the rotation of the molecule about its axis, making in all six degrees of freedom. If the rigid rod were removed, then three degrees of freedom for the translation of the first molecule, three degrees of freedom for the rotation of the first molecule and similarly for the second, making in all twelve degrees of freedom. So given a system of N particles with k degrees of freedom, we can express the positions of the N molecules as functions of k generalized coordinates, i.e.

$$\mathbf{r}_i = \mathbf{r}_i(t, q_1, ..., q_k), i = 1, 2, ..., N$$

If $\mathbf{F}_i$ is the force on the ith molecule, then the equations of motion can be formulated in the form of virtual work equals zero:

$$\sum_{i=1}^{N} \left(\mathbf{F}_i - \frac{d\mathbf{p}_i}{dt}, \delta \mathbf{r}_i \right) = 0$$

Note that

$$\delta \mathbf{r}_i = \sum_{j=1}^{k} \frac{\partial \mathbf{r}_i}{\partial q_j} \delta q_j$$

is a vector obtained by moving an infinitesimal distance on the k dimensional surface defined by the equations $\mathbf{r}_i = \mathbf{r}_i(t, q_1, ..., q_k), i = 1, 2, ..., N$. D'Alembert's principle states that the along the tangential directions to the surface, Newton's second law holds. Since the constraint forces are perpendicular to the surface, in order to validate Newton's law along the normal direction of the surface, we must add constraint forces to the external forces. D'Alembert's equation can be written in the form of k coupled differential equations

$$\sum_{i=1}^{N} \left(\frac{d\mathbf{p}_i}{dt}, \frac{\partial \mathbf{r}_i}{\partial q_j} \right) = \sum_{i=1}^{N} \left(\mathbf{F}_i, \frac{\partial \mathbf{r}_i}{\partial q_j} \right), j = 1, 2, ..., k$$

The momenta $\mathbf{p}_i$ of the particles can be evaluated as

$$\mathbf{p}_i = m_i \frac{d\mathbf{r}_i}{dt} = m_i \left(\frac{\partial \mathbf{r}_i}{\partial t} + \sum_{j=1}^{k} \frac{\partial \mathbf{r}_i}{\partial q_j} q_j' \right)$$

The rate of change of this momentum can also be evaluated as

$$\frac{d\mathbf{p}_i}{dt} = m_i \Big(\frac{\partial^2 \mathbf{r}_i}{\partial t^2} + 2 \sum_{j=1}^{k} \frac{\partial^2 \mathbf{r}_i}{\partial q_j \partial t} q_j'$$

$$+ \sum_{j,m=1}^{k} \frac{\partial^2 \mathbf{r}_i}{\partial q_j \partial q_m} q_j' q_m' + \sum_{j=1}^{k} \frac{\partial \mathbf{r}_i}{\partial q_j} q_j'' \Big)$$

Further, if the forces on the particle are derivable from a potential $V(\mathbf{r}_1, ..., \mathbf{r}_N)$, i.e., $\mathbf{F}_i = -\nabla_{\mathbf{r}_i} V$, then

$$\sum_{i=1}^{N} (\mathbf{F}_i, \frac{\partial \mathbf{r}_i}{\partial q_j}) = -\frac{\partial V}{\partial q_j}$$

as follows from the chain rule for partial derivatives. Substituting these into D' Alembert's differential equation gives us coupled ordinary differential equations for the generalized coordinates $\{q_i\}$. As a final example of constrained motion, we have the classic example of a pendulum. This consists of a bob tied to a string hung from a nail on the wall. If there were no thread, we would require two coordinates to describe the motion of the bob on the wall. The constraint force that keeps the pendulum in angular orbit is the tension in the string and this constraint reduces the number of degrees of freedom from two to one angle.

An example of constrained motion: Consider a particle moving on a two dimensional surface defined by the equation $z = f(x, y)$. At time t, the position of the particle is $(x(t), y(t), f(x(t), y(t)))$. The constraint force acts normal to the surface and in order to obtain the equations of motion, this constraint force must be eliminated. Let N denote the constraint force. The unit normal to the surface is given by

$$grad(z - f(x, y))/|grad(z - f(x, y))| = (-f_x, -f_y, 1)/\sqrt{1 + f_x^2 + f_y^2}$$

where

$$f_x = \frac{\partial f}{\partial x}, f_y = \frac{\partial f}{\partial y}$$

Thus, the equations of motion are

$$M \frac{d^2}{dt^2} (x(t), y(t), f(x(t), y(t))) = N'(-f_x, -f_y, 1) - Mg(0, 0, 1)$$

where

$$N' = N/\sqrt{1 + f_x^2 + f_y^2}$$

These are three ordinary differential equations from which N' can be eliminated to obtain two ordinary differential equations for the variables $(x(t), y(t))$. We leave the job of performing the computations to the student.

1.8   Euler-Lagrange equations of motion and variational calculus

Variational calculus deals with problems of extremizing functionals of curves. In ordinary calculus, we are used to dealing with problems of extremizing a functions of a finite number of variables $f(x_1, ..., x_n)$. The optimal values of the $x_i's$ that extremize this function are obtained as solutions to the system $\frac{\partial f}{\partial x_i} = 0, i = 1, 2, ..., n$. However suppose that F depends on a curve drawn between two fixed points (t_1, x_1) and (t_2, x_2). We can express F as $F[\xi(t) : t_1 \leq t \leq t_2]$. The curve is specified by a continuous infinity of points $(t, \xi(t)), t_1 \leq t \leq t_2$. In order to extremize this functional, i.e., determine the optimal curve that maximizes or minimizes the functional F, we must have the concept of partial derivative of F with respect to $\xi(t)$ for each $t \in [t_1, t_2]$. This is hard to define. An alternate route taught to us by Euler and Lagrange is to assume that $t \to \xi*(t)$ is the optimal curve that extremizes the functional. Then any small perturbation of this optimal curve will have the form $t \to \xi^*(t) + \epsilon\eta(t)$ and we can regard F as a function of the single real parameter ϵ:

$$\psi(\epsilon) = F[\xi^*(t) + \epsilon\eta(t) : t_1 \leq t \leq t_2]$$

The perturbation $\eta(t)$ must be such that it leaves the endpoints fixed since we are interested in extremizing the functional F over all curves that start from a fixed point (t_1, x_1) and terminate at the end point (t_2, x_2). This condition implies that $\eta(t_1) = \eta(t_2) = 0$. It is clear that ψ must attain an extremum at $\epsilon = 0$ for all such perturbations $t \to \eta(t)$, i.e., $\psi'(0) = 0$. In many problems one can derive from this condition a differential equation for the optimal trajectory $t \to \xi^*(t)$. The typical example involves extremizing the functional

$$S[x] = \int_{t_1}^{t_2} L(t, x(t), x'(t), ..., x^{(n)}(t))dt$$

subject to the terminal conditions that $x, x', x'', ..., x^{(n-1)}$ are fixed at $t = t_1$ and at $t = t_2$.

$$S[x + \epsilon\eta] = \int_{t_1}^{t_2} L(t, x(t) + \epsilon\eta(t), x'(t) + \epsilon\eta'(t), ..., x^{(n)}(t) + \epsilon\eta^{(n)}(t))dt$$

We are choosing the perturbing function η is such that $\eta(t), \eta'(t), ..., \eta^{(n-1)}(t)$ are zero at $t = t_1$ and at $t = t_2$. These correspond to the boundary conditions that $x(t), x'(t), ..., x^{(n-1)}(t)$ are fixed at $t = t_1, t = t_2$.

$$\frac{d}{d\epsilon}S[x + \epsilon\eta]|_{\epsilon=0}$$

$$= \int_{t_1}^{t_2} (\frac{\partial L}{\partial x}\eta + \frac{\partial L}{\partial x'}\eta' + ... + \frac{\partial L}{\partial x^{(n)}}\eta^{(n)})dt$$

Integrating by parts with the above mentioned boundary conditions gives us

$$\int_{t_1}^{t_2} (\frac{\partial L}{\partial x} - \frac{d}{dt}\frac{\partial L}{\partial x'} + \frac{d^2}{dt^2}\frac{\partial L}{\partial x''} + ... + (-1)^n\frac{\partial^n}{\partial t^n}\frac{\partial L}{\partial x^{(n)}})\eta dt = 0$$

Since η in the open interval (t_1, t_2) is arbitrary, we obtain the Euler-Lagrange equations in the form

$$\sum_{k=0}^{n} (-1)^k \frac{d^k}{dt^k} \frac{\partial L}{\partial x^{(k)}} = 0$$

A special case of maximum importance in mechanics is when $n = 1$, i.e. $L = L(t, x, x')$ depends only on time, $x(t)$ and its first derivative $x'(t)$. Then, the Euler-Lagrange equations simplify to

$$\frac{\partial L}{\partial x} - \frac{d}{dt} \frac{\partial L}{\partial x'} = 0$$

Suppose for example we take $L(t, x, x') = \frac{mx'^2}{2} - V(t, x)$, where $V(t, x)$ is the potential field in which the particle moves. The Euler-Lagrange equations for this function are obtained as follows:

$$\frac{\partial L}{\partial x'} = mx', \frac{d}{dt} \frac{\partial L}{\partial x'} = mx''$$

$$\frac{\partial L}{\partial x} = -\frac{\partial V}{\partial x}$$

yielding

$$mx'' = -\frac{\partial V}{\partial x}$$

which is precisely Newton's equation of motion in the potential V. Thus, Newton's laws of motion arise by extremizing the functional $\int_{t_1}^{t_2} (mx'^2/2 - V(t, x))dt$ in the time interval $[t_1, t_2]$ subject to the boundary conditions that $x(t_1)$ and $x(t_2)$ are fixed. This holds even for a system of interacting particles. For the system, we take

$$L(t, \mathbf{r}_1, \mathbf{r}_2, ..., \mathbf{r}_N, \mathbf{v}_1, ..., \mathbf{v}_N) = \sum_{i=1}^{N} \frac{1}{2} m_i |\mathbf{v}_i|^2 - V(t, \mathbf{r}_1, ..., \mathbf{r}_N)$$

Euler-Lagrange equations obtained by minimizing $S = \int_{t_1}^{t_2} L(t, \mathbf{r}_1, ..., \mathbf{r}_N, \mathbf{v}_1, ..., \mathbf{v}_N)dt$ with $\mathbf{v}_i(t) = \frac{d\mathbf{r}_i(t)}{dt}$ and terminal conditions that $\mathbf{r}_i(t_1)$ and $\mathbf{r}_i(t_2)$ are fixed are

$$\frac{d}{dt} \frac{\partial L}{\partial \mathbf{v}_i} = \frac{\partial L}{\partial \mathbf{r}_i}, i = 1, 2, ..., N$$

and these give

$$m_i \mathbf{v}_i'' = -\frac{\partial V}{\partial \mathbf{r}_i}, i = 1, 2, ..., N$$

which are precisely Newton's equations of motion. We give below three examples illustrating the Lagrangian method. The method consists of three steps: (1) Choose a set of generalized coordinates, (2) Express the kinetic and potential energies of the system in terms of these generalized coordinates, (3) Write down the Euler-Lagrange equations.

Example 1: Simple pendulum: Let L denote the length of the string and m the mass of the bob. Let θ denote the angle of deflection from the vertical. Then, the speed of the bob is $L\theta'$ and hence its kinetic energy is $\frac{mL^2\theta'^2}{2}$. The potential energy of the bob with respect to the point of suspension as the reference is $-mgL.cos\theta$. The Lagrangian function is the difference of these:

$$L(\theta, \theta') = \frac{mL^2\theta'^2}{2} + mgL.cos\theta$$

$$\frac{\partial L}{\partial \theta} = -mgL.sin\theta, \qquad \frac{\partial L}{\partial \theta'} = mL^2\theta'$$

and the Euler-Lagrange equation

$$\frac{d}{dt}\frac{\partial L}{\partial \theta'} = \frac{\partial L}{\partial \theta}$$

reads

$$\theta'' = -(g/L)sin\theta$$

Example 2: Particle moving on an inclined plane. Let α be the angle of the plane and x the distance of the particle along the slope from the bottom of the plane. The kinetic energy of the particle is given by $mx'^2/2$ and the potential energy by $mgx.sin\alpha$. The Lagrangian of the particle equals

$$L(x, x') = mx'^2/2 - mgx.sin\alpha$$

and Euler-Lagrange equation is

$$mx'' = -mg.sin\alpha$$

which is well known.

Example 3: Double pendulum. This consists of a bob of mass m_1 attached to a string of length L_1 whose other end is tied to a nail on the wall. Another bob of mass m_2 is attached to the first bob by a string of length L_2. This double pendulum oscillates in a plane and we want to determine the equations of motion. Let θ_1 denote the angle made by the top string with the vertical and θ_2 the angle made by the bottom string with the vertical. The position of the top bob in cartesian coordinates is given by $(L_1 sin\theta_1, L_1 cos\theta_1)$ and that of the bottom bob is given by $(L_1 sin\theta_1 + L_2 sin\theta_2, L_1 cos\theta_1 + L_2 cos\theta_2)$. The velocity of the top bob is then

$$\mathbf{v}_1 = \frac{d}{dt}(L_1 sin\theta_1, L_1 cos\theta_1) = L_1(cos(\theta_1)\theta_1', -sin(\theta_1)\theta_1')$$

and hence the kinetic energy of the top bob is

$$T_1 = \frac{m_1}{2}v_1^2 = \frac{m_1 L_1^2}{2}\theta_1'^2$$

The velocity of the bottom bob is

$$\mathbf{v}_2 = \frac{d}{dt}(L_1 sin\theta_1 + L_2 sin\theta_2, L_1 cos\theta_1 + L_2 cos\theta_2)$$

$$= (L_1 cos(\theta_1)\theta_1' + L_2 cos(\theta_2)\theta_2', -L_1 sin(\theta_1)\theta_1' - L_2 sin(\theta_2)\theta_2')$$

and hence the kinetic energy of the bottom bob is

$$T_2 = \frac{m_2 v_2^2}{2} = \frac{m_2}{2}(L_1^2\theta_1'^2 + L_2^2\theta_2'^2 + 2L_1 L_2\theta_1'\theta_2' cos(\theta_1 - \theta_2))$$

The potential energy of the top bob is

$$V_1 = -m_1 g L_1 cos(\theta_1)$$

and the potential energy of the bottom bob is

$$V_2 = -m_2 g(L_1 cos(\theta_1) + L_2 cos(\theta_2))$$

Our reference is chosen as the suspension point of the top string. The Lagrangian of the system is then

$$L(\theta_1, \theta_2, \theta_1', \theta_2') = T_1 + T_2 - V_1 - V_2$$

and this can be used to write down the equations of motion. As these examples illustrate, the Lagrangian approach directly furnishes us with the equations of motion avoiding the need to eliminate unknown forces of constraint like tension and normal reaction. If we were to write Newton's equations directly, constraint forces would creep into the equations and to eliminate them, additional algebra is required.

D'Alembert's Justification of the Lagrangian approach: Given a system whose dynamics is described by k generalized coordinates, we write down the expressions for the kinetic and potential energies in terms of these coordinates and set the Lagrangian equal to their difference. For this Lagrangian, we write down the Euler-Lagrange equations of motion and show that they coincide with D-Alembert's equations. This justifies the Lagrangian approach to all problems of classical mechanics. If $\mathbf{r}_i(t, q_1, ..., q_k), i = 1, 2, ..., N$ denote the positions of the particles, then the velocity vectors are given by

$$\frac{d\mathbf{r}_i}{dt} = \frac{\partial\mathbf{r}_i}{\partial t} + \sum_{i=1}^{k}\frac{\partial\mathbf{r}_i}{\partial q_k}q_k'$$

and the kinetic energy is given by

$$T(q_1, ..., q_k, q_1', ..., q_k') = \sum_{i=1}^{N}\frac{m_i}{2}|\frac{d\mathbf{r}_i}{dt}|^2$$

It is easy to show that T has the form

$$T_0(t, q_1, ..., q_k) + \sum_{j=1}^{k} a_j(t, q_1, ..., q_k) q_j' + \frac{1}{2} \sum_{j,m=1}^{k} b_{jm}(t, q_1, ..., q_m) q_j' q_m'$$

where

$$T_0(t, q_1, ..., q_k) = \sum_{i=1}^{N} \frac{m_i}{2} \left| \frac{\partial \mathbf{r}_i}{\partial t} \right|^2$$

$$a_j(t, q_1, ..., q_k) = \sum_{i=1}^{N} m_i \left(\frac{\partial \mathbf{r}_i}{\partial t}, \frac{\partial \mathbf{r}_i}{\partial q_j} \right)$$

$$b_{jm}(t, q_1, ..., q_k) = \sum_{i=1}^{N} m_i \left(\frac{\partial \mathbf{r}_i}{\partial q_j}, \frac{\partial \mathbf{r}_i}{\partial q_m} \right)$$

The potential $V(\mathbf{r}_1, ..., \mathbf{r}_N)$ is calculable directly from the physics of the arrangement. This furnishes us with V as a function of the $q_i's$:

$$V(\mathbf{r}_1(t, q_1, ..., q_k), ..., \mathbf{r}_N(t, q_1, ..., q_k))$$

1.9   Hamilton's equations of motion

Assume that a system having k degrees of freedom is described by the Lagrangian $L(q_1, ..., q_k, q_1', ..., q_k', t)$. Define

$$p_i = \frac{\partial L}{\partial q_i'}, i = 1, 2, ..., k$$

We can solve these equations for $q_1', ..., q_k'$ thereby expressing these variables as functions of $q_1, ..., q_k, p_1, ..., p_k, t$. Now introduce the function

$$H = \sum_{i=1}^{k} p_i q_i' - L$$

We can assume that H has been expressed as a function of $q_1, ..., q_k, p_1, ..., p_k, t$. We compute its total differential. On the one hand,

$$dH = \sum_{i=1}^{k} \left(\frac{\partial H}{\partial q_i} dq_i + \frac{\partial H}{\partial p_i} dp_i \right) + \frac{\partial H}{\partial t} dt$$

while on the other hand,

$$dH = d(\sum_i p_i q_i' - L) = \sum_i (q_i' dp_i + p_i dq_i') - dL$$

$$= \sum_i (q_i' dp_i + p_i dq_i') - \frac{\partial L}{\partial t} dt - \sum_{i=1}^{k} (\frac{\partial L}{\partial q_i} dq_i + \frac{\partial L}{\partial q_i'} dq_i')$$

Now make use of our definition of p_i:

$$p_i = \frac{\partial L}{\partial q_i'}$$

and also make use of the Euler-Lagrange equations

$$p_i' = \frac{d}{dt} \frac{\partial L}{\partial q_i'} = \frac{\partial L}{\partial q_i}$$

This gives

$$dH = \sum (q_i' dp_i + p_i dq_i') - \frac{\partial L}{\partial t} dt - \sum_i (p_i' dq_i + p_i dq_i') = \sum_i (q_i' dp_i - p_i' dq_i) - \frac{\partial L}{\partial t} dt$$

Comparing this expression for the total differential of H with that obtained earlier gives us

$$q_i' = \frac{\partial H}{\partial p_i}, p_i' = -\frac{\partial H}{\partial q_i}, \frac{\partial L}{\partial t} = -\frac{\partial H}{\partial t}$$

The first $2k$ equations define the dynamics of the variables $\{q_i, p_i : i = 1, 2, ..., k\}$. These form a set of coupled first order ordinary differential equations and completely specify the dynamics of the system. They are called the Hamilton equations of motion. A convenient way to express these is to define the state vector

$$\mathbf{x} = \left(\; q_1, ..., q_k, p_1, ..., p_k \; \right)$$

and the symplectic $2k \times 2k$ matrix

$$\mathbf{J} = \begin{pmatrix} \mathbf{0} & \mathbf{I} \\ -\mathbf{I} & \mathbf{0} \end{pmatrix}$$

where each identity is a $k \times k$ identity. Then, the Hamiltonian equations can be expressed as

$$\frac{d\mathbf{x}}{dt} = \mathbf{J} \nabla_{\mathbf{x}} H$$

The state of the system at all times is completely determined by the Hamiltonian equations and the state $\mathbf{x}(t_0)$ at any given time t_0. The transformation from the Lagrangian function to the Hamiltonian function is also called the Legendre transformation. The Lagrangian is expressed as a function of the generalized coordinates $\{q_i\}$, the generalized velocities $\{q_i'\}$ and time t while the Hamiltonian is expressed as a function of the generalized coordinates, $\{q_i\}$, the generalized momenta $\{p_i\}$ and time. In general, suppose we are given a function $F(x_1, ..., x_k, y_1, ..., y_k)$, we can define $z_i = \frac{\partial F}{\partial y_i}$ and solve these equations for $y_1, ..., y_k$ in terms of $x_1, ..., x_k, z_1, ..., z_k$ and define $G(x_1, ..., x_k, z_1, ..., z_k) = \sum z_i y_i - F$. This transformation from the independent variables $\{(x_i, y_i)\}$ to the independent variables $\{(x_i, z_i)\}$ accompanied by the corresponding transformation from the function F to the function G is called the Legendre transformation. The Lagrangian-Hamiltonian transformation is a special instance of this. Another important example comes from thermodynamics. The differential change in the internal energy $dU = TdS - pdV$ is expressed in terms of the independent variables V, S. We therefore write $U(S, V)$. To effect the Legendre transformation, we use $-p = \frac{\partial U}{\partial V}$ and define $H = U - (-p)V = U + pV$. H is called the enthalpy of the system. The independent variables for describing H are p, S. Its total differential is given by $dH = dU + pdV + Vdp = TdS + Vdp$. We next evaluate the total time rate of change of the Hamiltonian of the system:

$$\frac{dH}{dt} = \frac{\partial H}{\partial t} + \sum \frac{\partial H}{\partial q_i} q_i' + \frac{\partial H}{\partial p_i} p_i'$$

$$= \frac{\partial H}{\partial t} + \sum \frac{\partial H}{\partial q_i} \frac{\partial H}{\partial p_i} - \frac{\partial H}{\partial p_i} \frac{\partial H}{\partial q_i}$$

$$= \frac{\partial H}{\partial t} = -\frac{\partial L}{\partial t}$$

When the Lagrangian does not depend explicitly on time, the Hamiltonian also does not depend explicitly on time and then $\frac{dH}{dt} = 0$, i.e., the Hamiltonian is a conserved quantity. It turns out that H corresponds to the energy of the system and hence we have the law of conservation of energy in its most general form. For example, consider a system of N particles moving in a potential field $V(q_1, ..., q_k)$. The Lagrangian of the system is

$$L = \frac{1}{2} \sum_{i,j=1}^{k} a_{ij}(q_1, ..., q_k) q_i' q_j' - V(q_1, ..., q_k)$$

where $((a_{ij}))$ can be taken to be a symmetrix matrix. The generalized momenta are given by

$$p_i = \frac{\partial L}{\partial q_i'} = \sum_{j=1}^{k} a_{ij}(\mathbf{q}) q_j'$$

If $b_{ij}(\mathbf{q})$ denotes the inverse of the matrix $a_{ij}(\mathbf{q})$, then the above equations can be inverted to give

$$q_i' = \sum_{j=1}^{k} b_{ij}(\mathbf{q}) p_j$$

$$H = \sum b_{ij} p_i p_j - L$$

Noting that

$$\sum_{i,j} a_{ij} q_i' q_j' = \sum_{i,j,m,r} a_{ij} b_{im} p_m b_{jr} p_r = \sum_r b_{mr} p_r$$

we have

$$H = \frac{1}{2} \sum b_{ij} p_i p_j + V$$

In intuitive terms, $((b_{ij}))$ can be regarded as a mass matrix. If the generalized coordinates correspond to ordinary cartesian coordinates, so that $k = 3N, (q_1, ..., q_k) = (x_1, y_1, z_1, ..., x_N, y_N, z_N)$, then

$$L = \frac{1}{2} \sum m_i (x_i'^2 + y_i'^2 + z_i'^2) - V(x_1, y_1, z_1, ..., x_N, y_N, z_N)$$

then the inverse mass matrix $((a_{ij}))$ is $[1/m_1, 1/m_1, 1/m_1, .., 1/m_N, 1/m_N, 1/m_N]$, then the mass matrix $((b_{ij}))$ has the form $[m_1, m_1, m_1, ..., m_N, m_N, m_N]$ and the generalized momenta are given by $p_{x_i} = m_i x_i', p_{y_1} = m_i y_i', p_{z_i} = m_i z_i'$. Then,

$$H = \frac{1}{2} \sum \frac{1}{m_i} (p_{x_i}^2 + p_{y_i}^2 + p_{z_i}^2) + V$$

The $\{(p_{x_i}, p_{y_i}, p_{z_i})\}$ are the generalized momenta. As another example, consider a simple pendulum. Its Lagrangian function is given by

$$L(\theta, \theta') = \frac{mL^2 \theta'^2}{2} + mgL\cos\theta$$

The angle θ is the only generalized coordinate and the corresponding generalized momentum equals

$$p_\theta = \frac{\partial L}{\partial \theta'} = mL^2 \theta'$$

Thus,

$$H(\theta, p_\theta) = mL^2 \theta'^2 - mL^2 \theta'^2/2 - mgL\cos\theta$$

$$= mL^2 \theta'^2/2 - mgL\cos\theta = p_\theta^2/2mL^2 - mgL\cos\theta$$

A final remark can be made by saying that the Legendre transformation corresponding to the transition from the Lagrangian to the Hamiltonian system corresponds to changing our description of the system from a system of k coupled second order ordinary differential equations to a system of $2k$ coupled first order ordinary differential equations. Newton's equations of motion are naturally second order in time and the Hamiltonian approach transforms these to a system of first order equations by increasing the number of variables two fold.

Lagrangian approach for fields: Let S be a surface in N dimensional space and let $\phi_1(\mathbf{x}), ...,$ $\phi_m(\mathbf{x})$ be well behaved functions defined on the volume V enclosed by S. Each point $\mathbf{x} \in V$ is specified by N coordinates $(x_1, ..., x_N)$. Suppose L is a function of these fields and their first order partial derivatives and the point, i.e., $L(\phi_1, ..., \phi_m, \phi_{1,i}, ..., \phi_{m,i}, i = 1, 2, ..., N, \mathbf{x})$. We want to extremize the functional

$$S[\phi_1, ..., \phi_m] = \int_N L(\phi_1(\mathbf{x}), ..., \phi_m(\mathbf{x}), \phi_{1,i}(\mathbf{x}), ..., \phi_{m,i}(\mathbf{x}), \mathbf{x}) d^N\mathbf{x}$$

subject to the boundary condition that $\phi_1, ..., \phi_m$ are specified on S. This problem is easily seen to be a generalization of the Lagrangian problem earlier for $N = 1$. There, the volume V was replaced by an interval $[t_1, t_2]$ in $\mathbb{R}$ with boundary being the two endpoints $\{t_1, t_2\}$. To carry out the variation, we assume that $\phi_1, ..., \phi_m$ are the optimal extremizing functions. Then, consider $S[\phi_1 + \epsilon\eta_1, ..., \phi_m + \epsilon\eta_m]$ with ϵ a real parameter and $\eta_1(\mathbf{x}), ..., \eta_m(\mathbf{x})$ are perturbing functions that vanish on the boundary S to guarantee that the perturbed functions $\phi_k + \epsilon\eta_k, k = 1, 2, ..., m$ also satisfy the boundary conditions. The condition of optimality is defined by the fact that the function

$$\psi(\epsilon) = S[\phi_1 + \epsilon\eta_1, ..., \phi_m + \epsilon\eta_m]$$

attains an extremum when $\epsilon = 0$ no matter what the perturbing functions $\eta_k, k = 1, 2, ..., m$ are, i.e.,

$$\frac{d}{d\epsilon}\psi(\epsilon)|_{\epsilon=0} = 0$$

1.10   Vibrating string and vibrating membrane

These problems are discussed in the section on waves.

1.11   Poisson brackets

Let $u(q_1, ..., q_n, p_1, ..., p_n)$ and $v(q_1, ..., q_n, p_1, ..., p_n)$ be two observables, i.e., functions of canonical coordinates and momenta. Their Poisson bracket is defined as

$$\{u, v\} = \sum_{i=1}^{n} \frac{\partial u}{\partial q_i}\frac{\partial v}{\partial p_i} - \frac{\partial u}{\partial p_i}\frac{\partial v}{\partial q_i}$$

It satisfies three important properties: (a) $\{au + bv, w\} = a\{u, w\} + b\{v, w\}$ where u, v, w are observables and a, b are two constants, (b) $\{u, v\} = -\{v, u\}$ and (c) the Jacobi identity:

$$\{u, \{v, w\}\} + \{v, \{w, u\}\} + \{w, \{u, v\}\} = 0$$

These properties enable one to compare the Poisson bracket with the Lie bracket that appears in quantum mechanics. If U, V are two linear operators on Hilbert space, their Lie bracket is defined by $[U, V] = UV - VU$. The Lie bracket satisfies all the three properties satisfied by the Poisson bracket. The Hamiltonian equations of motion can be expressed using the Poisson bracket as

$$\frac{dq_i}{dt} = \{q_i, H\}, \frac{dp_i}{dt} = \{p_i, H\}$$

More generally, let u be any observable. It can also depend explicitly on time, i.e., $u(t, q_1, ..., q_n, p_1, ..., p_n)$. The rate of change of this observable is given by

$$\frac{du}{dt} = \frac{\partial u}{\partial t} + \sum_{i=1}^{n} \frac{\partial u}{\partial q_i} \frac{dq_i}{dt} + \frac{\partial u}{\partial p_i} \frac{dp_i}{dt}$$

$$= \frac{\partial u}{\partial t} + \sum_{i=1}^{n} \frac{\partial u}{\partial q_i} \frac{\partial H}{\partial p_i} - \frac{\partial u}{\partial p_i} \frac{\partial H}{\partial q_i}$$

$$= \frac{\partial u}{\partial t} + \{u, H\}$$

In particular, taking $u = H$ gives

$$\frac{dH}{dt} = \frac{\partial H}{\partial t}$$

so that if H does not depend explicitly on time, it must be a constant of the motion.

Canonical transformations: Suppose the Hamiltonian of a system is not a function of any of the coordinates, i.e., $H = H(p_1, ..., p_n)$. Then, Hamilton's equations imply $\frac{dp_i}{dt} = 0$ and hence the $p_i's$ are constants of the motion. We set $p_i(t) = \alpha_i$. Hamilton's equations for the $q_i's$ are $\frac{dq_i}{dt} = \omega_i = \frac{\partial H(\alpha_1,...,\alpha_n)}{\partial \alpha_i}$. The $\omega_i's$ are evidently constants of the motion. The solution for the $q_i's$ is $q_i(t) = \omega_i t + \beta_i$. Rarely does it happen that all the coordinates are cyclic but it often happens that one or more of the coordinates is cyclic. By an appropriate transformation of the generalized coordinates and momenta accompanied with a corresponding transformation of the Hamiltonian, some coordinates become cyclic thereby permitting an easy description of the motion. For example, consider motion in a central force defined by the potential $V(r)$ in two dimensions. The Lagrangian is given by

$$L(r, r', \theta, \theta') = \frac{m}{2}(r'^2 + r^2\theta'^2) - V(r)$$

and the generalized momenta are

$$p_r = mr', p_\theta = mr^2\theta'$$

so that the Hamiltonian is

$$H(r, \theta, p_r, p_\theta) = p_r^2/2m + p_\theta^2/2mr^2 + V(r)$$

This Hamiltonian is evidently cyclic in the coordinate θ and this cyclicity implies conservation of angular momentum. On the other hand, if we used cartesian coordinates, the Hamiltonian would assume the form

$$H(x, y, p_x, p_y) = p_x^2/2m + p_y^2/2m + V(\sqrt{x^2 + y^2})$$

which would not be cyclic in any coordinate. This example shows that by adopting an appropriate transformation of the coordinates, we can arrive at equations of motion that are easily integrable. This transformation however involves only transformations of the coordinates, i.e., transformations of the form $Q_i = Q_i(q_1, ..., q_n, t)$. Such transformations are known as point transformations. One can consider more generally, transformations of both the coordinates and momenta, i.e., transformations having the general form

$$Q_i = Q_i(q_1, ..., q_n, p_1, ..., p_n, t) \qquad P_i = P_i(q_1, ..., q_n, p_1, ..., p_n, t)$$

This transformation will be called a canonical transformation if there exists a function $K(Q_1, ..., Q_n, P_1, ..., P_n, t)$ such that the equations of motion of the transformed coordinates and momenta are given by

$$\frac{dQ_i}{dt} = \frac{\partial K}{\partial P_i} \qquad \frac{dP_i}{dt} = -\frac{\partial K}{\partial Q_i}$$

The original Lagrangian is given by

$$L = \sum p_i q_i' - H$$

and hence the original equations of motion can be formulated in variational form as

$$\delta \int_{t_1}^{t_2} (\sum p_i q_i' - H) dt = 0$$

If the above transformation is canonical with K as the new Hamiltonian, then the transformed Lagrangian is given by

$$L' = \sum P_i Q_i' - K$$

and hence the new equations of motion are obtained from the variational principle

$$\delta \int_{t_1}^{t_2} (\sum P_i Q_i' - K) dt = 0$$

These equations of motion must coincide with the old equations by our definition of canonical transformation and one way in which this can happen is that the new Lagrangian should differ from a constant multiple of the old one by a total time derivative, i.e.,

$$\lambda(\sum p_i q_i' - H) = \sum P_i Q_i' - K + \frac{dF}{dt}$$

However we can always change the scale of the original coordinates and Hamiltonian in the following way. Define $\tilde{p}_i = ap_i, \tilde{q}_i = bq_i, \tilde{H} = cH$. Then,

$$\frac{\tilde{q}_i}{dt} = b\frac{dq_i}{dt} = b\frac{\partial H}{\partial p_i} = \frac{ab}{c}\frac{\partial \tilde{H}}{\partial \tilde{p}_i}$$

$$\frac{d\tilde{p}_i}{dt} = a\frac{dp_i}{dt} = -a\frac{\partial H}{\partial q_i} = -\frac{ab}{c}\frac{\partial \tilde{H}}{\partial \tilde{q}_i}$$

We choose the constants a, b, c so that $c = \lambda, ab = c$. Then, $(\tilde{q}_i, \tilde{p}_i, \tilde{H})$ form a canonical set and

$$\lambda(\sum p_i q_i' - H) = \sum \tilde{p}_i \tilde{q}_i' - \tilde{H}$$

which means that without loss of generality, we can assume the constant multiplier λ to equal unity. Thus, the condition for the transition from the triplet (q_i, p_i, H) to the triplet (Q_i, P_i, K) to be canonical is that there exist a function $F(q_1, ..., q_n, p_1, ..., p_n, t)$ such that

$$\sum p_i q_i' - H = \sum P_i Q_i' - K + \frac{dF}{dt}$$

The function F is called the generating function of the canonical transformation. We can instead of regarding $(q_i, p_i), i = 1, 2, ..., n$ as the independent variables on which F depends, equivalently regard any one of $\{(q_i, P_i), i = 1, 2, ..., n\}$, $\{(Q_i, p_i), i = 1, 2, ..., n\}$, $\{(Q_i, P_i), i = 1, 2, ..., n\}$, $\{(p_i, P_i), i = 1, 2, ..., n\}$ as the independent variables. For example taking F to be a function of q_i, Q_i, i.e., $F(q, Q, t)$ using the abbreviated notation q in place of q_i and Q in place of Q_i, we get

$$\sum p_i q_i' - H = \sum P_i Q_i' - K + \frac{\partial F}{\partial t} + \sum \frac{\partial F}{\partial q_i} q_i' + \frac{\partial F}{\partial Q_i} Q_i'$$

and comparing coefficients gives us

$$p_i = \frac{\partial F}{\partial q_i} \qquad P_i = -\frac{\partial F}{\partial Q_i}, \qquad K = H + \frac{\partial F}{\partial t}$$

Suppose on the other hand, we take our generating function as $F = F_2(q, P, t) - \sum_i Q_i P_i$. Substituting this into the canonical condition gives us

$$\sum p_i q_i' - H = \sum P_i Q_i' - K + \frac{d}{dt}(F_2 - \sum Q_i P_i)$$

so that

$$\sum p_i q_i' - H = -\sum Q_i P_i' - K + \frac{\partial F_2}{\partial t} + \sum \frac{\partial F_2}{\partial q_i} q_i' + \frac{\partial F_2}{\partial P_i} P_i'$$

yielding

$$p_i = \frac{\partial F_2}{\partial q_i} \qquad Q_i = \frac{\partial F_2}{\partial P_i} \qquad K = H + \frac{\partial F_2}{\partial t}$$

We can also consider generating functions expressed as functions of p, Q and p, P. For example writing the canonical condition as

$$\sum p_i q_i' - H = \frac{d}{dt}\left(\sum p_i q_i\right) - \sum q_i p_i' - H = \sum P_i Q_i' - K + \frac{dF}{dt}$$

so that

$$-\sum q_i p_i' - H = \sum P_i Q_i' - K + \frac{d}{dt}\left(F - \sum p_i q_i\right)$$

we set

$$F = \sum p_i q_i + F_3(p, Q, t)$$

and get

$$-\sum q_i p_i' - H = \sum P_i Q_i' - K + \frac{dF_3}{dt}$$

yielding

$$-q_i = \frac{\partial F_3}{\partial p_i} \qquad P_i = \frac{\partial F_3}{\partial Q_i} \qquad K = H + \frac{\partial F_3}{\partial t}$$

Similarly, the canonical condition can be expressed as

$$\frac{d}{dt}\left(\sum_i (p_i q_i - P_i Q_i)\right) - \sum q_i p_i' - H = -\sum Q_i P_i' - K + \frac{dF}{dt}$$

or

$$-\sum q_i p_i' - H = -\sum Q_i P_i' - K + \frac{d}{dt}\left(F - \sum_i (p_i q_i - P_i Q_i)\right)$$

so that taking

$$F = F_4(p, P, t) + \sum_i (p_i q_i - P_i Q_i)$$

gives us

$$-q_i = \frac{\partial F_4}{\partial p_i} \qquad Q_i = \frac{\partial F_4}{\partial P_i} \qquad K = H + \frac{\partial F_4}{\partial t}$$

There is another approach to canonical transformations, namely, the symplectic approach. Consider a canonical transformation $Q_i(q, p), P_i(q, p)$ which for simplicity, we assume to be time independent. The time independence implies that the Hamiltonian is an invariant (The fact that the functions Q_i, P_i do not depend explicitly on time implies that the generating function F does not depend explicitly on time and that implies $K = H + \frac{\partial F}{\partial t} = H$). The equations of motion of the transformed coordinates and momenta are given by

$$Q_i' = \frac{\partial Q_i}{\partial q_j} q_j' + \frac{\partial Q_i}{\partial p_j} p_j'$$

$$= \frac{\partial Q_i}{\partial q_j}\frac{\partial H}{\partial p_j} - \frac{\partial Q_i}{\partial p_j}\frac{\partial H}{\partial q_j}$$

Similarly,

$$P_i' = \frac{\partial P_i}{\partial q_j}q_j' + \frac{\partial P_i}{\partial p_j}p_j'$$

$$= \frac{\partial P_i}{\partial q_j}\frac{\partial H}{\partial p_j} - \frac{\partial P_i}{\partial p_j}\frac{\partial H}{\partial q_j}$$

Of course, these equations are simply the Poisson bracket relations for the rate of change of an observable. Summation over the repeated index j is assumed (the Einstein convention). Regarding H as a function of Q_i, P_i gives us

$$\frac{\partial H}{\partial Q_i} = \frac{\partial H}{\partial q_j}\frac{\partial q_j}{\partial Q_i} + \frac{\partial H}{\partial p_j}\frac{\partial p_j}{\partial Q_i}$$

$$\frac{\partial H}{\partial P_i} = \frac{\partial H}{\partial q_j}\frac{\partial q_j}{\partial P_i} + \frac{\partial H}{\partial p_j}\frac{\partial p_j}{\partial P_i}$$

The condition for the transformation from the q_i, p_i set to the Q_i, P_i set to be canonical is that

$$Q_i' = \frac{\partial H}{\partial P_i} \qquad P_i' = -\frac{\partial H}{\partial Q_i}$$

Substituting the above equations gives

$$\frac{\partial Q_i}{\partial q_j}\frac{\partial H}{\partial p_j} - \frac{\partial Q_i}{\partial p_j}\frac{\partial H}{\partial q_j}$$

$$= \frac{\partial H}{\partial P_i} = \frac{\partial H}{\partial q_j}\frac{\partial q_j}{\partial P_i} + \frac{\partial H}{\partial p_j}\frac{\partial p_j}{\partial P_i}$$

and

$$\frac{\partial P_i}{\partial q_j}\frac{\partial H}{\partial p_j} - \frac{\partial P_i}{\partial p_j}\frac{\partial H}{\partial q_j}$$

$$-\frac{\partial H}{\partial Q_i} = -\left(\frac{\partial H}{\partial q_j}\frac{\partial q_j}{\partial Q_i} + \frac{\partial H}{\partial p_j}\frac{\partial p_j}{\partial Q_i}\right)$$

These equations can be rearranged as

$$\frac{\partial H}{\partial q_j}\left(\frac{\partial q_j}{\partial P_i} + \frac{\partial Q_i}{\partial p_j}\right) + \frac{\partial H}{\partial p_j}\left(\frac{\partial p_j}{\partial P_i} - \frac{\partial Q_i}{\partial q_j}\right) = 0$$

and

$$\frac{\partial H}{\partial q_j}\left(\frac{\partial q_j}{\partial Q_i} - \frac{\partial P_i}{\partial p_j}\right) + \frac{\partial H}{\partial p_j}\left(\frac{\partial p_j}{\partial Q_i} + \frac{\partial P_i}{\partial q_j}\right) = 0$$

We can infer that the following necessary and sufficient conditions for the transformation to be canonical:

$$\frac{\partial q_j}{\partial P_i} = -frac\partial Q_i\partial p_j$$

$$\frac{\partial p_j}{\partial P_i} = \frac{\partial Q_i}{\partial q_j}$$

$$\frac{\partial q_j}{\partial Q_i} = \frac{\partial P_i}{\partial p_j}$$

$$\frac{\partial p_j}{\partial Q_i} = -\frac{\partial P_i}{\partial q_j}$$

These equations are known as the direct conditions in comparison to the generator formalism which are the indirect conditions.

Hamilton-Jacobi theory: Here, the generating function $F(q, P, t)$ is selected so that the transformed Hamiltonian $K = H + \frac{\partial F}{\partial t}$ is zero. Then, the new coordinates and momenta Q_i, P_i become constants of the motion and by inverting the transformation, we obtain the solution to the equations of motion. The generating function F is also denoted by S. We have

$$\sum p_i q_i' - H = \sum P_i Q_i' - K + \frac{dS}{dt} = \sum P_i Q_i' - K + \sum \frac{\partial S}{\partial q_i} q_i' + \frac{\partial S}{\partial P_i} P_i' + \frac{\partial S}{\partial t}$$

from which we deduce that

$$p_i = \frac{\partial S}{\partial q_i}$$

Thus, $S(q, P, t)$ is a function of the old coordinates q and the new momenta P which are constants of the motion and satisfies the partial differential equation

$$H(q_1, ..., q_n, \frac{\partial S}{\partial q_1}, ..., \frac{\partial S}{\partial q_n}, t) + \frac{\partial S}{\partial t} = 0$$

The solution to this partial differential equation S will involve $n + 1$ independent constants of which n can be taken as the new momenta P_i. The other constant is an additive constant and does not play any role since the canonical transformations linking the old to the new coordinates and momenta involve derivatives of the generating function and the additive constant disappears on taking the derivatives. Having obtained the generating function, we use the equations

$$p_i = \frac{\partial S(q, P)}{\partial q_i}, Q_i = \frac{\partial S(q, P)}{\partial P_i}$$

to solve for q_i, p_i in terms of the constants P_i, Q_i. The P_i appear as constants in S and the Q_i can be taken as arbitrary. The final solution gives the time functions $q_i(t), p_i(t)$ in terms of the $2n$ constants Q_i, P_i which are obtained from the initial conditions for the problem. S is called

Hamilton's principal function and the above partial differential equation satisfied by S is called the Hamilton-Jacobi equation. The solution to the Hamilton-Jacobi partial differential equation is completely equivalent to the solution of the Euler-Lagrange's differential equation or equivalently, to the Hamilton canonical differential equations of motion. As an example, consider the one dimensional Harmonic oscillator. The Hamiltonian is given by

$$H(q,p) = \frac{p^2}{2m} + \frac{m\omega^2 q^2}{2}$$

The Hamilton-Jacobi equation for this problem is given by

$$\frac{1}{2m}\left(\frac{\partial S}{\partial q}\right)^2 + \frac{m\omega^2 q^2}{2} + \frac{\partial S}{\partial t} = 0$$

We write

$$S(q,t) = W(q) - \alpha t$$

Substituting this into the partial differential equation gives

$$\frac{1}{2m}W'(q)^2 + \frac{m\omega^2 q^2}{2} - \alpha = 0$$

or

$$W'(q) = \sqrt{2m(\alpha - m\omega^2 q^2/2)} = m\omega\sqrt{2\alpha/m\omega^2 - q^2}$$

The integral

$$\int \sqrt{a^2 - q^2}\, dq$$

evaluates to

$$\frac{a^2}{2}sin^{-1}(q/a) + \frac{q}{2}\sqrt{a^2 - q^2}$$

This gives

$$W(q) = (\alpha/\omega)sin^{-1}(q\omega\sqrt{m/2\alpha}) + (m\omega q/2)\sqrt{2\alpha/m\omega^2 - q^2}$$

We take the constant α as our new momentum P. Then,

$$p = \frac{\partial S}{\partial q} = m\omega\sqrt{2\alpha/m\omega^2 - q^2}$$

$$Q = \beta = \frac{\partial S}{\partial P} = \frac{\partial S}{\partial \alpha}$$

$$= \frac{1}{\omega}sin^{-1}(q\omega\sqrt{m/2\alpha}) - t$$

Inverting this equation gives

$$q(t) = \frac{1}{\omega}\sqrt{2\alpha/m}sin(\omega(t + Q))$$

Plugging this expression into the expression for p gives

$$p(t) = \sqrt{2\alpha m}\cos(\omega(t+Q))$$

We are now in a position to obtain the physical meanings of the constants $Q, P = \alpha$. The expression for $q(t)$ shows that $-Q$ is the initial delay, or equivalently, $\phi = \omega Q$ is the initial phase. Since

$$E = p^2/2m + m\omega^2 q^2/2 = \alpha$$

it follows that α is the total energy of the oscillator. This compeletes the solution to the oscillator using the Hamilton-Jacobi method.

1.12   Rigid body motion

Motion of a spinning top. The analysis of the rigid body motion starts with the introduction of the Euler angles into the description of rotations. Any rotation of space can be expressed as $R_z(\phi)R_x(\theta)R_z(\psi)$, where $R_n(\alpha)$ denotes rotation about the direction $\mathbf{n}$ by the angle α. Here, ψ, θ, ϕ are termed as the Euler angles. Suppose we are given a heavy symmetric top with the bottom point fixed at the origin and the top rotating about this pivot. We choose three mutually orthogonal axes on the top, one of them, the z' axis coinciding with the top's axis and the two remaining x', y' axes perpendicular to the z' axis. The axes (x', y', z') constitute the body set of axes in contrast to the (x, y, z) axes which is the space set. After time t, assume that a rotation described by the Euler angles $\theta(t), \phi(t), \psi(t)$ has been applied to the space set yielding the body set of axes. This means that the top has undergone a rotation

$$R(\phi(t), \theta(t), \psi(t)) = R_z(\phi(t))R_x(\theta(t))R_z(\psi(t))$$

Any point $\mathbf{r}$ at time $t = 0$ fixed to the top undergoes the above rotation taking it to the new point

$$\mathbf{r}(t) = R_z(\phi(t))R_x(\theta(t))R_z(\psi(t))\mathbf{r}$$

For convenience we set

$$R(t) = R(\phi(t), \theta(t), \psi(t))$$

$R(t)$ is a rotation matrix valued function of time and the problem of describing the top's motion amounts to determining this function or equivalently the Euler angles as functions of time. The velocity of the point $\mathbf{r}(t)$ fixed to the top is given by

$$\mathbf{v}(t) = \frac{dR(t)}{dt}\mathbf{r} = \Omega(t)\mathbf{r}(t)$$

where

$$\Omega(t) = \frac{dR(t)}{dt}R(t)^{-1}$$

is the angular velocity tensor. $\Omega(t)$ is an antisymmetric matrix having the form

$$\Omega(t) = \begin{pmatrix} 0 & -\omega_z(t) & \omega_y(t) \\ \omega_z(t) & 0 & -\omega_x(t) \\ -\omega_y(t) & \omega_x(t) & 0 \end{pmatrix}$$

Given any rigid body, let its angular velocity pseudovector be given by $\omega(t) = (\omega_x(t), \omega_y(t), \omega_z(t))$. The velocity of any point $\mathbf{r}$ on the body undergoing rotation along with it is given by the usual formula

$$\mathbf{v}(t) = \omega(t) \times \mathbf{r}(t)$$

If we denote the mass of the ith particle on the rigid body by m_i, then the kinetic energy of the body is given by

$$T = \sum \frac{m_i}{2} |\omega \times \mathbf{r}_i|^2 = \frac{1}{2} \sum m_i(\omega^2 r_i^2 - (\omega.\mathbf{r}_i)^2)$$

$$= \frac{1}{2} \sum m_i((\omega_x^2 + \omega_y^2 + \omega_z^2)(x_i^2 + y_i^2 + z_i^2) - (\omega_x x_i + \omega_y y_i + \omega_z z_i)^2)$$

$$= \frac{1}{2} \sum m_i(\omega_x^2(y_i^2 + z_i^2) + \omega_y^2(z_i^2 + x_i^2) + \omega_z^2(x_i^2 + y_i^2) - 2\omega_x \omega_y x_i y_i - 2\omega_y \omega_z y_i z_i - 2\omega_z z_i x_i)$$

$$= \frac{1}{2}\omega^T I \omega$$

where I is the moment of inertia tensor of the body defined by the symmetric matrix

$$I = \begin{pmatrix} I_{xx} & I_{xy} & I_{xz} \\ I_{yz} & I_{yy} & I_{yz} \\ I_{zx} & I_{zy} & I_{zz} \end{pmatrix}$$

where

$$I_{xx} = \sum m_i(y_i^2 + z_i^2) \qquad I_{yy} = \sum m_i(z_i^2 + x_i^2) \qquad I_{zz} = \sum m_i(x_i^2 + y_i^2)$$

$$I_{xy} = I_{yx} = -\sum m_i x_i y_i \qquad I_{yz} = I_{zy} = -\sum m_i y_i z_i \qquad I_{zx} = I_{xz} = -\sum m_i z_i x_i$$

This tensor has been defined to the space set of axis. The difficulty with expressing the kinetic energy of the top in terms of the moment of inertia computed relative to the space set is that this moment of inertia varies with time. We would rather have our expression for the kinetic energy in terms of the moment of inertia tensor relative to the body set of axis which is invariant. Suppose that we apply the rotation R to the space set of axes thereby yielding the body set of axes. We want an expression for the moment of inertial of the body with respect to the body set in terms of the moment of inertia relative to the space set. Applying the rotation R to the axes is equivalent to applying the rotation $R^{-1} = R^T$ to the particles. The coordinates of a point relative to the body set are thus given by

$$x' = R_{xx}x + R_{yx}y + R_{zx}z \qquad y' = R_{xy}x + R_{yy}y + R_{zy}z \qquad z' = R_{xz}x + R_{yz}y + R_{zz}z$$

or equivalently, $\mathbf{r}' = R^{-1}\mathbf{r}$. Before proceeding further, we look at a nicer way to write down the moment of inertia tensor. The kinetic energy is given by

$$T = \frac{1}{2}\sum m|\omega \times \mathbf{r}|^2 = \frac{1}{2}\sum m(\omega^2 r^2 - (\omega.\mathbf{r})^2)$$

$$= \frac{1}{2}\sum m(\omega_i\omega_i r^2 - \omega_i\omega_j x_i x_j) = \frac{1}{2}\sum m(r^2\delta_{ij}\omega_i\omega_j - \omega_i\omega_j x_i x_j)$$

where summation over repeated indices is understood. It follows that if we define

$$I_{ij} = \sum m(r^2\delta_{ij} - x_i x_j)$$

then

$$T = \frac{1}{2}\sum I_{ij}\omega_i\omega_j = \frac{1}{2}\omega^T I\omega$$

Thus, $I = ((I_{ij}))$ is the moment of inertia tensor. If $\mathbf{r}$ is regarded as a column vector having components x_1, x_2, x_3 (where of course x_1 corresponds to x, x_2 corresponds to y and x_3 corresponds to z), then

$$I = \sum n(r^2\mathcal{I} - \mathbf{r}\mathbf{r}^T)$$

where $\mathcal{I}$ stands for the 3×3 identity matrix. The moment of inertia tensor I' with respect to the body set of axes is obtained by replacing $\mathbf{r}$ with $\mathbf{r}' = R^{-1}\mathbf{r}$. Thus,

$$I' = \sum m(r'^2\mathcal{I} - \mathbf{r}'\mathbf{r}'^T) = \sum m(r^2\mathcal{I} - R^{-1}\mathbf{r}\mathbf{r}^T R) = R^{-1}IR$$

This is the fundamental formula relating the moment of inertia tensors relative to two orthogonal systems. Now, the kinetic energy can be expressed as

$$T = \frac{1}{2}\omega^T I\omega = \frac{1}{2}\omega^T RI'R^{-1}\omega = \frac{1}{2}\omega'^T I'\omega'$$

where

$$\omega' = R^{-1}\omega$$

is the angular velocity pseudo-vector relative to the body set of axes. First we shall determine ω, the angular velocity pseudo-vector relative to the space set of axes. This can be obtained as entries of the antisymmetric matrix $\Omega = \frac{dR}{dt}R^{-1}$. Observe that

$$\frac{dR}{dt} = \frac{\partial R}{\partial \phi}\phi' + \frac{\partial R}{\partial \theta}\theta' + \frac{\partial R}{\partial \psi}\psi'$$

$$= R'_z(\phi)R_x(\theta)R_z(\psi)\phi' + R_z(\phi)R'_x(\theta)R_z(\psi)\theta' + R_z(\phi)R_x(\theta)R'_z(\psi)\psi'$$

Also

$$R^{-1} = R_z(-\psi)R_x(-\theta)R_z(-\phi)$$

giving

$$\Omega = \frac{dR}{dt}R^{-1}$$

$$= \phi' R'_z(\phi)R_z(-\phi) + \theta' R_z(\phi)R'_x(\theta)R_x(-\theta)R_z(-\phi) + \psi' R_z(\phi)R_x(\theta)R'_z(\psi)R_z(-\psi)R_x(-\theta)R_z(-\phi)$$

These matrix products have to be evaluated in order to determine the angular velocity vector. We use the expressions

$$R_z(\phi) = \begin{pmatrix} cos\phi & -sin\phi & 0 \\ sin\phi & cos\phi & 0 \\ 0 & 0 & 1 \end{pmatrix}$$

$$R_x(\theta) = \begin{pmatrix} 1 & 0 & 0 \\ 0 & cos\theta & -sin\theta \\ 0 & sin\theta & cos\theta \end{pmatrix}$$

$$R'_z(\phi)R_z(-\phi) = \begin{pmatrix} -sin\phi & -cos\phi & 0 \\ cos\phi & -sin\phi & 0 \\ 0 & 0 & 0 \end{pmatrix} \begin{pmatrix} cos\phi & sin\phi & 0 \\ -sin\phi & cos\phi & 0 \\ 0 & 0 & 1 \end{pmatrix}$$

$$= \begin{pmatrix} 0 & -1 & 0 \\ 1 & 0 & 0 \\ 0 & 0 & 0 \end{pmatrix}$$

$$R_z(\phi)R'_x(\theta)R_x(-\theta)R_z(-\phi)$$

$$= \begin{pmatrix} cos\phi & -sin\phi & 0 \\ sin\phi & cos\phi & 0 \\ 0 & 0 & 1 \end{pmatrix} \begin{pmatrix} 0 & 0 & 0 \\ 0 & -sin\theta & -cos\theta \\ 0 & cos\theta & -sin\theta \end{pmatrix}$$

$$\cdot \begin{pmatrix} 1 & 0 & 0 \\ 0 & cos\theta & sin\theta \\ 0 & -sin\theta & cos\theta \end{pmatrix} \begin{pmatrix} 1 & 0 & 0 \\ cos\phi & sin\phi & 0 \\ -sin\phi & cos\phi & 0 \\ 0 & 0 & 1 \end{pmatrix}$$

$$= \begin{pmatrix} 0 & sin\phi sin\theta & sin\phi cos\theta \\ 0 & -sin\theta cos\phi & -cos\theta cos\phi \\ 0 & cos\theta & -sin\theta \end{pmatrix} \cdot \begin{pmatrix} cos\phi & sin\phi & 0 \\ -cos\theta sin\phi & cos\theta cos\phi & sin\theta \\ sin\theta sin\phi & -sin\theta cos\phi & cos\theta \end{pmatrix}$$

$$= \begin{pmatrix} 0 & 0 & sin\phi \\ 0 & 0 & -cos\phi \\ -sin\phi & cos\phi & 0 \end{pmatrix}$$

Finally,

$$R_z(\phi)R_x(\theta)R'_z(\psi)R_z(-\psi)R_x(-\theta)R_z(-\phi)$$

$$= \begin{pmatrix} cos\phi & -sin\phi & 0 \\ sin\phi & cos\phi & 0 \\ 0 & 0 & 1 \end{pmatrix} \begin{pmatrix} 1 & 0 & 0 \\ 0 & cos\theta & -sin\theta \\ 0 & sin\theta & cos\theta \end{pmatrix} \cdot \begin{pmatrix} -sin\psi & -cos\psi & 0 \\ cos\psi & -sin\psi & 0 \\ 0 & 0 & 0 \end{pmatrix} \begin{pmatrix} cos\psi & sin\psi & 0 \\ -sin\psi & cos\psi & 0 \\ 0 & 0 & 1 \end{pmatrix}$$

$$\cdot \begin{pmatrix} 1 & 0 & 0 \\ 0 & cos\theta & sin\theta \\ 0 & -sin\theta & cos\theta \end{pmatrix} \begin{pmatrix} cos\phi & sin\phi & 0 \\ -sin\phi & cos\phi & 0 \\ 0 & 0 & 1 \end{pmatrix}$$

$$= \begin{pmatrix} cos\phi & -sin\phi cos\theta & sin\phi sin\theta \\ sin\phi & cos\phi cos\theta & -cos\phi sin\theta \\ 0 & sin\theta & cos\theta \end{pmatrix} \cdot \begin{pmatrix} sin\phi cos\theta & -cos\phi cos\theta & -sin\theta \\ cos\phi & sin\phi & 0 \\ 0 & 0 & 0 \end{pmatrix}$$

$$= \begin{pmatrix} 0 & -cos\theta & -sin\theta cos\phi \\ cos\theta & 0 & -sin\theta sin\phi \\ sin\theta cos\phi & sin\theta sin\phi & 0 \end{pmatrix}$$

Thus,

$$\Omega = \phi' \begin{pmatrix} 0 & -1 & 0 \\ 1 & 0 & 0 \\ 0 & 0 & 0 \end{pmatrix} + \theta' \begin{pmatrix} 0 & 0 & sin\phi \\ 0 & 0 & -cos\phi \\ -sin\phi & cos\phi & 0 \end{pmatrix}$$

$$+ \psi' \begin{pmatrix} 0 & -cos\theta & -sin\theta cos\phi \\ cos\theta & 0 & -sin\theta sin\phi \\ sin\theta cos\phi & sin\theta sin\phi & 0 \end{pmatrix}$$

$$= \begin{pmatrix} 0 & -(\phi' + \psi' cos\theta) & \theta' sin\phi - \psi' sin\theta cos\phi \\ \phi' + \psi' cos\theta & 0 & -(\theta' cos\phi + \psi' sin\theta sin\phi) \\ -\theta' sin\phi + \psi' sin\theta cos\phi & \theta' cos\phi + \psi' sin\theta sin\phi & 0 \end{pmatrix}$$

We also have

$$\Omega(t) = \begin{pmatrix} 0 & -\omega_z(t) & \omega_y(t) \\ \omega_z(t) & 0 & -\omega_x(t) \\ -\omega_y(t) & \omega_x(t) & 0 \end{pmatrix}$$

From this expression, we obtain the components of the angular velocity pseudo-vector relative to the space set of axis, i.e., the frame at rest:

$$\omega_x = \theta' cos\phi + \psi' sin\theta sin\phi, \qquad \omega_y = \theta' sin\phi - \psi' sin\theta cos\phi, \qquad \omega_z = \phi' + \psi' cos\theta$$

To obtain the formulae for the angular velocity pseudo-vector relative to the body set of axes, we require an expression for $R^{-1} = R^T$. Noting that

$$R = R_z(\phi) R_x(\theta) R_z(\psi) = \begin{pmatrix} cos\phi & -sin\phi & 0 \\ sin\phi & cos\phi & 0 \\ 0 & 0 & 1 \end{pmatrix} \cdot \begin{pmatrix} 1 & 0 & 0 \\ 0 & cos\theta & -sin\theta \\ 0 & sin\theta & cos\theta \end{pmatrix} \cdot \begin{pmatrix} cos\psi & -sin\psi & 0 \\ sin\psi & cos\psi & 0 \\ 0 & 0 & 1 \end{pmatrix}$$

$$= \begin{pmatrix} cos\phi & -sin\phi cos\theta & sin\phi sin\theta \\ sin\phi & cos\phi cos\theta & -cos\phi sin\theta \\ 0 & sin\theta & cos\theta \end{pmatrix} \begin{pmatrix} cos\psi & -sin\psi & 0 \\ sin\psi & cos\psi & 0 \\ 0 & 0 & 1 \end{pmatrix}$$

$$= \begin{pmatrix} cos\phi cos\psi - sin\phi cos\theta sin\psi & -cos\phi sin\psi - sin\phi cos\psi cos\theta & sin\phi sin\theta \\ sin\phi cos\psi + cos\phi cos\theta sin\psi & -sin\phi sin\psi + cos\phi cos\theta cos\psi & -cos\phi sin\theta \\ sin\theta sin\psi & sin\theta cos\psi & cos\theta \end{pmatrix}$$

We now use the transformation formulas

$$\omega'_x = R_{11}\omega_x + R_{21}\omega_y + R_{31}\omega_z =$$

$$\omega'_y = R_{12}\omega_x + R_{22}\omega_y + R_{32}\omega_z =$$

$$\omega'_z = R_{13}\omega_x + R_{23}\omega_y + R_{33}\omega_z =$$

to get

$$\omega'_x = (cos\phi cos\psi - sin\phi cos\theta sin\psi)(\theta' cos\phi + \psi' sin\theta sin\phi)$$

$$+(sin\phi cos\psi + cos\phi cos\theta sin\psi)(\theta' sin\phi - \psi' sin\theta cos\phi)$$

$$+sin\theta sin\psi(\phi' + \psi' cos\theta)$$

$$= \phi' sin\theta sin\psi + \theta' cos\psi$$

$$\omega'_y = (-cos\phi sin\psi - sin\phi cos\psi cos\theta)(\theta' cos\phi + \psi' sin\theta sin\phi)$$

$$+(-sin\phi sin\psi + cos\phi cos\theta cos\psi)(\theta' sin\phi - \psi' sin\theta cos\phi)$$

$$+sin\theta cos\psi(\phi' + \psi' cos\theta)$$

$$= -\theta' sin\psi + \phi' sin\theta cos\psi$$

$$\omega'_z = sin\phi sin\theta(\theta' cos\phi + \psi' sin\theta sin\phi) - cos\phi sin\theta(\theta' sin\phi - \psi' sin\theta cos\phi)$$

$$+cos\theta(\phi' + \psi' cos\theta)$$

$$\phi' cos\theta + \psi'$$

The moment of inertia tensor with respect to the body set of axes has the form $diag[I_1, I_1, I_3]$. Thus, the kinetic energy of the top can be expressed as

$$T = I_1(\omega'^2_x + \omega'^2_y)/2 + I_3\omega'^2_z/2$$

$$= I_1((\phi' sin\theta sin\psi + \theta' cos\psi)^2 + (-\theta' sin\psi + \phi' sin\theta cos\psi)^2)/2 + I_3(\phi' cos\theta + \psi')^2/2$$

$$= I_1(\phi'^2 sin^2\theta + \theta'^2)/2 + I_3(\phi' cos\theta + \psi')^2/2$$

θ is clearly the angle made by the z axis of the top, i.e., the z' axis with the z axis of the space set as the expression $R = R_z(\phi)R_x(\theta)R_z(\psi)$ shows. Thus, if L is the distance of the centre of gravity of the top from the base point, the potential energy of the top can be expressed as $mgLcos\theta$ and the Lagrangian of the top is

$$L(\theta, \phi', \psi', \theta') = I_1(\phi'^2 sin^2\theta + \theta'^2)/2 + I_3(\phi' cos\theta + \psi')^2/2 - mgLcos\theta$$

We leave it as an exercise to write down the Euler-Lagrange equations of motion.

1.13   Waves

Sinusoidal waves: A sinusoidal plane wave of frequency ω propagating along the x axis has amplitude distribution in space-time given by

$$\psi(t, x) = A.cos(\omega t - kx + \delta)$$

At each time t, the wave pattern is sinusoidal in space with a wavelength of $\lambda = 2\pi/k$ and at each spatial point x, the wave pattern is sinusoidal in time with a frequency $\nu = \omega/2\pi$. To calculate the wave velocity, we simply note that the constant phase surface has equation $\omega t - kx = constt.$ Differentiating this gives $\omega dt - kdx = 0$ so that the phase velocity is given by $v_p = \frac{dx}{dt} = \omega/k = \nu\lambda$. More generally, we can consider a sinusoidal wave in three dimensions specified by the amplitude distribution

$$\psi(t, \mathbf{r}) = A.cos(\omega t - \mathbf{k}.\mathbf{r})$$

The constant phase surfaces are $\omega t - \mathbf{k}.\mathbf{r} = constt.$ Differentiating this relation gives $\omega dt - \mathbf{k}.d\mathbf{r} = 0$ from which $\mathbf{k}.\frac{d\mathbf{r}}{dt} = \omega$. If $\mathbf{n}$ is the unit vector along the direction $\mathbf{k}$, then this gives $\mathbf{n}.\mathbf{v}_p = \omega/k$. The maximum of $\mathbf{n}.\mathbf{v}_p$ occurs when the direction of $\mathbf{v}_p$ is taken as $\mathbf{n}$. This gives $v_p = \omega/k$. In other words, the wave propagates along the direction of the wave vector $\mathbf{k}$ with a speed of ω/k.

Types of waves: The type of waves that one comes across in physics and engineering broadly fall under two categories, transverse and longitudinal. In a transverse wave, the vibration or change in amplitude takes place in a direction perpendicular to the direction of propagation while in a longitudinal wave, the vibration takes place along the direction of wave propagation. An example of a transverse wave is a plucked string. Another example is a propagating electromagnetic wave. An example of a longitudinal wave is sound. Another example is the vibration of a rod by contraction and expansion of its length.

Wave equation in one dimension: Consider a signal pulse at the origin whose variation in time is defined by the function $f(t)$. This wave propagates along the x direction. Let $\psi(t, x)$ denote the amplitude at the point x as a function of time. The amplitude $\psi(t, x)$ at the point x at time t must coincide with the amplitude $\psi(t - x/c, 0)$ at the origin at the delayed time $t - x/c$. Here, c is the speed of wave propagation so that x/c equals the time taken for a pulse at the origin to reach the point x. Since $f(t) = \psi(t, 0)$, we get $\psi(t, x) = f(t - x/c)$ and hence $\psi(t, x)$ satisfies the one dimensional wave equation

$$\frac{\partial^2 \psi(t, x)}{\partial t^2} - c^2 \frac{\partial^2 \psi(t, x)}{\partial x^2} = 0$$

This can be checked by direct differentiation. Conversely suppose $\psi(t, x)$ satisfies the above partial differential equation. We can then show that ψ has the general form

$$\psi(t, x) = f(t - x/c) + g(t + x/c)$$

for some doubly differentiable functions f, g. To see this, we Fourier transform the wave amplitude with respect to time, i.e., define

$$\hat{\psi}(\omega, x) = \int_{-\infty}^{\infty} \psi(t, x) exp(-i\omega t) dt$$

The inverse Fourier equation is

$$\psi(t, x) = \frac{1}{2\pi} \int_{-\infty}^{\infty} \hat{\psi}(\omega, x) exp(i\omega t) d\omega$$

The Fourier transform of the wave equation reads

$$\frac{\partial^2 \hat{\psi}(\omega, x)}{\partial x^2} + (\omega^2/c^2)\hat{\psi}(\omega, x) = 0$$

which has general solution

$$\hat{\psi}(\omega, x) = A(\omega) exp(-i\omega x/c) + B(\omega) exp(i\omega x/c)$$

and hence

$$\psi(t, x) = \frac{1}{2\pi} \int_{-\infty}^{\infty} (A(\omega) exp(i\omega(t - x/c)) + B(\omega) exp(i\omega(t + x/c))) d\omega$$

since A, B are arbitrary integrable functions, we get the general solution to the wave equation in the above form.

Transverse vibrations of a stretched string Consider a taut string with ends connected at $x = 0$ and $x = L$. The string is stretched tightly between these endpoints. The line of the string coincides with the x axis. Suppose that this string is plucked at one or several points and allowed to vibrate. Let $u(t, x)$ denote the vertical displacement of an element of the string at a distance x from the origin. Let T be the tension in the string, and $tan\theta = \frac{\partial u}{\partial x}$. θ is the angle made by the tangent to a string element at the point (x, u). The horizontal component of the tension must be such that there is no net horizontal force, i.e. $T(t, x)cos(\theta(t, x)) = T(t, x + dx)cos(\theta(t, x + dx))$, which is the same as $\frac{\partial}{\partial x}(Tcos(\theta)) = 0$ and hence $Tcos\theta = F$ is a constant dependent only on time. We assume F to be independent of time too. The vertical component of the tension force on the element dx of the string is given by

$$T(t, x + dx).sin(\theta(t, x + dx)) - T(t, x).sin(\theta(t, x)) = dx\frac{\partial}{\partial x}(T.sin(\theta)) = dx\frac{\partial}{\partial x}(F.tan\theta)$$

$$= Fdx\frac{\partial^2 u}{\partial x^2}$$

The mass of the string element dx equals $\sigma\sqrt{1 + (\frac{\partial u}{\partial x})^2}dx$ and assuming small deflection, i.e., $|\frac{\partial u}{\partial x}| << 1$ gives the mass element as approximately $\sigma.dx$. This means that the equation of motion is

$$\sigma\frac{\partial^2 u}{\partial t^2} = F\frac{\partial^2 u}{\partial x^2}$$

or if $c = \sqrt{F/\sigma}$, we get the equation of motion as

$$\frac{\partial^2 u}{\partial t^2} - c^2 \frac{\partial^2 u}{\partial x^2}$$

In other words, to a very good degree of approximation, the deflection $u(t, x)$ of the string follows the one dimensional wave equation. One way to solve this wave equation with the boundary condition $u(t, 0) = u(L, 0) = 0$ corresponding to the fact that the two ends of the string are fixed is to employ the half-sine-wave Fourier expansion, i.e.,

$$u(t, x) = \sum_{n=1}^{\infty} u_n(t) sin(n\pi x/L)$$

Substituting this into the wave equation gives

$$\frac{d^2 u_n(t)}{dt^2} = (-n\pi c/L)^2 u_n(t), n = 1, 2, ...,$$

which has solutions

$$u_n(t) = A_n cos(n\pi ct/L) + B_n sin(n\pi ct/L)$$

Each value of n is called a mode of vibration. The general solution can also be expressed as linear combinations of forward propagating and backward propagating waves, i.e., $cos(n\pi(ct - x)/L), sin(n\pi(ct - x)/L)$ and $sin(n\pi(ct - x)/L), sin(n\pi(ct + x)/L)$.

Three dimensional wave equation and its solution: A wave propagating in three dimensions has the following form for the variation of amplitude as a function of time and space:

$$\psi(t, x, y, z) = f(t - (n_x x + n_y y + n_z z)/c) = f(t - \mathbf{n}.\mathbf{r}/c)$$

where $\mathbf{n}$ is a unit vector. This wave corresponds to a pulse $f(t)$ at the origin propagating in the form of plane waves along the direction $\mathbf{n}$. Elementary differentiation shows that ψ satisfies the three dimensional wave equation

$$\frac{\partial^2 \psi}{\partial t^2} = c^2 \nabla^2 \psi$$

We can show that the general solution to this partial differential equation is a superposition of plane waves propagating along different directions at the velocity c. To see this, we consider the spatial Fourier transform of a general solution ψ:

$$\hat{\psi}(t, \mathbf{k}) = \int \psi(t, \mathbf{r}) exp(-i\mathbf{k}.\mathbf{r}) d^3\mathbf{r}$$

The inverse Fourier relation is

$$\psi(t, \mathbf{r}) = (2\pi)^{-3} \int \hat{\psi}(t, \mathbf{k}) exp(i\mathbf{k}.\mathbf{r}) d^3\mathbf{k}$$

Plugging this into the wave equation shows that $\hat{\psi}(t, \mathbf{k})$ satisfies the following ordinary differential equation

$$\frac{\partial^2 \hat{\psi}(t, \mathbf{k})}{\partial t^2} = -k^2 c^2 \hat{\psi}(t, \mathbf{k})$$

giving the solution

$$\hat{\psi}(t, \mathbf{k}) = A(\mathbf{k})exp(ikct) + B(\mathbf{k})exp(-ikct)$$

The general solution is therefore given by

$$\psi(t, \mathbf{r}) = (2\pi)^{-3} \int (A(\mathbf{k}))exp(i(kct - \mathbf{k}.\mathbf{r})) + B(\mathbf{k})exp(-i(kct + \mathbf{k}.\mathbf{r})))d^3\mathbf{k}$$

Writing $\mathbf{k} = k\mathbf{n}$ with $\mathbf{n}$ as a unit vector, we can equivalently express the solution as

$$\psi(t, \mathbf{r}) = \int (A(\omega, \mathbf{n})exp(i\omega(t - \mathbf{n}.\mathbf{r}/c)) + B(\omega, \mathbf{n})exp(-i\omega(t + \mathbf{n}.\mathbf{r}/c)))d\omega d\Omega(\mathbf{n})$$

This proves our assertion that the general solution is a superposition of plane waves.

Standing waves: Consider a sinusoidal wave having a forward propagating component and a backward propagating component. Suppose both components have equal magnitudes. Then, the wave amplitude as a function of space and time has the form

$$\psi(t, x) = cos(\omega t - kx) + cos(\omega t + kx) = 2cos(\omega t)cos(kx)$$

This is called a standing wave. At all times, we have nodes (zero amplitude) at the points $x = (n + 1/2)\pi/k$ with n assuming integer values. Since both the forward and backward components satisfy the wave equation, the standing wave is also a solution to the wave equation.

Spherical waves: The three dimensional Laplace operator when acting on a function depending upon the radial coordinate r only, has the form $\frac{1}{r}\frac{d^2}{dr^2}r$. Thus, if the wave amplitude depends only on time and the radial coordinate, i.e., $\psi(t, r)$, then it satisfies the spherical wave equation

$$\frac{\partial^2 \psi(t, r)}{\partial t^2} - \frac{c^2}{r}\frac{\partial^2}{\partial r^2}(r\psi(t, r)) = 0$$

Assume the solution to have sinusoidal time dependence. This assumption does not reduce the generality because any reasonably well behaved function of time can be Fourier analyzed into sinusoids. Writing $\psi(t, r) = u(r)exp(i\omega t)$, we find that $u(r)$ satisfies

$$\frac{1}{r}(ru)'' + (\omega^2/c^2)u = 0$$

or

$$(ru)'' = -(\omega/c)^2 ru$$

which has the general solution

$$u(r) = Aexp(ikr)/r + Bexp(-ikr)/r$$

where $k = \omega/c$ and A, B are constants. Thus, the general solution the spherical wave equation is

$$\psi(t,r) = \int_0^\infty dk(A(k)exp(ik(ct+r))/r + B(k)exp(ik(-ct+r))/r)$$

which is fully equivalent to the general form

$$\psi(t,r) = f(t+r/c)/r + g(t-r/c)/r$$

where f, g are arbitrary twice differentiable functions. The first component represents a spherical wave converging to the origin and the second component a spherical wave diverging away from the origin.

Interference of light: Suppose we have two light beams of the same frequency ω. Assume that both of these propagate in the form of plane waves in the same or in different directions Let ω denote the frequency and $\mathbf{k}_1$ and $\mathbf{k}_2$ the wave vectors. Then, the complex amplitude distribution due to both the beams has the form

$$\psi(t,\mathbf{r}) = \psi_1(t,\mathbf{r}) + \psi_2(t,\mathbf{r})$$

where

$$\psi_1(t,\mathbf{r}) = A_1 exp(i(\omega t - \mathbf{k}_1.\mathbf{r})) \qquad \psi_2(t,\mathbf{r}) = A_2 exp(i(\omega t - \mathbf{k}_2.\mathbf{r}))$$

The wave amplitude ψ is obtained by superposing the amplitudes ψ_1 and ψ_2. The intensity of the total wave amplitude is the magnitude square of ψ. It is given by

$$|\psi(t,\mathbf{r})|^2 = A_1^2 + A_2^2 + 2A_1 A_2 cos((\mathbf{k}_1 - \mathbf{k}_2).\mathbf{r})$$

Intensity maxima occur on the planes defined by the equations

$$(\mathbf{k}_1 - \mathbf{k}_2).\mathbf{r} = 2n\pi$$

with n taking integral values. Intensity minima occur on the planes defined by the equations

$$(\mathbf{k}_1 - \mathbf{k}_2).\mathbf{r} = (2n+1)\pi$$

with n taking integral values. A classic example of the interference pattern is seen in the Young's double slit experiment. This experiment consists of a screen with two slits and a coherent source of light placed at an equal distance from the two slits. In front of this two slit screen at a distance L parallel to it is placed another screen. Two rays of light from the source enter propagate upto the two slits, pass through them an interfere on the screen in front. Since the light source is at an equal distance from both the slits, the phases of the light wave amplitude at the slits are equal. Thus, both slits can be regarded as sources of spherical waves. Draw a line from a point Q midway between the two slits perpendicular to the screen onto the screen in front and let P denote the point of intersection of this line on the screen. We shall evaluate the amplitude of the interference pattern produced by the rays emanating from the two slits at a point $\mathbf{r}$ on the screen that is located at a distance x from P in a direction parallel to the line joining the two slits. For this purpose, let the origin of our coordinate system be the point Q located midway between the two slits. Then, the coordinates of the slits are respectively $\mathbf{r}_{1,2} = (\pm d/2, 0, 0)$ d being the distance between the two slits. The point $\mathbf{r}$ on the screen has coordinates $(x, 0, L)$. The complex amplitude of the spherical wave at the point $\mathbf{r}$ due to the wave coming from the first slit is given by $A.exp(ik|\mathbf{r} - \mathbf{r}_1|)/|\mathbf{r} - \mathbf{r}_1|$ and the complex amplitude of the spherical wave at $\mathbf{r}$ at due to the wave coming from the second slit is given by $A.exp(ik|\mathbf{r} - \mathbf{r}_2|)/|\mathbf{r} - \mathbf{r}_2|$. The actual amplitude as a function of time is obtained by multiplying the complex amplitude by $exp(-ikct)$ and taking the real part. The total amplitude due to interference at $\mathbf{r}$ is the sum of these two amplitudes. This can be approximated to

$$\psi(\mathbf{r}) = \frac{A}{r}(exp(ik|\mathbf{r} - \mathbf{r}_1|) + exp(ik|\mathbf{r} - \mathbf{r}_2|))$$

The intensity of light is the modulus square of this. Since the variation of r can be neglected in comparison to phase changes, the intensity is proportional to

$$I(\mathbf{r}) = 2 + 2.cos(\delta(\mathbf{r}))$$

where

$$\delta(\mathbf{r}) = k(|\mathbf{r} - \mathbf{r}_1| - |\mathbf{r} - \mathbf{r}_2|)$$

is k times the path difference from the two slits to the point on the screen. Writing

$$|\mathbf{r} - \mathbf{r}_1| = \sqrt{L^2 + (x - d/2)^2} - \sqrt{L^2 + (x + d/2)^2} \approx \sqrt{L^2 - dx} - \sqrt{L^2 + dx}$$

$$\approx L(1 - dx/2L^2) - L(1 + dx/2L^2) = -dx/L$$

we get the fundamental equation of the Young's double slit experiment that the intensity as a function of x varies as $cos^2(kdx/2L)$. The intensity maxima occur when $kdx/2L = n\pi$ and the intensity minima occur when $dx/2L = (n + 1/2)\pi$ where n is any integer. The spacing between two successive maxima is thus given by $2\pi L/kd = L\lambda/d$.

Principle of laser: There are basically three types of processes in a laser. Consider a quantum system having two energy levels $E_1 > E_2$. Spontaneous emission of radiation of frequency $\nu = (E_2 - E_1)/h$ takes place if atoms occupying the energy level E_2 make a transition to the level E_1. The prefix spontaneous signifies that no external agent has triggered the transition. Stimulated emission of radiation takes place if radiation of frequency ν is incident upon the atomic system. Such a radiation triggers atoms occupying the energy level E_2 to make a transition to the energy level E_1 thereby leading to amplification of the incident radiation. The intensity of a beam of radiation caused by atoms making a transition from E_2 to E_1 is directly proportional to the number of atoms making a transition. The experimentally observed fact that in stimulated emission of radiation, the frequency of radiation caused by the transitions from E_2 to E_1 coincides with the incident frequency is known as the phenomenon of phase coherence. This means that stimulated emission of radiation occurs only if the incident frequency is $\nu = (E_2 - E_1)/h$. The typical operation of a laser however involves a quantum system having three energy levels $E_1 < E_2 < E_3$. A laser called the pump having frequency $(E_3 - E_1)/h$ incident upon this quantum system causes atoms occupying the level E_1 to make a transition to the highest level E_3. Almost immediately, these atoms make a transition from E_3 to E_2. No photon is however emitted in the process. The energy difference $E_3 - E_2$ associated with each such transition is released and appears either as heat or as vibrational energy in the system. The state E_2 is metastable. The fact that more atoms now occupy E_2 than E_1 is known as population inversion. Atoms can either make a direct transition from E_2 to E_1 causing radiation of frequency $(E_2 - E_1)/h$ to get released, or else, if there is a signal laser source of frequency $(E_2 - E_1)/h$ incident upon this system, then this signal is amplified by transitions from the metastable state E_2 to the state E_1. The stimulated emission is the typical way in which the laser is operated. The word laser derives its name from this phenomenon. Its full form is light amplified by stimulated emission of radiation. If however, the population of E_2 is still lesser than E_1 during the duration of the presence of the signal source, then rather than having stimulated emission, there would be stimulated absorption causing atoms from E_1 to absorb energy from the signal source and make a transition to E_2. This causes an attenuation of the incident beam from the signal source.

Einstein Coefficients: Let $E_1 < E_2$ and let N_i be the number of atoms in the state $E_i, i = 1, 2$. Let $u(\nu)$ denote the energy density of radiation, i.e., the total energy in the radiation per unit volume in the frequency interval $[\nu_1, \nu_2]$ is given by $\int_{\nu_1}^{\nu_2} u(\nu)d\nu$. The number of absorptions per unit volume per unit time is proportional to N_1 and also to $u(\nu)$ for $\nu = (E_2 - E_1)/h$. Thus, number can thus be written as $N_1 B_{12} u(\nu)$. B_{12} is a characteristic of the atomic system. The number of spontaneous transitions from E_2 to E_1 is $A_{21} N_2$. No frequency comes into picture. This rate is simply proportional to the number of atoms occupying the level E_2. The number of stimulated emissions per unit volume per unit time is given by $N_2 B_{21} u(\omega)$. At thermal equilibrium, the number of transitions from E_1 to E_2 must coincide with the number of transitions from E_2 to E_1. This means that

$$N_1 B_{12} u(\nu) = N_2 B_{21} u(\nu) + A_{21} N_2$$

Thus,

$$u(\nu) = \frac{N_2 A_{21}}{N_1 B_{12} - N_2 B_{21}} = \frac{A_{21}}{\frac{N_1}{N_2} B_{12} - B_{21}}$$

Now, the population of the energy levels obeys Boltzmann's statistics:

$$\frac{N_1}{N_2} = exp((E_2 - E_1)/kT) = exp(h\nu/kT)$$

Thus,

$$u(\nu) = \frac{A_{21}}{B_{12}exp(h\nu/kT) - B_{21}}$$

This formula for the energy spectral density of radiation has been obtained by a balancing principle. However Max Planck discovered that the energy density of black-body radiation is given by

$$u(\nu) = \frac{8\pi h\nu^3/c^3}{exp(h\nu/kT) - 1}$$

Comparing these two expressions, one gets

$$\frac{A_{12}}{B_{21}} = \frac{8\pi h\nu^3}{c^3}, B_{12} = B_{21}$$

These formulae connecting the coefficient for stimulated emission, spontaneous emission and stimulated absorption were discovered by Einstein. Agreement with Planck's formula of black-body radiation would not have been possible without postulating the phenomenon of stimulated emission.

Polarization of light:

Light is known to comprise of electric and magnetic fields propagating as a wave. Consider a plane light wave propagating along the z direction. Assume that the light wave has a definite frequency ω. Then, the electric field in the wave can be expressed as

$$\mathbf{E}(t,z) = Acos(\omega t - kz)\hat{x} + B.cos(\omega t - kz + \phi)\hat{y}$$

where $k = \omega/c$, c being the velocity of light in the medium. The electric field has only transverse components, i.e., x and y components in agreement with the Maxwell equation $\nabla.\mathbf{E} = 0$. Several cases can be considered regarding the nature of this transverse vibrations. The different cases correspond to different choices of ϕ. If $\phi = 0$ or π, we get $\frac{E_y}{E_x} = \pm\frac{B}{A}$, i.e., the electric field vibrations take place in a single plane. For simplicity we consider the field at $z = 0$ (we can replace this choice by any fixed z). Thus,

$$\mathbf{E}(t) = \mathbf{E}(t,0) = Acos(\omega t)\hat{x} + Bcos(\omega t + \phi)\hat{y}$$

The choice $\phi = 0$ gives $\frac{E_y}{E_x} = \frac{B}{A}$ showing that the electric field vector at $z = 0$ is always along the line in the xy plane passing through the origin and having a slope of $\frac{B}{A}$. Likewise, if $\phi = \pi$, the field is along the line passing through the origin and having a slope of $-\frac{B}{A}$. Such a wave is called a plane polarized light wave. The choice $\phi = \pi/2$ gives

$$E_x = Acos(\omega t), E_y = -Bsin(\omega t)$$

and the electric field vector thus satisfies the equation

$$\frac{E_x^2}{A^2} + \frac{E_y^2}{B^2} = 1$$

which is the equation of an ellipse in the xy plane having major and minor axes (A, B) or (B, A) depending upon whether $A > B$ or $B > A$. when $B = A$, we get the equation of a circle. When $A \neq B$ the we say that light is elliptically polarized since the electric field vector moves along the circumference of an ellipse. When $A = B$, we say that the light is circularly polarized for then the electric field vector moves along the circumference of a circle. Specifically taking $A = B = 1$ and $\phi = \pi/2$ gives

$$\mathbf{E}(t) = \hat{x}.cos(\omega t) - \hat{y}.sin(\omega t)$$

This equation shows that as t increases, the vector $\mathbf{E}(t)$ moves clockwise looking from the top, i.e., from the side $z > 0$ into the origin. We say call such a light wave a left circularly polarized wave. Now take $A = B = 1$ and $\phi = -\pi/2$. We get

$$\mathbf{E}(t) = \hat{x}.cos(\omega t) + \hat{y}.sin(\omega t)$$

This represents a right circularly polarized wave since the electric field vector looking from the region $z > 0$ moves counterclockwise. More generally, we can consider an arbitrary A, B, ϕ, i.e.,

$$E_x = A.cos(\omega t), \ E_y = B.cos(\omega t + \phi)$$

We eliminate the variable ωt between these two equations to obtain the curve along which the electric field vector travels. This is done as follows:

$$E_y = B.cos(\phi).cos(\omega t) - B.sin(\phi).sin(\omega t) = \frac{B}{A}.cos(\phi)E_x - B.sin(\phi)\sqrt{1 - E_x^2/A^2}$$

from which we get

$$\left(E_y - \frac{B}{A}.cos(\phi)E_x\right)^2 = B^2.sin^2(\phi)(1 - E_x^2/A^2)$$

or

$$E_y^2 + \frac{B^2}{A^2}cos^2(\phi)E_x^2 - 2\frac{B}{A}cos(\phi)E_x E_y = B^2.sin^2(\phi)(1 - E_x^2/A^2)$$

or

$$E_y^2 + \frac{B^2}{A^2}E_x^2 - 2\frac{B}{A}.cos(\phi)E_x E_y = 0$$

or

$$\frac{E_x^2}{A^2} + \frac{E_y^2}{B^2} - 2\frac{cos\phi}{AB}E_x E_y = 0$$

This is once again the equation of an ellipse but when $\phi \neq 0, \pi/2, \pi$, the major axis will be tilted, i.e., it will neither coincide with the x axis nor with the y axis.

An unpolarized wave can be plane polarized by reflection. Suppose a light wave is incident upon a medium 2 from medium 1. Let n_1 denote the refractive index of medium one and n_2 that of medium 2. When the angle of incidence equals the Brewster angle $tan^{-1}(n_2/n_1)$ and if the incident beam has electric field polarized in the plane of incidence, then using the Maxwell equations, one can show that the amplitude of the reflected wave is zero. So suppose the incident light is unpolarized. Then, the electric field vector in the incident wave can be decomposed into a component that is parallel to the plane of incidence and a component that is perpendicular to the plane of incidence. At the critical Brewster angle of incidence, the parallel component contributes to zero reflected amplitude while the perpendicular component has a non zero reflected amplitude. In effect, the reflected wave is perpendicularly polarized and thus we've created a linearly polarized wave. Now suppose we place alternately versions of the two media, i.e., a given strip of medium one followed by a given strip of medium two, followed by a given strip of medium one and so on. Unpolarized light is incident from medium one at the top at the Brewster angle. Then, the reflected light has only a perpendicularly polarized component and no parallely polarized component. All the parallel component thus goes into the transmitted wave. The perpendicular component in the transmitted wave is reduced in amplitude since part of it goes into the reflected wave. The incidence angle for the wave propagating from medium one to medium two in the third strip again takes place at the Brewster angle and hence the resulting transmitted wave has perpendicular component further attenuated. Thus after several reflections, the outcoming wave is nearly parallely polarized and we've thus created an almost linearly polarized wave as the outcoming wave.

Another way to create a linearly polarized wave is to use double refraction. Certain crystals are anisotropic, i.e., their complex refractive index or permittivity varies with the direction of polarization. As a consequence, a linearly polarized wave gets absorbed quickly if the polarization is along one direction and propagates through easily if the polarization is along the perpendicular direction. Thus if an unpolarized wave is incident upon the crystal, we can decompose the electric field vector in it into two components orthogonal to each other, one of which gets absorbed in the crystal and the other propagates through. The emergent wave is then linearly polarized. An example of such an anisotropic crystal is Tourmaline. It should be borne in mind that if the refractive indices of the anisotropic crystal are real, then no absorption can take place, for absorption to take place, like a metal, the crystal should have a complex index of refraction, i.e., non zero conductivity. If σ is the conductivity and ϵ the permittivity, then the effective conductivity at the frequency ω is given by $\sigma + i\omega\epsilon$ and the effective permittivity is $\epsilon - i\sigma/\omega$. The effective refractive index is equal to the complex permittivity divided by ϵ_0, the free space permittivity. Only if $\sigma \neq 0$, can absorption take place.

A Nicol prism has the following remarkable property. For a plane polarized light wave, if the plane of polarization is along one direction, the refractive index is ϵ_{r1} while if the plane of polarization is along the perpendicular direction, the refractive index is different ϵ_{r2}. Note that Nicol is a dielectric which means that its conductivity is nearly zero and hence the two refractive indices are real. These two refractive indices have the property that ϵ_{r1} exceeds the refractive index of air while ϵ_{r2} is smaller than that of air. If a ray propagates inside the prism and is plane polarized along the first direction, then there will exist a critical angle greater than which total internal reflection will occur. Note that such a critical angle exists for this polarization only because the refractive index ϵ_{r1} is more than that of air. For polarization along the second direction, there does not exist any critical angle. Thus, if unpolarized light is incident upon the prism so that the wave propagating inside the prism is also unpolarized, then one component will suffer total internal reflection at the other interface provided the incident angle has been chosen appropriately, i.e. such that the angle of incidence of the ray propagating inside the prism incident on the other edge exceeds the critical angle corresponding to the index ϵ_{r1}. The emergent wave will then have only the other component and will hence be plane polarized.

Frauhhoffer and Fresnel's diffraction theories: Consider an aperture S in the xy plane. Light from below is incident upon this aperture. We divide the aperture into little slots so that each slot acts as a secondary source. Assume that the frequency of the incident light is fixed, say ω. Let $E(x, y)$ be the complex amplitude of the light at the point (x, y) on the aperture. This means that the total amplitude due to light in the region δS of the aperture is $E(x, y)\delta S(x, y)$. Each little slot acts as a source of spherical waves that propagate outwards. The total amplitude of light at the point (X, Y, Z) above the plane is then given by a superposition of spherical waves:

$$A(X, Y, Z) = \int_S E(x, y) \frac{exp(-ik((X-x)^2 + (Y-y)^2 + Z^2)^{1/2})}{((X-x)^2 + (Y-y)^2 + Z^2)^{1/2}} dxdy$$

Assume that the aperture has its centre at the origin and that its dimensions are much smaller than the distance of the point (X, Y, Z) from it. We also assume that the wavelength is much smaller than the dimensions of the aperture. Then the variation of the phase factor in the above integral dominates the variation of the factor in the denominator and we get approximately

$$A(X, Y, Z) \approx \frac{1}{R} \int_S E(x, y) exp(-ikR(1 + (x^2 + y^2 - 2Xx - 2Yy)/R^2)^{1/2}) dxdy$$

$$\approx \frac{exp(-ikR)}{R} \int_S E(x, y) exp(-ik((x^2 + y^2)/2R^2 - (Xx + Yy)/R^2)) dxdy$$

where $R = \sqrt{X^2 + Y^2 + Z^2}$. If X, Y are very much larger than the dimensions of the aperture, then the term $(x^2 + y^2)/R^2$ is negligible in comparison with $(Xx + Yy)/R^2$ and we get the Fraunhoffer diffraction formula which with neglect of a constant phase and amplitude factor reads

$$A(X, Y, Z) = \int_S E(x, y) exp(ik(Xx + Yy)/R^2) dxdy$$

If however, X, Y are large but not very much larger than the aperture dimensions, then we get the Fresnel diffraction formula:

$$A(X, Y, Z) = \int_S E(x, y) exp(ik((Xx + Yy)/R^2 - (x^2 + y^2)/2R^2)) dxdy$$

Holography The wave coming from the object has the form $O(x, y, t) = a(x, y)cos(\phi(x, y) - \omega t)$ when incident on the photographic plate which is on the $z = 0$ plane. This is the result of superposition of waves from each point on the object which acts as a scatterer of the incident wave. The object wavefield superposes with a reference wave field $r(x, y, z, t) = A\cos(\mathbf{k.r} - \omega t)$. Since both the object wave and the reference waves have the same frequencies, the source of light must be common. For example a laser source shines on the object generating the object wave and also shines on a mirror generating the reference wave. Assume that the light coming from the mirror propagates along the direction $(sin\theta, 0, cos\theta)$. Then the superposition of the object and reference waves generates the following amplitude distribution on the photographic plate:

$$u(x, y, t) = a(x, y)cos(\phi(x, y) - \omega t) + A.cos(kxsin\theta - \omega t)$$

The intensity distribution on the plate is the time averaged value of $u(x, y, t)^2$, i.e.,

$$I(x, y) = < u(x, y, t)^2 > = a(x, y)^2/2 + A^2/2 + A.a(x, y).cos(\phi(x, y) - kx.sin\theta)$$

This intensity distribution is recorded on the plate by means of chemicals. Suppose we shine light from the reference source upon this photographic plate which has recorded the above intensity distribution. Then the amplitude of the output at $z = 0$ will be proportional to $I(x, y)r(x, y, 0, t)$:

$$I(x, y)r(x, y, 0, t) = A.cos(kx.sin\theta - \omega t).(A^2/2 + a(x, y)^2/2$$

$$+A.a(x, y).cos(\phi(x, y) - kx.sin\theta))$$

$$= (A^2 + a(x, y)^2)A.cos(kx.sin\theta - \omega t)/2 + (A^2.a(x, y)/2)cos(\phi(x, y) - \omega t)$$

$$+(A^2.a(x, y)/2).cos(2kx.sin\theta - \phi(x, y) - \omega t)$$

The first term is a wave traveling in the direction of the reference source. The second is a scaled replica of the object amplitude pattern and the third is a conjugated object pattern, i.e., a real image of the object but its direction makes an angle $sin^{-1}(2sin\theta)$ with the x axis. The second term is a virtual image of the object and the third one is a real image of the object which can be filtered out. Equivalently, the third one can be filtered out by applying a spatial filter that filters out spatial frequencies $2k$ along the x direction. Thus, a virtual image of the object can be produced shining light from the reference source upon the recorded photographic plate.

Optical activity and specific rotation: Sugar solution is an optically active medium in that left and right circularly polarized waves travel with different velocities through it. Consider a linearly polarized wave propagating through sugar solution. The wave amplitude is given by $\mathbf{E} = 2E_0\hat{x}.cos(kz - \omega t)$. It can be resolved into right and left circularly polarized waves:

$$\mathbf{E} = \mathbf{E}_r + \mathbf{E}_l$$

where

$$\mathbf{E}_r = E_0(\hat{x}.cos(kz - \omega t) - \hat{y}.sin(kz - \omega t))$$

and

$$\mathbf{E}_l = E_0(\hat{x}.cos(kz - \omega t) + \hat{y}.sin(kz - \omega t))$$

The two beams when they propagate through the solution respectively become

$$\mathbf{E}'_r = E_0(\hat{x}.cos(k_r z - \omega t) - \hat{y}.sin(k_r z - \omega t))$$

and

$$\mathbf{E}'_l = E_0(\hat{x}.cos(k_l z - \omega t) + \hat{y}.sin(k_l z - \omega t))$$

The total wave field in the solution is $\mathbf{E}'_r + \mathbf{E}'_l$. Its x component is given by

$$E'_x = E_0(cos(k_r z - \omega t) + cos(k_l z - \omega t))$$

and its y component is given by

$$E'_y = E_0(sin(k_l z - \omega t) - sin(k_r z - \omega t))$$

Using standard trigonometric formulas, these can be expressed as

$$E'_x = 2E_0 cos((k_l - k_r)z/2)cos((k_l + k_r)z/2 - \omega t)$$

$$E'_y = 2E_0 sin((k_l - k_r)z/2)cos((k_l + k_r)z/2 - \omega t)$$

If $k_l = k_r$, then the resulting wave inside the solution is linearly polarized along the x direction. Since however k_l and k_r differ slightly, the plane of polarization rotates slowly at the rate of $(k_l - k_r)/2$ radians per metre.

Solved Problems, Study Projects, Suggestions and Remarks

Study Project 1

A seminar surveying numerical integration of the Newtonian equations of motion could be given. The seminar may start with the Newtonian equations of motion in differential form, say first in one dimension in a potential field $V(x)$

$$m\frac{d^2x(t)}{dt^2} = -V'(x(t))$$

how one constructs the energy integral for this equation,

$$\frac{m}{2}(\frac{dx(t)}{dt})^2 + V(x(t)) = E$$

The integration of this first order differential equation for a variety of choices of the potential $V(x)$

$$t - t_0 = \int_{x_0}^{x} \frac{dx}{\sqrt{2(E - V(x))/m}}$$

Numerical integration of the energy integral using

$$x[n+1] = x[n] + \Delta.\sqrt{2(E - V(x[n]))/m}$$

Numerical integration can be illustrated for the harmonic oscillator case for which $V(x) = \frac{m\omega^2 x^2}{2}$, $V(x) = cx^n$ (power law potential) and many other useful and commonly occurring potentials. Some time can be given to integration of the original Newtonian differential equation of motion by adopting an appropriate Taylor expansion for the discretization. This means that we use the following approximations:

$$x(t+\Delta) \approx x(t) + v(t)\Delta + v'(t)\Delta^2/2 = x(t) + v(t)\Delta - V'(x(t))\Delta^2/2m$$

$$v(t+\Delta) \approx v(t) + v'(t)\Delta + v''(t)\Delta^2/2 = v(t) - V'(x(t))\Delta/m - V''(x(t))v(t)\Delta^2/2m$$

and extend this to include higher derivatives in the Taylor expansion. Some time could be devoted also to two dimensional problems of Newtonian mechanics in a potential field $V(x,y)$. This amounts to solving the following differential equations numerically:

$$\frac{d^2x}{dt^2} = -\frac{\partial V(x,y)}{\partial x}$$

$$\frac{d^2y}{dt^2} = -\frac{\partial V(x,y)}{\partial y}$$

One of the obvious integrals is the energy integral which is

$$\frac{1}{2}\left(\left(\frac{dx}{dt}\right)^2 + \left(\frac{dy}{dt}\right)^2\right) + V(x,y) = E$$

No other first integral for this system is known for a general potential V. The use of numerical techniques for the integration of the original equations of motion could be illustrated. For example, if one adopts a second order Taylor expansion,

$$x(t+\Delta) \approx x(t) + \Delta.v_x(t) - \frac{\Delta^2}{2}\frac{\partial V(x(t), y(t))}{\partial x}$$

$$y(t+\Delta) \approx y(t) + \Delta.v_y(t) - \frac{\Delta^2}{2}\frac{\partial V(x(t), y(t))}{\partial y}$$

$$v_x(t+\Delta) \approx v_x(t) - \Delta.\frac{\partial V(x(t), y(t))}{\partial x} - \frac{\Delta^2}{2}\left(v_x(t)\frac{\partial^2 V(x(t), y(t))}{\partial x^2} + v_y(t)\frac{\partial^2 V(x(t), y(t))}{\partial x \partial y}\right)$$

$$v_y(t+\Delta) \approx v_y(t) - \Delta.\frac{\partial V(x(t), y(t))}{\partial y} - \frac{\Delta^2}{2}\left(v_x(t)\frac{\partial^2 V(x(t), y(t))}{\partial x \partial y} + v_y(t)\frac{\partial^2 V(x(t), y(t))}{\partial y^2}\right)$$

Study project 2

Tools for the analysis of trajectories of a dynamical system can also be dealt with while discussing the mechanics of Newton. These include Fourier analysis of the particle's trajectory over a finite duration of time and from the dominant frequencies, to draw conclusions regarding the parameters of the system. A classic example of this could include the anharmonic oscillator whose equation of motion is given by

$$\frac{d^2x}{dt^2} = -kx - \epsilon.x^3$$

This is solved using canonical perturbation theory, i.e. by expanding the solution in powers of the parameter ϵ and then performing a Fourier analysis of the resulting position process.

Study project 3

The use of variational calculus in the solution of problems of mechanics can be discussed in a seminar and how corresponding algorithms can be implemented on the digital computer. The seminar can be confined to one dimensional problems. Specifically, the following approach can be adopted: If a particle of mass m moves in one dimension in the potential field $V(x)$, then its action integral between points $(t_a, x_a = x(t_a))$ and $(t_b, x_b = x(t_b))$ is given by

$$S[t_a, x_a, t_b, x_b, x(.)] = \int_{t_a}^{t_b} \frac{m}{2}(\frac{dx(t)}{dt})^2 - V(x(t)))dt$$

Minimization of this action integral keeping the two end points x_a, x_b fixed yields the true classical trajectory of the particle. This can be seen by perturbing the trajectory from $x(t)$ to $x(t) + \eta(t), t_a \leq t \leq t_b$ with the endpoints fixed so that $\eta(t_a) = \eta(t_b) = 0$. The corresponding perturbation in the action upto first order in η is given by

$$S[x(.) + \eta(.)] - S[x(.)] = \int_{t_a}^{t_b} (m.x'(t)\eta'(t) - V'(x(t))\eta(t))dt$$

$$= -\int_{t_a}^{t_b} (mx''(t) + V'(x(t)))\eta(t)dt$$

on integrating the first term by parts. $S[.]$ has an extremum for the trajectory $x(.)$ between the prescribed endpoints for the prescribed times when and only when the above variation vanishes for all perturbations $\eta(.)$ of the trajectory that leave the endpoints fixed. This happens when and only when the trajectory $x(.)$ satisfies the differential equation

$$mx''(t) + V'(x(t)) = 0, t_a \leq t \leq t_b$$

Discrete formulation of the Lagrangian variational principle: This amounts to discretizing the time interval $[0, T]$ and replacing the action integral $\int_0^T L(x(t), x'(t), t)dt$ by the discrete sum $\sum_{n=1}^{N-1} L(x[n], (x[n] - x[n-1])/\Delta, n\Delta)\Delta$ where $T = N\Delta$ and optimizing this function with respect to $x[1], ..., x[N-1]$ with $x[0], x[N]$ kept fixed. For example, for the Lagrangian $\frac{m}{2}x'(t)^2 + V(x(t))$ we would use

$$\Phi(x[1], ..., x[N-1]) = \sum_{n=1}^{N} m(x[n] - x[n-1])^2/2\Delta - V(x[n])\Delta$$

as the function to be optimized.

Study project 4

Present the theory of the Hamilton-Jacobi equation in classical mechanics. If (q_i, p_i) are canonical position-momentum pairs and $H(\{q_i\}, \{p_i\})$ is the corresponding Hamiltonian, then the equations of motion are

$$\frac{dq_i}{dt} = \frac{\partial H}{\partial p_i}, \frac{dp_i}{dt} = -\frac{\partial H}{\partial q_i}$$

The Lagrangian of the system is related to the Hamiltonian via a Legendre transformation, i.e.

$$L = \sum p_i q_i' - H$$

The action principle reads

$$\delta \int_{t_1}^{t_2} (\sum p_i q_i' - H)dt = 0$$

We now transform to another set of canonical variables Q_i, P_i. The new Hamiltonian is K, i.e. the equations of motion of the same mechanical system appear in terms of the new variables as

$$\frac{dQ_i}{dt} = \frac{\partial K}{\partial P_i} \qquad \frac{dP_i}{dt} = -\frac{\partial K}{\partial Q_i}$$

The corresponding variational principle reads

$$\delta \int_{t_1}^{t_2} (\sum P_i Q_i' - K)dt = 0$$

For both of these equations to correspond to the same equations of motion, the difference between the two Lagrangians must be a total time derivative, i.e.,

$$\sum P_i Q_i' - K = \sum p_i q_i' - L + \frac{dF}{dt}$$

We can assume F to be a function of (q_i, Q_i, t) or (q_i, P_i, t) or (Q_i, p_i, t) or (Q_i, P_i, t). Taking our independent variables as (q_i, Q_i, t) gives

$$\sum P_i Q_i' - K = \sum p_i q_i' - H + \sum \frac{\partial F}{\partial q_i} q_i' + \frac{\partial F}{\partial Q_i} Q_i' + \frac{\partial F}{\partial t}$$

This means that

$$P_i = \frac{\partial F}{\partial Q_i}$$

$$p_i = -\frac{\partial F}{\partial q_i}$$

$$K = H - \frac{\partial F}{\partial t}$$

The above equation can also be expressed as

$$\frac{d}{dt}\left(\sum P_i Q_i - F\right) - \sum Q_i P_i' - K = \sum p_i q_i' - H$$

If we regard $G = \sum P_i Q_i - F$ as our new generating function, then we get

$$\frac{dG}{dt} - \sum Q_i P_i' - K = \sum p_i q_i' - H$$

Now we regard G as a function of q_i, P_i and express this equation in the following form:

$$\frac{\partial G}{\partial t} + \sum \frac{\partial G}{\partial q_i} q_i' + \sum \frac{\partial G}{\partial P_i} P_i' - \sum Q_i P_i' - K = \sum p_i q_i' - H$$

On equating coefficients we get from this,

$$K - H = \frac{\partial G}{\partial t}$$

$$\frac{\partial G}{\partial q_i} = p_i$$

$$\frac{\partial G}{\partial P_i} = Q_i$$

Now suppose we choose our generating function G so that the canonical variables Q_i, P_i are constants of the motion, i.e. $Q_i' = 0, P_i' = 0$. Then K must be a constant which can be set equal to zero. This gives

$$H + \frac{\partial G}{\partial t} = 0$$

or plugging the value of p_i given by the above equation gives

$$H(q_i, \frac{\partial G}{\partial q_i}) + \frac{\partial G}{\partial t} = 0$$

Note that in this partial differential equation, G is a function of the $q_i's$ and the constant $P_i's$. Having solved this equation for the function, G, we can determine q_i as functions of time via the equation

$$Q_i = \frac{\partial G(q_i, P_i, t)}{\partial P_i}$$

This is a set of algebraic equations for the variables $q_i(t)$ with P_i, Q_i as constants of the motion. Having thus determined $q_i(t)$, we can obtain $p_i(t)$ using the equation

$$p_i = \frac{\partial G(q_i, P_i)}{\partial q_i}$$

Thus the complete solution to the mechanical problem can be obtained once the generating function G of the canonical transformation that transforms the variables q_i, p_i into another set of canonical variables Q_i, P_i which are constants of the motion.

Study project 5

Describe the elastic collision of two particles showing the relationship between the laboratory and centre of mass reference frames. The description can start along the following lines. Let mass m_2 be initially at rest in the laboratory system and mass m_1 be moving with velocity v_1 along the x axis. After the collision, the mass m_1 moves with a velocity v_1' at an angle θ_1 to the x axis relative to the laboratory frame and mass m_2 moves with a velocity v_2' at an angle θ_2 relative to the laboratory frame. Momentum and energy are conserved since the collision is elastic. Using vector notation, these two conservation equations can be expressed as

$$m_1\mathbf{v}_1 = m_1\mathbf{v}_1' + m_2\mathbf{v}_2'$$

$$m_1 v_1^2 = m_1 v_1'^2 + m_2 v_2'^2$$

The velocity of the centre of mass relative to the laboratory before the collision is given by

$$\mathbf{V} = m_1\mathbf{v}_1/(m_1 + m_2)$$

The velocity of the centre of mass after the collision is given by

$$\mathbf{V}' = (m_1\mathbf{v}_1' + m_2\mathbf{v}_2')/(m_1 + m_2)$$

In view of the momentum conservation equation, $\mathbf{V}' = \mathbf{V}$, i.e. the velocity of the centre of mass remains invariant under the collision. Now we look at the collision relative to the centre of mass frame. The velocity of m_1 relative to the centre of mass before the collision is given by

$$\mathbf{u}_1 = \mathbf{v}_1 - \mathbf{V} = m_2\mathbf{v}_1/(m_1 + m_2)$$

and the velocity of m_2 relative to the centre of mass before the collision is given by

$$\mathbf{u}_2 = -\mathbf{V} = -m_1\mathbf{v}_1/(m_1 + m_2)$$

The momentum conservation equation is equivalent to $m_1\mathbf{u}_1 + m_2\mathbf{u}_2 = 0$ which is clearly satisfied by the above expressions for $\mathbf{u}_1, \mathbf{u}_2$. The velocities of m_1 and m_2 after the collision are respectively given by $\mathbf{u}_1' = \mathbf{v}_1' - \mathbf{V}$ and $\mathbf{u}_2' = \mathbf{v}_2' - \mathbf{V}$. The energy conservation equation can be expressed as

$$m_1|\mathbf{V} + \mathbf{u}_1|^2 + m_2|\mathbf{V} + \mathbf{u}_2|^2 = m_1|\mathbf{V} + \mathbf{u}_1'|^2 + m_2|\mathbf{V} + \mathbf{u}_2'|^2$$

which in view of the equations

$$m_1\mathbf{u}_1 + m_2\mathbf{u}_2 = 0 = m_1\mathbf{u}_1' + m_2\mathbf{u}_2'$$

(Momentum conservation) reduces to

$$m_1 u_1^2 + m_2 u_2^2 = m_1 u_1'^2 + m_2 u_2'^2$$

We now substitute $\mathbf{u}_2 = -m_1\mathbf{u}_1/m_2$, $\mathbf{u}_2' = -m_1\mathbf{u}_1'/m_2$ in this equation to get $u_1 = u_1'$, i.e., $|\mathbf{u}_1| = |\mathbf{u}_1'|$ and hence $|\mathbf{u}_2| = |\mathbf{u}_2'|$. This is the same as

$$|\mathbf{v}_1 - m_1\mathbf{v}_1/(m_1 + m_2)| = |\mathbf{v}_1' - (m_1\mathbf{v}_1' + m_2\mathbf{v}_2')/(m_1 + m_2)|$$

which is the same as

$$|\mathbf{v}_1| = |\mathbf{v}_1' - \mathbf{v}_2'|$$

i.e., the relative velocity before the collision of the two particles coincides with the relative velocity after the collision in magnitude. Let α_1, α_2 denote the angles made by the vectors $\mathbf{u}_1', \mathbf{u}_2'$ with the x axis.. The equation $\mathbf{v}_1' = \mathbf{V} + \mathbf{u}_1'$ reads

$$v_1' cos\theta_1 = V + u_1' cos\alpha_1$$

$$v_1' sin\theta_1 = u_1' sin\alpha_1$$

Since $u_1 = u_1'$, we get

$$tan(\theta_1) = \frac{u_1 sin\alpha_1}{V + u_1 cos\alpha_1}$$

θ_1 is the scattering angle for the first particle relative to the laboratory system and α_1 the scattering angle of the first particle relative to the centre of mass system. The two are connected by the above equation.

Study project 6

The perturbed Hamiltonian problem. Let $H(q_i, p_i) = H_0(q_i, p_i) + \epsilon.H_1(q_i, p_i)$ be the Hamiltonian of a mechanical system. The number of generalized coordinates and momenta is $2N$ which corresponds to $2N$ degrees of freedom in the coordinate space. The equations of motion are

$$\frac{dq_i}{dt} = \frac{\partial H_0}{\partial p_i} + \epsilon.\frac{\partial H_1}{\partial p_i}$$

$$\frac{dp_i}{dt} = -\frac{\partial H_0}{\partial q_i} - \epsilon.\frac{\partial H_1}{\partial q_i}$$

We make a perturbation expansion

$$q_i(t) = \sum_{n=0}^{\infty} \epsilon^n q_i^{(n)}(t)$$

$$p_i(t) = \sum_{n=0}^{\infty} \epsilon^n p_i^{(n)}(t)$$

Study project 7

Let us evaluate the scattering cross section for a collision between two molecules of masses M and m, when we assume that m moves in a central potential generated by the mass M. Let $V(r)$ denote the potential field produced by the mass M. The field is assumed to be repulsive for sufficiently small r. For large r, it may be attractive. For example, we can have a potential of the form $V(r) = \frac{A}{r^a} - \frac{B}{r^b}$ with $a > b$ and $A, B > 0$ For r close to zero, the first term is dominant implying a positive potential and hence a repulsive force. For large r, the second term is dominant implying an attractive potential. The equation of motion of m relative to M is given by the Lagrangian

$$L(r, \phi, r', \phi') = \frac{\mu}{2}(r'^2 + r^2\phi'^2) - V(r)$$

This Lagrangian leads to energy conservation as well as angular momentum conservation. The energy conservation equation reads

$$\frac{\mu}{2}(r'^2 + r^2\phi'^2) + V(r) = E$$

and the angular momentum conservation equation reads

$$r^2\phi' = \beta$$

where β equals angular momentum per unit mass. From these two, we obtain the differential equation for the trajectory:

$$\frac{\mu}{2}((dr/d\phi)^2 + r^2) + r^4 V(r)/\beta^2 = r^4 E/\beta^2$$

Letting $r = 1/u$ gives

$$\frac{\mu}{2}((du/d\phi)^2 + u^2) + V(1/u)/\beta^2 = E/\beta^2$$

The origin of our coordinate system is the centre of mass of the two particle system. Assume that the separation between the particles is infinite at $t = 0$ and that relative to M, m moves with an initial speed v with the line of motion being at a distance of s from the M. Then, the initial relative angular velocity is given by μvs. We can in fact visualize a stream of incident particles having intensity I particles crossing unit area per unit time. This is the incident flux. Then, the number of particles crossing the ring $s, s + ds$ per unit time is given by $I.2\pi s ds$. When $u = 0$, we have $r = \infty$ and then $\phi = \pi$. This corresponds to the starting of the incident particle's trajectory. When u is a maximum, r is a minimum and assume that this corresponds to $\phi = \phi_0$. Again when the particle reaches $r = \infty$, we have $u = 0$ and from symmetry, we can argue that $\phi = 2\phi_0 - \pi$. Let $u_m = 1/r_{min}$. Then, $du/d\phi = 0$ when $u = u_m, \phi = \phi_0$. This gives on substitution

$$\frac{\mu}{2}u_m^2 + V(1/u_m)/\beta^2 = E/\beta^2$$

This is a nonlinear equation for u_m and the solution depends on E and β, i.e. $u_m = u_m(E, \beta)$. From the differential equation for the orbit, we get

$$\int_0^{u_m} \left(\frac{\mu\beta^2}{2(E - V(1/u)) - \mu\beta^2 u^2}\right)^{1/2} du = \pi - \phi_0$$

Thus, the scattering angle is given by

$$\phi_{scatt} = 2\phi_0 - \pi = \pi - 2\int_0^{u_m} \left(\frac{\mu\beta^2}{2(E - V(1/u)) - \mu\beta^2 u^2}\right)^{1/2} du$$

This angle is a function of E and β. We write this nonlinear equation as $\phi_{scatt} = \phi_{scatt}(E, \beta)$. Note that $E = \frac{\mu}{2}v^2$ and $\beta = vs$. Thus,

$$\phi_{scatt} = \phi_{scatt}(\mu v^2/2, vs)$$

Inverting this equation gives us s as a function of ϕ_{scatt} and v. v is the initial relative speed, i.e. $|\mathbf{v}_1 - \mathbf{v}_2|$.

Study project 8

Differentiable manifolds

The theory of Lie groups plays a fundamental role in mechanics, since the Hamiltonian leads to a flow on a differentiable manifold which is a one parameter group of diffeomorphisms acting on the symplectic manifold. The Hamiltonian equations of motion can be described very nicely in the language of symplectic geometry for which we require the calculus of differentiable forms. Moreover, the Lagrangian can be viewed in a coordinate free fashion as a function on the tangent bundle such that the generalized coordinates define the point on the manifold and the generalized velocity the tangent vector. But to learn the theory of Lie groups, we must first master the theory of C^∞ manifolds. A separate lecture on C^∞ manifolds can be presented covering the following salient points. Let M be a topological space, i.e., a set with a topology τ on it. A topology consists of subsets of M called open sets, closed under arbitrary unions and finite intersections and containing M and the empty set. We consider the scheme by which coordinates are introduced in M. This amounts to defining local charts and an atlas. An atlas is a collection $\{(U_\alpha, \phi_\alpha) : \alpha \in \Gamma\}$ of local charts. A local chart (U_α, ϕ_α) consists of an open set U_α in M along with a mapping $\phi_\alpha : U_\alpha \to \mathbb{R}^n$. For each $p \in U_\alpha$, $\phi_\alpha(p)$ are the local coordinates of p in this local chart. ϕ_α is required to be homeomorphism from U_α onto $\phi_\alpha(U_\alpha)$. Note that a homeomorphism $T : (X_1, \tau_1) \to (X_2, \tau_2)$ from one topological space to another is a bijection which is continuous and has a continuous inverse. By continuity, we mean that if $O \in \tau_2$, i.e., O is any open subset of X_2, then $T^{-1}(O) \in \tau_1$, i.e., the inverse image of any open set is once again an open set. The C^∞ condition is defined as follows. Suppose $\alpha, \beta \in \Gamma, \alpha \neq \beta$ and suppose $U_\alpha \cap U_\beta$ is not empty. Then, any point p in this intersection will have two sets of coordinates, one with respect to the α chart and another with respect to the β chart. The map that carries one set of coordinates to another set is required to be C^∞, in other words,

$$\phi_\beta o \phi_\alpha^{-1} : \phi_\alpha(U_\alpha \cap U_\beta) \to \phi_\beta(U_\alpha \cap U_\beta)$$

is required to be C^∞, i.e., infinitely differentiable. Then, M becomes an n dimensional C^∞ manifold. Given a point $p \in M$, let U be a neighborhood of p, i.e. an open set containing p. A function $f : U \to \mathbb{R}$ is said to be defined in a neighborhood of p. In fact, one considers such a function to be represented by an equivalence class $[f]$. This class consists of all functions g defined in some neighborhood of p (the neighborhood can vary with the function chosen) such that g and f agree on a neighborhood of p. Suppose $f : U \to \mathbb{R}$. We say that f is a C^∞-function if the following happens. Let $\alpha \in \Gamma$ be such that $p \in U_\alpha$. Then, the map $f o \phi_\alpha^{-1} : \phi_\alpha(U \cap U_\alpha) \to \mathbb{R}$ is C^∞. In classical mechanics, C^∞ functions on the manifold play the role of observables. Tangent vectors and vector fields are important and they have to be introduced into this generalized framework of differentiable manifolds. Just as points on the manifold represent the generalized coordinates, a curve on the manifold which is a map $\gamma : I \to M$ where I is an interval of $\mathbb{R}$ represents the trajectory of the particle. The velocity vector at a given point represents a tangent vector. In our generalized framework, a tangent vector v at a point $p \in M$ is simply a linear map from the space $C^\infty(U)$ into $\mathbb{R}$, where U is a neighborhood of p that also satisfies the derivation property, i.e., if $f, g \in C^\infty(U)$, then $v(af + bg) = av(f) + bv(g)$ and $v(fg) = f(p)v(g) + g(p)v(f)$. In terms of local coordinates, if $(x_1, ..., x_n)$ are the local coordinates of p, then $v = \sum_{i=1}^{n} v_i \frac{\partial}{\partial x_i}$. $(v_1, ..., v_n)$ are termed as the local coordinates of v. The space of all tangent vectors at a point $p \in M$ is a vector space of dimension n. A basis for this space is $(\frac{\partial}{\partial x_i} : i = 1, 2, ..., n)$. This space is denoted by TM_p or TM_x where $x = (x_1(p), ..., x_n(p))$ are the local coordinates of p.

Study project 9

$X_1, ..., X_m$ are vector fields on C^∞ manifold M. $(X_1)_x, ..., (X_m)_x$ forms a basis for T_xM for each $x \in M$. Let $(\alpha_1, ..., \alpha_m)$ be an m tuple of non-negative integers. Consider the differential operator $X^{(\alpha)} = X_1^{\alpha_1}...X_m^{\alpha_m}$. This differential operator can be expressed in terms of local coordinates by setting

$$(X_i)_x = \sum_{i=1}^{n} a_{ik}(x)\frac{\partial}{\partial x_k}$$

Then for example,

$$(X_i X_j)_x = \sum_{k,r} a_{ik}\frac{\partial}{\partial x_k}a_{jr}\frac{\partial}{\partial x_r}$$

$$= \sum_{k,r} a_{ik} a_{jr} \frac{\partial^2}{\partial x_k \partial x_r} + \sum a_{ik} \frac{\partial a_{jr}}{\partial x_k} \frac{\partial}{\partial x_r}$$

Claim: $X^{(\alpha)}$ for different m tuples α are linearly independent differential operators over C^∞. Any differential operator D of order $\leq r$ can be expressed as $\sum_{|\alpha| \leq r} a_\alpha X^{(\alpha)}$.

D_r is spanned by the operators $X^{(\alpha)}, |\alpha| \leq r$. Note that differential operators on M form a module over $C^\infty(M)$. The module plays the role of the vector space and the space of functions chosen for forming linear combinations plays the role of the field. Any differential operator D in D_r can be expressed as $\sum_{|\alpha| \leq r} f_\alpha X^{(\alpha)}$ where the $f'_\alpha s$ are C^∞ functions. D_1 consists of vector fields. For the $X^{(\alpha)'s}$ which belong to D_1 are obtained by taking $\alpha_j = 1$ and $\alpha_k = 0$ for $k \neq j$. These operators are $X_1, ..., X_m$ and their linear span is the set of all vector fields on M.

Any vector field Z can be expressed as $Z = \sum_{j=1}^m c_j X_j$ where the $c'_j s$ are uniquely defined functions. We want to show that the $c'_j s$ are C^∞. This is obvious when we write

$$Z = \sum a_j \frac{\partial}{\partial x_j}$$

with the $a'_j s$ as C^∞ functions. Then,

$$\sum_k a_k \frac{\partial}{\partial x_k} = \sum_{j,k} c_j a_{jk} \frac{\partial}{\partial x_k}$$

so that

$$a_k = \sum_j c_j a_{jk}$$

Now linear independence of the $X'_k s$ implies that at each point $x \in M$, the matrix $((a_{jk}(x)))$ is invertible and hence if $((A_{jk}))$ denotes its inverse, then its entries $A'_{jk} s$ are C^∞ functions, since the $a'_{jk} s$ are C^∞ functions. Then,

$$c_j = \sum_k a_k A_{kj}$$

are also C^∞ functions. First of all we want to show that if $Z_1, ..., Z_l$ are l vector fields with $l \geq 1$, then $Z_1 ... Z_l \in D_l$. Let $Y_1, ..., Y_l$ be any l vector fields and let $F = Y_1 ... Y_l$. Let F' be obtained by interchanging two adjacent $Y's$ in this product, i.e., writing $F = Y_1 .... Y_i Y_{i+1} .... Y_l$, then $F' = Y_1 .... Y_{i+1} Y_i ... Y_l$. Then,

$$F - F' = Y_1 .... [Y_i, Y_{i+1}] ... Y_l$$

is in D_{l-1} by the induction hypothesis. Every permutation is a product of adjacent interchanges. Thus, if σ is any permutation of $1, 2, ..., l$, then

$$X_1 ... X_l - X_{\sigma 1} ... X_{\sigma l} \in D_{l-1}$$

If $1 \leq j_1 \leq j_2 \leq \ldots \leq j_l \leq m$, then $X_{j_1}\ldots X_{j_l} \in D_l$. Hence, if $(k_1, \ldots, k_m)$ is any permutation of $(1, 2, \ldots, m)$ and $\alpha_1 + \ldots + \alpha_m \leq l$, then

$$X_{k_1}^{\alpha_1}\ldots X_{k_m}^{\alpha_m} \in D_l$$

By the induction hypothesis, $Z_2\ldots Z_l \in D_{l-1}$, so

$$Z_2\ldots Z_l = \sum_{|\beta| \leq l-1} b_\beta X^{(\beta)}$$

Also

$$Z_1 = \sum_{j=1}^{m} c_j X_j$$

where c_j, b_β are C^∞ functions. Thus,

$$Z_1 Z_2\ldots Z_l = \sum_{j=1}^{m} c_j X_j \Big(\sum_{|\beta| \leq l-1} b_\beta X^{(\beta)} \Big)$$

$$= \sum_{j,\beta} c_j b_\beta X_j X^{(\beta)} + \sum_{j,\beta} c_j X_j(b_\beta) X^{(\beta)}$$

In view of the preceding statement, $Z_1\ldots Z_l \in D_l$. We've used this in that $X_j X^{(\beta)} \in D_l$ for all β occurring in the sum. For $\alpha = (\alpha_1, \ldots, \alpha_m)$, let

$$\partial^{(\alpha)} = \partial_1^{\alpha_1}\ldots\partial_m^{\alpha_m}$$

Then, taking the sequence $Z_1, \ldots, Z_l$ as $\partial_1, \ldots, \partial_1, \ldots, \partial_m, \ldots, \partial_m$, with ∂_i occurring α_i times, we get from what was just proved,

$$\partial^{(\alpha)} = \sum_{|\beta| \leq r} a(\alpha, \beta) X^{(\beta)}, |\alpha| \leq r$$

where the $a(\alpha, \beta)$ are C^∞ functions. Since the operators $\{\partial^{(\alpha)}, |\alpha| \leq r\}$ span $TM_x^{(r)}$, the space of all differential operators of order r at the point $x \in M$, it follows from the above formula that the operators $X^{(\beta)}, |\beta| \leq r$ also span this space. In fact, we know that the operators $\{\partial^{(\alpha)} : |\alpha| \leq r\}$ forms a basis for this space. Hence the dimension of $TM_x^{(r)}$ equals the number of tuples $(\alpha_1, \ldots, \alpha_m)$ for which $\alpha_i \geq 0$ and $\sum \alpha_i = |\alpha| \leq r$. Since this number exactly coincides with the number of operators in the set $\{X^{(\alpha)} : |\alpha| \leq r\}$, it follows that the latter set forms a basis for $TM_x^{(r)}$.

Study project 10

Definitions of flow on a Lie group and left invariant vector fields and the exponential map. Let G be a Lie group, i.e., a differentiable manifold with a group structure such that the group operations are differentiable. A vector field X on G has the coordinate representation

$$X_x = \sum_{i=1}^{n} X_i(x) \frac{\partial}{\partial x_i}$$

in terms of local coordinates. If f is a C^∞ function, then

$$X(f)(x) = X_x(f) = \sum_{i=1}^{n} X_i(x)\frac{\partial f(x)}{\partial x_i}$$

Define g−shift map $L_g : G \to G$ by $L_g(x) = gx$. If the Lie group G is n dimensional as a differentiable manifold, then we can by means of local co-ordinates, parameterize any point $x \in G$ as a vector in $\mathbb{R}^n$, i.e. $x = (x_1, ..., x_n)$, where $x_i \in \mathbb{R}$. The group composition operation is expressed as $x.y = \psi(x, y)$, where ψ is a differentiable function of two arguments in $\mathbb{R}^n$. Then, $L_g(x) = \psi(g, x)$. L_g induces a map L_g^* from the tangent space $T_x G$ to the tangent space $T_{gx}G$ according to the rule

$$(L_g^* X)_{gx}(f) = (L_g^* X f)(gx) = X(f o L_g)(x) = X_x(f o L_g)$$

In fact, L_g^*, by the above rule maps each vector field on G to another vector field on G. It maps the vector field X to the vector field $L_g^* X$. The vector field X is said to be left invariant if $L_g^* X = X$ for all $g \in G$, i.e., $X_{gx} = L_g^* X_x$, or equivalently,

$$X_x(f o L_g) = X_{gx}(f)$$

for all $g, x \in G$. In terms of components, we see that

$$\sum_{j} \frac{\partial \psi_i(x, y)}{\partial y_j} X_j(y) = X_i(\psi(x, y))$$

for all i, x, y. Let X be any vector field on G. In terms of local coordinates

$$X_x = \sum_{i} X_i(x)\frac{\partial}{\partial x_i}$$

Let $t \to \Phi(t, x)$ be the flow generated by X passing through the point x. In terms of local coordinates,

$$\frac{d\Phi_i(t, x)}{dt} = X_i(\Phi(t, x))$$

and $Phi(0, x) = x$. Let f be any smooth function on G. In terms of local coordinates, f is expressible as a function of n real variables $(x_1, ..., x_n)$. Then,

$$\frac{d}{dt}f(\Phi(t, x)) = \sum_{i} \frac{d\Phi_i(t, x)}{dt}\frac{\partial f}{\partial y_i}(y = \Phi(t, x))$$

$$= \sum_{i} X_i(y)\frac{\partial f(y)}{\partial y_i}\Big|_{y = \Phi(t, x)} = X(f)(\Phi(t, x))$$

For any vector field X, let $\Phi(t, x)$ be the flow generated by X passing through the point x. Define the map $exp(tX) : G \to G$ by $exp(tX)(x) = \Phi(t, x)$. Then, $exp(tX)$ is a diffeomorphism of G. Some authors prefer to define $exp(tX) \in G$ as $\Phi(t, e) = exp(tX)(e)$. We want to show that $exp(tX).exp(sX) = exp((t + s)X)$ as a relation between group elements whenever X is a left invariant vector field on G. In other words, we want to show that $\Phi(t, e)\Phi(s, e) = \Phi(t + s, e)$ for all $t, s \in \mathbb{R}$, or in terms of the group composition function ψ,

$$\psi(\Phi(t, e), \Phi(s, e)) = \Phi(t + s, e)$$

Let $\chi(t, s, e) = \psi(\Phi(t, e), \Phi(s, e))$. Then, this is equivalent to showing that χ satisfies the differential equation

$$\frac{\partial}{\partial t}\chi_i(t, s, e) = X_i(\chi(t, s, e))$$

in terms of local co-ordinates, for then in view of the identity, $\chi(0, s, e) = \Phi(s, e)$ and the uniqueness of the solutions to ordinary differential equations, it would follow that $\chi(t, s, e) = \Phi(t, s, e)$. Let x denote the first argument of ψ and y the second argument. Thus, $x.y = \Phi(x, y)$. Then,

$$\frac{\partial}{\partial t}\psi_i(\Phi(t, e), \Phi(s, e)) = \sum_j \frac{\partial \Phi_j(t, e)}{\partial t} \frac{\partial \psi_i(\Phi(t, e), \Phi(s, e))}{\partial y_j}$$

$$= \sum_j X_j(\Phi(t, e)) \frac{\partial \psi_i(\Phi(t, e), \Phi(s, e))}{\partial y_j}$$

But from left invariance of X, it follows that the above equals $X_i(\psi(\Phi(t, e), \Phi(s, e)))$. We've thus shown that

$$\frac{\partial}{\partial t}\chi_i(t, s, e) = X_i(\chi(t, s, e))$$

thereby proving the claim. Again, let X be any left invariant vector field on G and f any smooth function on G. We want to show that

$$\frac{d}{dt}f(x.exp(tX)) = \frac{d}{dt}f(x.\Phi(t, e)) = X(f)(x.exp(tX)) = X(f)(x\Phi(t, e))$$

In fact, this identity can be used as our definition of a left invariant vector field on G. First we prove the above statement. In terms of local co-ordinates,

$$\frac{d}{dt}f(x.\Phi(t, e)) = \frac{d}{dt}f(\psi(x, \Phi(t, e))) = \sum_{i,j} \frac{\partial \Phi_i(t, e)}{\partial t} \frac{\partial \psi_j(x, \Phi(t, e))}{\partial y_i} \frac{\partial f(z)}{\partial z_j}\Big|_{z=\psi(x,\Phi(t,e))}$$

$$= \sum_{i,j} X_i(\Phi(t, e)) \frac{\partial \psi_j(x, \Phi(t, e))}{\partial y_i} \frac{\partial f(z)}{\partial z_j}\Big|_{z=\psi(x,\Phi(t,e))}$$

But left invariance of X implies

$$\sum_i X_i(\Phi(t, e)) \frac{\partial \psi_j(x, \Phi(t, e))}{\partial y_i} = X_j(\psi(x, \Phi(t, e)))$$

Thus

$$\frac{d}{dt}f(x.\Phi(t,e)) = X(f)(x.Phi(t,e))$$

and this completes the proof. Suppose for example, G is a matrix Lie group and X is an element of $T_0(G)$, i.e., X is a matrix. we define a vector field $\tau(X)$ on G by the equation

$$\tau(X)(f)(x) = \frac{d}{dt}f(x.exp(tX))|_{t=0}$$

We want to show that $\tau(X)$ is a left invariant vector field on G. First of all, it is clear that $\tau(X)$ is a vector field by verifying the two required properties:

$$\tau(X)(cf+g) = c\tau(X)(f) + c\tau(X)(g)$$

Now, let $g \in G$. Then,

$$L_g^*\tau(X)(f)(gx) = \tau(X)(foL_g)(x) = \frac{d}{dt}foL_g(x.exp(tX))|_{t=0} = \frac{d}{dt}f(g.x.exp(tX))|_{t=0}$$

$$= \tau(X)(f)(gx)$$

proving that $L_g^*\tau(X) = \tau(X)$ and hence $\tau(X)$ is a left invariant vector field on G. In fact suppose X is any vector field on G and let $\Phi(t,x)$ be the flow generated by the vector field passing through the point $x \in G$, i.e., in terms of local coordinates,

$$\frac{d\Phi_i(t,x)}{dt} = X_i(\Phi(t,x)) \qquad \Phi(0,x) = x$$

Define another vector field $\tau(X)$ by the rule

$$\tau(X)(f)(x) = \frac{d}{dt}f(x.Phi(t,e))|_{t=0}$$

It is clear that $\tau(X)$ is a vector field. Linearity and the derivation property are easily verified. The nature of the vector field $\tau(X)$ is clearly dependent only on the nature of X in an arbitrarily small neighborhood of e. We want to show that $\tau(X)$ is a left invariant vector field on G. Indeed, for any $g \in G$,

$$L_g^*\tau(X)(f)(gx) = X(foL_g)(x) = \frac{d}{dt}f(g.x.exp(tX))|_{t=0} = \tau(X)(f)(gx)$$

In other words,

$$L_g^*\tau(X) = \tau(X)$$

proving left invariance of $\tau(X)$. We say that $\tau(X)_g$ is obtained by applying a left translation to the tangent vector X_e at T_eG by amount g. Let X be any vector field and consider the exponential map $exp(tX) : G \to G$ by the rule $x \to exp(tX)(x) = \Phi(t,x)$, i.e. $t \to \Phi(t,x)$ is the flow generated by the vector field X passing through the point x Then, $exp(tX)^*$ maps the tangent space T_xG linearly to the tangent space $T_{\Phi(t,x)}G = T_{exp(tX)x}G$. Taking $x = e$, we see that $exp(tX)^*$ maps T_eG linearly to $T_{\Phi(t,e)}G = T_{exp(tX)}G$. The mapping is defined by the formula

$$exp(tX)^*Y(f) = Y(foexp(tX))$$

or more precisely,

$$exp(tX)^*Y(f)(exp(tX)) = Y(foexp(tX))(e)$$

Here, Y is a vector in T_eG. More generally, suppose Y is a vector field. Then $exp(tX)^*Y$ is another vector field defined by the equation

$$(exp(tX)^*Y)_{exp(tX)x}(f) = (exp(tX)^*Y)(f)(exp(tX)x)$$

$$= Y(foexp(tX))(x) = Y_x((foexp(tX))$$

In terms of components,

$$(exp(tX)^*Y)_i(exp(tX)x) = \sum_j \frac{\partial \Phi_i(t,x)}{\partial x_j} Y_j(x)$$

Replacing x by $y = exp(-tX)x$ gives us

$$(exp(tX)^*Y)_i(x) = \sum_j \frac{\partial \Phi_i(t,y)}{\partial y_j} Y_j(y) = \sum_j \frac{\partial \Phi_i(t,\Phi(-t,x))}{\partial y_j} Y_j(\Phi(-t,x))$$

We then find that

$$\frac{d}{dt}(exp(tX)^*Y)_i(x)|_{t=0} = \sum_j \frac{\partial X_i(x)}{\partial x_j} Y_j(x) - Y_j(x) \frac{\partial X_i(x)}{\partial x_j}$$

$$= [Y,X]_i$$

or using coordinate free notation,

$$\frac{d}{dt}(exp(tX)^*Y)(x)_{t=0} = [Y,X]$$

Study project 11

Burnside's theorem. It finds applications in the quantum theory. Let $\mathcal{A}$ be an irreducible algebra over an algebraically closed field K. By an algebra, we mean a vector space over the field on which a product is defined which is associative and distributes with addition. Specifically, $(a,b) \to a+b$ from $\mathcal{A} \times \mathcal{A} \to \mathcal{A}$ is defined. This is commutative and associative and $(\alpha, a) \to \alpha.a$

from $K \times \mathcal{A} \to \mathcal{A}$ is defined such that $\alpha(a+b) = \alpha.a + \alpha.b, (\alpha+\beta).a = \alpha.a + \beta.a, 0.a = 0, \alpha.0 = 0$, where 0 is the zero element of the algebra. It satisfies $a + 0 = a$ for all $a \in \mathcal{A}$ and for any $a \in \mathcal{A}$, there is a $-a \in \mathcal{A}$ such that $a + (-a) = 0$. $(a,b) \to a.b$ from $\mathcal{A} \times \mathcal{A} \to \mathcal{A}$ is also defined. It satisfies $\alpha(a.b) = (\alpha.a).b = a.(\alpha.b)$ for all $\alpha \in K, a,b \in \mathcal{A}$, $a.(b+c) = a.b + a.c$ and $(a+b).c = a.c + b.c$. Associativity of the product holds, i.e., $a.(b.c) = (a.b).c$ for all $a, b, c \in \mathcal{A}$. The algebra is assumed to have a unit element 1, such that $a.1 = a = 1.a$ for all $a \in \mathcal{A}$. We are assuming $\mathcal{A}$ to be an irreducible matrix algebra, i.e., the elements of $\mathcal{A}$ are $n \times n$ matrices with the property that if W is a subspace of $\mathcal{K}^n$ invariant under all the $a's$ with $a \in \mathcal{A}$, then either $W = \{0\}$, or else, $W = K^n$. In other words, if W is a subspace such that $a(W) \subset W$ for all $a \in \mathcal{A}$, then $W = K^n$ or $\{0\}$. Burnside's theorem states that such an algebra must coincide with the algebra of all $n \times n$ matrices over K, i.e., if $\mathcal{A}$ an irreducible matrix algebra over an algebraically closed field, then $\mathcal{A} = K^{n \times n}$. In case, the field is not algebraically closed, the theorem gets modified as follows. Define $\mathcal{A}'$ to be the commutant of $\mathcal{A}$, i.e., the set of all matrices b with the property $ab = ba$ for all $a \in \mathcal{A}$. $\mathcal{A}'$ is called the commutant of $\mathcal{A}$. Now let $\mathcal{A}''$ denote the double commutant of $\mathcal{A}$, i.e., the set of all matrices c such that $cb = bc$ for all $b \in \mathcal{A}'$. Clearly, from the definitions, $\mathcal{A} \subset \mathcal{A}''$. Moreover, it is easy to see that $\mathcal{A}', \mathcal{A}''$ are also algebras. Burnside's theorem asserts that irreducibility of $\mathcal{A}$ implies that $\mathcal{A}'' = \mathcal{A}$, i.e., if c is any matrix over K that commutes with every b that commutes with every $a \in \mathcal{A}$, then $c \in \mathcal{A}$. In the special case when K is algebraically closed, we see that every $b \in \mathcal{A}'$ is of the form $b = \alpha.I$ for some $\alpha \in K$ when the algebra $\mathcal{A}$ is irreducible. This is Schur's lemma. In other words, $\mathcal{A}'$ consists of all scalar multiples of the identity and hence $\mathcal{A}''$ consists of all matrices in $K^{n \times n}$ and the result reduces to the previous one namely for irreducible algebras over an algebraically closed field, every endomorphism of the vector space is an element of the algebra.

Study project 12

Multilinear linear algebra related to Lie algebra theory

V is any vector space and V^* is its dual. Define $V_{r,s} = V^r \otimes V^{*s}$ where V^r is the tensor product of V with itself r times and V^{*s} is the tensor product of V^* with itself s times. Note that any element in $V_{r,s}$ can be viewed as an $r+s$-multilinear mapping on $V^* \times ... \times V^* \times V \times ... \times V$ where V^* appears r times and V appears s times. For example suppose $x_1, ..., x_r \in V$ and $f_1, ..., f_s \in V^*$. Then, the element $x_1 \otimes .... \otimes x_r \otimes f_1 \otimes ... \otimes f_s$ in $V_{r,s}$ acts on the point $(g_1, ..., g_r, y_1, ..., y_s) \in V^* \times ... \times V^* \times V \times ... \times V$ according to the rule

$$(x_1 \otimes ... \otimes x_r \otimes f_1 \otimes ... \otimes f_s)(g_1, ..., g_r, y_1, ..., y_s) = g_1(x_1)...g_r(x_r)f_1(y_1)...f_s(y_s)$$

Multilinearity is easy to verify. For $L \in End(V)$, define $\hat{L} \in End(V^*)$ by $(\hat{L}v*)(v) = -v * (Lv)$. Note that $End(V)$ can equivalently be viewed as the Lie algebra $gl(V)$ of the Lie group $GL(V)$ provided we introduce the Lie bracket between two endomorphisms as the commutator between the two endomorphisms. Now we can define $L_{r,s} \in End(V_{r,s}) = gL(V_{r,s})$ by the rule

$$L_{r,s}(x_1 \otimes ...x_r \otimes f_1 \otimes ... \otimes f_s) = \sum_{i=1}^{r} x_1 \otimes ... \otimes x_{i-1} \otimes Lx_i \otimes x_{i+1} \otimes ... \otimes x_r \otimes f_1 \otimes ... \otimes f_s$$

$$-\sum_{i=1}^{s} x_1 \otimes ... \otimes x_r \otimes f_1 \otimes ... \otimes f_{i-1} \otimes \hat{L}f_i \otimes f_{i+1} \otimes ... \otimes f_s$$

This is equivalent to defining

$$L_{r,0} = L \otimes I \otimes ... \otimes I + I \otimes L \otimes I \otimes ... \otimes I + ... + I \otimes ... \otimes I \otimes L$$

as a linear operator on $L_{r,0} = V^r$ and $L_{0,s} = (\hat{L})_{s,0}$ as a linear operator on $V_{0,s} = V^{*s}$ and setting $L_{r,s} = L_{r,0} \otimes I + I \otimes L_{0,s}$. Note how $L_{0,s}$ acts on V^{*s}:

$$L_{0,s}(f_1 \otimes ... \otimes f_s)(x_1, ..., x_s)$$

$$= -f_1(Lx_1)f_2(x_2)...f_s(x_s) - f_1(x_1)f_2(Lx_2)f_3(x_3)...f_s(x_s) - ... - f_1(x_1)...f_{s-1}(x_{s-1})f_s(Lx_s)$$

Note that $V_{1,1} = V \otimes V^*$. Define $Z : V \times V^* \to End(V)$ by $Z(v, v^*)(w) = v^*(w)v$. Then, Z is a bilinear map. From the basic definition of tensor products, it follows that there exists a linear map $\xi : V \otimes V^* \to End(V)$ such that $\xi(v \otimes v^*) = Z(v, v^*)$ for all $v \in V, v^* \in V^*$. Let $\{e_1, ..., e_m\}$ be a basis for V and $\{e_1^*, ..., e_m^*\}$ the dual basis for V^*. Let us determine the matrix of $\xi(e_l \otimes e_p^*) \in End(V)$ relative to the basis $\{e_1, ..., e_m\}$. We have

$$\xi(e_l \otimes e_p^*)(e_i) = e_p^*(e_i)e_l = \delta_{pi}e_l = \sum_q \delta_{pi}\delta_{lj}e_j$$

implying that the matrix of $\xi(e_l \otimes e_p^*)$ is $a_{ij} = \delta_{pi}\delta_{lj}$. Thus, ξ is a linear isomorphism of $V_{1,1}$ with $End(V)$. It is easy to see that $\xi o L_{1,1} o \xi^{-1} = ad(L)$. Indeed,

$$\xi o L_{1,1}(v \otimes v^*)(w) = \xi o(Lv \otimes v^* + v \otimes \hat{L}v^*)(w) = v^*(w)Lv - v^*(Lw)v$$

while on the other hand,

$$(ad(L)o\xi)(v \otimes v^*) = L\xi(v \otimes v^*) - \xi(v \otimes v^*)L$$

so that

$$(ad(L)o\xi)(v \otimes v^*)(w) = L\xi(v \otimes v^*)(w) - \xi(v \otimes v^*)Lw = Lv^*(w)v - v^*(Lw)v$$

$$= v^*(w)Lv - v^*(Lw)v$$

proving that

$$\xi o L_{1,1} = ad(L)o\xi$$

For $v \in V, v_1^*, v_2^* \in V^*$, define a bilinear map $y(v, v_1^*, v_2^*)$ from $V \times V \to V$ by the rule

$$y(v, v_1^*, v_2^*)(u_1, u_2) = v_1^*(u_1)v_2^*(u_2)v$$

Next define a linear map η from $V_{1,2}$ into the vector space B of bilinear maps from $V \times V \to V$ by the rule

$$\eta(v \otimes v_1^* \otimes v_2^*) = y(v, v_1^*, v_2^*)$$

η is a linear isomorphism of $V_{1,2}$ onto B. Since $dimV_{1,2} = dimB = (dimV)^3$, it suffices to show that η is a one-one map. $\{e_i \otimes e_j^* \otimes e_k^* : 1 \leq i,j,k \leq m\}$ is easily seen to be a basis for $V_{1,2}$. Suppose

$$\eta(\sum_{i,j,k} c(i,j,k)e_i \otimes e_j^* \otimes e_k^*) = 0$$

implies

$$\sum_{i,j,k} c(i,j,k)\eta(e_i \otimes e_j^* \otimes e_k^*) = 0$$

implies

$$\sum_{i,j,k} c(i,j,k)\eta(e_i \otimes e_j^* \otimes e_k^*)(v,w) = 0$$

for all $v,w \in V$ implies

$$\sum_{i,j,k} c(i,j,k)e_j^*(v)e_k^*(w)e_i = 0$$

for all $v,w \in V$. Taking $v = e_a, w = e_b$ gives $\sum_i c(i,a,b)e_i = 0$ for all a,b and hence $c(i,a,b) = 0$ for all i,a,b proving the claim. Let $\beta \in B$ and denote by V_β the algebra whose underlying vector space is V and whose multiplication law is given by $x.y = \beta(x,y)$. An element $L \in End(V)$ is a derivation of V_β, i.e.

$$L\beta(x,y) = \beta(Lx,y) + \beta(x,Ly)$$

for all $x,y \in V$ iff $L_{1,2}(\eta^{-1}(\beta)) = 0$. To see this, it suffices to show that for $\beta = \eta(v \otimes v_1^* \otimes v_2^*)$, we have

$$\eta(L_{1,2}(v \otimes v_1^* \otimes v_2^*))(x,y) = L\beta(x,y) - \beta(Lx,y) - \beta(x,Ly)$$

For linearity would imply that for any $t \in V_{1,2}$ and $\beta = \eta(t)$, we would have

$$\eta(L_{1,2}(t))(x,y) = L\beta(x,y) - \beta(Lx,y) - \beta(x,Ly)$$

so that $\eta(L_{1,2}\eta^{-1}(\beta)) = 0$ would imply $\eta(L_{1,2}(t)) = 0$ and hence that L is a derivation. The proof of the above identity follows by evaluating both sides.

$$\eta(L_{1,2}(v \otimes v_1^* \otimes v_2^*))(x,y) = \eta(Lv \otimes v_1^* \otimes v_2^* + v \otimes \hat{L}v_1^* \otimes v_2 + v \otimes v_1^* \otimes \hat{L}v_2^*)(x,y)$$

$$= v_1^*(x)v_2^*(y)Lv - v_1^*(Lx)v_2^*(y)v - v_1^*(x)v_2^*(Ly)v$$

On the other hand,

$$L\beta(x,y) = L\eta(v \otimes v_1^* \otimes v_2^*)(x,y) = v_1^*(x)v_2^*(y)Lv$$

$$\beta(Lx,y) = \eta(v \otimes v_1^* \otimes v_2^*)(Lx,y) = v_1^*(Lx)v_2^*(y)v$$

$$\beta(x,Ly) = \eta(v \otimes v_1^* \otimes v_2^*)(x,Ly) = v_1^*(x)v_2^*(Ly)v$$

thereby proving the validity of the identity.

Study project 13

Notion of integral manifold: Let M be an analytic manifold of dimension m. Consider a map $x \to \mathcal{L}_x$ from M, where for each $x \in M$, $\mathcal{L}_x$ is a subspace of $T_x M$, the tangent space to M at x. We are assuming that $dim \mathcal{L}_x = p$ where p is a fixed finite positive integer. Then, $\mathcal{L}$ is called a system of tangent spaces of rank p. A vector field X belongs to $\mathcal{L}$ on an open set U of M if $X_x \in \mathcal{L}_x$. We can visualize $\mathcal{L}$ as consisting of a subspace of dimension p attached to each point $x \in M$. Note that $T_x M$ is the underlying vector space at $x \in M$. $\mathcal{L}$ is said to be an analytic system of tangent spaces if for each $x \in M$, there is an open set U containing the point x and p analytic vector fields $X_1, ..., X_p$ on U such that $\{X_1(y), ..., X_p(y)\}$ is a basis for $\mathcal{L}_y$ at each point $y \in U$. $\mathcal{L}$ is said to be an involutive analytic system of in addition for every open subset U and analytic vector fields X, Y with $X_x, Y_x \in \mathcal{L}_x$ for all $x \in U$, we have $[X, Y]_x \in \mathcal{L}_x$ for all $x \in U$.

Study project 14

This project deals with an indirect definition of a differentiable mapping from one C^∞ manifold into another accompanied by the corresponding algebra homomorphism induced by such a mapping from the space of equivalence classes of C^∞ functions defined on a neighborhood of a point in the second manifold to those defined on the other. How the definition of a tangent vector can be made sharper by regarding it as acting on equivalence classes of C^∞ functions defined on a neighborhood of a point with two functions in a class coinciding on a neighborhood (perhaps smaller than the intersection of the domains) rather than as acting on individual functions.

Let M, N be C^∞ manifolds. Consider a continuous map $\pi : M \to N$. The map is said to be C^∞ if for any open set $U \subset N$ and $g \in C^\infty(U)$, $g o \pi \in C^\infty(\pi^{-1}(U))$. Note that π continuous implies $\pi^{-1}(U)$ is open in M. In terms of local coordinates, suppose $x \in M$ is represented by $(x_1, ..., x_n)$ and $\pi(x)$ by $(\pi_1(x_1, ..., x_n), ..., \pi_m(x_1, ..., x_n))$ with $n = dim M$ and $m = dim N$, then $g : U \to \mathbb{R}$ is represented by $g(y_1, ..., y_m)$ where $(y_1, ..., y_m)$ are local coordinates in $U \subset N$. This means that $g; \pi(x)$ is represented using our coordinates by the function $f(x_1, ..., x_n) = g(\pi_1(x_1, ..., x_n), ..., \pi_m(x_1, ..., x_n))$ and differentiability of this function f with respect to the x_i's for all differentiable functions g guarantees differentiability of the functions $\pi_j(x_1, ..., x_n)$. Indeed

$$\frac{\partial f}{\partial x_j} = \sum_i \frac{\partial \pi_i}{\partial x_j} \frac{\partial g}{\partial y_i}(\pi(x))$$

so that choosing $g(y_1, ..., y_m) = y_k$ gives

$$\frac{\partial f}{\partial x_j} = \frac{\partial \pi_k}{\partial x_j}$$

proving our claim. With the above notation, suppose g, g' are C^∞ around y and coincide in an open set U containing y where $y = \pi(x)$, then $g o \pi, g' o \pi$ are also C^∞ (g, g' are mappings on N while $g o \pi, g' o \pi$ are defined on N) around x and coincide in the open set $\pi^{-1}(U)$ containing x. This can be seen from the fact that $y = \pi(x) \in U$ implies $x \in \pi^{-1}(U)$ and if $z \in \pi^{-1}(U)$, then

$\pi(z) \in U$ so that $g(\pi(z)) = g'(\pi(z))$. Consider the map $g \to go\pi$ where g is a C^∞ map defined on an open set U in N so that $go\pi$ is a C^∞ map on the open set $\pi^{-1}(U)$ in M. The open set U can vary with the function g chosen but is restricted by the condition that it should contain $y = \pi(x)$ so that $\pi^{-1}(U)$ contains x. This map induces an algebra homomorphism from D_y into D_x, where D_y consists of equivalence classes of C^∞ functions defined on any neighborhood of y. Once again, this neighborhood can vary from function to function. By a neighborhood of y, we mean an open set containing y. Two C^∞ functions g, g' defined on neighborhoods U, U' of y are said to belong to the same equivalence class, or are simply said to be equivalent if there is a neighborhood V of y such that $V \subset U \cap U'$ and g restricted to V coincides with g' restricted to V. Thus each element of D_y is the collection of C^∞ functions defined on neighborhoods of y such that any two functions in the collection coincide on a neighborhood of y that is (obviously) contained in the intersection of the neighborhoods on which the two functions are defined. Similarly, one defines D_x in M. Suppose g, g' are two such functions, i.e., belonging to the same equivalence class in D_y with g defined on U and g' defined on U'. Then $go\pi$ is defined on the neighborhood $\pi^{-1}(U)$ of x and $g'o\pi$ is defined on the neighborhood $\pi^{-1}(U')$ of x and moreover, $go\pi$ restricted to $\pi^{-1}(V)$ coincides with $g'o\pi$ restricted to $\pi^{-1}(V)$ This means that $go\pi$ and $g'o\pi$ belong to the same equivalence class in D_x. If C denotes an equivalence class in D_y, then we let $Co\pi$ denote the equivalence class in D_x. It is easy to see that $C \to Co\pi$ is an algebra homomorphism. Indeed, suppose h, h' are two functions in some other equivalence class say C' in D_y. Then, h, h' are defined on neighborhoods W, W' of y in N and coincide on a neighborhood T of y where $T \subset W \cap W'$. It easily follows that $ag + bh$ and $ag' + bh'$ are C^∞ functions defined on neighborhoods $U \cap W$ and $U' \cap W'$ of y and coincide on the neighborhood $V \cap T$ of y. A similar statement holds for the functions gh and $g'h'$. We can define $aC + bC'$ as consisting of functions defined on a nhood of y and coinciding with $ag + bh$ on a nhood of y. similarly define CC'. Then the relations $(aC + bC')o\pi = aCo\pi + bC'o\pi$ and $(CC')o\pi = (Co\pi)(C'o\pi)$ are easily verified. These follow immediately from the equations $(ag + bg')o\pi = ago\pi + bg'o\pi$ and $(gg')0\pi = (go\pi)(g'o\pi)$. Let $X_x \in TM_x$. If $\pi : M \to N$ is a differentiable map, then we define $Y_y \in TM_y$ by the rule $Y_y(g) = X_x(go\pi)$ for g in $C^\infty(U)$. This definition can be made sharper by defining $Y_y(C)$ where C is an arbitrary member of the equivalence class D_y as $Y_y(C) = X_x(Co\pi)$. $Y_y(C)$. The operator Y_y can in this way be viewed as a derivation on the space D_y and likewise X_x as a derivation on the space D_x.

Study project 15

Suppose V is a finite dimensional vector space over $\mathbb{R}$ and let V_c denote its complexification. This can be defined along the following lines. Let $B = \{e_1, ..., e_n\}$ be a basis for V. Then, any vector $x \in V$, we can write $x = c_1 e_1 + ... + c_n e_n$ for some $c_1, ..., c_n \in \mathbb{R}$. For the space V_c we can take the same basis but consider formal linear combinations of the above type with the $c_j's$ as complex numbers. This is equivalent to considering vectors having the form $x + iy$ with $x, y \in V$. It is also equivalent to considering the elements of V_c as ordered doublets (x, y) with $x, y \in V$. The scalar multiplication law for such doublets reads $(a + ib)(x, y) = (ax - by, ay + bx)$ and the vector addition law reads $(x, y) + (x', y') = (x + x', y + y')$. We can regard V as a subspace of V_c, indeed, any vector in V is of the form $c_1 e_1 + ... + c_n e_n$ where $c_1, ..., c_n$ are complex numbers having zero imaginary part. Thus $x \in V$ implies $x \in V_c$. We regard V as an analytic manifold in the usual manner. Let S be the symmetric algebra over V_c, i.e., any element in S has the form $(\xi_0, \xi_1, \xi_2, ..., \xi_n, ...)$ where ξ_0 is a scalar, $\xi_1 \in V_c$ and for $k > 1$, ξ_k is spanned by $\sum_{\sigma \in S_k} e_{i_{\sigma 1}} \otimes .. \otimes e_{i_{\sigma k}}$, where $1 \le i_1 \le i_2 \le ... \le i_k \le n$. S_k is the symmetric group of order k, i.e., all permutations of $\{1, 2, ..., k\}$. The summation is over all such permutations. Equivalently, ξ_k can be regarded as a symmetric k-linear functional over $V_c^* \times ... \times V_c^*$, there being k terms in the cartesian product. For any $u \in V$, let $\partial(u)$ be the endomorphism of $C^\infty(V)$ defined by

$$\partial(u)f(x) = \frac{d}{dt}f(x + tu)|_{t=0}$$

with $x \in V$. It is clear that $\partial(u)f$ is a C^∞ function on V. In fact, if we use rectangular coordinate system, then

$$\partial(u)f(x) = (u, \nabla f(x))$$

$\partial(u)$ is easily seen to be a vector field. In terms of coordinates, it corresponds to the vector field $\sum_{i=1}^n u_i \frac{\partial}{\partial x_i}$, i.e., the vector field has the same components at each point $x \in V$. It can also be seen that the association $u \to \partial(u)$ extends uniquely to an isomorphism ∂ of S onto the subalgebra $Diff(V)$ consisting of all differential operators which are invariant under all translations of V. We define the action of ∂ on the symmetric element $\sum_{\sigma \in S_s} u_{\sigma 1} \otimes ... \otimes u_{\sigma s}$ to be $\partial(u_1, ..., u_s)$, where

$$\partial(u_1, ..., u_s)f(x) = \frac{\partial^s}{\partial t_1 ... \partial t_s}f(x + t_1 u_1 + ... + t_s u_s)|_{t_1 = ... = t_s = 0}$$

In terms of coordinates,

$$\partial(u_1, ..., u_s) = (u_1, \nabla)(u_2, \nabla)...(u_s, \nabla) = \sum_{i_1, ..., i_s = 1}^n u_{1,i_1} u_{2,i_2} ... u_{s,i_s} \frac{\partial^s}{\partial x_{i_1} ... \partial x_{i_s}}$$

This differential operator is evidently symmetric with respect to the vectors $u_1, ..., u_s$ and hence can be regarded as the image of the symmetric element $\sum_\sigma u_{\sigma 1} \otimes ... u_{\sigma s}$. In general, we can consider the image of any element of the symmetric algebra as the sum of the images of the different components and the resultant is a differential operator. The extension of ∂ from $u \to \partial(u)$ to the entire symmetric algebra is unique. In fact, let us denote the symmetric product of the vectors $u_1, ..., u_k$ by $S(u_1, ..., u_k)$. Then if ∂ is an isomorphism, we must have

$$\partial(S(u_1, ..., u_k)) = \partial(u_1)...\partial(u_k) = (u_1, \nabla)...(u_k, \nabla)$$

Now, let $\{u_1, ..., u_m\}$ be a basis for V and let dV be the m-form on V such that

$$dV(\partial(u_1), ..., \partial(u_m)) = 1$$

Then, dV is invariant under all translations. This can be seen from the fact that if $L_a x = x + a$ on V, then

$$L_a^* dV(\partial(u_1), ..., \partial(u_m)) = dV(L_a * \partial(u_1), ..., L_a * \partial(u_m)) = dV(\partial(u_1), ..., \partial(u_m))$$

since

$$(L_a^* \partial(u))_{x+a}(f) = \partial(u)_x(foL_a) = (u, \nabla)f(x+a) = \partial(u)_{x+a}(f)$$

so that $L_a^* \partial(u) = \partial(u)$. The measure associated with the m form can be seen to be the Lebesgue measure. Indeed, taking $u_1 = e_1, ..., u_m = e_m$ the standard basis, we have $\partial(u_i) = \frac{\partial}{\partial x_i}$ and since

$$dV(\frac{\partial}{\partial x_1}, ..., \frac{\partial}{\partial x_m}) = 1$$

the result follows. The next thing to note is that there is a unique automorphism $a \to a^*$ of the symmetric algebra S such that $u^* = -u$ for all and $u \in V$ and that $\partial(a^*)$ is the formal adjoint of $\partial(a)$ relative to dV for any $a \in S$. Such an adjoint will carry the symmetric element $S(v_1, ..., v_k)$ where $v_1, ..., v_k$ are vectors in V to the element $S(-v_1, ..., -v_k) = (-1)^k S(v_1, ..., v_k)$. The differential operator associated with this element is given by

$$(-1)^k \partial(v_1, ..., v_k) = (-1)^k \sum v_{1,i_1}...v_{k,i_k} \frac{\partial^k}{\partial x_{i_1}...\partial x_{i_k}}$$

It is known from integral calculus that this operator is the formal adjoint of $\partial(v_1, ..., v_k)$ with respect to the Lebesgue measure. Indeed,

$$\int f(x_1, ..., x_m)\partial(v_1, ..., v_k)g(x_1, ..., x_m)dx_1...dx_m$$

$$= (-1)^k \int g(x_1, ..., x_m)\partial(v_1, ..., v_k)f(x_1, ..., x_m)dx_1...dx_m$$

Study project 16

If D denotes a differential operator, then the notation $f(x; D)$ stands for $Df(x)$, the operator D acting on the function f evaluated at the point x. Let V be a finite dimensional vector space over the field $\mathbb{R}$ and let $\mathcal{M}$ denote the space of all endomorphisms of V. $\mathcal{M}$ is a vector space over $\mathbb{R}$ of dimension n^2 and is an analytic manifold in the usual manner. $GL(V)$, the space of all non-singular linear transformations on V is an open submanifold of $\mathcal{M}$ in the usual manner. The tangent space to $\mathcal{M}$ at any point is identified with $\mathcal{M}$ itself. Let $\}$ denote the Lie algebra of $GL(V)$. For $X \in \}$, let X^0 denote the value of X at the identity element I of $GL(V)$. $dim\mathcal{M} = n^2 = dim\}$, so it is clear that the map $X \to X^0$ from $\}$ to $\mathcal{M}$ is an isomorphism. Let f be a linear function on $\mathcal{M}$. For $A \in GL(V)$ and $Y \in \}$, we have

$$f(A; Y) = \frac{d}{dt}f(Aexp(tY^0))|_{t=0} = \frac{d}{dt}f(A + tAY^0)|_{t=0} = f(AY^0)$$

The last equation follows from the fact that since f is a linear function, it can be represented as $f(M) = Tr(\Lambda M)$ for some matrix Λ and hence

$$f(A + tAY^0) = Tr(\Lambda(A + tAY^0)) = Tr(\Lambda A) + t.Tr(\Lambda AY^0)$$

giving

$$\frac{d}{dt}f(A + tAY^0) = Tr(\Lambda AY^0) = f(AY^0)$$

Also

$$f(1; XY) = Yf(1; X) = \frac{d}{dt}Yf(exp(tX^0))|_{t=0} = \frac{\partial^2}{\partial t \partial s}f(exp(tX^0)exp(sY^0))|_{s=t=0}$$

$$= \frac{\partial^2}{\partial t \partial s}f((I + tX^0)(I + sY^0))|_{s=t=0} = \frac{\partial^2}{\partial t \partial s}f(I + tX^0 + sY^0 + tsX^0Y^0)|_{s=t=0}$$

$$= f(X^0Y^0)$$

Thus,

$$f(1; [X, Y]) = f(1; XY - YX) = f(1; XY) - f(1; YX)$$

$$= f(X^0Y^0) - f(Y^0X^0) = f(X^0Y^0 - Y^0X^0)$$

from which we conclude using the previous identity that

$$f([X, Y]^0) = f(X^0Y^0 - Y^0X^0)$$

implying that

$$[X, Y]^0 = X^0Y^0 - Y^0X^0$$

Study project 17

Deals with properties of the Lie algebra of a Lie subgroup of a Lie group regarding its relationship to how it is imbedded in the larger Lie algebra.

Let G_1, G_2 be two Lie groups with Lie algebras $\}_1, \}_2$ respectively. Let $\pi : G_1 \to G_2$ be an analytic homomorphism, by homomorphism, we mean that $\pi(gg') = \pi(g)\pi(g')$ for all $g, g' \in G_1$ and by analytic we mean that if we parameterize an element $g \in G_1$ using local coordinates as $g = (x_1, ..., x_n)$ and we also parameterize the elements in G_2 using local coordinates as $(y_1, ..., y_m)$, then in terms of coordinates, $\pi(g)$ has the form $(\pi_1(x_1, ..., x_n), ..., \pi_m(x_1, ..., x_n))$. By analytic, we meant that the functions $\pi_j(x_1, ..., x_n)$ have Taylor series expansions, i.e., if $\mathbf{x}, \mathbf{x}'$ are two points in G_1 and $\phi(\mathbf{x}, \mathbf{x}')$ denotes their composition, then $\pi_j(\phi(\mathbf{x}, \mathbf{x}')$ can be expanded as a power series in the variables $\mathbf{x}'$. The expansion will have the form

$$\pi_j(\mathbf{x}, \mathbf{x}') = \sum_{r_1,...,r_n=0}^{\infty} c_j(r_1, ..., r_n, \mathbf{x}))x_1'^{r_1}...x_n'^{r_n}$$

We claim that $\pi(expX) = expd\pi(X)$ for all $X \in \}_1$. Note that X can be regarded as a vector in $T_{e_1}G_1$ or equivalently as a left invariant vector field defined uniquely by its value at e_1.

If $t \rightarrow \xi_X(t)$ is the flow generated by the vector field X. Then in terms of components, we have the following differential equations for the flow:

$$\frac{d\xi_X(t)}{dt} = X(\xi_X(t))$$

or

$$\frac{d}{dt}\xi_{X,i}(t) = X_i(\xi_{X,1}(t), ..., \xi_{X,n}(t)), i = 1, 2, ..., n$$

The initial condition chosen for the flow is $\xi_X(0) = e_1$ where e_1 is expressed in terms of coordinates. $expX$ is the value of $\xi_X(t)$ at $t = 1$, i.e., $expX = \xi_X(1)$. $exptX = \xi_X(t)$.. To see the validity of this formula, we note first of all that if we define $\eta_X(t) = \pi(exp(tX))$, then $t \rightarrow \eta_X(t)$ is a homomorphism, i.e., $\eta_X(t)\eta_X(s) = \eta_X(t+s)$. This follows from the hypothesis that π is a homomorphism and that $t \rightarrow exp(tX)$ is also a homomorphism from $\mathbb{R}$ into G_1:

$$\eta_X(t)\eta_X(t') = \pi(exp(tX))\pi(exp(t'X) = \pi(exp(tX)exp(t'X))$$

$$= \pi(exp((t+t')X)) = \eta_X(t+t')$$

Using our coordinate notation, we have

$$\frac{d}{dt}\eta_{X,a}(t) = \frac{d}{dt}\pi_a(\xi_X(t)) = \sum_b \frac{\partial\pi_a(x)}{\partial x_b}\Big|_{x=\xi_X(t)}\frac{d\xi_{X,b}(t)}{dt}$$

Putting $t = 0$ in this equation gives

$$\frac{d}{dt}\eta_{X,a}(0) = \sum_b \frac{\partial\pi_a(e_1)}{\partial x_b}X_b(e_1) = d\pi(X)_a(e_2)$$

since $\pi(e_1) = e_2$. e_1 here is the identity element of G_1 at e_1 and e_2 is the identity element of G_2 at e_2. Thus, $\eta_X(t)$ is the flow generated by the element $d\pi(X) \in \}_2$, i.e.,

$$\eta_X = \xi_{d\pi(X)}$$

or equivalently,

$$\pi(exp(tX)) = exp(td\pi(X))$$

For $t = 1$, we get $\pi(expX) = exp(d\pi(X))$. Suppose X is an element of $\}_1$ such that $d\pi(X) = 0$. In other words, we have a left invariant vector field on G_1 that maps into the zero vector field on G_2 under the action of the homomorphism π. Then, it is clear from the equation $exp(t0) = e_2$ when 0 is the zero vector field in G_2 that $\pi(exp(tX)) = 0$ for all t. Conversely, if X is any left invariant vector field on G_1 such that $\pi(exp(tX)) = e_2$ for all $t \in \mathbb{R}$, i.e. $exp(tX)$ belongs to the kernel of π, then $exp(td\pi(X)) = e_2$ for all $t \in \mathbb{R}$ and differentiating the coordinate form of this equation at $t = 0$ gives us $d\pi(X) = 0$, i.e. X maps to the zero vector field in G_2.

Now suppose $\langle'$ is the Lie algebra of a Lie subgroup H of G. Define $i : H \to G$ as $i(h) = h$, i.e. i is the imbedding of H in G. i is clearly a homomorphism. For $X' \in \langle'$, we obtain from the above by taking $G_1 = H, G_2 = G$ and $\pi = i$ that

$$i(exp(tX')) = exp(tdi(X')), t \in \mathbb{R}$$

The subalgebra $\langle$ of G defined by the subgroup H is defined by the set of all left invariant vector fields on G having the form $di(X')$ with X' varying over $\langle'$. Note that the map i imbeds H in G while the map di imbeds the Lie algebra $\langle'$ of H in the Lie algebra $\}$ of G. Taking $X = di(X')$ in the above formula gives

$$exp(tX) = i(exp(tX'))$$

It follows from this formula that for all $X \in \langle$, $exp(tX) \in H$ because $exp(tX') \in H$ and $i(exp(tX')) = exp(tX')$.

Study project 18

This project deals with topological properties of the action defined by a group on a manifold. How the group action induces a mapping between the tangent spaces of the group and of the manifold and how the stabilizer subgroup defines a submanifold of the group, how the dimension of the stabilizer group is computed and what happens in the special case of transitive action.

Let G be a Lie group acting transitively and analytically on an analytic manifold M and let $x_0 \in M$. The stability subgroup $G_{x_0} = \{g \in G : gx_0 = x_0\}$ is a closed Lie subgroup of G. If G acts transitively on M, the map $\gamma : G \to M$ defined by $\gamma(g) = gx_0$ is an analytic submersion of G onto M. M is the quotient manifold of G relative to the map γ. By saying that G acts analytically on M, we mean that if local coordinates are introduced on G as well as on M, and the action of G on M is described in terms of these coordinates, i.e., $\psi(g, x) = gx$, then ψ has a Taylor expansion in the vicinity of any ordered pair (g, x). To prove the above result, we first of all note that $\gamma o l_g = t_g o \gamma$ where l_g denotes left translation on G, i.e., $l_g(g') = gg'$ and $t_g : M \to M$ is defined by $t_g(x) = gx$. The above identity follows from the fact that $\gamma o l_g(g') = \gamma(gg') = gg'x_0$ and $t_g o \gamma(g') = t_g(g'x_0) = gg'x_0$ for any $g' \in G$. We thus get

$$d\gamma.dl_g = dt_g.d\gamma$$

(Note that if M_1, M_2, M_3 are differentiable manifolds and $\phi_1 : M_1 \to M_2$ and $\phi_2 : M_2 \to M_3$ are differentiable maps, then $\phi_2 o \phi_1 : M_1 \to M_3$ is also a differentiable map. Now, $d\phi_1 : TM_1 \to TM_2$ is a linear map with $d\phi_1$ mapping TM_{1x} into $TM_{2\phi_1(x)}$ and $d\phi_2 : TM_2 \to TM_3$ is a linear map with $d\phi_2$ mapping TM_{2y} into $TM_{2\phi_2(y)}$. $d(\phi_2 o \phi_1)$ maps TM_1 linearly into TM_3 and if X is any vector field on M_1, i.e., $X \in TM_1$, then for any differentiable function f on M_1,

$$d(\phi_2 o \phi_1)(X)(f) = X(f o \phi_2 o \phi_1) = X((f o \phi_2) o \phi_1) = d\phi_1(X)(f o \phi_2) = d\phi_2 d\phi_1(X)(f)$$

Thus, we have the identity,

$$d(\phi_2 o \phi_1) = d\phi_2 d\phi_1$$

It follows from the above discussion therefore that

$$d\gamma dl_g = dt_g d\gamma$$

and hence

$$d\gamma dl_g(TG_e) = dt_g d\gamma(TG_e)$$

Now, $dl_g(TG_e) = TG_g$ and letting $V_g = d\gamma(TG_g)$, we thus get

$$V_g = dt_g(V_e)$$

Note that γ maps G onto M and hence $d\gamma$ maps TG_g into TM_{gx_0}. Thus, V_g is a subspace of TM_{gx_0} and V_e is therefore a subspace of TM_{x_0}. Thus the above equation should be more precisely written as

$$V_g = (dt_g)_{x_0}(V_e)$$

Note that $t_g : M \to M$ is a diffeomorphism and hence $dt_g : TM_x \to TM_{gx}$ is a bijective linear map. Thus,

$$dim V_g = dim(dt_g)_{x_0}(V_e) = dim V_e$$

which means that $dim V_g$ is independent of g. Call this dimension as k. Let $y \in G_{x_0}$. Then it follows from the fact that $dim(d\gamma(TG_g)) = k$ that there is an open set U containing y such that $\gamma(U)$ is a k dimensional regular submanifold of M and $\gamma : U \to \gamma(U)$ is a submersion. Thus, $\{z : z \in U : \gamma(z) = x_0\} = \gamma^{-1}(\{x_0\}) \cap U = G_{x_0} \cap U$ is a regularly imbedded submanifold of U. Suppose we introduce local coordinates on G and also on M. Then what does γ look like in terms of these coordinates ? It has the form $(\gamma_1(x_1, ..., x_n), ..., \gamma_k(x_1, ..., x_n), 0, 0, ..., 0)$ where $n = dim G$ and $m = dim M$. In this expression, the last $m - k$ entries are zeroes. The coordinates can be permuted so that the matrix $((\frac{\partial \gamma_i}{\partial x_j}))_{1 \leq i,j \leq k}$ is non-singular locally from the definition of submersion. If we parameterize $\gamma(U)$ by the first k coordinates of M, then the map $\gamma : U \to \gamma(U)$ has the form

$$(x_1, ..., x_n) \to (\gamma_1(x_1, ..., x_n), ..., \gamma_k(x_1, ..., x_n))$$

The set $G_{x_0} \cap U$ has the form

$$\{(x_1, ..., x_n) \in U : (\gamma_1(x_1, ..., x_n), ..., \gamma_k(x_1, ..., x_n)) = (x_{01}, ..., x_{0k})\}$$

where x_0 has the coordinate representation $(x_{01}, ..., x_{0k}, 0, ..., 0)$. Putting k constraints on the functions $\gamma_1, ..., \gamma_k$ leaves us with $n - k$ degrees of freedom. This means that the manifold $G_{x_0} \cap U$ has $n - k$ degrees of freedom and hence it is an $n - k$ dimensional submanifold of U. In particular, $G_{x_0} \cap U$ is a regularly imbedded submanifold of U. We've chosen $y \in G_{x_0}$ arbitrarily and U an open neighborhood of y. This means that G_{x_0} is a regular submanifold of G. Suppose that G acts transitively on M. Then, $\gamma(U)$ is open in M. A heuristic argument for this would run along the following lines. Let $(g, x) \to \psi(g, x)$ as earlier denote the action of G on M. Since the action is transitive, it follows that for fixed x_0 as g varies over the entire group G, the point $\psi(g, x_0)$ covers the entire M. In terms of coordinates, it must follow that as dg varies over a small open subset of the origin, $\psi(\phi(dg, g), x_0)$ will vary over an open neighborhood of $\psi(g, x_0)$ where $\phi(g, g')$ denotes the composition of g and g'. In fact this infinitesimal deviation from $\psi(g, x_0)$ can be represented as

$$\psi_i(\phi(dg, g), x_0) = \psi_i(g, x_0) + \sum_{j,k} \frac{\partial \psi_i(g, x_0)}{\partial g_j} \frac{\partial \phi_j(e, g)}{\partial e_k} dg_k$$

The coefficient matrix

$$a_{ik} = \sum_j \frac{\partial \psi_i(g, x_0)}{\partial g_j} \frac{\partial \phi_j(e, g)}{\partial e_k}$$

must be onto (i.e., full row rank) since otherwise transitivity will not hold good. In fact the finite transformation $\psi(\phi(h, g), x_0)$ can be built out of infinitesimal transformations since the finite transformation $\phi(h, g)$ can be built out of infinitesimal ones having the form

$$g_2 = \phi(dg_1, g), g_3 = \phi(dg_2, g_2), ..., g_n = \phi(dg_{n-1}, g_{n-1}) = \phi(h, g)$$

or in terms of our usual notation for composition, $g_2 = dg_1.g, g_3 = dg_2.g_2, ..., h.g = dg_{n-1}g_{n-1}$, i.e., $h.g = dg_{n-1}dg_{n-2}...dg_1.g$ or $h = dg_{n-1}.dg_{n-2}...dg_1$ which amounts to the equation

$$h = \phi(dg_{n-1}, \phi(dg_{n-2}, ..., \phi(dg_2, dg_1))...)$$

As $dg = (dg_i)$ varies over an open subset of G, $(\sum_j a_{ij} dg_j)$ will vary over an open subset of M since the matrix (a_{ij}) has full row rank and from this we can infer that as g varies over an open set $\psi(g, x_0)$ will vary over an open set, i.e., $\psi(U, x_0) = \gamma(U)$ is open in M. Hence $k = dim M$ for transitive actions which means that $(d\gamma)_g$ is surjective for all $g \in G$.

Study project 19

Further topological properties of group action. Let M be a G space with G, M both being locally compact and second countable. Let G act transitively on M, let $x_0 \in M$ and let G_{x_0} be the stability subgroup at x_0. Then the map $\pi : G/G_{x_0} \to M$ defined by $\pi(gG_{x_0}) = gx_0$ is well defined, is a homeomorphism and intertwines the actions of G on the two spaces. In particular, the map $g \to gx_0$ is an open map of G onto M.

Define $\beta : G \to G/G_{x_0}$ by $\beta(g) = g.G_{x_0}$. Define $\gamma : G \to M$ by $\gamma(g) = gx_0$. Since G acts transitively on M, γ is surjective. By definition, γ is continuous and $\pi o \beta(g) = \pi(gG_{x_0}) = gx_0 = \gamma(g)$, i.e., $\gamma = \pi o \beta$. G/G_{x_0} has the quotient topology, i.e., a set $U \in G/G_{x_0}$ is open iff $\beta^{-1}(U)$ is open in G. Let E be open in M. Then $\gamma^{-1}(E)$ is open in G since γ is continuous. Thus, $\beta^{-1}\pi^{-1}(U)$ is open in G. It follows that $\pi^{-1}(U)$ open in G/G_{x_0}, i.e., π is continuous. It is clear that π is one-one and intertwines the G actions on G/G_{x_0} and M. That π is one-one follows from the fact that $gG_{x_0} = g'G_{x_0}$ implies $g'^{-1}g \in G_{x_0}$ which in turn implies that $g'^{-1}gx_0 = x_0$ and hence $gx_0 = g'x_0$. That π intertwines the two actions follows from the identity

$$\pi(g'gG_{x_0}) = g'gx_0 = g'\pi(gG_{x_0}), g, g' \in G$$

which is the same as

$$\pi o g' = g' o \pi$$

for all $g' \in G$. To prove that π^{-1} is continuous, we must show that π is an open map, i.e., if U is an open subset of G/G_{x_0}, then $\pi(U)$ is an open subset of M. Suppose U is a non-empty open subset of G/G_{x_0}. Then, $\beta^{-1}(U)$ is a non-empty open subset of G. By second countability of G, there is a sequence $\{g_n\}$ in G such that $G = \bigcup_{n=1}^{\infty} g_n \beta^{-1}(U)$. It follows that

$$G/G_{x_0} = \beta(G) = \beta(\bigcup g_n \beta^{-1}(U)) = \bigcup g_n \beta^{-1}(U)G_{x_0}$$

It is clear that $\beta^{-1}(U)G_{x_0} = U$ for suppose we write $U = \{gG_{x_0} : g \in A\}$ with $A \subset G$. Then, $g' \in \beta^{-1}(U)$ iff $\beta(g') \in U$ iff $g'G_{x_0} = gG_{x_0}$ for some $g \in A$, iff $g^{-1}g' \in G_{x_0}$ for some $g \in A$ iff $g'G_{x_0} = gG_{x_0}$ for some $g \in A$. Thus, $g' \in \beta^{-1}(U)$ implies $g'G_{x_0} \in U$. Conversely, $g'G_{x_0} \in U$ implies $g'G_{x_0} = gG_{x_0}$ for some $g \in A$ implies $g' \in \beta^{-1}(U)$ implies $g'G_{x_0} \in \beta^{-1}(U)G_{x_0}$. Thus,

$$G/G_{x_0} = \bigcup g_n U$$

Now, let V be any compact subset of G/G_{x_0} with a non-empty interior. Then by the above argument, we can choose a sequence g_n in G such that

$$G/G_{x_0} = \bigcup g_n V$$

From this we get,

$$M = \pi(G/G_{x_0}) = \bigcup \pi(g_n V) = \bigcup g_n \pi(V)$$

Since π is continuous and V compact, $\pi(V)$ is also compact and hence by Baire's category theorem, some $g_n \pi(V)$ has a non-empty interior. Thus, $\pi(V)$ has a non-empty interior. Let $W = int\pi(V)$ and $V_1 = \pi^{-1}(W)$. Note that $W \subset \pi(V)$ and hence $V_1 = \pi^{-1}(W) \subset V$. Hence $Cl(V_1)$ is a closed subset of the compact set V and is therefore itself compact. Since π is one-one, it follows that π is a homeomorphism of V_1 onto W. In fact, π maps G/G_{x_0} onto M in a one-one fashion and hence, the equation $V_1 = \pi^{-1}(W)$ implies $\pi(V_1) = W$. To see the homeomorphism property, suppose E is open in W. Then, $E = W \cap E'$ where E' is open in M. Then, $\pi^{-1}(E) = \pi^{-1}(W) \cap \pi^{-1}(E')$ with $\pi^{-1}(W)$ and $\pi^{-1}(E')$ open in view of the continuity of π and $\pi^{-1}(W) = V_1$ implies that $\pi^{-1}(E)$ is open in V_1. Again, suppose C is closed in V_1. Then $C = V_1 \cap C'$ where C' is closed in G/G_{x_0}. Now, $\pi(C) = \pi(V_1) \cap \pi(C')$ since π is one-one. Also $Cl(C) \subset Cl(V_1) \cap C'$ and hence, Now

$$\pi(C) = \pi(V_1) \cap \pi(C') = \pi(V_1) \cap \pi(Cl(V_1)) \cap \pi(C')$$

$$= \pi(V_1) \cap \pi(Cl(V_1) \cap C') = W_1 \cap \pi(Cl(V_1) \cap C')$$

Now $Cl(V_1) \cap C'$ is a closed subset of the compact set $Cl(V_1)$ and is therefore compact and hence closed. Thus, $\pi(C)$ is closed in W_1 and this proves that $\pi : V_1 \rightarrow W = \pi(V_1)$ is a homeomorphism.

     Remark: If X is a Hausdorff space and C is a compact subset of X, then C is closed. This can be seen via the following argument. Let $x \in C^c$. For each $y \in C$ choose open sets U_y, V_y such that $y \in V_y, x \in U_y$ and $U_y \cap V_y = \phi$. Then, $\{V_y : y \in C\}$ is an open cover for C. By compactness of C, there is a finite subcover $\{V_{y_i} : i = 1, 2, ..., n\}$. Then, let $U = U_{y_1} \cap .. \cap U_{y_n}$. It is clear that $x \in U$ and since $U \cap V_{y_i} = \phi, i = 1, 2, ..., n$, it follows that $U \cap C = \phi$. This proves that C^c is open and hence C is closed.

Now, let E be any open subset of G/G_{x_0}. Then, $\beta^{-1}(E)$ is an open subset of G. Now, W is a non-empty open subset of M and since $\pi : G/G_{x_0} \to M$ is continuous and onto, $V_1 = \pi^{-1}(W)$ is a non-empty open subset of G/G_{x_0}. Thus, $\beta^{-1}(V_1)$ is a non-empty open subset of G. Let $g \in \beta^{-1}(E)$. Since $\beta^{-1}(E)$ is open, we can find an open set U_g in G such that $g \in U_g \subset \beta^{-1}(E)$. Let $g_0 \in \beta^{-1}(V_1)$. Then, $e \in g_0^{-1}\beta^{-1}(V_1)$ and hence $g \in gg_0^{-1}\beta^{-1}(V_1)$. Now, $gg_0^{-1}\beta^{-1}(V_1)$ is open and hence there exists an open set V_g in G such that $g \in V_g \subset gg_0^{-1}\beta^{-1}(V_1)$. Thus, $F_g = U_g \cap V_g$ is an open neighborhood of g contained in $\beta^{-1}(E)$ as well as in $gg_0^{-1}\beta^{-1}(V_1)$. Clearly we have

$$\beta^{-1}(E) = \cup_{g \in \beta^{-1}(E)} F_g = \cup_{g \in \beta^{-1}(E)} gg_0^{-1}(g_0 g^{-1} F_g)$$

Now $H_g = g_0 g^{-1} F_g$ is an open set in $\beta^{-1}(V_1)$ and the above implies that

$$\beta^{-1}(E) = \cup_{g \in I} g H_g$$

where $I = \beta^{-1}(E)g_0^{-1}$ is an index set. Thus,

$$\beta\beta^{-1}(E) = \beta(\cup_{g \in I} g H_g)$$

It is easy to see that $\beta\beta^{-1}(E) = E$. First note that E is of the form $\{gG_{x_0} : g \in A\}$ and hence $\beta^{-1}(E) = AG_{x_0}$ where A is some subset of G. Thus, $\beta\beta^{-1}(E)$ is given by $\{gG_{x_0} : g \in AG_{x_0}\}$. Obviously this coincides with $\{gG_{x_0} : g \in A\}$ since for any $h \in G_{x_0}$, we have $hG_{x_0} = G_{x_0}$. Also

$$\beta(gH_g) = \{ghG_{x_0} : h \in H_g\} = \{g\beta(h) : h \in H_g\} = g\beta(H_g)$$

Thus,

$$E = \cup_{g \in I} g\beta(H_g)$$

It is easy to see that $\beta(H_g)$ is an open subset of V_1. In fact, $\beta(H_g) == \{hG_{x_0} : h \in H_g\}$. Thus, $\beta^{-1}\beta(H_g) = H_g G_{x_0} = \cup_{h \in G_{x_0}} H_g h$ is a union of open sets and hence is itself an open subset of G. It follows that $\beta(H_g)$ is an open subset of V_1. Thus, E is a union of sets of the form $g.H$ where $g \in G$ and H is an open subset of V_1. Thus, $\pi(E)$ is a union of sets of the form $g\pi(H)$ with $g \in G$ and H open in V_1. Since $\pi : V_1 \to W$ is a homeomorphism, it follows that $\pi(H)$ is open and hence $\pi(E)$ is the union of open sets and hence is also open proving that π is an open map.

Study project 20

The Euclidean motion group consists of rotations and translations. This group can be viewed as the semidirect product of the rotation group and the translation group. With this intention, we describe the notion of semidirect product in its full generality.

Semidirect product

Let G be a group and N a normal subgroup of G, i.e., N is a group, $N \subset G$ and for every $g \in G$, $gNg^{-1} = N$. G/N is a group under the multiplication $aN.bN = abN$. This multiplication is well defined since $aN = a'N, bN = b'N$ imply $a^{-1}a', b^{-1}b' \in N$ imply $a' = an_1, b' = bn_2$ with $n_1, n_2 \in N$. Thus, $a'b' = an_1bn_2 = abb^{-1}n_1bn_2 = \in abN$ since normality of N implies $b^{-1}n_1b \in N$. Thus, $a'b'N = abN$. Suppose in addition to normality, N is also Abelian. Let H be any subgroup of G and suppose for every $g \in G$, there exist unique $h \in H, n \in N$ such that $g = hn$. Then, $H \cap N = (e)$, for if $g \in H \cap N$, then $g = g.e = e.g$ are two represenations of g having the form hn and hence $g = e$. If G is finite, then such a unique representation implies $o(G) = o(H)o(N)$. Suppose now that H acts on N by the rule $n \to hnh^{-1}$ with $n \in N$ and $h \in H$. Denote this action by τ_h. Clearly τ_h is an automorphism of N and $\tau_{hh'} = \tau_h \tau_{h'}$ for all $h, h' \in H$. Thus $h \to \tau_h$ is a homomorphism of the group H into the group $Aut(N)$ of all automorphisms of N. We can represent an element $g \in G$ by the ordered pair (h, n) with $h \in H$ and $n \in N$. Any element $g \in G$ can also be represented uniquely as $g = nh$ with $n \in N, h \in H$. In fact, the two representations are related by $hn = hnh^{-1}h = \tau_h(n)h$. Now, consider the product of two elements $g = nh, g' = n'h'$. $gg' = nhn'h' = nhn'h^{-1}hh'$. Representing $g = nh$ by the ordered pair (n, h), the product is thus given by $(n, h)(n', h') = (n\tau_h(n'), hh')$. Since the group N is Abelian, we represent the composition operation on N by $+$. Then, $(n, h).(n', h') = (n + \tau_h(n'), hh')$. More generally, we can assume that N is any Abelian group, that H is any group, that $\tau : H \to Aut(N)$ is any homomorphism and define G to consist of all ordered pairs (n, h) with product of two elements given by $(n, h).(n', h') = (n + \tau_h(n'), hh')$. Then G is called the semidirect product of the groups N and H. It is not difficult to verify that G so defined is indeed a group under this product. For example, to verify associativity of this product, we note that

$$((n, h).(n', h')).(n'', h'') = (n + \tau_h(n'), hh').(n'', h'') = (n + \tau_h(n') + \tau_{hh'}(n''), hh'h'')$$

$$= (n + \tau_h(n') + \tau_h \tau_{h'}(n''), hh'h'')$$

while on the other hand,

$$(n, h).((n', h').(n'', h'')) = (n, h).(n' + \tau_{h'}(n''), h'h'') = (n + \tau_h(n' + \tau_{h'}(n'')), hh'h'')$$

$$= (n + \tau_h(n') + \tau_h \tau_{h'}(n''), hh'h'')$$

proving associativity. The inverse of any element is calculated easily. Assume that (n', h') is the inverse of (n, h). Then,

$$(n, h).(n', h') = (e, e)$$

or

$$(n + \tau_h(n'), hh') = (0, e)$$

so that

$$h' = h^{-1}, n' = \tau_h^{-1}(-n) = -\tau_h^{-1}(n) = -\tau_{h^{-1}}(n)$$

Thus,

$$(n,h)^{-1} = (-\tau_h^{-1}(n), h^{-1})$$

Note that $\tilde{N} = \{(n,e) : n \in N\}$ is an Abelian normal subgroup of G, that $\tilde{H} = \{(0,h) : h \in H\}$ is a subgroup of G, that any element $g \in G$ has the unique expression $g = \tilde{n}\tilde{h}$ with $\tilde{n} \in \tilde{N}, \tilde{h} \in \tilde{H}$ and that

$$(m,h)(n,e)(m,h)^{-1} = (m,h)(n,e)(-\tau_{h^{-1}}(m), h^{-1}) = (m + \tau_h(n), h).(-\tau_{h^{-1}}(m), h^{-1})$$

$$= (m + \tau_h(n) - \tau_h\tau_{h^{-1}}(m), e) = (\tau_h(n), e)$$

This means that G is the semidirect product of the groups $\tilde{N}$ and $\tilde{H}$ in the sense defined earlier.

An example of the semidirect product: Let G denote the group of all Euclidean motions, i.e., an element $g \in G$ consists of a pair (A, v) where $A \in SO(3)$, i.e., A is a proper rotation and $v \in \mathbb{R}^3$ is a vector representing a translation. The action of g on a point $x \in \mathbb{R}^3$ is given by $gx = Ax + v$. There is a nice representation for G. This involves representing a point $x \in \mathbb{R}^3$ by the vector $\begin{pmatrix} x \\ 1 \end{pmatrix}$ and representing $g \in G$ by the block matrix

$$\begin{pmatrix} A & v \\ \mathbf{0} & 1 \end{pmatrix}$$

The action of G on $\mathbb{R}^3$ is then given by a matrix multiplication:

$$\begin{pmatrix} A & v \\ \mathbf{0} & 1 \end{pmatrix} \begin{pmatrix} x \\ 1 \end{pmatrix}$$

$$= \begin{pmatrix} Ax + v \\ 1 \end{pmatrix}$$

The composition of two element of G is obtained by acting on a point:

$$(A', v')(A, v)x = (A', v')(Ax + v) = (A'(Ax + v) + v') = A'Ax + A'v + v' = (A'A, A'v + v')x$$

To bring this into our formalism of semidirect products, we use the notation (v, A) in place of (A, v). v varies over the Abelian group $\mathbb{R}^3$ of all translations of space and A varies over $SO(3)$. In other words, $\mathbb{R}^3$ plays the role of N and $SO(3)$ the role of H. Then,

$$(v, A)(v', A') = (v + Av', AA')$$

shows that τ_A is simply multiplication of vectors in $\mathbb{R}^3$ by A. The automorphism and homomorphism properties are easy to verify:

$$\tau_A(v) = Av, \tau_A(v + v') = A(v + v') = Av + Av' = \tau_A(v) + \tau_A(v'),$$

$$\tau_{AA'}(v) = AA'v = \tau_A(A'v) = \tau_A\tau_{A'}(v)$$

In terms of our representation of G, we have

$$\begin{pmatrix} A & v \\ 0 & 1 \end{pmatrix}\begin{pmatrix} A' & v' \\ 0 & 1 \end{pmatrix}$$

$$= \begin{pmatrix} AA' & Av' + v \\ 0 & 1 \end{pmatrix}$$

The group G consisting of all matrices having the block form $\begin{pmatrix} A & v \\ 0 & 1 \end{pmatrix}$ with A varying over $SO(3)$ and v varying over $\mathbb{R}^3$ is a subgroup of $GL(\mathbb{R}^4)$. This group is isomorphic to the group of Euclidean motions and we may call this as the Euclidean motion group. The translation subgroup consists of all matrices having the above form with $A = I$, i.e. the group of all matrices having the block structure $\begin{pmatrix} I & v \\ 0 & 1 \end{pmatrix}$ with v varying over $\mathbb{R}^3$. The inverse of an element is given by

$$\begin{pmatrix} A & v \\ 0 & 1 \end{pmatrix} = \begin{pmatrix} A^{-1} & -A^{-1}v \\ 0 & 1 \end{pmatrix}$$

The translation subgroup forms a normal subgroup of G. This follows from the identity

$$\begin{pmatrix} A & v \\ 0 & 1 \end{pmatrix}\begin{pmatrix} I & w \\ 0 & 1 \end{pmatrix}\begin{pmatrix} A^{-1}7 - A^{-1}v \\ 0 & \qquad 1 \end{pmatrix}$$

$$= \begin{pmatrix} I & Aw \\ 0 & 1 \end{pmatrix}$$

First applying a translation w and then a rotation A amounts to applying the group element (A, Aw) since $x \to x + w \to A(x + w) = Ax + Aw$. This is seen via the equation

$$\begin{pmatrix} A & 0 \\ 0 & 1 \end{pmatrix}\begin{pmatrix} I & w \\ 0 & 1 \end{pmatrix}$$

$$= \begin{pmatrix} A & v \\ 0 & 1 \end{pmatrix}$$

This formula also shows that every $g \in G$ is uniquely expressible as a product hn with h a rotation and n a translation. Note that H here is the subgroup $\begin{pmatrix} A & 0 \\ 0 & 1 \end{pmatrix}$ with A varying over $SO(3)$.

Study project 21

This deals with how one obtains the irreducible representations of the semidirect product of two groups, one of which is Abelian.

Study project 22

This project deals with the analysis of the shapes of a bent beam. Suppose that the beam is bent so that at a given point on it, its central axis has a radius of curvature R. Take a point above it having a radius of curvature $R + r$. r is the distance of the point from the central axis. Let ϕ denote the angle subtended by the given beam segment at its centre of curvature. Then, the central portion of the segment has length $R\phi$ while the top portion of the segment that is at a distance r from the central segment has length $(R + r)\phi$. The longitudinal strain on this top portion is given by $((R + r)\phi - R\phi)/R\phi = r/R$ and the longitudinal stress is Yr/R where Y is Young's modulus. The moment of these stress forces about the central axis is therefore given by the integral $\int Yr^2 dA/R$, the integral being taken over the cross sectional area of the beam. Now, $I = \int r^2 dA$ is the moment of inertia (normalized) of the cross section of the beam. If τ is the torque of the external forces about the central axis at that point, we obtain the following equation for the shape of the beam $\frac{YI}{R} = \tau$. Note that I is a constant that depends only upon the cross sectional shape of the beam. R, the radius of curvature varies along the length of the beam and τ also varies along the beam length. As an application, consider a cantilever with one end fixed to the wall and a force F applied to the other end. Assume the deflection to be small. Let $y(x)$ denote the vertical deflection of the beam at a distance x from the fixed end. Let L denote the length of the beam. then, the torque of the external force about this point is $W(L - x)$ and the radius of curvature is approximately $y''(x)$. We thus obtain the following differential equation for the beam deflection $y''(x) = \frac{W}{YI}(L - x)$. This equation is to be solved subject to the conditions $y(0) = 0$ and $y'(0) = 0$ corresponding to the fact that the end $x = 0$ is fixed and the tangent to the beam at this point is horizontal.

Solved problem 1

Consider a bob of mass m attached to a string of upstretched length L_0. Assume that the bob swings as a pendulum. Let Y denote the Young's modulus of the string and A is cross sectional area. After time t, let $\theta(t)$ be the angle made by the pendulum string with the vertical and let $L(t)$ be the length of the string. Determine the equations of motion.

Ans:

The tension in the string is given by $T(t) = YA(L(t) - L_0)$. The potential energy of the string is $\frac{T(t)(L(t) - L_0)}{2} = YA(L(t) - L_0)^2/2$. The potential energy of the bob due to gravity is $-mgL(t)cos(\theta(t))$. The kinetic energy of the bob is given by $\frac{m}{2}(L'(t)^2 + L(t)^2\theta'(t)^2)$. The Lagrangian function of the pendulum is therefore

$$\mathcal{L}(L, \theta, L', \theta') = \frac{m}{2}(L'^2 + L^2\theta'^2) + mgLcos\theta - \frac{YA(L - L_0)^2}{2}$$

The Euler-Lagrange equations of motion can now be written down.

Study project 23

Dynamics on tori play an important role in mechanics especially in the action-angle variable formalism. This project deals with the torus T^2 and a certain dynamics on this torus generated by two irrationally related frequencies. We start with

Weyl's theorem: Let (a, b) be two real numbers such that a/b is irrational. Then, we claim that the map $t \to (exp(iat), exp(ibt))$ from $\mathbb{R}$ into T^2 is dense in T^2 where T^2 is the two torus defined by $\{(z_1, z_2) : z_1, z_2 \in \mathbb{C}, |z_1| = |z_2| = 1\}$. In other words, given any $(z_1, z_2) \in \mathbb{C}^2$ and any $\epsilon > 0$, there exists a real number t such that $|z_1 - exp(iat)| + |z_2 - exp(ibt)| < \epsilon$. To see this, we consider any continuous function $f(u, v)$ defined on $[0, 1) \times [0, 1)$. We may assume without loss of generality that f has period 1 in both the variables. It can thus be expanded as a Fourier series

$$f(u, v) = \sum_{n,m=-\infty}^{\infty} c(n, m) exp(2\pi i(nu + mv))$$

We now consider

$$\frac{1}{T} \int_0^T f(at, bt)dt = \sum_{n,m=-\infty}^{\infty} c(n, m) \frac{1}{T} \int_0^T exp(2\pi i(na + mb)t)dt$$

$$= c(0, 0) + \sum_{(n,m)\neq(0,0)}^{\infty} c(n, m) \frac{exp(2\pi i(na + mb)T) - 1}{2\pi iT(na + mb)}$$

$$\to c(0, 0)$$

as $T \to \infty$. In other words,

$$lim_{T\to\infty} \frac{1}{T} \int_0^T f(at, bt)dt = \int_0^1 \int_0^1 f(u, v)dudv$$

This formula implies that $\{(atmod1, btmod1) : t \in \mathbb{R}\}$ is a dense subset of $[0, 1) \times [0, 1)$. Thus, $\{(exp(2\pi iat), exp(2\pi ibt)) : t \in \mathbb{R}\}$ is dense in T^2, or replacing $2\pi t$ by t, we conclude that $\{(exp(iat), exp(ibt)) : t \in \mathbb{R}\}$ is a dense subset of T^2. We also note that the map $t \to (exp(iat), exp(ibt))$ from $\mathbb{R}$ into T^2 is one. This follows from the fact that the relations $exp(iat) = exp(iat')$, $exp(ibt) = exp(ibt')$ imply that $a(t - t') = 2n\pi, b(t - t') = 2m\pi$ for some integers n, m. If $t \neq t'$, it would follow by taking the ratio of these two equations that $a/b = n/m$ is a rational a contradiction to our hypothesis. Thus, $t = t'$ proving that the map is one-one. The map is also clearly differentiable. Denote this map by $\pi : \mathbb{R} \to T^2$. What can we say about the continuity of the map $\pi^{-1} : \pi(\mathbb{R}) \to \mathbb{R}$? We claim that this map is discontinuous and hence that π is not a homeomorphism of $\mathbb{R}$ onto $\pi(\mathbb{R})$. (Varadarajan, Lie groups, Lie algebras and their representations). To see this, we compute $|exp(iat) - exp(iat')| = 4sin^2(a(t - t')/2)$ and $|exp(ibt) - exp(ibt')| = 4sin^2(b(t - t')/2)$. To prove discontinuity of π^{-1}, it therefore suffices to choose to points t, t' such that $|t - t'|$ is not small but $sin(a(t - t')/2)$ and $sin(b(t - t')/2)$ are small in magnitude. Equivalently, we must show the existence of t that is not close to zero but

$sin(at/2)$ and $sin(bt/2)$ are close to zero. Let $0 < c < 2\pi$. Since $\{(at, bt) mod 2\pi : t \in \mathbb{R}\}$ is dense in $[0, 2\pi) \times [0, 2\pi)$, if follows that given $\epsilon > 0$ there exists a t such that $|at - 2n\pi| < \epsilon$ and $|bt - 2m\pi - c| < \epsilon$ for some integers n, m. Suppose $n = m = 0$. Then, $|t| < \epsilon/a, |t - c/b| < \epsilon/b$. These two imply $c/b < \epsilon(1/a + 1/b)$ which is a contradiction if c is chosen so that $c > \epsilon(1 + b/a)$. Let $c = 2\pi$. Note that $c = 2\pi$ is a permissible choice since we first choose t so that $|at - 2n\pi| < \epsilon$ and $|bt - 2m\pi - (2\pi - \epsilon/2)| < \epsilon/2$. The second of these implies $|bt - 2m\pi - 2\pi| < \epsilon$. Thus, either $n \neq 0$ or $m \neq 0$. If $n > 0$, then $t > 2\pi/a - \epsilon/a$. Writing $at = 2n\pi + \delta$, we have $|\delta| < \epsilon$ and $sin(at/2) = sin(n\pi + \delta/2) = (-1)^n sin(\delta/2)$ so that $|sin(at/2)| < |\delta|/2 < \epsilon/2$. Also $bt = 2m\pi + 2\pi + \delta'$ with $|\delta'| < \epsilon$. This gives $|sin(bt/2)| < |\delta'/2| < \epsilon/2$. This shows that $|sin(at/2)| + |sin(bt/2)| = O(\epsilon)$ while $t > 2\pi/a - \epsilon/a$ is not $O(\epsilon)$. Likewise, suppose $m > 0$. Then, $t > 2m\pi/b + 2\pi/b - \epsilon/b > 2\pi/b - \epsilon/b$ showing that t is not $O(\epsilon)$, while $sin(at/2)$ and $sin(bt/2)$ are $O(\epsilon)$. A similar argument works when we take $n < 0$ or $m < 0$. Taking $\epsilon = 1/n$, we've found a sequence $t_n, n = 1, 2, ...$, for which t_n does not converge to zero but $sin(at_n/2)$ and $sin(bt_n/2)$ both converge to zero. This proves that π^{-1} is not a homomorphism from $\pi(\mathbb{R})$ onto $\mathbb{R}$.

Chapter 2

Fluid Dynamics

2.1 Equation of continuity

The fundamental quantities of a fluid are density, velocity and pressure. These three are field functions, i.e., functions of position of the fluid particle and time. We use the following notations for these: Density: $\rho(\mathbf{x}, t)$, velocity $\mathbf{v}(\mathbf{x}, t)$, pressure $\rho(\mathbf{x}, t)$. The units of density are Kg/m^3 or mass per length cube. The units of velocity are m/s or length per time. The units of pressure are $Newton/m^2$ or Force per area. The quantity of fluid contained inside a volume V of space at time t is given by $M_t(V) = \int_V \rho(\mathbf{x}, t) d^3\mathbf{x}$. This is in general a function of time as the subscript t indicates. This follows from the fact that fluid can enter into the volume through the boundary surface S of V or can leave the volume V through the boundary surface S. The density ρ is a function of space and time in general corresponding to the fact that the fluid can be in general compressible. When we speak of fluids, we also include gases which are compressible. An incompressible fluid is that whose density is everywhere and at all times a constant. The first basic equation of fluid dynamics is conservation of mass. This states that the net rate at which fluid mass inside a given volume V increases with time equals the net rate at which fluid mass flows into the volume V through the boundary minus the rate at which fluid mass leaves the volume through the boundary. The quantity $\rho\mathbf{v}$ represents fluid flux at a point, i.e. the amount of fluid mass flowing across unit area per unit time. Its surface integral over a surface gives the flux of the fluid through the surface. For example, if S is a closed surface bounding the volume V, then

$$\int_S \rho\mathbf{v}.\mathbf{n} dS$$

is the net rate at which fluid leaves the volume. Here account has been taken of the fact fluid can enter or leave the volume. If at some point on the surface, the quantity $\mathbf{v}.\mathbf{n}$ is positive, then the quantity $\rho\mathbf{v}.\mathbf{n} dS$ at this point equals the rate at which fluid leaves the volume through the surface element dS. If at some other point $\mathbf{v}.\mathbf{n}$ is negative, then $-\rho\mathbf{v}.\mathbf{n} dS$ equals the rate at which fluid enters the volume through the surface element dS. Matter conservation implies

$$-\frac{\partial}{\partial t}\int_V \rho dV = \int \rho\mathbf{v}.\mathbf{n}dS = \int_V \nabla.(\rho\mathbf{v})dV$$

or equivalently,

$$\int_V (\frac{\partial \rho}{\partial t} + \nabla.(\rho\mathbf{v}))dV = 0$$

Since this must hold for all volumes V, it follows that

$$\frac{\partial \rho}{\partial t} + \nabla.(\rho\mathbf{v}) = 0$$

This is the fundamental equation of mass conservation. Note that $\frac{\partial \rho}{\partial t}$ equals the rate at which matter increases per unit volume at a point and $\nabla.(\rho\mathbf{v})$ equals the rate at which fluid mass flows out of a unit volume at the given point. The equation $\frac{\partial}{\partial t}\int_V \rho dV + \int_S \rho\mathbf{v}.\mathbf{n}dS = 0$ can be termed as a global matter conservation law and the above differential form as the local matter conservation law. If in addition, $g(t,\mathbf{r})$ denotes the rate at which matter is generated per unit volume, then the matter conservation law gets modified to

$$\frac{\partial \rho}{\partial t} + \nabla.(\rho\mathbf{v}) = g$$

In full notation, this reads

$$\frac{\partial \rho(t,x,y,z)}{\partial t} + \frac{\partial}{\partial x}(\rho(t,x,y,z)v_x(t,x,y,z)) +$$

$$\frac{\partial}{\partial y}(\rho(t,x,y,z)v_y(t,x,y,z)) + \frac{\partial}{\partial z}(\rho(t,x,y,z)v_z(t,x,y,z)) = 0$$

The divergence of a vector field in cylindrical coordinates is given by

$$\nabla.\mathbf{F} = \frac{1}{\rho}(\frac{\partial}{\partial \rho}(\rho F_\rho) + \frac{\partial}{\partial \phi}(F_\phi) + \frac{\partial}{\partial z}(\rho F_z))$$

$$= \frac{1}{\rho}\frac{\partial}{\partial \rho}(\rho F_\rho) + \frac{1}{\rho}\frac{\partial F_\phi}{\partial \phi} + \frac{\partial F_z}{\partial z}$$

In spherical polar coordinates, the divergence is given by

$$\nabla.\mathbf{F} = \frac{1}{r^2 sin\theta}(\frac{\partial}{\partial r}(r^2 sin\theta F_r) + \frac{\partial}{\partial \theta}(rsin\theta F_\theta) + \frac{\partial}{\partial \phi}(rF_\phi))$$

$$= \frac{1}{r^2}\frac{\partial}{\partial r}(r^2 F_r) + \frac{1}{r.sin\theta}\frac{\partial F_\theta}{\partial \theta} + \frac{1}{r.sin\theta}\frac{\partial F_\phi}{\partial \phi}$$

These equations can be used to write down the matter conservation equations in the two systems.Particularly important case consists of a fluid with cylindrically symmetric and spherically symmetric matter distributions, i.e., the density dependent upon ρ only and on r only. We use the symbol σ in place of the matter density in order not to confuse with the radial coordinate ρ of the cylindrical system. Then under cylindrical symmetry, the matter conservation equation reads

$$\frac{\partial \sigma(t,\rho)}{\partial t} + \frac{1}{\rho}\frac{\partial}{\partial \rho}(\rho\sigma(t,\rho)v_\rho(t,\rho)) = 0$$

Likewise, under spherical symmetry, the matter conservation equation reads

$$\frac{\partial \sigma(t,r)}{\partial t} + \frac{1}{r^2}\frac{\partial}{\partial r}(r^2\sigma(t,r)v_r(t,r)) = 0$$

2.2 Navier-Stokes Equation

The fluid momentum per unit volume at a given point of space and at a given time is given by $\rho(\mathbf{r},t)\mathbf{v}(\mathbf{r},t)$ which in our shorthand notation, is $\rho\mathbf{v}$. The total momentum of the fluid matter inside a volume V is the volume integral of this quantity, i.e., $\int_V \rho\mathbf{v}d^3\mathbf{r}$. Its rate of change with time is given by

$$\frac{d}{dt}\int \rho(\mathbf{r},t)\mathbf{v}(\mathbf{r},t)d^3\mathbf{r}$$

The rate at which momentum flows out of the boundary of the volume V is given by the surface integral of the momentum flux, i.e.

$$\int_S \rho\mathbf{v}\mathbf{v}.\mathbf{n}dS$$

The sum of the rate of increase of momentum inside V and the rate of flux of momentum out of V must equal the rate at which momentum is generated, i.e. the net force acting on the volume. This net force equals the sum of the pressure and gravitational forces, i.e. $-\int_S p\mathbf{n}dS + \int \rho\mathbf{g}d^3\mathbf{r}$. Thus, the momentum conservation equation reads

$$\frac{d}{dt}\int \rho\mathbf{v}d^3\mathbf{r} + \int_S \rho\mathbf{v}\mathbf{v}.\mathbf{n}dS = -\int_S p\mathbf{n}dS + \int \rho\mathbf{g}d^3\mathbf{r}$$

By use of Gauss, theorem, this can be transformed into

$$\int \frac{\partial}{\partial t}(\rho v_i)d^3\mathbf{r} + \int_V \sum_{j=1}^{3}\frac{\partial}{\partial x_j}(\rho v_i v_j)d^3\mathbf{r}$$

$$= -\int \frac{\partial p}{\partial x_i} d^3\mathbf{r} + \int \rho g_i d^3\mathbf{r}$$

Note that Gauss' divergence theorem implies that

$$\int_S p\mathbf{n}dS = \int \nabla p d^3\mathbf{r}$$

Thus, from the arbitrariness of the chosen volume V, we conclude that

$$\frac{\partial}{\partial t}(\rho v_i) + \sum_{j=1}^{3} \frac{\partial}{\partial x_j}(\rho v_i v_j) = -\frac{\partial p}{\partial x_i} + \rho g_i$$

This can be rearranged as

$$v_i\left(\frac{\partial \rho}{\partial t} + \sum_{j=1}^{3} \frac{\partial(\rho v_j)}{\partial x_j}\right)$$

$$+\rho\left(\frac{\partial v_i}{\partial t} + \sum_{j=1}^{3} v_j \frac{\partial v_i}{\partial x_j}\right)$$

$$= -\frac{\partial p}{\partial x_i} + \rho g_i$$

The first bracket on the left equals zero by mass conservation. Hence, we end up with the Navier-Stokes equation for a non-viscous (ideal) fluid:

$$\rho\left(\frac{\partial v_i}{\partial t} + \sum_{j=1}^{3} v_j \frac{\partial v_i}{\partial x_j}\right) = -\frac{\partial p}{\partial x_i} + \rho g_i$$

which in vector notation reads

$$\rho\left(\frac{\partial \mathbf{v}}{\partial t} + \mathbf{v}.\nabla\mathbf{v}\right) = -\nabla p + \rho \mathbf{g}$$

2.3   Viscosity

When there exists spatial gradients in the velocity field of a fluid, then there are shearing stresses. The stress tensor is a rank two tensor S_{ij}. S_{ij} gives the ith component of the force across a unit area perpendicular to the jth normal. If the unit normal to a surface S at a given point on it is denoted by (n_1, n_2, n_3), then the ith component of the force associated with the above stress tensor is given by the expression

$$F_i = \int_S \sum_{j=1}^{3} S_{ij} n_j dA = \int_S \mathbf{S}_i . \mathbf{n} dA$$

where

$$\mathbf{S}_i = (S_{i1}, S_{i2}, S)_{i3}) = S_{i1}\hat{x} + S_{i2}\hat{y} + S_{i3}\hat{z}$$

Using the divergence theorem,

$$F_i = \int_V \sum_{j=1}^{3} \frac{\partial S_{ij}}{\partial x_j} dV$$

the integration being over the volume V enclosed by the surface S. The most general form of a symmetric rank two tensor that is linear in the spatial gradients of the velocity is given by

$$S_{ij} = \eta \left(\frac{\partial v_i}{\partial x_j} + \frac{\partial v_j}{\partial x_i} \right) + \nu \delta_{ij} \sum_k \frac{\partial v_k}{\partial x_k}$$

$$= \eta \left(\frac{\partial v_i}{\partial x_j} + \frac{\partial v_j}{\partial x_i} \right) + \nu \delta_{ij} \nabla . \mathbf{v}$$

The ith component of the viscous force is

$$f_i = \sum_j \frac{\partial S_{ij}}{\partial x_i} = \eta \left(\nabla^2 v_i + \frac{\partial}{\partial x_i}(\nabla . \mathbf{v}) \right) + \nu \frac{\partial}{\partial x_i}(\nabla . \mathbf{v})$$

or in vectorial notation,

$$\mathbf{f} = \eta \nabla^2 \mathbf{v} + (\eta + \nu) \nabla (\nabla . \mathbf{v})$$

Navier-Stokes equation for the viscous fluid is thus given by

$$\rho \left(\frac{\partial \mathbf{v}}{\partial t} + \mathbf{v} . \nabla \mathbf{v} \right) = -\nabla p + \rho \mathbf{g} + \eta \nabla^2 \mathbf{v} + (\eta + \nu) \nabla (\nabla . \mathbf{v})$$

For an incompressible fluid, $\nabla . \mathbf{v} = 0$ and we get

$$\rho \left(\frac{\partial \mathbf{v}}{\partial t} + \mathbf{v} . \nabla \mathbf{v} \right) = -\nabla p + \rho \mathbf{g} + \eta \nabla^2 \mathbf{v}$$

2.4   Velocity potential

A fluid is said to irrotational if the curl of its velocity field vanishes. This is equivalent to saying that the line integral of the velocity field around any closed contour equals zero, i.e., for an irrotational fluid, $\nabla \times \mathbf{v} = \mathbf{0}$, or equivalently, $\int_\Gamma \mathbf{v}.d\mathbf{r} = 0$ for any closed contour Γ. When this happens $\mathbf{v} = -\nabla\phi$, where ϕ is a scalar function, i.e. an irrotational velocity field can be derived as the gradient of a scalar field. ϕ is in general a function of the spatial coordinates and time. For steady flows, $\mathbf{v}$ does not depend upon time and hence ϕ also is time-independent. If in addition to being irrotational, the fluid is incompressible, then $0 = -\nabla.\mathbf{v} = \nabla^2\phi$, i.e., ϕ is the solution to Laplace's equation. Thus any solution to Laplace's equation corresponds to an irrotational, incompressible flow. Now using the vector identity

$$(\nabla \times \mathbf{v}) \times \mathbf{v} = \mathbf{v}.\nabla\mathbf{v} - \frac{1}{2}\nabla v^2$$

Navier-Stokes equation can be expressed as

$$\rho(\Omega \times \mathbf{v} + \frac{1}{2}\nabla v^2 + \frac{\partial \mathbf{v}}{\partial t}) = -\nabla p + \eta\nabla^2\mathbf{v} + \rho\mathbf{g}$$

where $\Omega = \nabla \times \mathbf{v}$ is the vorticity. In particular, if the fluid is irrotational, Navier-Stokes equation reduces to

$$\rho(\frac{1}{2}\nabla v^2 + \frac{\partial \mathbf{v}}{\partial t}) = -\nabla p + \eta\nabla^2\mathbf{v} + \rho\mathbf{g}$$

2.5   Two dimensional potential flows

A two dimensional flow is specified by a velocity field having the form $v_x(x,y)\hat{x} + v_y(x,y)\hat{y}$, i.e., the velocity has only x and y as non-vanishing and these two components are functions of x and y only. The z component of the velocity field is zero. Suppose that the fluid is incompressible as well as irrotational. Incompressibility implies that

$$\frac{\partial v_x}{\partial x} + \frac{\partial v_y}{\partial y} = 0$$

and irrotationality implies that

$$\frac{\partial v_y}{\partial x} - \frac{\partial v_x}{\partial y} = 0$$

The second implies that $\mathbf{v} = -\nabla\phi$, i.e.,

$$v_x(x,y) = -\frac{\partial \phi(x,y)}{\partial x} \qquad v_y(x,y) = -\frac{\partial \phi(x,y)}{\partial y}$$

for some scalar function $\phi(x, y)$. Combining this with the incompressibility condition gives us $\nabla^2 \phi = 0$, i.e.,

$$\frac{\partial^2 \phi(x, y)}{\partial x^2} + \frac{\partial^2 \phi(x, y)}{\partial y^2} = 0$$

From the theory of analytic functions of a complex variable, we know that any solution to the two dimensional Laplace equation is the real part of a complex analytic function. Specifically, if $f(z) = f(x + iy)$ is an analytic function of the complex variable $z = x + iy$, then f satisfies Laplace's equation. In fact, $\frac{\partial f(z)}{\partial x} = f'(z)$ and $\frac{\partial f(z)}{\partial y} = if'(z)$ Repeating this gives $\frac{\partial^2 f(z)}{\partial x^2} = f''(z)$ and $\frac{\partial^2 f(z)}{\partial y^2} = -f''(z)$ so that

$$\nabla^2 f(z) = \frac{\partial^2 f(z)}{\partial x^2} + \frac{\partial^2 f(z)}{\partial y^2} = 0$$

Writing $f(z) = u(x, y) + iv(x, y)$, with u as the real part of $f(z)$ and v as the imaginary part, we get $\nabla^2 u = \nabla^2 v = 0$. We also know that

$$\frac{\partial u}{\partial x} + i\frac{\partial v}{\partial x} = \frac{\partial f(z)}{\partial x} = f'(z) = -i\frac{\partial f(z)}{\partial y} = -i\frac{\partial u}{\partial y} + \frac{\partial v}{\partial y}$$

and equating real and imaginary parts leads to the Cauchy-Riemann equations

$$\frac{\partial u}{\partial x} = \frac{\partial v}{\partial y} \qquad \frac{\partial v}{\partial x} = -\frac{\partial u}{\partial y}$$

The reverse is also true: If u, v form a Cauchy-Riemann pair, then $u + iv$ is an analytic function of z. Suppose that we have a two dimensional irrotational, incompressible fluid flow and that we have solved the Laplace equation for the velocity potential ϕ. Then, we solve the equation

$$\frac{\partial \phi}{\partial x} = \frac{\partial \psi}{\partial y}$$

to get

$$\psi(x, y) = \int_0^y \frac{\partial \phi(x, y)}{\partial x} dy + g(x)$$

where $g(x)$ is an arbitrary function of x. In order that the pair (ϕ, ψ) satisfy the other Cauchy-Riemann equation, we require that

$$-\frac{\partial \psi}{\partial x} = \frac{\partial \phi}{\partial y}$$

or

$$-\int_0^y \frac{\partial^2 \phi(x, y)}{\partial x^2} dy - g'(x) = \frac{\partial \phi}{\partial y}$$

or since $\nabla^2 \phi = 0$, it follows that

$$\frac{\partial \phi(x, y)}{\partial y} - \frac{\partial \phi(x, 0)}{\partial y} - g'(x) = \frac{\partial \phi(x, y)}{\partial y}$$

This means that we can take

$$g(x) = - \int_0^x \frac{\partial \phi(x,0)}{\partial y} dx + C$$

where C is some constant. The function ψ so determined is called the stream function of the flow. The velocity field of the irrotational, incompressible flow can be expressed in terms of the stream function as

$$v_x = -\frac{\partial \phi}{\partial x} = -\frac{\partial \psi}{\partial y}$$

$$v_y = -\frac{\partial \phi}{\partial y} = \frac{\partial \psi}{\partial x}$$

Since $\mathbf{v} = -\nabla\phi$, it follows that the velocity vector at any point is orthogonal to the curve $\phi = constt$ passing through that point. In other words, suppose we choose any constant C and look at the curve $\phi(x,y) = C$. Let (x,y) be any point on this curve. Then, the vector $\mathbf{v}(x,y)$ is orthogonal to the tangent of this curve at this point. The Cauchy-Riemann equations imply that

$$\nabla\phi.\nabla\psi = \frac{\partial \phi}{\partial x}\frac{\partial \psi}{\partial x} + \frac{\partial \phi}{\partial y}\frac{\partial \psi}{\partial y} = 0$$

This means that the vectors $\nabla\phi$ and $\nabla\psi$ are orthogonal to each other at every point in the plane. It follows that if C_1, C_2 are two constants, then the curves $\phi(x,y) = C_1, \psi(x,y) = C_2$ are orthogonal to each other at every point of their intersection. In view of the preceding discussion, it follows that the curve $\psi(x,y) = C$ is tangential to the velocity vector at any point on it. In other words, the fluid flows along the curves $\psi(x,y) = constt$. Thus, ψ is also called the stream function of the flow. Another way to view this is as follows. Let $f(z)$ be any analytic function of the complex variable z. Let $f(z) = \phi(x,y) + i\psi(x,y)$. Then consider the flow with velocity field $\mathbf{v} = -\nabla\phi$. Then, ψ is the stream function of the flow. For example, consider $f(z) = z$. We have $\phi = x, \psi = y$. This flow corresponds to the steady flow having $x = 0$ as a boundary, i.e., fluid flows steadily everywhere along the $\hat{y}$ direction. This can also be regarded as the flow between two vertical planes $x = a, x = b$. In general, if ψ is the stream function of a flow, then the flow corresponds to the flow of a fluid with a boundary wall being $\psi = C$, or equivalently, to the flow between two walls $\psi = C_1, \psi = C_2$. For example, take $f(z) = z^2$. Then, $\phi = x^2 - y^2, \psi = 2xy$. The equation $\psi = 0$ gives either $x = 0$ or $y = 0$. Thus, this flow corresponds to the flow in the first quadrant bounded by two walls, namely the positive x and y axes. Consider now $f(z) = iz^2$. Then, $\psi = x^2 - y^2$. The equation $\psi = 0$ yields two boundary walls $x = \pm y$ and we get a flow in the region $|y| \leq |x|$ or equivalently, in the region $|y| \geq |x|$. It is suggested that the reader make plots of the other streamlines. Now consider $f(z) = log(z)$. Then, $\phi = log(r), \psi = \theta$ where (r, θ) are the polar coordinates of the point (x,y), i.e. $x + iy = z = r.exp(i\theta)$. The stream lines $\theta = C$ are radial lines directed from the origin. The equipotentials $r = C$ are concentric circles with the origin as the centre. This flow corresponds either to a source or a sink at the origin, i.e., a point source at the origin releases a jet of fluid isotropically in all the directions (diverges) or flow from all directions gushes in (converges) isotropically inwards to the origin. Finally, consider

$f(z) = i\log(z)$. The equipotential lines are $\theta = C$ and the streamlines are $r = C$. This flow pattern corresponds to fluid flowing around the origin in circles. Note that the line integral of the velocity field around any closed curve enclosing the origin is non-zero. Thus, the velocity field is irrotational everywhere except at the origin. These examples make abundantly clear the importance of complex analytic function theory in the construction of irrotational, incompressible flows.

2.6   Bernoulli's equation

Consider a non-viscous fluid moving in an external force field defined by the potential V. The external force per unit mass is given by $\mathbf{F} = -\nabla V$ and Euler's equation for the flow is given by

$$\rho(\frac{\partial \mathbf{v}}{\partial t} + \mathbf{v}.\nabla\mathbf{v}) = -\nabla p - \rho\nabla V$$

In case the external force is gravitational, we set $V = gz$ and $-\nabla V = -g\hat{z}$. This equation can be expressed as

$$\rho(\frac{\partial \mathbf{v}}{\partial t} + (\nabla \times \mathbf{v}) \times \mathbf{v} + \frac{1}{2}\nabla v^2) = -\nabla p - \rho\nabla V$$

If we assume the fluid to be incompressible, then ρ is a constant and this equation can be written as

$$\frac{\partial \mathbf{v}}{\partial t} + (\Omega \times \mathbf{v}) \times \mathbf{v} + \nabla(p/\rho + V + \frac{v^2}{2}) = \mathbf{0}$$

where $\Omega = \nabla \times \mathbf{v}$ is the vorticity. Under steady state conditions, $\frac{\partial \mathbf{v}}{\partial t} = \mathbf{0}$ and if $\mathbf{n}$ denotes a unit vector along the direction of the velocity field at a given point, we get on forming the scalar product with $\mathbf{n}$ the equation

$$\mathbf{n}.\nabla(p/\rho + V + \frac{v^2}{2}) = 0$$

or

$$\frac{\partial}{\partial \mathbf{n}}(p/\rho + V + \frac{v^2}{2}) = 0$$

This equation states that the directional derivative of $p/\rho + V + \frac{v^2}{2}$ along a streamline for a steady incompressible fluid is zero. In other words, the quantity $p/\rho + V + \frac{v^2}{2}$ is a constant along each streamline. This is one version of Bernoulli's equation. Another version is obtained by making the assumption that the fluid is incompressible and irrotational, i.e., ρ is a constant and $\Omega = \mathbf{0}$. Then, $\mathbf{v} = -\nabla\phi$, where ϕ is the velocity potential and Euler's equation assumes the form

$$\nabla(-\frac{\partial \phi}{\partial t} + \frac{v^2}{2} + p/\rho + V) = \mathbf{0}$$

and hence the quantity

$$p/\rho + \frac{v^2}{2} + V - \frac{\partial \phi}{\partial t}$$

is a constant throughout the fluid volume. If in addition, the flow is steady, then $\frac{\partial \phi}{\partial t} = 0$ and Bernoulli's equation assumes the form

$$p/\rho + \frac{v^2}{2} + V = constt.$$

at all points of the fluid. If there is no external force, then $V = 0$ and we get $p/\rho + \frac{v^2}{2} = constt.$. This equation states that the pressure is lower at points where the fluid velocity is greater. This is precisely the reason why an aircraft wing develops a lift. The top curve of the wing is longer than the bottom curve. So when the wing moves forward, in the same time, air covers a longer distance at the top of the wing than at the bottom. This means that the air speed is higher at the top of the wing than at the bottom and hence the pressure is lower at the top of the wing than the bottom causing the wing to get lifted up. Another illustration of Bernoulli's principle is obtained by taking two sheets of paper, holding them vertically parallel to each other and blowing air in between them. One would expect the sheets to get spread apart on blowing air in this fashion. However what actually happens is that the two sheets come closer together on blowing air between them. This is because Bernoulli's principle is at work. When air is blown between the sheets, the pressure in between them becomes lower than the pressure outside them causing the sheets to come together. For more examples see the Feynman lectures on physics, volume II.

2.7   Equation for vorticity

Consider Navier-Stokes equation for a viscous fluid in the form

$$\rho\left(\frac{\partial \mathbf{v}}{\partial t} + \Omega \times \mathbf{v} + \frac{1}{2}\nabla v^2\right) = -\nabla p + \eta \nabla^2 \mathbf{v}$$

Assuming the fluid to be incompressible, we get on taking curl,

$$\frac{\partial \Omega}{\partial t} + \nabla \times (\Omega \times \mathbf{v}) - \nu \nabla^2 \Omega = 0$$

where $\nu = \eta/\rho$ is the modified viscosity coefficient. If terms quadratic in the velocity are neglected, i.e., only terms linear in the viscosity are retained, then this reduces to the diffusion equation

$$\frac{\partial \Omega}{\partial t} - \nu \nabla^2 \Omega = 0$$

This equation has solutions that correspond to a decay of Ω. This means that an initial circulation set up in a viscous fluid eventually decays to zero due to viscous damping. The exact solution to the above equation is given by

$$\Omega(\mathbf{r}, t) = \int \Omega(\mathbf{r}', \mathbf{0})(4\pi Dt)^{-1/2} exp(-|\mathbf{r} - \mathbf{r}'|^2/4\nu t)d^3\mathbf{r}'$$

This implies that

$$lim_{t\to\infty}\Omega(\mathbf{r}, t) = \mathbf{0}$$

The term $\nabla \times (\Omega \times \mathbf{v})$ represents convection. Thus if no approximations are made, vorticity propagates in the form of convection and diffusion. For a more detailed discussion of the convection term, we refer to the Feynman lectures, vol.II. We shall now look at the form assumed by the vorticity equation for a two dimensional flow. The equation $\nabla.\mathbf{v} = 0$ in two dimensions reads

$$\frac{\partial v_x(t, x, y)}{\partial x} + \frac{\partial v_y(t, x, y)}{\partial y} = 0$$

and hence there is a function $\psi(t, s, y)$ such that

$$v_x = \frac{\partial \psi}{\partial y}, v_y = -\frac{\partial \psi}{\partial x}$$

The normal to the curve $\psi = constt$ at any given time is $(\frac{\partial \psi}{\partial x}, \frac{\partial \psi}{\partial y})$ and the tangent to this curve is $(\frac{\partial \psi}{\partial y}, -\frac{\partial \psi}{\partial x})$ which is the velocity. Thus, the curves $\psi = constt.$ represent the streamlines. Note that we are not assuming that the flow is a potential flow, i.e., the curl of the velocity field is not necessarily zero. The incompressibility condition itself implies the presence of a stream function.

8. Energy conservation equation for a fluid

Let ϵ denote the internal (thermal) energy per unit mass of the fluid. The total energy per unit mass is given by $\frac{1}{2}\rho v^2 + \rho\epsilon$. We want the use the Euler equation and the basic equations of thermodynamics to determine its rate of change. Let w denote the enthalpy per unit mass, i.e., $w = \epsilon + pV = \epsilon + p/\rho$ where $V = 1/\rho$ is the volume per unit mass. The thermodynamical equation $dw = Tds + Vdp = TdS + dp/\rho$ where s is the entropy per unit mass yields $\nabla w = T\nabla s + (\nabla p)/\rho$. If the flow is adiabatic, then no heat is added to any fluid particle and $ds = 0$. For such a fluid, $\nabla w = (\nabla p)/\rho$. Note that the adiabatic condition can also be expressed as $\frac{Ds}{Dt} = 0$, or equivalently as

$$\frac{\partial s}{\partial t} + \mathbf{v}.\nabla s = 0$$

Combining this with the mass conservation equation

$$\frac{\partial \rho}{\partial t} + \nabla.(\rho\mathbf{v}) = 0$$

gives

$$\frac{\partial}{\partial t}(\rho s) = s\frac{\partial \rho}{\partial t} + \rho\frac{\partial s}{\partial t}$$

$$= -s\nabla.(\rho \mathbf{v}) - \rho\mathbf{v}.\nabla s = -\nabla.(\rho s\mathbf{v})$$

or

$$\frac{\partial(\rho s)}{\partial t} + \nabla.(\rho s\mathbf{v}) = 0$$

This equation along with the Euler and mass conservation equation completely defines an adiabatic fluid. Since $\nabla w = (\nabla p)/\rho$ for an adiabatic fluid, Euler's equation for an adiabatic fluid assumes the form

$$\mathbf{v}.\nabla\mathbf{v} + \frac{\partial \mathbf{v}}{\partial t} = -\nabla w$$

from which writing $\mathbf{v}.\nabla\mathbf{v} = (\nabla \times \mathbf{v}) \times \mathbf{v} + \frac{1}{2}\nabla v^2$, we deduce

$$\nabla \times (\Omega \times \mathbf{v}) + \frac{\partial \Omega}{\partial t} = \mathbf{0}$$

where $\Omega = \nabla \times \mathbf{v}$ is the vorticity. This equation holds for any adiabatic fluid even if the fluid is compressible. Now we drop the condition of adiabaticity and derive an equation for the rate of energy increase.

$$\frac{\partial}{\partial t}(\frac{1}{2}\rho v^2) = \frac{v^2}{2}\frac{\partial \rho}{\partial t} + \rho\mathbf{v}.\frac{\partial \mathbf{v}}{\partial t}$$

$$= -\frac{v^2}{2}\nabla.(\rho\mathbf{v}) - \rho\mathbf{v}.(\mathbf{v}.\nabla\mathbf{v} + (\nabla p)/\rho)$$

$$= -\frac{v^2}{2}\nabla.(\rho\mathbf{v}) - \rho\mathbf{v}.((\Omega \times \mathbf{v}) \times \mathbf{v} + \frac{1}{2}\nabla v^2 + (\nabla p)/\rho)$$

$$= -\frac{v^2}{2}\nabla.(\rho\mathbf{v}) - \rho\mathbf{v}.\nabla(v^2/2 + w) + \rho T\mathbf{v}.\nabla s$$

since

$$(\nabla p)/\rho = \nabla w - T\nabla s$$

Next,

$$\frac{\partial}{\partial t}(\rho\epsilon) = \rho\frac{\partial \epsilon}{\partial t} + \epsilon\frac{\partial \rho}{\partial t}$$

By energy conservation, we have

$$d\epsilon = Tds - pdV = Tds + pd\rho/\rho^2$$

so that

$$\frac{\partial}{\partial t}(\rho\epsilon) = \rho T\frac{\partial s}{\partial t} + (\epsilon + \frac{p}{\rho})\frac{\partial \rho}{\partial t}$$

$$= \rho T \frac{\partial s}{\partial t} + w \frac{\partial \rho}{\partial t}$$

$$= -\rho T \mathbf{v}.\nabla s - w \nabla.(\rho \mathbf{v})$$

combining the above two equations gives

$$\frac{\partial}{\partial t}(\frac{1}{2}\rho v^2 + \rho\epsilon)$$

$$= -\frac{v^2}{2}\nabla.(\rho \mathbf{v}) - \rho \mathbf{v}.\nabla(v^2/2 + w) + \rho T \mathbf{v}.\nabla s - \rho T \mathbf{v}.\nabla s - w\nabla.(\rho \mathbf{v})$$

$$= -\nabla.(\rho \mathbf{v}(w + v^2/2))$$

This is the required energy conservation equation.

9. Kelvin's theorem on conservation of circulation

Let C denote a closed contour that moves along with the fluid. The circulation of the velocity field around this contour is given by

$$\Gamma = \int_C \mathbf{v}.d\mathbf{r}$$

We want to evaluate the rate of change of Γ and show that it equals zero. To do so, we use

$$\frac{d\Gamma}{dt} = \int_C \frac{d\mathbf{v}}{dt}.d\mathbf{r} + \int_C \mathbf{v}.\frac{dd\mathbf{r}}{dt}$$

Now,

$$\frac{dd\mathbf{r}}{dt} = d\frac{d\mathbf{r}}{dt} = d\mathbf{v}$$

Also

$$\frac{d\mathbf{v}}{dt} = -(\nabla p)/\rho = -\nabla w$$

assuming adiabaticity and hence

$$\frac{d\Gamma}{dt} = -\int_C \nabla w.d\mathbf{r} + \int_C \mathbf{v}.d\mathbf{v} = \int_C d(-w + v^2/2) = 0$$

This result is known as Kelvin's vorticity theorem.

2.8 Some examples from fluid dynamics

The first problem deals with determining the shape of the surface of an incompressible fluid subject to a gravitational field contained in a cylindrical vessel rotating at constant angular velocity Ω about its axis. The velocity field is two dimensional and given by $v_x = -\Omega y, v_y = \Omega x, v_z = 0$. This follows from the obvious relation between angular velocity and velocity:

$$\mathbf{v} = \Omega \times \mathbf{r} = \Omega\hat{z}(x\hat{x} + y\hat{y} + z\hat{z}) = \Omega(x\hat{y} - y\hat{x})$$

Then,

$$\mathbf{v}.\nabla\mathbf{v} = (-\Omega y\hat{x} + \Omega x\hat{y}).\nabla(-\Omega y\hat{x} + \Omega x\hat{y})$$

$$= -\Omega y\frac{\partial}{\partial x}(-\Omega y\hat{x} + \Omega x\hat{y}) + \Omega x\frac{\partial}{\partial y}(-\Omega y\hat{x} + \Omega x\hat{y})$$

$$-\Omega^2(x\hat{x} + y\hat{y})$$

The steady state Navier-Stokes equation is

$$\mathbf{v}.\nabla\mathbf{v} = -(\nabla p)/\rho - g\hat{z}$$

which in component form are

$$\Omega^2 x = -\frac{1}{\rho}\frac{\partial p}{\partial x} \qquad \Omega^2 y = -\frac{1}{\rho}\frac{\partial p}{\partial y}$$

$$= -\frac{1}{\rho}\frac{\partial p}{\partial z} + g = 0$$

The solution to these equations is given by

$$\frac{p}{\rho} = \frac{1}{2}\Omega^2(x^2 + y^2) - gz + C$$

and the free surface $p = constt$ is a paraboloid. The second problem deals with the flow pattern for a fluid in which a sphere of radius R moves with a velocity $\mathbf{u}$ which may possibly be time dependent and to determine the pressure on the surface of the sphere. We assume that the flow is irrotational and incompressible. Irrotationality implies that the velocity field is derivable from a velocity potential function ϕ: $\mathbf{v} = \nabla\phi$. Combining this with incompressibility gives $\nabla^2\phi = 0$. ϕ is a scalar function involving only one vector $\mathbf{u}$. The only scalar function that we can form out of this vector that satisfies Laplace's equation is $C\mathbf{u}.\mathbf{r}/r^3 = -C\mathbf{u}.\nabla(1/r)$ where C is a numerical constant. In fact, suppose we assume that $\phi = -\mathbf{A}.\nabla(1/r)$ where $\mathbf{A}$ is some vector, we can determine $\mathbf{A}$ by matching the boundary conditions for the velocity at the spherical surface. At the spherical surface $r = R$, we must have $\mathbf{v}.\hat{r} = \mathbf{u}.\hat{r}$. Now,

$$\mathbf{v} = \nabla(\mathbf{A}.\mathbf{r}/r^3) = (r^3\mathbf{A} - 3\mathbf{A}.\mathbf{r}r^2\mathbf{r})/r^6$$

so that

$$\mathbf{v}.\hat{r}|_{r=R} = (\mathbf{A}.\hat{r})/R^3 - 3(\mathbf{A}.\hat{r})/R^3 = -2\mathbf{A}.\hat{r}/R^3$$

Equating this to $\mathbf{u}.\hat{r}$ gives us

$$\mathbf{A} = -R^3\mathbf{u}/2$$

and hence

$$\phi = -\frac{R^3\mathbf{u}.\mathbf{r}}{2r^3}$$

The gradient of this gives the velocity field. To calculate the pressure distribution on the spherical surface, we use the equation

$$p = p_0 - \frac{\rho v^2}{2} - \rho\frac{\partial\phi}{\partial t}$$

where the partial derivative $\frac{\partial\phi}{\partial t}$ is calculated relative to the moving sphere. We use

$$\frac{\partial\phi}{\partial t}\Big|_{sphere} = \frac{\partial\phi}{\partial\mathbf{u}}.\frac{d\mathbf{u}}{dt} + \mathbf{u}.\nabla\phi$$

at $r = R$. The third problem is a two dimensional fluid flow problem in which we have an infinitely long cylinder of radius R moving perpendicularly to its axis with a velocity $\mathbf{u}$. The solution to the two dimensional Laplace equation for the velocity potential involving $\mathbf{u}$ as the only vector is $\phi = C\mathbf{u}.\nabla log(\rho) = C\mathbf{u}.\hat{\rho}/\rho$. This can be seen from the fact that

$$\nabla^2 log(\rho) = \frac{1}{\rho}\frac{d}{d\rho}\rho\frac{dlog(\rho)}{d\rho} = 0, \rho > 0$$

Note that we are assuming that at the given time, the axis of the cylinder is the z axis, i.e., the centre of the cross section of the cylinder coincides with the origin. We are dealing with the flow for $\rho > R$. We can write $\phi = C\mathbf{u}.\rho/\rho^2$. The velocity field is given by the gradient of this:

$$\mathbf{v} = C\nabla(\mathbf{u}.\rho/\rho^2) = C(\rho^2\mathbf{u} - 2(\mathbf{u}.\rho)\rho)/\rho^4$$

The velocity on the surface of the cylinder is given by setting $\rho = R\hat{\rho}$. Thus,

$$\mathbf{v}|_{\rho=R\hat{\rho}} = C\mathbf{u}/R^2 - 2C(\mathbf{u}.\hat{\rho})\hat{\rho}/R^2$$

The normal component of this velocity is given by

$$\mathbf{v}.\hat{\rho} = C\mathbf{u}.\hat{\rho}/R^2 - 2C\mathbf{u}.\hat{\rho}/R^2 = -C\mathbf{u}.\hat{\rho}/R^2$$

This must equal $\mathbf{u}.\hat{\rho}$ since the velocity of fluid relative to the cylinder on its surface cannot have any normal component. Thus, $C = -R^2$ and we get for the velocity field

$$\mathbf{v} = -R^2\mathbf{u}/\rho^2 + 2R^2(\mathbf{u}.\hat{\rho})\hat{\rho}/\rho^2$$

Using Bernoulli's equation, we can determine once again the pressure distribution on the surface of the cylinder.

2.9   Gravity waves

Consider the flow of a non-viscous incompressible fluid under the uniform gravitational force. Neglecting the nonlinear term in the Navier-Stokes equation gives us the linearized equation

$$\frac{\partial \mathbf{v}}{\partial t} = -\nabla p/\rho - g\hat{z}$$

Setting $\mathbf{v} = \nabla\phi$ with ϕ the velocity potential gives

$$\nabla(p/\rho + gz + \frac{\partial \phi}{\partial t}) = 0$$

so that

$$p/\rho + gz + \frac{\partial \phi}{\partial t} = Constt.$$

The constant can depend upon time but not on space. It can be absorbed into the velocity potential thereby simplifying the equation to

$$p/\rho + gz + \frac{\partial \phi}{\partial t} = 0$$

This is the approximate Bernoulli equation obtained after neglecting the nonlinear term. It is valid for small fluid velocities. Let $z = \xi(t, x, y)$ denote the free surface of the fluid. On this surface, the pressure p is a constant equal to p_0, the atmospheric pressure. Thus,

$$p_0/\rho + g\xi + \frac{\partial \phi}{\partial t} = 0$$

Now, $v_z = \frac{\partial \xi}{\partial t}$ is the z component of the velocity of a fluid particle on the surface. Taking the time derivative of the above equation gives

$$gv_z + \frac{\partial^2 \phi}{\partial t^2} = 0$$

at the surface $z = \xi(t, x, y)$. Now substitute $v_z = \frac{\partial \phi}{\partial z}$. We assume that the surface undulations are small so that surface can be taken as $z = 0$. Then, we deduce the following partial differential equation for ϕ at $z = 0$:

$$(g\frac{\partial \phi}{\partial z} + \frac{\partial^2 \phi}{\partial t^2})|_{z=0} = 0$$

There is another approach to gravity waves based on considering the global form of the mass conservation equation. As usual, let $z = \xi(t, x, y)$ denote the equation of the surface. We consider a rectangle $[x, x+dx] \times [y, y+dy]$ and the matter of fluid directly above this rectangle. The rate at which fluid enters this region from the sides of this column of fluid is given by $h(-\frac{\partial v_x}{\partial x} - \frac{\partial v_y}{\partial y})dxdy$

and the rate at which the volume of fluid inside this column increases with time is given by $\frac{\partial \xi}{\partial t} dx dy$. Thus, our matter conservation equation assumes the form

$$h\left(\frac{\partial v_x}{\partial x} + \frac{\partial v_y}{\partial y}\right) + \frac{\partial \xi}{\partial t} = 0$$

Here, h is the height of the fluid column and we are assuming v_x, v_y to be independent of z. The z component of the velocity is negligible. Bernoulli's equation with neglect of the nonlinear term reads

$$p/\rho + gz + \frac{\partial \phi}{\partial t} = 0$$

Taking $z = \xi(t, x, y)$ gives

$$p_0/\rho + g\xi + \frac{\partial \phi}{\partial t} = 0$$

Note that since ϕ is assumed to be independent of z, this equation holds for all x, y, t. Thus,

$$g\frac{\partial \xi}{\partial t} = -\frac{\partial^2 \phi}{\partial t^2}$$

Substituting this equation and $v_x = \frac{\partial \phi}{\partial x}, v_y = \frac{\partial \phi}{\partial y}$ into the previous one gives us

$$gh\left(\frac{\partial^2 \phi}{\partial x^2} + \frac{\partial^2 \phi}{\partial y^2}\right) - \frac{\partial^2 \phi}{\partial t^2} = 0$$

which is the wave equation for the velocity potential. Differentiating this equation with respect to t and using $\frac{\partial \phi}{\partial t} = -p_0/\rho - g\xi$ shows that $\xi(t, x, y)$ also satisfies the same two dimensional wave equation. The velocity of these waves is $\sqrt{gh}$.

Solved Problems, Study Projects, Suggestions and Remarks

Study project 1

A seminar can be given on the theory of boundary layers occurring in fluid dynamics. If the flow takes place parallel to the xz plane, so that the y direction is normal to the direction of flow, then the velocity components v_x, v_y in steady state are functions of x, y only. A boundary layer is formed close to the plane. Outside the boundary layer, there is potential flow, i.e. if $U(x)$ denotes the velocity at $|y| = \infty$, i.e. at infinite distance from the plane, then the Bernoulli equation $p + \frac{1}{2}\rho U(x)^2 = constant$ is satisfied. Note that the y component of the velocity has been neglected in comparison with the x component. This assumption will be justified later. This equation shows that the pressure cannot have any gradient in the y direction, i.e. $\frac{\partial p}{\partial y} = 0$. It also implies that the gradient of p along the x direction is given by $\frac{dp}{dx} = -U\frac{dU}{dx}$. The equations of motion that describe the boundary layer are the Navier Stokes equation and the equation of mass conservation. These are respectively

$$v_x\frac{\partial v_x}{\partial x} + v_y\frac{\partial v_x}{\partial y} - \nu\nabla^2 v_x = -\frac{dp}{dx} = -U\frac{dU}{dx}$$

$$v_x \frac{\partial v_y}{\partial x} + v_y \frac{\partial v_y}{\partial y} - \nu \nabla^2 v_y = 0$$

$$\frac{\partial v_x}{\partial x} + \frac{\partial v_y}{\partial y} = 0$$

The thickness of the boundary layer is small δ as compared to the dimensions of the surface along which the flow takes place. We let l denote the characteristic dimension of the surface and $R = lU_0/nu$ the Reynolds's number. Here U_0 is a number that is represents the order of magnitude of the velocity of the fluid at an infinite distance from the boundary layer. The gradients of the velocity along the x direction (parallel to the flow) are negligible as compared to the gradients along the y direction (normal to the flow) near the boundary layer. This means that $\frac{\partial^2 v_x}{\partial x^2}$ is negligible as compared to $\frac{\partial^2 v_x}{\partial y^2}$. For both the terms on the left side of the Navier Stokes equation to be comparable, we must have $v_x/v_y \approx l/\delta$. This means that v_y is negligible as compared to v_x. The fact that $\frac{\partial p}{\partial y}$ is negligible in comparison with $\frac{\partial p}{\partial x}$ can also be seen by comparing $\frac{\partial p}{\partial x}$ with $\nu \nabla^2 v_x \approx \nu \frac{\partial^2 v_x}{\partial y^2}$ and $\frac{\partial p}{\partial y}$ with $\nu \frac{\partial^2 v_x}{\partial y^2}$. This gives us the following estimate for the ratio of the pressure gradient along the two directions:

$$\frac{\partial p}{\partial x} \bigg/ \frac{\partial p}{\partial y} \approx l^2/\delta^2$$

which means that the pressure gradient along the y direction can be neglected in comparison to the pressure gradient along the x direction. The scaling is done as follows $x = lx', y = ay', v_x = U_0 v_x', v_y = bv_y'$ where a, b are numbers chosen so that the equations of motion appear in dimensionless form. x', y', v_x', v_y' are dimensionless quantities. We find that

$$v_x \frac{\partial v_x}{\partial x} = \frac{U_0^2}{l} v_x' \frac{\partial v_x'}{\partial x'}$$

$$v_y \frac{\partial v_x}{\partial y} = (U_0 b/a) v_y' \frac{\partial v_x'}{\partial y'}$$

$$\nabla^2 v_x = \frac{U_0}{a^2} \frac{\partial^2 v_x'}{\partial y'^2})$$

$$\nabla^2 v_y = (b/a^2) \frac{\partial v_y'}{\partial y'^2}$$

The first Navier Stokes equation is equivalent to

$$\frac{U_0^2}{l} v_x' \frac{\partial v_x'}{\partial x'} + \frac{U_0 b}{a} v_y' \frac{\partial v_x'}{\partial y'} - (\nu U_0)/a^2) \frac{\partial^2 v_x'}{\partial y'^2}$$

$$= (U_0^2/l) U' \frac{dU'}{dx'}$$

For this equation to acquire a dimensionless form, we must have

$$U_0 b/a = U_0^2/l = \nu U_0/a^2$$

Thus,

$$a = (\nu l/U_0)^{1/2} = \frac{l}{\sqrt{R}}$$

with $R = U_0 l/\nu$ as Reynolds's number. Thus, the first Navier Stokes equation reduces to

$$v'_x \frac{\partial v'_x}{\partial x'} + v'_y \frac{\partial v'_x}{\partial y'} - \frac{\partial^2 v'_x}{\partial y'^2} = -U' \frac{dU'}{dx'}$$

The second Navier Stokes equation is likewise in dimensionless form

$$v'_x \frac{\partial v'_y}{\partial x'} + v'_y \frac{\partial v'_y}{\partial y'} - \frac{\partial^2 v'_y}{\partial y'^2} = 0$$

and the mass conservation equation

$$\frac{\partial v'_x}{\partial x'} + \frac{\partial v'_y}{\partial y'} = 0$$

These are known as the Prandtl equations.

Study project 2

A basic seminar on the flow of a liquid inside a pipe or between two planes with a constant pressure gradient along the length can be delivered. It would be of interest to mechanical and civil engineers involved in the design of pipes. 7For example suppose that we are interested in the flow between the planes $x = 0$ and $x = w$. Assume that the flow takes place along the y direction. Thus the velocity field is $v_y(x)\hat{y}$. The flow takes place between the points $y = 0$ and $y = L$. The Navier Stokes equations boil down to

$$\eta v''_y(x) - \frac{\partial p}{\partial y} = 0$$

The pressure is assumed to depend only on y, i.e. it does not vary across the cross section of the pipe. We thus have situation in which a function of x equals a function of y. Thus, both sides must equal a constant, i.e. $\frac{\partial p}{\partial y} = $ constt. If Δp is the pressure difference between the two ends of the pipe, then this gradient is $-\frac{\Delta p}{L}$ and we get the following equation for the velocity field

$$\eta v''_y(x) = -\frac{\Delta p}{L}$$

This integrates to give

$$v_y(x) = -\frac{\Delta p}{2\eta L} x^2 + ax + b$$

The velocity must vanish at the two bounding planes, i.e. $v_y(0) = v_y(w) = 0$. This determines the constants a, b: $b = 0, a = \frac{\Delta p w}{2\eta L}$. Thus,

$$v_y(x) = \frac{\Delta p}{2\eta L} x(w - x)$$

The frictional force per unit area on the bounding planes can be evaluated as $\eta v_y'(0), \eta v_y'(w)$. The magnitudes of both of these quantities is given by

$$|\sigma_{xy}| = \frac{w\Delta p}{2\eta L}$$

The next kind of pipe is a rectangular pipe defined by the planes $x = 0, a, y = 0, b$ with a constant pressure gradient along its length L. The velocity field is given by $v_z(x, y)\hat{z}$ and the Navier-Stokes equation boils down to

$$\eta\left(\frac{\partial^2 v_z}{\partial x^2} + \frac{\partial^2 v_z}{\partial y^2}\right) = -\frac{\Delta p}{L}$$

The boundary conditions are $v_z(0, y) = v_z(a, y) = v_z(x, 0) = v_z(x, b) = 0$. The solution can be obtained as a Fourier series that satisfies the relevant boundary conditions:

$$v_z(x, y) = \sum_{n,m=1}^{\infty} c(n, m)sin(n\pi x/a)sin(m\pi y/b)$$

Plugging this into the partial differential equation gives us

$$\sum_{n,m=1}^{\infty} (n^2/a^2 + m^2/b^2)\pi^2 c(n, m)sin(n\pi x/a)sin(m\pi y/b) = \frac{\Delta p}{\eta L}$$

from which we obtain

$$(n^2/a^2 + m^2/b^2)\pi^2 c(n, m) = \frac{4\Delta p}{ab\eta L}\int_0^a \int_0^b sin(n\pi x/a)sin(m\pi y/b)dxdy$$

$$= \frac{4\Delta p}{ab\eta L}\frac{ab}{nm\pi^2}(1 - (-1)^n)(1 - (-1)^m)$$

This gives $c(n, m) = 0$ if either n or m is even and for both odd n, m, we have

$$c(2r + 1, 2s + 1) = \frac{4\Delta p}{\eta L}\frac{(2r + 1)(2s + 1)}{(2r + 1)^2/a^2 + (2s + 1)^2/b^2}$$

and we obtain the following Fourier series for the velocity field inside the pipe

$$v_z(x, y) = \sum_{r,s=0}^{\infty} c(2r + 1, 2s + 1)sin((2r + 1)\pi x/a)sin((2s + 1)\pi y/b)$$

The flow can be simulated on a digital computer by plotting the a graph showing the variation of the velocity field along the cross section.

Study project 3

A seminar on the physical meaning of the divergence operation acting on vector fields would be of use to the researcher in electrodynamics and fluid dynamics. If $\mathbf{F}$ is a vector field and S is a surface bounding a volume V, then $\Phi_F = \int_S \mathbf{F}.\mathbf{n}dS$ gives us the flux of the vector field through the surface S. If $\mathbf{F}$ represents the velocity field, then Φ_F gives the amount of fluid flowing out of the volume V per unit time. If $\mathbf{F}$ is the electric field, then Φ_F gives the number of electric field lines cutting through the surface bounding the volume. The divergence theorem states that Φ_F equals the volume integral of $div\mathbf{F}$ over V, i.e.

$$\Phi_F = \int_V div\mathbf{F}dv$$

So, $div\mathbf{F}$ can be interpreted in the case of fluid flow as the amount of fluid leaving unit volume per unit time. If δV is a volume element containing the point $\mathbf{r}$, then the limit $lim_{\delta V \to 0} \frac{\int_V div\mathbf{F}dv}{\delta V} = div\mathbf{F}(\mathbf{r})$.

Study project 4

Basic equations of magnetohydrodynamics: Variables are $\mathbf{v}, \mathbf{E}, \mathbf{B}, \rho$, the velocity field, the electric field, the magnetic field and the density field. These are all functions of position and time, i.e.

$$\mathbf{v}(\mathbf{r}, t), \rho(\mathbf{r}, t), \mathbf{E}(\mathbf{r}, t), \mathbf{B}(\mathbf{r}, t)$$

The conductivity of the fluid is taken as a constant σ. The current density is given by

$$\mathbf{J} = \sigma(\mathbf{E} + \mathbf{v} \times \mathbf{B})$$

The equations of motion are matter conservation, momentum equation and the Maxwell equations. Written in full, these are

$$\frac{\partial \rho}{\partial t} + \nabla.(\rho\mathbf{v}) = 0$$

$$\rho(\frac{\partial \mathbf{v}}{\partial t} + \mathbf{v}.\nabla\mathbf{v}) = -\nabla p + \nu\nabla^2\mathbf{v} + \frac{\rho e}{m}(\mathbf{E} + \mathbf{v} \times \mathbf{B})$$

$$\nabla.\mathbf{E} = \rho e/m\epsilon$$

$$\nabla.\mathbf{B} = 0$$

$$\nabla \times \mathbf{E} = -\frac{\partial \mathbf{B}}{\partial t}$$

$$\nabla \times \mathbf{B} = \mu\sigma(\mathbf{E} + \mathbf{v} \times \mathbf{B}) + \mu\epsilon\frac{\partial \mathbf{E}}{\partial t}$$

Assume the fluid to be incompressible, i.e., ρ is a constant. Then, the matter conservation equation reduces to $\nabla.\mathbf{v} = 0$. The momentum equation is

$$\rho(\frac{\partial \mathbf{v}}{\partial t} + (\nabla \times \mathbf{v}) \times \mathbf{v} + \frac{1}{2}\nabla v^2) = -\nabla p + \nu \nabla^2 \mathbf{v} + \frac{\rho e}{m}(\mathbf{E} + \mathbf{v} \times \mathbf{B})$$

Taking the curl of this equation gives us

$$\frac{\partial \Omega}{\partial t} + \nabla \times (\Omega \times \mathbf{v}) = \nu \nabla^2 \Omega + \frac{\rho e}{m}\nabla \times (\mathbf{E} + \mathbf{v} \times \mathbf{B})$$

Making use of the Faraday equation, this is the same as

$$\frac{\partial}{\partial t}(\Omega + \frac{\rho e}{m}\mathbf{B}) + \nabla \times ((\Omega + \frac{\rho e}{m}\mathbf{B}) \times \mathbf{v}) = \nu \nabla^2 \Omega$$

where $\Omega = \nabla \times \mathbf{v}$ is the vorticity of the flow. When the fluid has no viscosity, i.e., $\nu = 0$, this reduces to the equation

$$\frac{\partial}{\partial t}(\Omega + \frac{\rho e}{m}\mathbf{B}) + \nabla \times (\Omega + \frac{\rho e}{m}\mathbf{B}) = \mathbf{0}$$

One solution to this equation is $\Omega + \frac{\rho e}{m}\mathbf{B} = \mathbf{0}$. This means that the vorticity follows the magnetic field. Taking curl of the Ampere equation and using the Maxwell equation $\nabla.\mathbf{B} = 0$, gives

$$\nabla^2 \mathbf{B} - \mu\sigma\frac{\partial \mathbf{B}}{\partial t} + \mu\sigma\nabla \times (\mathbf{v} \times \mathbf{B}) - \mu\epsilon\frac{\partial^2 \mathbf{B}}{\partial t^2} = 0$$

This furnishes us with two vector equations for the variables $\mathbf{v}, \mathbf{B}$ which can be used to analyze wave phenomena in the plasma. Assume that the magnetic field has a constant component $\mathbf{B}_0$ which is strong and a weaker temporally and spatially dependent component: $\mathbf{B}(\mathbf{r}, t) = \mathbf{B}_0 + \delta\mathbf{B}(\mathbf{r}, t)$. At equilibrium, i.e., only when the constant magnetic field component is present, there is no fluid velocity, i.e. $\mathbf{v} = 0$. The constant magnetic field combined with zero velocity field satisfy the MHD equations. Assume that the perturbing magnetic field $\delta\mathbf{B}(\mathbf{r}, t)$ causes a small perturbation to the velocity field $\delta\mathbf{v}(\mathbf{r}, t)$. Substituting these into the equations we get on neglecting second order terms, the linearized equation

$$\frac{\partial}{\partial t}(\delta\Omega + \frac{\rho e}{m}\delta\mathbf{B}) + \frac{\rho e}{m}\nabla \times (\mathbf{B}_0 \times \delta\mathbf{v}) = \nu \nabla^2 \delta\Omega$$

$$\nabla^2 \delta\mathbf{B} - \mu\sigma\frac{\partial \delta\mathbf{B}}{\partial t} + \mu\sigma\nabla \times (\delta\mathbf{v} \times \mathbf{B}_0) - \mu\epsilon\frac{\partial^2 \delta\mathbf{B}}{\partial t^2} = 0$$

We want to obtain the dispersion equation for waves propagated in the plasma due to a small perturbation either in the velocity field or in the magnetic field or in both. To this end, we set

$$\delta\mathbf{B}(\mathbf{r}, t) = \delta\mathbf{B}_0 exp(i(\omega t - \mathbf{k}.\mathbf{r}))$$

$$\delta\mathbf{v}(\mathbf{r}, t) = \delta\mathbf{v}_0 exp(i(\omega t - \mathbf{k}.\mathbf{r}))$$

Then,

$$\delta\Omega = \nabla \times \delta\mathbf{v} = -i\mathbf{k} \times \delta\mathbf{v}_0 exp(i(\omega t - \mathbf{k}.\mathbf{r}))$$

$$\delta\Omega + \frac{\rho e}{m}\delta\mathbf{B} = (-i\mathbf{k} \times \delta\mathbf{v}_0 + \frac{\rho e}{m}\delta\mathbf{B}_0)exp(i(\omega t - \mathbf{k}.\mathbf{r}))$$

$$\frac{\partial}{\partial t}(\delta\Omega + \frac{\rho e}{m}\mathbf{B}) = (\omega\mathbf{k} \times \delta\mathbf{v}_0 + i\omega\frac{\rho e}{m}\delta\mathbf{B}_0)exp(i(\omega t - \mathbf{k}.\mathbf{r}))$$

$$\nabla \times (\mathbf{B}_0 \times \delta\mathbf{v}) = -i\mathbf{k} \times (\mathbf{B}_0 \times \delta\mathbf{v}_0)exp(i(\omega t - \mathbf{k}.\mathbf{r}))$$

$$\nabla^2\Omega = ik^2\mathbf{k} \times \delta\mathbf{v}_0$$

The vorticity equation thus gives

$$\omega\mathbf{k} \times \delta\mathbf{v}_0 + i\omega\frac{\rho e}{m}\delta\mathbf{B}_0 = i\nu k^2\mathbf{k} \times \delta\mathbf{v}_0$$

Similarly, the magnetic field equation gives

$$-k^2\delta\mathbf{B}_0 - i\omega\mu\sigma\delta\mathbf{B}_0 + i\mu\sigma\mathbf{k} \times (\delta\mathbf{v}_0 \times \mathbf{B}_0) + \mu\epsilon\omega^2\delta\mathbf{B}_0 = 0$$

Study project 5

Alfven waves. The equations of magnetohydrodynamics are

$$\frac{\partial\rho_m}{\partial t} + div(\rho_m\mathbf{u}) = 0$$

$$\rho_m(\frac{\partial\mathbf{u}}{\partial t} + \mathbf{u}.\nabla\mathbf{u}) = -\nabla p + \mathbf{J} \times \mathbf{B}$$

$$\nabla p = v_s^2\nabla\rho_m$$

$$\nabla \times \mathbf{B} = \mu\mathbf{J}$$

$$\nabla \times \mathbf{E} = -\frac{\partial\mathbf{B}}{\partial t}$$

$$\mathbf{E} + \mathbf{u} \times \mathbf{B} = 0$$

The Navier-Stokes equation can be written in view of the other equations as

$$\rho_m(\frac{\partial\mathbf{u}}{\partial t} + \mathbf{u}.\nabla\mathbf{u}) = -v_s^2\nabla\rho_m + (\nabla \times \mathbf{B}) \times \mathbf{B}/\mu_0$$

Faraday's law also assumes the form

$$\nabla \times (\mathbf{u} \times \mathbf{B}) = \frac{\partial\mathbf{B}}{\partial t}$$

These two along with the mass conservation equation define our equations for $\rho_m, \mathbf{B}, \mathbf{u}$. We look at linearized versions of these equations, i.e. set

$$\mathbf{B} = \mathbf{B}_0 + \mathbf{B}_1$$

where $\mathbf{B}_0$ is a constant vector and $\mathbf{B}_1$ is a function of position and time. Also set

$$\rho_m = \rho_{m0} + \rho_{m1}$$

where ρ_{m0} is a constant and ρ_{m1} is a function of position and time. $\mathbf{B}_0$ and ρ_{m0} can respectively be regarded as the equilibrium magnetic field and the equilibrium matter density. At equilibrium, the velocity is zero. Thus, perturbation of the equilibrium gives $\mathbf{u} = \mathbf{u}_1$ which is a function of position and time. The linear matter conservation equation reads

$$\rho_{m0}\nabla.\mathbf{u}_1 + \frac{\partial \rho_{m1}}{\partial t} = 0$$

The linearized Navier-Stokes equation reads

$$\rho_{m0}\frac{\partial \mathbf{u}_1}{\partial t} + v_s^2 \nabla \rho_m + \mathbf{B}_0 \times (\nabla \times \mathbf{B}_1)/\mu_0 = 0$$

and finally, the linearized Faraday equation reads

$$\frac{\partial \mathbf{B}_1}{\partial t} - \nabla \times (\mathbf{u}_1 \times \mathbf{B}_0) = \mathbf{0}$$

To determine the dispersion relation corresponding to plane wave solutions to these three linearized equations, we set

$$\mathbf{u}_1 = \mathbf{a}exp(i(\omega t - \mathbf{k}.\mathbf{r}))$$
$$\mathbf{B}_1 = \mathbf{b}exp(i(\omega t - \mathbf{k}.\mathbf{r}))$$
$$\rho_{m1} = d.exp(i(\omega t - \mathbf{k}.\mathbf{r}))$$

The result is a set of homogeneous linear equations for the variables $\mathbf{a}, \mathbf{b}, d$. Setting equal to zero the determinant of the gives system yields the desired dispersion relation connecting the frequency ω and the wave vector $\mathbf{k}$. For the sake of completeness, we merely write down the linear equations first: The linearized matter conservation equation reads

$$-\rho_{m0}\mathbf{k}.\mathbf{a} + \omega d = 0$$

The linearized Navier-Stokes equation reads

$$i\rho_{m0}\omega\mathbf{a} - dv_s^2 k^2 - i\mathbf{B}_0 \times (\mathbf{k} \times \mathbf{b})/\mu_0 = 0$$

Finally, the linearized Faraday equation reads

$$\omega\mathbf{b} + \mathbf{k} \times (\mathbf{a} \times \mathbf{B}_0) = \mathbf{0}$$

These are seven linear homogeneous equations for the eight variables $d, \mathbf{a}, \mathbf{b}$ and hence the associated determinant is zero. This furnishes us with the desired dispersion relation. We shall not derive the dispersion relation here but leave it as an exercise to the student in linear algebra.

Chapter 3

ElectroMagnetics

3.1   Coulomb's law, Gauss Law and Electrostatics

The subject of electromagnetism deals with the analysis of how charge and current distributions in space produce electric and magnetic fields permeating the entire space. The starting point for any study of electromagnetism is the notion of charge. Charge is defined only in terms of the electric field that it produces. The ancient Greek discovery of amber which when rubbed is able to attract pieces of silk was the first direct discovery of charge. The modern explanation of this is based on the fact that amber when rubbed causes a polarization of the charge in each atom, i.e. a shift of the centre of positive and negative charges. In other words, each atom which was originally neutral now becomes a dipole that produces an electric field capable of polarizing any other material and hence attracting it. Coulomb discovered that if two charges Q_1 and Q_2 are placed at a distance d apart, they attract or repel each other with a force that acts along the line joining the two charges, is attractive if the two charges are of opposite sign and is repulsive if the two charges are of the same sign. Coulomb had already discovered that charges come in two types, positive and negative and that charges of the same sign attract each other while charges of opposite sign repel each other. Coulomb's law for the electrostatic force exerted by a charge Q_1 located at $\mathbf{r}_1$ on a charge Q_2 located at $\mathbf{r}_2$ can be expressed in vector notation as

$$\mathbf{F}_{12} = \frac{Q_1 Q_2 (\mathbf{r}_2 - \mathbf{r}_1)}{4\pi\epsilon_0 |\mathbf{r}_1 - \mathbf{r}_2|^3}$$

In this equation, ϵ_0 is a constant called the primitivity of free space. Inside a uniform dielectric medium, the same formula holds with ϵ_0 replaced by ϵ, the permitivity of the medium. The electrostatic force follows the principle of superposition, i.e., if charges $Q_1, ..., Q_n$ are located at $\mathbf{r}_1, ..., \mathbf{r}_n$, then the force exerted by this collection on a charge Q located at $\mathbf{r}$ is given by

$$\mathbf{F} = \sum_{i=1}^{n} \mathbf{F}_i$$

where $\mathbf{F}_i$, the force exerted by Q_i on Q is given by

$$\mathbf{F}_i = \sum_{i=1}^{n} \frac{Q_i Q(\mathbf{r} - \mathbf{r}_i)}{4\pi\epsilon_0|\mathbf{r} - \mathbf{r}_i|^3}$$

We cannot explain why the superposition principle holds in terms of simpler concepts. Superposition of the electrostatic force constitutes by itself a law of nature. Instead of a discrete distribution of charges, we can consider a continuous distribution with the charge density at $\mathbf{r}$ being $\rho(\mathbf{r})$. This amounts to saying that the total charge inside a volume V of space $\mathbb{R}^3$ is given by $\int_V \rho(\mathbf{r})d^3\mathbf{r}$. The electrostatic force on a charge Q located at $\mathbf{r}$ produced by this continuous charge distribution is given once again by combining Coulomb's law with the principle of superposition:

$$\mathbf{F} = \int \frac{Q\rho(\mathbf{r})(\mathbf{R} - \mathbf{r})}{4\pi\epsilon_0|\mathbf{R} - \mathbf{r}|^3}d^3\mathbf{r}$$

the integral being extended over all of space, i.e. $\mathbb{R}^3$ where charge is distributed. The electric field at a point $\mathbf{r}$ in space produced by a discrete or continuous distribution of charges is defined to be the electrostatic force exerted upon a unit positive charge located at $\mathbf{r}$ due to the charge distribution. Thus, the electric field at $\mathbf{R}$ produced by the combination of a discrete charge distribution consisting of charges Q_i at $\mathbf{r}_i, i = 1, 2, ..., n$ and a continuous charge distribution of density $\rho(\mathbf{r}), \mathbf{r} \in \mathbb{R}^3$ is given by the formula

$$\mathbf{E}(\mathbf{r}) = \int \frac{\rho(\mathbf{r})(\mathbf{R} - \mathbf{r})}{4\pi\epsilon_0|\mathbf{R} - \mathbf{r}|^3}d^3\mathbf{r} + \sum_i \frac{Q_i(\mathbf{R} - \mathbf{r})}{4\pi\epsilon_0|\mathbf{R} - \mathbf{r}|^3}$$

This formula can be specialized to several situations. Let S be any closed surface enclosing a charge Q located at the origin. The electric field at $\mathbf{r}$ due to this point charge is given by $\mathbf{E}(\mathbf{r}) = \frac{Q\hat{r}}{4\pi\epsilon_0 r^2}$. If $\mathbf{n}$ denotes the unit normal at a point on the surface, then we know from basic surface geometry that the total solid angle

$$\int \frac{\hat{r}.\mathbf{n}}{r^2}dS = \int \frac{dS\cos\theta}{r^2} = 4\pi$$

where θ is the angle between $\hat{r}$ and $\mathbf{n}$ on the surface. On the other hand, if the origin falls outside the surface, then the above integral is zero. Thus for the electric field generated by the point charge in question located inside S, one has

$$\int_S \mathbf{E}.\mathbf{n}dS = \frac{Q}{\epsilon_0}$$

Using superposition and this result with the above discussion regarding the origin falling outside the surface, we deduce Gauss' law: For any charge distribution, let $Q_{encl} = \int_V \rho dV$ denote the total charge enclosed by S, i.e., falling inside the volume V. If $\mathbf{E}(\mathbf{r})$ is the electric field produced by this distribution, then

$$\int \mathbf{E}.\mathbf{n}dS = \frac{Q_{encl}}{\epsilon_0}$$

Stated in words, the flux of the electric field out of a closed surface equals the net charge enclosed by the surface. Gauss' theorem of vector calculus states that the flux of a vector field out of a closed surface equals the volume integral of the divergence of the vector field over the volume V enclosed by the surface. Thus, Gauss' law can be expressed in the form of a volume integral equation:

$$\int_V \nabla.\mathbf{E}\,dV = \int_V \frac{\rho}{\epsilon_0}\,dV$$

and since the volume V can be chosen arbitrarily, we conclude that

$$\nabla.\mathbf{E} = \frac{\rho}{\epsilon_0}$$

3.2 Examples of computation of the electric field for various kinds of static charge distributions: Spherical charge, cylindrical charge, line charge, charged disc, infinite plane carrying uniform surface charge density

Consider a line charge located on the z axis between the points $z = a$ and $z = b$. The linear density of the charge is given by σ. Units of linear charge density are Coulombs per metre. The electric field produced by this line charge at (X, Y, Z) is given by the formula

$$\mathbf{E}(X, Y, Z) = \frac{\sigma}{4\pi\epsilon_0} \int_a^b \frac{(X, Y, Z - z)}{(X^2 + Y^2 + (Z - z)^2)^{3/2}}\,dz$$

We leave the evaluation of these three integrals to the reader. If the linear charge density is a function of z, say $\sigma(z)$, then the above formula for the electric field gets modified to

$$\mathbf{E}(X, Y, Z) = \frac{1}{4\pi\epsilon_0} \int_a^b \frac{\sigma(z)(X, Y, Z - z)}{(X^2 + Y^2 + (Z - z)^2)^{3/2}}\,dz$$

As another example, the electric field produced by a surface charge density $\sigma(x, y)$ on the xy plane (units of surface charge density are Coulombs per $metre^2$) at the point (X, Y, Z) in space is given by the expression

$$\mathbf{E}(X, Y, Z) = \frac{1}{4\pi\epsilon_0} \int_{-\infty}^{\infty} \int_{-\infty}^{\infty} \frac{\sigma(x, y)(X - x, Y - y, Z)}{((X - x)^2 + (Y - y)^2 + Z^2)^{3/2}}\,dx\,dy$$

when $\sigma(x, y)$ is a constant, we leave the evaluation of the three integrals to the reader. Frequently we are interested in obtaining expressions for the electric field in the cylindrical and spherical polar coordinate system. Suppose then that the volume charge density $\sigma(\rho, \phi, z)$ is specified in the cylindrical coordinate system. The cartesian components of the electric field are obtained by replacing, x, y, z with the corresponding cylindrical expressions. These components are given by

$$\mathbf{E}(\rho_0, \phi_0, z_0) =$$

$$\frac{1}{4\pi\epsilon_0}\int \frac{\sigma(\rho, \phi, z)(\rho_0 cos\phi_0 - \rho cos\phi, \rho_0 sin\phi_0 - \rho sin\phi, z_0 - z)}{(\rho_0 cos\phi_0 - \rho cos\phi)^2 + (\rho_0 sin\phi_0 - \rho sin\phi)^2 + (z - z_0)^2)^{3/2}}\rho d\rho d\phi dz$$

The cylindrical components of the field are given by

$$E_\rho = E_x cos\phi_0 + E_y sin\phi_0 \qquad E_\phi = -E_x sin\phi_0 + E_y cos\phi_0 \qquad E_z = E_z$$

We leave the simplifications of these expressions to the reader. Suppose that we have an infinitely long line charge coincident with the z axis with σ as the linear charge density. The electric field at (X, Y, Z) can be obtained by integration along the above lines. It leads to the result

$$\mathbf{E} = E_x(X, Y, Z)\hat{x} + E_y(X, Y, Z)\hat{y} + E_z(X, Y, Z)\hat{z}$$

where

$$E_x = \frac{\sigma X}{2\pi(X^2 + Y^2)^{3/2}}, \qquad E_y = \frac{\sigma Y}{2\pi(X^2 + Y^2)^{3/2}}, \qquad E_z = 0$$

or equivalently in terms of cylindrical coordinates,

$$\mathbf{E} = \frac{\sigma\hat{\rho}}{2\pi\rho}$$

We shall deduce this result directly, i.e., without integration later on using Gauss law. Consider now a charged disc of radius R located on the XY plane with origin coinciding with the centre of the disc and with σ as the surface charge density. The electric field at a point $(0, 0, d)$ on the Z axis with d positive is given by the integral (show this)

$$\mathbf{E}(0, 0, d) = E_z(0, 0, d)\hat{z}$$

where

$$E_z(0, 0, d) = \frac{\sigma}{4\pi\epsilon_0}\int_0^R 2\pi\rho d\rho \frac{d}{(\rho^2 + d^2)^{3/2}}$$

We leave the evaluation of this integral to the reader. Now consider a solid cylinder of radius R and length L with axis coinciding with the z axis and half the cylinder below the XY plane and the half above it. Assume that the volume density of charge is ρ_0, a constant. The electric field at a point (X, Y, Z) is given by the triple integral

$$\mathbf{E} = \frac{\rho_0}{4\pi\epsilon_0}\int \frac{\rho d\rho d\phi dz(X - \rho.cos\phi, Y - \rho.sin\phi, Z - z)}{((X - \rho.cos\phi)^2 + (Y - \rho.sin\phi)^2 + Z(-z)^2)^{3/2}}$$

It is easy to see that in terms of cylindrical coordinates, $E_\phi = 0$ and that E_ρ and E_z are functions only of ρ, z, not of ϕ. We leave it as an exercise to compute these components. Finally, consider a spherical charge distribution with charge density $\rho(r)$. The electric field at the point (X, Y, Z) due to this distribution is given by the triple integral

$$\mathbf{E} = \frac{1}{4\pi\epsilon_0} \int \frac{\rho(r) r^2 sin\theta dr d\theta d\phi}{((X - r.cos\phi.sin\theta)^2 + (Y - r.sin\phi.sin\theta)^2 + (Z - r.cos\theta)^2)^{3/2}}$$

where the integral is over the range $0 < r < \infty, 0 < \theta < \pi, 0 < \phi < 2\pi$. It is once again easily seen that in spherical polar coordinates, $\mathbf{E}$ has only an $\hat{r}$ component and that this component depends only upon r, the radial distance of the point from the origin. It can also be shown (take it as an exercise) that the field at r, θ, ϕ is given by $E_r(r)\hat{r}$ where $E_r(r)$ is the field at (r, θ, ϕ) produced by a point charge $Q(r)$ located at the origin with $Q(r) = \int_0^r \rho(r)4\pi r^2 dr$, the total charge contained within a sphere of radius r with centre coinciding with the origin.

3.3 Capacitors

Consider two parallel plates of area A having identical geometry and separated by a distance d. Assume that the bottom plate is on the xy plane so that the top plate is on the plane $z = d$. The medium between the plates is assumed to be vacuum. Neglecting edge effects, the electric field between the plates has only a z component and this component is a function of z only, i.e. $\mathbf{E} = E(z)\hat{z}$. Gauss' law gives $\frac{dE(z)}{dz} = 0$ implying that $E(z) = E_0$, a constant. By Gauss law, the surface charge density on the bottom plate is $\epsilon_0 E_0$ and that on the top plate is $-epsilon_0 E_0$. The total charge on the bottom plate is $Q = \epsilon_0 E_0 A$ and that on the top plate is $-Q$. The potential difference between the two plates is $V = E_0 d$. The ratio of the charge to potential difference is $C = Q/V = \epsilon_0 A/d$ which is a constant dependent only upon the plate area and the plate separation. Thus, the charge voltage relation for a capacitor is given by $Q = CV$ and the current flowing through the capacitor is $dQ/dt = CdV/dt$.

3.4 Electrostatic potential

Suppose $\mathbf{E}$ is the electric field generated by a static charge distribution $\rho(\mathbf{r})$. We can assume that this density has both a continuous component and a discrete component. Thus,

$$\rho(\mathbf{r}) = \rho_c(\mathbf{r}) + \sum_k Q_k \delta(\mathbf{r} - \mathbf{r}_k)$$

The electric field produced by this distribution is given by

$$\mathbf{E}(\mathbf{r}) = \frac{1}{4\pi\epsilon_0} \int \frac{\rho(\mathbf{r}')(\mathbf{r} - \mathbf{r}')}{|\mathbf{r} - \mathbf{r}'|^3} d^3\mathbf{r}'$$

Now,

$$\frac{\mathbf{r} - \mathbf{r}'}{|\mathbf{r} - \mathbf{r}'|^3} = -\nabla \frac{1}{|\mathbf{r} - \mathbf{r}'|}$$

Thus,

$$\mathbf{E}(\mathbf{r}) = -\frac{1}{4\pi\epsilon_0} \nabla \int \frac{\rho(\mathbf{r}')}{|\mathbf{r} - \mathbf{r}'|} d^3\mathbf{r}'$$

which implies that $\nabla \times \mathbf{E} = 0$, i.e. the electric field produced by a static charge distribution is irrotational. Any irrotational vector field can be obtained as the gradient of a scalar field. This is a general result in field theory. In fact, the above formula directly gives us a formula for the scalar function: $\mathbf{E} = -\nabla V$, where

$$V(\mathbf{r}) = \frac{1}{4\pi\epsilon_0} \int \frac{\rho(\mathbf{r}')}{|\mathbf{r} - \mathbf{r}'|} d^3\mathbf{r}'$$

We see that if Γ is any path going from the point $\mathbf{r}_1$ to the point $\mathbf{r}_2$, then

$$V(\mathbf{r}_1) - V(\mathbf{r}_2) = \int_\Gamma \mathbf{E}.d\mathbf{r}$$

In particular our reference potential is zero at distances infinitely separated from the charge distribution. Thus,

$$V(\mathbf{r}) = \int_\mathbf{r}^\infty \mathbf{E}.d\mathbf{r}$$

the integral being along any path starting from $\mathbf{r}$ and going upto ∞.

3.5   Poisson's equation for the potential

The complete equations of electrostatics are $\nabla.\mathbf{E} = \rho/\epsilon_0, \nabla \times \mathbf{E} = 0$. The latter implies the existence of a scalar function V called the potential such that $\mathbf{E} = -\nabla V$. Substituting this into the former gives $\nabla^2 V = -\rho/\epsilon_0$. This is called the Poisson equation. It contains all of electrostatics. Suppose we are interested in the solution for the potential in all of space. We note that the function $1/|\mathbf{r} - \mathbf{r}'|$ satisfies the equation

$$\nabla^2 |\mathbf{r} - \mathbf{r}'|^{-1} = -4\pi\delta(\mathbf{r} - \mathbf{r}')$$

Multiplying by $\rho(\mathbf{r}')$ and integrating with respect to $\mathbf{r}'$ gives

$$\nabla^2 \int \frac{\rho(\mathbf{r}')}{4\pi\epsilon_0 |\mathbf{r} - \mathbf{r}'|} dV' = -\rho(\mathbf{r})/\epsilon_0$$

Moreover, if the volume V where the charge is nonzero is a bounded region of space, the integral converges to zero as $r \to \infty$ and it follows that the solution to the Poisson equation is free space is the Coulomb integral. Thus, Coulomb's law is completely equivalent to the Poisson equation and to our definition of the electric field as the negative gradient of the potential function. The idea of using a potential function is very convenient because evaluating it involves the integral of just a single component whereas if one were to evaluate the electric field directly, three integrals for the three components would require to be evaluated.

3.6 Uniqueness theorem for the solution of the Poisson equation with Dirichlet and Neumann boundary conditions

Suppose that we are given a closed surface S enclosing a volume v containing charge. If the surface is a perfect conductor, then the potential on the surface is a constant which we can take as zero. Thus, the potential $V(\mathbf{r})$ for $\mathbf{r} \in v$ satisfies Poisson's equation $\nabla^2 V(\mathbf{r}) = -\rho(\mathbf{r})/\epsilon_0$ with the boundary condition that $V(\mathbf{r}) = 0$ for $\mathbf{r} \in S$. The notation for the boundary normally used is $S = \partial V$, since differential of volume is surface. The boundary condition corresponding to zero potential on the bounding surface is called the Dirichlet boundary condition and the corresponding electrostatics problem is called the Dirichlet problem. Another boundary condition often used is the Neumann boundary condition which corresponds to specifying the normal derivative $\frac{\partial V}{\partial \mathbf{n}}$ on the bounding surface. Since the normal component of the electric field on the surface of a conductor equals the surface charge density on the conductor, this latter condition corresponds to specifying the surface charge density on the conducting surface. We can also consider mixed boundary conditions, i.e., at some points on the surface, specify V while at the other points specify $\frac{\partial V}{\partial \mathbf{n}}$. The uniqueness theorem for the Poisson equation states that if $\nabla^2 V = \rho/\epsilon$ inside the volume and at each point on the surface, we specify either V or $\frac{\partial V}{\partial \mathbf{n}}$, then the potential in the volume is uniquely defined. To see this, suppose $\nabla^2 V_1 = \nabla^2 V_2 = \rho/\epsilon$ in the volume and at each point on the bounding surface, either $V_1 = V_2$ or $\frac{\partial V_1}{\partial \mathbf{n}} = \frac{\partial V_2}{\partial \mathbf{n}}$, then define $V = V_1 - V_2$. We have $\nabla^2 V = 0$ inside the volume and $V\frac{\partial V}{\partial \mathbf{n}} = 0$ on the bounding surface. Now apply Green's theorem in the form

$$\int_v (V\nabla^2 V + |\nabla V|^2)dv = \int_s V\frac{\partial V}{\partial \mathbf{n}}dS$$

The right side is zero. Also, $\nabla^2 V = 0$. We therefore get

$$\int_v |\nabla V|^2 dv = 0$$

and it follows that $V = constt$ inside the volume v. At some point on the surface, if the potential is specified, then $V = V_1 - V_2 = 0$ at that point and we infer that the constant must be zero, i.e. $V_1 = V_2$ everywhere inside the volume v. This proves shows that the electric field is uniquely determined by two equations, Gauss' law and the irrotationality of the electric field. Gauss' law is assumed to be true universally, i.e., even in time varying scenarios while the irrotationality condition is modified to the Faraday-Lenz law according to which the circulation of the electric field around a closed contour equals the rate of change of magnetic flux through the surface bounded by the contour.

3.7 Green's functions

Suppose we solve the equation $\nabla^2 G(\mathbf{r}|\mathbf{r}') = \delta(\mathbf{r} - \mathbf{r}')$ for all $\mathbf{r}, \mathbf{r}' \in v$ with the boundary condition $G(\mathbf{r}|\mathbf{r}') = 0$ for all $\mathbf{r} \in s$. Here, s is a closed surface that bounds the volume v. Then G is called the Green's function for the electrostatics problem corresponding to a charge distribution inside v with s as a perfectly conducting surface. Using G, we can determine the solution to the boundary value problem $\nabla^2 V(\mathbf{r}) = -\rho(\mathbf{r})/\epsilon$ for $\mathbf{r} \in v$ with the boundary condition $V(\mathbf{r}) = \psi(\mathbf{r})$ for $\mathbf{r} \in s$ where ψ is any specified function on s. To see this, we start with the Green identity

$$\int_v (V(\mathbf{r})\nabla^2 G(\mathbf{r}|\mathbf{r}') - G(\mathbf{r}|\mathbf{r}')\nabla^2 V(\mathbf{r}))d^3\mathbf{r}$$

$$= \int_s (V(\mathbf{r})\frac{\partial G(\mathbf{r}|\mathbf{r}')}{\partial \mathbf{n}} - G(\mathbf{r}|\mathbf{r}')\frac{\partial V(\mathbf{r})}{\partial \mathbf{n}})dS(\mathbf{r})$$

In view of the Poisson's equations satisfied by V, G and the boundary condition on G, this identity assumes the form

$$\int_v (V(\mathbf{r})\delta(\mathbf{r} - \mathbf{r}') + G(\mathbf{r}|\mathbf{r}')\rho(\mathbf{r})/\epsilon)d^3\mathbf{r}$$

$$= \int_s V(\mathbf{r})\frac{\partial G(\mathbf{r}|\mathbf{r}')}{\partial \mathbf{n}}dS(\mathbf{r})$$

which simplifies to

$$V(\mathbf{r}') = -\frac{1}{\epsilon}\int G(\mathbf{r}|\mathbf{r}')\rho(\mathbf{r})d^3\mathbf{r} + \int_s V(\mathbf{r})\frac{\partial G(\mathbf{r}|\mathbf{r}')}{\partial \mathbf{n}}dS(\mathbf{r})$$

Since $V(\mathbf{r})$ for $\mathbf{r} \in s$ is known, this completely solves the problem of determining the potential in v in terms of the Green's function for the problem.

3.8 Dielectrics, polarization and electric displacement vector

Consider a medium that is polarizable, i.e., the atoms are modeled as a central positively charged nucleus with the negatively charged electrons orbiting around it such that the centre of the negative charge coincides with the positively charged centre. If an electric field is applied, the centers of negative and positive charge get separated thereby transforming each atom or a molecule into an infinitesimal dipole. Now imagine a closed surface S enclosing such polarizable medium, i.e., a dielectric. Let $\mathbf{P}(r)$ denote the dipole moment per unit volume at the point $\mathbf{r}$. We can imagine at a given point on the surface, a little cylinder of length d and cross sectional area A with a negative surface charge on the circular face of the cylinder within S and a positive surface charge on the circular face outside S. The length of the cylinder coincides in direction with the polarization vector. Let θ denote the angle between the normal to the surface at this point and the axis of the cylinder. Let $-\sigma$ denote the surface charge density on the inner surface and $+\sigma$ the surface charge density on the outer surface at that point. This charge is spread over an area $A/cos\theta$ and hence the total charge on the inner circular face of the cylinder is $-\sigma A/cos\theta$ and the total charge on the outer circular face of the cylinder is $+\sigma A/cos\theta$. The dipole moment of this little cylinder is thus $p = \sigma Ad/cos\theta$ and the dipole moment per unit volume at this location is $P = p/Ad = \sigma/cos\theta$. Thus, $Pcos\theta = \sigma$ which in vector notation translates to $\mathbf{P.n} = \sigma$. This is the formula for the surface polarization charge density just outside the surface. The total polarization charge enclosed by the surface is

$$Q_{pol} = -\int_S \sigma_{pol} dS = -\int_S \mathbf{P.n} dS = -\int_V div\mathbf{P} dV$$

This must equal the volume integral $\int_V \rho_{pol} dV$ of the polarization charge density over the volume. Thus

$$div\mathbf{P} = -\rho_{pol}$$

and Gauss' law acquires the form

$$div\mathbf{E} = (\rho_{free} + \rho_{pol})/\epsilon_0 = (\rho_{free} - div\mathbf{P})/\epsilon_0$$

where ρ_{free} is the volume density of free charge, i.e., external charge. Defining $\mathbf{D} = \epsilon_0\mathbf{E} + \mathbf{P}$, we get

$$div\mathbf{D} = \rho_{free}$$

which is what Gauss' law looks like for a dielectric. If the medium is linear, the polarization vector is proportional to the electric field, i.e. $\mathbf{P} = \epsilon_0\chi\mathbf{E}$ and we get

$$\mathbf{D} = \epsilon_0(1 + \chi)\mathbf{E} = \epsilon\mathbf{E}$$

where $\epsilon = \epsilon_0(1 + \chi)$ is called the permitivity of the medium.

3.9   Ampere's law and magnetostatics

Ampere's law states that if $\mathbf{H}$ is the magnetic field in space and Γ any closed contour, then $\int_\Gamma \mathbf{H.dr} = I$ where the left side equals the line integral of the magnetic field around the closed loop Γ and I equals the net current flowing through the loop. With $\mathbf{J}$ denoting the current density in space, i.e., the current flowing per unit area, $I = \int_S \mathbf{J}.\mathbf{n}dS$ where the surface S is any surface whose boundary is the closed loop Γ. Using Stokes' theorem, we can write Ampere's law as

$$\int_S \nabla \times \mathbf{H}.\mathbf{n}dS = \int_S \mathbf{J}.\mathbf{n}dS$$

and hence we deduce the differential form of Ampere's law:

$$\nabla \times \mathbf{H} = \mathbf{J}$$

Along with Ampere's law comes the no monopole condition. If μ_0 denotes the magnetic permeability of free space, we define another field quantity $\mathbf{B} = \mu_0\mathbf{H}$. $\mathbf{B}$ is called the magnetic flux density and the no monopole condition states that $\int_S \mathbf{B}.\mathbf{n}dS = 0$ where S is any closed surface. Using Gauss' law, this acquires the form $\int_V div\mathbf{B}dV = 0$ and since this holds for all volumes V, we must have $div\mathbf{B} = 0$. This equation in conjunction with Ampere's law determines the solution to all problems of magnetostatics, i.e., problems involving steady current.

3.10   Magnetic vector potential

The equation $div\mathbf{B} = 0$ implies that $\mathbf{B} = \nabla \times \mathbf{A}$ for some vector field $\mathbf{A}$. This can be proved easily by integration. $\mathbf{A}$ is called the magnetic vector potential. Now, substitute this into Ampere's law

$$\nabla \times \nabla \times \mathbf{A} = \mu_0\mathbf{J}$$

or

$$\nabla^2\mathbf{A} - \nabla(\nabla.\mathbf{A}) = \mu_0\mathbf{J}$$

Suppose $\mathbf{A}$ is altered to $\mathbf{A}' = \mathbf{A} + \nabla f$, where f is a scalar field. Then, the magnetic field corresponding to the potential $\mathbf{A}$ is the same:

$$\nabla \times \mathbf{A}' = \nabla \times (\mathbf{A} + \nabla f) = \nabla \times \mathbf{A}$$

We can select the function f so that $\nabla.\mathbf{A}' = \nabla.\mathbf{A} + \nabla^2 f = 0$. This means that

$$f(\mathbf{x}) = \frac{1}{4\pi} \int \frac{\nabla.\mathbf{A}(\mathbf{x}')}{|\mathbf{x} - \mathbf{x}'|} d^3x'$$

Thus, without loss of generality, we may assume that $\nabla.\mathbf{A} = 0$ and then $\mathbf{A}$ will satisfy the equation

$$\nabla^2\mathbf{A} = -\mu_0\mathbf{J}$$

which is amounts to three Poisson equations:

$$\nabla^2 A_x = -\mu_0 J_x, \nabla^2 A_y = -\mu_0 J_y, \nabla^2 A_z = -\mu_0 J_z$$

3.11 Derivation of the Biot-Savart law

The solution to Poisson's differential equation for the vector potential is based on the identity

$$\nabla^2 |\mathbf{x} - \mathbf{x}'|^{-1} = -4\pi\delta(\mathbf{x} - \mathbf{x}')$$

Multiplying both sides of this equation by $\mu_0 \mathbf{J}(\mathbf{x}')/4\pi$ and integrating over all $\mathbf{x}'$ results in the identity

$$\nabla^2 \frac{\mu_0}{4\pi} \int \frac{\mathbf{J}(\mathbf{x}')}{|\mathbf{x} - \mathbf{x}'|} d^3x' = -\mu_0 \mathbf{J}(\mathbf{x})$$

and hence we conclude that

$$\mathbf{A}(\mathbf{x}) = \frac{\mu_0}{4\pi} \int \frac{\mathbf{J}(\mathbf{x}')}{|\mathbf{x} - \mathbf{x}'|} d^3x'$$

For a thin wire carrying a current I, we make the replacement $\mathbf{J}d^3x' = I d\mathbf{x}'$ where $d\mathbf{x}'$ is the same as $d\mathbf{l}$, the length element along the wire. This follows from the fact that the volume element d^3x' equals the area element dS' times the length element $|d\mathbf{x}'|$ and the direction of $\mathbf{J}(\mathbf{x}')$ coincides with the direction of the length element $d\mathbf{x}'$. Thus, we end up with the following expression for the vector potential:

$$\mathbf{A}(\mathbf{x}) = \frac{\mu_0}{4\pi} \int_\Gamma \frac{I d\mathbf{x}'}{|\mathbf{x} - \mathbf{x}'|}$$

where the line integral is along the contour Γ of the wire. Taking the curl of the integral expression for the vector potential gives us the magnetic field:

$$\mathbf{B}(\mathbf{x}) = \nabla \times \mathbf{A}(\mathbf{x}) = \frac{\mu_0}{4\pi} \nabla \times \int \frac{\mathbf{J}(\mathbf{x}')}{|\mathbf{x} - \mathbf{x}'|} d^3x'$$

$$= \frac{\mu_0}{4\pi} \int \nabla(|\mathbf{x} - \mathbf{x}'|^{-1}) \times \mathbf{J}(\mathbf{x}') d^3x'$$

$$= \frac{\mu_0}{4\pi} \int \frac{\mathbf{J}(\mathbf{x}') \times (\mathbf{x} - \mathbf{x}')}{|\mathbf{x} - \mathbf{x}'|^3} d^3x'$$

The abbreviated notation for this formula is

$$\mathbf{B}_2 = \frac{\mu_0}{4\pi} \int \mathbf{J}_1 \times \mathbf{R}_{12} dV_1 / R_{12}^3$$

For the case of a thin wire, this reduces to

$$\mathbf{B}(\mathbf{x}) = \frac{\mu_0}{4\pi} \int_\Gamma I \frac{d\mathbf{x}' \times (\mathbf{x} - \mathbf{x}')}{|\mathbf{x} - \mathbf{x}'|^3}$$

or in terms of our abbreviated notation,

$$\mathbf{B}_2 = \frac{\mu_0}{4\pi} \int_\Gamma I \frac{d\mathbf{r}_1 \times \mathbf{r}_{12}}{r_{12}^3}$$

This equation is called the Biot-Savart law. It tells us how from the given current distribution, we can compute the magnetic field. For surface current densities, the Biot-Savart law acquires the following form. Let S be a surface on which a surface current $\mathbf{K}$ flows per unit length. The magnetic field produced by this current sheet can be expressed as

$$\mathbf{B}(\mathbf{x}) = \frac{\mu_0}{4\pi} \int_S \frac{\mathbf{K}(\mathbf{x}') \times (\mathbf{x} - \mathbf{x}')}{|\mathbf{x} - \mathbf{x}'|^3} dS(\mathbf{x}')$$

The corresponding expression for the magnetic vector potential is

$$\mathbf{A}(\mathbf{x}) = \frac{\mu_0}{4\pi} \int_S \frac{\mathbf{K}(\mathbf{x}')}{|\mathbf{x} - \mathbf{x}'|} dS(\mathbf{x}')$$

3.12   Examples of the application of the Biot-Savart law to the computation of the magnetic field due to static current distributions

Example 1: A thin straight wire of length L carrying a current I along its axis. Assume that the wire coincides with the z axis and stretches from $z = -L/2$ to $z = L/2$. The Biot-Savart law for the magnetic field reads

$$\mathbf{B}(x, y, z) = \frac{\mu_0 I}{4\pi} \int_{-L/2}^{L/2} dz' \hat{z} \times (x\hat{x} + y\hat{y} + (z - z')\hat{z})/(x^2 + y^2 + (z - z')^2)^{3/2}$$

$$= \frac{\mu_0 I}{4\pi} \int_{-L/2}^{L/2} dz'(-y\hat{x} + x\hat{y})/(x^2 + y^2 + (z - z')^2)^{3/2}$$

From the expression, it is clear that in terms of cylindrical components, the magnetic field has only a ϕ component which is given by the formula

$$B_\phi(\rho, \phi, z) = \frac{\mu_0 I \rho}{4\pi} \int_{-L/2}^{L/2} \frac{dz'}{(\rho^2 + (z - z')^2)^{3/2}}$$

We leave the job of evaluating this integral to the student. The special case when the wire length becomes infinite, i.e., $L \to \infty$. We then get

$$B_\phi(\rho, \phi, z) = \frac{\mu_0 I \rho}{4\pi} \int_{-\infty}^{\infty} \frac{dz'}{(\rho^2 + (z - z')^2)^{3/2}} = \frac{\mu_0 I}{2\pi\rho}$$

Example 2: A circular loop of wire of radius R carrying a current I. Assume that the loop lies on the xy plane with its centre coinciding with the origin. Application of the Biot-Savart law gives

$$\mathbf{B}(x,y,z) = \frac{\mu_0 I}{4\pi} \int_0^{2\pi} \frac{Rd\phi(-\hat{x}.sin\phi + \hat{y}.cos\phi) \times ((x - R.cos\phi)\hat{x} + (y - R.sin\phi)\hat{y} + z\hat{z})}{((x - Rcos\phi)^2 + (y - Rsin\phi)^2 + z^2)^{3/2}}$$

$$= \frac{\mu_0 IR}{4\pi} \int_0^{2\pi} (zcos\phi\hat{x} + zsin\phi\hat{y}$$

$$+\hat{z}(-(y - Rsin\phi)sin\phi - (x - Rcos\phi)cos\phi))d\phi/((x - Rcos\phi)^2 + (y - Rsin\phi)^2 + z^2)^{3/2}$$

In terms of cylindrical coordinates, it is clear that the magnetic field has only a $\hat{\rho}$ component and a $\hat{z}$ component. It has zero $\hat{\phi}$ component. From the above expression, we get

$$B_\rho = \frac{\mu_0 IRz}{4\pi} \int_0^{2\pi} \frac{d\phi}{((x - Rcos\phi)^2 + (y - Rsin\phi)^2 + z^2)^{3/2}}$$

$$= \frac{\mu_0 IRz}{4\pi} \int_0^{2\pi} \frac{d\phi}{(\rho^2 + R^2 + z^2 - 2\rho Rcos\phi)^{3/2}}$$

$$B_z = \frac{\mu_0 IR}{4\pi} \int_0^{2\pi} (-ysin\phi - xcos\phi + R)d\phi/((x - Rcos\phi)^2 + (y - Rsin\phi)^2 + z^2)^{3/2}$$

$$= \frac{\mu_0 IR}{4\pi} \int_0^{2\pi} \frac{R - \rho cos\phi}{(\rho^2 + R^2 + z^2 - 2\rho Rcos\phi)^{3/2}} d\phi$$

Consider the special case when the field is to be evaluated on the z axis, i.e., set $\rho = 0$ in the above expressions. On the z axis, it is meaningless to talk of the $\hat{\rho}$ component. The only nonvanishing component is the $\hat{z}$ component. It is given by

$$B_z(0,0,z) = \frac{\mu_0 IR}{4\pi} \int_0^{2\pi} \frac{Rd\phi}{(R^2 + z^2)^{3/2}} = \frac{\mu_0 IR^2}{2(R^2 + z^2)^{3/2}}$$

The field at the origin is obtained by setting $z = 0$. This gives

$$B_z(0,0,0) = \frac{\mu_0 I}{2R}$$

Example 3: A planar current sheet carrying current of uniform surface density. Assume that the sheet is laid along the xz plane with x varying from 0 to a. The sheet extends infinitely on both sides of z, i.e., the sheet is defined by the conditions $0 < x < a, y = 0, -\infty < z < \infty$. Assume that the surface current density is $K\hat{z}$. The current flowing in the infinitesimal strip between x' to $x' + dx'$ equals $K\,dx'$. The Biot-Savart law reads

$$\mathbf{B}(x,y,z) = \frac{\mu_0 K}{4\pi} \int_0^a \int_{-\infty}^{\infty} dx'dz'\hat{z} \times ((x-x')\hat{x} + y\hat{y} + (z-z')\hat{z})/((x-x')^2 + (y^2 + (z-z')^2)^{3/2}$$

$$= \frac{\mu_0 K}{4\pi} \int_0^a \int_{-\infty}^{\infty} (-y\hat{x} + (x-x')\hat{y})dx'dz'/((x-x')^2 + y^2 + (z-z')^2)^{3/2}$$

The only nonvanishing component is the $\hat{x}$ component which is given by

$$B_x(x,y,z) = -\frac{\mu_0 K y}{4\pi} \int_0^a \int_{-\infty}^{\infty} \frac{dx'dz'}{((x-x')^2 + y^2 + (z-z')^2)^{3/2}}$$

We leave it as an exercise the student to evaluate this double integral.

Example 4: Consider a block of current defined by the region $0 < x < a, 0 < y < b, 0 < z < c$. This block carries a current uniform density J_z along the z axis. Application of the Biot-Savart law gives

$$\mathbf{B}(x,y,z) = \frac{\mu_0 J_z}{4\pi} \int_0^a \int_0^b \int_0^c dx'dy'dz'\hat{z} \times ((x-x')\hat{x} + (y-y')\hat{y} + (z-z')\hat{z})$$

$$/((x-x')^2 + (y-y')^2 + (z-z')^2)^{3/2}$$

$$= \frac{\mu_0 J_z}{4\pi} \int_0^a \int_0^b \int_0^c dx'dy'dz'(-(y-y')\hat{x} + (x-x')\hat{y})/((x-x')^2 + (y-y')^2 + (z-z')^2)^{3/2}$$

or explicitly in terms of components,

$$B_x(x,y,z) = -\frac{\mu_0 J_z}{4\pi} \int_0^a \int_0^b \int_0^c \frac{y-y'}{((x-x')^2 + (y-y')^2 + (z-z')^2)^{3/2}}dx'dy'dz'$$

$$B_y(x,y,z) = \frac{\mu_0 J_z}{4\pi} \int_0^a \int_0^b \int_0^c \frac{x-x'}{((x-x')^2 + (y-y')^2 + (z-z')^2)^{3/2}}dx'dy'dz'$$

$B_z = 0.$

3.13   Field of a Magnetic dipole

A magnetic dipole by definition is simply an infinitesimally small current loop. Let the loop be specified in parametric form as $s \to \mathbf{r}(s)$ with $0 \le s \le 1$. Obviously the condition $\mathbf{r}(0) = \mathbf{r}(1)$ is satisfied for the loop to be closed. Let the loop carry a current I. We shall determine the field at a point $\mathbf{R}$ when $|\mathbf{R}|$ is very large compared to the dimensions of the loop. The vector potential is given by

$$\mathbf{A}(\mathbf{R}) = \frac{\mu_0 I}{4\pi} \int_\Gamma \frac{d\mathbf{r}}{|\mathbf{R} - \mathbf{r}|}$$

Since $R >> r$, we can make the approximation

$$|\mathbf{R} - \mathbf{r}|^{-1} = R^{-1}(1 + \mathbf{r}.\mathbf{R}/R^2) = R^{-1} + \mathbf{r}.\mathbf{R}/R^3$$

Then noting that since the loop is closed, $\int_\Gamma d\mathbf{r} = \mathbf{0}$, it follows that to a first order approximation,

$$\mathbf{A}(\mathbf{R}) = \frac{\mu_0 I}{4\pi R^3} \int_\Gamma d\mathbf{r}(\mathbf{r}.\mathbf{R})$$

Writing $\mathbf{r} = (x_1, x_2, x_3)$ and $\mathbf{R} = (X_1, X_2, X_3)$ in terms of cartesian coordinates, we have

$$\int d\mathbf{r}(\mathbf{r}.\mathbf{R}) = \sum_{i,j=1}^{3} e_i \int x_j X_j dx_i = \sum e_i X_j \int x_j dx_i$$

where e_1, e_2, e_3 are the unit vectors along the three cartesian axes. Also

$$\int x_j dx_i = -\int x_i dx_j = \frac{1}{2} \int (x_j dx_i - x_i dx_j) = aji$$

where aji is the area tensor of the loop. aji is an antisymmetric tensor and hence has only three independent components, $a23, a31, a12$. In terms of this tensor, we can write

$$\mathbf{A} = \frac{\mu_0 I}{4\pi R^3}(e_1(X_2 a_{21} + X_3 a_{31}) + e_2(X_1 a_{12} + X_3 a_{32}) + e_3(X_1 a_{13} + X_2 a_{23}))$$

Introduce the notation $a_1 = a_{23}, a_2 = a_{31}, a_3 = a_{12}$. Then,

$$\mathbf{A} = \frac{\mu_0 I}{4\pi R^3}(e_1(a_2 X_3 - a_3 X_2) + e_2(a_3 X_1 - a_1 X_3) + e_3(a_1 X_2 - a_2 X_1)) = \frac{\mu_0 I \mathbf{a} \times \mathbf{R}}{4\pi R^3}$$

Where $\mathbf{a} = a_1 e_1 + a_2 e_2 + a_3 e_3$. We leave it as an exercise to compute the magnetic field corresponding to this vector potential by forming the curl. It is a usual convention to set $\mathbf{m} = I\mathbf{a}$ and call $\mathbf{m}$ the magnetic dipole moment of the loop. In terms of this,

$$\mathbf{A} = \frac{\mu_0 \mathbf{m} \times \mathbf{R}}{4\pi R^3}$$

If the current loop is planar and falls on the xy plane, then $\mathbf{m} = m_3 e_3$. A permanent magnet can be viewed as a collection of such dipoles oriented along the same direction. The above formula thus yields an expression for the magnetic field produced by a permanent magnet.

3.14   Magnetization vector

Consider a medium consisting of magnetic dipoles. Let $\mathbf{M}$ be the number of magnetic dipoles per unit volume. This is in general a function of the position $\mathbf{r}$, thus we write $\mathbf{M}(\mathbf{r})$. According to the discussion of the previous section, we can express the magnetic vector potential produced by this dipole density as

$$\mathbf{A}(\mathbf{r}) = \frac{\mu_0}{4\pi} \int \frac{\mathbf{M}(\mathbf{R}) \times (\mathbf{r} - \mathbf{R})}{|\mathbf{r} - \mathbf{R}|^3} dV(\mathbf{R})$$

$$= \frac{\mu_0}{4\pi} \int \mathbf{M}(\mathbf{R}) \times \nabla_R(|\mathbf{r} - \mathbf{R}|^{-1}) dV(\mathbf{R})$$

$$= \frac{\mu_0}{4\pi} \int -\nabla_R \times \frac{\mathbf{M}(\mathbf{R})}{|\mathbf{r} - \mathbf{R}|} dV(\mathbf{R}) + \frac{\mu_0}{4\pi} \int \frac{\nabla_R \times \mathbf{M}(\mathbf{R})}{|\mathbf{r} - \mathbf{R}|} dV(\mathbf{R})$$

If the integration is over infinite space, then the first integral can be transformed into a surface integral which vanishes provided the dipole density does not increase in magnitude unboundedly. From the second integral, we deduce that the current density produced by this magnetic dipole distribution is given by $\mathbf{J}_m = \nabla \times \mathbf{M}$. If $\mathbf{J}_e$ denotes the current due to external sources, then the total current is given by the sum $\mathbf{J}_e + \nabla \times \mathbf{M}$. Hence, Ampere's law assumes the form

$$\nabla \times \mathbf{B}/\mu_0 = \mathbf{J}_e + \nabla \times \mathbf{M}$$

or

$$\nabla \times (\mathbf{B}/\mu_0 - \mathbf{M}) = \mathbf{J}_e$$

This equation can be expressed as

$$\nabla \times \mathbf{H} = \mathbf{J}_e$$

where

$$\mathbf{B} = \mu_0(\mathbf{H} + \mathbf{M})$$

This is the fundamental equation connecting the two vectors $\mathbf{B}$ and $\mathbf{H}$. $\mathbf{B}$ is a fundamental field quantity and $\mathbf{H}$ is a quantity depending upon the magnetic properties of the material. For a linear material, $\mathbf{M}$ is proportional to $\mathbf{H}$, i.e., $\mathbf{M} = \mu_0\chi_m\mathbf{H}$ and then $\mathbf{B} = \mu_0(1 + \chi_m)\mathbf{H} = \mu\mathbf{H}$ where $\mu = \mu_0(1 + \chi_m)$ is the magnetic permeability of the medium. χ_m is called the magnetic susceptibility of the medium.

3.15   Faraday's law of induction

This law states that the rate of change of magnetic flux through a closed loop equals the emf or voltage drop around the loop. If $\mathbf{E}$ is the electric field in space and Γ a closed loop, then the line integral $V = \int_\Gamma \mathbf{E}.d\mathbf{r}$ of the electric field around the loop is the emf around the loop. If $\mathbf{B}$ is the magnetic field in space and S any surface having Γ as its boundary, then $\Phi_B = \int_S \mathbf{B}.\mathbf{n}dS$ is the magnetic flux through S. Faraday's law can then be stated as

$$\int_\Gamma \mathbf{E}.\mathbf{r} = -\frac{d\Phi_B}{dt}$$

Transforming the line integral on left using Stokes theorem gives

$$\int_S \nabla \times \mathbf{E}.\mathbf{n}dS = -\int_S \frac{\partial \mathbf{B}}{\partial t}.\mathbf{n}dS$$

Since S can be chose arbitrarily, we deduce the differential form of Faraday's law

$$\nabla \times \mathbf{E} = -\frac{\partial \mathbf{B}}{\partial t}$$

The negative sign appearing in Faraday's law means that the voltage drop around the loop follows the left hand rule, i.e. if the flux increases in one direction, then the voltage drop is counterclockwise with respect to the direction of the increase of the magnetic field.

3.16 Examples of application of Faraday's law

1. Consider a spatially uniform magnetic field $B_0(t)\mathbf{n}$ where $\mathbf{n}$ is a fixed unit vector in space. Let S denote a planar surface having unit normal $\mathbf{m}$. Let Γ be the closed loop that bounds the surface Γ. Then, the magnitude of the voltage induced around the loop is given by $B_0'(t)\mathbf{n}.\mathbf{m}$. if α is the angle between the magnetic field and the surface normal, then $\mathbf{n}.\mathbf{m} = cos\alpha$ and the voltage drop is $B_0'(t)cos\alpha$.

2. Suppose a closed loop Γ moves in space such that a point $\mathbf{r}$ on this loop has at a certain instant of time a velocity $\mathbf{v}$. The magnetic flux through this loop is given by $\int_S \mathbf{B}.d\mathbf{S}$. A small line element $d\mathbf{r}$ on the loop sweeps an area $\mathbf{v}dt \times d\mathbf{r}$ in the time interval dt. Thus, the rate of change of magnetic flux thorough the loop is given by

$$\frac{d\Phi_B}{dt} = \int_S \frac{\partial \mathbf{B}}{\partial t}.d\mathbf{S} + \int_\Gamma \mathbf{B}.\mathbf{v} \times d\mathbf{r}$$

If the magnetic field does not change, then only the second term contributes to the emf around Γ which according to Faraday's law is the negative rate of change of the flux. Thus,

$$\int_\Gamma \mathbf{E}.d\mathbf{r} = \int_\Gamma \mathbf{v} \times \mathbf{B}.d\mathbf{r}$$

This means that relative to the loop, the electric field is given by $\mathbf{E} = \mathbf{v} \times \mathbf{B}$. This phenomenon involving emf produced due to motion is called motional emf.

3. Assume that we have a rectangular loop with side lengths a, b. One side of the rectangle having length b is slideable. This means that a varies with time. If the side moves with an instantaneous velocity $v(t)$, then a as a function of time is given by $a(t) = a(0) + \int_0^t v(t)dt$. The instantaneous area of the loop equals $a(t)b$. Let B_0 be a constant magnetic field perpendicular to the rectangular plane. Then, the instantaneous flux through the loop is given by $a(t)bB_0$ and its rate of change gives the emf around the loop as $V = bB_0a'(t) = bv(t)B_0$.

4. Consider a rectangular loop with side-lengths a, b. One side of length a is coincident with the x axis and stretches from the origin to $(a, 0, 0)$. There is hinge on this line that allows the loop to oscillate. Let $\theta(t)$ denote the instantaneous angle made by the plane of the loop with the xz plane and let the magnetic field be $B_0\hat{y}$. When the loop makes an angle $\theta(t)$ with the xz plane, the angle between the magnetic field and the normal to the plane of the loop is $\theta(t)$. At this instant the flux of the magnetic field through the loop is given by $B_0 ab\cos\theta(t)$ and its rate of change $-B_0 ab\frac{d\theta}{dt}\sin(\theta)$ gives the emf induced around the loop.

5. Consider a loop in the xy plane which is described by the equation $r(\phi), 0 \leq \phi < 2\pi$. Assume that a conducting rod with one end pivoted at the origin sweeps along the loop such that after time t, the rod makes an angle $\phi(t)$ with the x axis. Let $B_0\hat{z}$ denote the magnetic field assumed to be a constant. We want to determine the emf induced in the loop formed by the section of the loop and the conducting rod. After time t, the sectorial area subtended by the rod is given by the integral $\frac{1}{2}\int_0^\phi r^2 d\phi$ and the flux of the magnetic field through this sector is obtained by multiplying B_0 with this area. The rate of change of this flux is given by $\frac{B_0}{2}r^2(\phi(t))\frac{d\phi(t)}{dt}$ and this is the emf induced around the loop.

6. Consider again a rectangular loop whose plane coincides with the xy plane and which falls vertically downwards. The side of the loop having length a is parallel to the x axis while the side of the loop having length b is parallel to the z axis. After time t assume that the loop has fallen a distance $\xi(t)$ downwards so that the coordinates of the four points of the loop are given by $(0, 0, -\xi(t)), (0, 0, -\xi(t) - b), (a, 0, -\xi(t))$ and $(a, 0, -\xi(t) - b)$. Let $B_0(x, z)\hat{y}$ be the magnetic field in space. The emf induced in the loop is motional. To calculate it, we need to evaluate the integral $\int_\Gamma \mathbf{v} \times \mathbf{B}.d\mathbf{r}$ around the loop. The contribution to the integral comes from the top and bottom sides only. The integral is seen to evaluate to

$$\int_0^a \frac{d\xi(t)}{dt} B_0(x, -\xi(t))dx - \int_0^a \frac{d\xi(t)}{dt} B_0(x, -\xi(t) - b)dx$$

and this is precisely the induced emf around the loop.

7. Lorentz force: This states that if a charge q is located at a point $\mathbf{r}$ and is moving with velocity $\mathbf{v}$ in an electromagnetic field $\mathbf{E}(\mathbf{r}, t), \mathbf{B}(\mathbf{r}, t)$, then the force on the charge is given by

$$\mathbf{F} = q(\mathbf{E}(\mathbf{r}, t) + \mathbf{v} \times \mathbf{B}(\mathbf{r}, t))$$

Thus, the equation of motion of the charge is given according to Newton's law of motion by

$$m\frac{d^2\mathbf{r}(t)}{dt^2} = q(\mathbf{E}(\mathbf{r}(t),t) + \frac{d\mathbf{r}(t)}{dt} \times \mathbf{B}(\mathbf{r}(t),t))$$

When the electric and magnetic fields are constant in space and time, it is easy to obtain the solution to this system. Formally, this is achieved by arranging the equations of motion in matrix form:

$$\frac{dx}{dt} = v_x, \frac{dy}{dt} = v_y, \frac{dz}{dt} = v_z,$$

$$\frac{dv_x}{dt} = \frac{q}{m}E_x + \frac{q}{m}(v_y B_z - v_z B_y)$$

$$\frac{dv_y}{dt} = \frac{q}{m}E_y + \frac{q}{m}(v_z B_x - v_x B_z)$$

$$\frac{dv_z}{dt} = \frac{q}{m}E_z + \frac{q}{m}(v_x B_y - v_y B_x)$$

which arranges as

$$\frac{d}{dt}\begin{pmatrix} x \\ y \\ z \\ v_x \\ v_y \\ v_z \end{pmatrix}$$

$$= \frac{q}{m}\begin{pmatrix} 0 \\ 0 \\ 0 \\ E_x \\ E_y \\ E_z \end{pmatrix}$$

$$+ \begin{pmatrix} 0 & 0 & 0 & 1 & 0 & 0 \\ 0 & 0 & 0 & 0 & 1 & 0 \\ 0 & 0 & 0 & 0 & 0 & 1 \\ 0 & 0 & 0 & 0 & \frac{qB_z}{m} & -\frac{qB_y}{m} \\ 0 & 0 & 0 & -\frac{qB_z}{m} & 0 & \frac{qB_x}{m} \\ 0 & 0 & 0 & \frac{qB_y}{m} & -\frac{qB_x}{m} & 0 \end{pmatrix}\begin{pmatrix} x \\ y \\ z \\ v_x \\ v_y \\ v_z \end{pmatrix}$$

This is of the form

$$\frac{d\mathbf{z}(t)}{dt} = \mathbf{a} + \mathbf{C}\mathbf{z}$$

Where $\mathbf{a}$ is a constant vector and $\mathbf{C}$ is a constant matrix. When the electromagnetic field depends upon time but not on position, then $\mathbf{a}$ and $\mathbf{C}$ become functions of time. The general solution in the time independent case is given by

$$\mathbf{z}(t) = exp(t\mathbf{C})\mathbf{z}(0) + \int_0^t exp((t-s)\mathbf{C})\mathbf{a}(s)ds$$

In the time dependent case, we first determine the solution to the equation

$$\frac{\partial \Phi(t,s)}{\partial t} = \mathbf{C}(t)\Phi(t,s), t \geq s$$

with the initial condition $\Phi(s,s) = \mathbf{I}$. $\Phi(t,s)$ is called the state transition matrix. Then, the general solution to the state equations is given by

$$\mathbf{z}(t) = \Phi(t,0)\mathbf{z}(0) + \int_0^t \Phi(t,s)\mathbf{a}(s)ds$$

For constant electromagnetic fields, it can be shown via elementary integration that the trajectories are spirals.

3.17   Displacement current, Maxwell's equations and electromagnetic waves

Start with Ampere's law in the form $\nabla \times \mathbf{B} = \mu_0\mathbf{J}$. Taking the divergence of this equation gives $\nabla.\mathbf{J} = 0$. However this equation violates the equation of charge conservation in the case of non-steady currents, for the general equation of charge conservation states that

$$\nabla.\mathbf{J} + \frac{\partial \rho}{\partial t} = 0$$

Using Gauss' law in the form $\nabla.(\epsilon_0\mathbf{E}) = \rho$, the equation of charge conservation can be expressed as

$$\nabla.(\mathbf{J} + \epsilon\frac{\partial \mathbf{E}}{\partial t}) = 0$$

This suggests that Ampere's law for general non-steady field problems should be modified to

$$\nabla \times \mathbf{B} = \mu_0\mathbf{J} + \mu_0\epsilon_0\frac{\partial \mathbf{E}}{\partial t}$$

Taking the divergence of this equation and using Gauss' law then yields the correct equation of charge conservation. $\epsilon_0\frac{\partial \mathbf{E}}{\partial t}$ is called the displacement current or equivalently, Maxwell's displacement current correction term. It has to be added to the ordinary current density on the right side of Ampere's law to get agreement with the charge conservation law. The complete set of Maxwell's equations can now be written down:

$$\nabla.\mathbf{E} = \rho/\epsilon_0, \quad \nabla \times \mathbf{E} = -\frac{\partial \mathbf{B}}{\partial t}$$

$$\nabla \times \mathbf{B} = \mu_0 \mathbf{J} + \mu_0 \epsilon_0 \frac{\partial \mathbf{E}}{\partial t}, \quad \nabla.\mathbf{B} = 0$$

Consider now a medium that is polarizable as well as magnetizable. Let ρ_f denote the free or external charge density and $\mathbf{J}_c$ the external or conduction current. If there are N electrons per unit volume and each electron at a given point suffers a displacement $\mathbf{x}$ relative to the positive nucleus, then the dipole moment per unit volume at that point is given by $\mathbf{P} = -Ne\mathbf{x}$. The rate of change of this polarization vector is given by $\frac{\partial \mathbf{P}}{\partial t} = -Ne\frac{d\mathbf{x}}{dt} = -Ne\mathbf{v}$. Now, $-Ne$ is the polarization charge density and hence $-Ne\mathbf{v} = \frac{\partial \mathbf{P}}{\partial t}$ is the polarization current density. The magnetization current density is given by $\mathbf{J}_m = \nabla \times \mathbf{M}$. Thus, Maxwell's equations for such a medium can be expressed as

$$\epsilon_0 \nabla.\mathbf{E} = \rho_f - \nabla.\mathbf{P}$$

$$\nabla \times \mathbf{E} = -\frac{\partial \mathbf{B}}{\partial t}$$

$$\nabla \times \mathbf{B} = \mu_0 \left(\mathbf{J}_c + \nabla \times \mathbf{M} + \frac{\partial \mathbf{P}}{\partial t}\right) + \mu_0 \epsilon_0 \frac{\partial \mathbf{E}}{\partial t}$$

$$\nabla.\mathbf{B} = 0$$

The first equation is rearranged as

$$\nabla.(\epsilon_0 \mathbf{E} + \mathbf{P}) = \rho_f$$

or

$$\nabla.\mathbf{D} = \rho_f$$

The third equation is rearranged as

$$\nabla \times (\mathbf{B}/\mu_0 - \mathbf{M}) = \mathbf{J}_c + \frac{\partial}{\partial t}(\epsilon_0 \mathbf{E} + \mathbf{P})$$

or equivalently,

$$\nabla \times \mathbf{H} = \mathbf{J}_c + \frac{\partial \mathbf{D}}{\partial t}$$

The striking experimental confirmation of the Maxwell equations is that the Maxwell equations predict the existence of electromagnetic waves propagating in vacuum at the speed of light and this has been verified experimentally and in fact forms the basis of radio and television. Taking the curl of Faraday's law gives

$$\nabla \times (\nabla \times \mathbf{E}) = -\frac{\partial}{\partial t}\nabla \times \mathbf{B}$$

which can be rearranged as

$$\nabla(\nabla.\mathbf{E}) - \nabla^2\mathbf{E} = -\frac{\partial}{\partial t}\nabla \times \mathbf{B}$$

In free space, $\rho = 0, \mathbf{J} = \mathbf{0}$, so Gauss' law simplifies to $\nabla.\mathbf{E} = 0$ and Ampere's law simplifies to $\nabla \times \mathbf{B} = \mu_0\epsilon_0\frac{\partial \mathbf{E}}{\partial t}$. Thus, in free space, we get

$$\nabla^2\mathbf{E} - \mu_0\epsilon_0\frac{\partial^2\mathbf{E}}{\partial t^2} = \mathbf{0}$$

The value of $\mu_0\epsilon_0$ turns out to be $\frac{1}{c^2}$ where $c = 3 \times 10^8 m/s$ is the speed of light in vacuum. Thus, the electric field in vacuum propagates as a wave traveling at the speed of light. Similarly, we can show that the magnetic field also propagates as a wave in vacuum traveling at the speed of light. To see this start with Ampere's law in the absence of currents:

$$\nabla \times \mathbf{B} = \mu_0\epsilon_0\frac{\partial \mathbf{E}}{\partial t}$$

Take the curl of this equation to get

$$\nabla \times (\nabla \times \mathbf{B})) = \mu_0\epsilon_0\frac{\partial}{\partial t}\nabla \times \mathbf{E}$$

Using the standard vector identity and Faraday's law this reads

$$\nabla(\nabla.\mathbf{B}) - \nabla^2\mathbf{B} = -\mu_0\epsilon_0\frac{\partial^2\mathbf{B}}{\partial t^2}$$

Since $\nabla.\mathbf{B} = 0$, we obtain the wave equation for the magnetic field in free space:

$$\nabla^2\mathbf{B} - \mu_0\epsilon_0\frac{\partial^2\mathbf{B}}{\partial t^2} = \mathbf{0}$$

Consider a plane wave solution to these equations, i.e.,

$$\mathbf{E}(\mathbf{r}, t) = Re(\mathbf{E}_0 exp(i(\omega t - \mathbf{k}.\mathbf{r})))$$

ω is the wave frequency and $\mathbf{k}$ is the wave vector. $\lambda = 2\pi/|\mathbf{k}|$ is the wavelength. The direction of propagation is $\mathbf{n} = \mathbf{k}/|\mathbf{k}|$. Substituting this into the wave equation for the electric field gives

$$\omega^2 - k^2c^2 = 0$$

which means $\omega = \pm kc$. This is called the dispersion relation. The normal way of expressing this equation is to set $\nu = \omega/2\pi$, the wave frequency in cycles per second and $\lambda = 2\pi/k$, the wavelength in metres. Then, the dispersion relation reads $\nu\lambda = c$. We shall now deduce very general equations connecting the wave vector and the electric and magnetic field amplitudes of a plane wave using Maxwell's equations in free space. Let

$$\mathbf{E}(\mathbf{r}, t) = Re(\mathbf{E}_0 exp(i(\omega t - \mathbf{k}.\mathbf{r}))), \qquad \mathbf{B}(\mathbf{r}, t) = Re(\mathbf{B}_0 exp(i(\omega t - \mathbf{k}.\mathbf{r})))$$

The equation $\nabla.\mathbf{E} = 0$ translates to

$$-Re(i\mathbf{k}.\mathbf{E}_0 exp(i(\omega t - \mathbf{k}.\mathbf{r}))) = 0$$

If this equation is to hold for all times and at all points of space, then we must have

$$\mathbf{k}.\mathbf{E}_0 = 0$$

That is, the electric field vector must be perpendicular to the direction of propagation. Likewise plugging in the expression for the magnetic field into the equation $\nabla.\mathbf{B} = 0$ gives

$$-Re(i\mathbf{k}.\mathbf{B}_0 exp(i(\omega t - \mathbf{k}.\mathbf{r}))) = 0$$

and hence

$$\mathbf{k}.\mathbf{B}_0 = 0$$

That is, the magnetic field vector is also perpendicular to the direction of wave propagation. Again, plugging the expressions for the electric and magnetic field into Faraday's law

$$\nabla \times \mathbf{E} = -\frac{\partial \mathbf{B}}{\partial t}$$

gives

$$Re(i\mathbf{k} \times \mathbf{E}_0 exp(i(\omega t - \mathbf{k}.\mathbf{r}))) = Re(i\omega \mathbf{B}_0 exp(i(\omega t - \mathbf{k}.\mathbf{r})))$$

from which we deduce

$$\mathbf{k} \times \mathbf{E}_0/\omega = \mathbf{B}_0$$

Note that this equation implies $\mathbf{k}.\mathbf{B}_0 = 0$ an equation that was obtained using another Maxwell equation. This equation is more conveniently expressed as

$$\mathbf{n} \times \mathbf{E}_0/c = \mathbf{B}_0$$

since $\omega/k = c$. Finally, Ampere's law

$$\nabla \times \mathbf{B} = \mu_0\epsilon_0 \frac{\partial \mathbf{E}}{\partial t}$$

gives

$$-Re(i\mathbf{k} \times \mathbf{B}_0 exp(i(\omega t - \mathbf{k}.\mathbf{r}))) = \mu_0\epsilon_0 Re(i\omega \mathbf{E}_0 exp(i(\omega t - \mathbf{k}.\mathbf{r})))$$

from which we deduce

$$-c^2\mathbf{k} \times \mathbf{B}_0/\omega = \mathbf{E}_0$$

This is better expressed as

$$\mathbf{E}_0 = -c\mathbf{n} \times \mathbf{B}_0$$

This equation implies $\mathbf{k}.\mathbf{E}_0 = 0$ which was derived using another Maxwell equation. Thus, the vectors $(\mathbf{E}_0, \mathbf{B}_0, \mathbf{n})$ form a right handed orthogonal system of axes.

3.18 Complete solution to the Maxwell equations in terms of retarded potentials

To obtain the complete solution to the Maxwell equations, we start with the last one $\nabla.\mathbf{B} = 0$ which implies that $\mathbf{B} = \nabla \times \mathbf{A}$ for some vector field $\mathbf{A}$ called the magnetic vector potential. Now plug this into Faraday's equation

$$\nabla \times \mathbf{E} = -\frac{\partial \mathbf{B}}{\partial t} = -\frac{\partial}{\partial t}\nabla \times \mathbf{A}$$

$$= -\nabla \times \frac{\partial \mathbf{A}}{\partial t}$$

which can be rearranged as

$$\nabla \times \left(\mathbf{E} + \frac{\partial \mathbf{A}}{\partial t}\right) = \mathbf{0}$$

Thus, there exists a scalar field V, called the scalar potential such that

$$\mathbf{E} + \frac{\partial \mathbf{A}}{\partial t} = -\nabla V$$

or

$$\mathbf{E} = -\nabla V - \frac{\partial \mathbf{A}}{\partial t}$$

Now observe that if V is replaced by $V' = V - \frac{\partial f}{\partial t}$ and $\mathbf{A}$ by $\mathbf{A}' = \mathbf{A} + \nabla f$ where $f(\mathbf{r}, t)$ is any scalar function, the electric and magnetic fields do not change. Indeed, we need only verify that

$$-\nabla V' - \frac{\partial \mathbf{A}'}{\partial t} = -\nabla V - \frac{\partial \mathbf{A}}{\partial t}$$

and

$$\nabla \times \mathbf{A}' = \nabla \times \mathbf{A}$$

Now, the scalar function f can be selected so that $\nabla.\mathbf{A}' + \mu_0\epsilon_0\frac{\partial V'}{\partial t} = 0$. This means that f should satisfy

$$\nabla^2 f - \mu_0\epsilon_0\frac{\partial^2 f}{\partial t^2} = -\left(\nabla.\mathbf{A} + \mu_0\epsilon_0\frac{\partial V}{\partial t}\right)$$

Having done so we work completely with the transformed potentials $V', \mathbf{A}'$. Without loss of generality we may thus assume that the potentials $V, \mathbf{A}$ have been obtained after such a gauge transformation and hence satisfy the gauge condition $\nabla.\mathbf{A} + \mu_0\epsilon_0\frac{\partial V}{\partial t} = 0$. Substituting the expression for the electric field into Gauss' law gives us

$$\nabla.\left(-\nabla V - \frac{\partial \mathbf{A}}{\partial t}\right) = \rho/\epsilon_0$$

or

$$\nabla^2 V + \frac{\partial}{\partial t}(\nabla.\mathbf{A}) = -\rho/\epsilon_0$$

Using the gauge condition, this becomes

$$\nabla^2 V - \frac{1}{c^2}\frac{\partial^2 V}{\partial t^2} = -\rho/\epsilon_0$$

In other words, the scalar potential satisfies the wave equation with source with the source being proportional to the charge density. Substituting for the magnetic field $\mathbf{B} = \nabla \times \mathbf{A}$ and the electric field $\mathbf{E} = -\nabla V - \frac{\partial \mathbf{A}}{\partial t}$ into Ampere's law with the displacement current correction term gives

$$\nabla \times (\nabla \times \mathbf{A}) = \mu\mathbf{J} - \epsilon_0\mu_0\frac{\partial}{\partial t}(\nabla V + \frac{\partial \mathbf{A}}{\partial t})$$

or

$$\nabla^2 \mathbf{A} - \frac{1}{c^2}\frac{\partial^2 \mathbf{A}}{\partial t^2} - \nabla(\nabla.\mathbf{A} + \frac{1}{c^2}\frac{\partial V}{\partial t}) = -\mu\mathbf{J}$$

In view of the gauge condition this reduces to

$$\nabla^2 \mathbf{A} - \frac{1}{c^2}\frac{\partial^2 \mathbf{A}}{\partial t^2} = -\mu\mathbf{J}$$

Thus both the electric potential and the magnetic vector potential satisfy the wave equation with source. The solutions to these equations are of the Coulomb type with a retardation term arising due to the extra term $\frac{\partial^2}{\partial t^2}$ added to the Laplacian:

$$V(t,\mathbf{r}) = \frac{1}{4\pi\epsilon_0} \int \frac{\rho(t - |\mathbf{r} - \mathbf{r}'|/c, \mathbf{r}')}{|\mathbf{r} - \mathbf{r}'|} dV'$$

$$\mathbf{A}(t,\mathbf{r}) = \frac{\mu_0}{4\pi} \int \frac{\mathbf{J}(t - |\mathbf{r} - \mathbf{r}'|/c, \mathbf{r}')}{|\mathbf{r} - \mathbf{r}'|} dV'$$

The proofs of these solutions are based on the theory of the Fourier transform and we skip them here. For a physical proof, the student is referred to the Feynman lectures on physics Volume II.

3.19   Solved problems, study projects, suggestions, remarks

Study Project 1

Numerical solution to the Maxwell equations and study of wave propagation on the digital computer.

Maxwell's equations in differential form are

$$\nabla.\mathbf{E} = \frac{\rho}{\epsilon} \qquad \nabla \times \mathbf{E} = -\frac{\partial \mathbf{B}}{\partial t}$$

$$\nabla.\mathbf{H} = 0 \qquad \nabla \times \mathbf{H} = \mathbf{J} + \epsilon\frac{\partial \mathbf{E}}{\partial t}$$

where $\mathbf{E}(t,\mathbf{x})$ and $\mathbf{H}(t,\mathbf{x})$ are respectively the electric and magnetic field in space as functions of time. ρ is the charge density and $\mathbf{J}$ is the current density field. When the charge density is zero and the current density is also zero, we can derive the wave propagation equations for the fields. In one dimension, the wave propagation equation reads

$$\frac{\partial^2 f(t,z)}{\partial t^2} - c^2\frac{\partial^2 f(t,z)}{\partial z^2} = 0$$

The general solution to this is given by

$$f(t,z) = F(t - z/c) + G(t + z/c)$$

The former represents a wave propagating in the positive z direction and the latter a wave propagating in the negative z direction. Consider just a forward propagating wave i.e. $f(t,z) = F(t - z/c)$. We evaluate

$$\frac{\partial f(t,z)}{\partial t} = F'(t - z/c)$$

$$\frac{\partial f(t,z)}{\partial z} = -\frac{1}{c}F'(t - z/c)$$

Thus, f satisfies the advection equation

$$\frac{\partial f}{\partial t} + c\frac{\partial f}{\partial z} = 0$$

To solve this equation numerically, we discretize space and time. Let $f[n,j] = f(n\tau, j\Delta)$. Then, the discretized advection equation can be expressed in the form

$$\frac{f[n+1,j] - f[n,j]}{\tau} + c\frac{f[n,j+1] - f[n,j]}{\Delta} = 0$$

or

$$f[n+1,j] = (1 + \frac{\tau c}{\Delta})f[n,j] - \frac{c\tau}{\Delta}f[n,j+1]$$

Let us try to solve this equation by assuming a sinusoidal dependence of the amplitude upon the spatial index, i.e.

$$f[n,j] = g[n]exp(ij\Delta) \qquad i = \sqrt{-1}$$

Substituting this into the difference equation gives us

$$g[n+1] = (1 + \frac{\tau c}{\Delta})g[n] - \frac{c\tau}{\Delta}exp(i\Delta)g[n]$$

$$= g[n](1 + \frac{\tau c}{\Delta}(1 - exp(i\Delta))) = ag[n]$$

say. We next compute $|a|$:

$$a = 1 + \frac{\tau c}{\Delta}(1 - cos(\Delta)) - \frac{ic\tau}{\Delta}sin(\Delta)$$

so that

$$|a|^2 = (1 + \frac{\tau c}{\Delta}(1 - cos(\Delta)))^2 + (\frac{c\tau}{\Delta})^2 sin^2(\Delta)$$

$$= 1 + \frac{2c^2\tau^2}{\Delta^2} + \frac{2c\tau}{\Delta}(1 - cos(\Delta))$$

This quantity is greater than unity implying that the amplitude of the wave increases exponentially with time. So the advection equation needs to be modified. Likewise, one can discretize the wave equation to get

$$\frac{f[n+1,k] - 2f[n,k] + f[n-1,k]}{\tau^2} - c^2 \frac{f[n,k+1] - 2f[n,k] + f[n,k-1]}{\Delta^2} = 0$$

We need to investigate stablity issues here too. The general three dimensional wave equation can also be formulated in discrete terms.

Study Project 2

A seminar could be given on the basic problems of electrostatics. This seminar should include a survey of the techniques used to calculate the electric field produced by different static charge distributions using Coulomb's law and how numerical evaluation of the field can be achieved by dividing the charge distribution into pixels and replacing the continuous integral by a discrete Riemann sum. Specifically, the seminar can begin with Coulomb's finding that the force of repulsion between two charges Q_1 and Q_2 separated by a distance d is given by

$$F(Q_1, Q_2, d) = \frac{Q_1 Q_2}{4\pi\epsilon_0 d^2}$$

The meaning of the term permitivity should be stressed upon highlighting the fact that when the medium changes from vacuum to some other medium, the constant ϵ_0 gets replaced by ϵ_0 times the refractive index of the medium also called the dielectric constant. The fact that the electrostatic force varies inversely as the square of the distance between the two charges is to be highlighted as an experimental result. Tests by Henri Cavendish may be mentioned here which showed that rather than the second power of 2 appearing in the denominator, the power r is specified by the tolerance limit $2 - \delta < r < 2 + \delta$. It should be mentioned how the entire structure of matter would be different,i.e. the world would appear differently if the electrostatic law of attraction and repulsion had a different power law, i.e. a law of the form $\frac{Q_1 Q_2}{d^r}$ with r different from two. The notion of the electric field lines of force produced by a system of charges and numerical techniques of plotting these field lines can be illustrated in the seminar. This section of the talk could begin with the principle of superposition for the electric field, i.e. if $\mathbf{E}(\mathbf{r}|\mathbf{R}, Q)$ is the field produced by a charge Q located at $\mathbf{R}$, then the electric field produced by a system of charges $Q_1, ..., Q_n$ located at the points $\mathbf{R}_1, ..., \mathbf{R}_n$ is given by the sum $\mathbf{E}(\mathbf{r}) = \sum_{i=1}^{n} \mathbf{E}(\mathbf{r}|\mathbf{R}_i, Q_i)$. How Coulomb's law and superposition are used to derive the electric field produced by discrete and continuous charge distributions:

$$\mathbf{E}(\mathbf{r}) = \sum_{i=1}^{n} \frac{Q_i(\mathbf{r} - \mathbf{R}_i)}{4\pi\epsilon_0|\mathbf{r} - \mathbf{R}_i|^3}$$

for a discrete distribution and

$$\mathbf{E}(\mathbf{r}) = \int \frac{\rho(\mathbf{R})(\mathbf{r} - \mathbf{R})}{4\pi\epsilon_0|\mathbf{r} - \mathbf{R}|^3}$$

for a continuous charge distribution having density $\rho(\mathbf{r})$. Some problems highlighting the correlations in the electric field components at two different points in space when there are random fluctuations in the positions of the charges or in the values of the charges themselves can also be illustrated via simple computations. The signal processing problems appearing in electrostatics should be stressed upon. For example, given the electric field in space, how the field lines are plotted by numerically solving the differential equations

$$\frac{dx}{E_x(x, y, z)} = \frac{dy}{E_y(x, y, z)} = \frac{dz}{E_z(x, y, z)}$$

which on discretization give us

$$x_{k+1} = x_k + \Delta.E_x(x_k, y_k, z_k)$$

$$y_{k+1} = y_k + \Delta.E_y(x_k, y_k, z_k)$$

$$z_{k+1} = z_k + \Delta.E_z(x_k, y_k, z_k)$$

This could be illustrated by plotting the field lines corresponding to a point charge, of a spherical mass of charge having uniform density, of a solid cylinder of charge having radius R and infinite length, of a circular disc of charge having uniform surface charge density and many other problems. The method of discretization for computing the field produced by a volume charge distribution can be illustrated by means of familiar examples. This involves replacing the continuous integral for the electric field by the discrete sum

$$\mathbf{E}(\mathbf{r}) \approx \frac{1}{4\pi\epsilon_0} \sum_{k,m,n} \Delta^3 \frac{\rho(k\delta, m\delta, n\delta)(X - k\delta, Y - m\delta, Z - n\delta)}{((X - k\delta)^2 + (Y - m\delta)^2 + (Z - n\delta)^2)^{3/2}}$$

The TMS card could also be programmed to evaluate the electric field given the charge density in discretized format.

The computation of the field integrals is considerably simplified by the introduction of the electrostatic potential. Numerical techniques for computing the potential and then differentiating this potential numerically, i.e. by adopting the technique of finite differences can be used. For example, the potential due to a flat disc of radius R carrying uniform surface charge density σ with the disc lying in the xy plane and having centre coinciding with the origin at the point (X, Y, Z) is given by the formula

$$V(X, Y, Z) = \frac{\sigma}{4\pi\epsilon_0} \int_0^R \int_0^{2\pi} ((X - \rho.cos\phi)^2 + (Y - \rho.sin\phi)^2 + Z^2)^{-1/2} \rho.d\rho.d\phi$$

$$= \frac{\sigma}{4\pi\epsilon_0} \int_0^R \int_{)}^R \int_0^{2\pi} (X^2 + Y^2 + Z^2 - 2\rho(X.cos\phi + Y.sin\phi))^{-1/2} \rho.d\rho.d\phi$$

This is computed in practice by replacing an integral of the form $\int_0^R \int_0^{2\pi} f(\rho, \phi) d\rho d\phi$ by the discrete sum $\sum_{k=0}^{N-1} \sum_{m=0}^{M-1} f(k\Delta, m\mu)\Delta\mu$ with $\Delta = R/N, \mu = 2\pi/M$. Note that one could alternately define the potential in terms of cylindrical coordinates, i.e. at $X = \rho'.cos\phi', Y = \rho'.sin\phi', Z$ and evaluate the potential appropriately.

TMS programmes for the solution of Laplace's equation with appropriate boundary conditions can be developed. For example, consider a curve Γ in the plane. We want to solve Laplace's equation

$$\frac{\partial^2 V(x, y)}{\partial x^2} + \frac{\partial^2 V(x, y)}{\partial y^2} = 0$$

for all (x, y) falling inside the region S enclosed by the curve and with V assuming specified values on the curve, i.e. if the curve is specified in parametric form as $\gamma(s) = \gamma_1(s)\hat{x} + \gamma_2(s)\hat{y}$, then $V(\gamma_1(s), \gamma_2(s)) = \psi(s)$, with ψ a specified function. We discretize the region S using a grid and let $(\gamma(1, i), \gamma(2, i)), i = 1, 2, ..., m$ denote the points on the grid closest to the curve Γ. The curve can then be regarded as being defined by these boundary points of the grid. The value of ψ at the boundary grid point (x_i, y_i) can be taken as ψ_i, the value of ψ at the point on the curve closest to this grid point. The discretized version of Laplace's equation is given by

$$V[k + 1, m] - 4V[k, m] + V[k - 1, m] + V[k, m + 1] + V[k, m - 1] = 0$$

where $V[k, m] = V(k\Delta, m\Delta)$. Whenever one of the points occurring in the above equation corresponds to a boundary point, the corresponding value of the potential is replaced by the known value ψ at that point. In this manner, we obtain a sequence of linear equation for the values of the potential at the grid points whose solution solves the boundary value problem. The generalization of this two dimensional boundary value problem to the case when the dielectric constant depends on the spatial coordinate could also be mentioned. The corresponding equation is given by

$$\frac{\partial}{\partial x}(\epsilon(x, y)\frac{\partial V(x, y)}{\partial x}) + \frac{\partial}{\partial y}(\epsilon(x, y)\frac{\partial V(x, y)}{\partial y}) = 0$$

The discretization of this equation gives us

$$\epsilon[k + 1, m](V[k + 2, m] - V[k + 1, m]) - \epsilon[k, m](V[k + 1, m] - V[k, m])$$

$$+\epsilon[k, m+1](V[k, m+2] - V[k, m+1]) - \epsilon[k, m](V[k, m+1] - V[k, m]) = 0$$

With once again the value of V at a boundary point being replaced by the corresponding value of ψ. An iterative method to solve the above equation proceeds by computing the potentials at the $(i+1)^{th}$ stage using the rule

$$V^{(i+1)}[k, m] = \frac{V^{(i)}[k+1, m] + V^{(i)}[k-1, m] + V^{(i)}[k, m+1] + V^{(i)}[k, m-1]}{4}$$

where (k, m) correspond only to the interior points of the grid. Eventual convergence to the true potential is almost certain if the grid is chosen to be sufficiently dense.

Study project 3

Solution to problems of electrostatics in a medium having spatially varying dielectric constant. When the dielectric constant is a constant, it is known that the potential obeys the Poisson equation, i.e. $\nabla^2 V(\mathbf{r}) = -\rho(\mathbf{r})/\epsilon$. The solution to this equation is the Coulomb potential

$$V(\mathbf{r}) = \frac{1}{4\pi\epsilon} \int \frac{\rho(\mathbf{r}')}{|\mathbf{r} - \mathbf{r}'|} dV'$$

This follows from the identity $\nabla^2 \frac{1}{|\mathbf{r}-\mathbf{r}'|} = -4\pi\delta(\mathbf{r} - \mathbf{r}')$. When however the dielectric constant varies with space, the electric displacement vector is given by $\mathbf{D}(\mathbf{r}) = -\epsilon(\mathbf{r})\nabla V(\mathbf{r})$ and Gauss' law $\nabla.\mathbf{D} = \rho$ reads as

$$\nabla.(\epsilon\nabla V) = -\rho$$

This can be recast as

$$\nabla^2 V + (\nabla log\epsilon, \nabla V) = -\rho/\epsilon$$

The second term on the left is a perturbation to the Laplacian operator. The equation is still linear in V and the second term can be regarded as a perturbation to the first. One can consider this perturbation to be of the first order of smallness. We make a perturbation series expansion of the potential

$$V = V_0 + V_1 + V_2 + ...$$

where V_n is of the nth degree of smallness. Then, equating terms of equal degrees of smallness gives us

$$\nabla^2 V_0 = -\rho/\epsilon$$

$$\nabla^2 V_n + (\nabla log(\epsilon), \nabla V_{n-1}) = 0$$

and one can solve this equation recursively for the correction terms of various orders $\{V_n\}$. Alternately, following a system theoretic viewpoint, one can regard ϵ as a spatial input signal and V as a spatial output signal and develop a Volterra representation for the output in terms of the former.

This standpoint is to be investigated in conjunction with computer simulations. Specifically, we find that

$$V_0(\mathbf{r}) = \frac{1}{4\pi} \int \frac{\rho(\mathbf{R})}{\epsilon(\mathbf{R})|\mathbf{r} - \mathbf{R}|} d^3\mathbf{R}$$

$$V_n(\mathbf{r}) = \frac{1}{4\pi} \int \frac{(\nabla log(\epsilon(\mathbf{R})), \nabla V_{n-1}(\mathbf{R}))}{|\mathbf{r_R}|} d^3\mathbf{R}$$

Study project 4

This project deals with deriving the equations of macroscopic electromagnetism. Suppose that we have a discrete charge distribution specified by charges $e_1, ..., e_k$ located at points $\mathbf{r}_1, ..., \mathbf{r}_k$. The charge density is given by $\rho_{mic}(t, \mathbf{x}) = \sum_{j=1}^{k} e_j \delta(\mathbf{r} - \mathbf{r}_j)$. We can consider the general case in which the positions $\mathbf{r}_j, j = 1, 2, ..., k$ depend upon time. If $\mathbf{v}_j(t) = \frac{d\mathbf{r}_j(t)}{dt}$ is the velocity of the jth charge, then the current density is given by $\mathbf{J}_{mic}(t, \mathbf{x}) = \sum e_j \mathbf{v}_j(t) \delta(\mathbf{x} - \mathbf{r}_j(t))$. These quantities are called respectively the microscopic charge and current densities. This situation can correspond to atoms located at the sites of a crystal lattice with each atom having the classical model of a nucleus with a positive charge and electrons orbiting around it. The $e_j's$ then include both the charge of the nucleus as well as the electronic charge. Let $\mathbf{e}(t, \mathbf{x})$ and $\mathbf{b}(t, \mathbf{x})$ denote the electric and magnetic fields produced by these charges. These fields satisfy the Maxwell equations

$$\nabla.\mathbf{e} = \rho_{mic}/\epsilon$$

$$\nabla \times \mathbf{e} = -\frac{\partial \mathbf{b}}{\partial t}$$

$$\nabla \times \mathbf{b} = \mathbf{J}_{mic}(t, \mathbf{x}) + \epsilon \frac{\partial \mathbf{e}}{\partial t}$$

$$\nabla.\mathbf{b} = 0$$

The fields at the macroscopic level are obtained by averaging out the discrete charge and current densities with respect to a weighting function. For example, in the quantum mechanical picture, an electron is regarded as a smeared out charge distribution with the charge density $-e|\psi(t, \mathbf{x})|^2$ where $\psi(t, \mathbf{x})$ is the electronic wave function. The macroscopic fields are thus obtained by averaging the microscopic fields with respect to a weighting function:

$$\rho_{mac}(t, \mathbf{x}) = \int \rho_{mic}(t, \mathbf{x} - \mathbf{x}')f(\mathbf{x}')d^3\mathbf{x}'$$

$$\mathbf{J}_{mac}(t, \mathbf{x}) = \int \mathbf{J}_{mic}(t, \mathbf{x} - \mathbf{x}')f(\mathbf{x}')d^3\mathbf{x}'$$

$$\mathbf{E}(t, \mathbf{x}) = \int \mathbf{e}(t, \mathbf{x} - \mathbf{x}')f(\mathbf{x}')d^3\mathbf{x}'$$

$$\mathbf{B}(t,\mathbf{x}) = \int \mathbf{b}(t,\mathbf{x}-\mathbf{x}')f(\mathbf{x}')d^3\mathbf{x}'$$

It is easy to see that by averaging the microscopic Maxwell equations, we arrive at the macroscopic Maxwell equations:

$$\nabla.\mathbf{E} = \rho_{max}/\epsilon$$

$$\nabla \times \mathbf{E} = -\frac{\partial \mathbf{B}}{\partial t}$$

$$\nabla \times \mathbf{B} = \mathbf{J}_{max} + \epsilon\frac{\partial \mathbf{E}}{\partial t}$$

$$\nabla.\mathbf{B} = 0$$

To obtain the macroscopic Maxwell equations, we average the microscopic Maxwell equations with respect to the weight function f: We derive the first one leaving the derivations of the other macroscopic equations to the interested student.

$$\nabla.\mathbf{E}(t,\mathbf{x}) = \nabla.\int \mathbf{e}(t,\mathbf{x}-\mathbf{x}')f(\mathbf{x}')d\mathbf{x}'$$

$$= \int f(\mathbf{x}')\nabla.\mathbf{e}(t,\mathbf{x}-\mathbf{x}')d\mathbf{x}' = \int f(\mathbf{x}')\rho_{mic}(t,\mathbf{x}-\mathbf{x}')d\mathbf{x}'$$

$$= \rho_{mac}(t,\mathbf{x})/\epsilon$$

Solved Problem 1

Develop the multipole expansion for the electrostatic potential function.

Ans: The first step is to expand the Green's function $\frac{1}{|\mathbf{x}-\mathbf{x}'|}$ in terms of the Legendre polynomials: Assuming that $r' < r$ where $r = |\mathbf{x}|$ and $r' = |\mathbf{x}'|$, we have with γ denoting the angle between the vectors $\mathbf{x}$, the expansion

$$\frac{1}{|\mathbf{x}-\mathbf{x}'|} = \frac{1}{r^2+r'^2-2rr'\cos\gamma} = \sum_{l=0}^{\infty}\frac{r'^l}{r^{l+1}}P_l(\cos\gamma)$$

We then make use of the identity

$$P_l(\cos\gamma) = \sum_{m=-l}^{l}\frac{4\pi}{2l+1}Y_{lm}(\theta,\phi)Y_{lm}(\theta',\phi')$$

where (θ,ϕ) is the direction of the vector $\mathbf{x}$ and (θ',ϕ') is the direction of the vector $\mathbf{x}'$. This gives us the following expansion for the potential due to a charge distribution corresponding to the density $\rho(\mathbf{x})$:

$$V(\mathbf{x}) = \frac{1}{4\pi\epsilon}\int \frac{\rho(\mathbf{x}')}{|\mathbf{x}-\mathbf{x}'|}d^3\mathbf{x}'$$

$$= \frac{1}{\epsilon} \sum_{l=0}^{\infty} \sum_{m=-l}^{l} \frac{Y_{lm}(\theta, \phi)}{(2l+1)} \int \rho(\mathbf{x}') r'^{l} Y_{lm}(\theta', \phi') d^3\mathbf{x}'$$

$$= \sum_{l=0}^{\infty} \sum_{m=-l}^{l} \frac{q_{lm}}{r^{l+1}} Y_{lm}(\theta, \phi)$$

where

$$q_{lm} = \frac{1}{\epsilon(2l+1)} \int \rho(\mathbf{x}') r'^{l} Y_{lm}(\theta', \phi') d^3\mathbf{x}'$$

the integral being over the entire region where charge is distributed.

Solved problem 2

Derive the expression for the gradient operator in general curvilinear coordinates. Specialize to orthogonal curvilinear coordinates

Ans

$$\nabla f = \sum_{i=1}^{3} \frac{\partial f}{\partial q_i} \nabla q_i$$

In case that the system is orthogonal, the unit vector along the q_i direction is given by $\mathbf{e}_{q_i} = \nabla q_i / |\nabla q_i|$ and we get

$$\nabla f = \sum_{i=1}^{3} \frac{\mathbf{e}_{q_i}}{H_i} \frac{\partial f}{\partial q_i}$$

where $H_i = \frac{1}{|\nabla q_i|}, i = 1, 2, 3$ are the Lame coefficients.

Question 1

Explain using the Maxwell equations why the electric field lines need not be closed while the magnetic field lines are always closed. Your explanation should contain the fact that electric monopoles exist in nature while magnetic monopoles do not exist in nature.

Unsolved problem 1

Prove the following vector identity using (a) cartesian coordinates, (b) cylindrical coordinates and (c) spherical polar coordinates: $\nabla \times (\nabla \times \mathbf{A}) = \nabla(\nabla.\mathbf{A}) - \nabla^2 \mathbf{A}$. Here, $\mathbf{A}(\mathbf{r})$ is an arbitrary vector field. Your can use the following expressions for the gradient, divergence operator in arbitrary orthogonal curvilinear coordinate system:

$$\nabla = \frac{\mathbf{e}_1}{H_1} \frac{\partial}{\partial q_1} + \frac{\mathbf{e}_1}{H_2} \frac{\partial}{\partial q_2} + \frac{\mathbf{e}_3}{H_3} \frac{\partial}{\partial q_3}$$

$$\nabla.\mathbf{f} = \frac{1}{H_1H_2H_3}\left(\frac{\partial}{\partial q_1}(f_1H_2H_3) + \frac{\partial}{\partial q_2}(f_2H_3H_1) + \frac{\partial}{\partial q_3}(f_3H_1H_2)\right)$$

Study project 5

A short seminar on electromagnetic potentials can be given. This seminar should introduce the concept of scalar and vector potential and the operations of gradient and curl and should clearly bring out the reason for introducing the potentials. How to solve for the potentials in terms of the fields should also be explained here. The notion of retarded potentials arising naturally in the solution of the Maxwell equations and how these can be used in obtaining expressions for the fields produced by accelerating charges should be brought out. The seminar should begin characteristically by introducing the Maxwell equations in differential form, i.e.,

$$\nabla.\mathbf{E} = \rho/\epsilon_0, \ \nabla \times \mathbf{E} = -\frac{\partial \mathbf{B}}{\partial t}$$

$$\nabla \times \mathbf{B} = \mu_0\epsilon_0\frac{\partial \mathbf{E}}{\partial t} + \mu_0\mathbf{J}, \ \nabla.\mathbf{B} = 0$$

It should then be mentioned that the equation $\nabla.\mathbf{B} = 0$ implies that $\mathbf{B}$ is the curl of a vector field, i.e. $\mathbf{B} = \nabla \times \mathbf{A}$ where $\mathbf{A}$ is some vector field. The proof of this fact as well as the computation of $\mathbf{A}$ from $\mathbf{B}$ should be given. This can proceed along the following lines. Write down the equation $\mathbf{B} = \nabla \times \mathbf{A}$ in component form:

$$\frac{\partial A_z}{\partial y} - \frac{\partial A_y}{\partial z} = B_x$$

$$\frac{\partial A_x}{\partial z} - \frac{\partial A_z}{\partial x} = B_y$$

$$\frac{\partial A_y}{\partial x} - \frac{\partial A_x}{\partial y} = B_z$$

The first equation integrates to give

$$A_z = \int_0^y B_x dy + \int_0^y \frac{\partial A_y}{\partial z}dy + \psi(x, z)$$

where ψ is arbitrary. Here, A_y is assumed to be known. The third equation integrates to give

$$A_x = \int_0^y \frac{\partial A_y}{\partial x}dy - \int_0^y B_z dy + \phi(x, z)$$

where ϕ is arbitrary. Substituting for A_x, A_z from here into the second equation gives

$$\int_0^y \frac{\partial^2 A_y}{\partial x \partial z}dy - \int_0^y \frac{\partial B_z}{\partial z}dy + \frac{\partial \phi}{\partial z}$$

$$-\int_0^y \frac{\partial B_x}{\partial x}dy - \int_0^y \frac{\partial^2 A_y}{\partial x \partial z}dy - \frac{\partial \psi}{\partial x} = 0$$

Since $\nabla.\mathbf{B} = 0$, this is the same as

$$\int_0^y \frac{\partial B_y}{\partial y}dy + \frac{\partial \phi}{\partial z} - \frac{\partial \psi}{\partial x} = B_y$$

This gives

$$-B_y(x,0,z) + \frac{\partial \phi}{\partial z} - \frac{\partial \psi}{\partial x} = 0$$

Assume that ψ has been chosen arbitrarily. Then, to meet the above constraint, ϕ must be chosen so that

$$\phi(x,z) = \int_0^z \frac{\partial \psi}{\partial x}dz + \int_0^z B_y(x,0,z)dz + \eta(x)$$

where η is any function of x alone. This shows how one can explicitly construct the magnetic vector potential. Having done so, one plugs this into the second Maxwell equation to get

$$\nabla \times \left(\mathbf{E} + \frac{\partial \mathbf{A}}{\partial t}\right) = \mathbf{0}$$

Then explanation should be provided for the fact that if the curl of a vector field is zero, then the vector field must be the gradient of a scalar field. The proof would proceed along the following lines. Suppose $\nabla \times \mathbf{F} = \mathbf{0}$. Express this in component form

$$\frac{\partial F_z}{\partial y} - \frac{\partial F_y}{\partial z} = 0$$

$$\frac{\partial F_x}{\partial z} - \frac{\partial F_z}{\partial x} = 0$$

$$\frac{\partial F_y}{\partial x} - \frac{\partial F_x}{\partial y} = 0$$

Integration of the first equation gives us

$$F_z = \int_0^y \frac{\partial F_y}{\partial z}dy + \psi(x,z)$$

Similarly, integration of the last equation gives

$$F_x = \int_0^y \frac{\partial F_y}{\partial x}dy + \phi(x,z)$$

Substitute these two expressions into the second equation to get

$$\frac{\partial \phi}{\partial z} - \frac{\partial \psi}{\partial x} = 0$$

which implies that

$$\psi(x,z) = \int_0^x \frac{\partial \phi}{\partial z} dx + \eta(z)$$

Thus,

$$F_x = \frac{\partial}{\partial x}\left(\int_0^y F_y dy + \int_0^x \phi(x,z)dx + \int_0^z \eta(z)dz\right)$$

$$F_y = \frac{\partial}{\partial y}\left(\int_0^y F_y dy + \int_0^x \phi(x,z)dx + \int_0^z \eta(z)dz\right)$$

$$F_z = \frac{\partial}{\partial z}\left(\int_0^y F_y dy + \int_0^z \psi(x,z)dz\right) = \frac{\partial}{\partial z}\left(\int_0^y F_y dy + \int_0^x \phi dx + \int_0^z \eta(z)dz\right)$$

This shows that $\mathbf{F} = \nabla V$ where

$$V(x,y,z) = \int_0^y F_y dy + \int_0^x \phi(x,z)dx + \int_0^z \eta(z)dz$$

Now $\int_0^x \phi(x,z)dx + \int_0^z \eta(z)dz$ is an arbitrary function of x and z. Thus, we can equivalently write the general form of V as

$$V = \int_0^y F_y dy + \theta(x,z)$$

where θ is an arbitrary function of x and z. Taking for $\mathbf{F}$ $\mathbf{E} + \frac{\partial \mathbf{A}}{\partial t}$, we obtain

$$\mathbf{E} + \frac{\partial \mathbf{A}}{\partial t} = -\nabla V$$

V is called the electric potential and $\mathbf{A}$ the magnetic vector potential.

Study project 6

A seminar on basic aspects of transmission line theory can be delivered. This discussion should first of all focus around propagation of voltage and current signals along a two wire line having definite values of capacitance, inductance, conductance and resistance per unit length. Specifically, if these values are defined by the functions $C(t,z), L(t,z), G(t,z), R(t,z)$ where allowance is made for the fact that these quantities can vary with time and distance of the point from the source end, then how application of the Kirchhoff's laws lead to the following equations for the current and voltage $i(t,z)$ and $v(t,z)$ along the line should be explained clearly:

$$i(t,z) - i(t,z+dz) = G(t,z).dz.v(t,z) + \frac{\partial}{\partial t}(C(t,z).dz.v(t,z))$$

$$v(t,z) - v(t,z+dz) = R(t,z).dz.i(t,z) + \frac{\partial}{\partial t}(L(t,z).dz.i(t,z))$$

using which one deduces the following fundamental partial differential equations for the transmission line:

$$-\frac{\partial i}{\partial z} = Gv + \frac{\partial(Cv)}{\partial t}$$

$$-\frac{\partial v}{\partial z} = Ri + \frac{\partial(Li)}{\partial t}$$

The special case when G, L, R, C are constant along the line should be considered and the process by which one of the variables i or v is eliminated to arrive at damped wave equations for the current and voltage sources should be considered.

Study project 7

A seminar on numerical evaluation of the potential of a charge distribution using basis functions will act as a starting point for the engineer interested in applications of finite element techniques in electromagnetic theory. If $\rho(\mathbf{r})$ is the charge density in space, the potential is the Coulomb integral

$$V(\mathbf{r}) = \int \frac{\rho(\mathbf{r}')}{|\mathbf{r} - \mathbf{r}'|} dV'$$

We expand the density in terms of a basis set $\{\phi_1(\mathbf{r}), ..., \phi_N(\mathbf{r})\}$ as follows:

$$\rho(\mathbf{r}) \approx \sum_{k=1}^{N} c_k \phi_k(\mathbf{r})$$

The coefficients $\{c_k\}$ are chosen so as to minimize the integral of the error squared:

$$min_{c_1,...,c_N} \int \left(\rho(\mathbf{r}) - \sum_{k=1}^{N} c_k \phi_k(\mathbf{r})\right)^2 dV$$

Setting the derivative of this with respect to the c_j' to zero yields the normal equations:

$$\sum_{k=1}^{N} c_k \int \phi_k(\mathbf{r})\phi_m(\mathbf{r})dV = \int \rho(\mathbf{r})\phi_m(\mathbf{r})dV, m = 1, 2, ..., N$$

Having determined the $c_m's$ in this way, we obtain the following approximate expansion for the potential:

$$V(\mathbf{r}) \approx \sum_{k=1}^{N} c_k \int \frac{\phi_k(\mathbf{r}')}{|\mathbf{r} - \mathbf{r}'|} dV'$$

Note the advantage with this method. We can once and for all evaluate the functions

$$\psi_k(\mathbf{r}) = \int \frac{\rho(\mathbf{r}')}{|\mathbf{r} - \mathbf{r}'|} dV'$$

for the given basis function. These functions are independent of the charge density. Now given a new charge density $\rho(\mathbf{r})$, we compute the coefficients $\{c_k\}$ that determine the best mean square approximation and plug it into the expression for the potential:

$$V(\mathbf{r}) = \sum c_k \psi_k(\mathbf{r})$$

Study project 8

Give a seminar on the reflection of an electromagnetic wave at the interface of two dielectric media at normal incidence. Assume that the first medium has permitivity ϵ_1 and permeability μ_1 and the second medium has permitivity ϵ_2 and permeability μ_2. The interface is assumed to be the xy plane. Medium one is the region $z > 0$ and medium two is the region $z < 0$. the wave is thus incident from the top. The incident electric field phasor is given by

$$\mathbf{E}_i(z) = E_{i0}\hat{x}.exp(-j\beta_1 z)$$

where $\beta_1 = \omega\sqrt{\epsilon_1\mu_1}$ is the wave number or propagation constant for plane waves in medium one. The reflected electric field is given by

$$\mathbf{E}_r(z) = E_{r0}\hat{x}.exp(j\beta_1 z)$$

and the transmitted electric field by

$$\mathbf{E}_t(z) = E_{t0}\hat{x}.exp(-j\beta_2 z)$$

The magnetic field in the incident wave is given by

$$-j\omega\mu_1\mathbf{H}_i = curl\mathbf{E}_i = -j\beta_1 E_{i0}\hat{z} \times \hat{x}.exp(-j\beta_1 z)$$

or

$$\mathbf{H}_i(z) = \frac{\beta_1}{\omega\mu_1}E_{i0}\hat{y}.exp(-j\beta_1 z)$$

Here, $\frac{\beta_1}{\omega\mu_1} = \left(\frac{\epsilon_1}{\mu_1}\right)^{1/2}$ or in terms of the characteristic impedance of medium one $\eta_1 = \left(\frac{\mu_1}{\epsilon_1}\right)^{1/2}$, we can write the incident magnetic field as

$$\mathbf{H}_i(z) = \frac{E_{i0}}{\eta_1}\hat{y}.exp(-j\beta_1 z)$$

Similarly the reflected magnetic field is given by

$$\mathbf{H}_r(z) = -\frac{E_{r0}}{\eta_1}\hat{y}.exp(j\beta_1 z)$$

and the transmitted magnetic field by

$$\mathbf{H}_t(z) = \frac{E_{t0}}{\eta_2}\hat{y}exp(-j\beta_2 z)$$

where $\eta_2 = (\frac{\mu_2}{\epsilon_2})^{1/2}$. $\beta_2 = \omega\sqrt{\epsilon_2\mu_2}$ is the propagation constant or wave vector in medium two. Equating the tangential electric and magnetic fields on both sides of the boundary gives us

$$E_{i0} + E_{r0} = E_{t0}$$

$$E_{i0}/\eta_1 - E_{r0}/\eta_1 = E_{t0}/\eta_2$$

from which, we deduce the reflection and transmission coefficients $R = \frac{E_{r0}}{E_{i0}}, T = \frac{E_{t0}}{E_{i0}}$.

Solved problem 3

Compute of the electric field and potential due to a spherical charge distribution with density given by $\rho(r) = \rho_1$, for $0 \leq r \leq a, \rho(r) = \rho_2$ for $a \leq r \leq R$ and $\rho(r) = 0$ for $r > R$.

Gauss' law assumes the form

$$\frac{1}{r^2}\frac{d}{dr}(r^2 D(r)) = \rho_1, 0 \leq r \leq a$$

$$\frac{1}{r^2}\frac{d}{dr}(r^2 D(r)) = \rho_2, a \leq r \leq R$$

$$\frac{1}{r^2}\frac{d}{dr}(r^2 D(r)) = 0, r > R$$

where $D(r)$ is the radial electric displacement vector. While delivering the project presentation, these equations can be explained on the blackboard. First of all, the expression for the divergence of a vector field in spherical polar coordinates should be derived:

$$\nabla.\mathbf{F} = \frac{1}{r^2}\frac{\partial}{\partial r}(r^2 F_r) + \frac{1}{r.sin(\theta)}\frac{\partial}{\partial\theta}(F_\theta sin(\theta)) + \frac{1}{r.sin(\theta)}\frac{\partial}{\partial\phi}(F_\phi)$$

This expression should be derived using the chain rule of differential calculus. If needed, the physical meaning of the divergence of a vector field can also be explained. The integration of the above equations is to be explained next leading to

$$D(r) = \rho_1 r/3, 0 \leq r \leq a$$

$$D(r) = \rho_2 r/3 + c_2/r^2, a \leq r \leq R$$

$$D(r) = c_3/r^2, r > R$$

A term c_1/r^2 should also appear in the first of these equgations but we take $c_1 = 0$ since the field must be finite at $r = 0$. c_2 and c_3 are chosen to match the boundary conditions that

$D(a-) = D(a+), D(R-) = D(R+)$. Numerical integration of these equations will involve dividing their radial line into intervals of length Δ with $M\Delta = a, N\Delta = R$ where $M < N$ are positive integers. The differential equations transform into difference equations

$$\frac{1}{\Delta}(D[k+1] - D[k]) + \frac{2}{k\Delta}D[k] = \rho_1$$

for $1 \leq k \leq M$, the same expression equal to ρ_2 for $M \leq k \leq N$ and $= 0$ for $k > N$. From the symmetry of the problem $D[1] = 0$, i.e. the electric displacement at the origin vanishes. This means that one can successively solve these difference equations for the $D[k]'s$. The equations should be solved and compared with the solution obtained by actual integration.

Solved Problem 4

Derive the form of the energy density of the electromagnetic field from theoretical principles Also compute the energy in the electrostatic field produced by a spherical charge distribution.

Hint: We start with the general expression for the total energy in the field produced by a charge distribution having density $\rho(\mathbf{r})$:

$$W = \frac{1}{8\pi\epsilon_0} \int \frac{\rho(\mathbf{r})\rho(\mathbf{r}')}{|\mathbf{r} - \mathbf{r}'|} dV\, dV'$$

as a generalization of the expression for the energy density produced by system of point charges:

$$W = \frac{1}{8\pi\epsilon_0} \sum_{i \neq j} \frac{Q_i Q_j}{|\mathbf{r}_i - \mathbf{r}_j|}$$

Then using the Coulomb expression for the potential, one casts W in the form

$$W = \frac{1}{2} \int \rho(\mathbf{r}) V(\mathbf{r}) d^3\mathbf{r}$$

and using Poisson's equation $\nabla^2 V = -\rho/\epsilon_0$,

$$W = -\frac{\epsilon_0}{2} \int V\nabla^2 V d^3\mathbf{r} = -\frac{\epsilon_0}{2} \int \nabla.(V\nabla V)d^3\mathbf{r} + \frac{\epsilon_0}{2} \int |\nabla V|^2 d^3\mathbf{r}$$

The first integral transforms by Gauss' theorem into a surface integral over a surface of infinite radius and hence vanishes. The second is the same as $\int \frac{\epsilon_0}{2}E^2(\mathbf{r})d^3\mathbf{r}$, yielding for the energy density the expression $U_E(\mathbf{r}) = \frac{\epsilon_0}{2}E(\mathbf{r})^2$. We then look at the expression for the energy density in a magnetic field. Argue that the energy density can be expressed as $\frac{1}{2}\int \mathbf{J}.\mathbf{A}dV$ and simplify using Ampere's equation $\nabla \times \mathbf{B} = \mu\mathbf{J}$ to arrive at the expression $U_B = \frac{\mu_0}{2}B(\mathbf{r})^2$. The total energy of the electromagnetic field is the sum of the electrostatic and magnetostatics energies. This total sum yields a formula that is also valid in the dynamic situation, i.e. when the fields are time varying.

For a spherical charge distribution, $\rho(r)$, the electric field is radially directed outward and the field at a distance r from the origin depends only upon the charge enclosed by a sphere of radius r. This expression can be obtained using Gauss' law. Plug this into the equation for the energy density to obtain the answer.

Study project 9

Physicists specializing in applications of Lie groups to problems of mechanics and electrodynamics use the apparatus of differentiable manifolds. A manifold M is a topological space with coordinates defined via a chart $(U_i, \phi_i), i = 1, ..., N$. Each U_i is an open subset of M and M equals the union of all the $U_i's$. For each i, $\phi_i : U_i \rightarrow \phi(U_i) \subset \mathbb{R}^n$ is a homeomorphism, i.e. a one-one onto continuous map with a continuous inverse. The maps ϕ_i have a differentiable structure, i.e. if x is any point in $\phi_i(U_i) \cap \phi_j(U_j)$, then the map $x \rightarrow \phi_i^{-1} o \phi_j(x)$ is differentiable, i.e., each component is differentiable. The wedge product and the Hodge star operator form an important component in the theory of differentiable manifolds. These operations can be used to cast the Maxwell equations in general differentiable geometric notation. Specifically one considers space time $\mathbb{R}^4$ as a differentiable manifold, with the tangent vector space at a given point (t, x, y, z) being spanned by

$$\left(\frac{\partial}{\partial t}, \frac{\partial}{\partial x}, \frac{\partial}{\partial y}, \frac{\partial}{\partial z}\right)$$

Denote this tangent space by $T_{(t,x,y,z)}\mathbb{R}^4$. The dual to this space is the space of linear functions spanned by the dual bases

$$(dt, dx, dy, dz)$$

where

$$(c_0 dt + c_1 dx + c_2 dy + c_3 dz)(a_0 \frac{\partial}{\partial t} + a_1 \frac{\partial}{\partial x} + a_2 \frac{\partial}{\partial y} + a_3 \frac{\partial}{\partial z})$$

$$= c_0 a_0 + c_1 a_1 + c_2 a_2 + c_3 a_3$$

Let (V, A_x, A_y, A_z) be the electromagnetic four potential. This is to be regarded as a contravariant vector. We form the one form

$$A = -V dt + A_x dx + A_y dy + A_z dz$$

and this is to be regarded as a covariant vector. Its differential is given by

$$dA = -dV \wedge dt + dA_x \wedge dx + dA_y \wedge dy + dA_z \wedge dz$$

$$= -(\frac{\partial V}{\partial x} dx + \frac{\partial V}{\partial y} dy + \frac{\partial V}{\partial z} dz) \wedge dt$$

$$+(\frac{\partial A_x}{\partial t} dt + \frac{\partial A_x}{\partial y} dy + \frac{\partial A_x}{\partial z} dz) \wedge dx$$

$$+(\frac{\partial A_y}{\partial t} dt + \frac{\partial A_y}{\partial x} dx + \frac{\partial A_y}{\partial z} dz) \wedge dy$$

$$+(\frac{\partial A_z}{\partial t}dt + \frac{\partial A_z}{\partial x}dx + \frac{\partial A_z}{\partial y}dy) \wedge dz$$

$$= (\frac{\partial V}{\partial x} + \frac{\partial A_x}{\partial t})dt \wedge dx + ...$$

A full expansion of this shows the appearance of the electric and magnetic fields in the differential of A. The Maxwell equations can be formulated using this differential form notation. The seminar should include a thorough discussion of differential forms on manifolds and should clearly explain how the Maxwell equations are to be formulated using this new language.

Study project 10

A short seminar dealing with the following boundary value problem of electromagnetic theory can be presented. The potential $V(x,y)$ in the region $0 < x < a, y > 0$ is to be evaluated. The bottom boundary $y = 0, 0 < x < a$ is at a constant potential V_0 while the boundaries $x = 0, y > 0$ and $x = a, y > 0$ are at potentials zero. $V(x,y)$ satisfies the Laplace equation

$$\frac{\partial^2 V(x,y)}{\partial x^2} + \frac{\partial^2 V(x,y)}{\partial y^2} = 0$$

The first step in the seminar would be to solve this equation by the separation of variables technique. This means setting $V(x,y) = X(x)Y(y)$ and casting the equation as

$$\frac{X''(x)}{X(x)} = -\frac{Y''(y)}{Y(y)}$$

Both sides must equal a constant since the left side is a function of x only and the right side of y only. Thus $X(x) = A\cos(\alpha x) + B\sin(\alpha x)$ and $Y(y) = A'exp(\alpha y) + B'exp(-\alpha y)$. The boundary conditions at $x = 0$ and $x = a$ imply that $\alpha = n\pi/a, n = 1, 2, ..., A = 0$ and $A' = 0$.

$$V(x,y) = \sum_{n=1}^{\infty} c_n exp(-n\pi y/a)sin(n\pi x/a)$$

The constants $\{c_n\}$ are determined from the boundary condition at $y = 0$. These give

$$c_n = \frac{2}{a} \int_0^a V_0 sin(n\pi x/a)dx$$

A plot of this solution should be made using MATLAB. Then, other boundary conditions at $y = 0$, i.e. $V(x,0) = V_0(x)$ should be looked at and computer plots of the potential variation should be made. Finally, the same boundary value problem should be solved by means of discretizing the Laplace equation. The Laplace equation discretizes to give the difference equation

$$2V[k,n] - 2V[k-1,n] + V[k-2,n] - 2V[k,n-1] + V[k,n-2] = 0$$

with $k, n \geq 2$ and the boundary condition $V[0, n] = V[N, n] = 0, V[k, L] = 0, V[k, 0] = V_0$ where $a = N\Delta, L >> N$. Algorithms for solving this difference equation should be devised by the student and presented in a seminar.

Study project 11

A short seminar on the computation of the magnetic field for a static current distribution and the numerical evaluation of this field for the field produced by a thin wire carrying current I. The numerical computation of the mutual inductance between two wire loops can also be dealt with here. The starting point of the seminar is the Maxwell equations

$$\nabla.\mathbf{B} = 0, \nabla \times \mathbf{B} = \mu\mathbf{J}$$

It should be stressed in the seminar that the second of these equations is Ampere's law and is valid only for static situations, i.e. situations in which no time variation is present. The reason for this should be given, namely that forming the divergence of this equation leads to $\nabla.\mathbf{J} = 0$ which is the charge conservation equation when the charge density in space does not vary with time. This can be called the static charge conservation equation. The physical meaning of the first equation should also be given as corresponding to the absence of magnetic monopoles. The first equation leads us to the fact that the magnetic field is the curl of another vector field which is called the magnetic vector potential: $\mathbf{B} = \nabla \times \mathbf{A}$. The proof of this fact may be indicated by deriving this from basic integration of the three components of the equation. Plugging this into Ampere's law then leads to

$$\nabla \times (\nabla \times \mathbf{A}) = \mu\mathbf{J}$$

or

$$\nabla^2\mathbf{A} - \nabla(\nabla.\mathbf{A}) = -\mu\mathbf{J}$$

Then, the gauge condition can be stated. Specifically, if $\mathbf{A}$ is changed to $\mathbf{A} + \nabla f$, where f is an arbitrary scalar field, then the magnetic field will not change:

$$\nabla \times (\mathbf{A} + \nabla f) = \nabla \times \mathbf{A}$$

Now, f can be chosen, so that $\nabla.(\mathbf{A} + \nabla f) = 0$. This amounts to choosing f as the solution to Poisson's equation: $\nabla^2 f = -\nabla.\mathbf{A}$. That is the new potential $\mathbf{A}'$ satisfies the Gauge condition $\nabla.\mathbf{A}' = 0$ and $\mathbf{A}'$ becomes the solution to the Poisson equation

$$\nabla^2\mathbf{A}' = -\mu\mathbf{J}$$

the solution to which is given by

$$\mathbf{A}'(\mathbf{r}) = \frac{\mu}{4\pi} \int \frac{\mathbf{J}(\mathbf{r}')}{|\mathbf{r} - \mathbf{r}'|} dV'$$

The magnetic field is found by forming the curl of this equation:

$$\mathbf{B} = \frac{\mu}{4\pi} \int \mathbf{J}(\mathbf{r}') \times \frac{(\mathbf{r} - \mathbf{r}')}{|\mathbf{r} - \mathbf{r}'|^3} dV'$$

This is the Biot-Savart law. For a current carrying wire, the volume differential $\mathbf{J}'dV'$ is replaced by the line differential $I d\mathbf{r}'$ and one obtains the classical Biot-Savart law

$$\mathbf{B}(\mathbf{r}) = \frac{\mu I}{4\pi} \int_\Gamma d\mathbf{r}' \times \frac{(\mathbf{r} - \mathbf{r}')}{|\mathbf{r} - \mathbf{r}'|^3}$$

with the line integral being carried out over the curve Γ. The computation of the field defined by this expression by numerical integration involves discretizing the curve into points $\mathbf{r}'_k, k = 1, 2, ..., N$ and replacing the line integral by the discrete sum

$$\mathbf{B}(\mathbf{r}) \approx \frac{\mu I}{4\pi} \sum_{k=0}^{N-1} (\mathbf{r}'_{k+1} - \mathbf{r}'_k) \times \frac{(\mathbf{r} - \mathbf{r}'_k)}{|\mathbf{r} - \mathbf{r}'_k|^3}$$

The seminar can deal with displaying the field line plots for wires of various shapes obtained using this formula.

Study project 12

An interesting seminar can be delivered on the boundary conditions for the electric and magnetic fields that one comes across in the theory of electromagnetic fields and how one applies these boundary conditions to practical problems. The derivation of the boundary conditions from the first principles, i.e. application of the Maxwell equations with appropriate surface and line integrals should be explained. For example, Gauss' law in the form $\nabla.\mathbf{D} = \rho$ when transformed to integral form by the aid of Gauss' theorem reads $\int_S \mathbf{D}.\mathbf{n}dS = Q_{encl}$, with Q_{encl} as the charge enclosed by the volume. If one takes a surface separating two media one and two such that at the vicinity of a point P on the surface, the normal component of $\mathbf{D}$ on side one is D_{n1} and the normal component of $\mathbf{D}$ on side two is D_{n2}, then application of Gauss' law to a small cylindrical pill-box one side of which falls in region one and the other side of which falls in region two gives the boundary condition $D_{n2} - D_{n1} = \sigma$ where σ is the surface charge density at P. This derivation should be explained with the aid of a clear diagram on the blackboard or using transparencies by showing the pill box and the surface charge and how the net charge enclosed by the pillbox equals $\sigma.\delta S$ and how the surface integral of $\mathbf{D}$ over the surface equals $(D_{n2} - D_{n1})\delta S$ with reasons provided for the vanishing of the contribution to the surface integral from the sides of the cylindrical surface in the limit of the cylinder length tending to zero. The next boundary condition to be derived is that on the tangential component of the electric field. One starts with the Faraday-Lenz law of induction in integral form $\int_\Gamma \mathbf{E}.d\mathbf{r} = -\frac{d\Phi_B}{dt}$, where Γ is a closed contour enclosing a surface S and Φ_B is the flux of the magnetic field through S. Now take a surface partitioning space into two regions one and two and a rectangular contour with lengths parallel to the surface and falling exactly on the two sides of the surface touching it. The breadth of the rectangular contour is perpendicular to the surface and its length Δ can be neglected. If E_{t1} and E_{t2} denote the tangential components of the electric fields on the two sides of the interface parallel to the rectangle length, then application of the Faraday-Lenz law to this contour gives $E_{t1} - E_{t2} = 0$. The flux of the magnetic field through the rectangle is zero since the rectangle in the limit of the breadth converging to zero has zero flux. The length of the contour can be rotated in any direction keeping it parallel to the surface. Thus one ends up with the continuity of the tangential component of the electric field. It should be mentioned here that there are exactly two linearly independent vectors tangential to the plane of the interface at the point P. If $\mathbf{n}$ denotes the unit normal to the surface at P, then this condition on the tangential component can be expressed in the form $\mathbf{n} \times (\mathbf{E}_1 - \mathbf{E}_2) = \mathbf{0}$. Again for the magnetic field, one has two boundary conditions corresponding to the normal and the tangential parts. The equation $\nabla.\mathbf{B} = 0$ translates to continuity of the normal component of the magnetic field.

Study project 13

A special seminar can be given on the origin of the attenuation constant while studying the propagation of plane waves inside a conductor. How the attenuation arises from the Maxwell equations should be explained from first principles and a derivation of the skin effect should be obtained. The dissipation of power in a wave propagating inside a conductor should also be clarified. The derivations should start by deriving the fundamental damped wave equations

$$\nabla^2 \mathbf{F} - \mu\sigma \frac{\partial \mathbf{F}}{\partial t} - \mu\epsilon \frac{\partial^2 \mathbf{F}}{\partial t^2} = \mathbf{0}$$

where $\mathbf{F}$ equals either the electric field or the magnetic field. The dispersion relation

$$\gamma^2 - j\mu\sigma\omega + \omega^2\mu\epsilon = 0$$

should then be obtained by plugging in the expression for the plane wave into the damped wave equation. Forming square roots the real and imaginary parts of γ should be extracted. Note that the form of the plane wave is given by

$$\mathbf{F} = \mathbf{F}_0 exp(-\gamma\mathbf{n}.\mathbf{r} + j\omega t)$$

γ is called the propagation constant and if one sets $\gamma = \alpha + j\beta$, α becomes the attenuation constant and $2\pi/\beta$ the wavelength of the plane waves.

Study project 14

A seminar can be given on the plotting of equipotential lines for a given charge distribution. For example, if there are just two charges in the configuration, say Q located at $(0, 0, -d)$ and $-Q$ located at $(0, 0, d)$, the equipotential surfaces are given by

$$\frac{1}{(x^2 + y^2 + (z + d)^2)^{1/2}} - \frac{1}{(x^2 + y^2 + (z - d)^2)^{1/2}} = C$$

with C a constant. The shapes of these surfaces for different values of C can be shown. One can also consider as an example the equipotential surface produced by a discrete charge distribution $Q_1, ..., Q_n$ located at $\mathbf{r}_1, ..., \mathbf{r}_n$ respectively. The surfaces are given by

$$\sum_{\alpha=1}^{n} \frac{Q_\alpha}{|\mathbf{r} - \mathbf{r}_\alpha|} = C$$

These surfaces cannot be obtained in closed form only approximate computer plots of these surfaces can be shown. If $f(\mathbf{r})$ is a function of the position vector $\mathbf{r}$ and we've found a point $\mathbf{r}_0$ on the surface $f(\mathbf{r}) = 0$ and we want to determine approximately the neighbouring points on this surface, then we set $\mathbf{r} = \mathbf{r}_0 + \delta\mathbf{r}$ and obtain approximately

$$0 = f(\mathbf{r}) = f(\mathbf{r}_0) + \delta\mathbf{r}.\nabla f(\mathbf{r}_0) = \delta\mathbf{r}.\nabla f(\mathbf{r}_0)$$

The neighbouring points $\mathbf{r}$ on this surface will therefore all fall on the plane $(\mathbf{r} - \mathbf{r}_0).\nabla f(\mathbf{r}_0) = 0$ and this scheme can be used to obtain approximate computer plots of the equipotential surfaces.

Study project 15

A seminar can be given on formulae giving the power flow in a transmission line corresponding to sinusoidal signal excitation. The voltage phasor $V(z)$ and current phasor $I(z)$ are obtained as solutions to the governing differential equations of the line as

$$V(z) = V_+ exp(-j\beta z) + V_- exp(j\beta z)$$

$$I(z) = \frac{1}{Z_0}(V_+ exp(-j\beta z) - V_- exp(j\beta z))$$

where Z_0 is the characteristic impedance of the line. The proof of these formulae starting from the basic transmission line differential equations should be explained in a clear fashion. The exact voltage and current down the line are expressed by the formulae

$$V(z,t) = Re(V(z)exp(j\omega t)) \qquad I(z,t) = Re(I(z)exp(j\omega t))$$

For a shorted line, $V(d,t) = 0$ and if the source voltage is $A.cos(\omega t)$, we have the additional condition $V_+ + V_- = V(0) = A$. These equations can be used to solve for $V(z,t)$ and $I(z,t)$ and hence determine the average power flow across any section $P(z) = \frac{1}{T}\int_0^T V(z,t)I(z,t)dt$. These issues can be discussed in a seminar.

Study project 16

A seminar on the computation of the magnetic field energy stored in a system of conductors carrying current. The idea is either use symmetry combined with Ampere's law $\int_\Gamma \mathbf{H}.d\mathbf{r} = I$, to evaluate the magnetic field where Γ is a closed contour enclosing the current I and then use the standard formula $W_B = \frac{\mu}{2} \int H^2 dV$ for the magnetic field energy. The evaluation of the integral can sometimes be hard. In such cases, numerical techniques can be adopted. The derivation of the expression of the magnetic field energy density should also be covered in the discussion.

Study project 17

A seminar describing the operation of a cathode ray oscilloscope would be of use to beginners in a basic laboratory course on electronic circuits. The construction of the CRO is first defined by showing the placement of the fluorescent screen in behind which we have an electron gun. To the right and left sides of the electron beam, we have x deflection plates and to the top and bottom of the beam, we have the y deflection plates. When the electron beam emitted from the gun and no voltages are applied to the plates, the beam goes straight and hits the screen causing a bright spot in the screen's centre. We normally apply a ramp voltage to the x plates of the CRO. This creates an electric field in the horizontal direction that varies proportionally to time for a given duration, then falls down to zero and repeats. In mathematical notation, the horizontal field between the plates can be expressed in the form $E_x(t) = \alpha.t$ for $0 \le t \le \tau$ with a period of τ. Equivalently, if $p_x(t) = \alpha.t$ for $0 \le t \le \tau$ and $p_x(t) = 0$ for $t < 0$ and for $t > \tau$, then the horizontal electric field between the plates can be expressed as $E_x(t) = \sum_{n=0}^{\infty} p_x(t - n\tau)$. The presence of such an electric field causes the beam to traverse from one end of the screen to the other end, return back and repeat. If λ denotes the time base, i.e. the time per centimetre setting on the CRO, then a deflection of x centimetre along the x direction corresponds to a time duration of $\lambda.x$ seconds. The beam takes a time τ to traverse from one end to the other end of the screen. Thus, if L denotes the screen length, we must have the equation $\lambda.L = \tau$ or $L = \frac{\tau}{\lambda}$ in centimetres. Now we apply the voltage waveform that we wish to see to the y or vertical plates. Suppose, the waveform is a sinusoid $v(t) = A.sin(\omega.t)$. By the time the beam traverses from one end to the other end of the screen, we want an integral number of cycles of the voltage waveform to be completed. Otherwise, when the beam returns to its starting end the waveform on the screen would begin with a different phase and we would observe on the screen a superposition of sinsoidal waveforms with different starting phases, i.e. a stable sinusoidal waveform would not be observed on the screen. So assume that $\omega = 2\pi/T$ and that $\tau = NT$, where N is a positive integer. The deflection along the x direction is proportional to the x component of the electric field, i.e. $x(t) = \beta.p_x(t) = \beta\alpha t$. It is clear from this expression that $\lambda = (\beta\alpha)^{-1}$. The y deflection after time t is given by

$$y(t) = A.sin(\omega.t) = A.sin(\omega.x/\beta\alpha) = A.sin(\omega\lambda.x)$$

and we find setting $x = L$ that $\omega\lambda.L = \omega.\tau = \frac{2\pi\tau}{T} = 2\pi n$. This means that an integer number of cycles of the y waveform are completed in the time that the beam sweeps from the right to the left.

Study project 18

A seminar on Faraday's law and its applications to practical problems can be delivered to an audience interested in exploring applications of the electromagnetic field equations. The seminar can begin by explaining Faraday's experimental findings wherein he moved a bar magnet inside a wire loop and measured a current through the loop. Actually, he did not measure the exact value of the current but rather noted the deflection of the galvanometer needle observing that the deflection had a greater amplitude when the magnet was moved more vigorously. He concluded that a voltage was developed through the loop when the magnetic field through the loop varied with time. In other words, a changing magnetic field produces an electric field. He quantified this observation in the form of an integral equation: Let S denote the cross sectional area of the loop and let $\mathbf{B}$ denote the magnetic field through the loop. The flux of the magnetic field through the loop is given by the surface integral $\Phi_B = \int_S \mathbf{B}.\mathbf{n}dA$. The magnetic field varies with the spatial location as well as with time, thus $\mathbf{B} = \mathbf{B}(\mathbf{r}, t)$. The flux through the area bounded by the loop is computed as follows. Assume that the loop is planar and falls in the xy plane. The flux of the magnetic field through the area bounded by the loop is then given by the double integral

$$\Phi_B(t) = \int\int_S B_z(x, y, 0, t)dxdy = \int\int_S B_z(\rho, \phi, 0, t)\rho.d\rho.d\phi$$

If the loop is non-planar, its flux can be computed along the following lines. Let $\mathbf{r}(u, v)$ denote an arbitrary point on the surface bounded by the loop. This point is specified by giving two parameters, (u, v) since a surface is two dimensional and to specify any p dimensional surface, one requires p independent parameters. Displacements of the free parameters u, v by du and dv respectively cause the point $\mathbf{r}(u, v)$ on the surface to get displaced by the vectors $\frac{\partial \mathbf{r}}{\partial u}du$ and $\frac{\partial \mathbf{r}}{\partial v}dv$ respectively. The infinitesimal area bounded by these two infinitesimal displacement vectors is given by the magnitude of their cross product. This follows from the following observation: Let $\mathbf{a}, \mathbf{b}$ be two vectors in three dimensional space drawn from the origin. Complete the parallelogram having two adjacent sides defined by these two vectors. The area of the resulting parallelogram is then given by $|\mathbf{a}||\mathbf{b}|sin\alpha$, where α is the angle between these two vectors. But from the definition of the cross product, $|\mathbf{a} \times \mathbf{b}| = |\mathbf{a}||\mathbf{b}|sin\alpha$. Thus, the area of the infinitesimal parallelogram is given by

$$dS(u, v) = |\frac{\partial \mathbf{r}}{\partial u} \times \frac{\partial \mathbf{r}}{\partial v}|dudv$$

The direction of the normal to the surface at the point defined by the parameter pair (u, v) is the same as $\frac{\partial \mathbf{r}}{\partial u} \times \frac{\partial \mathbf{r}}{\partial v}$. Thus, if $\mathbf{B}(\mathbf{r}, t)$ is the magnetic field, we have at the point (u, v),

$$\mathbf{B}.\mathbf{n}dS = \mathbf{B}(\mathbf{r}(u, v), t).\frac{\partial \mathbf{r}(u, v)}{\partial u} \times \frac{\partial \mathbf{r}(u, v)}{\partial v}dudv$$

and the magnetic flux as a function of time can be expressed as the integral of this quantity:

$$\Phi_B(t) = \int \mathbf{B}(\mathbf{r}(u, v), t).\frac{\partial \mathbf{r}(u, v)}{\partial u} \times \frac{\partial \mathbf{r}(u, v)}{\partial v}dudv$$

Faraday' law states that the magnitude of the voltage induced through the loop equals $|\frac{d\Phi_B(t)}{dt}|$ and the direction is governed by the left hand rule.

Study project 19

Illustrating the computation of potential difference between two points due to a surface charge distribution, one can present as a preliminary step to the computation of the capacitance of a parallel plate capacitor, the computation of the potential difference between two points due to a constant surface charge density on a plane. The computation can be based on direct integration as well as numerical evaluation. For example, consider charge distributed uniformly with a surface density of σ on the plane $y = 0, 0 < x < a, -L < z < L$. The potential at the point $(x, y, 0)$ due to this surface charge density is given by

$$V(x, y, 0) = \frac{\sigma}{4\pi\epsilon} \int_0^a \int_{-L}^{L} \frac{dX\,dZ}{\sqrt{(x - X)^2 + y^2 + Z^2}}$$

This potential is important while evaluating the capacitance between two rectangular plates. One divides the length of the plates into intervals of constant width w and assumes constant surface charge density in each section. Call these as $\sigma_1, ..., \sigma_k$. With d denoting the distance between the two plates, one can using the above integral formula, evaluate the potential difference between each pair of opposite sections to the charge on each of the sections. For example the potential difference between the first pair will have the form $\sigma_1 V_{11} + \sigma_2 V_{12} + \sigma_3 V_{13} + ... + \sigma_k V_{1k}$. This is set equal to V_0, the potential difference between the plates. In general, the potential difference between the ith pair will have the form $\sigma_1 V_{i1} + \sigma_2 V_{i2} + ... + \sigma_k V_{ik}$. This is also set equal to V_0 since the potential difference between the plates is a constant. Thus we end up with the system of equations

$$V_0 = \sum_{j=1}^{k} \sigma_j V_{ij}, i = 1, 2, ..., k$$

which can be solved for the charge densities $\{\sigma_j\}$ in the different sections. These are summed up, multiplied by the area of each section and divided by V_0 to obtain a corrected version of the formula for the capacitance.

Study project 20

Illustrating the computation of potential difference between two points due to a surface charge distribution, one can present as a preliminary step to the computation of the capacitance of a parallel plate capacitor, the computation of the potential difference between two points due to a constant surface charge density on a plane. The computation can be based on direct integration as well as numerical evaluation. For example, consider charge distributed uniformly with a surface density of σ on the plane $y = 0, 0 < x < a, -L < z < L$. The potential at the point $(x, y, 0)$ due to this surface charge density is given by

$$V(x, y, 0) = \frac{\sigma}{4\pi\epsilon} \int_0^a \int_{-L}^{L} \frac{dX\,dZ}{\sqrt{(x - X)^2 + y^2 + Z^2}}$$

This potential is important while evaluating the capacitance between two rectangular plates. One divides the length of the plates into intervals of constant width w and assumes constant surface charge density in each section. Call these as $\sigma_1, ..., \sigma_k$. With d denoting the distance between the two plates, one can using the above integral formula, evaluate the potential difference between each pair of opposite sections to the charge on each of the sections. For example the potential difference between the first pair will have the form $\sigma_1 V_{11} + \sigma_2 V_{12} + \sigma_3 V_{13} + ... + \sigma_k V_{1k}$. This is set equal to V_0, the potential difference between the plates. In general, the potential difference between the ith pair will have the form $\sigma_1 V_{i1} + \sigma_2 V_{i2} + ... + \sigma_k V_{ik}$. This is also set equal to V_0 since the potential difference between the plates is a constant. Thus we end up with the system of equations

$$V_0 = \sum_{j=1}^{k} \sigma_j V_{ij}, i = 1, 2, ..., k$$

which can be solved for the charge densities $\{\sigma_j\}$ in the different sections. These are summed up, multiplied by the area of each section and divided by V_0 to obtain a corrected version of the formula for the capacitance.

Study project 21

Explain how one can study using first order perturbation theory the scattered wave field in the Maxwell equations when the permitivity has a small spatio-temporally varying term. The problem can be viewed as scattering of an incident wave field in a medium with varying permitivity.

Ans: Analysis of bilinear systems using perturbation theory has been done in great detail by experts in control engineering and signal processing. However, most of the bilinear systems considered in the literature deals only with abstract systems defined either via difference or differential equations. Very few applications to practical problems have been considered. In this paper, we look at the problem of obtaining expansions for the electric and magnetic fields in a medium in which the permitivity and permeability vary with time but the time varying components are small in amplitude so that they can be regarded as first order perturbations to the Maxwell equations in free space. Such problems arise for example in the ionosphere where a propagating electromagnetic wave tends to ionize the medium slightly and one has to describe the propagation in a medium having a complex refractive index. We are able to obtain a Volterra expansion for the electromagnetic field. We also look at the problem of a field dependent refractive index, a situation that is more common and use perturbation theory to analyze the effect of the non-linearity.

Description of a general abstract bilinear system

By the term bilinear system, one means that the output of the system couples to the driving input bilinearly and contributes to the forcing term. A common example is: Motion of a forced spring-mass system with damping with the spring constant and damping constant having a time varying component. The equation of motion is given by

$$\frac{d^2x(t)}{dt^2} + 2(\gamma_0 + \gamma_1(t))\frac{dx(t)}{dt} + (k_0 + k_1(t))x(t) = F(t)$$

Here, $F(t)$ is the input, γ_0 is the constant part of the damping coefficient and $\gamma_1(t)$ is the time varying part of the damping coefficient. k_0 is the constant part of the spring constant and $k_1(t)$ is the time varying part of the spring constant. If one assumes that the processes $\{\xi_1(t)\}$ and $\{k_1(t)\}$ are within our control, then this system has three inputs or equivalently, a single vector input $[F(t), \xi_1(t), k_1(t)]^T$ and a single output $x(t)$. One can generalize this idea and consider a linear differential equation with time varying coefficients and regard the time varying component of each coefficient as an input to the system. Thus, the output $x(t)$ satisfies the differential equation

$$\frac{d^n x(t)}{dt^n} + (a_1 + \xi_1(t))\frac{d^{n-1}x(t)}{dt^{n-1}} + ... + (a_{n-1} + \xi_{n-1}(t))\frac{dx(t)}{dt} + (a_n + \xi_n(t))x(t) = F(t)$$

In this system, the vector $[\xi_1(t), ..., \xi_n(t), F(t)]^T$ is the vector valued input and $x(t)$ is the output. This defines a bilinear system since the coupling terms $\xi_j(t)\frac{d^{n-j}x(t)}{dt^{n-j}}, j = 1, 2, ..., n$ are bilinear in the input components and in the output. When one implements a system defined by a differential equation on a digital computer, a difference equation results after discretization and a bilinear system in continuous time transforms on discretization to a bilinear system in discrete time. The general equation has the form

$$x[n] + (a_1 + \xi_1[n])x[n-1] + ... + (a_{p-1} + \xi_{p-1}[n])x[n-p+1] + (a_p + \xi_p[n])x[n-p] = F[n]$$

Bilinear systems occur in a variety of other physical models like RLC circuits with R, L, C being controllable with time flow. By introducing state variables, the general bilinear system can be cast in the form of a system of first order linear differential or difference equations with time varying coefficients, i.e.,

$$\frac{d\mathbf{X}(t)}{dt} = \mathbf{A}_0\mathbf{X}(t) + \sum_{k=1}^{m} \xi_k(t)\mathbf{A}_k\mathbf{X}(t) = \mathbf{F}(t)$$

where the inputs are $\xi_1(t), ..., \xi_k(t), \mathbf{F}(t)$ and the output is a linear transformation of the state vector: $y(t) = \mathbf{c}^T\mathbf{X}(t)$. The analogous discrete time system is defined by the difference equation

$$\mathbf{X}[n] = \mathbf{A}_0\mathbf{X}[n] + \sum_{k=1}^{m} \xi_k[n]\mathbf{A}_k\mathbf{X}[n] + \mathbf{F}[n]$$

and $y[n] = \mathbf{c}^T\mathbf{X}[n]$.

Examples Perturbation Theoretic Analysis

Consider first the following elementary bilinear system

$$\frac{dx(t)}{dt} = \epsilon u(t)x(t) + ax(t) + u(t)$$

The parameter ϵ has been introduced into the bilinear coupling term in order to understand that it can be regarded as a small perturbation to the otherwise linear system described by the differential equation

$$\frac{dx(t)}{dt} = ax(t) + u(t)$$

To solve for the output, we assume that it can be expanded in powers of the small parameter ϵ, i.e.

$$x(t) = x(t, \epsilon) = \sum_{m=0}^{\infty} \epsilon^m x_m(t)$$

Substituting this into the differential equation and equating coefficients of equal powers of ϵ on both sides gives us the sequence of differential equations

$$\frac{dx_0(t)}{dt} = ax_0(t) + u(t) \qquad \frac{dx_m(t)}{dt} = u(t)x_{m-1}(t) + ax_m(t), m \geq 1$$

The solutions to the various correction terms $x_m(t), m = 0, 1, 2, ...$, are obtained sequentially:

$$x_0(t) = \int_0^t exp(-a(t - \tau))u(\tau)d\tau$$

$$x_m(t) = \int_0^t exp(-a(t - \tau))x_{m-1}(\tau)u(\tau)d\tau, m \geq 1$$

We can easily evaluate the first few terms:

$$x_1(t) = \int_0^t exp(-a(t - \tau_1))x_0(\tau_1)u(\tau_1)d\tau_1$$

$$= \int_{0<\tau_2<\tau_1<t} exp(-a(t - \tau_1))exp(-a(\tau_1 - \tau_2))u(\tau_1)u(\tau_2)d\tau_2 d\tau_1$$

$$= \int_{0<\tau_2<\tau_1<t} exp(-a(t - \tau_2))u(\tau_1)u(\tau_2)d\tau_2 d\tau_1$$

In general,

$$x_m(t) = \int_{0<\tau_{m+1}<...<\tau_1<t} exp(-a(t - \tau_{m+1}))u(\tau_1)...u(\tau_{m+1})d\tau_{m+1}...d\tau_1$$

This results in a Volterra expansion for the output. A general Volterra system is defined via the input output equation

$$x(t) = \sum_{m=1}^{\infty} \int_{-\infty}^{\infty} ... \int_{-\infty}^{\infty} h(t; \tau_1, ..., \tau_m)u(\tau_1)...u(\tau_m)d\tau_1...d\tau_m$$

Now, consider the following more general bilinear system

$$\frac{d\mathbf{X}(t)}{dt} - \mathbf{A}\mathbf{X}(t) - \epsilon\xi(t)\mathbf{B}\mathbf{X}(t) = \mathbf{F}(t)$$

This is of the more general state variable type considered earlier. The parameter ϵ has been introduced to facilitate perturbation theoretic analysis. We make a Taylor expansion of the state vector:

$$\mathbf{X}(t) = \mathbf{X}(t, \epsilon) = \sum_{m=0}^{\infty} \epsilon^m \mathbf{X}_m(t)$$

Substituting into the differential equation and equating coefficients of same powers of ϵ gives us once again a system of recursive state equations

$$\frac{d\mathbf{X}_0(t)}{dt} - \mathbf{A}\mathbf{X}_0(t) = \mathbf{F}(t)$$

$$\frac{d\mathbf{X}_m(t)}{dt} - \mathbf{A}\mathbf{X}_m(t) = \xi(t)\mathbf{B}\mathbf{X}_{m-1}(t), m \geq 1$$

which can be solved to obtain a perturbation series.

Maxwell's equations in a time varying medium as a bilinear system

We now look at the Maxwell equations in a medium having no external charge or current density but with the permitivity and permeability functions of position and time. We denote these quantities as $\epsilon(t, \mathbf{r})$ and $\mu(t, \mathbf{r})$. The equations are

$$\nabla.(\epsilon\mathbf{E}) = 0 \qquad \nabla \times \mathbf{E} = -\frac{\partial}{\partial t}(\mu\mathbf{H})$$

$$\nabla \times \mathbf{H} = \frac{\partial}{\partial t}(\epsilon\mathbf{E}) \qquad \nabla.(\mu\mathbf{H}) = 0$$

Take the curl of the second equation to get

$$\nabla(\nabla.\mathbf{E}) - \nabla^2\mathbf{E} = -\frac{\partial}{\partial t}(\nabla \times (\mu\mathbf{H}))$$

Now, the first Maxwell equation gives

$$\nabla\epsilon.\mathbf{E} + \epsilon\nabla.\mathbf{E} = 0$$

so that

$$\nabla.\mathbf{E} = -(\nabla(log(\epsilon)), \mathbf{E})$$

We also use

$$\nabla \times (\mu\mathbf{H}) = \nabla\mu \times \mathbf{H} + \mu\nabla \times \mathbf{H} = \nabla\mu \times \mathbf{H} + \mu\frac{\partial}{\partial t}(\epsilon\mathbf{E})$$

Thus, we end up with a modified wave equation for the electric field:

$$\nabla^2 \mathbf{E} - \frac{\partial}{\partial t}\left(\nabla \mu \times \mathbf{H} + \mu \frac{\partial}{\partial t}(\epsilon \mathbf{E})\right)$$

or

$$\nabla^2 \mathbf{E} + \nabla(\nabla(log\epsilon), \mathbf{E}) - \mu \frac{\partial^2}{\partial t^2}(\epsilon \mathbf{E}) - \frac{\partial}{\partial t}(\nabla \mu \times \mathbf{H}) = 0$$

When μ is a constant, we end up with a modified wave equation for the electric field alone which clearly depicts a bilinear coupling between the field and functions derived from the permitivity:

$$\nabla^2 \mathbf{E} + \nabla(\nabla(log(\epsilon), \mathbf{E}) - \mu \frac{\partial^2}{\partial t^2}(\epsilon \mathbf{E}) = 0$$

We shall analyze approximate solutions to this equation using perturbation theory. Assume that $\epsilon(t, \mathbf{r}) = \epsilon_0(1 + \chi(t, \mathbf{r}))$ where $\chi(t, \mathbf{r})$ is small in comparison with unity. Then,

$$log(\epsilon) = log(\epsilon_0) + log(1 + \chi) \approx log(\epsilon_0) + \chi$$

$$\nabla(log(\epsilon)) \approx \nabla\chi$$

and the above equation further approximates to

$$\nabla^2 \mathbf{E} + \nabla(\nabla\chi, \mathbf{E}) - \mu\epsilon_0 \frac{\partial^2 \mathbf{E}}{\partial t^2} - \mu\epsilon_0 \frac{\partial^2(\chi\mathbf{E})}{\partial t^2} = 0$$

This can be rearranged as

$$\nabla^2 \mathbf{E} - \frac{1}{c^2}\frac{\partial^2 \mathbf{E}}{\partial t^2} = -\nabla(\nabla\chi, \mathbf{E}) + \mu\epsilon_0 \frac{\partial^2}{\partial t^2}(\chi\mathbf{E})$$

The right side can be regarded as a perturbation to the wave equation satisfied by the electric field caused by varying refractive index of the medium. We want to determine the scattering of a wave caused by this perturbation to an incident propagating wave. So assume that the incident wave is given by

$$\mathbf{E}_1(t, \mathbf{r}) = Re(\mathbf{A}exp(i(\omega_0 t - \mathbf{k}_0.\mathbf{r}))$$

and the total wave field after scattering by the medium is given by

$$\mathbf{E}(t, \mathbf{r}) = \mathbf{E}_1(t, \mathbf{r}) + \mathbf{E}_2(t, \mathbf{r})$$

where $\mathbf{E}_2$ represents the scattered component. Then regarding the terms involving χ on the right side as being first order perturbations, we find on equating first order terms on both sides, the equation

$$\nabla^2 \mathbf{E}_2 - \frac{1}{c^2}\frac{\partial^2 \mathbf{E}_2}{\partial t^2} = -\nabla(\nabla\chi, \mathbf{E}_1) + \mu\epsilon_0 \frac{\partial^2}{\partial t^2}(\chi\mathbf{E}_1)$$

To develop the scattered wave field $\mathbf{E}_2$ into Fourier components, we first develop the right side of the above equation into Fourier modes. The first step is to Fourier analyze the susceptibility $\chi(t, \mathbf{r})$:

$$\chi(t, \mathbf{r}) = \int_{\mathbb{R}^4} \hat{\chi}(\omega, \mathbf{k}) exp(i(\omega t - \mathbf{k}.\mathbf{r})) d\omega d\mathbf{k}$$

Then, the various terms on the right side evaluate as follows:

$$(\nabla\chi, \mathbf{E}_1) = \int \hat{\chi}(\omega, \mathbf{k})(-i\mathbf{k}) exp(i(\omega t - \mathbf{k}.\mathbf{r})) d\omega d\mathbf{k}.(\frac{1}{2}\mathbf{A} exp(i(\omega_0 t - \mathbf{k}_0.\mathbf{r})) +$$

$$\frac{1}{2}\mathbf{A}^* exp(-i(\omega_0 t - \mathbf{k}_0.\mathbf{r})))$$

$$= -\frac{i}{2} \int \hat{\chi}(\omega, \mathbf{k})\mathbf{A}.\mathbf{k} exp(i(\omega + \omega_0)t - (\mathbf{k} + \mathbf{k}_0).\mathbf{r})) d\omega d\mathbf{k}$$

$$- \frac{i}{2} \int \hat{\chi}(\omega, \mathbf{k})\mathbf{A}^*.\mathbf{k} exp(i(\omega - \omega_0)t - (\mathbf{k} - \mathbf{k}_0).\mathbf{r})) d\omega d\mathbf{k}$$

$$\nabla(\nabla\chi, \mathbf{E}_1) = -\frac{1}{2} \int (\mathbf{k} + \mathbf{k}_0)\hat{\chi}(\omega, \mathbf{k})\mathbf{A}.\mathbf{k} exp(i(\omega + \omega_0)t - (\mathbf{k} + \mathbf{k}_0, \mathbf{r})) d\omega d\mathbf{k}$$

$$- \frac{1}{2} \int (\mathbf{k} - \mathbf{k}_0)\mathbf{A}^*.\mathbf{k}\hat{\chi}(\omega, \mathbf{k}) exp(i(\omega - \omega_0)t - (\mathbf{k} - \mathbf{k}_0, \mathbf{r})) d\omega d\mathbf{k}$$

Next,

$$\chi\mathbf{E}_1 = \int \hat{\chi}(\omega, \mathbf{k}) exp(i(\omega t - \mathbf{k}.\mathbf{r})) d\omega d\mathbf{k} Re(\mathbf{A} exp(i(\omega_0 t - \mathbf{k}_0.\mathbf{r})))$$

$$= \frac{\mathbf{A}}{2} \int \hat{\chi}(\omega, \mathbf{k}) exp(i(\omega + \omega_0)t - (\mathbf{k} + \mathbf{k}_0, \mathbf{r})) d\omega d\mathbf{k}$$

$$+ \frac{\mathbf{A}^*}{2} \int \hat{\chi}(\omega, \mathbf{k}) exp(i(\omega - \omega_0)t - (\mathbf{k} - \mathbf{k}_0, \mathbf{r})) d\omega d\mathbf{k}$$

Its second time derivative needs to be computed:

$$\frac{\partial^2}{\partial t^2})(\chi\mathbf{E}_1) = -\frac{\mathbf{A}}{2} \int \hat{\chi}(\omega, \mathbf{k})(\omega + \omega_0)^2 exp(i(\omega + \omega_0)t - (\mathbf{k} + \mathbf{k}_0, \mathbf{r}) d\omega d\mathbf{k}$$

$$- \frac{\mathbf{A}^*}{2} \int \hat{\chi}(\omega, \mathbf{k})(\omega - \omega_0)^2 exp(i(\omega - \omega_0)t - (\mathbf{k} - \mathbf{k}_0, \mathbf{r})) d\omega d\mathbf{k}$$

Combining the above results, it follows that the differential equation for the scattered field thus assumes the form

$$\nabla^2\mathbf{E}_2 - \frac{1}{c^2}\frac{\partial^2\mathbf{E}_2}{\partial t^2} =$$

$$\frac{1}{2} \int (\mathbf{k} + \mathbf{k}_0)\hat{\chi}(\omega, \mathbf{k})\mathbf{A}.\mathbf{k} exp(i(\omega + \omega_0)t - (\mathbf{k} + \mathbf{k}_0, \mathbf{r})) d\omega d\mathbf{k}$$

$$+\frac{1}{2}\int (\mathbf{k} - \mathbf{k}_0)\mathbf{A}^*.\mathbf{k}\hat{\chi}(\omega, \mathbf{k})exp(i(\omega - \omega_0)t - (\mathbf{k} - \mathbf{k}_0, \mathbf{r}))d\omega d\mathbf{k}$$

$$+\frac{\mathbf{A}}{2c^2}\int \hat{\chi}(\omega, \mathbf{k})(\omega + \omega_0)^2 exp(i(\omega + \omega_0)t - (\mathbf{k} + \mathbf{k}_0, \mathbf{r})d\omega d\mathbf{k}$$

$$+\frac{\mathbf{A}^*}{2c^2}\int \hat{\chi}(\omega, \mathbf{k})(\omega - \omega_0)^2 exp(i(\omega - \omega_0)t - (\mathbf{k} - \mathbf{k}_0, \mathbf{r}))d\omega d\mathbf{k}$$

Since we are assuming that the susceptibility function $\chi(t, \mathbf{r})$ comprises the purely varying part, it does not have any d.c. component. Thus, we can for all practical purposes assume that ω cannot assume the value zero. Thus the tricky situation of resonance occurring in the scattering process is avoided. Now we can evaluate the amplitude of each Fourier component in the scattered wave field by direct substitution into the wave equation. Specifically, the solution to the equation

$$\nabla^2\psi - \frac{1}{c^2}\frac{\partial^2\psi}{\partial t^2} = B.exp(i(\omega t - \mathbf{k}.\mathbf{r}))$$

has the form

$$\psi(t, \mathbf{r}) = C.exp(i(\omega t - \mathbf{k}.\mathbf{r}))$$

where

$$C = \frac{B}{\omega^2/c^2 - k^2}$$

Superposition implies that the solution to the equation

$$\nabla^2\psi - \frac{1}{c^2}\frac{\partial^2\psi}{\partial t^2} = \int B(\omega, \mathbf{k})exp(i(\omega t - \mathbf{k}.\mathbf{r}))d\omega d\mathbf{k}$$

is given by

$$\psi(t, \mathbf{r}) = \int \frac{B(\omega, \mathbf{k})}{\omega^2/c^2 - k^2}exp(i(\omega t - \mathbf{k}.\mathbf{r}))d\omega d\mathbf{k}$$

Applying this to the preceding expression, we can derive the Fourier expansion for the scattered wave field:

$$\mathbf{E}_2(t, \mathbf{r}) =$$

$$\frac{1}{2}\int \frac{(\mathbf{k} + \mathbf{k}_0)}{(\omega + \omega_0)^2/c^2 - |\mathbf{k} + \mathbf{k}_0|^2}\hat{\chi}(\omega, \mathbf{k})\mathbf{A}.\mathbf{k}exp(i(\omega + \omega_0)t - (\mathbf{k} + \mathbf{k}_0, \mathbf{r}))d\omega d\mathbf{k}$$

$$+\frac{1}{2}\int \frac{(\mathbf{k} - \mathbf{k}_0)}{(\omega - \omega_0)^2/c^2 - |\mathbf{k} - \mathbf{k}_0|^2}\hat{\chi}(\omega, \mathbf{k})\mathbf{A}^*.\mathbf{k}exp(i(\omega - \omega_0)t - (\mathbf{k} - \mathbf{k}_0, \mathbf{r}))d\omega d\mathbf{k}$$

$$+\frac{\mathbf{A}}{2c^2}\int \frac{\hat{\chi}(\omega, \mathbf{k})}{(\omega + \omega_0)^2/c^2 - |\mathbf{k} + \mathbf{k}_0|^2}(\omega + \omega_0)^2 exp(i(\omega + \omega_0)t - (\mathbf{k} + \mathbf{k}_0, \mathbf{r}))d\omega d\mathbf{k}$$

$$+\frac{\mathbf{A}^*}{2c^2}\int \frac{\hat{\chi}(\omega, \mathbf{k})}{(\omega - \omega_0)^2/c^2 - |\mathbf{k} - \mathbf{k}_0|^2}(\omega - \omega_0)^2 exp(i(\omega - \omega_0)t - (\mathbf{k} - \mathbf{k}_0, \mathbf{r}))d\omega d\mathbf{k}$$

Study project 22

Curve fitting in the Maxwell equations. Suppose that we want to determine the permitivity of space from measurements of the charge density and the electrostatic potential. The exact equation is given by

$$V(\mathbf{R}) = \sum_{m=1}^{N} \frac{Q_m}{4\pi\epsilon|\mathbf{R} - \mathbf{r}_m|}$$

where the charge distribution is assumed to be discrete. ϵ is obtained by taking measurements of the potential $V_1, ..., V_N$ at the points $\mathbf{R}_1, ..., \mathbf{R}_N$ and minimizing $\sum_{j=1}^{N}(V_j - V(\mathbf{R}_j))^2$ with respect to the parameter ϵ, i.e.

$$\frac{d}{d\epsilon} \sum_{j=1}^{N}(V_j - V(\mathbf{R}_j))^2 = 0$$

We leave the problem of writing down the exact normal equations for this problem

Study project 23

Consider a conductor of varying cross section. Let $\mathbf{E}$ denote the electric field in it. It is assumed to be a constant across each cross section of the conductor. The potential difference between its two ends is given by the line integral $\int \mathbf{E}.d\mathbf{l}$ with the line integral being along the length of the conductor. The current flowing through the conductor is $\sigma \int \mathbf{E}.d\mathbf{S}$ with the surface integral being along the cross section. The ratio of the two gives the resistance of the conductor:

$$R = \frac{\int \mathbf{E}.d\mathbf{l}}{\sigma \int_S \mathbf{E}.d\mathbf{S}}$$

From measurements of the resistance, we want to estimate the electric field inside the conductor. How can this problem be formulated as a least squares problem ? From the above expression, it is clear that

$$R = \int_\Gamma \frac{dl}{\int_S \sigma dS}$$

a general formula valid even in the case when the conductivity σ varies across the cross section and along the length of the conductor. This formula can be taken as the starting point for estimating the conductivity σ within the conductor from measurements of the total resistance between two points in along the conductor length. We choose three coordinates u, v, l where u, v are along the cross section of the conductor, i.e., they specify where the point is on the cross section and l is taken along the length of the conductor. Then, the conductivity σ is defined as a function of these three coordinates $\sigma(u, v, l)$, this can be called the "conductivity field". The resistance of the conductor is then expressible as

$$R(L, S) = \int_0^L \frac{dl}{\int_S \sigma(u, v, l)\,du\,dv}$$

We let L and S vary, measure $R(L, S)$ for each pair and then try to estimate from this data $\sigma(u, v, l)$.

Study project 24

Suppose that there is a force field $\mathbf{F}(\mathbf{r}) = F_x(x, y)\hat{x} + F_y(x, y)\hat{y}$ in the plane. We want to estimate this force field by measuring the change in the kinetic energy of a particle moving along a path that can be varied.

Study project 25

Integration can be used to determine the electrostatic force between two charge distributions. If $\rho(\mathbf{r})$ denotes the density of the first distribution and $\rho(\mathbf{r}')$ the density of the second distribution, the force exerted by the first distribution on the second is given by the formula

$$\mathbf{F} = \frac{1}{4\pi} \int \frac{\rho(\mathbf{r})\rho(\mathbf{r}')(\mathbf{r}' - \mathbf{r})}{|\mathbf{r} - \mathbf{r}'|^{3/2}}\,dV\,dV'$$

When the charge distributions are symmetric, this integral can be evaluated or at least, simplified considerably. For example, consider a ring of charge of radius a having uniform linear charge density σ over it. The ring is placed in the xy plane with its centre coinciding with the origin. Likewise, we have another ring of radius b having uniform linear charge density σ'. This ring has its plane parallel to the xy plane and its centre is at $(0, 0, d)$. The force between the two rings is directed along the $\hat{z}$ axis. The z component of the electric field due to the first ring is given by

$$E_z(x, y, z) = \int \frac{\sigma.a.z.d\phi'}{((x - a.cos\phi')^2 + (y - a.sin\phi')^2 + z^2)^{3/2}}$$

or in terms of cylindrical coordinates

$$E_z(\rho, \phi, z) = \int_0^{2\pi} \frac{\sigma.a.z.d\phi'}{(\rho^2 + a^2 + z^2 - 2a\rho.cos(\phi' - \phi))^{3/2}}$$

The result is clearly independent of ϕ. It coincides with the expression

$$E_z(\rho, z) = \int_0^{2\pi} \frac{\sigma.a.z.d\phi'}{(\rho^2 + a^2 + z^2 - 2a.\rho.cos\phi')^{3/2}}$$

The force exerted by the first ring on the second has only a z component which is given by

$$F_z = \int_0^{2\pi} \sigma'.b.d\phi' E_z(b,d) = 2\pi b \sigma'.E_z(b,d)$$

One can develop a spherical harmonic expansion for the force. TMS assembly language programmes can be developed for computing and tabulating the Legendre polynomials which are important in providing us with an expansion for the potentials and fields. The Legendre polynomial $P_n(x)$ satisfies the differential equation $(1-x^2)f''(x) - 2xf'(x) + n(n+1)f(x) = 0$. Numerical recipes for solving this differential equation via a discretization scheme can be devised. An alternate way is to directly solve the eigenvalue problems $(1-x^2)f''(x) - 2xf'(x) - Ef(x) = 0$ in the range $[-1,1]$ subject to the boundary conditions $f(0) = f(1) = 0$. This differential equation can be transformed into a matrix eigenvalue problem and the eigenvalues of the matrix determined by means of a programme that computes the determinant and makes a search for the minimum. When the above differential equation is transformed into a matrix eigenvalue problem of the form $(\mathbf{A} - c\mathbf{I})\mathbf{f} = 0$, we need to evaluate $det(\mathbf{A} - c\mathbf{I})$ for various choices of c and locate the minimum. The computation of the determinant of a matrix $\mathbf{A}$ is based upon the expression

$$det(\mathbf{A}) = \sum_\sigma sgn(\sigma) A(1,\sigma 1) A(2,\sigma 2)...A(n,\sigma n)$$

the sum being over all permutations σ of the indices $\{1,2,...,n\}$. The idea is to first of all tabulate all the permutations of $\{1,2,...,n\}$ and store each permuted sequence in a string of consecutive n locations in the memory. Each permutation is extracted by reading the consecutive memory locations and the corresponding product $A(1,\sigma 1)...A(n,\sigma n)$ is evaluated. Actually, we along with each permutation σ, we also have to store its signature $sgn(\sigma)$ in a memory location adjacent to the n consecutive locations in which the permutation $\sigma 1,...,\sigma n$ has been stored.

Study project 26

Finite element methods for the solution of the waveguide problem. $\Pi(x,y)$ represents the axially directed electric field or magnetic field vector in the waveguide. For the TE mode, Π will correspond to the z component of the magnetic field vector and satisfies the Neumann boundary condition $\frac{\partial \Pi}{\partial \mathbf{n}} = 0$ where $\mathbf{n}$ is the unit normal on the walls of the waveguide. For the TM mode Π will correspond to the z component (axial component) of the electric field and the boundary condition that is appropriate is $\Pi = 0$ on the walls of the waveguide. (These conditions represent the vanishing of the tangential component of the electric field and the normal component of the magnetic field at the boundary of the waveguide). Π satisfies the Helmholtz equation which is $\nabla^2 \Pi + k^2 \Pi = 0$ within the guide. In terms of the components,

$$\frac{\partial^2 \Pi(x,y)}{\partial x^2} + k^2 \Pi(x,y) = 0$$

for (x, y) varying within the waveguide. The Helmholtz equation can be derived from a variational principle that corresponds to the minimization of the functional

$$\mathcal{F} = \int_S (|\nabla \phi|^2 - k^2 \phi^2) dx dy$$

Carrying out the variation gives us

$$\delta \mathcal{F} = 2 \int_S (\nabla \phi, \delta \nabla \phi) dx dy + k^2 \int_S \phi \delta \phi dx dy = 0$$

We integrate the first term by parts and use to get

$$2 \int_S \nabla . (\delta \phi \nabla \phi) dx dy - 2 \int_S \delta \phi \nabla^2 \phi dx dy - 2k^2 \int_S \phi \delta \phi dx dy = 0$$

The first integral transforms to a line integral around a contour that encircles any one cross section of the waveguide. This line integral vanishes if we assume the Dirichlet boundary condition that ϕ equals zero on the waveguide boundary. Noting that the variation $\delta \phi$ within the guide volume can be arbitrary, the vanishing of the remaining term implies that ϕ satisfies the Helmholtz equation

$$\nabla^2 \phi + k^2 \phi = 0$$

The finite element method consists in dividing the cross sectional surface area of the guide into smaller polygonal regions and within each polygonal region, we assume that ϕ is a linear function of its values at the polygonal vertices, i.e. $\phi(x, y) = \sum_{i=1}^n \phi_i \alpha_i(x, y)$. This value of ϕ is substituted into the expression for the functional to be minimized leading to the following expression

$$\mathcal{F} = F(\phi_1, ..., \phi_n) = \int_S (|\sum_{i=1}^n \phi_i \nabla \alpha_i|^2 - k^2 (\sum \phi_i \alpha_i)^2) dx dy$$

$$= \sum_{i,j=1}^n \phi_i \phi_j \int_S (\nabla \alpha_i, \nabla \alpha_j) dx dy - k^2 \sum_{i,j=1}^n \phi_i \phi_j \int_S \alpha_i \alpha_j dx dy$$

The optimal minimizing equations are

$$\frac{\partial F}{\partial \phi_i} = 0, i = 1, 2, ..., n$$

and these lead to the following generalized eigenvalue equations for the parameters $\{\phi_i\}$

$$\sum_{j=1}^n S_{ij} \phi_j = k^2 \sum_{j=1}^n T_{ij} \phi_j$$

where

$$S_{ij} = \int_S (\nabla \alpha_i, \nabla \alpha_j) dx dy$$

$$T_{ij} = \int_S \alpha_i \alpha_j dx dy$$

Note that the generalized eigenvalue k^2 is not determined apriori. The variational principle being adopted corresponds to the minimization of $\int_S (\nabla \phi, \nabla \phi) dx dy$ subject to the constraint that the energy $\int_S \phi^2 dx dy$ is kept fixed. An example of this method is provided by dividing the region S into triangular elements. Consider a particular triangle. Denote its three vertices by $(x_i, y_i), i = 1, 2, 3$. Within the triangle, we assume that $\phi(x, y)$ varies linearly with (x, y), i.e.,

$$\phi(x, y) = a + bx + cy$$

The value of ϕ at the vertices is given, i.e. $\phi(x_i, y_i) = \phi_i, i = 1, 2, 3$. This gives us three linear equations for the three variables a, b, c:

$$a + bx_i + cy_i = \phi_i, i = 1, 2, 3$$

In matrix form,

$$\begin{pmatrix} 1 & x_1 & y_1 \\ 1 & x_2 & y_2 \\ 1 & x_3 & y_3 \end{pmatrix} \begin{pmatrix} a \\ b \\ c \end{pmatrix}$$

$$= \begin{pmatrix} \phi_1 \\ \phi_2 \\ \phi_3 \end{pmatrix}$$

The resulting interpolation can be shown to be given by the expression

$$\phi(x, y) = \frac{1}{2A} \sum_{j=1}^{3} (a_j + b_j x + c_j y) \phi_j$$

where

$$a_j = x_{j+1} y_{j+2} - x_{j+2} y_{j+1}$$

$$b_j = y_{j+2} - y_{j+1}$$

$$c_j = x_{j+1} - x_{j+2}$$

with the indices being taken modulo 3. A represents the triangle area. To see this, we apply Cramer's rule to the above matrix equation and successively solve for a, b, c:

$$a = \frac{\det \begin{pmatrix} \phi_1 & x_1 & y_1 \\ \phi_2 & x_2 & y_2 \\ \phi_3 & x_3 & y_3 \end{pmatrix}}{\det \begin{pmatrix} 1 & x_1 & y_1 \\ 1 & x_2 & y_2 \\ 1 & x_3 & y_3 \end{pmatrix}}$$

$$b = \frac{det \begin{pmatrix} 1 & \phi_1 & y_1 \\ 1 & \phi_2 & y_2 \\ 1 & \phi_3 & y_3 \end{pmatrix}}{det \begin{pmatrix} 1 & x_1 & y_1 \\ 1 & x_2 & y_2 \\ 1 & x_3 & y_3 \end{pmatrix}}$$

$$c = \frac{det \begin{pmatrix} 1 & x_1 & \phi_1 \\ 1 & x_2 & \phi_2 \\ 1 & x_3 & \phi_3 \end{pmatrix}}{det \begin{pmatrix} 1 & x_1 & y_1 \\ 1 & x_2 & y_2 \\ 1 & x_3 & y_3 \end{pmatrix}}$$

The denominator determinant here equals $2A$ where A is the triangle area. This can be seen by noting that

$$det \begin{pmatrix} 1 & x_1 & y_1 \\ 1 & x_2 & y_2 \\ 1 & x_3 & y_3 \end{pmatrix}$$

$$= det \begin{pmatrix} 0 & x_1 - x_2 & y_1 - y_2 \\ 0 & x_2 - x_3 & y_2 - y_3 \\ 1 & x_3 & y_3 \end{pmatrix}$$

$$= (x_1 - x_2)(y_2 - y_3) - (y_1 - y_2)(x_2 - x_3)$$

This is simply the cross product of the vectors $(x_1 - x_2, y_1 - y_2)$ and $(x_3 - x_2, y_3 - y_2)$ or equivalently of $\mathbf{r}_1 - \mathbf{r}_2$ and $\mathbf{r}_3 - \mathbf{r}_2$. From elementary vector algebra, this cross product is the area of the parallelogram subtended by the vectors $\mathbf{r}_1 - \mathbf{r}_2$ and $\mathbf{r}_3 - \mathbf{r}_2$ which is twice the triangle area.

Study project 27

Finite element formulation of problems of electrostatics. The potential inside a volume v bounded by a surface S on which the potential V is prescribed satisfies Poisson's equation $\nabla^2 V(\mathbf{r}) = -\rho(\mathbf{r})/\epsilon$ with $V(\mathbf{r})$ on the surface s a definite function $\psi(\mathbf{r})$. Consider the problem of minimizing the functional

$$S(V) = \frac{\epsilon}{2} \int_v |\nabla V|^2 d^3\mathbf{r} - \int \rho.V d^3\mathbf{r}$$

The variational principle $\delta S(V) = 0$ leads to the equation

$$\epsilon \int <\nabla V, \delta \nabla V> d^3\mathbf{r} - \int \rho \delta V d^3\mathbf{r} = 0$$

Since $\delta \nabla V = \nabla \delta V$, we get on integration by parts,

$$\epsilon \int \nabla.(\delta V \nabla V) d^3\mathbf{r} - \epsilon \int \nabla^2 V \delta V d^3\mathbf{r} - \int \rho \delta V d^3\mathbf{r} = 0$$

The first integral here transforms to a surface integral which evaluates to zero since the variation δV on the surface s is zero. If the above equation is therefore to be valid for arbitrary variations δV within v, we must then have the validity of the Poisson equation:

$$\nabla^2 V = -\rho/\epsilon$$

within v. Thus minimization of the above functional subject to the prescribed boundary conditions on V leads to the solution of the electrostatics problem. We adopt a finite element approach to the problem by considering an expansion of the potential inside v in terms of test functions $\phi_i(\mathbf{r}), i = 1, 2, ..., N$:

$$V(\mathbf{r}) = \sum_{i=1}^{N} c_i \phi_i(\mathbf{r})$$

We assume that the boundary conditions are that $V = 0$ on the boundary s. This means automatically that the above expansion satisfies the boundary conditions. Then,

$$\frac{\epsilon}{2} \int_v |\nabla V|^2 d^3\mathbf{r} = \frac{\epsilon}{2} \sum_{i,j=1}^{N} c_i c_j \int_v (\nabla \phi_i(\mathbf{r}), \nabla \phi_j(\mathbf{r}) d^3\mathbf{r}$$

$$\int_v ho(\mathbf{r}) V(\mathbf{r}) d^3\mathbf{r} = \sum_{i=1}^{N} c_i \int_v \rho(\mathbf{r}) \phi_i(\mathbf{r}) d^3\mathbf{r}$$

or equivalently letting

$$a_{ij} = \epsilon \int_v (\nabla \phi_i(\mathbf{r}) hi_j(\mathbf{r}) d^3\mathbf{r}$$

and

$$b_i = \int_v \rho(\mathbf{r}) \phi_i(\mathbf{r}) d^3\mathbf{r}$$

the approximate solution to the electrostatics problem can be found by minimizing

$$F(c_1, ..., c_N) = \frac{1}{2} \sum_{i,j=1}^{N} a_{ij} c_i c_j - \sum_{i=1}^{N} b_i c_i$$

The optimal equations $\frac{\partial F}{\partial c_i} = 0$ yield the following linear equations for $\{c_i\}$:

$$\sum_{j=1}^{N} a_{ij} c_j = b_i, i = 1, 2, ..., N$$

Study project 28

Partial differential equations arising in electromagnetic theory

The first problem involves computing the fields as a function of time in free space when the fields at time $t = 0$ are specified. In continuous time, this evolution of the field is described by the two Maxwell equations

$$\frac{\partial \mathbf{B}}{\partial t} = -\nabla \times \mathbf{E}$$

$$\frac{\partial \mathbf{E}}{\partial t} = c^2 \nabla \times \mathbf{B}$$

The spatio-temporal discretization of these two equations can be achieved via a Taylor expansion. This aspect should be discussed. It leads to linear state variable equations. The other problem is that given the charge and current densities, we want to evaluate the electromagnetic potentials $V, \mathbf{A}$. These are obtained as solutions to the wave equation with source:

$$\Delta V - \frac{1}{c^2}\frac{\partial^2 V}{\partial t^2} = -\rho/\epsilon$$

$$\Delta \mathbf{A} - \frac{1}{c^2}\frac{\partial^2 \mathbf{A}}{\partial t^2} = -\mathbf{J}/\epsilon$$

Direct spatio temporal discretization of these equations will lead to discrete time state equations for the potentials. The other option is to consider the actual solution to these equations in terms of the retarded potentials

$$V(t, \mathbf{r}) = \int \frac{\rho(\mathbf{r}', t - |\mathbf{r} - \mathbf{r}'|/c)}{4\pi\epsilon|\mathbf{r} - \mathbf{r}'|} d^3\mathbf{r}'$$

$$\mathbf{A}(t, \mathbf{r}) = \int \frac{\mathbf{J}(\mathbf{r}', t - |\mathbf{r} - \mathbf{r}'|/c)}{4\pi\epsilon c^2|\mathbf{r} - \mathbf{r}'|} d^3\mathbf{r}'$$

and to discretize these integrals. The electromagnetic field itself is obtained by differentiating these equations with respect to space and time. The numerical computation of these fields should also be discussed in the lecture for accelerating charged particles and for sinusoidal current sources. For sinusoidal sources, we have the following frequency domain version of the above equations:

$$V(\mathbf{r}) = \int \rho(\mathbf{r}') \frac{exp(-ik|\mathbf{r} - \mathbf{r}'|)}{4\pi\epsilon|\mathbf{r} - \mathbf{r}'|} d^3\mathbf{r}'$$

$$\mathbf{A}(\mathbf{r}) = \int \mathbf{J}(\mathbf{r}') \frac{exp(-ik|\mathbf{r} - \mathbf{r}'|)}{4\pi\epsilon c^2|\mathbf{r} - \mathbf{r}'|} d^3\mathbf{r}'$$

where the charge density is given by the equation $Re(\rho(\mathbf{r})exp(i\omega t))$ and the current density by the equation $Re(\mathbf{J}(\mathbf{r})exp(i\omega t))$. The above equations define the potentials in phasor form, i.e., the true potentials are given by $Re(V(\mathbf{r})exp(i\omega t))$ and $Re(\mathbf{A}(\mathbf{r})exp(i\omega t))$. The fields are given by

$$\mathbf{B} = \nabla \times \mathbf{A}$$

$$i\omega\epsilon\mu\mathbf{E} = \nabla \times \mathbf{B} = \nabla(\nabla.\mathbf{A})) - \nabla^2\mathbf{A}$$
$$= \nabla(\nabla.\mathbf{A}) + k^2\mathbf{A}$$

or

$$\mathbf{E} = -\frac{ic^2}{\omega}\nabla(\nabla.\mathbf{A}) + k^2\mathbf{A}$$

Once the potentials $V(t,\mathbf{r})$ and $\mathbf{A}(t,\mathbf{r})$ have been computed by discretizing the integrals involving the retarded potentials, we can approximate the above partial derivatives defining the fields in terms of the potentials by discretizing the partial derivatives. Specifically, the following method can be adopted: Let $\mathbf{F}(t,\mathbf{r})$ be a vector field. Its discretized version is

$$\mathbf{F}[n,k,m,l] = \mathbf{F}(n\tau, k\Delta, m\Delta, l\Delta)$$

where τ is the time step size and Δ is the spatial step size. Its partial derivatives can be approximated by finite differences with respect to the integer variables n, k, m, l. For example, $\frac{\partial \mathbf{F}}{\partial t}$ is approximated by

$$\frac{\mathbf{F}[n+1,k,m,l] - \mathbf{F}[n,k,m,l]}{\tau}$$

This is to be interpreted as the vector

$$\hat{x}\frac{F_x[n+1,k,l,m] - F_x[n,k,l,m]}{\tau} + \hat{y}\frac{F_y[n+1,k,l,m] - F_y[n,k,l,m]}{\tau}$$

$$+\hat{z}\frac{F_z[n+1,k,l,m] - F_z[n,k,l,m]}{\tau}$$

A similar argument goes for the partial derivatives with respect to the spatial variables. In the talk, you should also explain how the wave equation with source term can be solved numerically using the multivariate discrete Fourier transform.

Study project 29

Determine the shape of a region V of space filled with charge of uniform density from measurements on the electric field at different points. Assume that the region to start with is a cuboid with faces $x = 0, a, y = 0, b, z = 0, c$ and that we are taking measurements of the electric field at points $(x,y,z) \in E$ where E is a region of space. The electric field can be expressed as

$$E_x(x,y,z|a,b,c) = \frac{\rho}{4\pi\epsilon}\int_0^a\int_0^b\int_0^c \frac{x-X}{((x-X)^2+(y-Y)^2+(z-Z)^2)^{3/2}}dX\,dY\,dZ$$

$$E_y(x,y,z|a,b,c) = \frac{\rho}{4\pi\epsilon}\int_0^a\int_0^b\int_0^c \frac{y-Y}{((x-X)^2+(y-Y)^2+(z-Z)^2)^{3/2}}dX\,dY\,dZ$$

$$E_z(x,y,z|a,b,c) = \frac{\rho}{4\pi\epsilon}\int_0^a\int_0^b\int_0^c \frac{z-Z}{((x-X)^2+(y-Y)^2+(z-Z)^2)^{3/2}}dX\,dY\,dZ$$

We take measurements of the field at the points $(x_i, y_i, z_i), i = 1, 2, ..., N$ and from these measurements, try to estimate a, b, c. By adopting a discretization scheme, we have obtain the following approximate equations

$$E_x(x, y, z | a, b, c) \approx \frac{\rho}{4\pi\epsilon} \sum_{k,m,n=0}^{N-1} \frac{x - ka/N}{((x - ka/N)^2 + (y - mb/N)^2 + (z - nc/N)^2))^{3/2}} abc/N^3$$

$$E_y(x, y, z | a, b, c) \approx \frac{\rho}{4\pi\epsilon} \sum_{k,m,n=0}^{N-1} \frac{y - mb/N}{((x - ka/N)^2 + (y - mb/N)^2 + (z - nc/N)^2))^{3/2}} abc/N^3$$

$$E_z(x, y, z | a, b, c) \approx \frac{\rho}{4\pi\epsilon} \sum_{k,m,n=0}^{N-1} \frac{z - nc/N}{((x - ka/N)^2 + (y - mb/N)^2 + (z - nc/N)^2)^{3/2}} abc/N^3$$

These can be regarded as equations for the unknowns a, b, c in terms of the electric field values at different points. A least squares approach can then be formulated for determining these parameters.

Study project 30

Implementation on the 8085 microprocessor of the computation of the radiation fields produced by a two dimensional sinusoidal current density source. The vector potential is given according to a previous problem by the formula

$$\mathbf{A}(\rho, \phi) = \frac{i\mu}{4} \int_0^\infty \int_0^{2\pi} \mathbf{J}(\rho', \phi') H_0(k\sqrt{\rho^2 + \rho'^2 - 2\rho\rho'.cos(\phi - \phi')})\rho' d\rho' d\phi'$$

The variable ρ' varies over a length a, where a is the order of magnitude of the linear dimensions of the current source. The radiation zone approximation amounts to looking at the fields at distances ρ much larger than both the wavelength of the radiation as well as the linear dimensions of the region, i.e. $k\rho >> 1$ and $\rho >> a$. For $|x| >> 1$, we have the following asymptotic approximation for the Hankel function:

$$H_0(x) \approx \sqrt{\frac{2}{\pi x}} exp(-i\pi/4) exp(ix)$$

Since $k\rho >> 1$ and $\rho >> \rho'$, we have $k|\rho - \rho'| >> 1$ and hence we obtain the approximation

$$H_0(k|\rho - \rho'|) \approx \sqrt{\frac{2}{\pi k|\rho - \rho'|}} exp(-i\pi/4) exp(ik|\rho - \rho'|)$$

Now $\rho >> \rho'$ implies that

$$|\rho - \rho'| \approx \rho - \rho'.cos(\phi - \phi')$$

so that we obtain the following radiation zone approximation to the distant magnetic vector potential:

$$\mathbf{A}(\rho,\phi) \approx \frac{i\mu}{4}\sqrt{\frac{2}{\pi k \rho}}exp(-i\pi/4)\int \mathbf{J}(\rho',\phi')exp(-ik(\rho-\rho'.cos(\phi-\phi')))\rho'd\rho'd\phi'$$

This can be expressed as

$$\mathbf{A}(\rho,\phi) \approx C.\frac{exp(-ik\rho)}{\sqrt{\rho}}\int \mathbf{J}(\rho',\phi')exp(ik\rho'.cos(\phi-\phi'))\rho'd\rho'd\phi'$$

where C is a constant. It is an interesting exercise to compute the radiation zone electric and magnetic fields and hence the Poynting vector and the total power radiated by the current source. To implement this formula and plot the radiation fields for a specified current density phasor $\mathbf{J}(\rho,\phi)$, we've got to discretize the above double integral. This amounts to replacing the integral by the finite sum

$$\sum_{n=0}^{N-1}\sum_{m=0}^{K-1}\mathbf{J}(n\Delta,m\delta)exp(ikn\Delta.cos(\phi-m\delta))n\Delta^2\delta$$

where $N\Delta$ is of the order of the linear dimensions of the source current field and $\delta = 2\pi/K$.

Study project 31

The potential field in space due to a discrete distribution of charges with Q_k located at $\mathbf{r}_k = (x_k, y_k, z_k), k = 1, 2, ..., n$ is given by the Coulomb formula

$$V(x,y,z) = \sum_{k=1}^{n}\frac{Q_k}{\sqrt{(x-x_k)^2+(y-y_k)^2+(z-z_k)^2}}$$

Assume that the positions fluctuate randomly with the position of the kth charge being $\mathbf{r}_k + \delta\mathbf{r}_k$, where $\delta\mathbf{r}_k$ is small, i.e. of the first order of smallness. Then, upto first order fluctuations, the fluctuation in the potential is given by

$$\delta V(x,y,z) = \sum_{k=1}^{n}\frac{\partial V}{\partial x_k}\delta x_k + \frac{\partial V}{\partial y_k}\delta y_k + \frac{\partial V}{\partial z_k}\delta z_k$$

$$= \sum_{k=1}^{n}E_{xk}\delta x_k + E_{yk}\delta y_k + E_{zk}\delta z_k$$

where

$$E_{xk} = \frac{\partial V}{\partial x_k} = \frac{Q_k(x-x_k)}{|\mathbf{r}-\mathbf{r}_k|^3}$$

is the x-component of the electric field at (x, y, z) produced by the charge Q_k. Grouping the fluctuations $\delta x_k, \delta y_k, \delta z_k, k = 1, 2, ..., n$ into a single vector $\delta \mathbf{X}$, this formula can be expressed as

$$\delta V = \sum_{k=1}^{n} a_k \delta X_k$$

with the $a_k's$ being position dependent. The fluctuation in the potential at some other point $\mathbf{r}'$ is given by

$$\delta V' = \sum_{k=1}^{n} a_k' \delta X_k$$

The variances and correlations of the potential fluctuations are given by

$$\mathbb{E}(\delta V)^2 = \sum_{k,m=1}^{n} a_k a_m r_{km}$$

$$\mathbb{E}((\delta V).\delta V') = \sum_{k,m=1}^{n} a_k a_m' r_{km}$$

where $r_{km} = \mathbb{E}(\delta X_k \delta X_m)$. In the special case when the positions are uncorrelated i.e.

$$\mathbb{E}\delta x_k^2 = \sigma_{xk}^2, \mathbb{E}\delta y_k^2 = \sigma_{yk}^2, \mathbb{E}\delta z_k^2 = \sigma_{zk}$$

$$\mathbb{E}(\delta x_k \delta x_m) = 0 = \mathbb{E}(\delta y_k \delta y_m) = \mathbb{E}(\delta z_k \delta z_m)$$

for $k \neq m$ and

$$\mathbb{E}(\delta x_k \delta y_m) = 0 = \mathbb{E}(\delta y_k \delta z_m) = \mathbb{E}(\delta z_k \delta x_m)$$

for all k, m, we have the nice formula

$$\mathbb{E}(\delta V)^2 = \sum_{k=1}^{n} \sigma_{xk}^2 E_{xk}^2 + \sigma_{yk}^2 E_{yk}^2 + \sigma_{zk}^2 E_{zk}^2$$

$$\mathbb{E}((\delta V)(\delta V')) = \sum_{k=1}^{n} \sigma_{xk}'^2 E_{xk} E_{xk}' + \sigma_{yk}'^2 E_{xk}' E_{yk}' + \sigma_{zk}^2 E_{zk} E_{zk}'$$

Study project 32

A charge Q has radius R that is fixed. Q is however unknown and is distributed according to the Gibbs distribution, i.e. its density is proportional to $exp(-\beta W(Q))$, where $W(Q)$ is the energy in the electrostatic field produced by the charge Q. Determine the moments of the charge Q, in particular its mean and variance.

The electric field in space is given by $E(r) = \frac{Q}{4\pi\epsilon r^2}$ if $r > R$ and by $E(r) = Qr/4\pi\epsilon R^3$ if $r \leq R$. The field energy is found by integrating $\epsilon E^2/2$ over all of space:

$$W(Q) = \frac{\epsilon}{2}\int_0^R E(r)^2 4\pi r^2 dr + \frac{\epsilon}{2}\int_R^\infty E(r)^2 4\pi r^2 dr$$

This has the form $W(Q) = \alpha.Q^2/R$ where α is a numerical constant. We leave the evaluation of α to the interested student. Thus, the Gibbs distribution implies that the density of Q has the form $f(Q) = C.exp(-\alpha Q^2/R)$. This is a normal distribution with a mean of zero and a variance of $R/2\alpha$. The average charge $< Q >$ equals zero and the variance $< Q^2 >$ equals $R/2\alpha$. The higher moments can be computed from the general formula

$$\frac{1}{\sqrt{2\pi}}\int_{-\infty}^\infty x^{2n}exp(-x^2/2)dx = \frac{2}{\sqrt{2\pi}}\int_0^\infty x^{2n}exp(-x^2/2)dx$$

$$= (\frac{2}{\pi})^{1/2}\int_0^\infty (2y)^n exp(-y)dy/\sqrt{2y} = (\frac{2}{\pi})^{1/2}2^{n-1/2}\Gamma(n+1/2)$$

Study project 33

Consider a one dimensional electrostatics problem in which we have a linear charge density $\sigma(x)$ along the line. Let $V(x)$ denote the potential produced by this charge distribution. Assuming $\sigma(x)$ to be random with a mean of zero and an autocorrelation function of $\mathbb{E}(\sigma(x)\sigma(y)) = R_\sigma(x,y)$, determine the autocorrelation $\mathbb{E}(V(x)V(y))$ of the potential distribution function. Also assuming the linear charge density $\{\sigma(x)\}$ to be a normal one dimensional random field, determine the conditional density of $V(x)$ given $\sigma(x)$. This is of course, equivalent to determining the conditional mean $\mathbb{E}(V(x)|\sigma(x))$ and the conditional variance $\mathbb{E}(V(x)^2|\sigma(x))-\mathbb{E}(V(x)|\sigma(X))^2$. These two latter statistical quantities can be determined by using the fact that if X,Y is a pair of jointly normal random variables with mean vector (μ_x, μ_y) and covariance matrix

$$\begin{pmatrix} \sigma_x^2 & \sigma_{xy} \\ \sigma_{xy} & \sigma_y^2 \end{pmatrix}$$

then $\mathbb{E}(X|Y) = \mu_x + \frac{\sigma_{xy}}{\sigma_y^2}(Y - \mu_y)$ and $Var(X|Y) = \sigma_x^2 - \sigma_{xy}^2/\sigma_y^2$. Thus to evaluate the conditional density of $V(x)$ given $\sigma(x)$, we need only evaluate the correlations $\mathbb{E}(V(x)), \mathbb{E}(V(x)^2), \mathbb{E}(V(x)\sigma(x))$ and $\mathbb{E}\sigma(x)^2$. The Poisson equation $V''(x) = -\sigma(x)/\epsilon$ has the solution

$$V(x) = -\int_{-\infty}^x dy \int_{-\infty}^y \sigma(z)dz/\epsilon = -\int_{-\infty<z<y<x} \sigma(z)dzdy/\epsilon$$

$$= -\int_{-\infty}^x (x - z)\sigma(z)dz/\epsilon$$

Assuming that $V(x) \to 0$ as $x \to -\infty$. This shows that the potential field $\{V(x)\}_{x \in \mathbb{R}}$ is also a Gaussian field. Its mean is zero and its autocorrelation is given by

$$R_{VV}(x, y) = \mathbb{E}(V(x)V(y)) = \int_{-\infty}^{x} \int_{-\infty}^{y} (x - u)(y - v) R_\sigma(u, v) du dv$$

$R_{VV}(x, x)$ equals the variance of $V(x)$ and

$$R_{V\sigma}(x, y) = \mathbb{E}(V(x)\sigma(y)) = - \int_{-\infty}^{x} (x - z) \mathbb{E}(\sigma(z)\sigma(y)) dz/\epsilon$$

$$= - \int_{-\infty}^{x} (x - z) R_\sigma(z, y) dz/\epsilon$$

Study project 34

Consider the problem of determining the electric field produced by an electrostatic potential $V(t, \mathbf{r})$ and a magnetic vector potential $\mathbf{A}(t, \mathbf{r})$. The required relationship is given by the equation

$$\mathbf{E}(t, \mathbf{r}) = -\nabla V(t, \mathbf{r}) - \frac{\partial \mathbf{A}(t, \mathbf{r})}{\partial t}$$

The Lorentz Gauge condition yields the scalar potential in terms of the vector potential:

$$-\frac{1}{c^2} \frac{\partial V(t, \mathbf{r})}{\partial t} = \nabla . \mathbf{A}$$

or equivalently,

$$V = -c^2 \int_0^t dt \nabla . \mathbf{A}$$

The electric field can thus be expressed as

$$\mathbf{E} = -\frac{\partial \mathbf{A}}{\partial t} + c^2 \int_0^t \nabla(\nabla . \mathbf{A}) dt$$

Assume that the sources of current have sinusoidal variation with a frequency of ω. Then, the electric field as given by the above equation has the following phasor representation:

$$\mathbf{E} = -j\omega \mathbf{A} + (c^2/j\omega)\nabla(\nabla . \mathbf{A})$$

Now assume that $\mathbf{A}(\mathbf{r})$ is a random Gaussian field. This will happen when the sources of current, i.e. current density is a random Gaussian field with respect to the spatial variables. Let the correlation between the field components be given by the equation

$$\mathbb{E}(A_\alpha(\mathbf{r})A_\beta(\mathbf{r}')^*) = R_{\alpha\beta}(\mathbf{r}, \mathbf{r}')$$

We want now to calculate the correlations between the electric field phasor components at two points in space. Note that the above formula can be expressed as

$$E_\alpha = -j\omega A_\alpha + (c^2/j\omega)\frac{\partial}{\partial x_\alpha}(\nabla.\mathbf{A})$$

$$= -j\omega A_\alpha + (c^2/j\omega)\sum_{\beta=1}^{3}\frac{\partial^2 A_\beta}{\partial x_\alpha x_\beta}$$

The field correlations at two different points are then evaluated as follows:

$$< E_\alpha(\mathbf{r}).E_\beta(\mathbf{r}')^* >$$

$$= \omega^2 < A_\alpha(\mathbf{r})A_\beta(\mathbf{r}') > +(c^4/\omega^2)\sum_{\rho,\sigma=1}^{3}\frac{\partial^4}{\partial x_\alpha \partial x_\rho' \partial x_\beta \partial x_\sigma'} < A_\rho(\mathbf{r})A_\sigma(\mathbf{r}') >$$

$$-c^2\sum_{\rho=1}^{3}\frac{\partial^2}{\partial x_\beta \partial x_\rho} < A_\alpha(\mathbf{r})A_\rho(\mathbf{r}') >$$

$$-c^2\sum_{\rho=1}^{3}\frac{\partial^2}{\partial x_\alpha' \partial x_\rho'} < A_\rho(\mathbf{r}')A_\beta(\mathbf{r}) >$$

Study project 35

In studying the motion of a plasma, we are sometimes interested in the individual trajectories of the charged particles in a time varying electromagnetic field where the electric field has a sinusoidal variation but is constant in space and the magnetic field is constant in space as well as in time. The equation of motion of the particle has the form

$$m\frac{d\mathbf{v}}{dt} = q(\mathbf{E}_0 exp(i\omega t) + \mathbf{v} \times \mathbf{B})$$

where $\mathbf{E}_0$ and $\mathbf{B}_0$ are constant vectors. We decompose the velocity into a harmonically time varying component with frequency ω_c and a component that may vary with time but does not have the frequency ω_c (all terms with frequency ω_c have been absorbed into the first component). Thus, $\mathbf{v}(t) = \mathbf{v}_m(t) + \mathbf{v}_e exp(i\omega t)$, where $\mathbf{v}_e$ is a constant vector, it can be complex in general. Plugging this into the equation of motion gives

$$m(\frac{d\mathbf{v}_m}{dt} + i\omega\mathbf{v}_e exp(i\omega t)) = q(\mathbf{E}_0 exp(i\omega t) + \mathbf{v}_m \times \mathbf{B} + \mathbf{v}_e \times \mathbf{B}exp(i\omega t))$$

Equate terms on both sides that do not vary with the frequency ω and terms that do vary separately. This gives us two vector equations:

$$m\frac{d\mathbf{v}_m}{dt} = q\mathbf{v}_m \times \mathbf{B}$$

$$i\omega m\mathbf{v}_e = q\mathbf{E}_0 + q\mathbf{v}_e \times \mathbf{B}$$

The first equation corresponds to circular motion with frequency $\omega_c = qB/m$ about the direction of the magnetic field. This component is $\mathbf{v}_m(t)$. The second equation is an algebraic equation for the vector $\mathbf{v}_e$. It can be solved by decomposing $\mathbf{v}_e$ into a component parallel to $\mathbf{B}$ and a component perpendicular to $\mathbf{B}$: $\mathbf{v}_e = \mathbf{v}_{e,par} + \mathbf{v}_{e,perp}$. Plugging this into the equation gives

$$i\omega m(\mathbf{v}_{e,par} + \mathbf{v}_{e,perp}) = q(\mathbf{E}_{0,par} + \mathbf{E}_{0,perp} + \mathbf{v}_{e,perp} \times \mathbf{B})$$

Equating the parallel component on both sides gives

$$i\omega m\mathbf{v}_{e,par} = q\mathbf{E}_{0,par}$$

or

$$\mathbf{v}_{e,par} = -\frac{iq}{m\omega}\mathbf{E}_{0,par}$$

Equating the perpendicular component gives

$$i\omega m\mathbf{v}_{e,perp} = q\mathbf{E}_{0,perp} + q\mathbf{v}_{e,perp} \times \mathbf{B}$$

Take cross product on both sides with $\mathbf{B}$ to get

$$i\omega m\mathbf{B} \times \mathbf{v}_{e,perp} = q\mathbf{B} \times \mathbf{E}_{0,perp} + qB^2\mathbf{v}_{e,perp}$$

or

$$\mathbf{v}_{e,perp} \times \mathbf{B} = (iq/m\omega)\mathbf{B} \times \mathbf{E}_{0,perp} + (iq/m\omega)B^2\mathbf{v}_{e,perp}$$

Substitute this into the previous equation to get

$$i\omega m\mathbf{v}_{e,perp} = q\mathbf{E}_{0,perp} + (iq^2/m\omega)\mathbf{B} \times \mathbf{E}_{0,perp} + (iq^2/m\omega)B^2\mathbf{v}_{e,perp}$$

so that

$$\mathbf{v}_{e,perp} = \frac{q\mathbf{E}_{0,perp} + (iq^2/m\omega)\mathbf{B} \times \mathbf{E}_{0,perp}}{i\omega m(1 - q^2B^2/m^2\omega^2)}$$

It is clear that the expression for $\mathbf{v}_{e,perp}$ displays resonance. When $\omega = qB/m$, the cyclotron frequency, there is a sudden blow up in the amplitude of $\mathbf{v}_{e,perp}$.

Study project 36

Finite element methods can be used to solve the scattering problem occurring in electromagnetic field theory. Consider a surface S bounding a volume V. Assume that an incident wave field $\mathbf{E}_i, \mathbf{H}_i$ impinges upon the surface from the outside. Currents and charges are induced on the surface. Let $\mathbf{J}_s, \rho_s$ denote the surface current density and surface charge density respectively induced on the surface. These induced surface currents and charges produce a reflected component $\mathbf{E}_r, \mathbf{H}_r$ outside S and a transmitted component $\mathbf{E}_t, \mathbf{H}_t$ in the region enclosed by S. The field outside S

is given by the sum of the incident and reflected fields. To calculate the reflected component, we note that the vector potential outside S is given by the expression

$$\mathbf{A}(\mathbf{r}) = \frac{\mu_0}{4\pi} \int_S \mathbf{J}_s(\mathbf{r}') \frac{exp(-jk|\mathbf{r} - \mathbf{r}'|)}{|\mathbf{r} - \mathbf{r}'|} dS'$$

The fields are assumed to be of a definite frequency $\omega = kc$. We can use the Green's function

$$G(\mathbf{r}|\mathbf{r}') = \frac{\mu_0}{4\pi} exp(-jk|\mathbf{r} - \mathbf{r}'|)/|\mathbf{r} - \mathbf{r}'|$$

Then,

$$\mathbf{A}(\mathbf{r}) = \int \mathbf{J}_s(\mathbf{r}') G(\mathbf{r}|\mathbf{r}') dS'$$

The Lorentz gauge condition $\nabla . \mathbf{A} = -(j\omega/c^2)V$ gives $V = (jc^2/\omega)\nabla . \mathbf{A}$, so that

$$V(\mathbf{r}) = (jc^2/\omega) \int \mathbf{J})_s(\mathbf{r}') . \nabla G(\mathbf{r}|\mathbf{r}') dS'$$

and

$$\mathbf{E}(\mathbf{r}) = -\nabla V(\mathbf{r}) - j\omega \mathbf{A}(\mathbf{r}) = -(jc^2/\omega)\nabla \int \mathbf{J}_s(\mathbf{r}') . \nabla G(\mathbf{r}|\mathbf{r}') dS' - i\omega \int \mathbf{J}_s(\mathbf{r}') G(\mathbf{r}|\mathbf{r}') dS'$$

This can be expressed in component form as

$$E_{r,a}(\mathbf{r}) = \sum_{b=1}^{3} \int_S J_{sb}(\mathbf{r}') H_{ab}(\mathbf{r}|\mathbf{r}') dS'$$

where the functions $H_{ik}(\mathbf{r}|\mathbf{r}')$ are expressible in terms of $G(\mathbf{r}|\mathbf{r}')$. We're assuming that the surface is a conducting surface. The region to the exterior of S is free space having permitivity ϵ_0 and permeability μ_0 and the region to the interior of S is a dielectric having permitivity ϵ and permeability μ. The field inside the surface S can likewise be computed using the same method but with μ_0 replaced by μ and ϵ_0 replaced by ϵ, i.e. c^2 replaced by $1/\mu\epsilon$. The resulting functions H_{ik} get replaced by functions H'_{ik}. Thus, the field inside the surface has the form

$$\mathbf{E}_{t,a}(\mathbf{r}) = \sum_{b=1}^{3} \int_S J_{sb}(\mathbf{r}') H_{ab}(\mathbf{r}|\mathbf{r}') dS'$$

An integral equation for the surface current can be obtained by equating the tangential components of the electric field on the surface. Specifically, if $\mathbf{n}$ denotes the unit normal at a point $\mathbf{r}$ on the surface, then the boundary condition to be applied is

$$\mathbf{n} \times (\mathbf{E}_i(\mathbf{r} + \mathbf{E}_r(\mathbf{r})) = \mathbf{n} \times \mathbf{E}_t(\mathbf{r})$$

The terms $\mathbf{E}_r(\mathbf{r})$ and $\mathbf{E}_t(\mathbf{r})$ are linear functions of the surface current and the source $\mathbf{E}_i(\mathbf{r})$ is known to us. The solution to this integral equation gives us the surface current density. This integral equation is also known as Pocklington's integral equation.

Study project 37

This project deals with the numerical simulation of charged particle motions and plotting of field lines.

Basic Equations describing charged particle motions

Equation of motion of a charge in an electrostatic field

$$\frac{d^2x(t)}{dt^2} = \frac{e}{m}E(x(t))$$

Electric field is derivable from a potential function: $E(x) = -V'(x)$. Then, equation of motion can be integrated to give

$$\frac{m}{2}(\frac{dx}{dt})^2 + eV(x) = C$$

where C is a constant equal to the total energy of the particle. The equation of motion can be cast as a first order differential equation

$$\frac{d}{dt}\begin{pmatrix} x(t) \\ v(t) \end{pmatrix} = \begin{pmatrix} v(t) \\ eE(x(t)) \end{pmatrix}$$

One can more generally consider a system of n coupled first order ordinary differential equations of the form

$$\frac{d\xi_i(t)}{dt} = F_i(\xi_1(t), ..., \xi_n(t)), i = 1, 2, ..., n$$

which in vectorial form can be expressed as

$$\frac{d\xi(t)}{dt} = \mathbf{F}(\xi(t))$$

where $\xi(t)$ is a vector in $\mathbb{R}^n$ and $\mathbf{F}$ is a smooth map from $\mathbb{R}^n \rightarrow \mathbb{R}^n$. The numerical simulation is based on the Taylor expansion

$$\xi(t + \Delta) = \sum_{n=0}^{\infty} \Delta^n \xi^{(n)}(t)/n!$$

For a charged particle moving in an electromagnetic field, the equations of motion are

$$\frac{d\mathbf{r}}{dt} = \mathbf{v} \qquad \frac{d\mathbf{v}}{dt} = \frac{e}{m}(\mathbf{E}(t, \mathbf{r}(t)) + \mathbf{v}(t) \times \mathbf{B}(t, \mathbf{r}(t)))$$

which in component form correspond to the following system of six coupled first order ordinary differential equations

$$\frac{dx}{dt} = v_x, \; \frac{dy}{dt} = v_y, \; \frac{dz}{dt} = v_z$$

$$\frac{dv_x}{dt} = \frac{e}{m}(E_x + v_y B_z - v_z B_y), \; \frac{dv_y}{dt} = \frac{e}{m}(E_y + v_z B_x - v_x B_z), \; \frac{dv_z}{dt} = \frac{e}{m}(E_z + v_x B_y - v_y B_x)$$

The use of Taylor expansion in simulating this motion shall be discussed in the talk. Another important ordinary differential equation that one comes across in electromagnetic field theory is that involving the plotting of field lines. If the electric field in space is given in cartesian form as

$$\mathbf{E}(x, y, z) = E_x(x, y, z)\hat{x} + E_y(x, y, z)\hat{y} + E_z(x, y, z)\hat{z}$$

then the field lines are obtained as solutions to the following system of three coupled ordinary differential equations:

$$\frac{dx}{E_x(x, y, z)} = \frac{dy}{E_y(x, y, z)} = \frac{dz}{E_z(x, y, z)}$$

This can be viewed as the system

$$\frac{dx}{d\lambda} = E_x(x, y, z)$$

$$\frac{dy}{d\lambda} = E_y(x, y, z)$$

$$\frac{dz}{d\lambda} = E_z(x, y, z)$$

The field lines are plotted by replacing this system by a system of difference equations

$$x_{n+1} = x_n + \Delta.E_x(x_n, y_n, z_n)$$

$$y_{n+1} = y_n + \Delta.E_y(x_n, y_n, z_n)$$

$$z_{n+1} = z_n + \Delta.E_z(x_n, y_n, z_n)$$

This corresponds to a first order Taylor approximation. In the talk, second order Taylor approximations will should be discussed. We can look at field line plotting in other coordinate systems like the cylindrical and spherical polar systems or more generally in an arbitrary curvilinear system. The ordinary differential equations for the field lines for each system are derived from the expression of the line element. In the cylindrical system, for example, the line element is given by

$$d\mathbf{r} = \hat{\rho}d\rho + \hat{\phi}\rho d\phi + \hat{z}dz$$

while in the spherical polar system, the line element is given by

$$d\mathbf{r} = \hat{r}dr + \hat{\theta}rd\theta + \hat{\phi}.r.sin\theta$$

The corresponding differential equations are

$$\frac{d\rho}{E_\rho} = \frac{\rho d\phi}{E_\phi} = \frac{dz}{E_z}$$

and

$$\frac{dr}{E_r} = \frac{r d\theta}{E_\theta} = \frac{r sin\theta d\phi}{E_\phi}$$

Examples of field plotting that should be discussed in the lecture are field produced by (1) two equal and opposite charges Q and $-Q$ separated by a distance of d metres, (2) a disc of radius R carrying uniform surface charge density σ,(3) A solid cylinder of charge of radius R and length L with the volume charge density being ρ, (4) Truncated multipole expansion of a charge distribution.

Study project 38

Computing the energy of a distribution of dipoles in an electric field and numerically evaluating this energy. One way to compute is to note that the field energy is given by

$$U = \frac{1}{2} \int \mathbf{D}.\mathbf{E}dV$$

Setting $\mathbf{D} = \epsilon_0 \mathbf{E} + \mathbf{P}$ gives

$$U = U_{field} + U_{dipoles}$$

where

$$U_{field} = \frac{\epsilon_0}{2} \int E^2 dv$$

is the energy of the field and

$$U_{dipoles} = \frac{1}{2} \int \mathbf{P}.\mathbf{E}dv$$

gives the total energy due to interaction between the field with the dipoles and the dipoles with themselves. We want to derive this formula by direct method. If a dipole $\mathbf{p}$ is placed in a field $\mathbf{E}$, there will be an expansion or a contraction of the dipole and this causes the dipole to acquire an internal energy. The stretching $\mathbf{x}$ is proportional to the field $\mathbf{x} = \alpha\mathbf{E}$ and hence the internal energy is given by $q \int \mathbf{E}.d\mathbf{x}$ where q is the charge on the dipole. This evaluates to

$$q \int \mathbf{E}.\alpha d\mathbf{E} = q\alpha|\mathbf{E}|^2/2 = q\mathbf{x}.\mathbf{E}/2 = \mathbf{p}.\mathbf{E}/2$$

This is the internal energy stored in the dipole. We now look at the energy el interaction between the dipole and the external field. If $-q, q$ are the two charges separated by the vector $\mathbf{d}$, then

$\mathbf{p} = q\mathbf{d}$ is its dipole moment and assuming the two charges to be located at $-\mathbf{d}/2$ and $\mathbf{d}/2$, we get for the field-dipole interaction energy the expression $-qV(-\mathbf{d}/2) + qV(\mathbf{d}/2)$. Using a Taylor expansion upto first order for the potential, this expression for the interaction energy approximates to $q\mathbf{d}.\nabla V(0) = -\mathbf{p}.\mathbf{E}$, where the field is evaluated at the origin. We next look at the interaction energy of each pair of dipoles. Assume that dipoles $\mathbf{p}_1, ..., \mathbf{p}_N$ are located at the points $\mathbf{r}_1, ..., \mathbf{r}_N$. There is an external field which we denote by $\mathbf{E}_{ext}$ as well as an internal field $\mathbf{E}_{int}$ which is produced by the dipoles. The energy stored in the dipoles due to stretching or contracting is given by $W_1 = \frac{1}{2}\sum_i \mathbf{p}_i.(\mathbf{E}_{ext}(i) + \mathbf{E}_{int}(i))$. The energy stored due to the interaction of each dipole with the field at its site is $-\sum_i \mathbf{p}_i.(\mathbf{E}_{ext}(i) + \mathbf{E}_{int}(i))$. However if we see carefully, this is the sum of two components: $-\sum_i \mathbf{p}_i.\mathbf{E}_{ext}(i)$ which represents the interaction energy of the dipoles with the external field and $-\sum_i \mathbf{p}_i.\mathbf{E}_{int}(i)$ which represents the interaction energy of each dipole with the field due to the other dipoles. This last term can be seen to be the sum of interaction energies between each dipole pair. A double counting is taking place here. We are counting in this term a component of involving the interaction of the ith dipole with the field due to the jth dipole and the interaction of the jth dipole with the field due to the ith dipole. Effectively the pairwise interaction energy is being counted twice. In order to avoid this, we divide the last term by a factor of two. Thus The total energy of the system of the dipoles in the external field with the energy of the external field excluded is given by

$$W = W_1 + W_2 + W_3$$

where $W_1 = \frac{1}{2}\sum \mathbf{p}_i.(\mathbf{E}_{ext}(i) + \mathbf{E}_{int}(i))$ is the energy stored in the dipoles due to stretching or contracting. $W_2 = -\sum_i \mathbf{p}_i.\mathbf{E}_{ext}(i)$ is the energy of interaction of the system of dipoles with the external field and $W_3 = -\frac{1}{2}\sum_i \mathbf{p}_i.\mathbf{E}_{int}(i)$ is the energy of mutual interaction of the dipoles. The sum equals

$$W = -\frac{1}{2}\sum_i \mathbf{p}_i.\mathbf{E}_{ext}(i)$$

Study project 39

Assembly language implementation of the computation of electric field and potential fields in space due to a specified charge density, plotting of electric field lines of force and equipotentials. The potential due to a distribution of charge having density $\rho(x, y, z)$ in the region V of space is given by the equation

$$V(X, Y, Z) = \frac{1}{4\pi\epsilon_0} \int_V \frac{\rho(x, y, z)}{\sqrt{(X - x)^2 + (Y - y)^2 + (Z - z)^2}} dx\,dy\,dz$$

The electric field is given by the gradient of this function

$$\mathbf{E}(X, Y, Z) = -\nabla V = -\hat{x}\frac{\partial V}{\partial X} - \hat{y}\frac{\partial V}{\partial y} - \hat{z}\frac{\partial V}{\partial Z}$$

The project will involve programming a TMS card to compute these two field quantities and plot the field lines and equipotential surfaces. The types of charge distribution that we shall consider are (1) spherically symmetric charge distributions $\rho(r)$, i.e., the charge density depends only upon the distance from the origin (2) Cylindrically symmetric charge distribution $\rho(\rho)$, i.e. the charge density depends only upon the distance from the $\hat{z}$ axis, (3) Fields produced by a disc of radius R having uniform surface charge density, (4) Fields produced by a thin wire coinciding with the $\hat{z}$ axis and having a specified linear charge density $\sigma(z)$ along its length. The programs will involve discretization of the integral, approximation of the integral by a finite sum over pixels into which the volume V has been partitioned.

Study project 40

We want to investigate the motion of a charged particle in a static magnetic field having a small gradient. This is important in plasma physics. If $\mathbf{B}(\mathbf{r})$ is the magnetic field, we make a Taylor expansion about the origin and retain only first order correction terms, i.e. terms involving only the first order derivatives of $\mathbf{B}$ with respect to spatial location. Thus,

$$\mathbf{B}(\mathbf{r}) \approx \mathbf{B}(0) + \mathbf{r}.\nabla\mathbf{B}(0)$$

$$= (B_x(0) + x\frac{\partial B_x(0)}{\partial x} + y\frac{\partial B_x(0)}{\partial y} + z\frac{\partial B_x(0)}{\partial z})\hat{x}$$

$$+ (B_y(0) + x\frac{\partial B_y(0)}{\partial x} + y\frac{\partial B_y(0)}{\partial y} + z\frac{\partial B_y(0)}{\partial z})\hat{y}$$

$$+ (B_z(0) + x\frac{\partial B_z(0)}{\partial x} + y\frac{\partial B_z(0)}{\partial y} + z\frac{\partial B_z(0)}{\partial z})\hat{z}$$

The constant part $\mathbf{B}(0)$ can be regarded as being of zeroth order of smallness and the spatially varying part $\mathbf{r}.\nabla\mathbf{B}(0)$ as being of the first order of smallness. In the equation of motion

$$m\frac{d\mathbf{v}}{dt} = q\mathbf{v} \times \mathbf{B}$$

we substitute $\mathbf{v} = \mathbf{v}_0(t) + \mathbf{v}_1(t)$, where $\mathbf{v}_0(t)$ is of zeroth order of smallness and $\mathbf{v}_1(t)$ is of the first order of smallness. Then equating terms of equal orders of smallness, we find that

$$\frac{d\mathbf{v}_0(t)}{dt} = \frac{q}{m}\mathbf{v}_0(t) \times \mathbf{B}(0)$$

$$\frac{d\mathbf{v}_1(t)}{dt} = \frac{q}{m}\mathbf{v}_0(t) \times (\mathbf{r}_0(t).\nabla)\mathbf{B}(0)$$

The first equation implies that the unperturbed position $\mathbf{r}_0(t) = \int_0^t \mathbf{v}_0(\tau)d\tau$ is a circular orbit around the constant component $\mathbf{B}(0)$ of the magnetic field vector. The second equation can then be integrated directly to obtain the perturbation $\mathbf{v}_1(t)$.

Study project 41

Study the motion of an electric dipole in an external electric field via computer simulations. The external field can vary in space as well as in time. Assume that this field has the form $\mathbf{E}(t, \mathbf{r})$. Let $\mathbf{p}$ denote the dipole moment of the dipole whose motion is being investigated. The energy of interaction between the dipole and the electric field has the form $U(t, \mathbf{r}) = -\mathbf{p}.\mathbf{E}(t, \mathbf{r})$. Its negative gradient gives the force on the dipole: $\mathbf{F}(t, \mathbf{r}) = -\nabla U(t, \mathbf{r})$. In terms of components, this is

$$F_x = p_x \frac{\partial E_x}{\partial x} + p_y \frac{\partial E_y}{\partial x} + p_z \frac{\partial E_z}{\partial x}$$

$$F_y = p_x \frac{\partial E_x}{\partial y} + p_y \frac{\partial E_y}{\partial y} + p_z \frac{\partial E_z}{\partial y}$$

$$F_z = p_x \frac{\partial E_x}{\partial z} + p_y \frac{\partial E_y}{\partial z} + p_z \frac{\partial E_z}{\partial z}$$

With m denoting the mass of the dipole, its equation of motion in the external field is given by

$$m \frac{d^2 \mathbf{r}}{dt^2} = \mathbf{F}(t, \mathbf{r})$$

This equation is simulated on the digital computer using the process of discretization.

Study project 42

The study of partial derivatives is very important in problems of mathematical physics. Specifically, the kinds of equations one comes across in physical modeling are (1) the wave equation in one dimension $\frac{\partial^2 u(t,x)}{\partial t^2} - c^2 \frac{\partial^2 u(t,x)}{\partial x^2} = 0$. This describes the motion of a vibrating string. (2) the wave equation for a circular membrane

$$\frac{\partial^2 u(t, r, \phi)}{\partial t^2} = c^2 \left(\frac{1}{r} \frac{\partial}{\partial r} \left(r \frac{\partial u}{\partial r} \right) + \frac{1}{r^2} \frac{\partial^2 u}{\partial \phi^2} \right)$$

and the three dimensional wave equation in spherical polar coordinates:

$$\frac{\partial^2 u(t, r, \theta, \phi)}{\partial t^2} = c^2 \left(\frac{1}{r} \frac{\partial^2}{\partial r^2} (ru) + \frac{1}{r^2 sin\theta} \frac{\partial}{\partial \theta} \left(sin\theta \frac{\partial u}{\partial \theta} \right) + \frac{1}{r^2 sin^2(\theta)} \frac{\partial^2 u}{\partial \phi^2} \right)$$

The last equation arises when we try to solve the wave equation for the propagation of an electromagnetic wave under various kinds of symmetry conditions, like spherical symmetry, azimuthal symmetry etc. Usually, one adopts the separation of variables method leading to the use of special functions like spherical harmonics and Bessel's functions. However one can equivalently look at a moment method for solving these equations. This involves assuming that the wave can be expanded as a linear combination of a collection of functions $\psi_n(r, \theta, \phi)$ (considering the last example), i.e. $u(t, r, \theta, \phi) = \sum_{n=1}^{N} c_n(t)\psi_n(r, \theta, \phi)$. or in short, $u(t, \mathbf{x}) = \sum_{n=1}^{N} c_n(t)\psi_n(\mathbf{x})$. Let Δ denote the Laplacian operator. Then, plugging this into the wave equation gives us

$$\sum_{n=1}^{N} c_n''(t)\psi_n(\mathbf{x}) = \sum_{n=1}^{N} c_n(t)\Delta\psi_n(\mathbf{x})$$

This equation can in practice never be satisfied since the functions ψ_n may not be appropriate for the problem. However, assuming that these are satisfied, we can derive from these an auxiliary set of equations for the coefficients $\{c_n(t)\}$ that satisfy a set of linear coupled ordinary differential equations. These follow by multiplying both sides by $\psi_m(\mathbf{x})$ and integrating both sides over the domain D where the wave amplitude is assumed to be dominant. This results in

$$\sum_{n=1}^{N} c_n''(t) \int_{D} \psi_m(\mathbf{x})\psi_n(\mathbf{x})d\mathbf{x} = \sum_{n=1}^{N} c_n(t) \int_{D} \psi_m(\mathbf{x})\Delta\psi_n(\mathbf{x})d\mathbf{x}$$

Introduce the $N \times N$ matrix $\mathbf{A}$ having entries

$$a_{mn} = \int_{D} \psi_m(\mathbf{x})\psi_n(\mathbf{x})d\mathbf{x}$$

and the matrix $\mathbf{B}$ having entries

$$b_{mn} = \int_{D} \psi_m(\mathbf{x})\Delta\psi_n(\mathbf{x})d\mathbf{x}$$

Then the dynamics for the coefficient $\mathbf{c}(t) = [c_1(t), ..., c_N(t)]^T$ assumes the form

$$\mathbf{A}\mathbf{c}''(t) = \mathbf{B}\mathbf{c}(t)$$

or

$$\mathbf{c}''(t) = \mathbf{A}^{-1}\mathbf{B}\mathbf{c}(t)$$

which can be easily solved using eigen-techniques.

Study project 43

Before introducing the Maxwell equations, a thorough grinding the theory of the partial derivative must be introduced. This can be presented as a separate seminar. If $u(x, y)$ is a function of two variables, then its partial derivatives at the point (x, y) are defined by the limits

$$\frac{\partial u(x, y)}{\partial x} = lim_{\epsilon \to 0} \frac{u(x + \epsilon, y) - u(x, y)}{\epsilon}$$

$$\frac{\partial u(x, y)}{\partial y} = lim_{\epsilon \to 0} \frac{u(x, y + \epsilon, y) - u(x, y)}{\epsilon}$$

The limits should not depend on whether ϵ approaches zero from the right or from the left. Suppose that we have a unit vector (n_x, n_y) in the plane and we wish to compute the derivative of u at the point (x, y) along this direction. We're assuming that both the partial derivatives of u exist at (x, y). The derivative of u at (x, y) along this direction is given by

$$\frac{\partial u(x, y)}{\partial \mathbf{n}} = lim_{\epsilon \to 0} \frac{u(x + \epsilon n_x, y + \epsilon n_y) - u(x, y)}{\epsilon}$$

This is the same as

$$lim_{\epsilon \to 0} \frac{u(x + \epsilon n_x, y + \epsilon n_y) - u(x, y + \epsilon n_y) + u(x, y + \epsilon n_y) - u(x, y)}{\epsilon}$$

$$= lim_{\epsilon \to 0} n_x \frac{u(x + \epsilon n_x, y + \epsilon n_y) - u(x, y + \epsilon n_y)}{n_x \epsilon}$$

$$+ n_y \frac{u(x, y + \epsilon n_y) - u(x, y)}{n_y \epsilon}$$

Assuming continuity of the partial derivatives at (x, y) this limit is easily seen to evaluate to

$$n_x \frac{\partial u(x, y)}{\partial x} + n_y \frac{\partial u(x, y)}{\partial y}$$

We can introduce unit vectors along the two axis $\hat{x}, \hat{y}$ and define a vector field in the plane by the rule

$$\mathbf{F}(x, y) = \frac{\partial u}{\partial x} \hat{x} + \frac{\partial u}{\partial y} \hat{y}$$

This vector field is denoted by ∇u. ∇ is also termed as the gradient operator in the plane. It has the representation

$$\nabla = \hat{x} \frac{\partial}{\partial x} + \hat{y} \frac{\partial}{\partial y}$$

The chain rule for partial derivatives: If x, y are functions of u, v and we have a function $w(x, y)$, then we want to express the partial derivative of w with respect to the pair u, v. Note that when we talk about the partial derivative pair $(\frac{\partial}{\partial x}, \frac{\partial}{\partial y})$, we are assuming that the first partial derivative is taken keeping y constant and the second partial derivative is taken keeping x constant. When we regard w as a function of u, v. To make our notation precise, we set $\psi(u, v) = w(x(u, v), y(u, v))$. When we talk about taking the partial derivative of ψ with respect to u (v), then it means that v (u) is being held as constant. With this understanding, we want to show that

$$\frac{\partial \psi}{\partial u} = \frac{\partial x}{\partial u} \frac{\partial w}{\partial x} + \frac{\partial y}{\partial v} \frac{\partial \psi}{\partial y}$$

Study project 44

The solution to two dimensional electromagnetic radiation problems using the Hankel function expansion can be dealt with in a seminar. A two dimensional sinusoidal current distribution is specified by a current density $\mathbf{J}(\rho, \phi)exp(j\omega t)$. The current density is independent of the z coordinate. The magnetic vector potential phasor is then given by

$$\mathbf{A} = \frac{\mu}{4\pi} \int \mathbf{J}(\rho', \phi') \frac{exp(-ik\sqrt{\rho^2 + \rho'^2 - 2\rho\rho' cos(\phi - \phi') + z^2})}{\sqrt{\rho^2 + \rho'^2 - 2\rho\rho' cos(\phi - \phi') + z^2}} \rho' d\rho' d\phi' dz$$

This expression follows from the following vector identity: If ρ denotes the radial vector of the point from the z axis, then the position vector is given by $\rho + z\hat{z}$. Note ρ can be expressed as $\rho\hat{\rho} = \rho.cos\phi\hat{x} + \rho sin\phi\hat{y}$. Thus the position vector is given by the following cylindrical to cartesian formula

$$\mathbf{r} = \rho.cos(\phi)\hat{x} + \rho.sin(\phi)\hat{y} + z\hat{z}$$

We are trying to evaluate the field at this point. The position vector of a general point in the current distribution is given by an analogous formula:

$$\mathbf{r'} = \rho'.cos(\phi')\hat{x} + \rho'.sin(\phi')\hat{y} + z'\hat{z}$$

Then,

$$|\mathbf{r} - \mathbf{r'}|^2 = (\rho.cos\phi - \rho'.cos\phi')^2 + (\rho.sin\phi - \rho'sin\phi')^2 + (z - z')^2 = |\rho - \rho'|^2 + (z - z')^2$$

since

$$|\rho - \rho'|^2 = |(\rho.cos\phi\hat{x} + \rho.sin\phi\hat{y}) - (\rho'.cos\phi'\hat{x} + \rho'sin\phi'\hat{y})|^2$$

$$= (\rho.cos\phi - \rho'.cos\phi')^2 + (\rho.sin\phi - \rho'.sin\phi')^2 = \rho^2 + \rho'^2 - 2\rho\rho' cos(\phi - \phi')$$

The standard three dimensional formula for the vector potential using retarded potentials is given by

$$\mathbf{A(r)} = \frac{\mu}{4\pi} \int \mathbf{J(r')} \frac{exp(-ik|\mathbf{r} - \mathbf{r'}|)}{|\mathbf{r} - \mathbf{r'}|} dV'$$

The volume element $dV' = \rho' d\rho' d\phi' dz'$ and this gives

$$\mathbf{A(r)} = \frac{\mu}{4\pi} \int \mathbf{J}(\rho', \phi') \frac{exp(-ik\sqrt{\rho^2 + \rho'^2 - 2\rho\rho' cos(\phi - \phi') + (z - z')^2})}{\sqrt{\rho^2 + \rho'^2 - 2\rho\rho'.cos(\phi - \phi') + (z - z')^2}} \rho' d\rho' d\phi' dz'$$

We define the function

$$H_0(x) = -\frac{i}{\pi} \int_{-\infty}^{\infty} \frac{exp(i\sqrt{x^2 + z^2})}{\sqrt{x^2 + z^2}} dz$$

This function is called the Hankel function of zeroth order. It has a nice physical interpretation. Consider a unit two dimensional current element at the origin. The current density for this can be expressed as

$$\mathbf{J(r)} = \hat{z}\delta(\rho) = \hat{z}\delta(x)\delta(y)$$

This is to be interpreted as the current phasor. The vector potential due to this element is given by

$$\mathbf{A}(x,y,z) = \frac{\mu}{4\pi}\int \mathbf{J}(x',y',z')\frac{exp(-ik\sqrt{(x-x')^2+(y-y')^2+(z-z')^2})}{\sqrt{(x-x')^2+(y-y')^2+(z-z')^2}}dx'dy'dz'$$

$$= \hat{z}\frac{\mu}{4\pi}\int \frac{exp(-ik\sqrt{x^2+y^2+(z-z')^2})}{\sqrt{x^2+y^2+(z-z')^2}}dz'$$

$$= \hat{z}\frac{\mu}{4\pi}\int \frac{exp(-ik\sqrt{\rho^2+z'^2})}{\sqrt{\rho^2+z'^2}}dz'$$

$$= \hat{z}\frac{i\mu}{4}H_0(k\rho)$$

The magnetic vector potential has just a z component and it depends only upon the distance from the $\hat{z}$ axis. The vector potential due to the two dimensional current distribution can now be expressed in terms of the Hankel function as

$$\mathbf{A}(\rho,\phi) = \frac{i\mu}{4}\int_0^\infty \int_0^{2\pi} \mathbf{J}(\rho',\phi')H_0(k\sqrt{\rho^2+\rho'^2-2\rho.\rho'.cos(\phi-\phi')})\rho'd\rho'd\phi'$$

The differential equation satisfied by the function $H_0(x)$ can be derived using the Helmholtz equation satisfied the kernel that defines the retarded potential. The function $G(\mathbf{r},\mathbf{r}') = \frac{exp(-ik|\mathbf{r}-\mathbf{r}'|)}{|\mathbf{r}-\mathbf{r}'|}$ satisfies the equation

$$(\nabla^2+k^2)G(\mathbf{r},\mathbf{r}') = -4\pi\delta(\mathbf{r}-\mathbf{r}') = -4\pi\delta(x-x')\delta(y-y')\delta(z'-z')$$

Integrating both sides with respect to z' from $-\infty$ to ∞ gives

$$(\frac{\partial^2}{\partial x^2}+\frac{\partial^2}{\partial y^2}+k^2)\int G(\mathbf{r},\mathbf{r}')dz'$$

$$= -4\pi\delta(x-x')\delta(y-y')$$

Note that

$$\int \frac{\partial^2}{\partial z^2}G(\mathbf{r},\mathbf{r}')dz' = 0 = \frac{\partial^2}{\partial z^2}\int G(\mathbf{r},\mathbf{r}')dz' = 0$$

since the integral is independent of z. The integral depends on z through $z-z'$ and its integral from $-\infty$ to ∞ will therefore be independent of z. However, it is easy to see that

$$\int G(\mathbf{r},\mathbf{r}')dz' = \int \frac{exp(-ik\sqrt{(x-x')^2+(y-y')^2+(z-z')^2})}{\sqrt{(x-x')^2+(y-y')^2+(z-z')^2}}dz'$$

$$= i\pi H_0(k\sqrt{(x-x')^2+(y-y')^2}) = i\pi H_0(k|\rho-\rho'|)$$

Plugging this into the preceding equation gives

$$(\frac{\partial^2}{\partial x^2} + \frac{\partial^2}{\partial y^2} + k^2)H_0(k\sqrt{(x-x')^2 + (y-y')^2}) = 4i\delta(x-x')\delta(y-y')$$

Transforming this equation into cylindrical coordinates by letting $x - x' = \rho.cos\phi, y - y' = \rho.sin\phi$ and noting that $H_0(k\sqrt{(x-x')^2 + (y-y')^2})$ is independent of ϕ gives us

$$(\frac{1}{\rho}\frac{d}{d\rho}\rho\frac{d}{d\rho} + k^2)H_0(k\rho) = 4i\delta(\rho)$$

In particular, for $\rho > 0$ we have

$$\frac{1}{\rho}\frac{d}{d\rho}H_0(k\rho) + \frac{d^2}{d\rho^2}H_0(k\rho) + k^2 H_0(k\rho) = 0$$

which is equivalent to

$$xH_0''(x) + H_0'(x) + xH_0(x) = 0$$

This is precisely the Bessel differential equation of zeroth order. The Bessel equation of order zero has two linearly independent solutions, one of which is nonsingular at the origin and the other is singular at the origin. A specific complex linear combination of these two functions yields $H_0(x)$, that is $H_0(x) = J_0(x) + iN_0(x)$, where J_0 is the zeroth order Bessel function that is nonsingular at the origin and N_0 is the zeroth order Bessel function that is singular at the origin. Clearly, $H_0(x)$ is singular at the origin.

Study project 45

Method of moments directly applied to the Maxwell equations to solve for the electric and magnetic fields in terms of the charge and current densities. For sinusoidal sources, the potential $V(\mathbf{r})$ satisfies the Helmholtz equation with source

$$(\nabla^2 + \omega^2/c^2)V(\mathbf{r}) = -\rho(\mathbf{r})/\epsilon$$

and the vector potential satisfies three similar equations

$$(\nabla^2 + \omega^2/c^2)\mathbf{A}(\mathbf{r}) = -\mu\mathbf{J}(\mathbf{r})$$

we expand the charge and current densities and the scalar and vector potential as linear combinations of test functions, i.e.,

$$\rho(\mathbf{r}) = \sum_{k=1}^{n} \rho_k\phi_k(\mathbf{r})$$

$$V(\mathbf{r}) = \sum_{k=1}^{n} V_k\phi_k(\mathbf{r})$$

Plugging this into the differential equations gives

$$\sum_k V_k \nabla^2 \phi_k(\mathbf{r}) + (\omega^2/c^2) \sum_k V_k \phi_k(\mathbf{r}) = -\sum_k \rho_k \phi_k(\mathbf{r})/\epsilon$$

This equation can in general not be satisfied exactly. So we derive a set of linear equations for the $V_k's$ from this that can be solved and which yield an approximate solution for the potential. Multiply both sides by $\phi_m(\mathbf{r})$ and integrate over the volume:

$$\sum_k V_k \int \phi_m(\mathbf{r}) \nabla^2 \phi_k(\mathbf{r}) d^3\mathbf{r} + (\omega^2/c^2) \sum_k V_k \int \phi_m(\mathbf{r}) \phi_k(\mathbf{r}) d^3\mathbf{r}$$

$$= -\sum_k \rho_k \int \phi_m(\mathbf{r}) \int \phi_k(\mathbf{r}) d^3\mathbf{r}/\epsilon$$

This is a system of n linear equations for the variables $V_1, ..., V_n$ that can be solved. The $\rho_k's$ are obtained from the equations

$$\int \rho(\mathbf{r}\phi_m(\mathbf{r}) d^3\mathbf{r} = \sum_k \rho_k \int \phi_m(\mathbf{r}) \phi_k(\mathbf{r}) d^3\mathbf{r}$$

if the region is such that V vanishes on the boundary, then the test functions ϕ_k must be chosen to meet the same constraint.

Study project 46

How to generate magnetic fields from time varying electric fields ? In the absence of current sources, Ampere's law with the displacement current correction term reads

$$\nabla \times \mathbf{B} = \frac{1}{c^2} \frac{\partial \mathbf{E}}{\partial t}$$

This law is in fact the reciprocal of Faraday's law of induction which relates the spatial derivatives of the electric field to the time derivative of the magnetic field. To see how this can be used to generate a magnetic field, we mount the iron core of a transformer on a wooden base. Assume that the core is square with a cross sectional area of A and each side having length L. We then place two metallic plates on either side of the core and apply a sinusoidally varying voltage source $V(t) = V_0 sin(\omega t)$ to these plates. If d is the distance between the plates, then the electric field between the plates is approximately given by $E(t) = \frac{V_0}{d} sin(\omega t)$. Its time derivative equals $\frac{V_0 \omega}{d} cos(\omega t)$. The flux of this field through the square area enclosed by the core is $\frac{V_0 \omega L^2}{d} cos(\omega t)$ and this divided by c^2 must equal the line integral of B around the four sides of the core, i.e. $4LB$. This gives $B = \frac{V_0 \omega L}{4dc^2} cos(\omega t)$. Measurement of B in the core by cutting a small gap in it and placing a magnetic meter in the gap would thus verify the truth of the third Maxwell equation apart from suggesting a method to produce magnetic fields without the use of wire windings.

Solved problem 5: A region in space consists of a volume V, a surface S and a curve Γ. The volume V contains charge of volume density ρ_v, the surface S contains charge of surface density ρ_s and finally the line Γ has charge of linear density ρ_L. Determine the total charge.

Ans:

$$Q = \int_V \rho_v dV + \int_S \rho_s dS + \int_L \rho_L dl$$

If the volume V is bounded by the surfaces $z = f_1(x,y)$ and $z = f_2(x,y), -A < x < A, -B < y < B$, the surface S specified by the equation $z = f(x,y), -a < x < a, -b < y < b$ and the line by the parametric equation $(x(s),y(s),z(s)), 0 < s < 1$, then

$$Q = \int_{-A}^{A} \int_{-B}^{B} dx dy \int_{f_1(x,y)}^{f_2(x,y)} \rho_v(x,y,z)dz + \int_{-a}^{a} \int_{-b}^{b} \rho_s(x,y,f(x,y))\sqrt{1 + f_x^2 + f_y^2}dx dy$$

$$+ \int_0^1 \sigma_L(x(s),y(s),z(s))\sqrt{x'(s)^2 + y'(s)^2 + z'(s)^2}ds$$

Solved problem 6

If a vector $\mathbf{a}$ having cartesian components a_x, a_y, a_z is changed by an infinitesimal amount $d\mathbf{a}$, then by how much does the magnitude of the vector change upto first order in $d\mathbf{a}$?

Ans:

$$|\mathbf{a}| = \sqrt{a_x^2 + a_y^2 + a_z^2}$$

Thus,

$$d|\mathbf{a}| = \frac{a_x da_x + a_y da_y + a_z da_z}{\sqrt{a_x^2 + a_y^2 + a_z^2}} = \frac{\mathbf{a}.d\mathbf{a}}{|\mathbf{a}|}$$

Solved problem 7

Given two vectors $\mathbf{a}$ and $\mathbf{b}$. Determine the first order change in the angle between these two vectors when the vectors themselves undergo infinitesimal changes $d\mathbf{a}$ and $d\mathbf{b}$?

Ans: Let α denote the angle between the two vectors. Then,

$$\mathbf{a}.\mathbf{b} = |\mathbf{a}||\mathbf{b}|cos\alpha$$

Now observe that

$$\mathbf{a}.\mathbf{b} = a_x b_x + a_y b_y + a_z b_z$$

so that

$$d(\mathbf{a}.\mathbf{b}) = a_x db_x + a_y db_y + a_z db_z + b_x da_x + b_y da_y + b_z da_z = \mathbf{a}.d\mathbf{b} + \mathbf{b}.d\mathbf{a}$$

Using the previous formula, we thus find

$$d(\mathbf{a}.\mathbf{b}) = \mathbf{a}.d\mathbf{b} + \mathbf{b}.d\mathbf{a} = d(|\mathbf{a}||\mathbf{b}|cos\alpha)$$

$$= (d|\mathbf{a}|)(|\mathbf{b}|cos\alpha) + (d\mathbf{b}|)(|\mathbf{a}|cos\alpha) + |\mathbf{a}||\mathbf{b}|d(cos\alpha)$$

$$= \frac{\mathbf{a}.d\mathbf{a}}{|\mathbf{a}|}|\mathbf{b}|cos\alpha + \frac{\mathbf{b}.d\mathbf{b}}{|\mathbf{b}|}|\mathbf{a}|cos\alpha$$

$$-|\mathbf{a}||\mathbf{b}|sin(\alpha)d\alpha$$

This equation can be solved for $d\alpha$ in terms of $d\mathbf{a}, d\mathbf{b}$.

Solved problem 8

Explain the different polarizations of a vector field at a given point starting from first principles.

Ans: Let $\tilde{\mathbf{A}}(t) = Re((A_x\hat{x} + jA_y\hat{y})exp(j\omega t))$ be a phasor at a given point. Here, A_x and A_y are complex numbers. Let their phases be denoted respectively by ϕ_x, ϕ_y. Then, the x component of the field at the point is given by

$$\tilde{A}_x(t) = |A_x|cos(\omega t + \phi_x) \qquad \tilde{A}_y(t) = |A_y|cos(\omega t + \phi_y)$$

If $\phi_x = \phi_y$, we obtain the equation

$$\frac{\tilde{A}_y(t)}{\tilde{A}_x(t)} = \frac{|A_y|}{|A_x|}$$

and this corresponds to linear polarization. The direction of the field in this polarization state always makes an angle $\beta = tan^{-1}(|A_y|/|A_x|)$ with the x axis. If $\phi_y = \phi_x \pm \pi/2$ and $|A_x| = |A_y| = R$, then we clearly have

$$\tilde{A}_x(t)^2 + \tilde{A}_y(t)^2 = R^2$$

and this corresponds to circular polarization. In this case, the field vector rotates in a circle of radius R. For all other cases, we have elliptic polarization. The equation of the curve along which the field vector moves can be obtained by eliminating the time variable. To determine this equation, we note that

$$\tilde{A}_x(t) = |A_y|cos(cos^{-1}(\tilde{A}_y(t)/|A_y|) - \phi_y + \phi_x)$$

$$= \tilde{A}_y(t)cos(\phi_x - \phi_y) + |A_y|\sqrt{1 - \tilde{A}_y(t)^2/|A_y|^2}sin(\phi_y - \phi_x)$$

so that

$$(\tilde{A}_x(t) - \tilde{A}_y(t)cos(\phi_x - \phi_y))^2 = (|A_y|^2 - \tilde{A}_y(t)^2)sin^2(\phi_y - \phi_x)$$

or

$$\tilde{A}_x(t)^2 + \tilde{A}_y(t)^2 - 2\tilde{A}_x(t)\tilde{A}_y(t)cos(\phi_x - \phi_y) = |A_y|^2 sin^2(\phi_y - \phi_x)$$

This is the equation of an ellipse.

Solved problem 9

Let $\mathbf{A} = A_x(x,y)\hat{x} + A_y(x,y)\hat{y}$. Note that this defines a vector field in the plane. Consider the square with vertices $(0,0), (a,0), (a,a), (0,a)$. Determine an expression for the line integral of $\mathbf{A}$ around this square counterclockwise.

Ans:

$$\int_\Gamma \mathbf{A}.d\mathbf{r} = \int_0^a A_x(x,0)dx + \int_0^a A_y(a,y)dy - \int_0^a A_x(x,a)dx - \int_0^a A_y(0,y)dy$$

Solved problem 10

Explain how using a digital computer, you would generate a plot of the electric field line generated by two charges of Q_1 and Q_2 coulombs respectively located at $(0,0)$ and $(a,0)$ in the plane. The field line is assumed to pass through the point (b,c). Describe by means of a MATLAB programme.

Ans: The electric field in the plane has components

$$E_x(x,y) = \frac{Q_1 x}{(x^2 + y^2)^{3/2}} + \frac{Q_2(x-a)}{((x-a)^2 + y^2)^{3/2}}$$

$$E_y(x,y) = \frac{Q_1 y}{(x^2 + y^2)^{3/2}} + \frac{Q_2 y}{((x-a)^2 + y^2)^{3/2}}$$

Let $(X(s), Y(s)), s \geq 0$ denote the field line passing through the point (b,c). This line satisfies the differential equation

$$\frac{dX(s)}{ds} = E_x(X(s), Y(s)) \qquad \frac{dY(s)}{ds} = E_y(X(s), Y(s)), s \geq 0$$

with the initial condition $X(0) = b, Y(0) = c$. To plot this using a digital computer programme, we discretize the parameter s with a step size δ. Letting $X_n = X(n\delta), Y_n = Y(n\delta)$, the above differential equation approximates to

$$\frac{Xn+1 - X_n}{\delta} = E_x(X_n, Y_n) \qquad \frac{Y_{n+1} - Y_n}{\delta} = E_y(X_n, Y_n), n = 0, 1, 2, ...$$

with initial condition $X_0 = b, Y_0 = c$. A MATLAB programme to plot this would run along the following lines.

X(1)=b; Y(1)=c;

for n=1: N

X(n+1)=X(n)+delta*Ex(X(n),Y(n));

Y(n+1)=Y(n)+delta*Ey(X(n),Y(n));

end;

plot(X,Y);

Here, we have to substitute for the expressions Ex and Ey from the above discussion.

Solved problem 11

Charges Q Coulomb each are located at the vertices of a regular n sided polygon circumscribed by a circle of radius R. Determine the electrostatic force on each charge.

Ans: The angle subtended by each side of the polygon at the centre of the circle is given by $\alpha = 2\pi/n$. So, we can assume the positions of the charges to be given by $(R.cos(k\alpha), R.sin(k\alpha)), k = 0, 1, ..., n-1$. The force on the charge at $(R, 0)$ is given by applying Coulomb's law combined with superposition

$$\mathbf{F} = \sum_{k=0}^{n-1} \frac{Q^2(R.cos(k\alpha) - R, R.sin(k\alpha))}{((R.cos(k\alpha) - R)^2 + R^2 sin^2(k\alpha))^{3/2}}$$

$$= \frac{Q^2}{R^2} \sum_{k=0}^{n-1} \frac{(cos(k\alpha) - 1, sin(k\alpha))}{(2 - 2.cos(k\alpha))^{3/2}}$$

$$= \frac{Q^2}{8R^2} \sum_{k=0}^{n-1} \frac{(cos(k\alpha) - 1, sin(k\alpha))}{|sin^3(k\alpha)|}$$

Solved problem 12

Point charges each of Q coulombs are located at the vertices of a regular tetrahedron of edge length a. Determine the electrostatic force on each charge.

Ans: We can assume the base of the tetrahedron to fall in the xy plane. The base is an equilateral triangle with vertices at $(0,0,0), (a,0,0), (a/2, a\sqrt{3}/2, 0)$. The top vertex of the tetrahedron is directly above the centre of the equilateral base triangle. The coordinates of this centre are clearly given by $(a/2, (a/2)tan(\pi/6), 0) = (a/2, a/(2\sqrt{3}), 0)$. Thus, the top vertex of the tetrahedron has coordinates $(a/2, a/(2\sqrt{3}), z)$. z is determined from the condition that the distance of this vertex from the origin equals a, i.e.

$$a^2/4 + a^2/12 + z^2 = a^2$$

or

$$z^2 = a^2(1 - 1/4 - 1/12) = a^2(12 - 3 - 1)/12 = 2a^2/3$$

so that the top vertex is $(a/2, a/2\sqrt{3}, a\sqrt{2/3})$. We use the following notations for the four vertices of the tetrahedron:

$$\mathbf{A} = (0,0,0) \qquad \mathbf{B} = (a,0,0) \qquad \mathbf{C} = (a/2, a\sqrt{3}/2, 0) \qquad \mathbf{D} = (a/2, a/2\sqrt{3}, a\sqrt{2/3})$$

The force on the charge at $\mathbf{A}$ is given by Coulomb's law combined with superposition:

$$\mathbf{F_A} = Q^2\left(\frac{\mathbf{A} - \mathbf{B}}{|\mathbf{A} - \mathbf{B}|^3} + \frac{\mathbf{A} - \mathbf{C}}{|\mathbf{A} - \mathbf{C}|^3} + \frac{\mathbf{A} - \mathbf{D}}{|\mathbf{A} - \mathbf{D}|^3}\right)$$

Since

$$|\mathbf{A} - \mathbf{B}| = |\mathbf{A} - \mathbf{C}| = |\mathbf{A} - \mathbf{D}| = a$$

this force can be expressed as

$$\mathbf{F_A} = \frac{Q^2}{a^3}(3\mathbf{A} - \mathbf{B} - \mathbf{C} - \mathbf{D})$$

We leave the evaluation of this vector to the student.

Solved problem 13

Given the charge density in space $\rho = \rho_0(1 - |x|/a), |x| \leq a$ and $\rho = 0$ when $|x| > a$, determine the electric field vector in space. Assume that space has permitivity ϵ_0.

Ans: Gauss', law reads $\frac{dE_x(x)}{dx} = \rho(x)/\epsilon_0$. Since $\rho(x) = 0$ when $x < -a$ and $x > a$, for $x < -a$, $E_x(x) = E_0$, a constant and for $x > a$, $E_x(x) = E_0'$, another constant. From the symmetry of the problem $E_0' = -E_0$. Also for $|x| < a$, $E_x(x) = c + \int_{-a}^{x} \rho(x)dx/\epsilon_0$ where c is a constant. Matching the field at $x = -a$ and $x = a$ gives

$$c = E_0, c + \int_{-a}^{a} \rho(x)dx/\epsilon_0 = E_0' = -E_0$$

This gives

$$c = E_0 = -\frac{1}{2}\int_{-a}^{a} \rho(x)dx/\epsilon_0$$

The rest is routine evaluation of the integral. We leave the details to the student.

Solved problem 14

The charge density in space is given by $\rho(r) = \rho_1$ if $0 < r < a$, $\rho(r) = \rho_2$ if $a < r < b$ and finally $\rho(r) = 0$ if $b < r < \infty$. Determine the electric field and potential everywhere in space.

Ans: The electric field is radial $E(r)$ depending only upon the radial distance from the origin. It satisfies Gauss' law $\frac{1}{r^2}(r^2 E(r))' = -\rho/\epsilon_0$. We find in the region $0 < r < a$, the equation $(r^2 E(r))' = -\rho_1 r^2/\epsilon_0$. This integrates to give $r^2 E(r) = -\rho_1 r^3/3\epsilon_0 + c_1$, or $E(r) = -\rho_1 r/3\epsilon_0$ ($c_1 = 0$ since the field must be finite at the origin). Next we find $(r^2 E(r))' = -\rho_2 r^2/\epsilon_0$ in the region $a < r < b$. This integrates to give $E(r) = -\rho_2 r/3\epsilon_0 + c_2/r^2$ for $a < r < b$. Finally $(r^2 E(r))' = 0$ for $r > b$ and this integrates to give $E(r) = c_3/r^2$ for $r > b$. Continuity of the field at $r = a$ gives

$$-\rho_1 a/3\epsilon_0 = -\rho_2 a/3\epsilon_0 + c_2/a^2$$

and continuity of the field at $r = b$ gives

$$-\rho_2 b/3\epsilon_0 + c_2/b^2 = c_3/b^2$$

Thus,

$$c_2 = \frac{(\rho_2 - \rho_1)a^3}{3\epsilon_0}$$

$$c_3 = \frac{(\rho_2 - \rho_1)a^3}{3\epsilon_0} - \frac{\rho_2 b^3}{3\epsilon_0}$$

This determines the field everywhere. The potential can be determined using the formula

$$V(r) = \int_r^\infty E(r)dr$$

Specifically, for $r > b$,

$$V(r) = \int_r^\infty \frac{c_3}{r^2}dr = \frac{c_3}{r}$$

for $a < r < b$,

$$V(r) = \int_r^b E(r)dr + \int_b^\infty E(r)dr = \int_r^b (-\rho_2 r/3\epsilon_0 + c_2/r^2)dr + \int_b^\infty c_3 dr/r^2$$

$$= \rho_2(r^2 - b^2)/6\epsilon_0 + c_3/b$$

and for $r < a$,

$$V(r) = \int_r^a E(r)dr + \int_a^b E(r)dr + \int_b^\infty E(r)dr$$

We leave the evaluation of this integral as an exercise.

Solved problem 15

Determine the equation of a ray of light propagating in a medium whose refractive index varies as $n(x)$.

Ans: Assume that the ray propagates in the xz plane. Let θ_0 be the initial angle of the ray with the x axis and $\theta(x)$ be the angle between the ray tangent and the x axis at the point $(x, 0, z(x))$ where $z(x)$ is the equation of the ray trajectory. Then, Snell's law implies that $n(x)sin(\theta(x)) = n_0.sin(\theta_0)$ where n_0 is $n(0)$, the refractive index at $(0,0,0)$, we're assuming that the ray starts at the origin. Noting that $\frac{dz}{dx} = tan(\theta(x))$, we obtain the following differential equation for the ray

$$\frac{dz}{dx} = \frac{sin(\theta(x))}{\sqrt{1 - sin^2(\theta(x))}} = \frac{n_0 sin(\theta_0)}{n(x)\sqrt{1 - n_0^2 sin(\theta_0)^2/n(x)^2}}$$

or

$$\frac{dz(x)}{dx} = \frac{n_0 sin(\theta_0)}{\sqrt{n(x)^2 - n_0^2 sin^2(\theta_0)}}$$

Integration of both sides gives the equation of the ray trajectory.

Solved problem 16

The rectangular region $-a < x < a, -b < y < b, z = 0$ is covered by charge of constant surface density σ. Determine an expression for the electric field at the point $(0, 0, h)$.

Ans: $E(0, 0, h) = E_z\hat{z}$ where

$$E_z = \frac{\sigma}{4\pi\epsilon} \int_{-a}^{a} \int_{-b}^{b} \frac{h}{(x^2 + y^2 + h^2)^{3/2}} dxdy$$

Note that setting $y = \sqrt{h^2 + x^2} tan\phi$, the above integral transforms to

$$E_z = \frac{\sigma}{4\pi\epsilon} \int_{-a}^{a} \frac{\sigma h}{(h^2 + x^2)} \int_{-tan^{-1}(b/\sqrt{h^2+x^2})}^{tan^{-1}(b/\sqrt{(h^2+x^2)})} cos\phi d\phi$$

Solved problem 17

Determine the relationships between the polarization vector field, the electric displacement vector field and the electric field.

Ans: If $\mathbf{p}$ is an infinitesimal dipole located at the origin, the potential due to this is given by

$$V(\mathbf{r}) = \frac{\mathbf{p}.\mathbf{r}}{4\pi\epsilon r^3} = -\frac{1}{4\pi\epsilon}\mathbf{p}.\nabla(1/r)$$

Thus, if a volume V bounded by the surface S is filled with a volume density of dipoles $\mathbf{P}$ dipoles per $metre^3$, the electrostatic potential is given by

$$V(\mathbf{r}) = -\frac{1}{4\pi\epsilon} \int_{V} \mathbf{P}(\mathbf{r}').\nabla(1/|\mathbf{r} - \mathbf{r}'|)dV'$$

$$= \frac{1}{4\pi\epsilon} \int \mathbf{P}(\mathbf{r}').\nabla'(1/|\mathbf{r}-\mathbf{r}'|)dV'$$

The integral by Gauss' theorem is

$$\int \nabla'.(\mathbf{P}'/|\mathbf{r}-\mathbf{r}'|)dV' - \int \nabla'.P'/|\mathbf{r}-\mathbf{r}'|dV'$$

$$= \int \mathbf{P}'.\mathbf{n}'dS'/|\mathbf{r}-\mathbf{r}'| - \int \nabla'.P'/|\mathbf{r}-\mathbf{r}'|dV'$$

Thus the volume density of polarization charge equals $\rho_p = -\nabla.P$ and the surface density of polarization charge equals $\sigma_p = P.n$. Gauss' law now reads $\epsilon_0 \nabla.E = \rho_f + \rho_p = \rho_f - \nabla.P$ so that $\nabla.(\epsilon_0 E + P) = \rho_f$ where ρ_f is the free or external charge density. The vector field $D = \epsilon_0 E + P$ is called the electric displacement vector field. If the medium is linear, the polarization vector is directly proportional to the electric field and hence the displacement is also proportional to the electric field. We write $D = \epsilon E$ where ϵ is called the permitivity of the medium. $\epsilon_r = \epsilon/\epsilon_0$ is called the refractive index of the medium. Thus, $D = \epsilon_0 \epsilon_r E, P = (\epsilon - \epsilon_0)E = \epsilon_0(\epsilon_r - 1)E$. The total amount of polarization charge in V is $Q_p = \int_V \rho_p dV = -\int_V \nabla.P dV = -\int_S P.n dS$. This polarization charge can be attributed to a surface density $-P.n$ of polarization charge on the inner side of the surface S.

Solved problem 18

A rod placed along the x axis extending from $x = 0$ upto $x = L$ is polarized. The polarization vector is given by $P_x = ax^2 + b$. Determine the surface density of polarization charge at $x = 0, L$ and also the volume density of polarization charge at these two ends.

Ans: $divP = \frac{dP_x}{dx} = 2ax$. Using the relation $divP = -\rho_p$, gives the volume density of the polarization charge at $x = 0, L$ as 0 and $-2aL$ respectively. The surface density of polarization charge at $x = 0$ inside the rod is $\sigma_p = -P.n$. With $n = \hat{x}$ this evaluates to $-b$. The same formula for the surface polarization charge density at $x = L$ with $n = -\hat{x}$ gives $\sigma_p = aL^2 + b$.

Solved problem 19

The x axis has a uniform linear charge density σ and a charge Q is placed at (x_0, y_0, z_0). Determine the electric field everywhere in space.

Ans: The electric field due to an infinite line charge has only a radial component with the radial distance being taken along the perpendicular to the line from the point in question. This gives for the electric field produced by the line charge, the expression

$$\mathbf{E}_1(x,y,z) = \frac{\sigma}{2\pi\epsilon} \frac{y\hat{z} - z\hat{y}}{y^2 + z^2}$$

The electric field produced by the point charge Q is given by the expression

$$\mathbf{E}_2(x,y,z) = \frac{Q}{4\pi\epsilon} \frac{(x-x_0)\hat{x} + (y-y_0)\hat{y} + (z-z_0)\hat{z}}{((x-x_0)^2 + (y-y_0)^2 + (z-z_0)^2)^{3/2}}$$

The total electric field in space due to the line charge and the point charge is the superposition of these two fields:

$$\mathbf{E}(x,y,z) = \mathbf{E}_1(x,y,z) + \mathbf{E}_2(x,y,z) =$$

$$\frac{\sigma}{2\pi\epsilon} \frac{y\hat{z} - z\hat{y}}{y^2 + z^2} + \frac{Q}{4\pi\epsilon} \frac{(x-x_0)\hat{x} + (y-y_0)\hat{y} + (z-z_0)\hat{z}}{((x-x_0)^2 + (y-y_0)^2 + (z-z_0)^2)^{3/2}}$$

Summing up the individual components gives us

$$E_x(x,y,z) = \frac{Q}{4\pi\epsilon} \frac{x-x_0}{((x-x_0)^2 + (y-y_0)^2 + (z-z_0)^2)^{3/2}}$$

$$E_y(x,y,z) = -\frac{\sigma}{2\pi\epsilon} \frac{z}{y^2 + z^2} + \frac{Q}{4\pi\epsilon} \frac{y-y_0}{((x-x_0)^2 + (y-y_0)^2 + (z-z_0)^2)^{3/2}}$$

$$E_z(x,y,z) = \frac{\sigma}{2\pi\epsilon} \frac{y}{y^2 + z^2} + \frac{Q}{4\pi\epsilon} \frac{z-z_0}{((x-x_0)^2 + (y-y_0)^2 + (z-z_0)^2)^{3/2}}$$

Solved problem 20

This problem deals with the measurement of the potential difference between two points using a balance. One pan of the balance is replaced by a capacitor plate, the other plate of the capacitor being attached to the ground below the attached capacitor plate. The potential difference V is applied between the two plates. Let d be the separation between the two plates. d can vary as the balance oscillates. The capacitance between the two plates is given by $C = \epsilon_0 S/d$ and the electrostatic energy between the plates is given by $U = CV^2/2 = \epsilon_0 SV^2/2d$. The force between the plates is attractive and has a magnitude $F = \left|\frac{\partial U}{\partial d}\right| = \epsilon_0 SV^2/2d^2$. This force causes the side on which the capacitor plate is attached to be pulled towards the ground with this force. The force can be balanced by placing weights on the other side of the balance. Knowing these weights $F = Mg$, we determine the potential difference from the equation $Mg = \epsilon_0 SV^2/2d^2$. Of course d must be measured when equilibrium is achieved, d being the distance between the two plates of the capacitor.

Study project 47

This project deals with the description of the boundary conditions for the fields on the boundary of a conductor. Fields at the surface of and within a conductor: The fields inside a perfect conductor are zero. Thus, if $\mathbf{D}$ is the electric displacement vector just at the surface of a perfect conductor and $\mathbf{n}$ the unit normal to the surface directed from within to without and Σ is the surface charge density on the conductor, then application of Gauss' law gives the following boundary condition: $\mathbf{n}.\mathbf{D} = \Sigma$. In a perfect conductor, the charges are highly mobile so that even a small electric field causes charges to flow to the surfaces and produce a neutralizing electric field. Similarly, the magnetic field inside a perfect conductor is zero and if $\mathbf{H}$ denotes the magnetic field on the surface outside and $\mathbf{K}$ the surface current density, then the following boundary condition $\mathbf{n} \times \mathbf{H} = \mathbf{K}$ is satisfied. If the conductor is imperfect and $\mathbf{B}_c$ and $\mathbf{E}_c$ denote the magnetic and electric fields inside the conductor, then the Maxwell equations result in the following boundary conditions $\mathbf{n}.(\mathbf{B} - \mathbf{B}_c) = 0$ and $\mathbf{n} \times (\mathbf{E} - \mathbf{E}_c) = \mathbf{0}$. For a perfect conductor $\mathbf{B}_c$ and $\mathbf{E}_c$ are zero and we get $\mathbf{n}.\mathbf{B} = 0$ and $\mathbf{n} \times \mathbf{E} = \mathbf{0}$. Note that the boundary condition $\mathbf{n}.\mathbf{D} = \Sigma$ is obtained as a consequence of the integral form of Gauss' law $\int_S \mathbf{D}.\mathbf{n}dS = Q$ where S is a closed surface enclosing a charge $Q/$ The boundary condition $\mathbf{n} \times \mathbf{H} = \mathbf{K}$ is obtained as a consequence of the integral form of Ampere's law $\int_\Gamma \mathbf{H}.\mathbf{dl} = \mathbf{I} + \frac{\partial \Phi_E}{\partial t}$ where Γ is a closed contour enclosing a current I and Φ_E is the flux of the electric field through the surface S enclosed by the contour Γ. Note that the boundary condition $\mathbf{n}.\mathbf{D} = \Sigma$ is obtained from the integral form of Gauss' law $\int_S \mathbf{D}.\mathbf{n}dS = Q$ where S is a closed surface enclosing a charge Q. The boundary condition $\mathbf{n} \times \mathbf{H} = \mathbf{K}$ is obtained from the integral version of Ampere's law with the displacement current correction term and the boundary condition $\mathbf{n} \times (\mathbf{E} - \mathbf{E}_c) = \mathbf{0}$ is obtained from the integral version of Faraday's law, while $\mathbf{n}.(\mathbf{B} - \mathbf{B}_c) = 0$ is obtained from the integral version of the no magnetic monopole condition. If the conductivity is high but not infinite, then there cannot be any surface current since such a surface current would imply an infinite current density and this combined with the finite conductivity would imply an infinite electric field. Thus, the boundary condition $\mathbf{n} \times \mathbf{H} = \mathbf{K}$ is replaced by the condition $\mathbf{n} \times (\mathbf{H} - \mathbf{H}_c) = \mathbf{0}$. There is a thin transitional layer of width called the skin depth (that is calculable from the Maxwell equations) and to determine the fields outside the conductor, the effect of the fields in this transitional region must be determined. The method is the successive approximation scheme wherein we first determine the fields outside assuming boundary conditions corresponding to a perfect conductor. Then, applying the boundary conditions for a finite conductor, we determine the fields inside the conductor. Then, using these fields inside the conductor, we determine the correction to the fields outside by calculating the current densities and charge distribution produced by the fields inside. The new fields outside are used to with the finite conductivity boundary conditions to determine the correction to the fields inside the conductor. The iterative process goes on.

Study project 48

Determination of the skin depth of a good conductor. The variation in the fields is maximum along a direction normal to the conducting surface. Let $\mathbf{n}$ denote this direction and ξ the coordinate along this direction. Then the curl operator can be replaced by $\mathbf{n} \times \frac{\partial}{\partial \xi}$ on neglect of the variation of the fields along directions parallel to the surface. Let $\mathbf{H}_c$ and $\mathbf{E}_c$ denote the magnetic and electric fields inside the conductor. The displacement current is neglected in comparison with the conduction current. The Maxwell equation $\nabla \times \mathbf{H}_c = \sigma \mathbf{E}_c$ approximates to

$$\mathbf{n} \times \frac{\partial \mathbf{H}_c}{\partial \xi} = \sigma \mathbf{E}_c$$

while the Maxwell equation $\nabla \times \mathbf{E}_c = i\mu\omega \mathbf{H}_c$ approximates to

$$\mathbf{n} \times \frac{\partial \mathbf{E}_c}{\partial \xi} = i\mu\omega \mathbf{H}_c$$

From the first equation

$$\frac{\partial \mathbf{E}_c}{\partial \xi} = \frac{1}{\sigma} \frac{\partial^2}{\partial \xi^2}(\mathbf{n} \times \mathbf{H}_c)$$

in which we've used the fact that the normal $\mathbf{n}$ does not vary with small changes in the coordinate ξ along the normal to the conducting surface. From the second equation, we however have

$$i\mu\omega \mathbf{n} \times \mathbf{H}_c = \mathbf{n} \times (\mathbf{n} \times \frac{\partial \mathbf{E}_c}{\partial \xi}$$

$$= -\frac{\partial \mathbf{E}_c}{\partial \xi}$$

since $\mathbf{n}.\frac{\partial}{\partial \xi}\mathbf{E}_c = 0$ in accordance with the equation $\nabla.\mathbf{E}_c = 0$ since there is no volume density of charge inside the conducting body. Combining the above two equations gives

$$\frac{\partial^2}{\partial \xi^2}(\mathbf{n} \times \mathbf{H}_c) + i\mu\omega\sigma \mathbf{n} \times \mathbf{H}_c = 0$$

from which we conclude that $\mathbf{n} \times \mathbf{H}_c$ varies with ξ as $exp(\sqrt{\mu\omega\sigma/2}(-1+i)\xi)$ inside the conductor, with the understanding that ξ increases as one penetrates more and more inside the conductor.

Solved Problem 21

Suppose the polarization vector $\mathbf{P}$ is specified and the electric field $\mathbf{E.n}$ along a direction $\mathbf{n}$ is specified. From this data, we want to determine the electric field and the electric displacement everywhere. Assume the medium to be linear.

Ans: Let χ denote the susceptibility of the medium. Then, $\mathbf{P} = \epsilon_0\chi\mathbf{E}$ and in particular, $\mathbf{P.n} = \epsilon_0\chi\mathbf{E.n}$ from which we get $\chi = \frac{\mathbf{P.n}}{\epsilon_0\mathbf{E.n}}$. The electric field is now given by

$$\mathbf{E} = \frac{\mathbf{P}}{\epsilon_0\chi} = \frac{\mathbf{E.n}}{\mathbf{P.n}}\mathbf{P}$$

and the electric displacement is given by

$$\mathbf{D} = \epsilon_0(1 + \chi)\mathbf{E} = \epsilon_0(1 + \frac{\mathbf{P.n}}{\epsilon_0\mathbf{E.n}})\frac{\mathbf{E.n}}{\mathbf{P.n}}\mathbf{P}$$

$$= (\epsilon_0 \frac{\mathbf{E.n}}{\mathbf{P.n}} + 1)\mathbf{P}$$

Solved problem 22

Let charges $Q, \alpha Q$ be located a distance d apart. Assume that a charge Q' is located on the line joining the two charges. What should Q' be and at what distance from Q should it be located for the entire system comprising the three charges to be in electrostatic equilibrium.

Ans: Assume that Q' is located at a distance a from Q and at a distance $d - a$ from αQ. Equating the net forces on $Q, Q', \alpha Q$ to zero gives

$$\frac{QQ'}{a^2} + \frac{\alpha Q^2}{d^2} = 0$$

$$\frac{QQ'}{a^2} - \frac{\alpha QQ'}{(d - a)^2} = 0$$

$$\frac{\alpha QQ'}{(d - a)^2} + \frac{\alpha Q^2}{d^2} = 0$$

These are three equations for two variables Q', a. From the second equation

$$(d - a)^2 = \alpha a^2$$

so that

$$d - a = \pm a\sqrt{\alpha}$$

If we assume $\alpha > 1$, then the positive sign must be chosen to ensure that $d > a$. Thus, $d = a(1 + \sqrt{\alpha})$. From the first equation,

$$Q' = -Q\frac{\alpha a^2}{d^2} = -Q/(1 + \sqrt{\alpha})^2$$

while from the third equation

$$Q' = -\alpha Q\frac{(d - a)^2}{d^2} = -\alpha^2 Q a^2/d^2 = -\alpha^2 Q/(1 + \sqrt{\alpha})^2$$

So we cannot get agreement unless $\alpha = 1$, i.e., unless the two charges $Q, \alpha Q$ are equal. In this case, the charge $Q' = -Q/4$ and must be located in between the two charges $Q, \alpha Q$.

Study Project 49

Fourier techniques in electrodynamics:

Example 1: Solution to the Poisson equation of electrostatics

$$\nabla^2 V(\mathbf{x}) = -\rho(\mathbf{x})/\epsilon$$

Spatial Fourier transform is computed:

$$\hat{\rho}(\mathbf{k}) = \int \rho(\mathbf{x}) exp(-i\mathbf{k}.\mathbf{x}) d^3 x$$

$$\hat{V}(\mathbf{k}) = \int V(\mathbf{x}) exp(-i\mathbf{k}.\mathbf{x}) d^3 x$$

$$\int \nabla^2 V(\mathbf{x}) exp(-i\mathbf{k}.\mathbf{x}) d^3 x = -k^2 \hat{V}(\mathbf{k})$$

Poisson equation in the spatial frequency domain reads

$$k^2 \hat{V}(\mathbf{k}) = -\hat{\rho}(\mathbf{k})/\epsilon$$

so that

$$\hat{V}(\mathbf{k}) = (2\pi)^{-3} \frac{1}{\epsilon} \int \frac{\hat{\rho}(\mathbf{k})}{k^2} exp(i\mathbf{k}.\mathbf{x}) d^3 k$$

The integral can be evaluated numerically by replacing k^2 with $k^2 + \delta$ where $\delta > 0$. This avoids the problem of having to face a singularity at $k = 0$.

Example 2: Solution to the three dimensional wave equation with source

$$\nabla^2 \psi(t, \mathbf{x}) - \frac{1}{c^2} \frac{\partial^2 \psi(t, \mathbf{x})}{\partial t^2} = s(t, \mathbf{x})$$

Let

$$\hat{\psi}(\omega, \mathbf{k}) = \int \psi(t, \mathbf{x}) exp(-i(\omega t - \mathbf{k}.\mathbf{x})) dt d^3 x$$

$$\hat{s}(\omega, \mathbf{k}) = \int s(t, \mathbf{x}) exp(-i(\omega t - \mathbf{k}.\mathbf{x})) dt d^3 x$$

Fourier transforming the wave equation with source gives

$$\hat{\psi}(\omega, \mathbf{k}) = \frac{\hat{s}(\omega, \mathbf{k})}{\omega^2/c^2 - k^2}$$

The solution is obtained by inverse Fourier transforming this equation.

Example 3: Computation of the time averaged Poynting vector in the Fourier domain.

The vector potential is given by

$$\mathbf{A}(t,\mathbf{r}) = \int \hat{\mathbf{A}}(\mathbf{k}) exp(i(kct - \mathbf{k}.\mathbf{r}))d^3k + \int \hat{\mathbf{A}}(\mathbf{k})^* exp(-i(kct - \mathbf{k}.\mathbf{r}))d^3k$$

The Gauge condition is chosen so that $\phi = 0$, $\nabla.\mathbf{A} = 0$. The second condition implies $\mathbf{k}.\hat{\mathbf{A}}(\mathbf{k}) = 0$. The electric field is given by

$$\mathbf{E}(t,\mathbf{r}) = -\frac{\partial \mathbf{A}(t,\mathbf{r})}{\partial t} = -ic\int k\hat{\mathbf{A}}(\mathbf{k})exp(i(kct - \mathbf{k}.\mathbf{r}))d^3k + ic\int k\hat{\mathbf{A}}(\mathbf{k})^* exp(-i(kct - \mathbf{k}.\mathbf{r}))d^3k$$

and the magnetic field is given by

$$\mathbf{B}(t,\mathbf{r}) = \nabla \times \mathbf{A}(t,\mathbf{r}) = -i\int \mathbf{k} \times \hat{\mathbf{A}}(\mathbf{k})exp(i(kct - \mathbf{k}.\mathbf{r}))d^3k + i\int \mathbf{k} \times \mathbf{A}(\mathbf{k})^* exp(-i(kct - \mathbf{k}.\mathbf{r}))d^3k$$

Note that our Fourier expression for the potentials has both the positive and negative frequency components. The component $exp(i(kct - \mathbf{k}.\mathbf{r}))$ has frequency $ck = c|\mathbf{k}| = \omega(\mathbf{k})$ while the component $exp(-i(kct - \mathbf{k}.\mathbf{r}))$ has frequency $-ck = -c|\mathbf{k}| = -\omega(\mathbf{k})$. Both the positive and negative frequency components are needed to ensure reality of the potentials and fields. The Poynting vector is given by

$$\mathbf{S}(t,\mathbf{r}) = \frac{1}{\mu_0}\mathbf{E}(t,\mathbf{r}) \times \mathbf{B}(t,\mathbf{r})$$

$$= -\frac{c}{\mu_0}(\int k(\hat{\mathbf{A}}(\mathbf{k})exp(i(kct - \mathbf{k}.\mathbf{r})) - \hat{\mathbf{A}}(\mathbf{k})^* exp(-i(kct - \mathbf{k}.\mathbf{r})))d^3k)$$

$$\times(\int (\mathbf{k} \times \hat{\mathbf{A}}(\mathbf{k})exp(i(kct - \mathbf{k}.\mathbf{r})) - \mathbf{k} \times \hat{\mathbf{A}}(\mathbf{k})^* exp(-i(kct - \mathbf{k}.\mathbf{r})))d^3k$$

$$= -\frac{c}{\mu_0}(\int k\hat{\mathbf{A}}(\mathbf{k}) \times (\mathbf{k}' \times \hat{\mathbf{A}}(\mathbf{k}'))exp(i(k + k')ct - (\mathbf{k} + \mathbf{k}').\mathbf{r})d^3k d^3k'$$

$$- \int k\hat{\mathbf{A}}(\mathbf{k}) \times (\mathbf{k}' \times \hat{\mathbf{A}}(\mathbf{k}')^*)exp(i(k - k')ct - i(\mathbf{k} - \mathbf{k}').\mathbf{r})d^3k d^3k'$$

$$- \int k\hat{\mathbf{A}}(\mathbf{k})^* \times (\mathbf{k}' \times \hat{\mathbf{A}}(\mathbf{k}'))exp(-i(k - k')ct + i(\mathbf{k} - \mathbf{k}').\mathbf{r})d^3k d^3k'$$

$$+ \int k\hat{\mathbf{A}}(\mathbf{k})^* \times (\mathbf{k}' \times \hat{\mathbf{A}}(\mathbf{k}')^*)exp(-i(k + k')ct + i(\mathbf{k} + \mathbf{k}').\mathbf{r})d^3k d^3k')$$

Averaging this expression over time gives

$$< \mathbf{S}(t,\mathbf{r}) >= lim_{T\to\infty}\frac{1}{T}\int_0^T \mathbf{S}(t,\mathbf{r})dt$$

$$= \frac{c}{\mu_0}\int k\mathbf{A}(\mathbf{k}) \times (\mathbf{k}' \times \hat{\mathbf{A}}(\mathbf{k}')^*)exp(-i(\mathbf{k} - \mathbf{k}').\mathbf{r}))\delta_{k,k'}d^3k d^3k'$$

$$+\frac{c}{\mu_0}\int k\mathbf{A(k)}^* \times (\mathbf{k'} \times \hat{\mathbf{A}}(\mathbf{k'}))exp(i(\mathbf{k} - \mathbf{k'}).\mathbf{r})\delta_{k,k'}d^3kd^3k'$$

Suppose we are dealing with plane waves of a definite frequency. Then, the differentials d^3k and d^3k' are replaced by $dS(\mathbf{k})$ and $dS(\mathbf{k'})$ which are surface differentials on the surface $|\mathbf{k}| = \omega/c$. The expression for the time averaged Poynting vector is then

$$< \mathbf{S}(t,\mathbf{r}) >= \frac{c}{\mu_0}\int k\hat{\mathbf{A}}(\mathbf{k}) \times (\mathbf{k'} \times \hat{\mathbf{A}}(\mathbf{k'})^*)exp(-i(\mathbf{k} - \mathbf{k'}).\mathbf{r})dS(\mathbf{k})dS(\mathbf{k'})$$

$$+\frac{c}{\mu_0}\int k\hat{\mathbf{A}}(\mathbf{k})^* \times (\mathbf{k'} \times \hat{\mathbf{A}}(\mathbf{k'}))exp(i(\mathbf{k} - \mathbf{k'}).\mathbf{r})dS(\mathbf{k})dS(\mathbf{k'})$$

We can look at the case when the wave consists of just a single wave vector. Then we get the following expression for the Poynting vector:

$$< \mathbf{S}(t,\mathbf{r}) >= \frac{c}{\mu_0}k(\mathbf{A(k)} \times (\mathbf{k} \times \mathbf{A(k)}^*) + \mathbf{A(k)}^* \times (\mathbf{k} \times \mathbf{A(k)}))$$

This is independent of the spatial location. Another important case concerns a random wave with Fourier amplitudes uncorrelated for different wave-vectors. Assuming waves of definite frequency, we obtain the following integral expression for the Poynting vector:

$$< \mathbf{S}(t,\mathbf{r}) >= \frac{c}{\mu_0}\int k(\mathbf{A(k)} \times (\mathbf{k'} \times \mathbf{A(k')}^*) + \mathbf{A(k)}^* \times (\mathbf{k'} \times \mathbf{A(k')}))dS(\mathbf{k})dS(\mathbf{k'})$$

The uncorrelatedness assumption implies that

$$< \mathbf{A(k)}.\mathbf{A(k')}^* >= P(\mathbf{k})\delta_S(\mathbf{k} - \mathbf{k'})$$

where $\delta_S(\mathbf{k} - \mathbf{k'})$ is the delta function on the surface of constant frequency, i.e.,

$$\int f(\mathbf{k})\delta_S(\mathbf{k} - \mathbf{k'})dS(\mathbf{k}) = f(\mathbf{k'})$$

with $\mathbf{k}, \mathbf{k'}$ satisfying $|\mathbf{k}| = |\mathbf{k'}| = \omega$. Then,

$$< \mathbf{S}(t,\mathbf{r}) >= \frac{2\omega}{\mu_0}\int P(\mathbf{k})kdS(\mathbf{k})$$

We've used $kc = \omega$ which is a constant for the integral.

Chapter 4

Relativity

4.1 Introduction

There are certain basic principles that are violated by the Newtonian equations of motion. According to Newtonian physics, a particle can acquire an arbitrarily large velocity. Our practical intuition however states that no speed can exceed the speed of light, for if we travel in a spacecraft that moves with a speed greater than that of light forward with a mirror placed in front of us, then light will never be able to reach the mirror from our face and hence our image will disappear. Intuition tells us that this cannot happen. An experiment was performed by Michelson and Morley to detect the presence of the ether. It was believed during the time of this experiment that light needs a medium to travel and hence we cannot have total vacuum in space. This medium was called ether. If there is a light source in one reference frame S that is moving at a speed v relative to the other frame S', then the speed of light in the second frame will according to Newtonian mechanics be given by the vectorial sum of the velocity of light and the velocity of the frame S relative to S'. Based on this assumption, Michelson and Morley constructed an interference device that could detect the motion of the earth through the ether medium. Ether would have to be postulated in agreement with the Newtonian law of velocity addition if there was a shift in the interference pattern when the interferometer was rotated by ninety degrees, so that in the former position, the interferometer was moving along the direction of the earth's motion and in the latter position, the interferometer was moving perpendicular to the direction of the earth's motion. However no such shift was observed and it was subsequently concluded by Einstein that there was no ether. However, even before Einstein gave his postulates of relativity, H.A.Lorentz postulated that there could be an ether but the rods of which the interferometer was made contracted during motion along the rod's length. This contraction according to Lorentz was caused by stresses in the rod due to the electron-ion bonding. However, Einstein threw away all these ideas by introducing the postulates of special relativity.

4.2   Breakdown of classical mechanics

The definitions and postulates of classical mechanics break down when the velocities involved approach the velocity of light, for at these high speeds the postulates do not yield predictions that are in conformity with experimental facts. Although special relativity alters the formal structure of classical mechanics, the alteration is not as violent as the alteration provided by the quantum theory. There are examples where the alteration of the classical mechanical results provided by special relativity is not as drastic as that provided by the quantum theory and likewise there are instances where quantum mechanics does not alter the description as violently as special relativity. For example when we deal with the motion of macroscopic bodies like a tennis ball, the quantum mechanical alteration is not so significant, i.e., the quantum mechanical results for a macroscopic body nearly agree with the corresponding classical results obtained by allowing Planck's constant to tend to zero. If however the tennis ball moves with a speed close to that of the speed of light, the alteration provided by the special relativity is drastic. Analogously, while dealing with the motion of microscopic bodies like the motion of an electron around the nucleus, the alteration provided by the quantum theory is drastic, indeed the correct quantum mechanical description of the motion has a probabilistic flavour which differs drastically from the classical mechanical description. However the relativistic effects for an electron are negligible, indeed the non-relativistic Schrodinger equation gives a reasonably accurate picture of the electron without having to introduce Dirac's relativistic wave equation.

4.3   Galilean transformations, Inertial frames and Michelson experiment

An inertial frame is one in which Newton's law $\mathbf{F} = m\mathbf{a}$ is valid. If S is an inertial frame and S' is a frame that accelerates relative to S, then fictitious forces need to be introduced whilst writing the equations of motion in S'. For example, if S' rotates relative to S, then the fictitious forces to be introduced are the centrifugal and coriolis forces while if S' accelerates linearly relative to S, the fictitious forces are mass times the acceleration of S' relative to S with an appropriate sign. Suppose S is a frame and S' another frame that moves relative to S with a uniform velocity $\mathbf{v}$. Assume that at time $t = 0$, the origins of the two frames are coincident. Then, after a time t has elapsed, the position vectors in the two frames are related by the equation $\mathbf{r}' = \mathbf{r} - \mathbf{v}t$. If a constant force such as gravity $m\mathbf{g}$ is the only force, then this force is the same in the unprimed and primed frames and since

$$\frac{d^2\mathbf{r}'}{dt^2} = \frac{d^2}{dt^2}(\mathbf{r} - \mathbf{v}t) = \frac{d^2\mathbf{r}}{dt^2}$$

it follows that the equations of motion in both the frames have the same form, i.e., Newton's law goes unaltered. Suppose that the forces in S are derivable from a potential that depends only on the relative positions of the particles, i.e., if $\mathbf{r}_1, ..., \mathbf{r}_n$ are the particle positions, the potential from which the forces in S are derived has the form $V(\mathbf{r}_i - \mathbf{r}_j : 1 \leq i < j \leq n)$, then the forces in S' will have the same form, i.e., the form of the equations of motion in the two frames is the same. To see this explicitly, write down the equations of motion in S as

$$m_i \frac{d^2 \mathbf{r}_i}{dt^2} = \mathbf{F}_i(\mathbf{r}_1, ..., \mathbf{r}_n)$$

where

$$\mathbf{F}_i = -\nabla_{\mathbf{r}_i} V(\mathbf{r}_k - \mathbf{r}_j : 1 \leq k < j \leq n)$$

The potential in S' is given by replacing $\mathbf{r}_i$ with $\mathbf{r}'_i + \mathbf{v}t$. We see that $\mathbf{r}_i - \mathbf{r}_j - \mathbf{r}'_i - \mathbf{r}'_j$ and hence the potential has the same form:

$$V(\mathbf{r}_i - \mathbf{r}_j : 1 \leq i < j \leq n) = V(\mathbf{r}'_i - \mathbf{r}'_j : 1 \leq i < j \leq n)$$

and hence the forces in the primed frame are

$$\mathbf{F}'_i = -\nabla'_i V(\mathbf{r}'_k - \mathbf{r}'_j : 1 \leq k < j \leq n) = \mathbf{F}_i(\mathbf{r}'_1, ..., \mathbf{r}'_n)$$

In other words, the equations of motion in S' are

$$m_i \frac{d^2 \mathbf{r}'_i}{dt^2} = \mathbf{F}_i(\mathbf{r}'_1, ..., \mathbf{r}'_n)$$

which have the same structure as the equations of motion in S. It therefore appears that using the Galilean law of transformation $\mathbf{r}' = \mathbf{r} - \mathbf{v}t$ between two frames having a constant relative velocity leads to no anomaly as far as the Newtonian equations of motion go. However, this law of transformation breaks down when one tries to explain the constancy of the speed of light propagation. The velocities of a particle in the two frames are related via the equation

$$\mathbf{u}' = \frac{d\mathbf{r}'}{dt} = \frac{d\mathbf{r}}{dt} - \mathbf{v} = \mathbf{u} - \mathbf{v}$$

If we apply this to light, then the velocity of light in S' should be $\mathbf{c}' = \mathbf{c} - \mathbf{v}$ which contradicts Michelson's experiments. Michelson's experiments were based on the interference of light to detect change in the velocity of light on passing from one inertial frame to another. We shall not describe the details of the experiment here but just mention that Michelson's apparatus consisted of a half silvered mirror and two other mirrors and a screen. Two light paths were generated using this apparatus. When the apparatus was oriented in one direction then owing to the earth's movement, there was a certain interference pattern generated on the screen, i.e. a certain relative delay was generated between the two interfering light beams owing to the earth's movement while when the apparatus was oriented in an orthogonal direction, the relative delay was different thereby producing a shift in the theoretical interference pattern if one were to assume the Galilean law of velocity addition. However such a shift was never observed thereby contradicting the correctness of the application of the Galilean law of velocity addition and the non-constancy of the velocity of light. Hence, one had to assume that the motion of the earth could not have any effect on the velocity of light and hence Galilean law of velocity addition had to be revised. Formally the Galilean transformation between two frames S, S' with S' moving with a velocity $\mathbf{v}$ relative to S is defined by $\mathbf{r}' = \mathbf{r} - \mathbf{v}t, t' = t$. Time is thus regarded as absolute. The corresponding velocity addition rule is obtained by differentiation: $\frac{d\mathbf{r}'}{dt} = \frac{d\mathbf{r}}{dt} - \mathbf{v}$. More generally, a Galilean transformation consists of such a transformation along with rotations, reflections, translations and time inversion, i.e., the Galilean group is generated by the above transformations along with the transformations $\mathbf{r}' = \mathbf{A}\mathbf{r}$ where $\mathbf{A}$ is a real matrix with $AA^T = A^T A = I$, i.e., A is an element of $O(3)$, i.e., a proper rotation or the product of a proper rotation and a reflection of the form $\mathbf{r}' = -\mathbf{r}$, and the transformation $\mathbf{r}' = \mathbf{r}, t' = -t$ a time reversal and $\mathbf{r}' = \mathbf{r} + \mathbf{a}$ where $\mathbf{a}$ is a three vector. Thus, the general element of the Galilean group can be represented as $\mathbf{r}' = \mathbf{A}\mathbf{r} - \mathbf{v}t + \mathbf{a}, t' = t$ or $\mathbf{r}' = \mathbf{A}\mathbf{r} - \mathbf{V}t + \mathbf{a}, t' = -t$ where $\mathbf{A}\mathbf{A}^T = \mathbf{I}$. Proper Galilean transformations consist of proper rotations and boosts, i.e., transformations of the form $\mathbf{r}' = \mathbf{A}\mathbf{r} - \mathbf{v}t + \mathbf{a}, t' = t$ where $\mathbf{A}$ is a 3×3 real matrix with $AA^T = A^T A = I$ and $det A = 1$. Proper Galilean transformations form a subgroup of the general Galilean group of transformations. Galilean transformations fail to preserve the velocity of light and consequently fail to preserve the Maxwell equations. Indeed, Maxwell's equations in free space lead to the wave equation for the electric and magnetic field components and the wave equation is not invariant under the Galilean transformation $\mathbf{r}' = \mathbf{r} - \mathbf{v}t, t' = t$.

4.4 Einstein's postulates of relativity and the Lorentz transformation

The two postulates are (1) the laws of physics in two frames S, S' moving relative to each other with a constant velocity have the same form, (2) the speed of light is the same in any two frames S, S' moving relative to each other with a constant velocity. The first postulate is also called the equivalence principle. Newton's laws of motion satisfy the equivalence principle if one agrees that the Galilean transformation is the right transformation between two frames moving relative to each other with a constant velocity. If the Galilean law is correct then Maxwell's equations will not satisfy the equivalence principle. One way to get over this problem is to reject the Galilean transformation and come up with another transformation that guarantees that Maxwell's equations satisfy the equivalence principle. The new transformation introduced will then prevent the Newtonian laws of motion from satisfying the equivalence principle. Consequently Newton's laws must then be modified so that the modified laws satisfy the equivalence principle with respect to the new transformation. The new transformation must be such that it involves the speed of light as a parameter in such a matter that if this parameter was made infinite, the new transformation would lead to the Galilean law of transformation. The constancy of the velocity of light in two frames S, S' moving relative to each other with the same velocity implies that

$x'^2 + y'^2 + z'^2 - c^2 t'^2 = x^2 + y^2 + z^2 - c^2 t^2$. If S' moves relative to S with a speed v along the z axis, then we can deduce from this invariance the Lorentz transformation

$$x' = x, x' = y, z' = \gamma(z - vt), t' = \gamma(t - vx/c^2)$$

where $\gamma = (1 - \beta^2)^{-1/2}, \beta = v/c$. When $\beta \to \infty$, we have $\gamma \to 1$ and this transformation reduces to the Galilean transformation: $x' = x, y' = y, z' = z - vt, t' = t$. The invariance $x^2 + y^2 + z^2 - c^2 t^2 = x'^2 + y'^2 + z'^2 - c^2 t'^2$ alone does not imply the Lorentz transformation. We must put the additional restriction that the transformation from the unprimed system to the primed coordinate system must be linear. This linearity condition can be shown to be a consequence of the fact that if S is an inertial frame then so is S', i.e., if no fictitious forces act in S, then no fictitious forces can act in S', or in other words, uniform motion in S transforms to uniform motion in S' (this is a consequence of the second Einstein postulate that the laws of physics in two frames moving relative to each other with a constant velocity have the same form). Just as reversing the sign of $\mathbf{v}$ in a Galilean transformation results in inversion of the transformation, the same holds for the Lorentz transformation. Change of the sign of $\mathbf{v}$ in the Lorentz transformation that carries S coordinates to S' coordinates results in an inverse transformation that carries S' coordinates to S coordinates. This is obvious since S' moves relative to S with a velocity $-\mathbf{v}$. The above Lorentz transformation can be generalized to be valid for uniform motion along any direction. Suppose S' moves relative to S with a velocity $\mathbf{v}$. Let $\mathbf{r}_{\parallel}$ and $\mathbf{r}'_{\parallel}$ denote the components of $\mathbf{r}$ and $\mathbf{r}'$ along the direction parallel to $\mathbf{v}$. Likewise let $\mathbf{r}_{\perp}$ and $\mathbf{r}'_{\perp}$ denote the components of $\mathbf{r}$ and $\mathbf{r}'$ on the plane perpendicular to $\mathbf{v}$, i.e.,

$$\mathbf{r}_{\parallel} = \mathbf{r}.\mathbf{v}\mathbf{v}/v^2 \qquad \mathbf{r}_{\perp} = \mathbf{r} - \mathbf{r}_{\parallel}$$

with similar expressions with $\mathbf{r}$ replaced by $\mathbf{r}'$. Then the Lorentz transformation reads

$$\mathbf{r}'_{\parallel} = \gamma(\mathbf{r}_{\parallel} - \mathbf{v}t) \qquad \mathbf{r}'_{\perp} = \mathbf{r}_{\perp}, t' = \gamma(t - \mathbf{v}.\mathbf{r}/c^2)$$

Adding the first two equations gives

$$\mathbf{r}' = \mathbf{r}'_{\parallel} + \mathbf{r}'_{\perp} = \gamma((\mathbf{r}.\mathbf{v})\mathbf{v}/v^2 - \mathbf{v}t) + \mathbf{r} - (\mathbf{r}.\mathbf{v})\mathbf{v}/v^2$$

$$= \mathbf{r} + (\gamma - 1)(\mathbf{r}.\mathbf{v})\mathbf{v}/v^2 - \gamma\mathbf{v}t$$

One way to express the Lorentz transformation is as a linear transformation on four dimensional space where the four coordinates are given by $x_1 = x, x_2 = y, x_3 = z, x_4 = ict$. The fourth coordinate is complex. The fact that the speed of light is preserved is expressed by the statement that the quadratic form $x_1^2 + x_2^2 + x_3^2 + x_4^2$ is an invariant, i.e.,

$$\sum_{i=1}^{4} x_i^2 = \sum_{i=1}^{4} x_i'^2$$

That the transformation on this four dimensional space must be linear corresponds to the fact that if S is inertial, so must be S' which moves with a uniform velocity relative to S. For uniform motion with speed v along the z axis, the Lorentz transformation reads

$$x_1' = x_1, x_2' = x_2, x_3' = \gamma(x_3 - vt) = \gamma(x_3 + i\beta x_4), x_4' = ict' = ic\gamma(t - vx_3/c^2) = \gamma(x_4 - i\beta x_3)$$

with $\beta = v/c$ or in terms of matrices,

$$\mathbf{x}' = \mathbf{L}\mathbf{x}$$

where

$$\mathbf{L} = \begin{pmatrix} 1 & 0 & 0 & 0 \\ 0 & 1 & 0 & 0 \\ 0 & 0 & \gamma & i\beta\gamma \\ 0 & 0 & -i\beta\gamma & \gamma \end{pmatrix}$$

This can be expressed as a rotation of four dimensional space with the rotation taking place only in the $x_3 - x_4$ plane:

$$\mathbf{L} = \begin{pmatrix} 1 & 0 & 0 & 0 \\ 0 & 1 & 0 & 0 \\ 0 & 0 & cos\phi & sin\phi \\ 0 & 0 & -sin\phi & cos\phi \end{pmatrix}$$

where

$$cos\phi = \gamma, sin\phi = i\beta\gamma$$

or equivalently,

$$tan\phi = i\beta$$

The rotation angle is thus imaginary. We set $\phi = i\psi$ and noting that

$$tan\phi = tan(i\psi) = \frac{exp(-\psi) - exp(\psi)}{i(exp(\psi) + exp(-\psi))} = i.tanh(\psi)$$

we get

$$\beta = tanh^{-1}(\psi)$$

$$cos\phi = cos(i\psi) = cosh(\psi) \qquad sin(\phi) = sin(i\psi) = i.sinh(\psi)$$

The matrix for the general Lorentz boost along any given direction is obtained using

$$\mathbf{r}' = \mathbf{r} + (\gamma - 1)(\mathbf{r}.\mathbf{v})\mathbf{v}/v^2 - \gamma\mathbf{v}t = \mathbf{r} + (\gamma - 1)\mathbf{r}.\beta\beta/\beta^2 + i\gamma\beta x_4$$

and

$$x_4' = ict' = ic\gamma(t - \mathbf{v}.\mathbf{r}/c^2) = \gamma(x_4 - i\beta.\mathbf{r})$$

or in matrix notation $\mathbf{x}' = \mathbf{L}\mathbf{x}$ with

$$\mathbf{L} = \begin{pmatrix} 1 + (\gamma - 1)\beta_1^2/\beta^2 & (\gamma - 1)\beta_1\beta_2/\beta^2 & (\gamma - 1)\beta_1\beta_3/\beta^2 & i\gamma\beta_1 \\ (\gamma - 1)\beta_1\beta_2/\beta^2 & 1 + (\gamma - 1)\beta_2^2/\beta^2 & (\gamma - 1)\beta_2\beta_3/\beta^2 & i\gamma\beta_2 \\ (\gamma - 1)\beta_1\beta_3/\beta^2 & (\gamma - 1)\beta_2\beta_3/\beta^2 & 1 + (\gamma - 1)\beta_3^2/\beta^2 & i\gamma\beta_3 \\ -i\gamma\beta_1 & -i\beta_2\gamma & -i\beta_3\gamma & \gamma \end{pmatrix}$$

This can be cast more elegantly as a block structured matrix:

$$\mathbf{L} = \begin{pmatrix} \mathbf{I} + (\gamma - 1)\beta\beta^T/\beta^2 & i\gamma\beta \\ -i\gamma\beta^T & \gamma \end{pmatrix}$$

In general, a Lorentz transformation on four dimensional space-time, is a linear transformation that preserves the speed of light, i.e., a linear transformation $\mathbf{x}' = \mathbf{L}\mathbf{x}$ for which $\mathbf{x}'^T\mathbf{x}' = \mathbf{x}^T\mathbf{x}$. A necessary and sufficient condition for a matrix $\mathbf{L}$ to be such that this condition is satisfied is that $\mathbf{L}^T\mathbf{L} = \mathbf{L}\mathbf{L}^T = \mathbf{I}$, i.e., $\mathbf{L}$ is a complex orthogonal matrix. This condition implies that $(det L)^2 = 1$ and hence $det L = \pm 1$. The set of Lorentz transformations for which $det L = 1$ are continuously deformable to the identity. In other words, such transformations, called the proper Lorentz transformations belong to the connected component of the identity. The equation $LL^T = I$ implies in particular that $L_{44}^2 + \sum_{i=1}^{3} L_{4i}^2 = 1$. Also

$$x_4' = ict' = L_{41}x_1 + L_{42}x_2 + L_{43}x_3 + L_{44}x_4 = \sum_{i=1}^{3} L_{4i}x_i + L_{44}ict$$

Since $t, t', x_i, i = 1, 2, 3$ are real variables, it follows that L_{44} is real while $L_{4i}, i = 1, 2, 3$ are pure imaginary. Thus, $L_{4i}^2 < 0, i = 1, 2, 3$ and the equation $L_{44} = 1 - \sum_{i=1}^{3} L_{4i}^2$ then implies that $L_{44}^2 \geq 1$. Thus, either $L_{44} \geq 1$ or $L_{44} < -1$. The set of proper Lorentz transformations L for which $L_{44} \geq 1$ are called restricted Lorentz transformations. The transformations L for which $L_{44} \leq -1$ are transformations involving time reversal. It is a well known fact that every restricted Lorentz transformation L can be expressed as $L = RP$ where R is a rotation of the spatial coordinates and P is a boost, i.e., a Lorentz transformation corresponding to uniform motion along some direction. To see this, suppose that L is a restricted Lorentz transformation. We shall determine the velocity of the primed frame associated with L. This is obtained by setting $x_i' = 0, i = 1, 2, 3$. Since $L^T = L^{-1}$ we have

$$x_j = \sum_{i=1}^{3} L_{ij}x_i' + L_{4j}x_4' = L_{4j}x_4'$$

$$x_4 = \sum_{i=1}^{3} L_{i4}x_i' + L_{44}x_4' = L_{44}x_4'$$

Thus, for the point fixed to the origin of the primed frame, we have $t = L_{44}t', x_j = iL_{4j}ct', j = 1, 2, 3$, so that $\frac{dx_j}{dt} = icL_{4j}/L_{44}, j = 1, 2, 3$, or

$$\beta_j = \frac{1}{c}\frac{dx_j}{dt} = iL_{4j}/L_{44}, j = 1, 2, 3$$

This is the velocity of the origin of the primed frame relative to the unprimed frame. We now define the matrix $P(\beta)$ as the Lorentz boost associated with the velocity $\beta = (\beta_1, \beta_2, \beta_3)$. Then, it remains to show that $R = LP(-\beta)$ is a spatial rotation. For simplicity of notation, set $P = P(-\beta)$. Then,

$$R_{41} = L_{41}P_{11} + L_{42}P_{21} + L_{43}P_{31} + L_{44}P_{41}$$

where

$$P_{11} = 1 + (\gamma - 1)\beta_1^2/\beta^2, P_{21} = (\gamma - 1)\beta_1\beta_2/\beta^2, P_{31} = (\gamma - 1)\beta_1\beta_3/\beta^2, P_{41} = i\gamma\beta_1$$

Since further,

$$L_{4j} = -iL_{44}\beta_j, j = 1, 2, 3$$

we get

$$R_{41} = L_{44}(-i\beta_1(1 + (\gamma - 1)\beta_1^2/\beta^2) - i\beta_2(\beta_1\beta_2/\beta^2) - i\beta_3(\beta_1\beta_3/\beta^2) + i\gamma\beta_1)$$

$$iL_{44}\beta_1(-1 - (\gamma - 1)(\beta_1^2 + \beta^2 + \beta_3^2)/\beta^2 + \gamma) = 0$$

Similarly, $R_{42} = R_{43} = 0$. Thus, R has the following block matrix form

$$R = \begin{pmatrix} R_1 & \mathbf{p} \\ \mathbf{0}^T & q \end{pmatrix}$$

where R_1 is a 3×3 matrix, $\mathbf{p}$ is a 3×1 column vector, $\mathbf{0}^T$ is a 1×3 row vector and q is a scalar Since L and P are Lorentz transformations, so is their product R, i.e., $RR^T = I$. In particular, R is non-singular and hence $q \neq 0$. The equation $RR^T = I$ gives

$$\begin{pmatrix} R_1 R_1^T + \mathbf{p}\mathbf{p}^T & q\mathbf{p} \\ q\mathbf{p}^T & q^2 \end{pmatrix} = I$$

from which it follows that $\mathbf{p} = \mathbf{0}$ and $R_1 R_1^T = I$. Thus, $det R_1 = \pm 1$ and hence the condition $det R = det(L)det(P) = 1$ implies $q = \pm 1$. Now L is a restricted Lorentz transformation by hypothesis and P is also a restricted Lorentz transformation. Hence their product R is also restricted. Thus, $R_{44} \geq 1$ and hence, $R_{44} = 1$. Thus,

$$R = \begin{pmatrix} R_1 & \mathbf{0} \\ \mathbf{0}^T & 1 \end{pmatrix}$$

The condition $det R = 1$ now implies $det R_1 = 1$ and hence R_1 is a rotation. This proves our claim.

4.5 Thomas precession

Let O_1 denote the laboratory frame, O_2 a frame moving with a velocity $c\beta\hat{z}$ relative to O_1 and O_3 a frame moving with a velocity $c\Delta\beta$ relative to O_2. We assume that the only non-vanishing components of $\Delta\beta$ are the y and z components, i.e., $\Delta\beta = \Delta\beta_2\hat{y} + \Delta\beta_3\hat{z}$. The transformation connecting the space-time coordinates in O_2 to the space-time coordinates in O_1 is a Lorentz boost and the transformation connecting the space-time coordinates in O_3 to those in O_2 is another Lorentz boost. The overall transformation connecting the space-time coordinates in O_3 to those in O_1 is thus a composition of two Lorentz boosts. Call the first transformation as $\mathbf{L}_1$. Then,

$$
\mathbf{L}_1 = \begin{pmatrix} 1 & 0 & 0 & 0 \\ 0 & 1 & 0 & 0 \\ 0 & 0 & \gamma & i\beta\gamma \\ 0 & 0 & -i\beta\gamma & \gamma \end{pmatrix}
$$

Call the second transformation as $\mathbf{L}_2$. Then,

$$
\mathbf{L}_2 = \begin{pmatrix} 1 & 0 & 0 & 0 \\ 0 & 1+(\gamma(\Delta\beta)-1)(\Delta\beta_2)^2/(\Delta\beta)^2 & (\gamma(\Delta\beta)-1)\Delta\beta_2\Delta\beta_3/(\Delta\beta)2 & i\gamma(\Delta\beta)\Delta\beta_2 \\ 0 & (\gamma(\Delta\beta)-1)\Delta\beta_2\Delta\beta_3/(\Delta\beta)^2 & 1+(\gamma(\Delta\beta)-1)(\Delta\beta_3)^2/(\Delta\beta)^2 & i\gamma(\Delta\beta)\Delta\beta_3 \\ 0 & -i\gamma(\Delta\beta)\Delta\beta_2 & -i\gamma(\Delta\beta)\Delta\beta_3 & \gamma(\Delta\beta) \end{pmatrix}
$$

$$
\gamma(\Delta\beta) = (1-(\Delta\beta)^2)^{-1/2} = 1+(\Delta\beta)^2/2
$$

Thus,

$$
(\gamma(\Delta\beta)-1)(\Delta\beta_2)^2/(\Delta\beta)^2 = (\Delta\beta_2)^2 = 0
$$

and similarly for other terms having the same form. Likewise,

$$
i\gamma(\Delta\beta)\Delta\beta_2 = i\Delta\beta_2
$$

and upto first order, we get

$$
\mathbf{L}_2 = \begin{pmatrix} 1 & 0 & 0 & 0 \\ 0 & 1 & 0 & i\Delta\beta_2 \\ 0 & 0 & 1 & i\Delta\beta_3 \\ 0 & -i\Delta\beta_2 & -i\Delta\beta_3 & 1 \end{pmatrix}
$$

The overall transformation connecting the coordinates of O_3 to those of O_1 is given by

$$
\mathbf{L}_2\mathbf{L}_1 =
$$

$$
\begin{pmatrix} 1 & 0 & 0 & 0 \\ 0 & 1 & 0 & i\Delta\beta_2 \\ 0 & 0 & 1 & i\Delta\beta_3 \\ 0 & -i\Delta\beta_2 & -i\Delta\beta_3 & 1 \end{pmatrix} \cdot \begin{pmatrix} 1 & 0 & 0 & 0 \\ 0 & 1 & 0 & 0 \\ 0 & 0 & \gamma & i\beta\gamma \\ 0 & 0 & -i\beta\gamma & \gamma \end{pmatrix}
$$

$$
= \begin{pmatrix} 1 & 0 & 0 & 0 \\ 0 & 1 & \gamma\beta\Delta\beta_2 & i\gamma\Delta\beta_2 \\ 0 & 0 & \gamma+\gamma\beta\Delta\beta_3 & i\gamma\beta+i\gamma\Delta\beta_3 \\ 0 & -i\Delta\beta_2 & -i\gamma\Delta\beta_3 - i\beta\gamma & \gamma+\gamma\beta\Delta\beta_3 \end{pmatrix}
$$

This is a new Lorentz transformation and hence has the form $L = RP$ where P is a boost and R a rotation. The velocity components of the boost part P are given by

$$\beta_j' = iL_{4j}/L_{44}, j = 1, 2, 3$$

These evaluate to

$$\beta_1' = 0$$

$$\beta_2' = \Delta\beta_2/\gamma(1 + \beta\Delta\beta_3)$$

$$= \Delta\beta_2/\gamma, \beta_3' = (\gamma\beta + \gamma\Delta\beta_3)/\gamma(1 + \beta\Delta\beta_3) = (\beta + \Delta\beta_3)(1 - \Delta\beta_3) = \beta + \Delta\beta_3(1 - \beta)$$

These components can be used to reconstruct the matrix P and hence determine the rotation R. We do not provide the computations here but just state that R differs from the identity by nonvanishing first order linear terms. Physically this means that the combination of two boosts along different directions produces a rotation of the final coordinates. This phenomenon is known as the Thomas precession. It is a purely relativistic phenomenon.

4.6   Length contraction and time dilation

Consider two frames S, S'. S' moves relative to S with a constant velocity along the x axis. At $t = 0$, the two frames are coincident. The Lorentz transformation connecting the space-time coordinates of the two frames is given by

$$x' = \gamma(x - vt), t' = \gamma(t - vx/c^2), y' = y, z' = z$$

and the inverse transformation by

$$x = \gamma(x' + vt'), t = \gamma(t + vx'/c^2)$$

Consider a rigid rod attached to the S' frame. Assume that the rod lies along the x' axis with its ends at $x' = 0$ and $x' = L_0$. L_0 is the length of the rod in the S' frame, i.e., in a frame relative to which the rod is at rest. Now we take measurements of the two ends of the rod in S at the same time t. If x_1', x_2' are the x coordinates of the rod at this instant (t in the S frame, note that time t in the S frame corresponds to two different times in the S' frame for the two ends of the rod), then $x_2' - x_1' = L$. Taking the difference of the equations

$$x_1' = \gamma(x_1 - vt), x_2' = \gamma(x_2 - vt)$$

gives

$$x_2' - x_1' = \gamma(x_2 - x_1)$$

so that at time t, the length of the rod as measured in the S frame (relative to which the rod is moving with a constant velocity v) is given by

$$L = x_2 - x_1 = (x_2' - x_1')/\gamma = L_0\sqrt{1 - v^2/c^2}$$

This is called the phenomenon of length contraction. Moving rods appear to be shorter than their true length. The next phenomenon is that of time dilation. In short, this theorem states that a clock in a moving frame is slower than that in the rest frame. Suppose a particle moving along the x axis with a velocity v. Relative to the moving frame of the clock, let the time coordinate be t'. Then, we have $t' = \gamma(t - vx/c^2)$ where (t, x) are the time and x coordinates relative to the rest frame. The infinitesimal change in the time recorded by the moving clock is

$$dt' = \gamma(dt - vdx/c^2) = \gamma(1 - \frac{v}{c^2}dx/dt)dt = \gamma(1 - v^2/c^2)dt = dt\sqrt{1 - v^2/c^2}$$

This formula is applicable even for non-uniform motion. Thus if τ is the time measured by a clock moving along a trajectory $\mathbf{r}(t)$ with velocity $\mathbf{v}(t)$, then

$$d\tau = dt\sqrt{1 - v^2/c^2}$$

and the time interval measured by the moving clock is given by

$$\int_1^2 d\tau = \tau(2) - \tau(1) = \int_{t_1}^{t_2} dt\sqrt{1 - v^2(t)/c^2}$$

This equation is sometimes expressed by saying that the proper time interval is given by

$$d\tau^2 = dt^2 - (dx^2 + dy^2 + dz^2)/c^2$$

In particular, the moving clock runs slower. There is a paradox associated with this phenomenon called the twin paradox. It runs as follows. Suppose an astronaut travels in a rocket a distance d upward with a uniform velocity v relative to the earth, then reverses his direction and returns back to the earth with a velocity v. The time dilation principle then states that if t is the time elapsed on the earth starting from the astronaut's departure upto his return, and τ is the corresponding time recorded by the astronaut, then $\tau = t\sqrt{1 - v^2/c^2}$, i.e., the astronaut has aged lesser than the observer on the earth. However, one could reverse the argument by considering the rocket as the rest frame of reference and the observer on the earth as the traveler. The observer on the earth travels relative to the astronaut with a velocity v in one direction a distance d and then in the reverse direction with the same speed. So by regarding the rocket as the rest frame, one could argue that the observer on the earth must have aged less than the astronaut. The fallacy in this argument is that the whole argument is centred around the notion of an inertial frame. The earth is regarded as an inertial frame while the rocket is a noninertial frame since it reverses its direction of motion at some stage. Thus, the two frames are not equivalent. The observer on the earth has a privileged reference frame in which there are no fictitious forces while the astronaut has a frame in which there are inertial forces produced by the acceleration of the rocket.

4.7   Addition of velocities

Let S' be a frame that moves with a velocity v relative to S along the x axis. Then, the coordinates in the two frames are related by

$$x' = \gamma(v)(x - vt), t' = \gamma(v)(t - vx/c^2)$$

Thus the velocity of a particle in the S' frame is given by

$$u' = \frac{dx'}{dt'} = \frac{\gamma(v)(dx - vdt)}{\gamma(v)(dt - vdx/c^2)} = \frac{dx - vdt}{dt - vdx/c^2} = \frac{dx/dt - v}{1 - \frac{v}{c^2}dx/dt} = \frac{u - v}{1 - uv/c^2}$$

where $u = dx/dt$ is the velocity of the particle in S and $u' = dx'/dt'$ is the velocity of the particle in S'. The inverse of this equation is

$$u = \frac{u' + v}{1 + u'v/c^2}$$

This is the formula for velocity addition. If in my frame, a particle is moving with a velocity u' and I run along the same direction with a velocity v relative to the ground, then the velocity of the particle relative to the ground will be not $u' + v$ but $\frac{u'+v}{1+u'v/c^2}$.

4.8   Mass energy equivalence and variation of mass with velocity

Let m_0 denote the rest mass of a particle, i.e., the mass of the particle relative to a reference frame in which it is at rest. The proper time, i.e., the time interval relative to the rest frame of the particle is given by $d\tau = dt\sqrt{1 - v^2/c^2}$ where $\mathbf{v}$ is the velocity of the particle relative to an inertial frame S. Clearly like rest mass, $d\tau$ is a Lorentz invariant, i.e., a scalar. It remains the same in two frames having a uniform relative velocity. The four tuple (x, y, z, t) is a four vector, i.e., transforms from one inertial frame to another in accordance with the Lorentz transformation. It is clear that its differential (dx, dy, dz, dt) is also a four vector. The ratio of a four vector and a scalar is clearly also a four vector. Hence $(dx/d\tau, dy/d\tau, dz/d\tau, dt/d\tau)$ is a four vector. Taking another derivative relative to τ of this quantity will also preserve the four vector character. That is, $\left(\frac{d^2x}{d\tau^2}, \frac{d^2y}{d\tau^2}, \frac{d^2z}{d\tau^2}, \frac{d^2t}{d\tau^2}\right)$ is also a four vector. The equations of motion in special relativity should be Lorentz covariant, i.e., should be valid in all frames moving relative to each other with a constant velocity. For this to be the case, the equations of motion should have the form

$$(\frac{d^2x}{d\tau^2}, \frac{d^2y}{d\tau^2}, \frac{d^2z}{d\tau^2}, \frac{d^2t}{d\tau^2}) = (F_x, F_y, F_z, F_t)$$
.

where (F_x, F_y, F_z, F_t) is a four vector called the force four vector. To get agreement with Newtonian mechanics in the limit $c \to \infty$, it is clear that (F_x, F_y, F_z) should in this limit become the usual force three vector of Newtonian mechanics. This is because in the limit $c \to \infty$, we have $d\tau \to dt$ and hence

$$(\frac{d^2 x}{d\tau^2}, \frac{d^2 y}{d\tau^2}, \frac{d^2 z}{d\tau^2}) \to (\frac{d^2 x}{dt^2}, \frac{d^2 y}{dt^2}, \frac{d^2 z}{dt^2})$$

is the usual Newtonian three acceleration. Newton's equations of motion

$$(\frac{d^2 x}{dt^2}, \frac{d^2 y}{dt^2}, \frac{d^2 z}{dt^2}) = (f_x, f_y, f_z)$$

with (f_x, f_y, f_z) the Newtonian force three vector are not Lorentz covariant. They are covariant only with respect to the Galilean transformations. We can also define a four momentum vector, i.e., another four vector by the equation

$$(P_x, P_y, P_z, P_t) = m_0 (\frac{dx}{d\tau}, \frac{dy}{d\tau}, \frac{dz}{d\tau}, \frac{dt}{d\tau})$$

namely, as the rest mass times the velocity four vector. The first three components coincide with the ordinary three momentum components of Newtonian mechanics in the limit $c \to \infty$. We write

$$\mathbf{V} = (\frac{dx}{d\tau}, \frac{dy}{d\tau}, \frac{dz}{d\tau}, \frac{dt}{d\tau})$$

the velocity four vector. Then, the momentum four vector can be written as $\mathbf{P} = m_0 \mathbf{V}$. If $\mathbf{F}$ denotes the force four vector, then the equation of motion is

$$m_0 \frac{d\mathbf{V}}{d\tau} = \mathbf{F}$$

Since $d\tau = \gamma dt$ and $\mathbf{V} = (\gamma v_x, \gamma v_y, \gamma v_z, \gamma)$ where $v_x = dx/dt, v_y = dy/dt, v_z = dz/dt$ the first three components of this equation can be expressed as

$$\gamma \frac{d}{dt}(\gamma m_0 \mathbf{v}) = (F_x, F_y, F_z)$$

We write $F_x = \gamma f_x, F_y = \gamma f_y, F_z = \gamma f_z$ so that this equation is the same as

$$\frac{d}{dt}(\gamma m_0 \mathbf{v}) = \mathbf{f}$$

Experiment confirms that this is the correct equation of motion when $\mathbf{f}$ is taken to as the usual Newtonian three force acting on the particle. Newton's equation of motion $\frac{d\mathbf{p}}{dt} = \mathbf{f}$ is therefore correct even in the special theory of relativity provided that we take the three momentum as $\mathbf{p} = \gamma m_0 \mathbf{v}$, or equivalently, $\mathbf{f} = m\mathbf{v}$, with $m = \gamma m_0$. This is to be interpreted by saying that momentum equals mass times three velocity with mass a function of the velocity in accordance with $m = \gamma m_0 = \frac{m_0}{\sqrt{1-v^2/c^2}}$. If $\mathbf{A} = (A_x, A_y, A_z, A_t)$ and $\mathbf{B} = (B_x, B_y, B_z, B_t)$ are two four vectors, then the quantities $\mathbf{A}^2 = A_x^2 + A_y^2 + A_z^2 - c^2 A_t^2$ and $\mathbf{B}^2 = B_x^2 + B_y^2 + B_z^2 - c^2 B_t^2$ are Lorentz invariants. But $(A_x + B_x, A_y + B_y, A_z + B_z, A_t + B_t)$ is also a Lorentz four vector and hence $(\mathbf{A} + \mathbf{B})^2 = (A_x + B_x)^2 + (A_y + B_y)^2 + (A_z + B_z)^2 - c^2 (A_t + B_t)^2$ is also a Lorentz invariant. It follows that

$$A_x B_x + A_y B_y + A_z B_z - c^2 A_t B_t = \mathbf{A}.\mathbf{B} = \frac{1}{2}((\mathbf{A} + \mathbf{B})^2 - \mathbf{A}^2 - \mathbf{B}^2)$$

is also a Lorentz invariant. We note that

$$\mathbf{V}^2 = \gamma^2(v_x^2 + v_y^2 + v_z^2 - c^2) = -c^2$$

and hence,

$$2\mathbf{V}.\frac{d\mathbf{V}}{dt} = \frac{d}{dt}\mathbf{V}^2 = 0$$

In other words, the four velocity and four acceleration are orthogonal to each other relative to the Lorentz metric. Since $\frac{d\mathbf{V}}{dt} = \mathbf{F}/m_0$, it follows that

$$\mathbf{F}.\mathbf{V} = 0$$

Since $\mathbf{V} = \gamma(\mathbf{v}, 1)$ and $\mathbf{F} = (\gamma\mathbf{f}, F_t)$, it follows that $\mathbf{f}.\mathbf{v} - c^2 F_t = 0$. But, $\mathbf{f}.\mathbf{v} = \frac{dE}{dt}$ is the rate at which the force does work on the particle. Thus, $F_t = \frac{1}{c^2}\frac{dE}{dt}$. In other words, the fourth component of the four force vector is proportional to the rate at which the particle energy increases. Now the equation of motion implies that $m_0 \frac{d\gamma}{dt} = F_t$. Combining this with the previous equation gives

$$\frac{d}{dt}(\gamma m_0 c^2 - E) = 0$$

so that if the particle velocity changes from v_1 to v_2 over a duration of time, then the change in the kinetic energy of the particle is given by $(\gamma(v_2) - \gamma(v_1))m_0 c^2$. Since $m = \gamma m_0$ is the instantaneous mass of the particle, we can write this equation as $\Delta E = \Delta m c^2$. This is the fundamental mass energy relation of Einstein because of which special relativity became famous. It states that the quantity of matter in a particle is directly proportional to its energy. It from the basis of extracting atomic energy and led to the development of the atomic bomb which were used recklessly by the Americans to destroy the Japanese cities Hiroshima and Nagasaki during the second world war.

4.9   Invariance of the Maxwell equations under Lorentz transformations

Maxwell's equations can be formulated using the four potential $(A_x, A_y, A_z, V/c)$ where (A_x, A_y, A_z) is the magnetic vector potential and V the scalar potential. The experimentally verified fact is that this forms a four vector relative to Lorentz transformations. For simplicity of notation, we set $A^1 = A_x, A^2 = A_y, A^3 = A_z, A^4 = V/c$. This is called the contravariant four potential vector. We also construct a covariant four potential vector $A_1 = A_x, A_2 = A_y, A_3 = A_z, A_4 = -V/c$. Any vector (A^1, A^2, A^3, A^4) that transforms according to the Lorentz transformation from one reference frame to another moving relative to this with a uniform velocity is called a contravariant four vector. Associated with such a vector, one can define a covariant four vector $A_1 = A^1, A_2 = A^2, A_3 = A^3, A_4 = -A^4$. Then the scalar product $\mathbf{A.B}$ between two four vectors (contravariant) can be expressed as $\sum_{i=1}^{4} A^i B_i$, or with an abbreviated notation as $A^i B_i$, provided that we choose our units so that $c = 1$. We define the Lorentz metric as $G = diag(1, 1, 1, -1)$. If $\mathbf{A}$ is a contravariant four vector in column format, then the corresponding covariant vector is $\mathbf{GA}$. The scalar product between two four vectors $\mathbf{A}$ and $\mathbf{B}$ is then $\mathbf{B'GA} = \mathbf{A'GB}$. Let $\mathbf{A}$ be a contravariant vector in column form and let $\mathbf{A}_c = \mathbf{GA}$ be the corresponding covariant vector. The transformed contravariant vector is given by $\mathbf{B} = \mathbf{LA}$. The corresponding law of transformation for the covariant vector is $\mathbf{B}_c = \mathbf{GLA} = \mathbf{GLGA}_c$. For simplicity of notation, we set $\mathbf{L}_c = \mathbf{GLG}$. Then, $\mathbf{B}_c = \mathbf{L}_c \mathbf{A}_c$. Now given two contravariant vectors $\mathbf{A}, \mathbf{B}$, we can form a contravariant tensor whose $(i, j)^{th}$ component equals $A^i B^j$. This can either be viewed as the column vector $\mathbf{A} \otimes \mathbf{B}$, or as the matrix $\mathbf{AB'}$. If it is viewed in the former way, its law of transformation is given by the matrix $\mathbf{L} \otimes \mathbf{L}$ while if it is viewed in the latter way, its law of transformation is given by $\mathbf{AB'} \to \mathbf{LAB'} = \mathbf{L'}$. Viewing the tensor in the latter way, we can construct more generally a tensor by taking a sum of such terms, i.e., a sum of $\mathbf{AB'}$ over different contravariant vectors $\mathbf{A}$ and $\mathbf{B}$. Denoting such a tensor by the 4×4 matrix $\mathbf{F}$, its transformation law is given by $\mathbf{F} \to \mathbf{LFL'}$. With A^i as the electromagnetic four potential, we define $F_{ij} = A_{j,i} - A_{i,j}$. F_{ij} is then a covariant tensor. Its components are

$$F_{4m} = A_{m,4} - A_{4,m} = A^m, 4 + A^4, m = \frac{\partial A^m}{\partial t} + \frac{\partial V}{\partial x^m} = -E_m, m = 1, 2, 3$$

where (E_1, E_2, E_3) is the electric field.

$$F_{mn} = A_{n,m} - A_{m,n} = A^n, m - A^m, n, m, n = 1, 2, 3$$

so that $F_{23} = B_1, F_{31} = B_2, F_{12} = B_3$ where (B_1, B_2, B_3) is the magnetic field. We now construct $\mathbf{F}_{contr} = \mathbf{GFG'}$. $\mathbf{F}$ is called the covariant field tensor and $\mathbf{F}_{contr}$ is called the contravariant field tensor. In terms of components, $F^{ij} = g^{ik} g^{jr} F_{kr}$ summation over r being implied. We evaluate this contravariant tensor. Before however doing so, we mention about its Lorentz transformation law. Let L have elements L_j^i. Then the law of transformation of a contravariant vector $B = LA$ can be expressed in matrix notation as $B^i = L_j^i A^j$ summation over j being implied. The transformation law of the corresponding covariant vector is given by $B_i = g_{ij} B^i = g_{ij} L_k^j A^k = g_{ij} L_k^j g^{kr} A_r$, where g^{ij} is the inverse matrix of g_{ij}. In our special relativistic theory, both g^{ij} and g_{ij} coincide and hence we could write this as $B_c = GLGA_c$ with L_c having matrix elements $(L_c)_{ir} = g_{ij} L_k^j g_{kr}$. The transformation law of F^{ij} is $F^{ij'} = L_k^i L_r^j F_{kr}$ which in matrix notation can be expressed as $F' = LFL^T$. The corresponding transformation law of F_{ij} is $F_{ij}' = (L_c)_i^k (L_c)_j^r F_{kr}$ where $(L_c)_i^j = g_{ir} L_m^r g^{jm} = g_{ir} g_{jm} L_m^r$. This the component form of the equation $L_c = GLG$. The transformation law of the covariant field tensor F_c is in matrix notation, $F_c' = L_c F L_c^T$. We now explicitly evaluate the components of the tensors F and F_c and show how the Maxwell equations in free space can be expressed in terms of this tensor. We shall choose our units so that $c = 1$. This means that the coordinate contravariant four vector is $(x, y, z, t) = (x^1, x^2, x^3, x^4)$ and the four potential contravariant four vector is (A_x, A_y, A_z, V) Now,

$$F^{4m} = g^{44} g^{mm} F_{4m} = -F_{4m} = E_m, m = 1, 2, 3$$

$$F^{12} = g^{11} g^{22} F_{12} = F_{12} = B_3, F^{23} = g^{22} g^{33} F_{23} = F_{23} = B_1, F^{31} = g^{33} g^{11} F_{31} = F_{31} = B_2$$

The Maxwell equation $\nabla.\mathbf{E} = 0$ when expressed in terms of the tensor F is

$$\sum_{m=1}^{3} F^{4m}_{,m} = 0$$

which is the same as

$$\sum_{m=1}^{4} F^{4m}_{,m} = 0$$

The Maxwell equation $\nabla \times \mathbf{B} = \frac{\partial \mathbf{E}}{\partial t}$ has three components. The first is

$$B_{3,2} - B_{2,3} = E_{1,4}$$

In terms of F, this reads

$$F_{12,2} - F_{31,3} = -F_{41,4}$$

or

$$F_{12,2} + F_{13,3} + F_{41,4} = 0$$

or

$$F^{12}_{,2} + F^{13}_{,3} + F^{14}_{,4} = 0$$

which is the same as

$$\sum_{m=1}^{4} \frac{\partial F^{1m}}{\partial x^m} = 0$$

with analogous equations for the other components. These two Maxwell equations can be combined to give

$$\sum_{m=1}^{4} \frac{\partial F^{ij}}{\partial x^j} = 0, i = 1, 2, 3, 4$$

or using Einstein's summation convention,

$$\frac{\partial F^{ij}}{\partial x^j} = 0$$

The Maxwell equation $\nabla \times \mathbf{E} = -\frac{\partial \mathbf{B}}{\partial t}$ has first component,

$$E_{3,2} - E_{2,3} = -B_{1,4}$$

which when expressed in terms of F reads

$$-F_{43,2} + F_{42,3} = -F_{23,4}$$

or

$$F_{34,2} + F_{42,3} + F_{23,4} = 0$$

Similarly for the other components. The Maxwell equation $\nabla.\mathbf{B} = 0$ is

$$B_{1,1} + B_{2,2} + B_{3,3} = 0$$

which when expressed in terms of F reads

$$F_{23,1} + F_{31,2} + F_{12,3} = 0$$

These two Maxwell equations combine to the single tensor equation

$$F_{ij,k} + F_{jk,i} + F_{ki,j} = 0$$

This equation is in fact a consequence of our definition of the field tensor in terms of the four potential: $F_{ij} = A_{j,i} - A_{i,j}$. For,

$$F_{ij,k} + F_{jk,i} + F_{ki,j} = (A_{j,i} - A_{i,j})_{,k} + (A_{k,j} - A_{j,k})_{,i} + (A_{i,k} - A_{k,i})_{,j}$$

$$= A_{j,ik} - A_{i,jk} + A_{k,ji} - A_{j,ki} + A_{i,kj} - A_{k,ij} = 0$$

in view of the interchangeability of the order in which partial derivatives are computed. Thus if one assumes the definition of the field tensor in terms of the four potential, then Maxwell's equations in free space reduce to the single tensor equation $F^{ij}_{\;,j} = 0$. To write down the Maxwell equations in the presence of charges and currents, we define the four current vector $(J_x, J_y, J_z, \rho) = (J^1, J^2, J^3, J^4)$. Then, the Maxwell equation $\nabla.\mathbf{E} = \rho/\epsilon$ reads

$$\sum_{m=1}^{4} \frac{\partial F^{4m}}{\partial x^m} = \rho/\epsilon$$

Similarly, the Maxwell equation $\nabla \times \mathbf{B} = \frac{\partial \mathbf{E}}{\partial t} + \mu \mathbf{J}$ has first component

$$F_{12,2} - F_{31,3} = -F_{41,4} + \mu J_1$$

or

$$\sum_{m=1}^{4} \frac{\partial F^{1m}}{\partial x^m} = \mu J^1$$

with similar expressions for the other components. These can be combined to the single tensor equation

$$F^{ij}_{,j} = \mu J^i$$

If this equation holds in a reference frame S, then it also holds is a frame S' that is a rotated version of S moving with a uniform velocity relative to S, for the space-time coordinates of S' are related to those of S via a Lorentz transformation and since F^{ij} is a contravariant tensor, $F^{ij}_{,j}$ is a contravariant vector and so is J^i. Thus, the validity of the Maxwell equation in S implies

$$L^k_i F^{ij}_{,j} = \mu L^k_i J^i$$

where L is the Lorentz transformation that carries the space time coordinates of S to those of S', i.e., $x^{i'} = L^i_k x^k$. But since $F^{ij}_{,j}$ is a contravariant vector (you should verify this), it follows that

$$L^k_i F^{ij}_{,j} = F^{k'j'}_{,j'}$$

This is an abbreviation for

$$\sum_{j,k=1}^{4} L^k_i \frac{\partial F^{ij}}{\partial x^j} = \sum_{j=1}^{4} \frac{\partial F^{k'j'}}{\partial x^{j'}}$$

Also

$$J^{k'} = L^k_i J^i$$

and hence we deduce that

$$F^{i'j'}_{,j'} = \mu J^{i'}$$

i.e. the Maxwell equations are also valid in S'. We shall now deduce the law of transformation of the components of the electric and magnetic field components under a Lorentz boost. Suppose S' moves relative to S along the x axis. Then, the space time coordinates and potentials in S and S' are related by the equations

$$x' = \gamma(x - vt),\, t' = \gamma(t - vx/c^2),\, y' = y,\, z' = z$$

$$A'_x = \gamma(A_x - vV/c^2),\, V' = \gamma(V - vA_x),\, A'_y = A_y,\, Az' = A_z$$

with the inverse transforms

$$x = \gamma(x' + vt'),\, t = \gamma(t' + vx'/c^2),\, y = y',\, z = z'$$

and

$$A_x = \gamma(A'_x + vV'/c^2),\, V = \gamma(V' + vA'_x),\, A_y = A'_y,\, A_z = A'_z$$

For the electric field components, we have

$$E'_x = -\frac{\partial V'}{\partial x'} - \frac{\partial A'_x}{\partial t'}$$

$$= -\frac{\partial x}{\partial x'}\frac{\partial V'}{\partial x} - \frac{\partial t}{\partial x'}\frac{\partial V'}{\partial t}$$

$$- \frac{\partial x}{\partial t'}\frac{\partial A'_x}{\partial x} - \frac{\partial t}{\partial t'}\frac{\partial A'_x}{\partial t}$$

$$= -\gamma\frac{\partial V'}{\partial x} - (\gamma v/c^2)\frac{\partial V'}{\partial t}$$

$$-\gamma v\frac{\partial A'_x}{\partial x} - \gamma\frac{\partial A'_x}{\partial t}$$

$$= -\gamma^2(V_{,x} - vA_{x,x}) - (\gamma^2 v/c^2)(V_{,t} - vA_{x,t}) - \gamma^2 v(A_{x,x} - (v/c^2)V_{,x}) - \gamma^2(A_{x,t} - (v/c^2)V_{,t})$$

$$= (-\gamma^2 + \gamma^2 v^2/c^2)V_{,x} + (\gamma^2 v - \gamma^2 v)A_{x,x} + (-\gamma^2 v/c^2 + \gamma^2 v/c^2)V_{,t} + (\gamma^2 v^2/c^2 - \gamma^2)A_{x,t}$$

$$= -V_{,x} - A_{x,t} = E_x$$

The x component of the electric field goes unchanged. In general, if S' moves with a uniform velocity with respect to S, then the component of the electric field along the direction of motion is unchanged. We use

$$\frac{\partial}{\partial t'} = \frac{\partial x}{\partial t'}\frac{\partial}{\partial x} + \frac{\partial t}{\partial t'}\frac{\partial}{\partial t}$$

$$= \gamma v\frac{\partial}{\partial x} + \gamma\frac{\partial}{\partial t}$$

to get

$$E'_y = -\frac{\partial V'}{\partial y'} - \frac{\partial A'_y}{\partial t'}$$

$$= -\frac{\partial V'}{\partial y} - \frac{\partial A_y}{\partial t'}$$

$$= -\gamma(V_{,y} - vA_{x,y}) - \gamma v A_{y,x} - \gamma A_{y,t}$$

$$= \gamma(-V_{,y} - A_{y,t}) - \gamma v(A_{y,x} - A_{x,y}) = \gamma E_y - \gamma v B_z = \gamma(E_y - vB_z)$$

A similar computation shows that

$$E'_z = \gamma(E_z + vB_y)$$

Similar transformations for the magnetic field give

$$B'_x = B_x,\ B'_y = \gamma(B_y + vE_z),\ B'_z = \gamma(B_z - vE_y)$$

We leave the derivations of these to the student. You will have to use the change of variable formula

$$\frac{\partial}{\partial x'} = \frac{\partial x}{\partial x'}\frac{\partial}{\partial x} + \frac{\partial t}{\partial x'}\frac{\partial}{\partial t}$$

$$= \gamma\frac{\partial}{\partial x} + (\gamma v/c^2)\frac{\partial}{\partial t}$$

4.10　Principle of equivalence and general relativity

The principle of equivalence states that any gravitational field over an infinitesimally small region of space and an infinitesimally small interval of time can be canceled out by a local transformation of space-time coordinates. If the gravitational field is globally constant, then the appropriate change of coordinates corresponds to the coordinates inside a freely falling elevator in the gravitational field. When the coordinates are transformed then the space time metric also gets altered appropriately. In an inertial frame, i.e., a frame in which there is no gravitational field, the space-time metric is Minkowskian, i.e., $ds^2 = dx^2 + dy^2 + dz^2 - c^2 dt^2$. Let S be such a frame. Let S be a frame that moves relative to S with a uniform velocity v along the x axis. The space-time coordinates in S' are related to those in S by a Lorentz transformation, i.e., $x = \gamma(x' + vt'), t = \gamma(t' + vx'/c^2), y = y', z = z'$. The resulting metric is invariant, i.e., $ds^2 = dx'^2 + dy'^2 + dz'^2 - c^2 dt'^2$. If however, S' is a frame that accelerates relative to S, then the space time metric will not remain invariant for then v becomes a function of time. More generally, we can consider S' to be any frame in which the coordinates are related to those of S by a nonlinear transformation. For simplicity of notation, let $(x^1, x^2, x^2, x^4) = (x, y, z, ict)$ and $(x^{1'}, x^{2'}, x^{3'}, x^{4'}) = (x', y', z', ict')$ where $x^i = f_i(x^{1'}, x^{2'}, x^{3'}, x^{4'}), i = 1, 2, 3, 4$. The metric in the inertial frame S can be expressed as $ds^2 = \sum_{i=1}^{4} dx^{i2}$. The metrical distance ds^2 is required to be an invariant in all reference frames in the general theory of relativity. This is to ensure that $(-ds^2/c^2)^{1/2}$ measures the proper time interval along a particle's trajectory, i.e., the time measured by a clock attached to the moving particle. Hence, the metric in S' (which is in general non-inertial) is given by

$$ds'^2 = ds^2 = \sum_i dx^{i2} = \sum_i (\sum_j \frac{\partial f_i}{\partial x^{j'}} dx^{j'})^2 = \sum_{j,k} g'_{jk} dx^{j'} dx^{k'}$$

where

$$g'_{jk} = \sum_i \frac{\partial f_i}{\partial x^{j'}} \frac{\partial f_i}{\partial x^{k'}}$$

g'_{jk} is called the metric in S'. It is in general, a function of the space-time coordinates in S'. The fact that S' accelerates relative to S means that S' is noninertial and hence there are fictitious forces like coriolis and centrifugal forces in S'. By applying the inverse transformation to $\{f_i\}$, we can pass over from the non-inertial frame S' to the inertial frame S where there are no fictitious forces. This situation can be compared to the cancelation of a globally constant gravitational field by taking measurements in a freely falling elevator or by the local cancelation of a non-uniform gravitational field by a local transformation of coordinates. This suggests that the gravitational field can be regarded as arising due to a local transformation of coordinates which reflects in the appearance of a metric tensor g_{ij} in the frame that differs from that of flat space time $diag(1, 1, 1, 1)$. In other words, in a frame in which there is a gravitational field, the space time metric will be given by $ds^2 = \sum_{i,j} g_{ij} dx^i dx^j$ where the $g'_{ij}s$ are in general functions of the $x^{i's}$. However the g_{ij} discussed above can be transformed to the identity by a global transformation of coordinates. We can more generally consider a metric $ds^2 = \sum_{i,j} g_{ij} dx^i dx^j$ that cannot be transformed into the Minkowskian form by a global transformation of coordinates, i.e., there does not exist any set of four smooth functions $x^i = f_i(y^1, y^2, y^3, y^4), i = 1, 2, 3, 4$ for which

$$\sum_{i,j} g_{ij} \frac{\partial f_i}{\partial x^k} \frac{\partial f_j}{\partial x^m} = \delta_{km}, k, m = 1, 2, 3, 4$$

over a finite region of space-time. At a given point **y** of space-time, we can always select functions f_i for which the above equality holds, but not at all points in a finite region of space-time. It was Einstein's genius to guess that gravitational fields produced by a finite distribution of matter cannot be canceled over a finite region of space-time and hence the metric in such a field is irreducible. A reducible metric is that produced by applying a coordinate transformation to an inertial frame. An inertial frame is one that is nonaccelerating relative to the distant stars and galaxies of our universe. In such a frame, gravitational field is absent. A frame that accelerates relative to an inertial frame has a gravitational field but this field is reducible, i.e., can be canceled out by a coordinate transformation that corresponds to going back to the original inertial frame. An irreducible metric leads to a space time having a nonzero curvature. The curvature tensor for a Riemannian manifold is a rank four tensor, i.e., has four indices and from it one can construct a scalar curvature, i.e., an invariant. Since an irreducible metric is related according to Einstein to the gravitational field produced by a definite quantity of matter and since an irreducible metric yields a nonzero curvature, Einstein made the remarkable guess that the curvature of space-time should be related to the energy momentum tensor of matter. He was thus able to arrive at field equations for gravitation. The predictions made using these field equations have been verified experimentally with remarkable accuracy and have without doubt established their validity in situations where the Newtonian inverse square law of gravitation breaks down.

4.11 Space-time metric and the geodesic equation

In general a four dimensional curved space-time (curvature is produced by matter distribution), can be viewed as a four dimensional manifold immersed in N dimensional Euclidean space. If the metric is reducible (i.e., corresponds to a definite matter distribution), then as mentioned in the previous section, N can be taken as four while if the metric is not reducible, then we've got to take $N > 4$. Suppose then that x^1, x^2, x^3, x^4 are the coordinates in our frame and $y^1, ..., y^N$ are the coordinates in the higher dimensional Euclidean space in which our space-time manifold is immersed. Then, $y^1, ..., y^N$ are functions of the $x^{i's}$, i.e., $y^i = y^i(x^1, x^2, x^3, x^4), i = 1, 2, ..., N$. The metric of our space-time is given by

$$ds^2 = \sum_{n=1}^{N} (dy^n)^2 = \sum_{i,j=1}^{4} g_{ij} dx^i dx^j$$

where

$$g_{ij} = \sum_{n=1}^{N} \frac{\partial y^n}{\partial x^i} \frac{\partial y^n}{\partial x^j}$$

Einstein's postulate was that the trajectory of motion of a particle moving in the gravitational field described by the metric tensor g_{ij} is obtained by minimizing the proper distance or proper time between two space time points, i.e., from the variational principle

$$\delta \int_1^2 \sqrt{g_{ij} dx^i dx^j} = 0$$

If we regard $x^{i's}$ as a function of a parameter λ describing the points on the trajectory, then the variational principle reads

$$\delta \int_{\lambda_1}^{\lambda_2} \sqrt{\sum_{i,j=1}^{4} g_{ij}(x^1(\lambda), x^2(\lambda), x^3(\lambda), x^4(\lambda)) \frac{dx^i(\lambda)}{d\lambda} \frac{dx^j(\lambda)}{d\lambda}} d\lambda = 0$$

The Lagrangian for this variational problem is the integrand and is given by

$$L(x^1, x^2, x^3, x^4, \frac{dx^1}{d\lambda}, \frac{dx^2}{d\lambda}, \frac{dx^3}{d\lambda}, \frac{dx^4}{d\lambda})$$

$$= \sqrt{\sum_{i,j=1}^{4} g_{ij}(x^1, x^2, x^3, x^4) \frac{dx^i}{d\lambda} \frac{dx^j}{d\lambda}}$$

We adopt the Einstein summation convention, i.e. use the notation $g_{ij} dx^i dx^j$ as an abbreviation for $\sum_{i,j} g_{ij} dx^i dx^j$. To write down the Euler-Lagrange equation, we set $x^{i'} = \frac{dx^i}{d\lambda}$ and compute

$$\frac{\partial L}{\partial x^{i'}} = \sum_j g_{ij} x^{j'} / L$$

Since $L = ds/d\lambda$, this can be written as

$$\frac{\partial L}{\partial x^{i'}} = \sum_j g_{ij} \frac{dx^j}{ds}$$

Also

$$\frac{\partial L}{\partial x^i} = \sum_{k,m} g_{km,i} x^{k'} x^{m'} / 2L = \frac{1}{2} \sum_{k,m} g_{km,i} \frac{dx^k}{ds} x^{m'}$$

Thus, the Euler-Lagrange equation

$$\frac{d}{d\lambda} \frac{\partial L}{\partial x^{i'}} = \frac{\partial L}{\partial x^i}$$

assumes the form

$$\frac{d}{d\lambda} \sum_j g_{ij} \frac{dx^j}{ds} = \frac{1}{2} \sum_{k,m} g_{km,i} \frac{dx^k}{ds} x^{m'}$$

or

$$\frac{d}{ds} \sum_j g_{ij} \frac{dx^j}{ds} = \frac{1}{2} \frac{d\lambda}{ds} \sum_{k,m} g_{km,i} \frac{dx^k}{ds} x^{m'}$$

and since

$$\frac{d\lambda}{ds} x^{m'} = \frac{dx^m}{ds}$$

it follows that

$$\frac{d}{ds} \sum_j g_{ij} \frac{dx^j}{ds} = \frac{1}{2} \sum_{k,m} g_{km,i} \frac{dx^k}{ds} \frac{dx^m}{ds}$$

Noting that

$$\frac{dg_{ij}}{ds} = g_{ij,k} \frac{dx^k}{ds}$$

we can write the geodesic equation as

$$\sum_j g_{ij} \frac{d^2 x^j}{ds^2} + \sum_{k,m} g_{im,k} \frac{dx^m}{ds} \frac{dx^k}{ds} - \frac{1}{2} \sum_{k,m} g_{km,i} \frac{dx^k}{ds} \frac{dx^m}{ds} = 0$$

or

$$\sum_j g_{ij} \frac{d^2 x^j}{ds^2} + \frac{1}{2} \sum_{k,m} (g_{im,k} + g_{ik,m} - g_{km,i}) \frac{dx^k}{ds} \frac{dx^m}{ds} = 0$$

Introduce the notation

$$\Gamma_{ikm} = \frac{1}{2}(g_{ik,m} + g_{im,k} - g_{km,i})$$

Then

$$g_{ij} \frac{dv^j}{ds} + \Gamma_{ikm} v^k v^m = 0$$

where

$$v^k = \frac{dx^k}{ds}$$

Since (dx^k) is a contravariant vector and ds is a scalar, (v^k) is a contravariant vector. By a contravariant vector (A^i), we mean that under a change of coordinates from $x^i \to \bar{x}^i$, its components change according to the rule

$$\bar{A}^i = A^j \frac{\partial \bar{x}^i}{\partial x^j}$$

Let g^{ij} be the matrix inverse of g_{ij}, i.e., $g^{ij} g_{jk} = \delta^i_k$. Then multiplying both sides of the geodesic equation derived above by g^{ir} gives

$$\frac{dv^r}{ds} + \Gamma^r_{km} v^k v^m = 0$$

where

$$\Gamma^{r}_{\;km} = g^{ri}\Gamma_{ikm}$$

Before closing this section, we mention the transformation law of the metric tensor. From the invariance of the proper time,

$$\bar{d}s^2 = \bar{g}_{ij}d\bar{x}^i d\bar{x}^j = \bar{g}_{ij}\frac{\partial \bar{x}^i}{\partial x^k}\frac{\partial \bar{x}^j}{\partial x^m}dx^k dx^m$$

$$= ds^2 = g_{km}dx^k dx^m$$

Thus,

$$g_{km} = \bar{g}_{ij}\frac{\partial \bar{x}^i}{\partial x^k}\frac{\partial \bar{x}^j}{\partial x^m}$$

with the inverse

$$\bar{g}_{ij} = g_{km}\frac{\partial x^k}{\partial \bar{x}^i}\frac{\partial x^m}{\partial \bar{x}^j}$$

4.12   Parallel displacement, Christoffel symbols and Covariant derivative

Consider a four dimensional curved space time imbedded in an N dimensional Euclidean space. The parametric equations that describe this curved space time are given by $y^n = y^n(x^1, x^2, x^3, x^4)$, $n = 1, 2, ..., N$ and the metric tensor for the curved space time is given by

$$g_{\mu\nu} = \sum_{n=1}^{N}\frac{\partial y^n}{\partial x^\mu}\frac{\partial y^n}{\partial x^\nu}$$

We consider a contravariant vector $A^\mu, \mu = 1, 2, 3, , 4$ in our curved space time at the point $x^\mu, \mu = 1, 2, 3, 4$. Its Euclidean components are given by $A^n = \sum_\mu A^\mu \frac{\partial y^n}{\partial x^\mu}$. From this equation, we get

$$\sum_{n=1}^{N} A^n \frac{\partial y^n}{\partial x^\nu} = g_{\nu\mu}A^\mu = A_\nu$$

namely the covariant components of the four vector. Parallely displacing the vector from the point x^μ to the point $x^\mu + \delta x^\mu$ on the curved surface will not alter the cartesian components of the vector. The change in the covariant components of the vector on performing such a parallel displacement is therefore given by

$$\delta A_\mu = \delta(\sum_{n=1}^{N} A^n \frac{\partial y^n}{\partial x^\mu}) = \sum_{n=1}^{N} A^n \delta(\frac{\partial y^n}{\partial x^\mu})$$

$$= \sum_{n=1}^{N} A^n \frac{\partial^2 y^n}{\partial x^\mu \partial x^\nu} \delta x^\nu$$

$$= A^\rho \frac{\partial y^n}{\partial x^\rho} \frac{\partial^2 y^n}{\partial x^\mu \partial x^\nu} \delta x^\nu$$

where summation over the repeated indices n, ρ, ν is understood. We now claim that

$$\frac{\partial y^n}{\partial x^\rho} \frac{\partial^2 y^n}{\partial x^\mu \partial x^\nu} = \Gamma_{\rho\mu\nu}$$

$$= \frac{1}{2}(g_{\rho\mu,\nu} + g_{\rho\nu,\mu} - g_{\mu\nu,\rho})$$

In fact, the right side equals

$$\frac{1}{2}((y^n_{,\rho}y^n_{,\mu})_{,\nu} + (y^n_{,\rho}y^n_{,\nu})_{,\mu} - (y^n_{,\mu}y^n_{,\nu})_{,\rho})$$

$$= \frac{1}{2}(y^n_{,\rho\nu}y^n_{,\mu} + y^n_{,\rho}y^n_{,\mu\nu} + y^n_{,\rho\mu}y^n_{,\nu} + y^n_{,\rho}y^n_{,\mu\nu} - y^n_{,\mu\rho}y^n_{,\nu} - y^n_{,\mu}y^n_{,\nu\rho})$$

$$= y^n_{,\rho}y^n_{,\mu\nu}$$

proving the claim. Thus, the formula for parallel displacement reads

$$\delta A_\mu = A^\rho \Gamma_{\rho\mu\nu} \delta x^\nu$$

A scalar does not change on parallel displacement. Thus, $\delta(A^\rho B_\rho) = 0$ for two vectors A^ρ, B^ρ. This means

$$0 = A^\rho \delta B_\rho + B_\rho \delta A^\rho = A^\rho \Gamma_{\mu\rho\nu} B^\mu \delta x^\nu + B_\rho \delta A^\rho$$

or

$$g_{\rho\mu} B^\mu \delta A^\rho + A^\rho \Gamma_{\mu\rho\nu} B^\mu \delta x^\nu = 0$$

or

$$B^\mu(g_{\rho\mu} \delta A^\rho + \Gamma_{\mu\rho\nu} A^\rho \delta x^\nu) = 0$$

since B^μ is arbitrary, it follows that

$$g_{\rho\mu} \delta A^\rho + \Gamma_{\mu\rho\nu} A^\rho \delta x^\nu = 0$$

so that

$$\delta A^\mu = -g^{\sigma\mu} \Gamma_{\sigma\rho\nu} A^\rho \delta x^\nu = -\Gamma^\mu_{\rho\nu} A^\rho \delta x^\nu$$

Γ_{ijk} are called the Christoffel symbols of the first kind and Γ^i_{jk} are called the Christoffel symbols of the second kind. Suppose A^μ is a contravariant vector field. This means that for each space-time point $\mathbf{x} = (x^\mu)$, $A^\mu(\mathbf{x})$ is a vector at $\mathbf{x}$. Parallely displacing this vector to the neighbouring point $x^\mu + dx^\mu$, we get the vector $A^\mu + \delta A^\mu = A^\mu - \Gamma^\mu_{\nu\sigma} A^\nu dx^\sigma$ at this displaced point. Also

$A^\mu + dA^\mu = A^\mu + A^\mu_{,\sigma}dx^\sigma$ is a vector at the displaced point. Taking the difference of two vectors at the same point yields a vector at that point. Hence,

$$DA^\mu = (A^\mu + dA^\mu) - (A^\mu + \delta A^\mu) = dA^\mu - \delta A^\mu$$

is a vector at $x^\mu + dx^\mu$. Clearly,

$$DA^\mu = (A^\mu_{,\sigma} + \Gamma^\mu_{\nu\sigma}A^\nu)dx^\sigma$$

We define

$$A^\mu_{:\sigma} = A^\mu_{,\sigma} + \Gamma^\mu_{\nu\sigma}A^\nu$$

Thus,

$$DA^\mu = A^\mu_{:\sigma}dx^\sigma$$

$A^\mu_{:\sigma}$ is called the covariant derivative of the vector field A^μ. It is a tensor in the sense that we shall define in the next section. There is another way to formulate the geodesic equation using the notion of parallel displacement and covariant derivative. Consider a curve in our curved space time defined by the parametric equation $s \to x^\mu(s)$. For convenience of notation in what will follows, we are parametrizing the curve by the quantity s where $ds = \sqrt{g_{\mu\nu}dx^\mu dx^\nu}$ is the infinitesimal distance parameter along the curve. The unit velocity or unit tangent vector to the curve at s is given by $v^\mu = \frac{dx^\mu}{ds}$. If we displace this vector parallely from s to $s + ds$ along the curve, we get the vector $v^\mu + \delta v^\mu$. We claim that the given curve is a geodesic curve iff this parallely displaced vector coincides with the actual unit vector at the displaced point $v^\mu + dv^\mu$. Indeed, when this happens, then $dv^\mu - \delta v^\mu = 0$ and this implies that

$$0 = \frac{dv^\mu}{ds} - \frac{\delta v^\mu}{ds} = \frac{dv^\mu}{ds} + \Gamma^\mu_{\nu\sigma}v^\nu v^\sigma$$

which is precisely the geodesic equation. This means that a geodesic is a curve for which the tangent vector along it is obtained by continuously displacing parallely the tangent vector at the initial point. In terms of the covariant derivative, the geodesic equation can be expressed as $v^\mu_{:\nu}v^\nu = 0$ since

$$\frac{dv^\mu}{ds} = v^\mu_{,\nu}\frac{dx^\mu}{ds} = v^\mu_{,\nu}v^\nu$$

4.13 Tensor fields and their transformation laws

A covariant four vector field A^μ is a four component quantity at each point of space-time that transforms like the four components dx^μ under a change of coordinates, i.e. if $x^{\mu'}$ is a new coordinate system, then the components of the vector in this new coordinate system are given by

$$A^{\mu'} = A^\nu \frac{\partial x^{\mu'}}{\partial x^\nu}$$

The left side is to be evaluated at $x^{\mu'}$ and the right side at x^μ. Note that x^μ and $x^{\mu'}$ are respectively the components of the same space-time point P in the two systems. A scalar field V is one that does not change under a change of coordinates. This is expressed by means of the equation $V'(x') = V(x)$, or simply as $V' = V$. Let A^μ be a contravariant vector field. The corresponding covariant vector field A_μ is defined by $A_\mu = g_{\mu\nu}A^\nu$ at each space-time point where $g_{\mu\nu}$ is the metric tensor of our curved space-time. Its transformation law can be deduced using the transformation law of the contravariant vector and that of the metric tensor:

$$A'_\mu = g'_{\mu\nu}A^{\nu'} = g_{\alpha\beta}\frac{\partial x^\alpha}{\partial x^{\mu'}}\frac{\partial x^\beta}{\partial x^{\nu'}}A^\rho\frac{\partial x^{\nu'}}{\partial x^\rho}$$

$$= g_{\alpha\beta}A^\rho\frac{\partial x^\alpha}{\partial x^{\mu'}}\frac{\partial x^\beta}{\partial x^{\nu'}}\frac{\partial x^{\nu'}}{\partial x^\rho}$$

$$= g_{\alpha\beta}A^\rho\frac{\partial x^\alpha}{\partial x^{\mu'}}\delta^\beta_\rho$$

$$= g_{\alpha\rho}A^\rho\frac{\partial x^\alpha}{\partial x^{\mu'}} = A_\alpha\frac{\partial x^\alpha}{\partial x^{\mu'}}$$

A covariant vector is thus one that transforms like the components of $\frac{\partial}{\partial x^\mu}$:

$$\frac{\partial}{\partial x^{\mu'}} = \frac{\partial x^\alpha}{\partial x^{\mu'}}\frac{\partial}{\partial x^\alpha}$$

Suppose we are given a set of p contravariant vector fields $A(i)^\mu, i = 1, ..., m$ and q covariant vector fields $B(j)_\mu, j = 1, 2, ..., q$. We can then form a field quantity having $p + q$ indices, namely,

$$T^{\mu_1...\mu_p}_{\nu_1...\nu_q} = A(1)^{\mu_1}...A(p)^{\mu_p}B(1)_{\nu_1}...B(q)_{\nu_q}$$

The transformation law of the field quantity is easily obtained:

$$T^{\mu'_1...\mu'_p}{}_{\nu'_1...\nu'_q} = A(1)^{\mu'_1}...A(p)^{\mu'_p}B(1)_{\nu'_1}...B(q)_{\nu'_q}$$

$$= A(1)^{\alpha_1}\frac{\partial x^{\mu'_1}}{\partial x^{\alpha_1}}...A(p)^{\alpha_p}\frac{\partial x^{\mu'_p}}{\partial x^{\alpha_p}}B(1)_{\beta_1}\frac{\partial x^{\beta_1}}{\partial x^{\nu'_1}}...B(q)_{\beta_q}\frac{\partial x^{\beta_q}}{\partial x^{\nu'_q}}$$

$$= T^{\alpha_1...\alpha_p}_{\beta_1...\beta_q}\frac{\partial x^{\mu'_1}}{\partial x^{\alpha_1}}...\frac{\partial x^{\mu'_p}}{\partial x^{\alpha_p}}\frac{\partial x^{\beta_1}}{\partial x^{\nu'_1}}...\frac{\partial x^{\beta_q}}{\partial x^{\nu'_q}}$$

The same transformation law will hold for any superposition of similar quantities. That is if for each $r = 1, ..., M$, we have p contravariant vector fields $A(i, r)^\mu, i = 1, 2, ..., p$ and q covariant vector fields $B(j, r)_\nu, j = 1, 2, ..., q$, then we can form the field quantity

$$T^{\mu_1...\mu_p}_{\nu_1...\nu_q} = \sum_{r=1}^{M} A(1, r)^{\mu_1}...A(p, r)^{\mu_p}B(1, r)_{\nu_1}...B(q, r)_{\nu_q}$$

Then T will have the same transformation law as above. Any field quantity indexed by p contravariant space-time indices and q covariant space-time indices that transforms according to the above law is called a (pq) tensor quantity, or simply a (pq) tensor. For example, if A^μ is a contravariant vector field, then $A^\mu_{:\nu}$ is a (11) tensor field, but $A^\mu_{,\nu}$ is not a tensor field. The fact that $A^\mu_{:\nu}$ is a (11) tensor follows from the fact that $DA^\mu = A^\mu_{:\nu}dx^\nu$ is the difference of two vectors at $x^\mu + dx^\mu$ and is therefore a vector at $x^\mu + dx^\mu$. Thus,

$$DA^{\mu'} = \frac{\partial x^{\mu'}}{\partial x^\alpha}DA^\alpha$$

This is the same as

$$A^{\mu'}_{:\nu'}dx^{\nu'} = \frac{\partial x^{\mu'}}{\partial x^\alpha}A^\alpha_{:\beta}dx^\beta$$

which in turn can be expressed as

$$A^{\mu'}_{:\nu'}\frac{\partial x^{\nu'}}{\partial x^\beta}dx^\beta = \frac{\partial x^{\mu'}}{\partial x^\alpha}A^\alpha_{:\beta}dx^\beta$$

Since the direction of the contravariant vector dx^β can be arbitrary, it follows that

$$A^{\mu'}_{:\nu'}\frac{\partial x^{\nu'}}{\partial x^\beta} = \frac{\partial x^{\mu'}}{\partial x^\alpha}A^\alpha_{:\beta}$$

and hence

$$A^{\mu'}_{:\nu'} = \frac{\partial x^{\mu'}}{\partial x^\alpha}\frac{\partial x^\beta}{\partial x^{\nu'}}A^\alpha_{:\beta}$$

proving that the field quantity $A^\mu_{:\nu}$ is a (11) tensor. On similar lines, it can be shown that $A_{\mu:\nu}$ is a (02) tensor. It can also be shown that if $T^{\mu_1...\mu_p}_{\nu_1...\nu_q}$ is a (pq) tensor, then

$$T_{\alpha_1...\alpha_p\nu_1...\nu_q} = g_{\mu_1\alpha_1}...g_{\mu_p\alpha_p}T^{\mu_1...\mu_p}_{\nu_1...\nu_q}$$

is a $(0, p+q)$ tensor. More generally, we may choose distinct indices $i_1, ..., i_r$ in $\{1, 2, ..., p\}$ and distinct indices $j_1, ..., j_s$ in $\{1, 2, ..., q\}$ and form

$$g_{\mu_{i1}\alpha_{i1}}...g_{\mu_{ir}\alpha_{ir}}g^{\nu_{j1}\beta_{j1}}...g^{\nu_{js}\beta_{js}}T^{\mu_1...\mu_p}_{\nu_1...\nu_q}$$

This is a tensor of the type $(p - r, q - s)$ having contravariant indices $\mu_a, a \in E$ and covariant indices $\nu_b, b \in F$ where E is the complement of the set $\{i_1, ..., i_r\}$ in $\{1, 2, ..., p\}$ and F is the complement of the set $\{j_1, ..., j_s\}$ in $\{1, 2, ..., q\}$. The reason why $A^\mu_{,\nu}$ is not a tensor is that it involves forming the difference of two vectors at different points, i.e. The difference of the vector A^μ at x^μ and the vector $A^\mu + dA^\mu$ at $x^\mu + dx^\mu$. Thus, dA^μ is not a vector and hence $A^\mu_{,\nu}$ is not a tensor. We can also verify this fact directly by writing down its transformation law:

$$A^{\mu'}_{,\nu'} = \frac{\partial}{\partial x^{\nu'}}\left(A^\alpha\frac{\partial x^{\mu'}}{\partial x^\alpha}\right)$$

$$\begin{aligned} \vdots \quad &= \frac{\partial x^\beta}{\partial x^{\nu'}} \frac{\partial}{\partial x^\beta}\left(A^\alpha \frac{\partial x^{\mu'}}{\partial x^\alpha}\right) \\ &= \frac{\partial x^\beta}{\partial x^{\nu'}} \frac{\partial x^{\mu'}}{\partial x^\alpha} A^\alpha_{,\beta} + A^\alpha \frac{\partial x^\beta}{\partial x^{\nu'}} \frac{\partial^2 x^{\mu'}}{\partial x^\alpha \partial x^\beta} \end{aligned}$$

One of the important ways to show that a certain field quantity is a tensor is the quotient theorem. A special case of this theorem is that if $T^{\mu_1,\dots,\mu_p}_{\nu_1\dots\nu_q}$ is a field quantity such that for all (r,s) tensors $A^{\nu_1\dots\nu_r}_{\mu_1\dots\mu_s}$ with $r \le q$ and $s \le p$, the field quantity

$$T^{\mu_1\dots\mu_p}_{\nu_1\dots\nu_q} A^{\nu_1\dots\nu_r}_{\mu_1\dots\mu_s}$$

is a $(p-s, q-r)$ tensor, then T is a (pq) tensor. For example if $T^\mu_{\nu\rho} A_\mu$ is a tensor for all covariant vector fields A_μ, then T is a tensor. Its proof is elementary and left as an exercise. The covariant derivative of tensors is obtained using the formula for the covariant derivative of a vector field. For example, we define the covariant derivative of the tensor $A_\mu B_\nu$ as

$$(A_\mu B_\nu)_{:\rho} = A_{\mu:\rho} B_\nu + A_\mu B_{\nu:\rho}$$

Obviously, this is a (03) tensor. since $A_{\mu:\rho}$ and $B_{\mu:\rho}$ are (02) tensors. Now using the formula for the covariant derivative of a covariant vector, we get

$$(A_\mu B_\nu)_{:\rho} = (A_{\mu,\rho} - \Gamma^\alpha_{\mu\rho} A_\alpha) B_\nu + A_\mu (B_{\nu,\rho} - \Gamma^\alpha_{\nu\rho} B_\alpha)$$

Writing $T_{\mu\nu} = A_\mu B_\nu$ gives us

$$T_{\mu\nu:\rho} = T_{\mu\nu,\rho} - \Gamma^\alpha_{\mu\rho} T_{\alpha\nu} - \Gamma^\alpha_{\nu\rho} T_{\mu\alpha}$$

This formula is extended to arbitrary (02) tensors, i.e., by assuming that the covariant derivative of tensors is additive. As another example, consider the (11) tensor $A^\mu B_\nu$. Its covariant derivative is defined by

$$(A^\mu B_\nu)_{:\rho} = A^\mu_{:\rho} B_\nu + A^\mu B_{\nu:\rho}$$
$$= (A^\mu_{,\rho} + \Gamma^\mu_{\rho\alpha} A^\alpha) B_\nu + A^\mu (B_{\nu,\rho} - \Gamma^\alpha_{\nu\rho} B_\alpha)$$

so that for any (11) tensor field T^μ_ν, the covariant derivative is given by

$$T^\mu_{\nu:\rho} = T^\mu_{\nu,\rho} + \Gamma^\mu_{\rho\alpha} T_{\alpha\nu} - \Gamma^\alpha_{\nu\rho} T^\mu_\alpha$$

The general principle of relativity states that the laws of nature must be expressed as tensor equations. If a tensor equation is valid in one frame, it will be valid in all frames. For example, if our law of nature is expressed as $T_{\mu\nu} = 0$ in the S frame where $T_{\mu\nu}$ is a tensor, the it follows that

$$T_{\mu'\nu'} = T_{\alpha\beta} \frac{\partial x^\alpha}{\partial x^{\mu'}} \frac{\partial x^\beta}{\partial x^{\nu'}} = 0$$

implying that the tensor equation also holds in the S' frame. Note the difference between the special and general principle of relativity. The special principle of relativity requires that the laws of physics be invariant only under the Lorentz transformation, i.e., they be valid in all frames moving relative to each other with a constant velocity. The general principle of relativity however requires that the laws of physics be invariant under all transformations of space time., i.e., they be valid in all reference frames and this can only be satisfied if the laws of physics are expressed as tensor equations. An example of a tensor equation is the geodesic equation: $v^\mu_{;\nu} v^\nu = 0$. Before concluding this section, we mention that the metric tensor $g_{\mu\nu}$ is a (02) tensor and its inverse $g^{\mu\nu}$ is a (20) tensor as follows from their transformation laws. Another important point is that if $T^{\mu\nu}$ is a tensor then $g_{\mu\nu} T^{\mu\nu}$ is a scalar ,i.e., a (00) tensor.

4.14   Riemann Christoffel curvature tensor

Let A_μ be a covariant vector field. It double partial derivatives obey the identity $A_{\mu,\rho\sigma} = A_{\mu,\sigma\rho}$. This is not true however for covariant derivatives. We shall evaluate $A_{\mu:\nu:\rho} - A_{\mu:\rho:\nu}$. First note that $A_{\mu:\nu}$ is a tensor and so

$$A_{\mu:\nu:\rho} = A_{\mu:\nu,\rho} - \Gamma^\alpha_{\mu\rho} A_{\alpha:\nu} - \Gamma^\alpha_{\nu\rho} A_{\mu:\alpha}$$

$$= (A_{\mu,\nu} - \Gamma^\beta_{\mu\nu} A_\beta)_{,\rho} - \Gamma^\alpha_{\mu\rho}(A_{\alpha,\nu} - \Gamma^\beta_{\alpha\nu} A_\beta)$$

$$- \Gamma^\alpha_{\nu\rho}(A_{\mu,\alpha} - \Gamma^\beta_{\mu\alpha} A_\beta)$$

$$= A_{\mu,\nu\rho} + (-\Gamma^\beta_{\mu\nu,\rho} + \Gamma^\alpha_{\mu\rho}\Gamma^\beta_{\alpha\nu} + \Gamma^\alpha_{\nu\rho}\Gamma^\beta_{\mu\alpha})A_\beta$$

$$- \Gamma^\beta_{\mu\nu} A_{\beta,\rho} - \Gamma^\alpha_{\mu\rho} A_{\alpha,\nu} - \Gamma^\alpha_{\nu\rho} A_{\mu,\alpha}$$

Thus,

$$A_{\mu:\nu:\rho} - A_{\mu:\rho:\nu} =$$

$$(-\Gamma^\beta_{\mu\nu,\rho} + \Gamma^\beta_{\mu\rho,\nu} + \Gamma^\alpha_{\mu\rho}\Gamma^\beta_{\alpha\nu} - \Gamma^\alpha_{\mu\nu}\Gamma^\beta_{\alpha\rho})A_\beta$$

$$= R^\alpha_{\mu\nu\rho} A_\alpha$$

where

$$R^\alpha_{\mu\nu\sigma} = \Gamma^\alpha_{\mu\sigma,\nu} - \Gamma^\alpha_{\mu\nu,\sigma} + \Gamma^\beta_{\mu\sigma}\Gamma^\alpha_{\beta\nu} - \Gamma^\beta_{\mu\nu}\Gamma^\alpha_{\beta\sigma}$$

$R^\alpha_{\mu\nu\sigma}$ by virtue of the quotient theorem is a tensor. It possesses certain symmetry properties. The first follows from the fact that the tensor quantity $A_{\mu:\nu:\rho} - A_{\mu:\rho:\nu}$ changes sign when the indices ν and ρ interchanged. It results in the symmetry property

$$R^\alpha_{\mu\nu\rho} = -R^\alpha_{\mu\rho\nu}$$

To obtain the second symmetry, we lower the raised index of the curvature tensor by defining

$$R_{\rho\mu\nu\sigma} = g_{\rho\alpha} R^\alpha_{\mu\nu\sigma}$$

$$= g_{\rho\alpha}\Gamma^{\alpha}_{\mu\sigma,\nu} - g_{\rho\alpha}\Gamma^{\alpha}_{\mu\nu,\sigma} + g_{\rho\alpha}\Gamma^{\beta}_{\mu\sigma}\Gamma^{\alpha}_{\beta\nu} - g_{\rho\alpha}\Gamma^{\beta}_{\mu\nu}\Gamma^{\alpha}_{\beta\sigma}$$

It can then be shown that

$$R_{\rho\mu\nu\sigma} = R_{\mu\rho\nu\sigma}$$

To see this, note that

$$R_{\mu\rho\nu\sigma}A^{\mu}B^{\rho} = R^{\mu}_{\rho\nu\sigma}A_{\mu}B^{\rho} = B^{\rho}(A_{\rho:\nu:\sigma} - A_{\rho:\sigma:\nu})$$

Also

$$R_{\mu\rho\nu\sigma}A^{\rho}B^{\mu} = R^{\mu}_{\rho\nu\sigma}A^{\rho}B_{\mu}$$

$$= A^{\rho}(B_{\rho:\nu:\sigma} - B_{\rho:\sigma:\nu})$$

The sum of these two gives

$$B^{\rho}(A_{\rho:\nu:\sigma} - A_{\rho:\sigma:\nu}) + A^{\rho}(B_{\rho:\nu:\sigma} - B_{\rho:\sigma:\nu})$$

$$= B_{\rho}(A^{\rho}_{:\nu:\sigma} - A^{\rho}_{:\sigma:\nu}) + A^{\rho}(B_{\rho:\nu:\sigma} - B_{\rho:\sigma:\nu})$$

Now,

$$0 = (B_{\rho}A^{\rho})_{:\nu:\sigma} = (B_{\rho:\nu}A^{\rho} + B_{\rho}A^{\rho}_{:\nu})_{:\sigma}$$

$$= B_{\rho:\nu:\sigma}A^{\rho} + B_{\rho}A^{\rho}_{:\nu:\sigma} + B_{\rho:\nu}A^{\rho}_{:\sigma} + B_{\rho:\sigma}A^{\rho}_{:\nu}$$

Interchanging the indices ν and σ and subtracting the two equations gives

$$A^{\rho}(B_{\rho:\nu:\sigma} - B_{\rho:\sigma:\nu}) + B_{\rho}(A^{\rho}_{:\nu:\sigma} - A^{\rho}_{:\sigma:\nu}) = 0$$

It follows that

$$R_{\mu\rho\nu\sigma}A^{\mu}B^{\rho} = -R_{\mu\rho\nu\sigma}A^{\rho}B^{\mu}$$

Since A^{μ} and B^{μ} are arbitrary vector fields, we conclude that

$$R_{\mu\rho\nu\sigma} = -R_{\rho\mu\nu\sigma}$$

Before deriving the other symmetry property, we give a convenient expansion for the curvature tensor that shows the separation into a component that depends linearly on the metric and a component that depends nonlinearly on the metric. Our starting point is the formula

$$R_{\rho\mu\nu\sigma} = g_{\rho\alpha}\Gamma^{\alpha}_{\mu\sigma,\nu} - g_{\rho\alpha}\Gamma^{\alpha}_{\mu\nu,\sigma}$$

$$+ g_{\rho\alpha}\Gamma^{\beta}_{\mu\sigma}\Gamma^{\alpha}_{\beta\nu} - g_{\rho\alpha}\Gamma^{\beta}_{\mu\nu}\Gamma^{\alpha}_{\beta\sigma}$$

Now,

$$g_{\rho\alpha}\Gamma^{\alpha}_{\mu\sigma,\nu} = g_{\rho\alpha}(g^{\alpha\beta}\Gamma_{\beta\mu\sigma})_{,\nu}$$

$$= g_{\rho\alpha}g^{\alpha\beta}\Gamma_{\beta\mu\sigma,\nu} + g_{\rho\alpha}g^{\alpha\beta}_{,\nu}\Gamma_{\beta\mu\sigma}$$

Now, if A is a matrix valued function of variables $x_1, ..., x_n$, then $d(A^{-1}) = -A^{-1}(dA)A^{-1}$. This follows from the equation $0 = dI = d(AA^{-1}) = (dA)A^{-1} + Ad(A^{-1})$. Using this formula, we get

$$g^{\alpha\beta}_{,\nu} = -g^{\alpha k}g_{kj,\nu}g^{j\beta}$$

Also,

$$g_{\rho\alpha}g^{\alpha\beta} = \delta^\beta_\rho$$

Thus,

$$g_{\rho\alpha}\Gamma^\alpha_{\mu\sigma,\nu} = \Gamma_{\rho\mu\sigma,\nu} - g_{\rho\alpha}g^{\alpha k}g^{\beta j}g_{kj,\nu}\Gamma_{\beta\mu\sigma}$$

$$= \Gamma_{\rho\mu\sigma,\nu} - g_{\rho j,\nu}\Gamma^j_{\mu\sigma}$$

Similarly,

$$g_{\rho\alpha}\Gamma^\alpha_{\mu\nu,\sigma} = \Gamma_{\rho\mu\nu,\sigma} - g_{\rho j,\sigma}\Gamma^j_{\mu\nu}$$

Combining this with the expression for the covariant curvature tensor

$$R_{\rho\mu\nu\sigma} = \Gamma_{\rho\mu\sigma,\nu} - \Gamma_{\rho\mu\nu,\sigma} - g_{\rho j,\nu}\Gamma^j_{\mu\sigma} + g_{\rho j,\sigma}\Gamma^j_{\mu\nu}$$

$$+g_{\rho\alpha}\Gamma^\beta_{\mu\sigma}\Gamma^\alpha_{\beta\nu} - g_{\rho\alpha}\Gamma^\beta_{\mu\nu}\Gamma^\alpha_{\beta\sigma}$$

The first two terms are linear in the metric tensor. Their total can be expressed as

$$R^{lin}_{\rho\mu\nu\sigma} = \Gamma_{\rho\mu\sigma,\nu} - \Gamma_{\rho\mu\nu,\sigma} = \frac{1}{2}(g_{\rho\mu,\sigma\nu} + g_{\rho\sigma,\mu\nu} - g_{\sigma\mu,\rho\nu} - g_{\rho\mu,\nu\sigma} - g_{\rho\nu,\mu\sigma} + g_{\mu\nu,\rho\sigma})$$

$$= \frac{1}{2}(g_{\rho\sigma,\mu\nu} - g_{\sigma\mu,\rho\nu} - g_{\rho\nu,\mu\sigma} + g_{\mu\nu,\rho\sigma})$$

It is clear that this expression is antisymmetric in the indices ρ and μ. Consider now the nonlinear part of the curvature tensor. Their total is

$$R^{nonlin}_{\rho\mu\nu\sigma} = -g_{\rho j,\nu}\Gamma^j_{\mu\sigma} + g_{\rho j,\sigma}\Gamma^j_{\mu\nu}$$

$$+g_{\rho\alpha}\Gamma^\beta_{\mu\sigma}\Gamma^\alpha_{\beta\nu} - g_{\rho\alpha}\Gamma^\beta_{\mu\nu}\Gamma^\alpha_{\beta\sigma}$$

$$= \Gamma^\beta_{\mu\sigma}(-g_{\rho\beta,\nu} + \Gamma_{\rho\beta\nu}) + \Gamma^\beta_{\mu\nu}(g_{\rho\beta,\sigma} - \Gamma_{\rho\beta\sigma})$$

$$= \frac{1}{2}\Gamma^\beta_{\mu\sigma}(g_{\rho\nu,\beta} - g_{\rho\beta,\nu} - g_{\nu\beta,\rho}) + \frac{1}{2}\Gamma^\beta_{\mu\nu}(g_{\rho\beta,\sigma} - g_{\rho\sigma,\beta} + g_{\sigma\beta,\rho})$$

$$= -\Gamma^\beta_{\mu\sigma}\Gamma_{\beta\rho\nu} + \Gamma^\beta_{\mu\nu}\Gamma_{\beta\rho\sigma}$$

We claim that this expression is also antisymmetric in the variables ρ, μ. Indeed,

$$R^{nonlin}_{\mu\rho\nu\sigma} = -\Gamma^\beta_{\rho\sigma}\Gamma_{\beta\mu\nu} + \Gamma^\beta_{\rho\nu}\Gamma_{\beta\mu\sigma}$$

$$= -\Gamma_{\beta\rho\sigma}\Gamma^\beta_{\mu\nu} + \Gamma_{\beta\rho\nu}\Gamma^\beta_{\mu\sigma} = -R^{nonlin}_{\rho\mu\nu\sigma}$$

proving the claim. Note that the Christoffel symbol $\Gamma_{\mu\nu\sigma}$ is symmetric in the last two indices and the Christoffel symbol $\Gamma^\mu_{\nu\sigma} = g^{\mu\rho}\Gamma_{\rho\nu\sigma}$ is symmetric in the bottom two indices. This furnishes us with another proof for the identity $R_{\rho\mu\nu\sigma} = -R_{\mu\rho\nu\sigma}$.

We now establish the important Bianchi identities for the curvature tensor. Let A_μ be a covariant vector field. Consider $A_{\mu:\nu:\rho:\sigma} = T_{\mu\nu:\rho:\sigma}$ where $T_{\mu\nu} = A_{\mu:\nu}$ is a (02) tensor. Now, consider the (02) tensor $B_\mu C_\nu$. We have

$$(B_\mu C_\nu)_{:\rho:\sigma} = (B_{\mu:\rho} C_\nu + B_\mu C_{\nu:\rho})_{:\sigma}$$

$$= B_{\mu:\rho:\sigma} C_\nu + B_\mu C_{\nu:\rho:\sigma} + B_{\mu:\rho} C_{\nu:\sigma} + B_{\mu:\sigma} C_{\nu:\rho}$$

Interchange the indices ρ and σ and subtract the resulting identity from this one to get

$$(B_\mu C_\nu)_{:\rho:\sigma} - (B_\mu C_\nu)_{:\sigma:\rho} =$$

$$(B_{\mu:\rho:\sigma} - B_{\mu:\sigma:\rho})C_\nu + B_\mu(C_{\nu:\rho:\sigma} - C_{\nu:\sigma:\rho})$$

$$= R^\alpha_{\mu\rho\sigma} B_\alpha C_\nu + R^\alpha_{\nu\rho\sigma} B_\mu C_\alpha$$

From this we deduce that for any (02) tensor $T_{\mu\nu}$,

$$T_{\mu\nu:\rho:\sigma} - T_{\mu\nu:\sigma:\rho} = R^\alpha_{\mu\rho\sigma} T_{\alpha\nu} + R^\alpha_{\nu\rho\sigma} T_{\mu\alpha}$$

In particular, this identity holds for the above defined (02) tensor. Thus,

$$A_{\mu:\nu:\rho:\sigma} - A_{\mu:\nu:\sigma:\rho} = R^\alpha_{\mu\rho\sigma} A_{\alpha:\nu} + R^\alpha_{\nu\rho\sigma} A_{\mu:\alpha}$$

Making cyclic permutations amongst the indices ν, ρ, σ gives us the equations

$$A_{\mu:\rho:\sigma:\nu} - A_{\mu:\rho:\nu:\sigma} = R^\alpha_{\mu\sigma\nu} A_{\alpha:\rho} + R^\alpha_{\rho\sigma\nu} A_{\mu:\alpha}$$

$$A_{\mu:\sigma:\nu:\rho} - A_{\mu:\sigma:\rho:\nu} = R^\alpha_{\mu\nu\rho} A_{\alpha:\sigma} + R^\alpha_{\sigma\nu\rho} A_{\mu:\alpha}$$

Adding the three equations gives

$$(A_{\mu:\nu:\rho} - A_{\mu:\rho:\nu})_{:\sigma} + (A_{\mu:\sigma:\nu} - A_{\mu:\nu:\sigma})_{:\rho}$$

$$+(A_{\mu:\rho:\sigma} - A_{\mu:\sigma:\rho})_{:\nu}$$

$$= R^\alpha_{\mu\rho\sigma} A_{\alpha:\nu} + R^\alpha_{\mu\sigma\nu} A_{\alpha:\rho} + R^\alpha_{\mu\nu\rho} A_{\alpha:\sigma}$$

$$+(R^\alpha_{\nu\rho\sigma} + R^\alpha_{\rho\sigma\nu} + R^\alpha_{\sigma\nu\rho}) A_{\mu:\alpha}$$

The left side equals

$$(R^\alpha_{\mu\nu\rho} A_\alpha)_{:\sigma} + (R^\alpha_{\mu\sigma\nu} A_\alpha)_{:\rho} + (R^\alpha_{\mu\rho\sigma} A_\alpha)_{:\nu}$$

$$= (R^\alpha_{\mu\nu\rho:\sigma} + R^\alpha_{\mu\sigma\nu:\rho} + R^\alpha_{\mu\rho\sigma:\nu}) A_\alpha$$

$$+R^{\alpha}_{\mu\nu\rho}A_{\alpha:\sigma} + R^{\alpha}_{\mu\sigma\nu}A_{\alpha:\rho} + R^{\alpha}_{\mu\rho\sigma}A_{\alpha:\nu}$$

After making appropriate cancelations, we thus get

$$(R^{\alpha}_{\mu\nu\rho:\sigma} + R^{\alpha}_{\mu\sigma\nu:\rho} + R^{\alpha}_{\mu\rho\sigma:\nu})A_{\alpha}$$

$$= (R^{\alpha}_{\nu\rho\sigma} + R^{\alpha}_{\rho\sigma\nu} + R^{\alpha}_{\sigma\nu\rho})A_{\mu:\alpha}$$

We now claim that the following identity holds for the curvature tensor

$$R^{\alpha}_{\nu\rho\sigma} + R^{\alpha}_{\rho\sigma\nu} + R^{\alpha}_{\sigma\nu\rho} = 0$$

This can be seen by using the identity that for any covariant vector A_{μ}, one has

$$A_{\mu:\nu} - A_{\nu:\mu} = A_{\mu,\nu} - A_{\nu,\mu}$$

Thus,

$$(R^{\alpha}_{\nu\rho\sigma} + R^{\alpha}_{\rho\sigma\nu} + R^{\alpha}_{\sigma\nu\rho})A_{\alpha}$$

$$= (A_{\nu:\rho:\sigma} - A_{\nu:\sigma:\rho} + A_{\rho:\sigma:\nu} - A_{\rho:\nu:\sigma} + A_{\sigma:\nu:\rho} - A_{\sigma:\rho:\nu})$$

$$= (A_{\nu:\rho} - A_{\rho:\nu})_{:\sigma} + (A_{\rho:\sigma} - A_{\sigma:\rho})_{:\nu} + (A_{\sigma:\nu} - A_{\nu:\sigma})_{:\rho}$$

Now let $T_{\mu\nu} = A_{\mu:\nu} - A_{\nu:\mu} = A_{\mu,\nu} - A_{\nu,\mu}$. Then, the above equals

$$T_{\nu\rho:\sigma} + T_{\rho\sigma:\nu} + T_{\sigma\nu:\rho}$$

$$= T_{\nu\rho,\sigma} - \Gamma^{\alpha}_{\nu\sigma}T_{\alpha\rho} - \Gamma^{\alpha}_{\rho\sigma}T_{\nu\alpha}$$

$$+T_{\rho\sigma,\nu} - \Gamma^{\alpha}_{\rho\nu}T_{\alpha\sigma} - \Gamma^{\alpha}_{\sigma\nu}T_{\rho\alpha}$$

$$+T_{\sigma\nu,\rho} - \Gamma^{\alpha}_{\sigma\rho}T_{\alpha\nu} - \Gamma^{\alpha}_{\nu\rho}T_{\sigma\alpha}$$

$$= (T_{\nu\rho,\sigma} + T_{\rho\sigma,\nu} + T_{\sigma\nu,\rho})$$

$$-(\Gamma^{\alpha}_{\nu\sigma}(T_{\alpha\rho} + T_{\rho\alpha}) + \Gamma^{\alpha}_{\rho\sigma}(T_{\nu\alpha} + T_{\alpha\nu}) + \Gamma^{\alpha}_{\rho\nu}(T_{\alpha\sigma} + T_{\sigma\alpha}) = 0$$

The first bracket vanishes from the definition of $T_{\mu\nu} = A_{\mu,\nu} - A_{\nu,\mu}$. The second bracket is zero since $T_{\mu\nu} = -T_{\nu\mu}$. This proves the stated identity. As a consequence, we obtain the Bianchi identity

$$R^{\alpha}_{\mu\nu\rho:\sigma} + R^{\alpha}_{\mu\rho\sigma:\nu} + R^{\alpha}_{\mu\sigma\nu:\rho} = 0$$

This will prove to be of fundamental importance in Einstein's formulation of the field equations of gravitation.

4.15 Ricci tensor and Einstein's field equations

The Ricci tensor is defined by contracting the curvature tensor with respect to its first and last variables:

$$R_{\mu\nu} = R^{\alpha}_{\mu\nu\alpha} = g^{\beta\alpha} R_{\beta\mu\nu\alpha}$$

$$\Gamma^{\alpha}_{\mu\alpha,\nu} - \Gamma^{\alpha}_{\mu\nu,\alpha} + \Gamma^{\beta}_{\mu\alpha}\Gamma^{\alpha}_{\beta\nu} - \Gamma^{\beta}_{\mu\nu}\Gamma^{\alpha}_{\beta\alpha}$$

To show that the Ricci tensor is symmetric, it suffices to show that the first term is symmetric. All the other three terms are easily seen to be symmetric. The first term is

$$\Gamma^{\alpha}_{\mu\alpha,\nu} = (g^{\alpha\beta}\Gamma_{\beta\mu\alpha,\nu}$$

$$g^{\alpha\beta}_{,\nu}\Gamma_{\beta\mu\alpha} + g^{\alpha\beta}\Gamma_{\beta\mu\alpha,\nu}$$

$$= -g^{\alpha i}g^{\beta j}g_{ij,\nu}\Gamma_{\beta\mu\alpha} + g^{\alpha\beta}\Gamma_{\beta\mu\alpha,\nu}$$

$$= -\frac{1}{2}g^{\alpha i}g^{\beta j}g_{ij,\nu}(g_{\beta\mu,\alpha} + g_{\beta\alpha,\mu} - g_{\mu\alpha,\beta})$$

$$+ \frac{1}{2}g^{\alpha\beta}(g_{\beta\mu,\alpha\nu} + g_{\beta\alpha,\mu\nu} - g_{\alpha\mu,\beta\nu})$$

$$= -\frac{1}{2}g^{\alpha i}g^{\beta j}g_{ij,\nu}g_{\alpha\beta,\mu} + \frac{1}{2}g^{\alpha\beta}g_{\alpha\beta,\mu\nu}$$

This expression is easily seen to be symmetric in the indices μ, ν proving symmetry of the Ricci tensor. We also define the scalar curvature $R = g^{\mu\nu}R_{\mu\nu} = R^{\mu}_{\mu}$. We now prove the following fundamental identity on which the Einstein field equations are based:

$$(R^{\mu\nu} - \frac{1}{R}g^{\mu\nu})_{:\nu} = 0$$

where

$$R^{\mu\nu} = g^{\mu\alpha}g^{\nu\beta}R_{\alpha\beta}$$

To prove this, we start with the Bianchi identity:

$$R^{\alpha}_{\mu\nu\rho:\sigma} + R^{\alpha}_{\mu\rho\sigma:\nu} + R^{\alpha}_{\mu\sigma\nu:\rho} = 0$$

It follows that

$$R^{\alpha}_{\mu\nu\alpha:\sigma} + R^{\alpha}_{\mu\alpha\sigma:\nu} + R^{\alpha}_{\mu\sigma\nu:\alpha} = 0$$

which is the same as

$$R_{\mu\nu:\sigma} + R^{\alpha}_{\mu\alpha\sigma:\nu} + R^{\alpha}_{\mu\sigma\nu:\alpha} = 0$$

It follows that

$$g^{\mu\nu}R_{\mu\nu:\sigma} + g^{\mu\nu}R^{\alpha}_{\mu\alpha\sigma:\nu} + g^{\mu\nu}R^{\alpha}_{\mu\sigma\nu:\alpha} = 0$$

Now,

$$g^{\mu\nu}R_{\mu\nu:\sigma} = R_{:\sigma}$$

$$g^{\mu\nu}R^{\alpha}_{\mu\alpha\sigma:\nu} = -g^{\mu\nu}R^{\alpha}_{\mu\sigma\alpha:\nu} = -g^{\mu\nu}R_{\mu\sigma:\nu}$$

$$-R^{\nu}_{\sigma:\nu}$$

$$g^{\mu\nu}R^{\alpha}_{\mu\sigma\nu:\alpha} = (g^{\mu\nu}g^{\alpha\beta}R_{\beta\mu\sigma\nu})_{:\alpha}$$

$$= -(g^{\alpha\beta}g^{\mu\nu}R_{\mu\beta\sigma\nu})_{:\alpha} = -(g^{\alpha\beta}R_{\beta\sigma})_{:\alpha} = -R^{\alpha}_{\sigma:\alpha}$$

We thus obtain the identity

$$2R^{\nu}_{\sigma:\nu} - R_{:\sigma} = 0$$

or

$$(R^{\nu}_{\sigma} - \frac{1}{2}R\delta^{\nu}_{\sigma})_{:\nu} = 0$$

or

$$(R^{\mu\nu} - \frac{1}{2}Rg^{\mu\nu})_{:\nu} = 0$$

which is the fundamental Equation of general relativity. A few remarks are in order. Since $R_{\mu\nu}$ is symmetric, we have $g^{\mu\alpha}R_{\mu\nu} = g^{\mu\alpha}R_{\nu\mu}$. This means that there can be no confusion when we write R^{σ}_{α} as to whether the first index was raised or the second index. Second important remark is that the covariant derivative of the metric tensor is zero. Hence it can be taken inside the bracketed expression when the covariant derivative of a tensor field is being computed. For example, $g^{\mu\nu}T_{\nu\rho:\sigma} = (g^{\mu\nu}T_{\nu\rho})_{:\sigma}$ and

$$g_{\mu\nu}T^{\nu}_{\rho\sigma:\alpha} = (g_{\mu\nu}T^{\nu}_{\rho\sigma})_{:\alpha}$$

The Einstein tensor is defined as $G_{\mu\nu} = R_{\mu\nu} - \frac{1}{2}Rg_{\mu\nu}$. Its divergence vanishes as just shown, i.e., $G^{\mu\nu}_{:\nu} = 0$ where

$$G^{\mu\nu} = g^{\mu\alpha}g^{\nu\beta}G_{\alpha\beta}$$

We also know that if $T^{\mu\nu}$ is the energy momentum tensor of matter, then $T^{\mu\nu}_{:\nu} = 0$, i.e., its covariant divergence vanishes. The vanishing divergence of the energy momentum tensor of matter signifies matter conservation. Let us look at the energy momentum tensor of an ideal fluid. Let v^{μ} denote the four velocity vector, ρ the density and p the pressure. The Navier Stokes equation of an ideal fluid as obtained from Newtonian mechanics reads in cartesian coordinates,

$$\sum_{j=1}^{3} \rho V_j V_{i,j} = -p, i$$

where V_i is the three velocity. and the matter conservation equation reads $\sum_{i=1}^{3}(\rho V_i)_{,i} + \rho_{,4} = 0$, where $x_4 = t$. If we take $V_4 = 1$, then the Navier-Stokes equation can be expressed as

$$\sum_{j=1}^{4} \rho V_j V_{i,j} = -p,i$$

and the matter conservation equation reads

$$\sum_{i=1}^{4} (\rho V_j)_{,j} = 0$$

This can be combined with the Navier-Stokes equation and expressed as

$$\sum_{j=1}^{4} (\rho V_i V_j)_{,j} + p_{,i} = 0$$

Our units are taken so that $c = 1$. This equation carries over to the general theory of relativity provided that we replace V_i by the four velocity vector and ordinary derivatives by covariant derivatives. In other words, we define the energy momentum tensor of the fluid as

$$T^{\mu\nu} = \rho v^{\mu} v^{\nu} + p g^{\mu\nu}$$

and then the equation of motion of the fluid along with matter conservation are contained in the single tensor equation

$$T^{\mu\nu}_{:\nu} = 0$$

A remark is in order. Suppose we don't take $c = 1$. Then, $x^4 = ct$ and $V_4 = 1$. In the cartesian system, the equation of motion of a Newtonian fluid reads

$$\sum_{j=1}^{4} \rho V_j V_{i,j} = -p_{,i}, i = 1, 2, 3$$

To this we augment the equation

$$\sum_{j=1}^{4} \rho V_j V_{4,j} = -p_{,4}$$

This equation is not valid, but if we take special relativistic corrections into account, then $V_4 = \gamma$ and the above equation is to be regarded as being a consequence of special relativity. To be precise, if we wish to write down the equations of motion and of matter conservation in special relativity, then we take $x^4 = ct$, $d\tau^2 = dt^2 - \frac{1}{c^2}(dx^2 + dy^2 + dz^2)$ and for the four velocity vector

$$(V^1, V^2, V^3, V^4) = \left(\frac{dx}{d\tau}, \frac{dy}{d\tau}, \frac{dz}{d\tau}, \frac{d(ct)}{d\tau}\right)$$

$$= \gamma(v_x, v_y, v_z, c)$$

Then set

$$T^{ij} = \rho V^i V^j + pg^{ij}$$

where $g^{ij} = diag[1, 1, 1, -1]$ corresponding to the metric $ds^2 = dx^2 + dy^2 + dz^2 - c^2 dt^2 = -c^2 d\tau^2$. The equation of motion and matter conservation equations are then expressed as

$$\sum_{j=1}^{4} T^{ij}_{,j} = 0$$

Since the covariant divergence of the energy momentum tensor $T^{\mu\nu}$ vanishes and that of the Einstein tensor $G^{\mu\nu}$ also vanishes, and since the Einstein tensor contains upto second order partial derivatives of the metric, Einstein postulated that the equations that determine the metric in the presence of matter should have the form

$$G^{\mu\nu} = KT^{\mu\nu}$$

or

$$R^{\mu\nu} - \frac{1}{2}Rg^{\mu\nu} = KT^{\mu\nu}$$

where K is a constant to be determined. To determine the constant, we consider the following metric:

$$ds^2 = dx^2 + dy^2 + dz^2 - (1 + 2U/c^2)c^2 dt^2$$

where U is some function of the space-time coordinates. Let us write down the geodesic equations for this metric. The action is

$$\int_1^2 ds = \int_1^2 \sqrt{x'^2 + y'^2 + z'^2 - (1 + 2U/c^2)c^2} dt$$

where $x' = dx/dt, y' = dy/dt, z' = dz/dt$. The Lagrangian for this problem is

$$L(x, y, z, x', y', x', t) = \sqrt{x'^2 + y'^2 + z'^2 - (c^2 + 2U(x, y, z, t))}$$

The first component of the equation of motion is

$$\frac{d}{dt}\frac{\partial L}{\partial x'} = \frac{\partial L}{\partial x}$$

which reads

$$\frac{d}{dt}\frac{dx}{ds} = -\frac{\partial U}{\partial x}/(ds/dt)$$

or

$$\frac{d^2 x}{ds^2} = -(dt/ds)^2 \frac{\partial U}{\partial x}$$

Now if $|U| << c^2$ and the velocities are small, then $(dt/ds)^2 \approx (d\tau/ds)^2 = -1/c^2$ and the above equation approximates to

$$\frac{d^2 x}{dt^2} = -\frac{\partial U}{\partial x}$$

with two similar equations for y, z. These correspond to the Newtonian equations of motion when the potential is U. Thus in the weak field limit, we have taking $x^2 = x, x^2 = y, x^3 = z, x^4 = ct$ the following approximate formula for the metric:

$$ds^2 = g_{\mu\nu}dx^\mu dx^\nu$$

with

$$g_{44} = -(1 + 2U/c^2), g_{ij} = \delta_{ij}, g_{i4} = 0, i, j = 1, 2, 3$$

where U is the gravitational potential. Let us now determine the approximate form of the Einstein tensor in this weak field limit. First observe that only the linear part of the curvature tensor will contribute terms, for nonlinear part will involve products of derivatives of U which can be neglected. The linear part of the curvature tensor is

$$R_{\rho\mu\nu\sigma} \approx R^{lin}_{\rho\mu\nu\sigma} = \frac{1}{2}(g_{\rho\sigma,\mu\nu} - g_{\sigma\mu,\rho\nu} - g_{\rho\nu,\mu\sigma} + g_{\mu\nu,\rho\sigma})$$

The approximate metric is

$$((g_{\mu\nu})) \approx diag[1, 1, 1, -1] \approx g^{\mu\nu}$$

Thus,

$$R_{\mu\nu} = g^{\rho\sigma} R_{\rho\mu\nu\sigma} \approx \sum_{i=1}^{3} R_{i\mu\nu i} - R_{4\mu\nu 4}$$

In particular,

$$R_{44} \approx \sum_{i=1}^{3} R_{i44i} - R_{4444}$$

If we assume that the metric is time independent, then the approximate value of this is

$$\frac{1}{2}\sum_{i=1}^{3} g_{44,ii} = \frac{1}{2}\nabla^2 g_{44} = -\frac{1}{c^2}\nabla^2 U$$

Also

$$R = g^{\rho\sigma} g^{\mu\nu} R_{\rho\mu\nu\sigma} \approx \sum_{i,j=1}^{3} R_{jiij} - \sum_{i=1}^{3} R_{4ii4} - \sum_{i=1}^{3} R_{i44i} + R_{4444}$$

This approximates to after considering only the linear component and using time independence to $\nabla^2 g_{44}$. Note that only the second and third terms here contribute. Thus,

$$G_{44} = R_{44} - \frac{1}{2}Rg_{44} \approx \frac{1}{2}\nabla^2 g_{44} + \frac{1}{2}\nabla^2 g_{44} = \nabla^2 g_{44}$$

$$= -\frac{2}{c^2}\nabla^2 U$$

$$G'^{44} \approx G_{44}$$

On the other hand,

$$KT^{44} = K\rho V^{42} \approx Kc^2\rho$$

Thus, the Einstein equation $G^{44} = KT^{44}$ approximates to

$$-\frac{2}{c^2}\nabla^2 U = Kc^2\rho$$

or

$$\nabla^2 U = -(Kc^4/2)\rho$$

For this to agree with Newton's law of gravitation $\nabla^2 U = 4\pi G\rho$, we've got to take $K = -8\pi G/c^4$. Thus by considering the weak field limit of the Einstein field equations, we've arrived at Einstein's law of gravitation:

$$R^{\mu\nu} - \frac{1}{2}Rg^{\mu\nu} = -\frac{8\pi G}{c^4}T^{\mu\nu}$$

4.16 Schwarzschild's solution for spherical matter distribution

It can be shown by direct computation that the metric

$$ds^2 = (1 - 2GM/rc^2)^{-1}dr^2 + r^2(d\theta^2 + sin^2(\theta)d\phi^2) - (1 - 2GM/rc^2)c^2 dt^2$$

satisfies the Einstein field equations in the absence of matter, i.e., $R_{\mu\nu} = 0$. To show this simply compute the various entries of the Ricci tensor using the metric

$$g_{11} = (1 - 2GM/rc^2)^{-1}, g_{22} = r^2, g_{33} = r^2 sin^2(\theta), g_{44} = -(1 - 2GM/rc^2)$$

where we've taken $x^1 = r, x^2 = \theta, x^2 = \phi, x^4 = ct$. Some authors prefer to use the proper time in place of the proper distance, i.e.,

$$d\tau^2 = (1 - 2GM/rc^2)dt^2 - \frac{1}{c^2}((1 - 2GM/rc^2)^{-1}dr^2 + r^2 d\theta^2 + r^2 sin^2(\theta)d\phi^2)$$

They define the coordinates $x^0 = t, x^1 = r, x^2 = \theta, x^3 = \phi$ and

$$g_{00} = 1 - 2GM/rc^2, g_{11} = -\frac{1}{c^2}(1 - 2GM/rc^2)^{-1}, g_{22} = -\frac{1}{c^2}r^2, g_{33} = -\frac{1}{c^2}r^2 sin^2(\theta)$$

By analogy with the weak field limit, it is easy to see that this solution corresponds to the space-time curvature produced by a spherical distribution of mass M in the region $r > 2GM/c^2$. We now look at the motion of a particle moving radially in this metric. The appropriate radial metric is obtained by choosing $d\theta = 0, d\phi = 0$. This gives

$$ds^2 = \alpha^{-1}dr^2 - \alpha c^2 dt^2$$

where

$$\alpha(r) = 1 - 2GM/rc^2$$

Equivalently,

$$d\tau^2 = \alpha dt^2 - \alpha^{-1}dr^2/c^2$$

or

$$d\tau = (\alpha t'^2 - \alpha^{-1}r'^2/c^2)^{1/2}$$

The Euler Lagrange equation

$$\frac{d}{d\tau}\frac{\partial L}{\partial t'} = \frac{\partial L}{\partial t}$$

gives

$$\alpha\frac{dt}{d\tau} = K$$

where K is a constant. Thus,

$$d\tau^2 = \alpha^2 dt^2/K^2 = \alpha dt^2 - \alpha^{-1}dr^2/c^2$$

This gives

$$(\alpha - \alpha^2/K^2)(dt/dr)^2 = \alpha^{-1}/c^2$$

so that

$$dr/dt = -c(\alpha - \alpha^2/K^2)^{1/2}\alpha^{1/2}$$

The minus sign corresponds to our assumption that the particle is falling freely towards the centre of the gravitating body, i.e., towards $r = 0$ and hence its radial velocity is negative. We are assuming that at time $t = 0$, the particle is located at $R > 2GM/c^2$ and that its velocity at this instant is zero. This determines the constant K as

$$\alpha(R) - \alpha(R)^2/K^2 = 0$$

or

$$K = \alpha(R)^{1/2} = (1 - 2GM/Rc^2)^{1/2}$$

The time taken to fall from R to the point $R_0 = 2GM/c^2$ is given by

$$\frac{1}{c}\int_{R_0}^{R} dr\,\alpha(r)^{-1/2}(\alpha(r) - \alpha(r)^2/K^2)^{-1/2}dr$$

$$= \frac{1}{c} \int_{R_0}^{R} dr (1 - 2GM/rc^2)^{-3/2} ((1 - 2GM/rc^2)^{-1} - (1 - 2GM/Rc^2)^{-1})^{-1/2} dr$$

$$= \frac{1}{c} \int_{R_0}^{R} dr (1 - 2GM/rc^2)^{-3/2} \left(\frac{rc^2}{rc^2 - 2GM} - \frac{Rc^2}{Rc^2 - 2GM} \right)^{-1/2} dr$$

$$= c \int_{R_0}^{R} dr \left(\frac{r}{rc^2 - 2GM} \right)^{3/2} (2GM(R - r)/(rc^2 - 2GM)(Rc^2 - 2GM))^{-1/2} dr$$

$$= c \int_{R_0}^{R} r^{3/2} (rc^2 - 2GM)^{-1} (Rc^2 - 2GM)^{-1/2} (2GM(R - r))^{-1/2} dr$$

It is clear from this expression that the particle takes infinite time to reach the point $r = 2GM/c^2$. Hence as perceived by an observer located at any fixed point $R' > 2GM/c^2$, the particle will never reach the critical radius $2GM/c^2$. We now compute the proper time taken by the particle to fall to the critical radius. The time is the time as measured by a clock attached to the moving particle. It will turn out to be finite. Start with the equation connecting coordinate time and proper time:

$$d\tau^2 = \alpha^2 dt^2 / K^2$$

or

$$d\tau = \alpha dt / K = - \frac{1 - 2GM/rc^2}{(1 - 2GM/Rc^2)^{1/2}} \frac{cr^{3/2}}{(Rc^2 - 2GM)^{1/2}} (2GM(R - r))^{-1/2} dr$$

$$= -\left(\frac{Rr}{2GM(R - r)} \right)^{1/2} dr$$

The integral of this expression from $r = R$ to $r = 2GM/c^2$ is clearly finite. This is one of the surprising predictions of general relativity that cannot be verified by experiment. An observer sitting on a particle moving radially towards the centre of a blackhole will see himself to reach the critical radius in finite time while an observer looking from outside will never be able to see the particle reach the critical radius. Note that an observer fixed at the point (R, θ, ϕ), measures the time interval as $(1 - 2GM/Rc^2)^{1/2} dt$ and when $R = \infty$, i.e., when the observer is at an infinite distance from the gravitating object, the time interval measured by him is dt, the coordinate time.

Consider next the radial motion of a light ray, in the Schwarzschild metric. The basic postulate of general relativity states that light travels along null geodesics, i.e., $ds^2 = 0$. This means that the equation of motion of a light ray traveling radially is given by

$$\alpha(r) dt^2 - \alpha(r)^{-1} dr^2 / c^2 = 0$$

i.e.

$$\frac{dr}{dt} = -c\alpha(r) = -c(1 - 2GM/rc^2)$$

the negative sign being taken assuming that the light ray is traveling radially towards the centre of the blackhole. The time taken by the light ray to travel from r_1 to r_2 radially inwards is given by

$$\frac{1}{c} \int_{r_2}^{r_1} dr (1 - 2GM/rc^2)^{-1} dr$$

There is a singularity in the integrand at $r = 2GM/c^2$ if $r_2 > 2GM/c^2 > r_1$. This means that the light from outside the critical radius of the blackhole will never be able to penetrate the critical radius. Likewise, light from within the critical radius of the blackhole will never be able to penetrate out of the critical radius. This is the reason why the gravitating sphere is called a blackhole. This effect makes sense only if the critical radius falls outside the matter distribution of the blackhole, i.e., if R is the blackhole radius, then $2GM/c^2 > R$. Consider next the propagation of a light ray from a distance r_1 to r_2 with $r_1 < r_2$. Assume that we transmit a pulse of light of coordinate time duration dt_1 at r_1 and that this pulse of light reaches r_2 where it has a duration dt_2. The equation of motion of the photon sent at the beginning of the pulse is given by

$$\int_{t_1}^{t_2} dt = \int_{r_1}^{r_2} (1 - 2GM/rc^2)^{-1} dr/c$$

and the equation of motion of the photon sent at the end of the pulse is given by

$$\int_{t_1+dt_1}^{t_2+dt_2} dt = \frac{1}{c} \int_{r_1}^{r_2} (1 - 2GM/rc^2)^1 dr/c$$

These two equations clearly imply that $dt_2 = dt_1$. As our intuition suggests, the coordinate time interval does not change. Let us now look at the frequency of the light pulse at the two points Assume that the pulse consists of one cycle so that the frequency at r_1 where the light was emitted is given by $\nu_1 = 1/d\tau_1$ and the frequency at r_2 where the light is received is given by $\nu_2 = 1/d\tau_2$. $d\tau_1$ is the time measured by a clock fixed at r_1. Thus,

$$d\tau_1 = (1 - 2GM/r_1 c^2)^{1/2} dt_1 = (1 - 2GM/r_1 c^2)^{1/2} dt$$

Similarly

$$d\tau_2 = (1 - 2GM/r_2 c^2)^{1/2} dt_2 = (1 - 2GM/r_2 c^2)^{1/2} dt$$

The ratio of the frequency of the light at r_1 to that at r_2 is then given by

$$\frac{\nu_1}{\nu_2} = \frac{d\tau_2}{d\tau_1} = \frac{(1 - 2GM/r_2 c^2)^{1/2}}{(1 - 2GM/r_1 c^2)^{1/2}}$$

Since $r_2 > r_1$, it follows that this ratio is greater than unity. This means that light in propagation from r_1 to r_2 suffers a red-shift. In other words, light when it propagates from a region of stronger gravitational field to a region of weaker gravitational field suffers a red shift. An example of this is the propagation of light from the sun's surface to the earth's surface. Another characteristic feature of the general theory of relativity which we have already observed is the slowing of clocks in a gravitational field. Suppose that the gravitational field is produced by a spherical distribution of matter for which we have the Schwarzschild solution. We now compare the time interval measured by a clock located at r_1 with the time interval measured by a clock located at $r_2 > r_1$. The field at r_1 is stronger than the field at r_2. if dt denotes the time interval measured at ∞, then the corresponding times $d\tau_1$ and $d\tau_2$ measured at r_1 and r_2 are respectively given by the equations

$$d\tau_1 = (1 - 2GM/r_1c^2)^{1/2}dt \qquad d\tau_2 = (1 - 2GM/r_2c^2)^{1/2}dt$$

so that

$$\frac{d\tau_1}{d\tau_2} = \frac{(1 - 2GM/r_1c^2)^{1/2}}{(1 - 2GM/r_2c^2)^{1/2}}$$

This ratio is smaller than unity meaning that $d\tau_1 < d\tau_2$, i.e., the clock runs slower in a stronger gravitational field.

Solved Problems, Study Projects, Suggestions and Remarks

Study Project 1

A seminar can be given on the basic foundations of Riemannian geometry and Einstein's general theory of relativity. The important points to be covered can include the following:

1. The idea of a vector field both covariant and contravariant as occurring in Riemannian geometry and used in Einstein's general theory of relativity. The explanations should include the law of transformation of the vector field under a change of the coordinate system.

2. The notion of a scalar field in the general theory of relativity. Reasons for considering the scalar field as functions invariant under the transformation of coordinates should be supplied. Examples of scalar and vector fields can be given. How scalar fields are constructed from vector fields can also be explained.

3. The quantity $A^{\mu}(x)B_{\mu}(x)$ is a scalar, i.e. an invariant where A^{μ} and B_{μ} are respectively contravariant and covariant vector fields. Explanations to be supplied.

4. Explain how you would calculate the distance between two neighbouring points on a surface. Assume that the equations of the surface are given by $x = x(u,v), y = y(u,v), z = z(u,v)$. Use the fact that distance in three dimensional Euclidean space is given by $ds^2 = dx^2 + dy^2 + dz^2$ and that

$$dx = \frac{\partial x}{\partial u}du + \frac{\partial x}{\partial v}dv$$

$$dy = \frac{\partial x}{\partial u}du + \frac{\partial y}{\partial v}dv$$

$$dz = \frac{\partial x}{\partial u}du + \frac{\partial z}{\partial v}dv$$

The form of the metric tensor on the curved surface is given by

$$ds^2 = dx^2 + dy^2 + dz^2 = (\frac{\partial x}{\partial u} + \frac{\partial x}{\partial v})^2 +$$

$$(\frac{\partial y}{\partial u} + \frac{\partial y}{\partial v})^2 + (\frac{\partial z}{\partial u} + \frac{\partial z}{\partial v})^2 +$$

$$= g_{uu}(u, v)du^2 + 2g_{uv}(u, v)dudv + g_{vv}(u, v)dv^2$$

where

$$g_{uu} = (\frac{\partial x}{\partial u})^2 + (\frac{\partial y}{\partial u})^2 + (\frac{\partial z}{\partial u})^2$$

$$g_{vv} = (\frac{\partial x}{\partial v})^2 + (\frac{\partial y}{\partial v})^2 + (\frac{\partial z}{\partial v})^2$$

$$g_{uv} = \frac{\partial x}{\partial u}\frac{\partial x}{\partial v} + \frac{\partial y}{\partial u}\frac{\partial y}{\partial v} + \frac{\partial z}{\partial u}\frac{\partial z}{\partial v}$$

Using this formula, we can for example compute the metric on the surface of a sphere of radius R as

$$ds^2 = R^2(d\theta^2 + sin^2(\theta)d\phi^2)$$

More generally, we can consider a p dimensional surface immersed in $\mathbb{R}^n$ defined by the equations

$$y^n = y^n(x^1, ..., x^p), n = 1, 2, ..., N$$

Then, the metric on the surface equals

$$ds^2 = \sum_{n=1}^{N}(dy^n)^2 = \sum_{n=1}^{N}(\sum_{\alpha=1}^{p}\frac{\partial y^n}{\partial x^\alpha}dx^\alpha)^2$$

$$= \sum_{\alpha,\beta=1}^{p} g_{\alpha\beta}(x^1, ..., x^p)dx^\alpha dx^\beta$$

where

$$g_{\alpha\beta}(x^1, ..., x^p) = \sum_{n=1}^{N}\frac{\partial y^n(x^1, ..., x^p)}{\partial x^\alpha}\frac{\partial y^n(x^1, ..., x^p)}{\partial x^\beta}$$

5. The notion of parallel displacement of a vector on a surface must be dealt with in any talk introducing the audience to the general theory of relativity. If the equations of the surface are given by $y^n(x^1, ..., x^p), n = 1, 2, ..., N$ (we're assuming a p dimensional curved surface immersed in n dimensional Euclidean space), then any vector on the surface will have cartesian components

$$A^n = \sum_{\mu=1}^{p} A^\mu \frac{\partial y^n}{\partial x^\mu}$$

From this we deduce that

$$\sum_{n=1}^{N} A^n \frac{\partial y^n}{\partial x^\nu} = \sum_{\mu} g_{\mu\nu} A^\mu = A_\nu$$

where the metric tensor of the surface is given by

$$g_{\mu\nu} = \sum_{n=1}^{N} \frac{\partial y^n}{\partial x^\mu} \frac{\partial y^n}{\partial x^\nu}$$

The curvilinear components A^μ define a contravariant vector while the components A_μ define a covariant vector field. During the process of parallel displacement, the cartesian components A^n of the vector do not undergo any change. Thus,

$$\delta A_\nu = \sum_{n=1}^{N} A^n \delta \frac{\partial y^n}{\partial x^\nu} = \sum_{n=1}^{N} A^n \frac{\partial^2 y^n}{\partial x^\nu \partial x^\mu} \delta x^\mu$$

$$= \sum_{\alpha,n,\mu} A^\alpha \frac{\partial y^n}{\partial x^\alpha} \frac{\partial^2 y^n}{\partial x^\mu \partial x^\nu} \delta x^\mu$$

This can be put in the form

$$\delta A^\alpha = \Gamma^\alpha_{\mu\nu} A^\nu \delta x^\mu$$

where

$$\Gamma_{\alpha\mu\nu} = \frac{1}{2}\left(\frac{\partial g_{\alpha\mu}}{\partial x^\nu} + \frac{\partial g_{\alpha\nu}}{\partial x^\mu} - \frac{\partial g_{\nu\mu}}{\partial x^\alpha}\right)$$

are the Christoffel symbols.

Study project 2

Covariant derivative in curved space-time:

One of the seminars introduced above discussed the equations of general relativity upto parallel displacement of vectors along a trajectory. Specifically, it was shown that if A^μ are the curvilinear components of a contravariant vector field and this vector is displaced by an amount dx^μ, the change in the components of the vector is given by the bilinear formula

$$\delta A^\mu = -\Gamma^\mu_{\alpha\beta} A^\alpha dx^\beta$$

The covariant derivative of this field is obtaining by subtracting the vector obtained by parallely displacing the vector $A^\mu(\mathbf{x})$ from $\mathbf{x}$ to $\mathbf{x} + d\mathbf{x}$ from the actual value of the vector field at $\mathbf{x} + d\mathbf{x}$. The that the parallely displaced vector $A^\mu + \delta A^\mu$ is a well defined vector at $\mathbf{x} + d\mathbf{x}$ and hence we are subtracting two vectors at $\mathbf{x} + d\mathbf{x}$ to obtain the covariant differential

$$DA^\mu = A^\mu + dA^\mu - (A^\mu + \delta A^\mu) = dA^\mu - \delta A^\mu$$

This if the displacement is along a curve parameterized as $\mathbf{x}(s)$, then we find for the covariant derivative of the vector field along the curve, the expression

$$\frac{DA^\mu}{ds} = \frac{dA^\mu}{ds} - \frac{\delta A^\mu}{ds} = \frac{dA^\mu}{ds} + \Gamma^\mu_{\alpha\beta}A^\alpha\frac{dx^\beta}{ds}$$

The covariant derivative of the vector field with respect to the νth coordinate is defined as

$$\frac{DA^\mu}{dx^\nu} = A^\mu_{:\nu} = \frac{\partial A^\mu}{\partial x^\nu} + \Gamma^\mu_{\alpha\nu}A^\alpha$$

The covariant derivative of the vector field along the curve is thus given in terms of these partial covariant derivatives by the formula

$$\frac{DA^\mu}{ds} = \frac{DA^\mu}{dx^\nu}\frac{dx^\nu}{ds}$$

It is interesting to look at the problem of displacing a vector parallely along a closed curve on a surface specified in parametric form. For example, one can take the unit sphere as our curved surface and a great circle as the curve on the surface and consider the problem of displacing a vector parallely on this surface. A vector on the surface should be tangential to the surface and hence will be specified only by its $\hat{\theta}$ and $\hat{\phi}$ components. MATLAB programs for numerically evaluating the parallel displacement of a vector along a curve can also be developed. Suppose A^μ is parallely displaced along a curve $x^\mu(s)$. If $A^\mu(s)$ denotes the components of the vector at the point $x^\mu(s)$ obtained by parallely displacing the vector along the curve, then it satisfies the obvious differential equation

$$\frac{dA^\mu}{ds} + \Gamma^\mu_{\alpha\beta}(\mathbf{x}(s))A^\alpha(s)\frac{dx^\beta(s)}{ds} = 0$$

which states that the covariant derivative of the vector along the curve is zero. This equation can be implemented numerically by discretization. For a surface specified as $z = f(x, y)$, we can compute the metric as a function of (x, y) as

$$ds^2 = (f_x dx + f_y dy)^2 + dx^2 + dy^2 = (1 + f_x^2)dx^2 + (1 + f_y^2)dy^2 + 2f_x f_y dx dy$$

where $f_x = \frac{\partial f}{\partial x}$ and $f_y = \frac{\partial f}{\partial y}$. Using these expressions, formulae for the Christoffel symbols are obtained and hence the equations of parallel displacement along a curve are formulated.

Study project 3

A seminar on the Lorentz transformation and its applications to the description of the kinematics in the special theory of relativity would be of interest to an audience interested in looking at relativistic corrections to Newtonian mechanics. Even control theorists will find it useful for many rocket and satellite control problems involve relativistic descriptions, the reason being that the motion is too rapid and relativistic description of the motion is needed. The seminar can begin with the postulate that the metric in two dimensional space time is preserved by the Lorentz transformation, i.e. $c^2t^2 - x^2 = c^2t'^2 - x'^2$. The Lorentz transformation is taken as

$$ct' = a_{00}ct + a_{01}x \qquad x' = a_{10}ct + a_{11}x$$

applying the equality condition for the metric with the condition that the primed frame is moving with a velocity v relative to the unprimed one leads to the equation $a_{10}/a_{11} = -v/c$ from which the Lorentz transformation first derived by Einstein rigorously is obtained:

$$t' = \gamma(v)(t - vx/c^2) \qquad x' = \gamma(v)(x - vt) \qquad \gamma(v) = (1 - v^2/c^2)^{-1/2}$$

The next important concept is that of proper time. If a particle moves such that at time t, its instantaneous speed is $\frac{dx}{dt}$, then a clock attached to it will read the passing interval of time as

$$dt' = d(\gamma(v)(t - vx/c^2)) = \gamma(v)(dt - vdx/c^2) = \gamma(v)(dt - v^2dt/c^2) = dt\sqrt{1 - v^2/c^2}$$

We set $dt' = d\tau$ and call it as the proper time. When the clock attached to the static inertial frame reads the time interval as $[t_1, t_2]$, then in this duration, the clock attached to the moving particle reads time

$$\tau_{12} = \int_{t_1}^{t_2} dt\sqrt{1 - v(t)^2/c^2}$$

The momentum of a particle is defined as $m_0\frac{d\mathbf{v}}{d\tau} = \gamma m_0\frac{d\mathbf{v}}{dt}$. Newton's equation of motion is thus replaced by

$$m_0\frac{d}{dt}\left(\gamma\frac{d\mathbf{v}}{dt}\right) = \mathbf{F}$$

where $\mathbf{F}$ is the force acting on the particle. In full notation,

$$\frac{d}{dt}\frac{\mathbf{v}}{\sqrt{1 - v^2/c^2}} = \mathbf{F}/m_0$$

This gives rise to a variety of nonlinear systems in engineering. The energy equation reads

$$m_0c^2(\gamma(\mathbf{v}_2) - \gamma(\mathbf{v}_1)) = \int_1^2 \mathbf{F}.\mathbf{v}dt$$

This is deduced as a first integral of the equation of motion. Many problems in mechanics can be formulated from a relativistic standpoint. The Lagrangian function for a particle moving in a potential V is given according to special relativity by

$$L = -m_0c^2\sqrt{1 - v^2/c^2} - U(\mathbf{r})$$

It is an easy check that this gives the correct equation

$$\frac{d}{dt}\frac{\partial L}{\partial \mathbf{v}} = \frac{\partial L}{\partial \mathbf{r}}$$

reads

$$\frac{d}{dt}\frac{m_0 \mathbf{v}}{\sqrt{1 - v^2/c^2}} = -\nabla U$$

The energy integral reads as

$$\frac{m_0 c^2}{\sqrt{1 - v^2/c^2}} + U = E$$

Relativistic corrections to the Keplerian orbit can be considered by applying this equation and performing a perturbation expansion. For example, relativistic motion in one dimension is approximated by the equation

$$\frac{m_0 v^2}{2} + 3m_0 v^4/4c^2 + U = E - m_0 c^2$$

The second term on the left is a perturbation to the first term arising from retaining upto the second term in the binomial expansion of the relativistic factor.

Study project 4

A seminar can be delivered on formulating Maxwell's equations in a curved space-time. This will enable one to describe the behaviour of electric and magnetic circuits in a space-craft moving very close to a massive gravitating object. The seminar can begin by introducing the covariant four potential $A_i, i = 0, 1, 2, 3$ and the antisymmetric field tensor $F_{ij} = A_{j:i} - A_{i:j} = A_{j,i} - A_{i,j}$. While describing this, one must introduce the covariant derivatives of tensor quantities. If A_i is a covariant vector, its covariant derivative is given by

$$A_{i:j} = A_{i,j} - \Gamma_{ij}^{k} A_k$$

This is in accordance with the principle that the covariant derivative of a scalar quantity must coincide with the ordinary derivative. If B^i is a contravariant vector, its covariant derivative is given by

$$B_{:j}^{i} = B_{,j}^{i} + \Gamma_{kj}^{i} B_k$$

The product rule implies that the covariant derivative of the scalar quantity $A_i B^i$ equals

$$(A_i B^i)_{:j} = A_{i:j} B^i + A_i B_{:j}^i = (A_{i,j} - \Gamma_{ij}^k A_k)B^i + A_i(B_{,j}^i + \Gamma_{kj}^i B^k)$$

$$= (A_i B^i)_{,j} - \Gamma_{ij}^k A_k B^i + \Gamma_{kj}^i A_i B^k = (A_i B^i)_{,j}$$

The covariant derivative of the metric tensor equals zero. This means that

$$F_{ij} = A_{j,i} - A_{i,j} - \Gamma_{ji}^k A_k + \Gamma_{ij}^k A_k = A_{j,i} - A_{i,j}$$

Specifically, if one chooses coordinates $x^0 = t, x^1 = x, x^2 = y, x_3 = z$ so that $A^0 = V, A^1 = A_x, A^2 = A_y, A^3 = A_z$ with $c = 1$, V the electric scalar potential and (A_x, A_y, A_z) the three components of the magnetic vector potential, then $A_0 = V, A_1 = -A_x, A_2 = -A_y, A_3 = -A_z$ and

$$F_{01} = -A_{x,t} - V_{,x} = -\frac{\partial V}{\partial x} - \frac{\partial A_x}{\partial t} = E_x$$

and likewise $F_{02} = E_y, F_{03} = E_z$. For the components of the magnetic field, we find

$$F_{12} = -A_{y,x} + A_{x,y} = -\frac{\partial A_y}{\partial x} + \frac{\partial A_x}{\partial y} = -B_z$$

and likewise $F_{23} = -B_x, F_{31} = -B_y$. The Maxwell equation $\nabla.\mathbf{E} = \rho/\epsilon$ can be expressed as

$$\sum_{r=1}^{3} \frac{\partial F_{0r}}{\partial x^r} = \rho/\epsilon$$

In flat space-time, we have corresponding to the above coordinates, $g_{00} = 1, g_{11} = g_{22} = g_{33} = -1$ and $g_{ij} = 0, i \neq j$. Thus, $F^{01} = g^{00}g^{11}F_{01} = -F_{01}$ and likewise, $F^{02} = -F_{02}, F^{03} = -F_{03}$ and the above Maxwell equation can be cast in the form

$$\sum_{r=1}^{3} \frac{\partial F^{0r}}{\partial x^r} = -\rho/\epsilon$$

We also note that $F^{23} = g^{22}g^{33}F_{23} = F_{23}$ and similarly, $F^{31} = F_{31}, F^{12} = F_{12}$. The Maxwell equation

$$\nabla \times \mathbf{E} = -\frac{\partial \mathbf{B}}{\partial t}$$

has first component

$$\frac{\partial F_{03}}{\partial x^2} - \frac{\partial F_{02}}{\partial x^3} = \frac{\partial F_{23}}{\partial x^0}$$

which is the same as

$$F_{03,2} + F_{20,3} + F_{32,0} = 0$$

Obtaining similar equations for the other components and combining the three gives us

$$F_{rs,0} + F_{s0,r} + F_{0r,s} = 0, r, s = 1, 2, 3$$

The Maxwell equation $\nabla.\mathbf{B} = 0$ becomes

$$F_{23,1} + F_{31,2} + F_{12,3} = 0$$

Thus, we can conclude that

$$F_{ij,k} + F_{jk,i} + F_{ki,j} = 0, i, j, k = 0, 1, 2, 3$$

Of course, these can also be deduced from the fact that the field tensor can be derived from the four potential. Indeed, we have

$$F_{ij,k} + F_{jk,i} + F_{ki,j} = (A_{j,i} - A_{i,j})_{,k} + (A_{k,j} - A_{j,k})_{,i} + (A_{i,k} - A_{k,i})_{,j}$$

$$= A_{j,ik} - A_{i,jk} + A_{k,ji} - A_{j,ki} + A_{i,kj} - A_{k,ij} = 0$$

Hence, once we assume that the field tensor can be derived from the four potential, this particular Maxwell equation is redundant. We now look at the Maxwell equation

$$\nabla \times \mathbf{B} = \mu \mathbf{J} + \frac{\partial \mathbf{E}}{\partial t}$$

The first component of this equation is

$$B_{z,y} - B_{y,z} = \mu J_x + E_{x,t}$$

which can be expressed in the form

$$F_{21,2} + F_{31,3} = \mu J_x + F_{01,0}$$

or

$$F^{12,2} + F^{10,0} + F^{13}_{,3} = -\mu J_x$$

with similar equations for the other components. Thus, we end up with the system

$$\sum_{\nu=0}^{3} \frac{\partial F^{\mu\nu}}{\partial x^\nu} = -\mu_0 J^\mu, \mu = 0, 1, 2, 3$$

where $J^0 = \rho, J^1 = J_x, J^2 = J_y, J^3 = J_z$. The entire set of Maxwell equations can be regarded as being equivalent to this single tensor equation once we agree that the field tensor is derivable from a four potential. In the general theory of relativity, ordinary derivatives of tensors get replaced by covariant derivatives of tensors. Thus, the above Maxwell equation is equivalent to the equation

$$F^{\alpha\beta}:\beta = -\mu_0.J^\alpha$$

Consider now a tensor of second rank $T^{\mu\nu}$. We want to express its covariant derivatives in terms of ordinary derivatives and the Christoffel symbols. We do this by postulating the product rule for a tensor of the form $A^\mu B^\nu$.

$$(A^\mu B^\nu)_{:\alpha} = A^\mu_{:\alpha} B^\nu + A^\mu B^\nu_{:\alpha}$$

$$= (A^\mu_{,\alpha} + \Gamma^\mu_{\alpha\beta} A^\beta) B^\nu + A^\mu (B^\nu_{,\alpha} + \Gamma^\nu_{\alpha\beta} B^\beta)$$

$$= (A^\mu B^\nu)_{,\alpha} + \Gamma^\mu_{\alpha\beta} A^\beta B^\nu + \Gamma^\nu_{\alpha\beta} A^\mu B^\beta$$

This equation can be generalized to an arbitrary tensor of rank two by noting that any such tensor is expressible as a finite sum of such product tensors. The general equation is given by

$$T^{\mu\nu}_{;\alpha} = T^{\mu\nu}_{,\alpha} + \Gamma^\mu_{\alpha\beta} T^{\beta\nu} + \Gamma^\nu_{\alpha\beta} T^{\mu\beta}$$

We can derive some useful formulas for the divergence of a rank two tensor by considering the case when $T^{\mu\nu}$ is a symmetric tensor and when it is an antisymmetric tensor. In the special theory of relativity, the divergence of a tensor field $T^{\mu\nu}$ is defined as $T^{\mu\nu}_{,\nu} = \frac{\partial T^{\mu\nu}}{\partial x^{\nu}}$. Consider a volume V in three dimensional space $\mathbb{R}^3$. If $T^{\mu\nu}$ is a tensor field satisfying $\frac{\partial T^{\mu\nu}}{\partial x^{\nu}} = 0$, then we get on integrating this equation over V,

$$\int_V \frac{\partial T^{\mu\nu}}{\partial x^{\nu}} dx_1 dx_2 dx_3 = 0$$

or

$$\frac{\partial}{\partial t} \int_V T^{\mu 0} dx_1 dx_2 dx_3 + \int \sum_{r=1}^{3} \frac{\partial T^{\mu r}}{\partial x^r} dx_1 dx_2 dx_3 = 0$$

Transforming the second volume integral into a surface integral by Gauss' theorem gives us

$$\frac{\partial}{\partial t} \int_V T^{\mu 0} dx_1 dx_2 dx_3 + \int \sum_{r=1}^{3} T^{\mu r} dS_r = 0$$

This is a global conservation law. When $\mu = 0$, we obtain the energy conservation equation

$$\frac{\partial}{\partial t} \int_V T^{00} dx_1 dx_2 dx_3 + \int \sum_{r=1}^{3} T^{0r} dS_r = 0$$

T^{00} is the energy density of the tensor field and T^{0r} is the amount of energy flowing across a unit area in the direction x^r per unit time. The above equation states that the net rate at which energy enters the volume V equals the net rate at which energy inside the volume increases. Similarly, when $\mu = r =, 1, 2, 3$, we have the momentum conservation equation

$$\frac{\partial}{\partial t} \int_V T^{r0} dx_1 dx_2 dx_3 + \int \sum_{s=1}^{3} T^{rs} dS_s = 0$$

T^{r0} equals the volume density of the rth component of the field momentum and T^{rs} the amount of rth component of momentum flowing across a unit area in the x^s direction. The above equation thus states that the net rate at which the rth component of momentum increases inside the volume equals the net rate at which the rth component of the momentum flows into V from its boundary S. For any tensor field $T^{\mu\nu}$, symmetric, or antisymmetric or neither, we have,

$$T^{\mu\nu}_{:\nu} = T^{\mu\nu}_{,\nu} + \Gamma^{\mu}_{\nu\beta} T^{\nu\beta} + \Gamma^{\nu}_{\nu\beta} T^{\mu\beta}$$

so that

$$T^{\mu\nu}_{:\nu} \sqrt{-g} = T^{\mu\nu}_{,\nu} \sqrt{-g} + \Gamma^{\mu}_{\nu\beta} T^{\nu\beta} \sqrt{-g} + T^{\mu\beta} \sqrt{-g}_{,\beta}$$

$$= (T^{\mu\nu} \sqrt{-g})_{,\nu} + \Gamma^{\mu}_{\nu\beta} T^{\nu\beta} \sqrt{-g}$$

When T is an antisymmetric tensor, $\Gamma^{\mu}_{\nu\beta}T^{\nu\beta} = 0$ and we get

$$T^{\mu\nu}_{:\nu}\sqrt{-g} = (T^{\mu\nu}\sqrt{-g})_{,\nu}$$

In view of the antisymmetry of the electromagnetic field tensor $F^{\mu\nu}$, the Maxwell equations can be expressed as

$$(F^{\alpha\beta}\sqrt{-g})_{,\beta} = -\mu_0 J^{\alpha}\sqrt{-g}$$

This equation leads to the charge conservation equation

$$(J^{\alpha}\sqrt{-g})_{,\alpha} = 0$$

which is the same as $J^{\alpha}_{:\alpha} = 0$. Writing $F_{\mu\nu} = A_{\nu,\mu} - A_{\mu,\nu}$ so that

$$F^{\alpha\beta} = g^{\mu\alpha}g^{\nu\beta}F_{\mu\nu} = g^{\mu\alpha}g^{\nu\beta}(A_{\nu,\mu} - A_{\mu,\nu})$$

the Maxwell equations are

$$(\sqrt{-g}g^{\mu\alpha}g^{\nu\beta}(A_{\nu,\mu} - A_{\mu,\nu}))_{,\beta} = -\mu_0 J^{\alpha}\sqrt{-g}$$

which is equivalent to

$$g^{\mu\alpha}g^{\nu\beta}(A_{\nu:\mu} - A_{\mu:\nu})_{:\beta} = -\mu_0 J^{\alpha}$$

Now

$$A_{\nu:\mu:\beta} - A_{\nu:\beta:\mu} = R^{\alpha}_{\nu\mu\beta}A_{\alpha}$$

where R is the curvature tensor. Thus, Maxwell equations are equivalent to

$$g^{\mu\alpha}g^{\nu\beta}(A_{\nu:\beta:\mu} + R^{\alpha}_{\nu\mu\beta}A_{\alpha} - A_{\mu:\nu})_{:\beta} = -\mu_0 J^{\alpha}$$

The Gauge condition imposed is $A^{\nu}_{:\nu} = 0$ which is the same as $g^{\nu\beta}A_{\nu:\beta} = 0$. With this condition imposed, the Maxwell equation simplifies to

$$g^{\mu\alpha}g^{\nu\beta}(R^{\alpha}_{\nu\mu\beta}A_{\alpha} - A_{\mu:\nu})_{:\beta} = -\mu_0 J^{\alpha}$$

This is the general relativistic modification for the D'Alembert equation with source obeyed by the scalar and vector potentials in special relativity.

Study project 5

Formula for the parallel translate of a vector around a little rectangular path in a Riemannian manifold endowed with a connection. Assume that we have a vector $X^i(a)$ located at a in terms of our local coordinate system. The first two coordinates of a are given by $x^1(a) = r, x^2(a) = s$. Choose a path C_1 that moves from a to a neighbouring point along the x^1 direction, i.e. this curve is given by $x^1(t) = r + t\delta r, x^2(t) = x^2(a) = s, x^3(t) = x^3(a).x^4(t) = x^4(a), 0 \leq t \leq 1$. Call the other end point of this new curve as b. Thus, b has coordinates $x^1(b) = r + \delta r, x^2(b) = s, x^3(b) =$

$x^3(a), x^4(b) = x^4(a)$. From b draw an infinitesimal curve C_2 parallel to the x_2 direction. This curve has parametric equation given by $x^1(t) = r + \delta r, x^2(t) = s + t\delta s, x^3(t) = x^3(a), x^4(t) = x^4(a), 0 \le t \le 1$. Call the other end point of this new curve as c. This point has coordinates $x^1(c) = r + \delta r, x^2(c) = s + \delta s, x^3(c) = x^3(a), x^4(c) = x^4(a)$. Now draw an infinitesimal curve C_3 from c along the negative x^1 direction. This curve has parametric equation given by $x^1(t) = r + \delta r - t\delta r, x^2(t) = s + \delta s, x^3(t) = x^3(a), x^4(t) = x^4(a), 0 \le t \le 1$. The end point d of this curve has coordinates $x^1(d) = r, x^2(d) = s + \delta s, x^3(d) = x^3(a), x^4(d) = x^4(a)$. Finally, draw a curve joining d and a along the x^2 direction. This curve C_4 has parametric equation given by $x^1(t) = r, x^2(t) = s + \delta s - t\delta s, x^3(t) = x^3(a), x^4(t) = x^4(a), 0 \le t \le 1$. The closed curve $C_1C_2C_3C_4$ defines a closed rectangular loop in the x^1x^2 plane. Parallel transport of $X^i(a)$ along C_1 gives

$$X^i(b) = X^i(a) - \int_0^1 (\Gamma^i_{j1}X^j)(r + t\delta r, s)\delta r dt$$

which to linear order in δr approximates to

$$X^i(b) = X^i(a) - \int_0^1 (\Gamma^i_{j1}X^j)(a) + \frac{\partial}{\partial x^1}(\Gamma^i_{j1}X^j)(a)t\delta r dt$$

$$= X^i(a) - \delta r(\Gamma^i_{j1}X^j(a) + \frac{\delta r}{2}\frac{\partial}{\partial x^1}(\Gamma^i_{j1}X^j)(a))$$

Parallel transport of $X^i(b)$ along C_2 to c gives

$$X^i(c) = X^i(b) - \int_0^1 (\Gamma^i_{j2}X^j)(r + \delta r, s + t\delta s)\delta s dt$$

$$= X^i(b) - \delta s \int_0^1 (\Gamma^i_{j2}(X^j)(a) + \delta r\frac{\partial(\Gamma^i_{j2}X^j)}{\partial x^1} + t\delta s\frac{\partial(\Gamma^i_{j2}X^j)(a)}{\partial x^2})dt$$

$$= X^i(b) - \delta s(\Gamma^i_{j2}X^j)(a) + \delta r\frac{\partial(\Gamma^i_{j2}X^j)}{\partial x^1} + \frac{\delta s}{2}\frac{\partial(\Gamma^i_{j2}X^j)(a)}{\partial x^2})$$

Similarly, parallel displacement of $X^i(c)$ along C_3 to d gives

$$X^i(d) = X^i(c) + \delta r \int_0^1 (\Gamma^i_{j1}X^j)(r + (1-t)\delta r, s + \delta s)dt$$

$$= X^i(c) + \delta r \int_0^1 ((\Gamma^i_{j1}X^j)(a) + (1-t)\delta r\frac{\partial(\Gamma^i_{j1}X^j(a)}{\partial x^1} + \delta s\frac{\partial(\Gamma^i_{j1}X^j)(a)}{\partial x^2})dt$$

$$= X^i(c) + \delta r(\Gamma^i_{j1}X^j(a) + \frac{\delta r}{2}\frac{\partial}{\partial x^1}(\Gamma^i_{j1}X^j(a)) + \delta s\frac{\partial}{\partial x^2}(\Gamma^i_{j1}X^j(a))$$

Finally, parallel transport of $X^i(c)$ along C_4 from d to a gives

$$X^i(a)' = X^i(d) + \delta s \int_0^1 (\Gamma^i_{j2}X^j)(r, s + (1-t)\delta s)dt$$

$$X^i(d) + \delta s(\Gamma^i_{j2}X^j(a) + \frac{\partial}{\partial x^2}(\Gamma^i_{j2}X^j)(a)\delta s)/2)$$

Thus,

$$X^i(a)' - X^i(a) = \delta X^i(a) = X^i(b) - X^i(a) + X^i(c) - X^i(b) + X^i(d) - X^i(c) + X^i(a)' - X^i(d)$$

$$= -\delta r(\Gamma^i_{j1}X^j(a) + \frac{\delta r}{2}\frac{\partial}{\partial x^1}(\Gamma^i_{j1}X^j)(a))$$

$$-\delta s(\Gamma^i_{j2}X^j)(a) + \delta r\frac{\partial(\Gamma^i_{j2}X^j)}{\partial x^1} + \frac{\delta s}{2}\frac{\partial(\Gamma^i_{j2}X^j)(a)}{\partial x^2})$$

$$+\delta r(\Gamma^i_{j1}X^j(a) + \frac{\delta r}{2}\frac{\partial}{\partial x^1}(\Gamma^i_{j1}X^j(a)) + \delta s\frac{\partial}{\partial x^2}(\Gamma^i_{j1}X^j(a))$$

$$+\delta s(\Gamma^i_{j2}X^j(a) + \frac{\partial}{\partial x^2}(\Gamma^i_{j2}X^j)(a)\delta s)/2)$$

$$= \frac{\delta r\delta s}{2}(-\frac{\partial}{\partial x^1}(\Gamma^i_{j2}X^j(a)) + \frac{\partial}{\partial x^2}(\Gamma^i_{j1}X^j)(a))$$

The partial derivatives of the vector field with respect to the two coordinates are evaluated using the condition that the covariant derivative of the vector field along the curve of parallel displacement equals zero. This means that

$$\frac{\partial X^j(a)}{\partial x_1} + \Gamma^j_{k1}X^k(a) = 0$$

for displacement in the x^1-direction and

$$\frac{\partial X^j(a)}{\partial x^2} + \Gamma^j_{k2}X^k(a) = 0$$

for displacement in the x^2-direction. Plugging these partial derivatives into the above expression gives

$$\delta X^i(a) = \frac{\delta r\delta s}{2}(-\frac{\partial \Gamma^i_{j2}}{\partial x^1}X^j(a) + \Gamma^i_{j2}\Gamma^j_{k1}X^k(a) + \frac{\partial \Gamma^i_{j1}}{\partial x^2}X^j(a) - \Gamma^i_{j1}\Gamma^j_{k2}X^k(a))$$

$$= \frac{\delta r\delta s}{2}(-\frac{\partial \Gamma^i_{j2}}{\partial x^1} + \Gamma^i_{k2}\Gamma^k_{j1} + \frac{\partial \Gamma^i_{j1}}{\partial x^2} - \Gamma^i_{k1}\Gamma^k_{j2})X^j(a)$$

This expression can be generalized to parallel displacement around any closed curve and furnishes us with an expression for the Riemann-Christoffel curvature tensor.

Chapter 5

Quantum mechanics

5.1   Wave function

Consider a system of n particles forming a system. In classical mechanics, the state of the system will be specified by specifying the positions of these particles as functions of time, i.e., the mappings $t \to \mathbf{r}_\alpha(t), \alpha = 1, 2, ..., n$ from $[0, \infty) \to \mathbb{R}^{3N}$. In the quantum theory, the particles do not move along definite trajectories and hence we cannot use position or momenta to specify the state of the system. The state of the system is rather specified by a wave function $\psi(t, \mathbf{r}_1, ..., \mathbf{r}_N)$ from which one can compute the probabilities for the particles at any time to occupy a given region of $3N$ dimensional space using Born's hypothesis:

$$P((\mathbf{r}_1(t), ..., \mathbf{r}_N(t)) \in E) = \int_E \psi(t, \mathbf{r}_1, ..., \mathbf{r}_N)|^2 d^3\mathbf{r}_1...d^3\mathbf{r}_N$$

The evolution of the wave function with time is given by Schrodinger's wave equation. This wave equation is derived from the energy or Hamiltonian function of classical mechanics by replacing positions and momenta with operators acting on Hilbert space. The normalization condition that the wave function must satisfy is that $\int |\psi|^2 d^3\mathbf{r}_1...d^3\mathbf{r}_N = 1$, corresponding to the fact that with probability one, the quantum particles will be found somewhere in space. If ψ and ϕ are two wave functions, and the particles of the system are in the state ψ, then the probability that a measurement of the state will show it to be in the state ϕ is given by $| < \phi, \psi > |^2$, where

$$< \phi, \psi >= \int \bar{\phi}(\mathbf{r}_1, ..., \mathbf{r}_N)\psi(\mathbf{r}_1, ..., \mathbf{r}_N)d^3\mathbf{r}_1...d^3\mathbf{r}_N$$

This is one of the postulates of quantum mechanics.

5.2   Mixed State of a quantum system

Suppose $\phi_1, ..., \phi_n$ are n orthonormal wave functions of a quantum system. By orthonormal, we mean $< \phi_i, \phi_j >= \delta_{ij}$. Then, the wave functions are mutually exclusive, i.e. if the system is in the state ϕ_i, there is zero probability for finding it in the state ϕ_j for $j \neq i$ since $< \phi_j, \phi_i >= 0$ for $j \neq i$. We can however conceive of a situation in which with probability p_i, the system is in the state ϕ_i. Under these circumstances, the probability that a measurement will show the system to be in the state ψ is given by Bayes formula: $\sum_i p_i | < \psi, \phi_i > |^2$. This probability can also be expressed as $\int \rho(\mathbf{x}, \mathbf{x}') \bar{\psi}(\mathbf{x}) \psi(\mathbf{x}') d\mathbf{x} d\mathbf{x}'$, or in operator theoretic notation, as $< \psi | \rho | \psi >$ where $\rho(\mathbf{x}, \mathbf{x}') = \sum p_i \phi_i(\mathbf{x}) \bar{\phi}_i(\mathbf{x}')$. ρ is called the density operator of the system. It is also called a mixed state of the system. Using the Bra-Ket notation, a wave function is represented by the ket vector $|\psi >$, its conjugate by the bra vector $< \psi|$ and the inner product of two wave functions $|\psi >$ and $|\phi >$ is viewed as the bra vector $< \phi|$ multiplied onto the ket vector $|\psi >$ to give $< \phi|\psi >$. More generally, suppose $L^2(\mathbb{R}^{3N})$ is our Hilbert space and $\phi_k, k = 1, 2, ...$ is a complete orthonormal basis for this space. Let $p_k, k = 1, 2, ...$ be non-negative real numbers adding to unity. Then we can define a mixed state

$$\rho(\mathbf{x}, \mathbf{x}') = \sum_{k=1}^{\infty} p_k \phi_k(\mathbf{x}) \bar{\phi}_k(\mathbf{x}')$$

Given any wave function $\psi \in L^2(\mathbb{R}^{3N})$, if the system is in the mixed state ρ, the probability that a measurement will show it to be in the pure state ψ is given by $\sum_{k=1}^{\infty} p_k | < \psi, \phi_k > |^2$ Two important properties that characterize a density operator ρ are that firstly it is non-negative definite in the sense that firstly if ϕ is any square integrable function, then $< \phi|\rho|\phi >\geq 0$. The second property is that $Tr(\rho) = 1$ in the sense that if $e_k, k = 1, 2, ...$ is any complete orthonormal basis for the Hilbert space, then $\sum_{k=1}^{\infty} < e_k|\rho|e_k >= 1$.

5.3   Observables

In the most general formalism of the quantum theory, the state space is any Hilbert space $\mathcal{H}$, a pure state is a vector $\psi \in \mathcal{H}$ having unit norm, i.e., $\| \psi \|= 1$ and a mixed state is a non-negative definite Hermitian operator on $\mathcal{H}$ having unit trace. Heisenberg also introduced the notion of an observable on the state space as a Hermitian linear operator X acting on the Hilbert space $\mathcal{H}$. If the Hilbert space is $L^2(\mathbb{R}^{3N})$ corresponding to our quantum system comprising N quantum particles, then a pure state ψ is simply a wave function whose magnitude square integrates to unity. An observable X can then be represented in position coordinates as a kernel $X(\mathbf{x}, \mathbf{y})$, namely a function of two variables $\mathbf{x}, \mathbf{y} \in \mathbb{R}^{3N}$ satisfying the conjugate symmetry property, i.e., $\bar{X}(\mathbf{y}, \mathbf{x}) = X(\mathbf{x}, \mathbf{y})$. This operator acts upon a vector $\psi \in L^2(\mathbb{R}^{3N})$ in accordance with the rule

$$X\psi(\mathbf{x}) = \int X(\mathbf{x}, \mathbf{y})\psi(\mathbf{y})dy$$

The inner product on this Hilbert space is the standard one:

$$< u, v >= \int \bar{u}(\mathbf{x})v(\mathbf{x})dx$$

It is easy to verify that X satisfies the Hermitian property:

$$< \phi|X|\psi >= \int \bar{\phi}(\mathbf{x})X(\mathbf{x},\mathbf{y})\psi(\mathbf{y})dxdy$$

$$= \int \bar{\phi}(\mathbf{x})\bar{X}(\mathbf{y},\mathbf{x})\psi(\mathbf{y})dxdy$$

$$= (\int \bar{\psi}(\mathbf{y})X(\mathbf{y},\mathbf{x})\phi(\mathbf{x})dxdy)^* =< \psi|X|\phi >^*$$

where bar and star are used interchangeably to represent conjugate of a complex number. Let X be an observable on the Hilbert space $\mathcal{H}$ of quantum states. Then by the spectral theorem, we can determine a complete orthonormal basis $e_k, k = 1, 2, \ldots$ for $\mathcal{H}$ and real numbers $c_k, k = 1, 2, \ldots$ such that $Xe_k = c_k e_k, k = 1, 2, \ldots.$ Using our ket notation, we have $X|e_k >= c_k|e_k >.$ $< e_k|$ denotes the conjugate transpose of the ket vector $|e_k >$ and the spectral theorem reads $X = \sum_k c_k|e_k >< e_k|.$ If $\mathcal{H}$ is $L^2(\mathbb{R}^{3N})$, then $e'_k s$ are wave functions $e_k(\mathbf{x})$ and the spectral resolution of X reads in kernel form

$$X(\mathbf{x},\mathbf{y}) = \sum c_k e_k(\mathbf{x})\bar{e}_k(\mathbf{y})$$

If X is an observable of a quantum system, then we can measure it. The possible outcomes of such a measurement are the eigenvalues $\{c_k\}$ of X. When the eigenvalue c_k shows up, then the state of the system no matter what it was prior to the measurement, collapses to the state $|e_k >$ where $|e_k >$ is the normalized eigenvector of X corresponding to the eigenvalue c_k. If $|\psi >$ was the state prior to the measurement, then the probability that c_k is the outcome is given by $p_k = | < e_k|\psi > |^2$. The average value of the observable X in the state $|\psi >$ is thus given by the following weighted sum:

$$< X >_\psi = \sum_k p_k| < e_k|\psi > |^2 =< \psi| \sum c_k|e_k >< e_k|\psi >=< \psi|X|\psi >$$

This is a fundamental equation. Suppose however that the system was in the mixed state ρ having the spectral resolution

$$\rho = \sum_k q_k|f_k >< f_k|$$

Then, the average of X in this mixed state turns out to be given by the intuitively obvious formula:

$$< X >_\rho = \sum_k q_k| < f_k|X|f_k >= Tr(\rho \sum q_k|f_k > |f_k|) = Tr(\rho X)$$

This equation states that we compute the average of X in each eigenstate of ρ, multiply by the probability of that eigenstate and sum the resultant.

5.4   Heisenberg's uncertainty principle

Let A, B be two observables and let ρ be a mixed or pure state. If the state is pure corresponding to the wave function ψ, we then set $\rho = |\psi><\psi|$. The averages of A, B in the state ρ are respectively given by $< A >= Tr(\rho A)$ and $< B >= Tr(\rho B)$. The dispersions of A, B in the state ρ are respectively $\sigma(A, \rho)^2 = Tr(\rho(A- < A >)^2)$ and $\sigma(B, \rho)^2 = Tr(\rho(B- < B >)^2)$. For simplicity, set $U = A- < A > I$ and $V = B- < B > I$. Then,

$$\sigma(A, \rho)^2 \sigma(B, \rho)^2 = Tr(\rho U^2)Tr(\rho V^2)$$

Since U, V are Hermitian, $(U - cV)^2$ is non-negative definite and hence $Tr(\rho(U - cV)^2) \geq 0$. Equality holds when and only when $\rho(U - cV) = 0$. Now,

$$0 \leq Tr(\rho(U - cV)^2) = Tr(\rho U^2) + c^2 Tr(\rho V^2) - 2c Tr(\rho UV)$$

This is valid for all real c. Hence, the discriminant of this quadratic function of c must be non-positive, i.e., $Tr(\rho UV)^2 \leq Tr(\rho U^2)Tr(\rho V^2)$. Also

$$ReTr(\rho UV) = \frac{1}{2}Tr(\rho(UV + VU)), ImTr(\rho UV) = \frac{1}{2i}Tr(\rho(UV - VU))$$

We can thus write the above inequality as

$$Tr(\rho U^2)Tr(\rho V^2) \geq \frac{1}{4}Tr(\rho[U, V]_+)^2 + \frac{1}{4}Tr(\rho[U, V])^2$$

and hence

$$\sigma(A, \rho)^2 \sigma(B, \rho)^2 \geq \frac{1}{4} < [A, B] >_\rho^2$$

This is called the Heisenberg uncertainty principle. It states that if A and B do not commute, then there exists a state in which both A and B cannot be measured simultaneously. That is, if in this state we try to measure A with accuracy, (i.e. low dispersion), then in this state, the dispersion of B will increase and vice-versa.

5.5   De-Broglie's and Planck's equations

De-Broglie's principle states that with every quantum particle, we can associate a wave. The momentum p of the particle and the wavelength λ of the associated wave are related by De-Broglie's equation $\lambda = h/p$. Equivalently, the wave vector of the associated wave k is given by $p = hk/2\pi$. Planck's quantum hypothesis states that if a quantum particle makes a transition from the energy state E_1 to the energy state E_2, then a photon of frequency $\nu = (E_1 - E_2)/h$ is emitted when $E_1 > E_2$ and if $E_1 < E_2$, then a photon of frequency $\nu = (E_2 - E_1)/h$ is absorbed.

A light wave having angular frequency ω and wave vector $\mathbf{k}k = \omega\mathbf{n}/c$ is described by the complex amplitude $exp(i(\mathbf{k.r} - \omega t))$. The De-Broglie and Planck's equations imply that the momentum of the associated photon is given by $\mathbf{p} = h\mathbf{k}/2\pi$ and the energy of the associated photon by $E = h\omega/2\pi$.

5.6 Bohr's model of the atom

Bohr proposed a classical model for the atom. This model can be described along the following lines. Let Ze denote the nuclear charge and $-e$ the charge of an orbiting electron. Bohr suggested that the electron stays in orbits of fixed radii and the angular momentum of the orbiting electron can assume values only in integral multiples of $h/2\pi$. Thus, if v is the velocity of the orbiting electron and a the radius of the orbit, then $mva = nh/2\pi$. On the other hand, Newton's law implies that the centripetal acceleration for the electron to stay in orbit must be provided by the nuclear attraction, i.e., $mv^2/a = Ze^2/a^2$, or $mv^2 = Ze^2/a$. Thus, the kinetic energy of the electron is $T = mv^2/2 = Ze^2/2a$ and the potential energy is given according to Coulomb's law by $-Ze^2/a$. The total energy of the electron is the sum of the kinetic and potential energies, i.e., $E = -Ze^2/2a$. The two equations $mva = nh/2\pi$ and $mv^2 = Ze^2/a$ imply $n^2h^2/4\pi^2ma^2 = Ze^2/a$, or, $a = n^2h^2/4\pi^2mZe^2$ and thus, $E = -2Z^2e^4\pi^2m/n^2h^2$. As n over positive integers, we get a sequence of allowable energy values E for the electron in the bound states. This spectrum agrees with experiment in the sense that the possible frequencies of emitted photon when the electron makes a transition from a higher to a lower energy state are $\frac{2Z^2e^4\pi^2m}{h^3}\left(\frac{1}{n^2} - \frac{1}{m^2}\right)$ with $n < m$ and n, m assuming positive integer values. These frequencies have been observed in experiments. Bohr' model is quasi classical. It violates the Maxwell equations which state that an accelerating electron must continuously emit electromagnetic radiation, lose energy and finally collapse to the centre. If one were to follow Maxwell's equations, then instability of matter would result. Bohr made the hypothesis, that radiation is emitted only when an electron makes a transition from one orbit to another and each orbit is characterized by a definite value of the angular momentum quantized in integer multiples of Planck's constant. Bohr's frequency condition states that if an electron makes a transition from an orbit having energy E_n to an orbit having energy $E_m < E_n$, then the frequency of the emitted radiation is given by $(E_n - E_m)/h$. Bohr's model for the atom is not a complete mathematical model. It combines a little bit of classical mechanics with a little bit of electrostatics and introduces a new thumb rule for quantization in order to prevent instability of matter. Despite its heuristic character, the Bohr model was able to explain the energy spectrum of the Hydrogen atom and hence can be called the start of modern quantum theory. The first complete mathematical model for the atom was provided by Erwyn Schrodinger.

5.7   Schrodinger's wave equation and wave mechanics

Consider a wave of frequency ω and wave vector $\mathbf{k}$. The complex amplitude in space-time is given by the equation $\psi(t, \mathbf{r}) = exp(i\mathbf{k}.\mathbf{r} - i\omega t)$. The rate of change of the amplitude at a given point in space is given by

$$\frac{\partial \psi}{\partial t} = -i\omega\psi$$

According to Planck, the energy of the associated particle is given by $E = h\omega/2\pi$. Hence, we can write

$$\frac{\partial \psi}{\partial t} = -i2\pi E\psi/h$$

or

$$\frac{ih}{2\pi}\frac{\partial \psi}{\partial t} = E\psi$$

Further, the spatial gradient of the complex amplitude is given by

$$\nabla \psi = i\mathbf{k}\psi$$

Thus,

$$\nabla^2 \psi = -k^2\psi$$

If we assume that the quantum particle associated with this wave has mass m, then the kinetic energy of the particle is $T = \frac{p^2}{2m}$, where according to De-Broglie, $\mathbf{p} = hk/2\pi$. Thus,

$$T\psi = (h^2k^2/8\pi^2m)\psi = -(h^2\nabla^2/8\pi^2m)\psi$$

We thus obtain the following wave equation for ψ:

$$\frac{ih}{2\pi}\psi = E\psi = (T + V)\psi = (-\frac{h^2\nabla^2}{8\pi^2m} + V)\psi$$

where V is the potential energy of the associated quantum particle. We assume that this equation is valid even if ψ does not have the above form, i.e., this equation describes the wave associated with any quantum particle moving in the potential field V. V is in general a function of time and space and the wave equation can be written as

$$(-\frac{h^2}{8\pi^2m}\nabla^2 + V(t, \mathbf{r}))\psi(t, \mathbf{r}) = \frac{ih}{2\pi}\frac{\partial \psi(t, \mathbf{r})}{\partial t}$$

This is the fundamental equation describing all quantum motions for a single particle. It was first written down by the Celebrated Austrian physicist Erwyn Schrodinger. Suppose V does not depend explicitly on time, i.e., $V = V(\mathbf{r})$. Then we can consider a wave corresponding to a definite energy E. The time dependence of the wave function is obtained from the equation

$$\frac{ih}{2\pi}\frac{\partial \psi}{\partial t} = E\psi$$

so that

$$\psi(t, \mathbf{r}) = exp(-2\pi iEt/h)\phi(\mathbf{r})$$

From Schrodinger's equation, we then derive the stationary Schrodinger equation for ϕ:

$$-\frac{h^2}{8\pi^2 m}\nabla^2\phi(\mathbf{r}) + V(\mathbf{r})\phi(\mathbf{r}) = E\phi(\mathbf{r})$$

According to Max Born $\int_V \psi(t,\mathbf{r})|^2 d^3\mathbf{r}$ should be interpreted as the probability of finding the quantum particle in the volume V of space at time t. This means that ψ and hence ϕ should be normalized so that

$$\int |\phi(\mathbf{r})|^2 d^3\mathbf{r} = \int |\psi(t,\mathbf{r})|^2 d^3\mathbf{r} = 1$$

The stationary Schrodinger equation does not produce a square integrable (i.e., normalizable) wave function ϕ for all values of the energy E. For a given potential, only for certain discrete levels E is the solution ϕ square integrable. These energy levels are called the bound states of motion. The stationary Schrodinger equation can be viewed as an eigen-equation for the operator

$$H = -\frac{h^2}{8\pi^2 m}\nabla^2 + V$$

The eigenvalues are discrete for most of the commonly occurring potentials. If $V = 0$, then no value of E produces a normalizable wave function. Such unnormalizable functions yield quantum states which describe a continuous flux of quantum particles and the magnitude square of the wave function then gives the number density of particles in space. After Schrodinger discovered his wave equation, he used it to obtain the energy levels of the quantum harmonic oscillator and of the Hydrogen atom. His equation provided the correct experimentally verified results.

5.8   Solution to the Schrodinger equation for particle in a box, harmonic oscillator and hydrogen atom

Particle in a box:

Harmonic oscillator: The motion is one dimensional and the potential energy is given by $V(x) = Kx^2/2$. The stationary Schrodinger equation is

$$\psi''(x) + \frac{8\pi^2 m}{h^2}(E - Kx^2/2)\psi(x) = 0$$

Let $u = ax$. Then, $\psi'(u) = a^{-1}\psi'(x), \psi''(u) = a^{-2}\psi''(x)$ and Schrodinger's equation gives

$$\psi''(u) + \frac{8\pi^2 m}{a^2 h^2}(E - Ku^2/2a^2)\psi(u) = 0$$

We choose a so that $4\pi^2 mK/h^2 a^4 = 1$. Let $b = 8\pi^2 mE/a^2 h^2$ for this choice of a. Then,

$$\psi''(u) + (b - u^2)\psi(u) = 0$$

Plugging in $\psi(u) = exp(-u^2/2)$ into this gives

$$-exp(-u^2/2) + u^2 exp(-u^2/2) + (b - u^2)exp(-u^2/2) = 0$$

or

$$(b - 1)exp(-u^2/2) = 0$$

This equation is asymptotically correct. So we try a solution of the form $\psi(u) = H(u)exp(-u^2/2)$. Then

$$\psi'(u) = (H'(u) - uH(u))exp(-u^2/2),$$

$$\psi''(u) = (H''(u) - 2uH'(u) + (u^2 - 1)H(u))exp(-u^2/2)$$

Plugging this into the differential equation gives

$$H''(u) - 2uH'(u) + (u^2 - 1)H(u) + (b - u^2)H(u) = 0$$

or

$$H''(u) - 2uH'(u) + (b - 1)H(u) = 0$$

Let

$$H(u) = u^r(a_0 + a_1 u + a_2 u^2 + ...) = \sum_{n=0}^{\infty} a_n u^{n+r}$$

Plugging this into the equation gives

$$\sum_{n=0}^{\infty}(n+r)(n+r-1)a_n u^{n+r-2} - 2\sum_{n=0}^{\infty}(n+r)a_n u^{n+r} + (b-1)\sum_{n=0}^{\infty} a_n u^{n+r} = 0$$

Equating coefficients of u^{r-2} to zero gives

$$r(r - 1)a_0 = 0$$

Equating coefficients of u^{r-1} to zero gives

$$r(r + 1)a_1 = 0$$

Equating coefficients of u^{n+r} with $n \geq 0$ gives

$$(n + r + 2)(n + r + 1)a_{n+2} - 2(n + r)a_n + (b - 1)a_n = 0, n \geq 0$$

or

$$a_{n+2} = \frac{2(n + r) - b + 1}{(n + r + 2)(n + r + 1)}, n \geq 0$$

The last equation implies that either $a_0 \neq 0$ or $a_1 \neq 0$ in order to get a non-zero solution. The first equation implies that either $a_0 = 0$ or else, $r \in \{0, 1\}$. The second equation implies that either $a_1 = 0$ or else $r = 0$. r cannot take the value -1 if we assume that $a_0 \neq 0$. If $a_0 = 0$, then $a_1 \neq 0$ and taking $r = -1$ is equivalent to assuming $a_0 \neq 0$ and $r = 0$. Also assuming $a_0 = 0$ and $r = 0$ (so that $a_1 \neq 0$) is equivalent to assuming $a_0 \neq 0$ and $r = 1$. Thus, without loss of generality, we may assume $a_0 \neq 0$ and $r \in \{0, 1\}$. Suppose $2(n + r) - b + 1 \neq 0$ for each $n = 0, 1, 2, \dots$. Then, it is clear that none of $a_0, a_2, a_4, \dots$ vanish and moreover, for large n,

$$a_{2n+2} = \frac{2(2n + r) - b + 1}{(2n + r + 2)(2n + r + 1)} a_{2n} \approx \frac{a_{2n}}{n}$$

Thus, for large n, $a_{2n} \approx a_0/n!$ and the asymptotic form of $H(u)$ is $a_0 \sum_{n=0}^{\infty} u^{2n}/n! = a_0 exp(u^2)$. This means that $\psi(u)$ grows as $exp(u/2)$ as $|u| \to \infty$, and is therefore not square integrable. To get a square integrable solution therefore, a_{2n} must be zero for some n. Assuming that $a_{2n+2} = 0$, we get $2(n+r) - b + 1 = 0$ and hence, $b = 2(2n+r) + 1$ which is either $4n+1$ or $4n+3$. with $n = 0, 1, 2, \dots$ Equivalently, $b = 2n+1$ with $n = 0, 1, \dots$ since $\{4n+1, 4n+3 : n = 0, 1, \dots\} = \{2n+1 : n = 0, 1, \dots\}$. This gives $E = (n + 1/2)h\omega/2\pi, n = 0, 1, \dots$ where $\omega = \sqrt{K/m}$.

5.9 Hydrogen atom (Quantum Mechanical Two body problem)

Consider two masses m_1 and m_2 moving under an interaction potential depending only on their relative positions. Choose a cartesian system and let $\mathbf{r}_1 = (x_1, y_1, z_1)$ be the position of m_1 and $\mathbf{r}_2 = (x_2, y_2, z_2)$ the position of m_2. The kinetic energy operator of the first particle is given by

$$T_1 = -\frac{h^2}{8\pi^2 m_1} \nabla_1^2 = -\frac{h^2}{8\pi^2 m} \left(\frac{\partial^2}{\partial x_1^2} + \frac{\partial^2}{\partial y_1^2} + \frac{\partial^2}{\partial z_1^2} \right)$$

and that of the second particle is given by

$$T_2 = -\frac{h^2}{8\pi^2 m_2} \nabla_2^2 = -\frac{h^2}{8\pi^2 m_2} \left(\frac{\partial^2}{\partial x_2^2} + \frac{\partial^2}{\partial y_2^2} + \frac{\partial^2}{\partial z_2^2} \right)$$

The potential field is given by $V(\mathbf{r}_1 - \mathbf{r}_2) = V(x_1 - x_2, y_1 - y_2, z_1 - z_2)$ and the stationary Schrodinger equation reads

$$(T_1 + T_2 + V)\psi = E\psi$$

where $\psi(x_1, y_1, z_1, x_2, y_2, z_2)$ is a function of both the particle coordinates. Let (X, Y, Z) denote the coordinates of the centre of mass of the two particles and (x, y, z) the relative position coordinates, i.e.,

$$X = \frac{m_1 x_1 + m_2 x_2}{M}, Y = \frac{m_1 y_1 + m_2 y_2}{M}, Z = \frac{m_1 z_1 + m_2 z_2}{M}, M = m_1 + m_2$$

$$x = x_1 - x_2, y = y_1 - y_2, z = z_1 - z_2$$

In terms of the centre of mass and relative coordinates, the Stationary Schrodinger equation assumes the form

$$-\frac{h^2}{8\pi^2 M}\left(\frac{\partial^2}{\partial X^2} + \frac{\partial^2}{\partial Y^2} + \frac{\partial^2}{\partial Z^2}\right)\psi$$

$$-\frac{h^2}{8\pi^2 m}\left(\frac{\partial^2}{\partial x^2} + \frac{\partial^2}{\partial y^2} + \frac{\partial^2}{\partial z^2}\right)\psi$$

$$+V(x,y,z)\psi = E\psi$$

where $\psi(X,Y,Z,x,y,z)$ has been expressed as a function of the CM and relative coordinates. We separate the variables (X,Y,Z) and (x,y,z) by setting

$$\psi(X,Y,Z,x,y,z) = \chi(X,Y,Z)\Phi(x,y,z)$$

Plugging this into the equation gives

$$-\frac{h^2}{8\pi^2 M\chi}(\chi_{XX} + \chi_{YY} + \chi_{ZZ}) = \frac{h^2}{8\pi^2 m\Phi}(\Phi_{xx} + \Phi)_{yy} + \Phi_{zz}) + (E - V)$$

where $m = \frac{m_1 m_2}{M}$ is the reduced mass of the system. The left side is a function of (X,Y,Z) only and the right side of (x,y,z) only. Hence both sides must equal a constant. Denote this constant by E_t. E_t is the translation energy of the two body system since it is an eigen value of the translational kinetic energy. The left side gives

$$\chi(X,Y,Z) = A.exp(i(k_{tX}X + k_{tY}Y + k_{tZ}Z))$$

where (k_{tX}, k_{tY}, k_{tZ}) satisfy

$$E_t = \frac{h^2}{8\pi^2 M}(k_{tX}^2 + k_{tY}^2 + k_{tZ}^2)$$

Setting $E - E_t = E_r$, the energy associated with relative motion of the two particles, we find that E_r is an eigenvalue of the energy of relative motion:

$$\frac{h^2}{8\pi^2 m}(\Phi_{xx} + \Phi_{yy} + \phi_{zz}) + (E_r - V)\Phi = 0$$

For most interaction potentials V, the possible values of E_r that allow bounded motion, i.e., square integrable Φ are discrete, i.e., countable. Denoting the possible values of E_r by $\epsilon_n, n = 1, 2, ...$, it follows that the total energy of the two body system is the sum of the translational energy and relative energy:

$$E = E(\mathbf{k}_t, n) = \frac{h^2 k_t^2}{8\pi^2 M} + \epsilon_n$$

$\mathbf{k}_t$ can assume any value in $\mathbb{R}^3$ corresponding to the situation that the two body system can undergo overall translation with arbitrary momentum/velocity in space. Consider the equation describing relative motion. Assume that in addition to being a function of only the relative positions, V is a function of only the distance between the two particles $r = \sqrt{x^2 + y^2 + z^2}$, i.e., $V = V(r)$. For simplicity of notation use the notation E for E_r. Expressing the Laplacian in spherical polar coordinates, the stationary Schrodinger equation reads

$$\frac{1}{r^2}\frac{\partial}{\partial r}\left(r^2\frac{\partial \Phi}{\partial r}\right) - \frac{4\pi^2 L^2}{h^2 r^2}\Phi + \frac{8\pi^2 m}{h^2}(E - V(r))\Phi = 0$$

where

$$L^2 = -\frac{h^2}{4\pi^2}\left(\frac{1}{sin\theta}\frac{\partial}{\partial\theta}sin\theta\frac{\partial}{\partial\theta} + \frac{1}{sin^2\theta}\frac{\partial^2}{\partial\phi^2}\right)$$

L^2 is the squared total angular momentum operator. It is known from the theory of ordinary differential equations that the eigenfunctions of L^2 are $Y_{lm}(\theta, \phi), m = -l, -l+1, ..., l, l = 0, 1, 2, ...$ $L_z = -\frac{ih}{2\pi}\frac{\partial}{\partial\phi}$ is the z-component of the angular momentum. Y_{lm} is a simultaneous eigenfunction of L^2, L_z with eigenvalues $l(l+1)h^2/4\pi^2$ and $mh/2\pi$:

$$L^2 Y_{lm} = (l(l+1)h^2/4\pi^2)Y_{lm}, \qquad L_z Y_{lm} = (mh/2\pi)Y_{lm}$$

Consider the eigenvalue equation

$$L^2 F(\theta, \phi) = \lambda F(\theta, \phi)$$

Substituting $F(\theta, \phi) = P(cos(\theta))exp(im\phi)$ results in an ordinary differential equation for $P(x)$ called Legendre's differential equation from which one deduces that λ assumes values $l(l+1)h^2/4\pi^2$. The overall wave function is assumed to have the separable form, i.e., $\Phi(r, \theta, \phi) = R(r)Y_{lm}(\theta, \phi)$. Plugging this into the stationary Schrodinger equation gives the following ordinary differential equation for $R(r)$:

$$\frac{1}{r^2}\frac{d}{dr}\left(r^2\frac{dR}{dr}\right) + \frac{8\pi^2 m}{h^2}\left(E - V(r) - \frac{l(l+1)h^2}{8\pi^2 mr^2}\right)R = 0$$

In particular when $V(r) = -e^2/r$, it can be shown using series method for the solution to ordinary differential equations that the possible values of E assume those given by Bohr's model, i.e., $-C/n^2$ where $n = l+1, l+2, ...$ We skip the details here

5.10 Heisenberg's matrix mechanics and its relation to Schrodinger's wave mechanics

Let $U_t = exp(-itH)$ denote the unitary evolution operator for a quantum system having Hamiltonian H. Schrodinger's equation in operator form reads $i\frac{\partial\psi_t}{\partial t} = H\psi_t$, where our units are chosen so that $h/2\pi = 1$. The solution to this equation is $\psi_t = U_t\psi_0$. U_t satisfies the operator equation $i\frac{dU_t}{dt} = HU_t$. Let X be any observable. Its average value at time t is given by

$$< X >_t = < \psi_t|X|\psi_t > = < U_t\psi_0|X|U_t\psi_0 > = < \psi_0|U_t * XU_t|\psi_0 >$$

We can equivalently regard the state as fixed ψ_0 and the observable X evolving in time according to the rule $X_t = U_t^* X U_t = exp(itH) X exp(-itH)$, where $X(0) = X$. Then X_t satisfies the operator equation

$$\frac{dX_t}{dt} = i(HX_t - X_t H)$$

This is called Heisenberg's equation of motion. The formalism in which the observables are constant in time and the wave functions time varying is called Schrodinger's wave mechanics while the formalism in which the wave functions are constant in time but the observables time varying is called Heisenberg's matrix mechanics. We have the formula:

$$< X >_t = < \psi_0 | X_t | \psi_0 > = < \psi_t | X | \psi_t >$$

Both the mechanics therefore yield the same evolution of the average value of an observable in the quantum state.

5.11 Tunneling problems

Consider the following quantum mechanical problem. The motion takes place in one dimension and the potential function is given by $V(x) = 0$ for $x < 0$ and for $x > L$. For $0 \leq x \leq L$ the potential is given by $V(x) = V_0$. The stationary Schrodinger equation corresponding to the energy value E is given by $\psi''(x) + \frac{8\pi^2 mE}{h^2}\psi(x) = 0$ for $x < 0$ and $x > L$ while for $0 \leq x \leq L$, $\psi''(x) + \frac{8\pi^2 m}{h^2}(E - V_0)\psi(x) = 0$. The solution is

$$\psi(x) = A_1.exp(-ikx) + B_1 exp(-ikx), x < 0$$

$$\psi(x) = A_2 exp(\alpha x) + B_2 exp(-\alpha x), 0 \leq x \leq L$$

$$\psi(x) = C.exp(ikx), x > L$$

where $\frac{h^2 k^2}{8\pi^2 m} = E$, $\frac{8\pi^2 m\alpha^2}{h^2} = V_0 - E$. We are assuming that $E < V_0$. This assumption means that a particle incident upon the potential barrier from the left cannot classically penetrate into the barrier. The wave function for $x < 0$ has been chosen so that we have both a beam of incident particles and a beam of reflected particles for $x < 0$ while for $x > L$, we have only a transmitted beam of particles and no reflected beam, since there is no barrier to the right of $x = L$. Quantum mechanically, we get nonzero values of A_2 and B_2 when the boundary conditions at $x = 0$ and $x = L$ are matched, i.e., $\psi(0-) = \psi(0+), \psi(L-) = \psi(L+)$. The nonzero values of A_2, B_2 imply that quantum mechanically, there is a nonzero probability of finding the particle in the range $0 \leq x \leq L$, i.e., within the barrier. Further C will be nonzero and that signifies that there is a beam of transmitted particles. Classically, no particle can tunnel from the region $x < 0$ to the region $x > L$ because $E < V_0$ implies that the kinetic energy of the particle in the region $0 < x < L$

is negative which is absurd. Thus tunneling is a purely quantum mechanical phenomenon. Before closing this section, we mention that the boundary conditions applied at $x = 0$ and $x = L$ give

$$A_1 + B_1 = A_2 + B_2, ik(A_1 - B_1) = \alpha(A_2 - B_2)$$

$$A_2 exp(\alpha L) + B_2 exp(-\alpha L) = C exp(ikL), \alpha(A_2 exp(\alpha L) - B_2 exp(-\alpha L)) = ikC exp(ikL)$$

Using these, we can compute the reflection coefficient $\frac{B_1}{A_1}$ and the transmission coefficient $\frac{C}{A_1}$. Their magnitude square gives respectively the ratio of the intensity of reflected particles to that of the incident particles and the intensity of transmitted particles to that of the incident particles.

5.12 Dirac equation for the electron

The non-relativistic formula for energy momentum relation for a free particle is $E = \frac{p^2}{2m}$. This leads us to the Schrodinger wave equation for the free particle $(E - p^2/2m)\psi = 0$ where E is replaced by the energy operator $i\frac{\partial}{\partial t}$ and $\mathbf{p}$ by the momentum operator $-i\nabla$. The units are chosen so that $h/2\pi = 1$. In the presence of the potential, we replace E by $E - V$ to get

$$(E - V - p^2/2m)\psi = 0$$

which is Schrodinger's wave equation. However to get a relativistically correct equation, we must use the relativistic formula for the energy momentum relation: $E^2 = c^2 p^2 + m^2 c^4$. Replacing E by the operator $i\frac{\partial}{\partial t}$ and $\mathbf{p}$ by $-i\nabla$ leads to the Klein-Gordon equation

$$(\frac{\partial^2}{\partial t^2} - c^2\nabla^2 + m^2 c^4)\psi = 0$$

This is not a correct evolution equation from the quantum mechanical standpoint since a correct quantum evolution equation should be first order in time. One should therefore use the equation

$$(E - c\sqrt{m^2 c^2 + p^2})\psi = 0$$

but this equation is difficult to solve since the momentum operator appears inside a square root sign and it is difficult to give meaning to the operator $\sqrt{m^2 c^2 - \nabla^2}$. Dirac suggested that we try to factorize the quadratic form $E^2 - c^2(p_x^2 + p_y^2 + p_z^2) - m^2 c^4$ into linear factors using 4×4 matrices. It is impossible to achieve such a factorization using scalars. Define the matrices

$$\alpha_k = \begin{pmatrix} 0 & \sigma_k \\ \sigma_k & 0 \end{pmatrix}, k = 1, 2, 3$$

and

$$\beta = \begin{pmatrix} I & 0 \\ 0 & -I \end{pmatrix}$$

where $\sigma_k, k = 1, 2, 3$ are the 2×2 Pauli spin matrices and the identity appearing in the definition of β is also 2×2. The Pauli spin matrices satisfy the anticommutation relations $\sigma_i \sigma_j + \sigma_j \sigma_i = 0$ for $i \neq j$ and also $\sigma_k^2 = I, k = 1, 2, 3$. This leads to the following anticommutation relations:

$$\alpha_k \alpha_m + \alpha_m \alpha_k = 0, \alpha_k \beta + \beta \alpha_k = 0, k = 1, 2, 3$$

From these, we easily deduce that

$$E^2 - c^2(p_x^2 + p_y^2 + p_z^2) - m^2 c^4$$

$$= (E + c(\alpha_x p_x + \alpha_y p_y + \alpha_z p_z) + \beta m c^2)(E - c(\alpha_x p_x + \alpha_y p_y + \alpha_z p_z) - \beta m c^2)$$

Hence, Dirac suggested the following relativistic wave equation for the free particle

$$(E - c(\alpha_x p_x + \alpha_y p_y + \alpha_z p_z) - \beta m c^2)\psi = 0$$

or plugging in the operators for E and $\mathbf{p}$,

$$(i\frac{\partial}{\partial t} + ic(\alpha_x \frac{\partial}{\partial x} + \alpha_y \frac{\partial}{\partial y} + \alpha_z \frac{\partial}{\partial z}) - \beta m c^2)\psi = 0$$

ψ will now be a four component wave function. Here, $\alpha_x = \alpha_1, \alpha_y = \alpha_2, \alpha_z = \alpha_3$. The beauty about Dirac's wave equation is that spin automatically creeps into in the form of 4×4 matrices. Dirac thus showed that spin is a relativistic effect. To account for spin in the non-relativistic Schrodinger equation, we need to add a term to the energy corresponding to the interaction between the magnetic field and the spin of the electron. If σ denotes the Pauli matrix vector $(\sigma_x, \sigma_y, \sigma_z)$, then the spin of the electron is represented by the operator $\sigma/2$ and the magnetic moment of the electron is $-e\sigma/4m$ where our units are chosen so that $h/2\pi = 1$. The interaction energy between the spin magnetic moment and the external magnetic field is given by $\frac{e}{4m}(\sigma_x B_x + \sigma_y B_y + \sigma_z B_z)$ and hence the non-relativistic Schrodinger equation with spin taken into account is given by

$$-\frac{h^2}{8\pi^2 m}\nabla^2 \psi + \frac{e}{4m}(\sigma, \mathbf{B})\psi = i\frac{\partial \psi}{\partial t}$$

Such a forced introduction of spin-magnetic field coupling effect into Schrodinger's theory is avoided in Dirac's theory. Spin arises naturally when we try to factorize the Klein-Gordon equation. Dirac's wave equation for the electron in an electromagnetic field is obtained by replacing E by $E + eV$ where V is the electric potential and $\mathbf{p}$ by $\mathbf{p} + e\mathbf{A}$. It can be solved for the Hydrogen atom where $V = -e/r, \mathbf{A} = \mathbf{0}$. We do not go into the details here. Neither do we go into the details of how Dirac used his relativistic wave equation for the electron to show the existence of the anti-particle of the electron, namely the positron having mass equal to that of the electron and positive charge. The arguments are based on symmetry and we refer the student to Dirac's celebrated book "Principles of Quantum Mechanics".

5.13 Solved Problems, Study Projects, Suggestions and Remarks

Study Project 1

Read up about the foundations of quantum mechanics and present the Heisenberg and Schrodinger theories succinctly. The points to be mentioned are as follows:

1. Notion of the wave function $\psi(x_1, ..., x_n)$ as describing the state of a system of quantum mechanical particles.

2. Max Born's interpretation of the wave function $P(V) = \int_V |\psi(x_1, ..., x_n)|^2 \, dx_1...dx_n$ as describing the probability that the system of particles will be found in the region V of configuration space.

3. History of the Schrodinger equation ? How Erwin Schrodinger in a series of papers derived the wave equation for the wave function and overthrew the semiclassical theory of Niels Bohr involving considering the electron to be revolving around the nucleus with quantized values of angular momentum. How Schrodinger formulated the problem of determining the energy levels of a quantum system as an eigenvalue problem for a partial differential operator.

4. How Schrodinger was able to obtain using his wave equation, the energy levels of a harmonic oscillator and that of the hydrogen atom.

5. Heisenberg's introduction of matrix mechanics prior to Schrodinger's discovery. The commutation relations between the position and momentum operators $QP - PQ = ih/2\pi$. Heisenberg's derivation of the quantum mechanical equations of motion for an observable

$$\frac{ih}{2\pi}\frac{dX}{dt} = HX - XH$$

6. The importance of quantum theory in modern electronics and communication engineering should be stressed as a final point: How the Schrodinger equation for a crystal leads to energy bands and how it is used in obtaining characteristic properties of semiconducting materials. Approximate numerical techniques that have been developed for solving the Schrodinger equation. The wave equation for a many electron atom may be mentioned here: If $\mathbf{r}_1, ..., \mathbf{r}_N$ are the position coordinates of the N electrons and $\mathbf{R}_1, ..., \mathbf{R}_M$ the positions of the M nuclei carrying charge Ze, then the Schrodinger equation to be solved for determining the characteristic energies of the system is

$$(-\frac{h^2}{8\pi^2}\sum_{j=1}^{M}\frac{\nabla_{\mathbf{R}_j}^2}{M_j} - \frac{h^2}{8\pi^2}\sum_{i=1}^{M}\frac{\nabla_{\mathbf{r}_i}^2}{m_i})\psi(\mathbf{r}_1, ..., \mathbf{r}_N, \mathbf{R}_1, ..., \mathbf{R}_M)$$

$$+(\sum_{1\leq i<j\leq M}\frac{Z^2 e^2}{|\mathbf{R}_i - \mathbf{R}_j|} - \sum_{i=1}^{N}\sum_{j=1}^{M}\frac{Ze^2}{|\mathbf{r}_i - \mathbf{R}_j|} + \sum_{1\leq i<j\leq N}\frac{e^2}{|\mathbf{r}_i - \mathbf{r}_j|})\psi(\mathbf{r}_1, ..., \mathbf{r}_N, \mathbf{R}_1, ..., \mathbf{R}_M)$$

$$= E\psi(\mathbf{r}_1, ..., \mathbf{r}_N, \mathbf{R}_1, ..., \mathbf{R}_M)$$

7. A lecture on the notion of measurement of an observable in the quantum theory and how measurement in a given state leads to a probability distribution amongst the values of the observable. Specifically, if ρ denotes the mixed state of the quantum system and $E(a, b]$ the spectral measure corresponding to the observable X assuming values in the interval $(a, b]$, i.e.

$$X = \int_{-\infty}^{\infty} x.E(dx)$$

then

$$P_{X,\rho}((a, b]) = Tr(\rho.E(a, b])$$

equals the probablity that the observable X will assume values in the range $(a, b]$ given that the state of the system is ρ.

8. A derivation of the Heisenberg uncertainty principle in quantum mechanics can be presented in a seminar along the following lines. Let X denote an observable and $|\psi>$ the state of the quantum system. Then, from the axiomatic rules of quantum mechanics, it is known that $\mu_X =< \psi|X|\psi >$ is the average value of X in the state $|\psi>$ and $\sigma_X^2 =< \psi|(X - \mu_X)^2|\psi >$ is the variance of X in the state $|\psi>$. The proof of this fact should be explained along the following lines. Let $|e_i>, i = 1, 2, ...$ denote a complete orthonormal basis of eigenvectors of the observable X and $a_i, i = 1, 2, ...$ the corresponding eigenvalues, i.e. $X|e_i >= a_i|e_i >$. If $|\psi>$ is the state of the system and a measurement on X is carried out, then $| < e_i|\psi > |^2$ is the probability that the value a_i shows up. From the principles of classical probability, the average value of X in the state $|\psi>$ is thus given by

$$< X >_\psi = \sum_i | < e_i|\psi > |^2 a_i =< \psi| \sum a_i|e_i >< e_i|\psi >=< \psi|X|\psi >$$

If the position representation is adopted, then $< x|\psi >= \psi(x)$ and $< x|X|y >= X(x, y)$ and

$$< \psi|X|\psi >= \int \bar{\psi}(x)X(x, y)\psi(y)dxdy$$

$X(x, y)$ is called the kernel of the operator X in the position representation and $\psi(x)$ is called the wave function of ψ in the position representation. The variance of X in the state $|\psi>$ is the average value of $(X - \mu_X)^2$ in this state, i.e. $< \psi|(X - \mu_X)^2|\psi >$. Now, let X, Y be two observables. Their variances in the quantum state $|\psi>$ are respectively given by $\sigma_X^2 =< \psi|(X - \mu_X)^2|\psi >$ and $\sigma_Y^2 =< \psi|(Y - \mu_Y)^2|\psi >$. These can be expressed as

$$\sigma_X^2 =\| (X - \mu_X)\psi >\|^2$$

$$\sigma_Y^2 = \| (Y - \mu_Y)\psi > \|^2$$

Thus, by the Cauchy-Schwarz inequality, we have

$$\sigma_X^2 \sigma_Y^2 \geq | < (X - \mu_X)\psi|(Y - \mu_Y)\psi > |^2 = | < \psi|(X - \mu_X)(Y - \mu_Y)|\psi > |^2$$

$$= (Re < \psi|(X - \mu_X)(Y - \mu_Y)|\psi >)^2 + (Im < \psi|(X - \mu_X)(Y - \mu_Y)|\psi > |^2$$

If U, V are two observables, then

$$Re(< \psi|UV|\psi >) = \frac{1}{2}(< \psi|UV|\psi > + < \psi|VU|\psi >) = \frac{1}{2} < \psi|[U, V]_+|\psi >$$

where $[U, V]_+ = UV + VU$ is the anticommutator of U and V. This fact should be derived on the board by explaining that

$$< \psi|UV|\psi >^* = < V\psi|U\psi > = < \psi|VU|\psi >$$

Likewise,

$$Im(< \psi|UV|\psi >= \frac{1}{2i} < \psi|UV - VU|\psi >= \frac{1}{2i} < \psi|[U, V]|\psi >$$

where $[U, V] = UV - VU$ is the anticommutator of U and V. Thus,

$$\sigma_X^2 \sigma_Y^2 \geq \frac{1}{4}| < \psi|[U, V]_+|\psi > |^2 + \frac{1}{4}| < \psi|[U, V]|\psi > |^2$$

This is the uncertainty principle in the form first derived by W.Pauli. In particular, we have

$$\sigma_X^2 \sigma_Y^2 \geq \frac{1}{4}| < \psi|[U, V]|\psi > |^2$$

if $[U, V] = c$ a constant, as in the case of position and momentum, then this boils down to

$$\sigma_X^2 \sigma_Y^2 \geq c^2/4$$

For example, if X is position and Y is momentum, then $[X, Y] = ih/2\pi$ and the product of the variances of the two in any state must exceed $h^2/16\pi^2$. Stress should be given on the fact implied by this inequality that if two observables do not commute then there exists a state in which both cannot be measured with perfect accuracy.

Study Project 2

Study the motion of the hydrogen atom in a magnetic field with spin taken into account. Assume that the hydrogen atom has mass M and momentum $\mathbf{P}$ and a spin magnetic moment of μ_B and a spin of $\mathbf{S} = S_1\hat{x} + S_2\hat{y} + S_3\hat{z}$. Assume that the motion of the atom takes place in a magnetic field $\mathbf{B}(\mathbf{Q}) = B_3(Q_3)\hat{z}$. Here,

$$S_3 = \frac{1}{2}\begin{pmatrix} 1 & 0 \\ 0 & -1 \end{pmatrix}$$

The Hamiltonian of the atom consists of two parts: The kinetic energy of the atom moving in the magnetic field and the interaction energy between the spin of the atom and the magnetic field:

$$H = \frac{\mathbf{P}^2}{2M} + 2\mu_B B_3(Q_3) S_3 = \begin{pmatrix} \frac{\mathbf{P}^2}{2M} + \mu_B B_3(Q_3) & 0 \\ 0 & \frac{\mathbf{P}^2}{2M} - \mu_B B_3(Q_3) \end{pmatrix}$$

Assume that the spin of the atom initially ($t = 0$) is in the state

$$|\phi> = \alpha|+> + \beta|->$$

where $|\alpha|^2 + |\beta|^2 = 1$. Here, $|+>$ denotes the state in which the particle has $+1/2$ for the $\hat{z}$ component of the spin, i.e. $S_3|+> = \frac{1}{2}|+>$ and $|->$ denotes the state in which the particle has $-1/2$ for the $\hat{z}$ component of the spin, i.e. $S_3|-> = -\frac{1}{2}|->$. In the state $|\phi>$, the probability of the particle having spin $+1/2$ for the z component is $|\alpha|^2$ and the probability of the particle having spin $-1/2$ for the z component is $|\beta|^2$. Assume also that the state of translatory motion of the atom is $|\psi_0>$. Then, the initial state of the overall particle with translatory motion and spin taken into account is given by $|\chi(0)> = |\phi> \otimes |\psi_0>$ and the corresponding density matrix is given by

$$W(0) = |\chi(0)> < \chi(0)| = (|\phi> \otimes |\psi_0>)(< \phi| \otimes < \psi_0|) = |\phi> < \phi| \otimes |\psi_0> < \psi_0|$$

Note that

$$|\chi(0)> = \alpha|+> \otimes |\psi_0> + \beta|-> \otimes |\psi_0>$$

After time t, the state of the system is given by

$$|\chi(t)> = exp(-2\pi i t(H_0 + 2\mu_B S_3 B_3(Q_3))/h)|\chi(0)>$$

$$= \alpha|+> \otimes exp(-2\pi i t(H_0 + \mu_B B_3(Q_3))/h)|\psi_0> + \beta|->$$

$$\otimes exp(-2\pi i t(H_0 - \mu_B B_3(Q_3))/h)|\psi_0>$$

$$= \alpha|+> \otimes |\psi_+(t)> + \beta|-> \otimes |\psi-(t)>$$

where $|\psi_+(t)>$ is the evolution of the translatory state of the atom after time t when the particle is in the upspin state and $|\psi-(t)>$ is the evolution of the translatory state of the atom after time t when the particle is in the downspin state. Here, $H_0 = \frac{\mathbf{P}^2}{2M}$. Note that actually, we should use the notation

$$H = \mathbf{I} \otimes \frac{\mathbf{P}^2}{2M} + 2\mu_B S_3 \otimes B_3(Q_3)$$

Thus,

$$H(|+> \otimes |\psi_0>) = |+> \otimes \frac{\mathbf{P}^2}{2M}|\psi_0> + |+> \otimes \mu_B B_3(Q_3)|\psi_0>$$

$$= |+> \otimes (\frac{\mathbf{P}^2}{2M} + \mu_B B_3(Q_3))|\psi_0>$$

and repetitive application gives

$$H^n(|+> \otimes|\psi_0>) = |+> \otimes(\frac{\mathbf{P}^2}{2M} + \mu_B B_3(Q_3))^n|\psi_0>$$

from which we deduce that

$$exp(-itH)(|+> \otimes|\psi_0>) = |+> \otimes exp(-it(\frac{\mathbf{P}^2}{2M} + \mu_B B_3(Q_3)))|\psi_0>$$

and likewise,

$$exp(-itH)(|-> \otimes|\psi_0>) = |-> \otimes exp(-it(\frac{\mathbf{P}^2}{2M} - \mu_B B_3(Q_3)))|\psi_0>$$

One project problem is to simulate the Schrodinger equation governing the motion of the hydrogen atom with spin in a magnetic field. The two component wave equation is to be solved:

$$-\frac{h^2}{8\pi^2 M}\nabla^2\psi_1(\mathbf{r},t) + \mu_B B_3(z)\psi_1(\mathbf{r},t) = \frac{ih}{2\pi}\frac{\partial\psi_1(\mathbf{r},t)}{\partial t}$$

$$-\frac{h^2}{8\pi^2 M}\nabla^2\psi_2(\mathbf{r},t) - \mu_B B_3(z)\psi_2(\mathbf{r},t) = \frac{ih}{2\pi}\frac{\partial\psi_2(\mathbf{r},t)}{\partial t}$$

These are to be solved numerically given the initial state

$$\begin{pmatrix} \psi_1(\mathbf{r},0) \\ \psi_2(\mathbf{r},0) \end{pmatrix}$$

Note that the product state

$$|\chi(0)>= \alpha|+> \otimes|\psi_0> +\beta|-> \otimes|\psi_0>$$

can be represented in our standard coordinate basis as

$$\begin{pmatrix} \alpha\psi_0(\mathbf{r}) \\ \beta\psi_0(\mathbf{r}) \end{pmatrix}$$

The measurement operator on the space corresponding to translatory motion in which the particle is found to belong to the interval $[x - \epsilon, x + \epsilon]$ is given by

$$\Lambda(x,\epsilon) = \int_{x-\epsilon}^{x+\epsilon} |x'> dx' <x'|$$

When viewed as an operator on the entire space that includes both translatory motion and internal spin, this operator is understood to be

$$\mathbf{I} \otimes \Lambda(x,\epsilon)$$

Suppose that the system is in the state $|\phi> \otimes |\psi>$. Then, the probability that a measurement of the position of the hydrogen atom will find it to be in the interval $[\dot{x} - \epsilon, x + \epsilon]$ is given by

$$Tr(|\phi> \otimes |\psi>< \phi| \otimes < \psi|(\mathbf{I} \otimes \Lambda)(x, \epsilon)))$$

$$=< \psi|\Lambda(x, \epsilon)|\psi> = \int_{x-\epsilon}^{x+\epsilon} | < x'|\psi > |^2 dx'$$

After time t, the state of the atom is

$$|\chi(t) >= \alpha|+> \otimes |\psi_+(t) > + \beta|-> \otimes |\psi_-(t) >$$

The probability of finding the position of the electron in the range $[x - \epsilon, x + \epsilon]$ is obtained as

$$|\alpha|^2 \int_{x-\epsilon}^{x+\epsilon} < x'|\psi_+(t) > |^2 dx' + |\beta|^2 \int_{x-\epsilon}^{x+\epsilon} | < x'|\psi_-(t) > |^2 dx'$$

These probabilities can be simulated as functions of time and spatial location.

Study project 3

A seminar on the computation of the energy levels of the Helium atom using perturbation theory could be included here. The lecture can begin by writing down explicitly the Hamiltonian of the Helium atom in atomic units. That is

$$H = \frac{1}{2}(p_1^2 + p_2^2) - 2(\frac{1}{r_1} + \frac{1}{r_2}) + \frac{1}{r_{12}}$$

Mention should be made as to why this problem cannot be treated using canonical perturbation theory. If one tried to use the product of hydrogenic wave functions with $Z = 2$, then the problem would be that if one electron is in a low energy state and another in a higher state, then the other electron would look at a nucleus that is screened heavily by the first one. This effect cannot be handled by perturbation theory. The idea is to use product of hydrogenic wave functions with the nuclear charge being a variable parameter. The 1s level with variable nuclear charge parameter has a wave function given by

$$u(r) = (4\pi)^{-1/2} 2\alpha^{3/2} exp(-\alpha r)$$

and the $2p$ level is given by

$$v_m(\mathbf{r}) = (2\sqrt{6})^{-1} \beta^{5/2} r exp(-\beta r/2) Y_{1m}(\hat{\mathbf{r}})$$

Symmetric and antisymmetric combinations of products of these wave functions can be formed and the variational principle applied to minimize the energy. A lecture on the variational principle as applied to quantum mechanical problems should of course precede this. If H is the energy operator and ψ any normalized wave function, then $< \psi|H|\psi >\geq E_0$ where E_0 is the ground state energy level of the atom. Given wave functions $\psi_1, ..., \psi_N$ one could form the combination

$$\psi = \sum_{j=1}^{N} c_j \psi_j$$

where the $c'_j s$ are selected so that

$$\sum_{j,k=1}^{N} \bar{c}_j c_k < \psi_j|\psi_k >= 1$$

and then subject to this constraint, minimize $< \psi|H|\psi >$ to obtain a value close to the ground state energy level. More generally, one could start with a wave function depending upon a number say p parameters $\psi(\mathbf{r}|\theta_1, ..., \theta_p)$ and minimize $< \psi|H|\psi >$ with respect to these parameters to arrive at a value close to the ground state energy of the quantum system. The trial function corresponding to the state in which both the electrons are in the $1s$ state (and hence by the Pauli exclusion principle have opposite spins) is given by $u(r_1)u(r_2)$ and the trial functions corresponding to the state in which one electron is in the $1s$ state and the other in the $2p$ state are respectively given by

$$\frac{1}{\sqrt{2}}(u(r_1)v_m(\mathbf{r}_2) + u(r_2)v_m(\mathbf{r}_1))$$

and

$$\frac{1}{\sqrt{2}}(u(r_1)v_m(\mathbf{r}_2) + u(r_2)v_m(\mathbf{r}_1))$$

In the former case, the two spins are oppositely aligned while in the latter case, they are aligned along the same direction.

Study project 4

A seminar presentation on the discrete space Schrodinger equation as a bilinear system from the systems theoretic standpoint can be given along the following lines.

Volterra systems are natural extensions of linear systems. They take into account quadratic, cubic and more generally, polynomial kind of nonlinearities in the system. The higher order Volterra terms are included in parallel with the linear terms in a block diagrammatic representation, where by the parallel connection of two or more systems, we mean summing the outputs of the individual systems. Physicists have developed perturbation theoretic analysis of the continuous time Schrodinger equation. This leads to a Volterra kind of representation of the wave function with the input being the perturbing potential. However while simulating a quantum mechanical system on the digital computer, we need to discretize the Schrodinger equation. The discrete time Schrodinger equation can also be brought into Volterra form by regarding the potential as an input. Such an analysis is based on the inherent bilinear character of the Schrodinger equation in which the potential is bilinearly coupled to the wave function. Bilinear systems in discrete time have been treated by research workers in signal processing.

Introduction: The physicists have treated the Schrodinger equation as a system defined by the Hamiltonian or energy operator. The Hamiltonian consists of a kinetic energy term which is a second order partial differential operator in the spatial variables along with a potential term which is a multiplication operator by a function of the position. The Hamiltonian operator is given by

$$H = -\frac{h^2}{8\pi^2 m}\frac{d^2}{dx^2} + V(x)$$

and the stationary Schrodinger equation involves obtaining the eigenvalues and eigenfunctions of the operator H, i.e. in solving the eigenequation

$$H\psi(x) = E\psi(x)$$

Physicists and mathematicians alike regard this as a linear operator equation. This formalism however obscures the fact that from a purely systems theoretic standpoint, the stationary Schrodinger equation does not define a linear system. If one regards the potential $\{V(x)\}$ as a function of the spatial variables as the input signal and the wave function $\{\psi(x)\}$ as the output signal, then the stationary Schrodinger equation defines a bilinear system. This is a special class of nonlinear systems and the inherent nonlinearity of the Schrodinger equation from the systems theoretic standpoint is obscured by the treatment given by the physicists and the mathematicians. There are papers that deal with the discrete two component Schrodinger equation which brings out the nonlinear aspect of the Schrodinger equation. These papers deal with the time dependent Schrodinger equation and consider a Riccatti representation of the same after an appropriate transformation. The treatment given here is however different. The Volterra representation of this equation treated as a bilinear system will show explicitly the nonlinear effect on the wave function produced by the bilinear coupling between the potential and the wave function. Before highlighting the systems theoretic aspect of the discrete Schrodinger equation, we shall cite some facts and references dealing with the theory of Volterra systems. A given nonlinear system, for example having the finite memory form

$$y(n) = F(x(n), x(n-1), ..., x(n-p))$$

can be represented as a Volterra series in the in the input process $\{x(n)\}$. This involves performing a Taylor expansion of the function F:

$$y(n) = \sum_{k_0,...,k_p=0}^{\infty} \frac{1}{k_0!...k_p!}\frac{\partial^{k_1+...+k_p}F(0,0,...,0)}{\partial x(n)^{k_0}...\partial x(n-p)^{k_p}}x(n)^{k_0}...x(n-p)^{k_p}$$

A truncation of this infinite series after a certain number of terms is a standard technique used to approximate the original nonlinear system. Most of the nonlinear effects of the original system are well taken care of to a good approximation if the number of terms chosen in the truncated series is sufficiently large. One of the techniques of transforming a polynomial type nonlinear difference equation into Volterra format is by the introduction of state variables. Starting with the Fredholm integral equation as our basic system, one can arrive at explicit expressions for the different Volterra kernels. Explicit computation of the second and third order kernels for certain Fredholm kind of kernels based on the Leonov-Shiryaev theorem have been dealt with several papers. These papers also talk about solution to the inverse scattering problem for the Schrodinger equation using transformation of the discrete time three term recursion equation into a two component wave propagation equation. Application of Volterra series to the quantum mechanical system derived from an optical communication problem has appeared in the literature. In this seminar, we present a Volterra representation of a general discretized Schrodinger equation based on decomposing the wave function relative to an appropriate function space.

The Time independent Schrodinger Equation

The motion of a quantum mechanical particle in one dimension in the potential field $V(x)$ is described by the Schrodinger equation

$$\phi''(x) + \frac{2m}{h^2}(E - V(x))\phi(x) = 0$$

$\phi(x)$ is the wave function corresponding to a stationary state of the system associated with the energy level E. This equation is in fact identical to the eigenvalue equation $H\phi = E\phi$ with H as the Hamiltonian operator

$$H = \frac{-h^2}{2m}\frac{d^2}{dq^2} + V(q)$$

When the system is in this state, the probability density of the position of the particle is given by $|\phi(x)|^2$. Assume that motion takes place in the interval $[a, b]$ which amounts to requiring that $V(x) = \infty$ for $x > b$ and for $x < a$. Then, the above partial differential equation is to be solved with $\phi \in L^2[a, b]$. We can however also conceive of the Hamiltonian operator acting on functions that do not belong to $L^2[a, b]$. Assume that S is a closed subspace of $L^2[a, b]$. For $\phi \in L^2[a, b]$, we have the decomposition $\phi = \phi_0 + \psi$ where $\phi_0 = P_S\phi$ is the orthogonal projection of ϕ onto S and $\psi = P_S^\perp\phi$ is the orthogonal projection of ϕ onto $S^\perp$. The Schrodinger equation can be expressed in the form

$$H\phi_0 + H\psi = E\phi_0 + E\psi$$

which is the same as

$$HP_S\phi + HP_S^\perp\phi = EP_S\phi + EP_S^\perp\phi$$

In case S is an invariant subspace for H, we have $P_S H = HP_S, P_S^\perp H = HP_S^\perp$ and the above eigen-equation in particular, gives

$$(H - E)P_S\phi + P_S^\perp(H - E)\phi = 0$$

from which we deduce that

$$P_S(H - E)P_S\phi = 0$$

or

$$P_S H P_S \phi_0 = E\phi_0$$

this means that the eigenvalues of H corresponding to states that fall in the invariant subspace H can be determined by a reduced eigen-equation. For example, if S is spanned by the functions $\{\phi_1, ..., \phi_N\}$, the above equation yields the finite dimensional eigen equation for $\phi_0 = \sum_{n=1}^{N} c_n\phi_n$:

$$\sum_{m=1}^{N} <\phi_m|H|\phi_n> c_n = Ec_m, \, m = 1, 2, ..., N$$

The required matrix elements are evaluated as

$$<\phi_m|H|\phi_n> = \int_{-\infty}^{\infty} \bar{\phi}_m(x)(-\frac{h^2}{2m}\phi_n''(x) + V(x)\phi_n(x))dx$$

$$= \frac{h^2}{2m}\int_{-\infty}^{\infty} \bar{\phi}_m'(x)\phi_n'(x)dx + \int_{-\infty}^{\infty} V(x)\bar{\phi}_m(x)\phi_n(x)dx$$

Sometimes we may be given a problem in which the projection of the state on $S^{\perp}$ is known, this means that the probability of finding the system in the states belonging to $S^{\perp}$ is known and we've got to determine the projection of the state on the subspace S, i.e. we want to determine the probability of finding the system in the states corresponding to the subspace S. We then substitute $\phi = \phi_0 + \psi$ where ψ is a known function into the time independent Schrodinger equation. Assuming that the energy E of the state has been measured, one ends up with the following differential equation for ϕ_0:

$$\phi_0''(x) + \psi''(x) + \frac{2m}{h^2}(E - V(x))(\phi_0(x) + \psi(x)) = 0$$

or after rearrangement,

$$\phi_0''(x) = -\psi''(x) - \frac{2mE}{h^2}\phi_0(x) + \frac{2m}{h^2}V(x)\phi_0(x) - \frac{2mE}{h^2}\psi(x) + \frac{2m}{h^2}V(x)\psi(x)$$

We can regard this as a system theoretic equation that defines the output $\phi(x)$ in terms of the input $V(x)$. The signal $\psi(x)$ is known. This is almost in the standard bilinear form except for the presence of the first term and the fourth term on the right. To transform this equation into standard bilinear form, we redefine our input $U(x) = V(x) - \frac{h^2}{2m}\psi''(x)/\psi(x) - E$. Then, the above differential equation boils down to

$$\phi_0''(x) = \frac{2m}{h^2}(U(x) + \frac{h^2}{2m}\psi''(x)/\psi(x))\phi_0(x) + \frac{2m}{h^2}\psi(x)U(x)$$

$$= \frac{2m}{h^2}U(x)\phi_0(x) + \frac{\psi''(x)}{\psi(x)}\phi_0(x) + \frac{2m}{h^2}\psi(x)U(x)$$

This is in the standard bilinear system form. We shall soon be adapting this technique to the discrete domain.

At this point, we make a digression to supply an example of how subspaces of the type $L^2[a,b]$ appear in the quantum theory. This example is taken from the theory of Legendre polynomials. Consider the motion of a particle in three dimensions in the radial potential $V(r)$. The time independent Schrodinger equation in polar variables reads

$$\frac{1}{r}\frac{\partial^2 r\psi}{\partial r^2} + \frac{1}{r^2 sin\theta}\frac{\partial}{\partial \theta}sin\theta\frac{\partial \psi}{\partial \theta} + \frac{1}{r^2 sin^2\theta}\frac{\partial^2 \psi}{\partial \phi^2} + \frac{2m}{h^2}(E - V(r))\psi = 0$$

Separation of variables leads to the following differential equation for the angular part of the wave function:

$$\frac{1}{sin\theta}\frac{\partial}{\partial \theta}sin\theta\frac{\partial Y(\theta,\phi)}{\partial \theta} + \frac{1}{sin^2\theta}\frac{\partial^2 Y(\theta,\phi)}{\partial \phi^2} = \lambda Y(\theta,\phi)$$

Here the point (θ,ϕ) varies over the unit circle. One can further separate out the variables into a θ part and a ϕ part. The ϕ part has the dependence $exp(im\phi)$ where m assumes integer values and the θ part leads to Legendre's differential equation

$$\frac{1}{sin\theta}\frac{d}{d\theta}sin\theta\frac{dP(\theta)}{d\theta} + (l(l+1) - \frac{m^2}{sin^2\theta})P(\theta) = 0$$

where l assumes integer values. This equation can be transformed into a differential equation in the variable $x = cos\theta$ over the interval $[-1,1]$. The differential equation is given by

$$(1 - x^2)P''(x) - 2xP'(x) + (l(l+1) - \frac{m^2}{1-x^2})P(x) = 0$$

It is called the modified Legendre equation. Its solutions are the modified Legendre polynomials $P_{lm}(x)$. Thus, the original Schrodinger equation for the wave function belonging to $L^2(\mathbb{R}^3)$ leads naturally after separation of variables to a differential equation whose eigenfunctions form an orthonormal basis for $L^2[-1,1]$ Note that we do not have a direct Schrodinger equation whose wave functions fall in this space, but rather an equation derived from the Schrodinger equation whose solutions fall in this space. The above Legendre differential equation can be transformed into a differential equation whose eigensolutions form an orthonormal basis for $L^2[a,b]$. This is achieved by transforming the variable from x to $y = \lambda x + \mu$ where $\mu - \lambda = a, \mu + \lambda = b$, i.e. $\mu = (a+b)/2, \lambda = (b-a)/2$. We leave the details of effecting this transformation to the reader. Now consider the original Schrodinger equation defined on $L^2[a,b]$ and with $\phi(x)$ denoting the wave function, set $\phi(x) = 1 + \phi_0)(x)$. The space of constant functions forms a one dimensional subspace of $L^2[a,b]$ and one may expect $\phi_0(x)$ to be orthogonal to this subspace. Plugging this into the Schrodinger equation leads us to the following differential equation for $\phi_0(x)$:

$$\phi_0''(x) + \frac{2m}{h^2}(E - V(x))(1 + \phi_0(x)) = 0$$

which is the same as

$$\phi_0''(x) + \frac{2mE}{h^2}\phi_0(x) - \frac{2mV(x)}{h^2}\phi_0(x) + \frac{2mE}{h^2} - \frac{2m}{h^2}V(x) = 0$$

Once again the decomposition of the wave function into a constant part and a part expected to be orthogonal to constant functions has been done in order to transform the Schrodinger equation into standard bilinear format. We next perform a discretization of this differential equation using the forward differencing technique. It leads us to the following difference equation:

$$\frac{\phi_0[k+2] - 2\phi_0[k+1] + \phi_0[k]}{\Delta^2} + \frac{2mE}{h^2}\phi_0[k] - \frac{2m}{h^2}V[k]\phi_0[k] + \frac{2mE}{h^2} - \frac{2m}{h^2}V[k]$$

This can be rearranged as

$$\phi_0[k+2] - 2\phi_0[k+1] + (1 - \frac{2m\Delta^2}{h^2}V[k])\phi_0[k] + \frac{2mE\Delta^2}{h^2} - \frac{2m\Delta^2}{h^2}V[k] = 0$$

We define the following:

$$b_1 = 2, b_2' = -1, c_{22} = \frac{2m\Delta^2}{h^2}, a_2 = \frac{2mE\Delta^2}{h^2}, -E + V[k] = V'[k]$$

We also use the notation $\phi[k]$ in place of $\phi_0[k]$. The above difference equation boils down to

$$\phi[k+2] = b_1\phi[k+1] + b_2'\phi[k] + a_2V'[k] + a_2V'[k] + c_{22}(V'[k] + E)\phi[k]$$

which can in turn be expressed as

$$\phi[k+2] = b_1\phi[k+1] + b_2\phi[k] + c_{22}\phi[k]V'[k] + a_2V'[k]$$

where $b_2 = b_2' + c_{22}E$. From this difference equation, by regarding the bilinear term $c_{22}\phi[k]V'[k]$ as a small perturbation, i.e. of the first order of smallness, we can in principle arrive at a Volterra representation for $\{\phi[k]\}$ in terms of $V'[k]$. Specifically, we let

$$\phi[k] = \phi_0[k] + \phi_1[k] + \phi_2[k] + ...$$

where $\phi_j[k]$ is of the jth order of smallness. Plugging this into the above equation and equating terms of same orders of smallness gives us the sequence of difference equations

$$\phi_0[k+2] = b_1\phi_0[k+1] + b_2\phi_0[k] + a_2V'[k]$$

$$\phi_n[k+2] = b_1\phi_n[k+1] + b_2\phi_n[k] + c_{22}\phi_{n-1}[k]V'[k], n \geq 1$$

This sequence of recursions can be solved iteratively and a Volterra series for ϕ set up.

Conclusions: In this talk, we presented the theory of the discrete Schrodinger equation from the viewpoint of the control theorist. We start with the one dimensional Schrodinger equation for a prescribed potential $V(x)$ and then discretize this by replacing second derivative in the Schrodinger equation by a second order difference operator. We next decompose the wave function into the sum of a constant term and a spatially varying term. This is done inorder to bring the Schrodinger equation into the standard form of a bilinear system discussed in the control theory literature. This results in a bilinear system in discrete time containing two difference terms for the output and a bilinear coupling between the wave function and the potential. The potential function is the input and the wave function is the output of the system. We then cast this difference equation in the state variable form and exploit this form to derive a Volterra representation for this system. Further work will focus upon two aspects, one application of this idea to specific potentials like the inverse square law appearing in the hydrogen atom and its perturbed versions, for example the hydrogen atom in an electric field. Two to exploit the Volterra representation to obtain formulas for the numerical computation of the perturbation in the eigenvalues of the energy operator caused by a perturbing potential. We shall also look at time dependent problems, i.e. situations in which the potential varies with time and we wish to numerically calculate the probabilities of transition from one stationary state to another or problems in which we wish to calculate the evolution of the average value of observables in time when the system is perturbed by a potential. The problem of motion in a random potential will be looked into after that.

Study project 5

A short seminar on the quantum mechanical description of the Helium atom can be presented along the following lines. The energy operator of the Helium atom is given by

$$H_0 = \frac{\mathbf{p}_1^2}{2m_e} + \frac{\mathbf{p}_2^2}{2m_e} - \frac{2e^2}{r_1} - \frac{2e^2}{r_2} + \frac{e^2}{r_{12}}$$

where $\mathbf{r}_1$ and $\mathbf{r}_2$ are respectively the positions of the first and the second electron, r_1, r_2 are their respective magnitudes and $\mathbf{p}_1, \mathbf{p}_2$ are respectively the momentum operators of the two electrons. These are respectively given by $\mathbf{p}_1 = -ih\nabla_1, \mathbf{p}_2 = -ih\nabla_2$, where ∇_1 and ∇_2 are respectively the gradient operators with respect to the position variables $\mathbf{r}_1$ and $\mathbf{r}_2$. If the spin of the electron is taken into account, then the energy is given by

$$H = H_0 + W$$

where W is the spin-orbit interaction term. If $\mathbf{L}_1, \mathbf{L}_2$ are respectively the angular momenta of the two electrons and $\mathbf{S}_1, \mathbf{S}_2$ are respectively the spin variables, then W has the form $W = c(\mathbf{L}_1.\mathbf{S}_1 + \mathbf{L}_2.\mathbf{S}_2$ where c is a constant. The state space of the Helium atom is

$$\mathcal{H} = \mathcal{H}_1 \otimes \mathcal{H}_2$$

where $\mathcal{H}_1$ is the state space of the first electron and $\mathcal{H}_2$ is the state space of the second electron. These two spaces can further be decomposed into the orbital part and the spin part:

$$\mathcal{H}_1 = \mathcal{H}_1^{orb} \otimes r_1^s$$

$$\mathcal{H}_2 = \mathcal{H}_2^{orb} \otimes r_2^s$$

where H_1^{orb} is the state space of the first electron on which only orbital operators corresponding to the first electron act (coordinate space of the first electron) and H_2^{orb} is the state space of the second electron on which only orbital operators corresponding to the second electron act (coordinate space of the second electron). Likewise, r_1^s is the state space of the first electron on which only spin observables of the first electron act and r_2^s is the state space of the second electron on which spin observables of the second electron act. The state space $\mathcal{H}$ is thus isomorphic to

$$(\mathcal{H}_1^{orb} \otimes \mathcal{H}_2^{orb}) \otimes (r_1^s \otimes r_2^s)$$

The space $\mathcal{H}_1^{orb} \otimes \mathcal{H}_2^{orb}$ can then be regarded as the orbital space of the entire two electron system and the space $r_1^s \otimes r_2^s$ as the spin space of the entire two electron system. However from the standpoint of the Pauli exclusion principle, it is more convenient to express the entire state space as direct sums of symmetric and antisymmetric spaces. Thus,

$$\mathcal{H}^{orb} = \mathcal{H}_1^{orb} \otimes \mathcal{H}_2^{orb} = \mathcal{H}_+^{orb} \oplus \mathcal{H}_-^{orb}$$

where $\mathcal{H}_+^{orb}$ is the symmetric orbital space and $\mathcal{H}_-^{orb}$ is the antisymmetric orbital space. The orbital space of the first electron is spanned by the vectors $|n_1, l_1, m_1 >$ where n_1 is the principal quantum number, l_1 is the angular momentum quantum number i.e. $l_1(l_1 + 1)$ is an eigenvalue of $\mathbf{L}_1^2$ m_1 is the eigenvalue of the z component of the angular momentum of the first electron. Likewise the orbital space of the second electron is spanned by the vectors $|n_2, l_2, m_2 >$. The space $\mathcal{H}_+^{orb}$ is spanned by the symmetric vectors

$$\frac{1}{\sqrt{2}}(|n_1, l_1, m_1 > \otimes |n_2, l_2, m_2 > + |n_2, l_2, m_2 > \otimes |n_1, l_1, m_1 >)$$

and the space $\mathcal{H}_-^{orb}$ is spanned by the vectors

$$\frac{1}{\sqrt{2}}(|n_1, l_1, m_1 > \otimes |n_2, l_2, m_2 > - |n_2, l_2, m_2 > \otimes |n_1, l_1, m_1 >)$$

Likewise the spin space of the system admits the decomposition

$$r^s = r^{s1} \otimes r^{s2} = r_+^s \otimes r_-^s$$

where r_+^s is spanned by the vectors $|+ > |+ >, |- > |- >$ and $\frac{1}{\sqrt{2}}(|+ - > + |- + >)$. r_-^s is spanned by the vector $\frac{1}{\sqrt{2}}(|+ - > - |- + >)$. The entire state space of the system can now be decomposed as follows

$$\mathcal{H} = \mathcal{H}^{orb} \otimes r^s = (H_+^{orb} \oplus H_-^{orb}) \otimes (r_+^s \oplus r_-^s)$$

$$= (H_+^{orb} \otimes r_+^s) \oplus (H_+^{orb} \otimes r_-^s) \oplus (H_-^{orb} \otimes r_+^s) \oplus (H_-^{orb} \otimes r_-^s)$$

The subspaces $H_+^{orb} \otimes r_+^s$ and $H_-^{orb} \otimes r_-^s$ are the symmetric subspaces while the subspaces $H_+^{orb} \otimes r_-^s$ and $H_-^{orb} \otimes r_+^s$ are the antisymmetric subspaces. One starts with the eigenfunctions of $H_{00} + W$ taken from the antisymmetric subspaces.Note that since this operator is separable into the sum of an operator that depends upon only the coordinates of the first electron and those that depend only upon the second electron, these eigenfunctions are easily constructed as

$$\frac{1}{2}(|n_1, l_1, m_1 > |n_2, l_2, m_2 > + |n_2, l_2, m_2 > |n_1, l_1, m_1 >)(|+->- |-+>)$$

and

$$\frac{1}{\sqrt{2}}(|n_1, l_1, m_1 > |n_2, l_2, m_2 > - |n_2, l_2, m_2 > |n_1, l_1, m_1 >)|++>$$

$$\frac{1}{2}(|n_1, l_1, m_1 > |n_2, l_2, m_2 > - |n_2, l_2, m_2 > |n_1, l_1, m_1 >)(|+->+ |-+>)$$

and

$$\frac{1}{2}(|n_1, l_1, m_1 > |n_2, l_2, m_2 > - |n_2, l_2, m_2 >)|-->$$

The correction to the energy levels of this unperturbed energy caused by the term corresponding to the interaction between the two electrons is then obtained using the standard results of quantum mechanical perturbation theory. These results can be presented in a seminar and a project can be carried out on the computer simulation of the wave functions and energy eigenvalues of the Helium energy operator. The student must first of all consult the important textbooks on the subject and then design computer programmes for the simulation of the wave equation.

Study project 6

One of the fundamental problems of quantum theory involves how angular momenta are composed in the quantum theory. The discussion can begin with the treatment of the quantum mechanical rotator for which the fundamental operators are $\mathbf{L}^2$ and L_z. Here, (L_x, L_y, L_z) are the three components of the angular momenta and $\mathbf{L}^2 = L_x^2 + L_y^2 + L_z^2$ equals the total angular momentum square. $\mathbf{L}^2, L_z$ form a complete set of commuting observables for the rotator for which the state space is $L^2(S^2)$ where S^2 is the surface of the unit sphere. Any wave function here can be specified as a function $F(\theta, \phi)$, with θ as the elevation angle and ϕ as the azimuth angle. The state space can be decomposed based on the irreducible representations of the rotation group as

$$L^2(S^2) = \bigoplus_{l=0}^{\infty} \mathcal{R}^l$$

where $\mathcal{R}^l$ is the eigenspace of $\mathbf{L}^2$ corresponding to the eigenvalue $l(l+1)$. The Lie algebra associated with the rotation group $SO(3)$ is generated by the operators $\{L_x, L_y, L_z\}$ and the appropriate commutation relations that define the structure of this Lie algebra are

$$[L_x, L_y] = iL_z \qquad [L_y, L_z] = iL_x \qquad [L_z, L_x] = iL_y$$

The space $\mathcal{R}^l$ has dimension $2l+1$ and is invariant under the operators of the Lie algebra. When spin is also taken into account, l can assume half integral values also, i.e. the direct sum is taken over $l = 0, 1/2, 1, 3/2, 2, ...$ If we have two elementary rotators with irreducible spaces $\mathcal{R}_1^{j_1}$ and $\mathcal{R}_2^{j_2}$, then the state vectors spanning the first space are denoted by the ket $|j_1, m_1 >$ with $m_1 = -j_1, -j_1+1, ..., j_1-1, j_1$ and the state vectors spanning the second space are denoted by the ket $|j_2, m_2 >$ with $m_2 = -j_2, -j_2 + 1, ..., j_2 - 1, j_2$. $j_1(j_1 + 1)$ is the squared angular momentum value of the first rotator while $j_2(j_2 + 1)$ is the squared angular momentum value of the second rotator. m_1 is the value of the z component of the angular momentum of the first rotator while m_2 is the value of the z component of the angular momentum of the second rotator. The state space of both the rotators is spanned by the vectors $|j_1, m_1 > \otimes |j_2, m_2 >$ with $|m_1| \leq j_1, |m_2| \leq j_2$. A convenient notation for this is $|j_1, m_1, j_2, m_2 >$. The state space itself is specified as the tensor product $\mathcal{R}_1^{j_1} \otimes \mathcal{R}_2^{j_2}$ and the operator of the ith component of the angular momentum in this space is given by

$$J_i = J_i^1 \otimes I + I \otimes J_i^2$$

where $J_i^1, i = 1, 2, 3$ are the components of the angular momentum for the first rotator while $J_i^2, i = 1, 2, 3$ are the components of the angular momentum for the second rotator. A complete set of commuting observables for the combined rotator is given by

$$\mathbf{J}^{12} \otimes I, J_3^1 \otimes I, I \otimes \mathbf{J}^{22}, I \otimes J_3^2$$

These operators act on the basis $\{|j_1, m_1, j_2, m_2 >: |m_1| \leq j_1, |m_2| \leq j_2\}$ in accordance with the following rules

$$\mathbf{J}^{12} \otimes I |j_1, m_1, j_2, m_2 >= j_1(j_1 + 1)|j_1, m_1, j_2, m_2 >$$

$$J_3^1 \otimes I |j_1, m_1, j_2, m_2 >= m_1 |j_1, m_1, j_2, m_2 >$$

$$I \otimes \mathbf{J}^{22} |j_1, m_1, j_2, m_2 >= j_2(j_2 + 1)|j_1, m_1, j_2, m_2 >$$

$$I \otimes J_3^2 |j_1, m_1, j_2, m_2 >= m_2 |j_1, m_1, j_2, m_2 >$$

When the Hamiltonian of the total system is invariant under rotations, generally the angular momenta of the individual rotators do not commute with it, rather the components of the total angular momenta $J_i, i = 1, 2, 3$ commute with it. It is therefore better to choose as our complete set of commuting observables the set

$$\mathbf{J}^{12} \otimes I, I \otimes \mathbf{J}^{22}, \mathbf{J}^2 = \sum_{i=1}^{3} J_i^2, J_3$$

The corresponding eigenbasis vectors are then denoted by $|j_1, j_2, j, m >$. It has the properties

$$\mathbf{J}^{12} \otimes I |j_1, j_2, j, m >= j_1(j_1 + 1)|j_1, j_2, j, m >$$

$$I \otimes \mathbf{J}^{22} |j_1, j_2, j, m >= j_2(j_2 + 1)|j_1, j_2, j, m >$$

$$\mathbf{J}^2 |j_1, j_2, j, m >= j(j + 1)|j_1, j_2, j, m >$$

$$J_3|j_1, j_2, j, m >= m|j_1, j_2, j, m >$$

Note that the numbers j_1, j_2 are fixed since our state space remains $\mathcal{R}_1^{j_1} \otimes \mathcal{R}_2^{j_2}$, it is only the basis for this space that is being selected differently in accordance with the overall rotational symmetry of the Hamiltonian. This symmetry corresponds to the conservation of the overall angular momentum of the system but not necessarily conservation of the individual components of the angular momentum. The quantum numbers j, m vary. One has therefore the expansion

$$|j_1, m_1, j_2, m_2 >= \sum_{j,m} |j_1, j_2, j, m >< j_1, j_2, j, m|j_1, m_1, j_2, m_2 >$$

The range of j, m and the Clebsch-Gordon coefficients $< j_1, j_2, j, m|j_1, m_1, j_2, m_2 >$ are to be determined from general considerations based on the selection rules of quantum mechanics.

Study project 7: The theory of linear estimation involves the following problem. Given a Hilbert space $\mathcal{H}$ and a subspace W of $\mathcal{H}$ and a vector $x \in \mathcal{H}$, we want to determine a vector $w \in W$ for which $\| x - w \|^2$ is a minimum. w turns out to be unique and is called the best approximation of x in W. The map that carries a vector $x \in \mathcal{H}$ to its best linear estimate $w \in W$ is linear, self adjoint and idempotent. In other words there is a map $P : \mathcal{H} \to W$ such that $P(cx + y) = cP(x) + P(y)$ for all $x, y \in \mathcal{H}$ and $c \in \mathbb{C}$ and $\| x - P(x) \| \leq \| x - v \|$ for all $v \in W$. P satisfies the property $P^2 = P$ (idempotence) and $< x, Py >=< Px, y >$ (self-adjointedness). The last property is the same as $P^* = P$. Also the range of P is W: $P(V) = W$. P can also be characterized by the orthogonality principle, i.e., $< x - P(x), v >= 0$ for all $v \in W$ and $x \in V$. In the quantum theory, the estimation problem can be cast in the following form. Given a wave function $\psi(\mathbf{r})$, we want to approximate it in the mean square sense by a linear combination of a set $\{\phi_1(\mathbf{r}), ..., \phi_m(\mathbf{r})\}$ of orthonormal wave functions, i.e.,

$$\int \bar{\phi}_i(\mathbf{r})\phi_j(\mathbf{r})d^3\mathbf{r} = \delta_{ij}$$

The minimum mean square approximant is given by

$$\hat{\psi}(\mathbf{r}) = \sum_{j=1}^{m} \phi_j(\mathbf{r}) < \phi_j, \psi >= P_W\psi(\mathbf{r})$$

where P_W denotes the orthogonal projection operator on the space W. If we make a measurement involving which state ϕ_j the system is in, then after the measurement has been made, where the $\phi_j's$ form a complete orthonormal set, then the state of the system after the measurement is given by the density operator

$$\rho(\mathbf{r}, \mathbf{r}') = \sum_{j=1}^{m} \phi_j(\mathbf{r})| < \phi_j|\psi > |^2\phi_j(\mathbf{r}')$$

When the $\phi'_j s$ are not complete but are orthonormal, then we've seen that

$$\hat{\psi} = \sum_j \phi_j < \phi_j, \psi >$$

is the best mean squared approximant of ψ in the space W. We now note that

$$< \psi, \hat{\psi} > = \sum_j | < \phi_j, \psi > |^2$$

gives the probability that the system on measurement will be found to be in one of the states ϕ_j.

Study project 8

A basic seminar on the notion of a qubit in quantum computation techniques is useful for the researcher in signal processing who wishes to start learning quantum algorithms. The state space for a single qubit quantum system is $\mathbb{C}^2$ and a basis is $\{|0 >, |1 >\}$ where $|0 >$ corresponds to the upspin state and $|1 >$ to the downspin state. The general state is given by a linear combination $a|0 > +b|1 >$. It is customary to represent the general state vector as $\begin{pmatrix} a \\ b \end{pmatrix}$, so that $|0 >$ is the state $\begin{pmatrix} 1 \\ 0 \end{pmatrix}$ and $|1 >$ is the state $\begin{pmatrix} 0 \\ 1 \end{pmatrix}$. Mention should be made of how probabilities are computed. That when the system is in the state $a|0 > +b|1 >$ and a measurement is made as to whether the state is $|0 >$ or $|1 >$, then with probability $|a|^2$, the state is $|0 >$ and with probability $|b|^2$, the state is $|1 >$. Multiple qubit states are introduced next. For example, letting $d = 2^n$, we can construct an n-qubit state in the space $\mathbb{C}^d$ as being spanned by the orthonormal basis $|x_1, x_2, ..., x_n >$ with each x_i assuming the value 0 or 1. For example the vector $|x_1, x_2, ..., x_n >$ can be represented by a column vector of size d having a one in the $1 + \sum_{i=1}^{n} 2^{i-1}x_i$ and zeroes as other entries. The quantum state $|x_1, x_2, ..., x_n >$ then corresponds to the classical bit string $.x_1, x_2, ..., x_n$. Alternately, $\mathbb{C}^d$ can be viewed as the tensor product of n copies of $\mathbb{C}^2$:

$$\mathbb{C}^d = \mathbb{C}^2 \otimes ... \otimes \mathbb{C}^2$$

and the state $|x_1, x_2, ..., x_n >$ as the tensor product $|x_1 > \otimes |x_2 > ... \otimes |x_n >$. A general state has the representation

$$|\psi > = \sum_{x_1, ..., x_n = 0}^{1} c(x_1, ..., x_n)|x_1, ..., x_n >$$

The notion of a measurement is introduced next. Suppose X is an observable, i.e. a Hermitian operator on the Hilbert space of quantum states. It has a spectral decomposition

$$X = \sum_{\alpha=1}^{m} c_\alpha E_\alpha$$

where $\{c_1, ..., c_m\}$ are distinct real numbers and $\{E_1, ..., E_m\}$ is a complete system of orthogonal projections on the Hilbert space, i.e. $\sum_{\alpha=1}^{m} E_\alpha = \mathbf{I}$ and $E_\alpha E_\beta = \mathbf{0}$ when $\alpha \neq \beta$. Then, if $|\psi>$ is the state of the quantum system and X is measured, the probability of the measurement showing up the value c_α equals $p_\alpha = |<\psi|E_\alpha|\psi>|^2$. When this measurement value shows up the state of the system collapses from the pure state $|\psi>$ to the mixed state $\frac{1}{p_\alpha} E_\alpha|\psi><\psi|E_\alpha$. In general, when no particular measurement outcome is noted and simply the observable X is measured, the state of the system following the measurement is given by the superposition

$$\rho = \sum_{\alpha=1}^{m} E_\alpha|\psi><\psi|E_\alpha$$

This holds good even for mixed states. If ρ is the starting state, then after measurement of the observable X, the state of the system becomes

$$E(\rho) = \sum_{\alpha=1}^{m} E_\alpha \rho E_\alpha$$

The concept of measurement can be extended to non-orthogonal families as follows: Let $\{M_\alpha : \alpha = 1, 2, ..., m\}$ be a family of operators on the Hilbert space of quantum states with the property

$$\sum_{\alpha=1}^{m} M_\alpha^* M_\alpha = \mathbf{I}$$

Suppose the starting state of the system is ρ. Assume that if the measurement M_α is carried out, the measurement outcome becomes α. Then, the probability of getting this outcome when the entire measurement is carried out is given by $p_\alpha = Tr(M_\alpha \rho M_\alpha^*)$. It is clear that the $p_\alpha's$ sum up to unity:

$$\sum_{\alpha=1}^{m} p_\alpha = \sum Tr(M_\alpha \rho M_\alpha^*) = \sum Tr(\rho M_\alpha^* M_\alpha)$$

$$= Tr(\rho \sum M_\alpha^* M_\alpha) = Tr(\rho) = 1$$

When the measurement yields the outcome α, the state of the system collapses to $\frac{M_\alpha \rho M_\alpha^*}{p_\alpha}$. When the entire measurement is carried, out, the state of the measurement system is the weighted average of these states, i.e.

$$M(\rho) = \sum_{\alpha=1}^{m} M_\alpha \rho M_\alpha^*$$

A gate acting on a qubit is simply a unitary operator in $\mathbb{C}^2$. An example is the not gate which flips the state $|x>$ to the state $|1 \oplus x>$ where $x = 0, 1$. In other words,

$$X|0> = |1> \qquad X|1> = |0>$$

Relative to the basis $|0>, |1>$,the not gate has matrix

$$[X]_B = \begin{pmatrix} 0 & 1 \\ 1 & 0 \end{pmatrix}$$

This is easily recognized to be the first Pauli spin matrix. Similarly, we have the phase gate which does not do anything to the state $|0>$ and alters the phase of the state $|1>$ by ϕ. Calling the phase gate as T_ϕ, we have

$$T_\phi|0>= |0> \qquad T_\phi|1>= exp(i\phi)|1>$$

Thus, relative to the basis $|0>, |1>$, the phase gate has matrix

$$[T_\phi]_B = \begin{pmatrix} 1 & 0 \\ 0 & exp(i\phi) \end{pmatrix}$$

Another important gate is the Hadamard gate. This transforms the state $|0>$ to the state $\frac{|0>+|1>}{\sqrt{2}}$ and the state $|1>$ to the state $\frac{|0>-|1>}{\sqrt{2}}$. relative to the standard basis $|0>, |1>$, the Hadamard gate has matrix

$$[H]_B = \begin{pmatrix} 1 & 1 \\ 1 & -1 \end{pmatrix}$$

There are also several gates acting on two qubit states. Note that qubit is the abbreviation for quantum bit. Suppose that U is a unitary operator acting on one qubit states, i.e. U is a one qubit gate. Then, a controlled U gate can be defined as follows. If the first qubit is a zero, then both qubits are left unchanged while if the first qubit is a one, then the first qubit is left unchanged while the operator U is applied to the second qubit. Thus, the controlled U operation can be described as $|0> |x> \longrightarrow |0> |x>$ and $|1> |x> \longrightarrow |1> U|x>$. This can be written in the general form as $|x> |y> \longrightarrow |x> U^x|y>$. Consider now the ordered basis $|0> |0>, |0> |1>, |1> |0>, |1> |1>$. The action of the controlled U gate on this basis is given by

$$|0> |0> \longrightarrow |0> |0> \qquad |0> |1> \longrightarrow |0> |1>$$

$$|1> |0> \longrightarrow |1> U|0> \qquad |1> |1> \longrightarrow |1> U|1>$$

Assuming that

$$U|0>= u_{00}|0> +u_{10}|1> \qquad U|1>= u_{01}|0> +u_{11}|1>$$

we can express the action of the controlled U operation on the standard basis as

$$|00> \longrightarrow |00> \qquad |01> \longrightarrow |01>$$

$$|10> \longrightarrow u_{00}|10> +u_{10}|11> \qquad |11> \longrightarrow u_{01}|10> +u_{11}|11>$$

The corresponding matrix is given by

$$\begin{pmatrix} 1 & 0 & 0 & 0 \\ 0 & 1 & 0 & 0 \\ 0 & 0 & u_{00} & u_{01} \\ 0 & 0 & u_{10} & u_{11} \end{pmatrix}$$

Study project 9

Seminar on the computation of the matrix elements of $r^{-\nu}$ for the Hydrogen atom energy eigenstates. This enables one to determine the first order shift in the energy level of the atom caused by the spin-orbit interaction terms. The energy of the Hydrogen atom in appropriate units is given by

$$K = \frac{\mathbf{p}^2}{2} - \frac{e^2}{r}$$

where $\mathbf{p}^2 = p_x^2 + p_y^2 + p_z^2$ is the magnitude square of the momentum and $r = \sqrt{x^2 + y^2 + z^2}$ is the distance of the electron from the nucleus. To obtain a recursion equation for the matrix elements, we need to evaluate $[r^{-\nu}[K, r], K]$. We state a relation in this regard:

$$[r^{-\nu}[r, K], K] + \frac{\nu}{2}[r^{-\nu-1}, K] = 2\nu r^{-\nu-1}K - (\nu + 1)r^{-\nu-3}\mathbf{L}^2$$

$$+(2\nu + 1)ar^{-\nu-2} + \frac{\nu(\nu + 1)(\nu + 2)}{4}r^{-\nu-3}$$

Prove this identity.

Hint: You can start with the equation

$$[r, K] = \frac{1}{2}[r, p_x^2 + p_y^2 + p_z^2] = \frac{1}{2}([r, p_x]p_x + p_x[r, p_x] + ...)$$

where ... are terms similar to the first two with p_x replaced by p_y and p_z and continue.

Study project 10

Explain how you would simulate the relativistic motion of a quantum mechanical free particle as well as a particle in an electromagnetic field. Your starting point can be the Klein-Gordon equation or equivalently, Einstein's energy-momentum relation.

Einstein's energy-momentum relation is $E^2 = c^2p^2 + m^2c^4$ where $p^2 = p_x^2 + p_y^2 + p_z^2$ is the total momentum squared of the particle. This relation can equivalently be expressed in the form

$$E = c\sqrt{p^2 + m^2c^2}$$

The Klein-Gordon equation is obtained by replacing the energy operator E by $\frac{ih}{2\pi}\frac{\partial}{\partial t}$ and the momentum operator $\mathbf{p}$ by $-\frac{ih}{2\pi}\nabla$ and equating the left and right sides of the equation operated on the wave function ψ. For simplicity of notation, we denote $\frac{h}{2\pi}$ by h. The resulting wave equation is

$$-h^2\frac{\partial^2\psi}{\partial t^2} = -c^2h^2\nabla^2\psi + m^2c^4\psi$$

which is the same as

$$\nabla^2\psi - \frac{1}{c^2}\frac{\partial^2\psi}{\partial t^2} - \frac{m^2c^2}{h^2}\psi = 0$$

This is the Klein-Gordon equation for a free particle. The Klein-Gordon equation for a charged particle e in an electromagnetic field is obtained by replacing the energy operator E by $E - e\phi$ and the momentum operator $\mathbf{p}$ by $\mathbf{p} - e\mathbf{A}$. This results in the wave equation

$$(E - e\phi)^2\psi = c^2(\mathbf{p} - e\mathbf{A})^2\psi + m^2c^4\psi$$

or in operator form

$$(ih\frac{\partial}{\partial t} - e\phi)^2\psi = -c^2(h\nabla + e\mathbf{A})^2\psi + m^2c^4\psi$$

We now make an expansion of the operators involved in the above relativistic wave equation.

$$(ih\frac{\partial}{\partial t} - e\phi)^2\psi = (-h^2\frac{\partial^2}{\partial t^2} - 2ieh\phi\frac{\partial}{\partial t} - ieh(\frac{\partial\phi}{\partial t}))\psi$$

$$(h\nabla + e\mathbf{A})^2\psi = h^2\nabla^2\psi + e^2A^2\psi + he\nabla.(\mathbf{A}\psi) + 2he\mathbf{A}.\nabla\psi$$

The partial derivatives with respect to space and time are discretized and implemented on the microprocessor.

Study project 11

Computer analysis of the quantum harmonic oscillator perturbed by a constant time varying force. The Hamiltonian as a function of time is given by

$$H(t) = \frac{P^2 + Q^2}{2} - K(t)Q$$

There does not exist any time independent Schrodinger equation for this system. The time varying Schrodinger equation is given by

$$i\frac{d\psi_t}{dt} = H(t)\psi_t$$

which reads as

$$i\frac{\partial\psi_t(Q)}{\partial t} = -\frac{1}{2}\frac{\partial^2\psi_t(Q)}{\partial Q^2} + \frac{Q^2}{2}\psi_t(Q) - K(t)Q\psi_t(Q)$$

Explicit solution to this partial differential equation gives the wave function of the system at time t. The Heisenberg form of the equations of motion are introduced via the creation and annihilation operators:

$$a = \frac{Q+iP}{\sqrt{2}} \qquad a^* = \frac{Q-iP}{\sqrt{2}}$$

Then,

$$aa^* = H_0 + 1/2 \qquad a^*a = H_0 - 1/2$$

where $H_0 = \frac{P^2+Q^2}{2}$ is the energy operator of the unperturbed Hamiltonian operator. The commutation relations are standard: $[a, a^*] = 1$. This gives $[a^n, a^*] = na^{n-1}$. The Hamiltonian operator of the perturbed system can be expressed in terms of this operator as

$$H(t) = a^*a + 1/2 - K(t)(a + a^*)$$

If $|0>$ is the ground state of the unperturbed oscillator, then $a|0>= 0$. The energy of this ground state is given by $1/2$. The wave function corresponding to the ground state is obtained by solving $a|0>= 0$ using the position representation: $a|0>= \frac{1}{2}(Q + \frac{d}{dQ})|0>= 0$. This gives

$$|0>= C.exp(-Q^2/2)$$

The normalization constant C is obtained using

$$1 =< 0|0 >= |C|^2 \int_{-\infty}^{\infty} exp(-Q^2)dQ = \sqrt{\pi/2}|C|^2$$

so that $C = (2/\pi)^{1/4}$. The state $|n>$ corresponding to the eigenvalue $n+1/2$ is given by applying the operator a^{*n} to the ground state:

$$|n>= \frac{a^{*n}|0>}{\sqrt{< 0|a^n a^{*n}|0 >}}$$

Let us expand the state ψ_t of the perturbed system at time t in terms of the eigenfunctions of the unperturbed oscillator: $\psi_t = \sum_{n=0}^{\infty} c_n(t)|n>$. Plugging this into the Schrodinger equation gives

$$i \sum_{n=0}^{\infty} c_n'(t)|n >= (H_0 - K(t)(a + a^*)) \sum_{n=0}^{\infty} c_n(t)|n >$$

$$= \sum_{n=0}^{\infty} c_n(t)(n + 1/2)|n > -K(t) \sum_{n=0}^{\infty} c_n(t)(a + a^*)|n >$$

We know that a lowers the energy of a state by unity and a^* raises the energy of the state by unity. Thus, $a|n >= A|n - 1 >$ where $|A|^2 =< n|a^*a|n >=< n|H_0 - 1/2|n >= n$, that is $a|n >= \sqrt{n}|n - 1 >$. Similarly, a^* raises the energy of the state by unity: $a^*|n >= B|n + 1 >$.

$|B|^2 = < n|aa^*|n > = < n|H_0 + 1/2|n > = (n + 1)$. Thus, $a^*|n > = \sqrt{n+1}|n + 1 >$. This gives us the following differential equations

$$ic'_n(t) = (n + 1/2)c_n(t) - K(t)(c_{n+1}(t)\sqrt{n+1} + c_{n-1}\sqrt{n})$$

We leave the problem of simulating this system of differential equations to the student.

Study project 12

This project deals with the computer simulation of finite state quantum systems.

Ordinary differential equations arising in the quantum theory

In the quantum theory, the state of a system is described by a vector in a Hilbert space. The Hilbert space for a finite state quantum system is finite dimensional. An example of a finite state quantum system is the ammonia molecule NH_3. This molecule has three Hydrogen atoms located in one plane in a triangle and a Nitrogen atom that can be either above the plane or below the plane. This is an example of a two state quantum system. Let $\mathbf{x}(t) = [x_1(t), x_2(t), ..., x_n(t)]^T$ be the state of an n dimensional quantum system at time t. Then, the $x_i(t)'s$ are complex numbers satisfying the property $\sum_{i=1}^{n} |x_i(t)|^2 = 1$. $|x_i(t)|^2$ equals the probability that at time t, the state of the system will be the ith standard coordinate vector $\mathbf{e}_i = [0, 0, ..., 0, 1, 0, ..., 0]^T$ with a one at the ith position. The dynamics of the state vector is described by the Schrodinger equation

$$\frac{ih}{2\pi}\frac{d\mathbf{x}(t)}{dt} = \mathbf{H}(t)\mathbf{x}(t)$$

where $\mathbf{H}(t)$ is a Hermitian matrix possibly dependent upon time and is called the Hamiltonian or the energy matrix. In terms of components,

$$\frac{ih}{2\pi}\frac{dx_i(t)}{dt} = \sum_{j=1}^{n} H_{ij}(t)x_j(t), i = 1, 2, ..., n$$

When $\mathbf{H}$ is a constant matrix, the solution to the Schrodinger equation can be expressed as

$$\mathbf{x}(t) = exp(-2\pi it\mathbf{H}/h)\mathbf{x}(0)$$

where $\mathbf{x}(0)$ is the initial state of the system. The matrix $\mathbf{U}(t) = exp(-2\pi it\mathbf{H}/h)$ is a unitary matrix and is called the evolution operator of the system. If $E_1, ..., E_n$ are the eigenvalues of the matrix $\mathbf{H}$ with eigenvectors $\mathbf{v}_1, ..., \mathbf{v}_n$, then we can write $\mathbf{H}\mathbf{v}_i = E_i\mathbf{v}_i, i = 1, 2, ..., n$. $\mathbf{v}_i$ are called the stationary states or energy eigenstates of the system. Since $\mathbf{H}$ is Hermitian, it can be shown (the spectral theorem) that $\{\mathbf{v}_1, ..., \mathbf{v}_n\}$ can be taken as an orthonormal basis for $\mathbb{C}^n$, i.e. $< \mathbf{v}_i, \mathbf{v}_j > = \delta_{ij}$. When the system is in the state $\mathbf{v}_i$, its energy is given by E_i. Any state can be expanded as a linear combination of the energy eigenstates. Specifically, $\mathbf{x}(0) = \sum_{i=1}^{N} c_i\mathbf{v}_i$. Then,

$$\mathbf{x}(t) = exp(-2\pi it\mathbf{H}/h)\mathbf{x}(0) = \sum_{k=1}^{n} c_k exp(-2\pi itE_k/h)\mathbf{v}_k$$

Note that the E_k are real numbers. This formula shows that no matter what the initial state is, the probability of finding the system after time t in the kth energy eigenstate remains a constant $|c_k|^2$. That is why the energy eigenstates are also called stationary states. Numerical techniques for the solution to the finite state Schrodinger equation will be discussed in the lecture. Both cases will be considered, i.e. (a) when the Hamiltonian does not explicitly depend upon time and (b) when the Hamiltonian explicitly depends upon time. The simplest kind of discretization involves replacing the Schrodinger equation with the discrete approximation:

$$\mathbf{x}[n+1] = \mathbf{x}[n] - \frac{2\pi i}{h}\Delta.\mathbf{H}[n]\mathbf{x}[n]$$

The trouble with this method however is that it does not define a unitary evolution and hence violates the law of total probability.

Study project 13

This project deals with the numerical simulation of certain important problems in the quantum theory. We've outlined only the theory and given hints about how the simulation is to be carried out.

One dimensional stationary Schrodinger equation

For a particle moving in a potential $V(x)$, the wave function corresponding to the energy level E satisfies the stationary Schrodinger equation

$$\psi''(x) + \frac{8\pi^2 m}{h^2}(E - V(x))\psi(x) = 0$$

This is an eigenvalue equation and the admissible values of E which lead to square integrable wave functions ψ, i.e. $\int_{-\infty}^{\infty} |\psi(x)|^2 dx < \infty$ form a discrete set. In the talk, we shall mention how the energy eigenvalues E can be computed by discretizing this ordinary differential equation into a linear matrix equation. After discretization, we get

$$\frac{\psi[k+1] - 2\psi[k] + \psi[k-1]}{\Delta^2} + \frac{8\pi^2 m}{h^2}E\psi[k] - \frac{8\pi^2 m}{h^2}V[k]\psi[k] = 0$$

where $k = -N, ..., N$ and $\psi[-N-1] = \psi[N+1] = 0$. The step size is Δ, so that $\psi[k] = \psi(k\Delta)$ and $V[k] = V(k\Delta)$. When cast as a matrix eigenvalue problem, the concerned matrix turns out to be tridiagonal. Tridiagonal matrices have been studied extensively in the literature. Elementary examples include particle in a box, harmonic oscillator and hydrogen atom. The other method of numerically a computing the eigenvalues is to choose a set of basis functions $\{\phi_1(x), ..., \phi_N(x)\}$ and assume that the wave function can be expanded as a linear combination of these basis functions, i.e.

$$\psi(x) = \sum_{k=1}^{N} c_k \phi_k(x)$$

where the constants $\{c_k\}$ are determined numerically. Substituting this into the Schrodinger equation gives

$$\sum_{k=1}^{N} c_k \phi_k''(x) + \frac{8\pi^2 m}{h^2}(E - V(x)) \sum_{k=1}^{N} c_k \phi_k(x) = 0$$

We now multiply this equation by $\phi_m(x)$ and integrate over $[-L, L]$ where L is assumed to be sufficiently large so that the probability of finding the particle outside this range is negligible. We obtain

$$\sum_{k=1}^{N} c_k \int_{-L}^{L} \phi_k''(x)\phi_m(x)dx + \frac{8\pi^2 mE}{h^2} \sum_{k=1}^{N} c_k \int_{-L}^{L} \phi_m(x)\phi_k(x)dx$$

$$-\frac{8\pi^2 m}{h^2} \sum_{k=1}^{N} c_k \int_{-L}^{L} V(x)\phi_k(x)\phi_m(x)dx = 0, m = 1, 2, ..., N$$

This is a system of N homogeneous linear equations in the variables $\{c_k\}$ and can be cast as a generalized eigenvalue problem:

$$(\mathbf{A} - E\mathbf{B})\mathbf{c} = 0$$

The possible energy eigenvalues are approximately given from the solution of the characteristic equation

$$det(\mathbf{A} - E\mathbf{B}) = 0$$

and the corresponding generalized eigenvectors $\mathbf{c}$ determine the wave function.

Three dimensional stationary Schrodinger equation

The discrete energy spectrum and stationary state wave functions of a particle moving in three dimensions in a potential field $V(x, y, z)$ are determined as the solution to the eigen-problem

$$\frac{\partial^2 \psi}{\partial x^2} + \frac{\partial^2 \psi}{\partial y^2} + \frac{\partial^2 \psi}{\partial z^2}$$

$$+\frac{8\pi^2 m}{h^2}(E - V(x, y, z))\psi(x, y, z) = 0$$

The talk should introduce you to a method by which this partial differential equation can be discretized over a box $[-L, L]^3$ and transformed into a generalized eigenvalue problem as in the one dimensional case. The method of moments applied to this three dimensional problem shall also be discussed in the talk.

Variational method

This technique is based on the principle that if one starts with an arbitrary wave function ψ of a stationary quantum mechanical problem with energy operator H and minimize the average energy $< \psi|H|\psi > = \int \bar{\psi}(x)H\psi(x)dx$ over all trial wave functions ψ, then one will end up with the ground state energy level. This is proved as follows. Let $\phi_n, n = 1, 2, ...$ be the normalized wave functions corresponding to energy levels $E_1 < E_2 < ....$ Expand the wave function $\psi = \sum_{n=1}^{\infty} c_n \phi_n$. Using orthogonality of the $\phi'_n s$ we get

$$< \psi|H|\psi >= \sum_{n=1}^{\infty} |c_n|^2 E_n \geq E_1 \sum_{n=1}^{\infty} |c_n|^2 = E_1$$

Equality is achieved when $\psi = \phi_1$. This proves the theorem. The variational method furnishes us with a powerful computational tool for determining the ground state energy level of a quantum mechanical system. This can be applied to the Helium atom as follows. The Helium atom has energy given by

$$H = -\frac{h^2}{8\pi^2 m}(\Delta_1 + \Delta_2) - \frac{2e^2}{r_1} - \frac{2e^2}{r_2} + \frac{e^2}{r_{12}}$$

In the talk, we shall discuss how by slightly perturbing the product of two ground state wave functions for the Helium ion, we can determine the ground state of the Helium atom approximately. The idea in brief is to use a trial wave function $\psi(x, \theta_1, ..., \theta_k)$ where the $\theta'_j s$ are parameters introduced. Given the parameters θ_j, the wave function is completely determined. We determine the ground state approximately by minimizing the quantity

$$F(\theta_1, ..., \theta_k) = \frac{\int_{-\infty}^{\infty} \bar{\psi}(x, \theta_1, ..., \theta_k) H \psi(x, \theta_1, ..., \theta_k) dx}{\int_{-\infty}^{\infty} |\psi(x, \theta_1, ..., \theta_k)|^2 dx}$$

with respect to these parameters. For example, if the particle is execting one dimensional motion in the potential $V(x)$, we get

$$F(\theta_1, ..., \theta_k) = \frac{\int_{-\infty}^{\infty} \bar{\psi}(x, \theta_1, ..., \theta_k)(-\frac{h^2}{8\pi^2 m} \frac{\partial^2 \psi(x,\theta_1,...,\theta_k)}{\partial x^2} + V(x)\psi(x, \theta_1, ..., \theta_k))dx}{\int_{-\infty}^{\infty} |\psi(x, \theta_1, ..., \theta_k)|^2 dx}$$

Motion in a radial potential

The talk should also discuss the motion of a particle in a radial potential $V(r)$. Laplace's operator in spherical polar coordinates is

$$\Delta\psi = \nabla^2 = \frac{1}{r}\frac{\partial^2}{\partial r^2}(r\psi) + \frac{1}{r^2 sin(\theta)}\frac{\partial}{\partial \theta}(sin\theta \frac{\partial\psi}{\partial\theta}) + \frac{1}{r^2 sin^2\theta}\frac{\partial^2\psi}{\partial\phi^2}$$

When this is substituted into the Schrodinger equation with radial potential and the variables separated, then the angular part turns out to be the spherical harmonic functions $Y_{lm}(\theta, \phi)$. The Laplace operator can be expressed as the sum of a radial part and an angular part:

$$\Delta = \frac{1}{r}\frac{\partial^2}{\partial r^2}r - \frac{4\pi^2 L^2}{h^2 r^2}$$

where

$$L^2 = -\frac{h^2}{4\pi^2}\left(\frac{1}{sin\theta}\frac{\partial}{\partial\theta}sin\theta\frac{\partial}{\partial\theta} + \frac{1}{sin^2\theta}\frac{\partial^2}{\partial\phi^2}\right)$$

Its eigenvalues are $l(l+1)$ where l takes values $0, 1, 2, ...$, and for each l, there are $2l+1$ linearly independent eigenfunctions $Y_{lm}, m = -l, -l+1, ..., l$. The eigenvalue equation for the angular part reads

$$L^2 Y_{lm}(\theta, \phi) = l(l+1)\frac{h^2}{4\pi^2}Y_{lm}(\theta, \phi)$$

The form of the spherical harmonic functions Y_{lm} is $Y_{lm}(\theta, \phi) = P_{lm}(cos\theta)exp(im\phi)$. The quantum number $mh/2\pi$ represents the value assumed by the z component of the angular momentum. L^2 is the operator of total angular momentum square. The wave function is assumed to have the form $R(r)Y_{lm}(\theta, \phi)$. When substituted into the stationary Schrodinger equation gives the following differential equation for the radial part $R(r)$:

$$\frac{1}{r}\frac{d^2}{dr^2}(rR(r)) - l(l+1)R(r)/r^2 + \frac{8\pi^2 m}{h^2}(E - V(r))R(r) = 0$$

The energy levels E will be dependent upon the angular momentum quantum number l, i.e. we can represent these as E_{nl} where n and l assume nonnegative integer values. The numerical solution to this differential equation enables one to determine the energy levels and this should be discussed in the lecture. In the special case when the potential V is Coulomb, i.e. $V(r) = -K/r$, the energy levels and wave functions depend on just one quantum number. For the general radial potential, we set $R(r) = R_{nl}(r)$ and for the Coulomb case, this is $R_n(r)$. When V is expanded in powers of $1/r$, the solutions are obtained by the series method and techniques for doing this will be discussed in the lecture.

Plotting of the free particle wave function, transformation from the position to momentum representation and vice-versa

The free particle Schrodinger equation reads

$$(ih/2\pi)\frac{\partial\psi(t, \mathbf{x})}{\partial t} = -\frac{h^2}{8\pi^2 m}\nabla^2\psi(t, \mathbf{x})$$

The stationary problem is

$$-\frac{h^2}{8\pi^2 m}\nabla^2\psi(\mathbf{x}) = E\psi(\mathbf{x})$$

Setting $E = h^2 k^2/8\pi^2 m$, this equation reduces to

$$\nabla^2\psi + k^2\psi = 0$$

whose solution is the plane wave $exp(i\mathbf{k}.\mathbf{x})$. The corresponding solution to the time dependent problem is given by $exp(i(\omega t - \mathbf{k}.\mathbf{x}))$, where $h\omega/2\pi = E = h^2 k^2/8\pi^2 m$ or equivalently, $\omega =$

$hk^2/4\pi m$. The general solution to the time dependent Schrodinger equation for the free particle is obtained by superposing solutions for different $\mathbf{k}'s$, i.e.,

$$\psi(t, \mathbf{x}) = \int A(\mathbf{k})exp(i(hk^2t/4\pi m - \mathbf{k}.\mathbf{x}))d^3\mathbf{k}$$

$|A(\mathbf{k})|^2 d^3\mathbf{k}$ is proportional to the number of particles with momenta in the range $h^3 d^3\mathbf{k}/8\pi^3$. To implement this in practical computations on the digital computer, we have to discretize the Fourier integral and implement it using the Discrete or Fast Fourier transform. This aspect shall be discussed in the lecture.

Numerical computation of the spherical harmonics

The ladder operators for angular momentum are defined by the equations $L_+ = L_x + iL_y$ and $L_- = L_x - iL_y$. The commutation relations for the angular momentum are $[L_x, L_y] = iL_z, [L_y, L_z] = iL_y$ and $[L_z, L_x] = iL_y$. It is known from the theory of angular momentum and spin that for each positive integer n, there exist $n \times n$ matrices satisfying these commutation relations. One approach is to assume general Hermitian matrices for each L_x, L_y, L_z and to substitute into these relations and numerically solve approximately for the matrix entries. The determination of the eigenfunctions of angular momentum can proceed using the theory of ladder operators. For given values of l, m, the $Y_{lm}'s$ satisfy $L_+ Y_{ll} = 0$ and $Y_{lm} = c_{lm} L_-^{l-m} Y_{ll}$. These two differential equations can be solved numerically by assuming $Y_{lm}(\theta, \phi)$ to have the form $F_{lm}(\theta)exp(im\phi)$. All the spherical harmonics can be obtained recursively from this formula. Numerical techniques for doing this shall be discussed in the talk. A comparison of the numerical solutions with the actual ones can be made by noting that $L_+L_- = L_x^2 + L_y^2 + L_z = L^2 - L_z^2 + L_z$ and $L_-L_+ = (L_x - iL_y)(L_x + iL_y) = L_x^2 + L_y^2 - L_z = L^2 - L_z^2 - L_z$. The matrix of L_z relative to the basis $\{Y_{lm} : |m| \leq l\}$ is given by $diag[-l, -l + 1, ..., l]$. The matrices of L_+ and L_- relative to this basis can also be determined as follows. Assume that $L_+ Y_{lm} = c_{lm} Y_{l,m+1}$. If $m = l$, we know that $c_{lm} = 0$. For $m < l$, taking norm gives us

$$< Y_{lm}|L_- L_+|Y_{lm} >= |c_{lm}|^2$$

or

$$|c_{lm}|^2 =< Y_{lm}|L^2 - L_z^2 - L_z|Y_{lm} >= l(l + 1) - m^2 - m$$

so that we can take

$$c_{lm} = \sqrt{l(l + 1) - m(m + 1)}$$

Thus, the only nonvanishing matrix elements of L_+ are

$$< Y_{l,m+1}|L_+|Y_{lm} >= \sqrt{l(l + 1) - m(m + 1)}$$

Similarly, taking $L_- Y_{lm} = d_{lm} Y_{l,m-1}$, we obtain the only nonvanishing matrix elements of L_- as $< Y_{l,m-1}|L_-|Y_{lm} >= d_{lm}$. We leave the calculation to the interested student.

Numerical computation of the Clebsch Gordon coefficients for the combination of angular momenta

Given two independent particles with spins j_1, j_2, we know that the states of the system are defined by the ket vectors $|j_1, j_2, m_1, m_2 >= |j_1, m_1 > |j_2, m_2 >$, where m_1 takes values in $\{-j_1, -j_2 + 1, ..., j_2\}$ and m_1 takes values in $\{-j_2, -j_2 + 1, ..., j_2\}$. $|j_1, j_2, m_1, m_2 >$ is the state in which the first particle has z component of angular momentum equal to m_1 while the second particle has z component of angular momentum equal to m_2. We can however also describe the state as $|j, m >$. In this state, the total squared angular momentum of the system of two particles equals $j(j + 1)$ and the z component of the angular momentum of the system equals m. Note that $m = m_1 + m_2$. For simplicity, we set $|j_1, j_2, m_1, m_2 >= |m_1, m_2 >$. We then want to find the unitary matrix that maps the states $\{|m_1, m_2 >: |m_1| \leq j_1, |m_2| \leq j_2\}$ to the states $\{|j, m >\}$. We know that for a j ranges over $|j_1 - j_2|, ..., j_1 + j_2$ and for a given value of j, m ranges over $-j, -j + 1, ..., j$. The algorithm for doing this should be discussed in the talk.

Scattering problems in the quantum theory and the Born approximation

Scattering theory deals the interaction of a beam of incident particles with a scattering centre defined by a radial potential. Before the interaction, the state of the system is one of a continuous set of eigenfunctions of a free particle. Thus, if the beam is incident along the z axis, the initial wave function can be taken as $exp(ikz)$. After the scattering, the asymptotic state of the system becomes

$$u(r, \theta, \phi) \rightarrow exp(ikz) + \frac{f(\theta, \phi)}{r} exp(ikr)$$

when $r \rightarrow \infty$. If μ denotes the relative mass of the scattering centre and the incident particle, Schrodinger's equation reads

$$-\frac{h^2}{2\mu}\nabla^2 u(r, \theta, \phi) + V(r)u(r, \theta, \phi) = Eu(r, \theta, \phi)$$

where $E = h^2 k^2/2\mu$ since energy must be conserved during the scattering process. Putting $u(\mathbf{r}) = exp(ikz) + v(\mathbf{r})$ with $v(\mathbf{r})$ as the perturbation due to the collision, we get on substituting into the Schrodinger equation,

$$(\nabla^2 + k^2)v = \frac{2\mu}{h^2}V(\mathbf{r})exp(ikz) + \frac{2\mu U(\mathbf{r})}{h^2}v(\mathbf{r})$$

Assuming that the interaction potential is weak, the product $U(\mathbf{r})v(\mathbf{r})$ can be neglected and we get

$$(\nabla^2 + k^2)v(\mathbf{r}) = \frac{2\mu}{h^2}V(\mathbf{r})exp(ikz)$$

the solution to which is given by

$$u(\mathbf{r}) = -\frac{1}{4\pi}\int \frac{2\mu}{h^2}V(\mathbf{r'})exp(ikz')|\mathbf{r} - \mathbf{r'}|^{-1}exp(ik|\mathbf{r} - \mathbf{r'}|)d^3\mathbf{r'}$$

An asymptotic analysis shows that

$$u(\mathbf{r}) \approx -\frac{1}{4\pi r} exp(ikr) \int \frac{2\mu}{h^2} V(\mathbf{r}')exp(i\mathbf{K}.\mathbf{r}')d^3\mathbf{r}'$$

where $\mathbf{K} = k\hat{z} - k\mathbf{r}/r$. $k\hat{z}$ is the momentum of the incident particle and $k\mathbf{r}/r$ is the momentum of the scattered particle after it has been scattered to ∞. The difference of these two is $\mathbf{K}$ which is the momentum transferred to the scattering centre (after multiplication by Planck's constant divided by 2π). This gives the probability distribution of the scattered particles as a function of the angle, i.e., the angular distribution. Thus, the scattered pattern is essentially determined by a Fourier transform of the interaction potential function. When implemented on a digital computer, we use the discrete Fourier transform. For a variety of scattering potentials, one can determine the asymptotic angular distribution of scattered particles. This aspect should be discussed in the talk.

Study project 14

Realization of quantum gates using harmonic oscillator perturbed by an anharmonic potential. The energy operator of a harmonic oscillator is given by the operator

$$H = \frac{1}{2}\left(-\frac{d^2}{dQ^2} + Q^2\right)$$

Its energy spectrum is $n + 1/2$ where $n = 0, 1, 2, ...$ We let $|n>, n = 0, 1, 2, ...$ denotes its energy eigenstates. Then, we can consider as our set of base states $|n>' = \sum_{m=0}^{\infty} c(n, m)|m>, n = 0, 1, 2, ...$ where $c(n, m)$ forms a unitary matrix. We can then study how these base states evolve under the unitary action of the harmonic oscillator with time. After time t, the unitary operator is given by $U_t = exp(-iHt)$. We can look at the matrix of U_t relative to these transformed base states. This matrix will vary as t varies. In this way one can generate a very large class of infinite unitary matrices. A quantum computer is essentially a device that can implement any unitary matrix. Modern research is focusing on how actual physical systems like the harmonic oscillator can be used to implement any unitary matrix and hence a quantum computer. The project will involve looking at this problem as well as the problem of how the harmonic oscillator perturbed by an arbitrary anharmonic potential can be used to simulate an arbitrary unitary gate. Part of the study will involve looking at the standard quantum gates like the Pauli spin gates, the CNOT gate, the Toffoli gate and the Fredkin gate. The anharmonic oscillator has the Hamiltonian

$$H = \frac{1}{2}\left(-\frac{d^2}{dQ^2} + Q^2\right) + \epsilon.Q^3$$

where ϵ is a controllable parameter.

Study project 15

Analyze the quantum mechanical motion of a charged particle in a weak electromagnetic field regarding the entire field as a perturbation to the free motion of the particle. The relevant Schrodinger equation reads choosing our units as $h/2\pi = 1$,

$$(-\frac{1}{2m}(\nabla + i.e.\mathbf{A})^2 - eV)\psi = E\psi$$

or

$$(\nabla + i.e.\mathbf{A})^2\psi + 2m(eV + E)\psi = 0$$

which expands to give

$$(\nabla^2 + 2i.e.\mathbf{A}.\nabla + i.e.\nabla.\mathbf{A} - e^2\mathbf{A}^2)\psi + 2m(eV + E)\psi = 0$$

We're assuming the electromagnetic field to be time independent, which means that we can talk about the stationary Schrodinger equation in the em field. The unperturbed equation reads

$$\nabla^2\psi + 2mE\psi = 0$$

whose solution is the plane wave

$$\psi(\mathbf{r}) = exp(i\mathbf{k}.\mathbf{r})$$

where $\mathbf{k} = \sqrt{2mE}\,\hat{\mathbf{n}}$, with $\mathbf{n}$ as a unit vector corresponding to the direction in which the incident beam of particles is flowing. We denote this wave function by $\psi_0(\mathbf{r})$. The value of the energy in this state is E. We now want to look at the perturbed state corresponding to this same value of E. Assume that $\psi = \psi_0 + \psi_1 + \psi_2$. We regard the potentials $\mathbf{A}, V$ as being of the first order of smallness and the quantity $\mathbf{A}^2$ as being of the second order of smallness. Then ψ_1 is the first order correction and ψ_2 is the second order correction to the wave function. We can choose $\nabla.\mathbf{A} = 0$. This is the Coulomb gauge. However for time independent problems, this coincides with the Lorentz gauge. The first order correction satisfies the equation

$$\nabla^2\psi_1 + 2i.e.\mathbf{A}.\nabla\psi_0 + 2mE\psi_1 + 2meV\psi_0 = 0$$

which can be rearranged to give

$$(\nabla^2 + k^2)\psi_1 = -2i.e.\mathbf{A}.\nabla\psi_0 - 2meV\psi_0$$

The right side of this equation evaluates to

$$2e\mathbf{A}.\mathbf{k}\psi_0 - 2meV\psi_0 = (2e\mathbf{k}.\mathbf{A}(\mathbf{r}) - 2meV(\mathbf{r}))exp(i\mathbf{k}.\mathbf{r})$$

It is seen that ψ_1 satisfies the Helmholtz equation with a source term dictated by the incident wave field. Its solution is given by

$$\psi_1(\mathbf{r}) \approx -\frac{1}{4\pi}\int\frac{exp(ik|\mathbf{r}-\mathbf{r}'|)}{|\mathbf{r}-\mathbf{r}'|}(2e\mathbf{k}.\mathbf{A}(\mathbf{r}') - 2meV(\mathbf{r}'))exp(i\mathbf{k}.\mathbf{r}')d^3\mathbf{r}'$$

At distances large from the field, this approximates to

$$\psi_1(\mathbf{r}) = -\frac{exp(-ikr)}{4\pi r} \int exp(ik(\hat{n} - \hat{r}).\mathbf{r}')(2e\mathbf{k}.\mathbf{A}(\mathbf{r}') - 2meV(\mathbf{r}'))d^3\mathbf{r}'$$

We can express this formula in terms the spatial Fourier transforms of $\mathbf{A}, V$:

$$\hat{V}(\mathbf{k}) = \int V(\mathbf{r})exp(i\mathbf{k}.\mathbf{r})d^3\mathbf{r}$$

$$\hat{\mathbf{A}}(\mathbf{k}) = \int \mathbf{A}(\mathbf{r})exp(i\mathbf{k}.\mathbf{r})d^3\mathbf{r}$$

Then,

$$\psi_1(r) = -\frac{exp(-ikr)}{4\pi r}(2e\mathbf{k}.\hat{A}(\hat{n} - \hat{r}) - 2me\hat{V}(\hat{n} - \hat{r}))$$

For determining ψ_2, we need to equate second order terms. This gives

$$(\nabla^2 + k^2)\psi_2 + 2i.e.\mathbf{A}.\nabla\psi_1 - e^2A^2\psi_0 + 2meV\psi_1 = 0$$

This equation can be solved for ψ_2 in terms of ψ_0 and ψ_1.

Study project 16

Group theory plays a fundamental role in quantum mechanics. A variety of interesting problems can be studied in the theory of groups that enable the student eventually to look at problems in the quantum theory that involve applications of groups and their representations. Some of these problems are listed below.

1. M is a C^∞ manifold, meaning that M is a topological space with a collection $\{U_\alpha : \alpha \in \Gamma\}$ of open sets covering M and for each $\alpha \in \Gamma$, we have a map $\phi_\alpha : U_\alpha \to \mathbb{R}^n$ such that ϕ_α is injective and $\phi_\alpha(U_\alpha) = V_\alpha$ is an open subset of $\mathbb{R}^n$. For any pair $\alpha, \beta \in \Gamma$, if $U_\alpha \cap U_\beta$ is non-empty, then

$$\phi_\alpha o\phi_\beta^{-1} : \phi_\beta(U_\alpha \cap U_\beta) \to \phi_\alpha(U_\alpha \cap U_\beta)$$

is infinitely differentiable with respect to each component. The collection $\{(U_\alpha, \phi_\alpha) : \alpha \in \Gamma\}$ of open sets along with coordinate maps is called an atlas. If $p \in M$ and $p \in U_\alpha$, then $\phi_\alpha(p)$ are called the coordinates of p relative to the local chart (U_α, ϕ_α). Note that if also $p \in U_\beta$ for some $\beta \neq \alpha$, then the coordinates of p relative to the chart (U_β, ϕ_β) are $\phi_\beta(p)$. By our definition of C^∞ manifold, the transformation that carries $\phi_\alpha(p)$, the coordinates of p relative to the first chart to $\phi_\beta(p)$, the coordinates of p relative to the second chart, is infinitely differentiable.

Let M be a C^∞ manifold and let $p \in M$ be a point. Let U be one of the $U'_\alpha s$ and ϕ the corresponding ϕ_α. Then $p \in U \subset M$, i.e., U is a neighborhood of p in M. A function f is said to be defined in a neighborhood of p, if there is an open subset V of M such that $f : V \to \mathbb{R}$ is defined and f is said to be locally C^∞, if the map $f : U \cap V \to \mathbb{R}$ is C^∞, that is, $f o \phi^{-1} : \phi(V \cap U) \to \mathbb{R}$ is infinitely differentiable. Note that $\phi(V \cap U)$ is an open nhood of $\phi(p)$ in $\mathbb{R}^n$.

2. Let $H(\mathbf{x})$ be the Hamiltonian of a quantum system. $\mathbf{x}$ are the canonical coordinates which define our representation for defining H. Schrodinger's equation reads $H(\mathbf{x}\psi((\mathbf{x}) = E\psi(\mathbf{x})$. Here, E is the energy level of the eigenstate $\psi(\mathbf{x})$. Suppose that H is invariant under a group G of transformations acting on the coordinates $\mathbf{x}$. This means that for any $g \in G$, $H(g^{-1}\mathbf{x}) = H(\mathbf{x})$ for all $\mathbf{x}$. What do we mean by $H(g^{-1}\mathbf{x})$? The group induces a representation in L^2 where by L^2, we mean all functions $f(\mathbf{x})$ defined on the coordinate space such that $\int |f(\mathbf{x})|^2 d\mathbf{x} < \infty$. Note that $H(\mathbf{x})$ acts on L^2 and in particular, the wave functions of the system belong to L^2. By $H(g^{-1}\mathbf{x})$, we mean $T_g H(\mathbf{x}) T_g^{-1}$ where T_g is the operator on L^2 induced by g, i.e., $T_g f(\mathbf{x}) = f(g^{-1}\mathbf{x})$. Thus, $H(g^{-1}\mathbf{x}) T_g f(\mathbf{x}) = T_g H(\mathbf{x}) f(\mathbf{x})$, or equivalently, $(Hf)(g^{-1}\mathbf{x}) = T_f Hf(\mathbf{x})$. If the Hamiltonian H is invariant under the group G, then $T_g H T_g^{-1} = H$ and hence, if ψ is an eigenfunction of H with eigenvalue E, then $T_g \psi$ is also an eigenfunction of H with eigenvalue E. In other words, if W_E is the eigenspace of L^2 with eigenvalue E, i.e.

$$W_E = \{\psi \in \ell^2 \colon H\psi = E\psi\}$$

then W_E is invariant under the operators $\{T_g : g \in G\}$. Thus, the restriction of the representation $g \to T_g$ to the space W_E defines a new representation of G. Usually, it turns out that this restriction defines an irreducible representation.

Example: Consider the motion of a particle in a central potential. The energy operator is given by

$$H(\mathbf{x}) = -\frac{h^2}{8\pi^2 m}\nabla^2 + V(r)$$

where

$$\nabla^2 = \frac{\partial^2}{\partial x_1^2} + \frac{\partial^2}{\partial x_2^2} + \frac{\partial^2}{\partial x_3^2}$$

$$V(r) = V(\sqrt{x_1^2 + x_2^2 + x_3^2})$$

If g is a rotation of $\mathbb{R}^3$, i.e., any 3×3 real matrix with $gg^T = I$, then it is easily verified that

$$\nabla^2 f(g^{-1}\mathbf{x}) = (\nabla^2 f)(g^{-1}\mathbf{x})$$

In other words,

$$\nabla^2 T_g = T_g \nabla^2$$

or equivalently,

$$T_g^{-1} \nabla^2 T_g = \nabla^2$$

Likewise,

$$V(r)f(g^{-1}\mathbf{x}) = V(|\mathbf{x}|)f(g^{-1}\mathbf{x}) = (Vf)(g^{-1}\mathbf{x})$$

since $|g\mathbf{x}| = |\mathbf{x}| = r$. Thus,

$$T_g^{-1} V T_g = V$$

Combining these two gives $T_g^{-1} H T_g = H$, i.e., the energy operator is invariant under the group of rotations. It is known in quantum mechanics that the energy levels are functions of two quantum numbers n, l where each l value determines an irreducible representation of the rotation group. Thus W_E defines a representation of the rotation group and for each $E = E_{nl}$, the representation is irreducible, We are assuming that there is no accidental degeneracy, i.e. if $(n', l') \neq (n, l)$, then $E_{nl} \neq E_{n'l'}$.

3. Tensor product of vector spaces: Let V be a finite dimensional vector space and $\otimes$ a tensor product of this space with itself. This means that if $dim V = n < \infty$, then $V \otimes V$ is a vector space having dimension n^2 and for each pair of vectors $(x, y) \in V \times V$, $x \otimes y$ is a vector in $V \otimes V$ such that the map $(x, y) \to x \otimes y$ is bilinear (linear in both its arguments) and if $B = \{e_1, ..., e_n\}$ is a basis for V, then $B \otimes B = \{e_i \otimes e_j : i, j = 1, 2, ..., n\}$ is a basis for $V \otimes V$. An example is provided by $V = \mathbb{F}^n$ where $\mathbb{F}$ is a field. The vector space V here consists of all $n \times 1$ column vectors with entries from $\mathbb{F}$. For any two vectors $[x_1, ..., x_n]^T, [y_1, ..., y_n]^T \in \mathbb{F}^n$, we can define their tensor product by the formula

$$[x_1 y_1, x_1 y_2, ..., x_1 y_n, x_2 y_1, x_2 y_2, ..., x_2 y_n, ..., x_n y_1, x_n y_2, ..., x_n y_n]^T$$

This tensor product is called the Kronecker tensor product. We leave it as an exercise to verify the product basis property. V^* is the dual space of V, i.e., the space of all linear functionals on V. If $B = \{e_1, ..., e_n\}$ is a basis for V, then the dual basis of B is $B^* = \{f_1, ..., f_n\}$ where $f_1, ..., f_n$ are linear functionals satisfying $f_i(e_j) = \delta_{ij}$. We can define the tensor product of V and V^* in the usual way If $e \in V, f \in V^*$, then $e \otimes f \in V \otimes V^*$ can be identified with a mapping from $V^* \times V \to \mathbb{F}$ linear in both its arguments, i.e, for $g \in V*$ and $x \in V$, we define $(e \otimes f)(g, x) = g(e)f(x)$. More generally, we can talk of the vector space

$$V_{r,s} = V \otimes ... \otimes V \otimes V^* \otimes .. \otimes V^*$$

where V appears r times and V^* appears s times in this tensor product. Given $L \in End(V)$ ($End(V)$ is the space of all linear transformations from V into itself. A linear transformation from V into itself is also called an endomorphism of V), we can define $L^* : V^* \to V^*$ by $(L^* v^*)(u) = v^*(Lw)$. Then, for any $L \in End(V)$, we can define $L_{r,s} \in End(V_{r,s})$ via the rule

$$L_{r,s}(u_1 \times ... \otimes u_r \otimes v_1^* \otimes ... \otimes v_s^*) = L u_1 \otimes ... L u_r \otimes L^* v_1^* \otimes ... \otimes L^* v_s^*$$

If however, L^* is defined by the rule $(L^* v^*)(u) = -v^*(Lu)$, then the definition of $L_{r,s}$ will undergo obvious modifications. This second definition of $L_{r,s}$ is important in the structure theory of Lie algebras.

4. If p is a prime, $Z_p = \{0, 1, 2, ..., p-1\}$ is a field under addition and multiplication defined modulo p. In particular, under addition modulo p, this defines an Abelian group. We want to

show that $Aut(\mathbb{Z}_p) = \mathbb{Z}_p^*$ where $\mathbb{Z}_p^* = \{1, 2, ..., p-1\}$ is the multiplicative group modulo p. To see this, let α be any automorphism of $\mathbb{Z}_p$. $\alpha(n) = n\alpha(1)$ implies α is completely determined by $\alpha(1)$. If $\alpha(1) = 0$, then it follows that $\alpha(n) = 0$ for all $n = 1, 2, ..., p-1$ and hence α cannot be an automorphism. Thus, $\alpha(1) \in \{1, 2, ..., p-1\}$. In fact, $\alpha(1)$ can assume any value j in $1, 2, ..., p-1$ since then $n \to nj \ (mod p)$ defines an automorphism of $\mathbb{Z}_p$. ((n+m)j=nj+mj and $n, m \in \{0, 1, ..., p-1\}$ implies $|n-m| \leq p-1$ and hence $nj = mj$ implies $(n-m)j = 0 mod p$ implies p divides j which is false., thus, as n ranges over $0, 1, ..., p-1$, $jn mod p$ ranges over the same set.). With $\alpha(n) = nj$ and $\beta(n) = nk$ where $k, j \in \{1, 2, ..., p-1\}$, we have $\beta(\alpha(n)) = \beta(nj) = njk$. So the group property of automorphisms follows. We've shown that all the automorphisms of $\mathbb{Z}_p$ are contained in $n \to nj$ with $j = 1, 2, ..., p-1$. We define a map $\alpha \to \alpha(1)$ from the group of all automorphisms into $\{1, 2, ..., p-1\}$. This defines an isomorphism from $Aut(\mathbb{Z}_p)$ onto $\mathbb{Z}_p^*$. Then, $\beta o \alpha(1) = \beta(\alpha(1)) = \beta(1)\alpha(1)$ shows that this map is a homomorphism of $Aut(\mathbb{Z}_p)$ into the multiplicative group $\mathbb{Z}_p^*$. Equivalently, if $\alpha(n) = jn, \beta(n) = kn$, then $\beta o \alpha(1) = jk = \beta(1)\alpha(1)$. Note that since we are dealing with Abelian groups, the order of multiplication does not matter.

5. $\mathbb{Z}_p^*$ is isomorphic to $\mathbb{Z}_{p-1}$. The first step in the proof is to show that there is an element $x \in \mathbb{Z}_{p-1}$ having order $p-1$. Assuming that this has been settled, the map $i \to x^i$ from the Abelian group $Z_{p-1} = \{0, 1, ..., p-2\}$ with composition defined as addition modulo p onto the Abelian group $Z_p^* = \{1, 2, ..., p-1\}$ with composition defined as multiplication modulo p is an isomorphism between the two groups.

6. Fundamental theorem of G-spaces: Let S be a set and let G be a finite group acting on S transitively. By saying that G acts on S, we mean that there is a map $(g, s) \to g.s$ from $G \times S \to S$ satisfying $e.s = s, g_2.(g_1.s) = (g_2 g_1).s$ for all $s \in S$ and $g_1, g_2 \in G$. By saying that the action is transitive, we mean that given any $s_1, s_2 \in S$, there is a $g \in G$ such that $s_2 = g.s_1$. For any $s \in S$, define $I_s = \{g \in G : g.s = s\}$. Clearly I_s is a group. It is called the isotropy group of s. We define a map $\psi : G/I_s \to S$ by the rule $\psi(g.I_s) = g.s$. This map is well defined since $g.I_s = g'.I_s$ implies $g'^{-1}g \in I_s$ implies $g'^{-1}g.s = s$ implies $g.s = g'.s$. The map is one since suppose $\psi(g.I_s) = \psi(g'.I_s)$. for some $g, g' \in G$. Then, $g.s = g'.s$ and hence $g'^{-1}g \in I_s$ which in turn implies $gI_s = g'I_s$. The map is onto in view of the transitivity of the G action. Further, ψ is a morphism in the sense that $g'.\psi(g.I_s) == \psi(g'.gI_s)$ for all $g, g' \in G$. In other words, $g.\psi = \psi.g$ for all $g \in G$. This follows from the identity $g'.\psi(g.I_s) = g'.g.s = \psi(g'g.I_s)$. In other words, studying the action of G on S is equivalent to studying the action of G on G/I_s. The two G spaces S and G/I_s are isomorphic in this sense.

7. If H and H' are conjugate subgroups of a group G, i.e., $xHx^{-1} = H'$ for some $x \in G$, then the G spaces G/H and G/H' are isomorphic G-spaces. The converse is also true. Let H, H' be any two subgroups of G and let $\psi : G/H \to G/H'$ be a G-space isomorphism, i.e., $\psi(g'.gH) = g'\psi(g.H)$ for all $g', g \in G$, ψ is one-one and onto. In particular, there is a $g \in G$ such that $g.\psi(H) = \psi(g.H) = H'$. Let us find what the isotropy group of H in G/H is. $x.H = H$ iff $\psi(x.H) = \psi(H)$ iff $x.\psi(H) = \psi(H)$ iff $x.g^{-1}H' = g^{-1}H'$ iff $gxg^{-1}H' = H'$, Thus, x belongs

to the isotropy group of H in G/H iff gxg^{-1} belongs to the isotropy group of H' in G/H'. Now suppose $y \in H$. Then, $yH = H$ so $gyg^{-1}H' = H'$, so $gyg^{-1} \in H'$, or $y \in g^{-1}H'g$. Conversely, suppose $y \in g^{-1}H'g$. Then, $gyg^{-1} \in H'$, so $gyg^{-1}H' = H'$, so $yH = H$ so $y \in H$. This proves that $H = g^{-1}H'g$ or equivalently, $gHg^{-1} = H'$, i.e., the subgroups H, H' are conjugate. Conversely, suppose $H' = gHg^{-1}$. Then, define $\psi : G/H \to G/H'$ by $\psi(x.H) = xg^{-1}H'$. First to show that ψ is well defined. $x.H = y.H$ implies $y^{-1}x \in H$ implies $gy^{-1}xg^{-1} \in H'$ implies $xg^{-1}H' = yg^{-1}H'$. Next to show that ψ is one. $xg^{-1}H' = yg^{-1}H'$ implies $gy^{-1}xg^{-1} \in H'$ implies $gy^{-1}xg^{-1} \in gHg^{-1}$ implies $y^{-1}x \in H$ implies $xH = yH$. Finally, ψ is onto because given any $y \in G$, define $x = yg$. Then, $yH' = \psi(x.H)$. Finally, to prove that ψ commutes with G-action. $g'\psi(x.H) = g'xg^{-1}H' = \psi(g'xH)$. We are done.

8. **Sylow's theorems:** Let p be a prime and $o(G) = p^s r$ where r is not divisible by p. There exists a subgroup H with $o(H) = p^s$. All such subgroups are conjugate and the number of such groups divides r. Moreover the number of such groups is congruent to 1 modulo p.

The basic step in the proof the Sylow theorems is to note that p does not divide $\binom{p^s r}{p^s}$ since

$$\binom{p^s r}{p^s} = \frac{p^s r(p^s r - 1)...(p^s r - p^s + 1)}{p^s(p^s - 1)...1}$$

Let E_k denote the set of elements in $p^s r - m, m = 0, 1, ..., p^s - 1$ that are divisible by p^k. Then,

$$E_s = \{p^s r\}, E_{s-k} = \{p^s r - jp^{s-k}, j = 0, 1, 2, ..., p^k - 1\}, k = 1, 2, ..., s - 1$$

If n_j denotes the number of elements in E_j, then we have $n_{s-k} = p^k, k = 0, 1, ..., s - 1$. Likewise, let F_k denote the set of elements in $p^s - m, m = 0, 1, ..., 1$ that are divisible by p^k. Then,

$$F_s = \{p^s\}, F_{s-k} = \{p^s - jp^{s-k}, j = 0, 1, ..., p^k - 1\}, k = 1, 2, ..., s - 1$$

Thus, if m_j denotes the number of elements in F_j, then $m_{s-k} = p^k, k = 0, 1, ..., s - 1$ and we have $m_k = n_k, k = 1, 2, ..., s$. Now, if n'_k denotes the number of elements in the set $\{p^s r - m, m = 0, 1, ..., p^s - 1\}$ that are divisible by p^k but not by p^{k+1}, then

$$n'_s = n_s, n'_{s-1} = n_{s-1} - n_s, ..., n'_1 = n_1 - n_2$$

Similarly, if m'_k denotes the number of elements in the set $\{p^s - m, m = 0, 1, ..., p^s - 1\}$ divisible by p^k but not by p^{k+1}, then $m'_s = m_s, m'_{s-1} = m_{s-1} - m_s, ..., m'_1 = m_1 - m_2$. This shows that $n_k = m_k, k = 1, 2, ..., s$, which means that all the powers of p cancel out from the numerator and the denominator proving that p cannot divide $\binom{\{p^s r\}}{\{p^s\}}$. Sylow's theorem in its more generalized setting is the following: If p is a prime and p^α divides $o(G)$, then G has a subgroup of order p^α. The first thing to note here is a generalization of the result just stated: If p^r divides m but p^{r+1} does not divide m, then p^r divides $\binom{p^\alpha m}{p^\alpha}$ but p^{r+1} does not divide $\binom{p^\alpha m}{p^\alpha}$. This follows from the fact that t is the maximum integer such that p^t divides $p^\alpha m - i$, then t is also the maximum integer such that p^t divides $p^\alpha - i$ where $i = 1, ..., p^\alpha - 1$. Thus, if p is a prime and k is the maximum integer for which p^k divides m, then k is also the maximum integer for which p^k divides $\binom{p^\alpha m}{p^\alpha}$. We shall show that if p is a prime and p^α divides $o(G)$, then G has a subgroup of order p^α. Let $o(G) = p^\alpha m$. Let $\mathcal{M}$ be the family of all subsets of G having p^α elements. Then, $o(\mathcal{M}) = \binom{p^\alpha m}{p^\alpha}$. Given $M_1, M_2 \in \mathcal{M}$, say that M_1 and M_2 are equivalent if there exists a $g \in G$ such that $M_1 = M_2 g$. This defines an equivalence relation on $\mathcal{M}$. Suppose p^{k+1} divides the size of every equivalence class. Then, p^{k+1} divides $o(M)$ which is a contradiction. Thus, there is at least one equivalence class such that p^{k+1} does not divide the number of elements in this equivalence class. Let one such class be denoted by $\{M_1, ..., M_n\}$, so that p^{k+1} does not divide n. Clearly, for each $g \in G$ and each $i = 1, 2, ..., n$, there is a $j = 1, 2, ..., n$ such that $M_i g = M_j$. In other words, G acts transitively on the space $\{M_1, ..., M_n\}$. Let $H = \{g \in G : M_1 g = M_1\}$. H is a subgroup of G. From the basic result on G spaces, the number of elements in G/H equals n, the size of the G space, i.e., $p^\alpha m = o(G) = no(H)$. Since p^{k+1} does not divide n and $p^{\alpha+k}$ divides $p^\alpha m = no(H)$, it must follows that p^α divides $o(H)$, so $o(H) \geq p^\alpha$. If $m \in M_1$ then for al $h \in H$, $m_1 h \in M_1$. Moreover, if $h, h' \in H$ and $h \neq h'$, then $m_1 h \neq m_1 h'$ since m_1 is an element of the group. Thus, M_1 has at least $o(H)$ distinct elements. However, M_1 is a subset of G containing p^α elements. Thus, $o(H) = p^\alpha$.

Remark: Let t be the maximum integer for which p^t divides $p^\alpha m - i$ where $i = 1, 2, ..., p^\alpha - 1$. We want to show that t is also the maximum integer for which p^t divides $p^\alpha - i$. Clearly $t \leq \alpha$ because $p^\alpha m - i$ is not divisible by p^α for any $i = 1, 2, ..., p^\alpha - 1$. Thus, p^t divides $p^\alpha m - i$ (by hypothesis) and also divides $p^\alpha m - p^\alpha$. Thus it divides their difference, namely $p^\alpha - i$. In a similar manner, we show that if t' is the largest integer for which $p^{t'}$ divides $p^\alpha - i$, then t' is also the largest integer for which $p^{t'}$ divides $p^\alpha m - i$.

Study project 17

Understand the spectral theorem for self adjoint linear operators on a finite dimensional inner product space over the complex field. Let V be such a space having dimension n. For any $T \in L(V)$, we define $T^* \in L(V)$ via the equation $< Tx, y >=< x, T^* y >$ holding for all $x, y \in V$. The existence of T^* is a consequence of the Riesz representation theorem according to which, if $f \in V^*$, then there is a unique $\xi \in V$ such that $f(x) =< x, \xi >$ for all $x \in V$. T is self adjoint if $T^* = T$, i.e., if $< Tx, y >=< x, Ty >$ for all $x, y \in V$. A subspace $W \subset V$ is said to be T-invariant if $T(W) \subset W$. It then follows that $W^\perp = \{x \in V :< x, y >= 0 \forall y \in V\}$ is also T-invariant. Since $x \in W^\perp, y \in W$ implies $x \in W^\perp, Ty \in W$ implies $< Tx, y >=< x, Ty >= 0$. The polynomial $p(t) = det(tI - T)$ has at least one complex root say c_1. Then, there is a nonzero vector $e_1 \in V$ such that $Te_1 = c_1 e_1$. By dividing e_1 by $< e_1, e_1 >^{1/2}$, we can assume that $< e_1, e_1 >= 1$. Thus $W = sp\{e_1\} = \{\lambda e_1 : \lambda \in \mathbb{C}\}$ is T invariant. Hence, $W^\perp$ is also T invariant. Also $dim W^\perp = n - 1$ since $V = W \oplus W^\perp$. If we make the induction hypothesis that every sulfadoxine operator on an inner product space of dimension $< n$ has a complete orthonormal basis of eigenvectors, then we can apply this hypothesis to the restriction of T to $W^\perp$, i.e., consider $T^{W^\perp}$. We can find by the hypothesis, orthonormal vectors $\{e_2, ..., e_n\}$ such that $Te_i = c_i.e_{.i}$ for some scalars $c_2, ..., c_n \in \mathbb{C}$.

Now $\{e_1, e_2, ..., e_n\}$ is an orthonormal basis for V and $Te_i = c_i.e._i, i = 1, 2, ..., n$ proving the digaonability of every Hermitian operator on a finite dimensional vector space by induction.

Study project 18

Time dependent problems in the quantum theory, implementation using TMS-ADSP cards as a part of implementation of quantum mechanical algorithms on DSP processors: Hamiltonian of the system is $H = H_0 + H'$. Eigenfunctions of unperturbed Hamiltonian $H_0 u_n = E_n u_n$. The state of the perturbed system is described by the dynamics

$$ih\frac{\partial\psi(t)}{\partial t} = (H_0 + H'(t))\psi(t)$$

Set

$$\psi(t) = \sum a_n(t)u_n(\mathbf{r})exp(-i\omega_n t)$$

where $\omega_n = E_n/h$. Plugging this into the differential equation gives

$$\sum iha'_n(t)u_n(\mathbf{r})exp(-i\omega_n t) + \sum a_n(t)E_n u_n(\mathbf{r})exp(-i.e._n t)$$

$$= \sum a_n(H_0 + H'(t))u_n(\mathbf{r})exp(-i\omega_n t)$$

which simplifies to

$$\sum iha'_n(t)u_n(\mathbf{r})exp(-i\omega_n t) = \sum a_n(t)H'(t)u_n(\mathbf{r})exp(-i\omega_n t)$$

which implies

$$iha'_n(t) = \sum_m a_m(t) < u_n|H'(t)|u_m > exp(i\omega_{nm}t)$$

This is a system of ordinary differential equations satisfied by the functions $a_n(t), n = 1, 2, ...$ The perturbation approximation involves replacing $H'(t)$ by $\lambda H'(t)$ and

$$a_n = a_n^{(0)} + \lambda a_n^{(1)} + \lambda^2 a_n^{(2)} + ... = \sum_{k=0}^{\infty} \lambda^k a_n^{(k)}$$

Substitution and equating coefficients of equal powers of λ leads to the sequence of differential equations

$$ih\frac{da_n^{(0)}}{dt} = 0$$

$$ih\frac{da_n^{(s+1)}}{dt} = \sum_m a_m^s < u_n|H'(t)|u_m > exp(i\omega_{nm}t), s = 0, 1, 2, ...$$

Assume that at time $t = 0$, the system in the state u_p. Then, $a_n^{(0)} = \delta_{np}$ and

$$a_n^{(s+1)}(t) = -\frac{i}{h}\sum_m \int_0^t a_m^{(s)}(t) < u_n|H'(t)|u_m > exp(i\omega_{nm}t)dt, s = 0, 1, 2, ..$$

In particular,

$$a_n^{(1)}(t) = -\frac{i}{h}\sum_m \int_0^t \delta_{mp}\int_0^t < u_n|H'(t)|u_m > exp(i\omega_{nm}t)dt$$

$$= -\frac{i}{h}\int_0^t < u_n|H'(t)|u_p > exp(i\omega_{np}t)dt$$

We can look at the problem of the electron of a Hydrogen atom perturbed by a time varying electric field. Then,

$$H_0 = -\frac{h^2}{2m}\Delta - \frac{e^2}{r} = -\frac{h^2}{2m}(\frac{\partial^2}{\partial x^2} + \frac{\partial^2}{\partial y^2} + \frac{\partial^2}{\partial z^2}) - \frac{e^2}{\sqrt{x^2 + y^2 + z^2}}$$

and

$$H'(t) = e(\mathbf{E}_0(t), \mathbf{r}) = e(E_{0x}(t)x + E_{0y}(t)y + E_{0z}(t)z)$$

Then,

$$< u_n|H'(t)|u_p >= \int u_n(\mathbf{r})H'(t)u_p(\mathbf{r})dV(\mathbf{r}) = e(\mathbf{E}_0(t), \int \bar{u}_n(\mathbf{r})\mathbf{r}u_p(\mathbf{r})dV(\mathbf{r}))$$

To implement the algorithm for computing the first order transition amplitudes $a_n^{(1)}(t)$ using the TMS processor, we need to evaluate the above integral numerically. Let

$$< \mathbf{r} >_{np}= \int \bar{u}_n(\mathbf{r})\mathbf{r}u_p(\mathbf{r})dV(\mathbf{r})$$

$$= \begin{pmatrix} \int x\bar{u}_p(x, y, z)u_n(x, y, z)dxdydz \\ \int y\bar{u}_p(x, y, z)u_n(x, y, z)dxdydz \\ \int z\bar{u}_p(x, y, z)u_n(x, y, z)dxdydz \end{pmatrix}$$

Space is discretized into pixels defined by the grid points $(k\Delta, r\Delta, q\Delta)$, $-N \leq k, m, n \leq N$. Then set $u_n[k, r, q] = u_n(k\Delta, r\Delta, q\Delta)$ and make the following approximations

$$\int x\bar{u}_n(x, y, z)u_p(x, y, z)dxdydz \approx \sum_{k,m,r=-N}^N k\Delta^4\bar{u}_n[k, m, r]u_p[k, m, r]$$

$$\int y\bar{u}_n(x, y, z)u_p(x, y, z)dxdydz \approx \sum_{k,m,r=-N}^N m\Delta^4\bar{u}_n[k, m, r]u_p[k, m, r]$$

$$\int z\bar{u}_n(x,y,z)u_p(x,y,z)dxdydz \approx \sum_{k,m,r=-N}^{N} r\Delta^4 \bar{u}_n[k,m,r]u_p[k,m,r]$$

The states of the unperturbed system themselves are computed using matrix eigenvalue techniques. For example, if we are dealing with a one dimensional quantum problem in a potential $V(x)$, the eigenvalue problem to be solved is

$$-\frac{h^2}{2m}\psi''(x) + V(x)\psi(x) = E\psi(x), x \in \mathbb{R}$$

Suppose we are discretize this equation over the interval $[-L, L]$ taking step size Δ and number of pixels $N = L/\Delta$. Then, we get a matrix eigenvalue problem. Fourier methods can also be introduced in the solution of this equation. For example, defining the Fourier transforms

$$\hat{V}(k) = \int_{-\infty}^{\infty} V(x)exp(-ikx)dx$$

$$\hat{\psi}(k) = \int_{-\infty}^{\infty} \psi(x)exp(-ikx)dx$$

we get on transforming the differential equation

$$\frac{h^2k^2}{2m}\hat{\psi}(k) + \frac{1}{2\pi}\int_{-\infty}^{\infty} \hat{V}(k-k')\hat{\psi}(k')dk' = E\hat{\psi}(k), k \in \mathbb{R}$$

This integral eigenequation can be discretized in the frequency domain as

$$\frac{h^2n^2\Delta_k^2}{2m}\hat{\psi}[n] + \frac{1}{2\pi}\sum_{r=0}^{N-1} \hat{V}[n-r]\hat{\psi}[r] = E\hat{\psi}[n], n = 0, 1, ..., N-1$$

where $\hat{V}[n]$ is computed as an FFT of $V[n]$. In fact, the continuous Fourier transform can be discretized as

$$\hat{V}(k) = \int_{-\infty}^{\infty} V(x)exp(-ikx)dx \approx \sum_{n=0}^{N-1} V(n\Delta)exp(-ikn\Delta)\Delta$$

Like space, the frequency space is also discretized with step size $\Delta_k = 2\pi/N\Delta$ to get

$$\hat{V}[r] = \hat{V}(r\Delta_k) = \sum_{n=0}^{N-1} V(n\Delta)exp(-2\pi inr/N) = \sum_{n=0}^{N-1} V[n]\Delta exp(-2\pi inr/N)$$

$\hat{\psi}[r]$ is interpreted as the FFT of the sequence $\{\psi[n], n = 0, 1, ..., N-1\}$ multiplied by Δ. Then, $\sum_{r=0}^{N-1} \hat{V}[n-r]\hat{\psi}[r]$ is the circular convolution of the DFT of the potential and that of the wave function. The differential equations that determine the evolution of the successive corrections to the transition probability amplitude can also be cast in matrix state variable format. Start with

$$ih\frac{da_n^{(s+1)}}{dt} = \sum_m a_m^s < u_n|H'(t)|u_m > exp(i\omega_{nm}t), s = 0, 1, 2, ...$$

Define the infinite matrix

$$\tilde{H}(t) = ((< u_n|H'(t)|u_m > exp(i\omega_{nm}t)))_{1 \leq n, m < \infty}$$

Also define the infinite dimensional state vectors

$$X^{(s)}(t) = ((a_n^{(s)}(t)))_{n=1}^{\infty}$$

Then, the above equations can be written as

$$ih\frac{dX^{(s+1)}(t)}{dt} = \tilde{H}(t)X^{(s)}(t)$$

The solution to which is

$$X^{(s+1)}(t) = -\frac{i}{h}\int_0^t \tilde{H}(t)X^{(s)}(t)dt$$

The discretized version of this equation is

$$X^{(s+1)}[k + 1] = X^{(s+1)}[k] - \frac{i\Delta}{h}\tilde{H}[k]X^{(s)}[k]$$

where $X^{(s)}[k] = X^{(s)}(k\Delta)$ and $\tilde{H}[k] = \tilde{H}(k\Delta)$. The recursion starts by setting $X^{(0)}[k] = ((\delta_{np}))_{n=1}^{\infty}$ and for $s \geq 1, X^{(s)}[0] = 0$. In practise, the state vector is truncated and so is the matrix $\tilde{H}(t)$ so that only the first N wave functions of the unperturbed system come into the picture. We then obtain a finite dimensional state variable problem. The perturbed problem can also be solved directly using discretization. Suppose that the Schrodinger equation is

$$ih\frac{\partial \psi_t}{\partial t} = H(t)\psi_t$$

Then, discretization this equation with a time step of Δ involves doing the following recursion:

$$\psi_{t+\Delta} = U(t, t + \Delta)\psi_t$$

where

$$U(t, t + \Delta) = exp(-i\Delta.H(t)/h)$$

The evaluation of the exponential of the above operator can be carried out via the approximation

$$exp(-i\Delta.H(t)/h) \approx I + \sum_{n=1}^{N}(-i\Delta/h)^n H(t)^n$$

To evaluate the powers of the operator $H(t)$, we need to do a lot of computation. In the special case when $H'(t)$ does not depend upon time, we have

$$a_n^{(1)}(t) = -\frac{i}{h} < u_n|H'|u_p > \int_0^t exp(i\omega_{np}t)dt = < u_n|H'|u_p > (1 - exp(i\omega_{np}t))/h\omega_{np}$$

The transition probability per unit time from the state p to the state n in the duration $[0, t]$ is given by

$$P_t(p \to n) = t^{-1}|a_n^{(1)}(t)|^2 = | < u_n|H'|u_p > |^2 \frac{sin^2(\omega_{np}t/2)}{t(h\omega_{np}/2)^2}$$

Study project 19

Some more quantum simulations.

TMS implementation of quantum Galilean transformation

1. Galilean transformation of space time is given by $\mathbf{r} \to \mathbf{r} - \mathbf{v}t, t \to t$. We want to see how the wave function changes under a Galilean transformation. Let

$$G(\mathbf{v}) = exp(\mathbf{v}.\mathbf{N})$$

be the unitary operator that induces a Galilean transformation of the position and momentum coordinates. Thus,

$$G(\mathbf{v})x_iG(\mathbf{v})^1 = x_i - v_it, i = 1, 2, 3$$

$$G(\mathbf{v})p_iG(\mathbf{v})^{-1} = p_i - mv_i, i = 1, 2, 3$$

where m is the mass of the quantum particle. Replacing $\mathbf{v}$ by the infinitesimal $\delta\mathbf{v}$ gives

$$(\mathbf{I} + \delta\mathbf{v}.\mathbf{N})x_i(\mathbf{I} - \delta\mathbf{v}.\mathbf{N}) = x_i - t\delta v_i$$

$$(\mathbf{I} + \delta\mathbf{v}.\mathbf{N})p_i(\mathbf{I} - \delta\mathbf{v}.\mathbf{N}) = p_i - m\delta v_i$$

Thus,

$$[\delta\mathbf{v}.\mathbf{N}, x_i] = -t\delta v_i \qquad [\delta\mathbf{v}.\mathbf{N}, p_i] = -m\delta v_i$$

Or

$$\sum_j \delta v_j[N_j, x_i] = -i\delta v_i \qquad \sum_j \delta v_j[N_j, p_i] = -m\delta v_i$$

Write

$$N_j = ap_j + bx_j$$

Then,

$$[N_j, x_i] = [ap_j + bx_j, x_i] = -iha\delta_{ij} \qquad [N_j, p_i] = [ap_j + bx_j, p_i] = ihb\delta_{ij}$$

and the above equations become

$$-iha\delta v_i = -t\delta v_i, ihb\delta v_i = -m\delta v_i$$

so that

$$a = -it/h, b = im/h$$

and

$$N(\mathbf{v}) = a\mathbf{p}.\mathbf{v} + b\mathbf{r}.\mathbf{v} = -\frac{it}{h}\mathbf{p}.\mathbf{v} + \frac{im}{h}\mathbf{r}.\mathbf{v}$$

In terms of operators,

$$G(bfv) = exp(-t\mathbf{v}.\nabla + \frac{im}{h}\mathbf{v}.\mathbf{r}))$$

Design TMS programmes to compute discretized versions of the operator $G(\mathbf{v})$ and act upon wave functions using this.

TMS implementation of Keplerian two body problem in Hamiltonian form.

The normalized Hamiltonian is

$$H(\mathbf{Q}, \mathbf{P}) = \frac{1}{2}\sum_{i=1}^{3} P_i^2 - \frac{1}{\sqrt{\sum_{i=1}^{3} Q_i^2}}$$

The Hamiltonian equations of motion are

$$\frac{dQ_i}{dt} = \frac{\partial H}{\partial P_i} = P_i$$

$$\frac{dP_i}{dt} = -\frac{\partial H}{\partial Q_i} = -\frac{Q_i}{(\sum Q_j^2)^{3/2}}$$

To implement these using TMS, we've got to use the Taylor approximations

$$Q_i(t + \Delta) = Q_i(t) + \Delta.Q_i'(t) + \frac{\Delta^2}{2}Q_i''(t)$$

$$= Q_i(t) + \Delta.P_i(t) - \frac{\Delta^2}{2}\frac{Q_i(t)}{(\sum Q_j(t)^2)^{3/2}}$$

$$P_i(t + \Delta) = P_i(t) + \Delta.P_i'(t) + \frac{\Delta^2}{2}P_i''(t)$$

with

$$P_i''(t) = -\frac{d}{dt}\frac{Q_i}{(\sum Q_j^2)^{3/2}} = -\frac{Q_i'(t)}{(\sum Q_j(t)^2)^{3/2}} + 3Q_i(t)\sum_j Q_j(t)Q_j'(t)/(\sum Q_k(t)^2)^{5/2}$$

These iterations are to be implemented using TMS.

TMS implementation of Gram Schmidt orthonormalization of wave functions. Let $\psi_i(\mathbf{dx}), i = 1, 2, ..., N$ be wave functions defined over space. We want to compute an orthonormal basis for the space spanned by these wave functions. This would enable one to compute probabilities of events related to these wave functions easily. The Gram Schmidt process involves performing the following operations

$$\phi_1 = \frac{\psi_1}{\parallel \psi_1 \parallel}, \phi_2 = \frac{\psi_2 - <\psi_2, \phi_1> \phi_1}{\parallel \psi_2 - <\psi_2, \phi_1> \phi_1 \parallel}$$

and in general,

$$\phi_{k+1} = \frac{\psi_{k+1} - \sum_{m=1}^{k} <\psi_{k+1}, \phi_m> \phi_m}{\parallel \psi_{k+1} - \sum_{m=1}^{k} <\psi_{k+1}, \phi_m> \phi_m \parallel}$$

The crucial point in these computations is that of the inner product

$$<u, v> = \int \bar{u}(\mathbf{x}) v(\mathbf{x}) d^3 x$$

This is achieved by choosing a cubical grid, discretizing the functions and replacing the Riemann integral by a Riemann sum. For example, the function $u(\mathbf{x})$ is replaced by $u[k, l, m] = u(k\Delta, l\Delta, m\Delta)$ with $0 \leq k, l, m \leq N - 1$ and $N\Delta = L$ where L is the length of the cubical box. Then, $<u, v>$ is replaced by $\Delta^3 \sum_{k,l,m=0}^{N-1} \bar{u}[k, l, m] v[k, l, m]$.

Study project 20

The following quantum mechanical problems can implemented on the TMS processor.

1. Particle in a box. A free particle moves in a one dimensional box of length L. The wave functions and energy levels are determined by discretization schemes for differential equations and the values obtained are compared with the exact analytical solution. The eigen-equation to be solved is

$$\psi''(x) + \frac{8\pi^2 mE}{h^2}\psi(x) = 0, 0 < x < L$$

with boundary conditions $\psi(0) = \psi(L) = 0$. This eigenvalue problem can be solved exactly, but we can obtain the approximate solution using TMS assembly language routines and compare with the exact analytic solution.

2. Flux of free quantum particles in space. The wave functions of the free Schrodinger equation are to be determined via numerical simulations and compared with experiment. The equation to be solved is

$$(\frac{\partial^2}{\partial x^2} + \frac{\partial^2}{\partial y^2} + \frac{\partial^2}{\partial z^2})\psi(x, y, z) + \frac{8\pi^2 mE}{h^2}\psi(x, y, z) = 0$$

for positive values of E. This partial differential equation can also be solved exactly but we can obtain the approximate solution using TMS assembly language routines and compare with the exact analytic solution.

3. Quantum mechanical rotator. Let $\mathbf{L}$ denote the angular momentum operator of a quantum mechanical rotator having moment of inertia I. The kinetic energy of the rotator is obtained from classical mechanics as $T = \frac{L^2}{2I}$. In quantum mechanics, the operator L^2 is given by

$$L^2 = -\frac{h^2}{4\pi^2}\left(\frac{1}{sin\theta}\frac{\partial}{\partial\theta}sin\theta\frac{\partial}{\partial\theta} + \frac{1}{sin^2\theta}\frac{\partial^2}{\partial\phi^2}\right)$$

The energy eigenvalues are obtained from the equation

$$L^2 Y(\theta,\phi) = 2IEY(\theta,\phi)$$

The solutions to this equation are known. They are expressed in terms of the solutions to Legendre's differential equation. We obtain these solutions using numerical routines implemented on the TMS processor. Specifically, the Legendre differential equation associated with the above eigenvalue problem is

$$\frac{d}{dx}(1-x^2)\frac{dP(x)}{dx} + (\lambda - \frac{m^2}{1-x^2})P(x) = 0, |x| \le 1$$

where m assumes integer values. The solutions to this eigenvalue problem may be obtained by approximating the differentials using finite differences.

4. Matrix mechanics of Heisenberg can be implemented using TMS routines. In Heisenberg's matrix mechanics, The equation of motion of an observable is given by $\frac{dX}{dt} = i(HX - XH)$ where H is the energy and X is an observable. If H is time independent, the solution is obtained as $X(t) = exp(itH)X(0)exp(-itH)$. If H is time dependent, then we need to use numerical methods to determine the temporal evolution of the observable X. We may determine the evolution of several observables X for a large class of energy operators that include free particle, harmonic oscillator and Hydrogen atom. Usually, the observables X, H are Hermitian operators on a Hilbert space like $L^2(\mathbb{R}^n)$. In order to obtain numerical methods for determining the solution to Heisenberg's equation, we need to choose a convenient orthonormal basis $\{\phi_1, \phi_2, ...\}$ for the Hilbert space $L^2(\mathbb{R}^n)$ and consider the matrices of the observables X, H relative to this basis. For example, if X is represented by an integral kernel $X\phi(\mathbf{x}) = \int X(\mathbf{x},\mathbf{y})\phi(\mathbf{y})d\mathbf{y}$, then we will represent X by a finite matrix having entries

$$X_{ab} = <\phi_a|X|\phi_b> = \int \bar{\phi}_a(\mathbf{x})X(\mathbf{x},\mathbf{y})\phi_b(\mathbf{y})d\mathbf{x}d\mathbf{y}$$

where $a, b = 1, 2, ..., N$ with N sufficiently large. N should be determined so that the squared error

$$\int_{L^2(\mathbb{R}^n)\times L^2(\mathbb{R}^n)} |X(\mathbf{x},\mathbf{y}) - \sum_{a,b=1}^{N} X_{ab}\phi_a(\mathbf{x})\phi_b(\mathbf{y})|^2 d\mathbf{x}d\mathbf{y}$$

is smaller than a prescribed threshold. Of course such a choice is possible only when $\int |X(\mathbf{x}, \mathbf{y})|^2$ $d\mathbf{x}d\mathbf{y} < \infty$, i.e., when X is Hilbert-Schmidt. This is not true for the position and momentum operators.

5. Determination of the energy eigenvalues of a Hermitian operator by the matrix approximation method. If H is a Hermitian operator on an infinite dimensional Hilbert space $\mathcal{H}$, then choose an orthonormal basis $\{e_1, e_2, ...\}$ for $\mathcal{H}$ and approximate H by its finite dimensional version $\tilde{H}_N = ((< e_a|H|e_b >))_{1 \leq a,b \leq N}$. Determine the eigenvalues of $\tilde{H}_N$ by standard numerical routines. This technique can be applied to the Hydrogen atom as well as to the harmonic oscillator.

Chapter 6

Circuit theory

6.1   Resistor, capacitor, inductor–voltage current relations

For a conductor, the relationship between the electric field and the current density is linear: $\mathbf{J} = \sigma\mathbf{E}$. σ is the conductivity of the conductor. If the conductor is a rod having cross sectional area A and length L, then the current through it is given by $I = JA$ and the voltage across it is given by $V = EL$. Thus, $\frac{V}{I} = \frac{J}{E}\frac{L}{A} = \frac{L}{\sigma A} = \rho L/A$, where $\rho = 1/\sigma$ is called the resistivity of the conductor. The ratio of V to I is called the resistance of the conductor and this equation shows that resistance is proportional to the length of the conductor and inversely proportional to the cross sectional area. A capacitor consists of two parallel plates of area A and separated by a distance d. Suppose a charge Q is uniformly distributed on one plate and a charge $-Q$ on the other. The surface charge densities on the plates are respectively $\sigma = Q/A$ and $-\sigma = -Q/A$ respectively. Neglecting edge effects, Gauss' law applied to a small pillbox gives for the electric field between the plates, $E = \sigma/\epsilon$, where ϵ is the permittivity of the dielectric between the two plates. The potential difference between the plates is thus $V = Ed = \sigma d/\epsilon = Qd/\epsilon A$ and the ratio $C = Q/V = \epsilon A/d$ is called the capacitance of this arrangement. The relation between voltage and current for this parallel plate capacitor is therefore given by $I = dQ/dt = CdV/dt$. An inductor in its most primitive form consists of a coil of N turns of wire wound around a iron core. Let μ denote the magnetic permeability of the core and let I be the current through the wire. Let A be the cross sectional area of the core. We make the approximation that the magnetic field B in the core is axial and uniform throughout and that the magnetic field outside the core is zero. Let L denote the length of the core. Then application of Ampere's law to a rectangular loop one length of which passes through the axis of the core and the other length outside the core gives $HL = NI$ or $H = NI/L$ or $B = \mu NI/L$. The flux of the magnetic field through the core is then given by $\Phi = BNA = \mu N^2 IA/L$ and this equation connecting flux linkage and current can be written as $\Phi = \mathcal{L}I$ where $\mathcal{L} = \mu AN^2/L$ is the inductance of the arrangement. By Faraday's law, the potential difference across the coil is given by $V = \frac{d\Phi}{dt} = \mathcal{L}\frac{dI}{dt}$. This is a summary of the voltage current relations for a resistor, capacitor and an inductor.

6.2   Kirchhoff's voltage and current relations

Kirchhoff's current relation states that the algebraic sum of the currents leaving a junction/node in a circuit is zero. Let the edges incident on a node have currents $I_1, ..., I_k$ directed away from the node. If one or more of the currents is directed into the node, then we may reverse the sign and assume that the current is directed away from the node. Then Kirchhoff's current equation states that $\sum_{m=1}^{k} I_m = 0$. This equation holds under the hypothesis that the node is so tiny that it cannot accumulate any charge. In case the node can accumulate charge, if $Q(t)$ denotes the charge in the node at time t, then the Kirchhoff current relation must get modified to $\sum_{m=1}^{k} I_m(t) = \frac{dQ(t)}{dt}$. These laws are derived from the general equation of local charge conservation $\nabla.\mathbf{J} + \frac{\partial \rho}{\partial t} = 0$ by integrating over a surface enclosing the node. Kirchhoff's voltage relation states that if a loop in a circuit consists of edges $e_1, ..., e_k$ and if V_i is the voltage across the edge e_i with the direction of V_i being all along the counterclockwise direction of the loop, then $\sum_{i=1}^{k} V_i = 0$, that is, the algebraic sum of the voltage drops around any loop is zero. This holds good provided we assume that no magnetic field lines pass through the loop. This is seen to be a consequence of Faraday's law in integral form $\int_\Gamma \mathbf{E}.dl = -\frac{d\Phi_B}{dt}$ with zero magnetic field. The Kirchhoff current and voltage relations completely describe the behaviour of all circuits when taken along with the individual voltage current equations for each circuit element.

6.3   Graph theoretic analysis of linear circuits

Any circuit is represented by its equivalent graph. Let $e_1, ..., e_n$ denote the edges of this graph and $v_1, ..., v_m$ its vertices or nodes. Across each edge e_i, we have either a resistor, or a capacitor or an inductor or voltage source or a current source or a controlled voltage or current source, with the control coming from the voltage or current across another edge. Let $v_i(t)$ denote the voltage across the edge e_i. A direction has been assigned to each edge so that the graph is a directed one and the direction of the voltage $v_i(t)$ follows the direction of the edge e_i. If a resistor is placed along e_i, then we have $v_i(t) = R_i i_i(t)$ where $i_i(t)$ is the current along e_i. In the Laplace domain, this equation reads $V_i(s) = R_i I_i(s)$. If a capacitor C_i is along e_i, then $i_i(t) = C_i \frac{dv_i(t)}{dt}$ which in the Laplace domain reads $I_i(s) = sC_i V_i(s) - C_i v_i(0)$. Finally, if the element across the edge e_i is an inductor, then the elemental equation for this edge reads $v_i(t) = L_i \frac{di_i(t)}{dt}$ which in the Laplace domain is $V_i(s) = sL_i I_i(s) - L_i i_i(0)$. If the element along e_i is a voltage source controlled by the voltage across edge e_j, then $v_i(t) = \alpha_{ij} v_j(t)$. Similarly for other kinds of controlled sources. In effect, all these elemental relations can be written down as a single matrix equation $\mathbf{V}_e(s) = \mathbf{Z}(s)\mathbf{I}_e(s) + \mathbf{E}_q(s)$

where $\mathbf{E}_q(s)$ contains terms from external voltage or current sources and initial conditions across capacitors and inductors. $\mathbf{V}_e(s) = V_1(s), ..., V_n(s)]^T, \mathbf{I}_e(s) = [I_1(s), ..., I_n(s)]^T$ are respectively the edge voltage and current vectors. We can now select a tree for this graph. A tree is a connected subgraph which contains all the nodes of the graph and no loop. The tree will have $m - 1$ edges (Edges of a tree are called branches). The edges of the original graph not contained in the tree are the chords. Each chord defines a loop of the graph called a fundamental loop. Likewise each edge of the tree defines a cutset called a fundamental cutset. By deleting a branch of the tree, the tree splits into two parts. The minimal set of edges of the original graph containing the deleted branch of the tree that must be deleted in order to separate out the original graph into two parts is the fundamental cutset associated with the branch. The algebraic sum of currents in a cutset is zero, since the edges of a cutset connect two disconnected components of the original graph and charge cannot be created in any component of the graph. Thus, one can formulate the Kirchhoff current equations in matrix form as $\mathbf{Q}_f\mathbf{I}_e = \mathbf{0}$, where $\mathbf{Q}_f$ is the fundamental cutset matrix. The rows of $\mathbf{Q}_f$ are indexed by the branches of the tree and columns by the edges of the entire graph. The first $m - 1$ columns of $\mathbf{Q}_f$ are indices corresponding to the branches of the tree while the remaining $n - m + 1$ columns are indices corresponding to the chords. If an edge e_i belongs to the fundamental cutset defined by the branch be e_j and has the same direction, then the entry corresponding to the row e_j and column e_i is $+1$. If the direction is opposite, then this entry is -1. If the edge e_i does not appear in the fundamental cutset defined by the branch e_i, then the corresponding entry is zero. It is easy to see that $\mathbf{Q}_f$ has the block form $[\mathbf{I}|\mathbf{Q}_{12}]$ where the identity is of size $m - 1 \times m - 1$. Similarly one can define the fundamental loop matrix $\mathbf{B}_f$. Its rows are indexed by the chords and columns by the edges, first branches and then chords. If e_j is a chord and if e_i is an edge which falls in the fundamental loop defined by the chord e_i and is along the direction of the loop dictated by e_j, then the entry having row indexed by e_j and column by e_i is a $+1$ while if the direction of e_i is opposite to that dictated by e_j, then the entry is -1. If e_i does not appear in the loop defined by the chord e_j, then the corresponding entry is zero. The Kirchhoff voltage equation that states that the algebraic sum of the voltages around any loop is zero, in matrix notation becomes $\mathbf{B}_f\mathbf{V}_e = \mathbf{0}$. It is clear that the f loop matrix has the structure $[B_{11}|\mathbf{I}]$. We arrange the edge current and voltage vector entries such that the first $m - 1$ indices are the branches and the last $k - m + 1$ indices are the chords. The equation $\mathbf{Q}_f\mathbf{B}_f^T = \mathbf{0}$ can be seen to be an easy consequence of the nature of any graph. Take the $(i, j)^{th}$ entry of the matrix $\mathbf{Q}_f\mathbf{B}_f^T$. It is given by $\sum_{r=1}^n Q_f(i, r)B_f(j, r)$. If $Q_f(i, r)B_f(j, r) = \pm 1$, then the rth edge appears in the cutset indexed by the branch e_i along the same direction and the rth edge also appears in the loop indexed by the chord e_j along the same direction. The rth edge is clearly a chord if $r \neq i$ and we cannot have two chords in the same f loop, i.e., we cannot have $B_f(j, r) = \pm 1$ unless $r = j$. Thus we have either $r \neq i$ and $r = j$ or $r = i$. Assume that $r = j$. Assume $Q_f(i, j) = 1$. Then, the chord e_j appears in the cutset indexed by the branch e_i. Also $B_f(j, j) = 1$ and hence $Q_f(i, j)B_f(j, j) = 1$. By the above discussion, the only value of r for which $r \neq j$ and $Q_f(i, r)B_f(j, r) = 1$ is given by $r = i$. The branch e_i must necessarily appear in the f loop corresponding to the chord e_j since the chord e_j appears in the cutset defined by the branch e_i and it is easily seen from a picture that $B_f(j, i) = -1$ since $Q_f(i, j) = 1$. Thus,

$\sum_r Q_f(i,r)B_f(j,r) = Q_f(i,i)B_f(j,i) + Q_f(i,j)B_f(j,j) = 0$. Other cases follow along similar lines. This proves that $Q_f B_f^T = 0$ or equivalently, $Q_{12} + B_{11}^T = 0$. Thus, $B_f = [-Q_{12}^T | I]$. We write

$$\mathbf{V}_e = \begin{pmatrix} \mathbf{V}_b \\ \mathbf{V}_c \end{pmatrix}$$

and similarly $\mathbf{I}_e = \begin{pmatrix} \mathbf{I}_b \\ \mathbf{I}_c \end{pmatrix}$. Then the Kirchhoff current equation $\mathbf{Q}_f \mathbf{I}_e = \mathbf{0}$ gives $\mathbf{I}_b + \mathbf{Q}_{12}\mathbf{I}_c = \mathbf{0}$ and the equation $\mathbf{B}_f \mathbf{V}_e = \mathbf{0}$ gives $-\mathbf{Q}_{12}^T \mathbf{V}_b + \mathbf{V}_c = \mathbf{0}$. Thus,

$$\mathbf{I}_e = \begin{pmatrix} -\mathbf{Q}_{12}\mathbf{I}_c \\ \mathbf{I}_c \end{pmatrix} = B_f^T I_c$$

$$\mathbf{V}_e = \begin{pmatrix} \mathbf{V}_b \\ \mathbf{Q}_{12}^T \mathbf{V}_b \end{pmatrix} = Q_f^T \mathbf{V}_b$$

The elemental relation can now be expressed as

$$V_e == ZB_f^T I_c + E_q$$

and since $B_f V_e = 0$, we get

$$B_f Z B_f^T I_c = -B_f E_q$$

so that

$$I_c = -(B_f Z B_f^T) B_f E_q$$

Alternately, the elemental relations can be inverted to express the edge current vector in terms of the edge voltage vector:

$$I_e(s) = Y(s)V_e(s) + J_q$$

or

$$I_e = Y Q_f^T V_b + J_q$$

where J_q is an equivalent current source vector. Then, KCL $Q_f I_e = 0$ gives

$$Q_f Y Q_f^T V_b = -Q_f J_q$$

so that

$$V_b = -(Q_f Y Q_f^T)^{-1} Q_f J_q$$

which completes the solution to the circuit. Note that from I_c, we can construct I_b and hence I_e and hence V_e. Similarly from V_b, we can construct V_c and hence V_e and then I_e.

6.4　Mutual inductance

Consider a pair of coils wound on opposite legs of a rectangular iron core. Assume that a current $i_1(t)$ flows through the first coil and a current $i_2(t)$ through the second coil. The flux produced by the current through the first coil links the second coil and if ϕ_{21} is the flux through the second coil due to the current in the first coil, we can write $\phi_{21} = M_{21}i_1$. If ϕ_{22} is the flux through the second coil due to the current in the second coil, then $\phi_{22} = L_2i_2$ Likewise, $\phi_{12} = M_{12}i_2$ and $\phi_{11} = L_1i_1$. It can be proved theoretically that $M_{21} = M_{12}$. In fact this is an elementary consequence of the Biot-Savart law. Depending on the relative directions of the two windings, M_{12} is positive or negative. This leads us to the dot convention which is as follows. If we mark a dot on the first coil at one end, say at A_1 with the other end being B_1 and we also mark a dot on the second coil at one end say A_2, the other end being B_2 and current through coil one flows from A_1 end to B_1 end, then $M_{21}\frac{di_1}{dt}$ will be the voltage induced in coil two due to current through the first coil directed from the A_2 end to the B_2 end where M_{21} is positive. On the other hand, if the dot in coil two is marked at the B_2 end, then the voltage induced in coil two due to changing current in coil one will be $M_{21}\frac{di_2}{dt}$ directed from the B_2 end to the A_2 end. Assuming dots in both coils are marked from the A end to the B end and currents are flowing from the A end to the B end in both the coils, we obtain the following expressions for the voltage drop across the two coils:

$$v_1 = L_1\frac{di_1}{dt} + M_{12}\frac{di_2}{dt}$$

$$v_2 = M_{21}\frac{di_1}{dt} + L_2\frac{di_2}{dt}$$

In terms of Laplace transforms,

$$V_1(s) = sL_1I_1(s) + sM_{12}I_2(s)$$

$$V_2(s) = sM_{21}I_1(s) + sL_2I_2(s)$$

or in matrix notation,

$$\begin{pmatrix} V_1(s) \\ V_2(s) \end{pmatrix} = \begin{pmatrix} sL_1 & sM_{12} \\ sM_{21} & sL_2 \end{pmatrix} \begin{pmatrix} I_1(s) \\ I_2(s) \end{pmatrix}$$

A linear circuit having mutual inductances can be analyzed using these equations.

6.5 Controlled sources

Consider a network whose graph consists of edges $e_1, ..., e_n$. We can pick a set of edges say $e_{i_1}, ..., e_{i_r}$ and another set of edges $e_{j_1}, ..., e_{j_s}$ disjoint from the previous set and place voltage sources across the former that are controlled by the voltages and currents through the remaining edges and likewise place current sources across the latter controlled by the voltages and currents through the remaining edges. Thus, for example, the voltage source across e_{i_m} will have the form $F_m(v_1, ..., v'_{i_m}, ..., v_n, i_1, ..., i'_{i_m}, ..., i_n)$ where prime signifies that particular variable is omitted. In most applications, a controlled source depends on only one voltage or one current variable. Thus, if the current through the edge e_j depends linearly upon the voltage across the edge e_i, we will have $i_j = \lambda v_i$. These equations will augment the other elemental relations and the controlling parameters λ will appear as entries in the matrix $Z(s)$ or in $Y(s)$. For example, if $I_j(s) = \lambda V_i(s)$, then $V_i(s) = I_j(s)/\lambda$ and the $(i,j)^{th}$ entry of the matrix $Z(s)$ will be $1/\lambda$.

6.6   Circuits with sinusoidal excitation

Consider a linear circuit. Suppose that we've chosen a tree and let $\mathbf{V}_b(s)$ denote the tree voltage vector. With $\mathbf{J}_q(s)$ as the equivalent current source, we've seen that $\mathbf{V}_b(s) = -(Q_f Y Q_f^T)^{-1} Q_f J_q(s)$. Write

$$\mathbf{H}(s) = -(Q_f Y(s) Q_f^T)^{-1} Q_f$$

Note that $\mathbf{H}(s)$ is a $m-1 \times n$ matrix where m is the number of vertices/nodes and n is the number of edges. In the time domain, this relationship between the branch voltage vector and the edge equivalent current source vector can be written as

$$v_p(t) = \sum_{r=1}^{n} \int_{-\infty}^{\infty} h_{pr}(\tau) J_r(t-\tau) d\tau, p = 1, 2, ..., m-1$$

where $e_1, ..., e_{m-1}$ are the tree edges. Now, some of the $J_r's$ will arise due to initial conditions and these terms will give rise to components in the output that decay away as $t \to \infty$. Let $J_1, ..., J_s$ denote the source terms that correspond to externally applied sources. Assume that these sources are sinusoidal, i.e., $J_a(t) = \sum_{l=1}^{N} A_{a,l} exp(i\omega_l t)$. Then, as $t \to \infty$, we get the following expressions for the branch voltages:

$$v_p(t) = \sum_{r=1}^{n} \int_{-\infty}^{\infty} h_{pr}(\tau) \sum_{l=1}^{N} A_{r,l} exp(i\omega_l(t-\tau)) d\tau$$

$$\sum_{r=1}^{n} \sum_{l=1}^{N} A_{r,l} H_{pr}(\omega_l) exp(i\omega_l t)$$

where $H_{pr}(\omega) = \int_{-\infty}^{\infty} h_{pr}(\tau) exp(-i\omega\tau) d\tau$. This analysis can be repeated for the chord currents and forms the basis of sinusoidal steady analysis.

6.7   Use of the Laplace transform in circuit analysis

Suppose we are given a linear circuit with time invariant elements i.e., C, L, R are time independent and the controlling parameters in controlled sources are also time independent. The using the elemental relations for the resistor, capacitor and inductor, we can write down the Kirchhoff voltage and current relations for the circuit. We may either perform a nodal analysis or a loop analysis. The net resulting equations can be reduced to a system of linear coupled first order differential equations with the number of differential equations not exceeding the number of capacitors plus the number of inductors. In summary, the differential equations governing the circuit can be expressed in the form

$$\frac{dx_i}{dt} = \sum_{j=1}^{d} a_{ij} x_j(t) + \sum_{j=1}^{p} b_{ij} u_j(t), i = 1, 2, ..., d$$

where the $x_i's$ are the capacitor voltages and the inductor currents and the $u_i's$ are the external inputs some of which may be voltages and the others currents. The initial conditions required are $x_i(0), i = 1, 2, ..., d$ namely, the initial capacitor voltages and the initial inductor currents. This system of linear equations can be solved by taking the Laplace transform. We use the fact that

$$\int_0^\infty \frac{dx(t)}{dt} exp(-st)dt = sX(s) - x(0)$$

Taking the Laplace transform of the above differential equation gives us

$$sX(s) - x(0) = AX(s) + BU(s)$$

so that

$$X(s) = (sI - A)^{-1}BU(s) + (sI - A)^{-1}x(0)$$

In the time domain, this equation reads

$$x(t) = exp(tA)x(0) + \int_0^t exp((t - \tau)A)bu(\tau)d\tau$$

Let A be diagonalized as $A = TDT^{-1}$ where $D = diag[c_1, ..., c_n]$ Writing $T = [t_1, ..., t_n]$ and $T^{-1} = [s_1, ..., s_n]^T$ we get

$$exp(tA) = Texp(tD)T^{-1} = \sum_{i=1}^{n} exp(c_i t)t_i s_i^T$$

Thus,

$$x(t) = \sum_{i=1}^{n} exp(c_i t)t_i s_i^T x(0) + \sum_{i=1}^{n} t_i \int_0^t exp(c_i(t - \tau))s_i^T bu(\tau)d\tau$$

If there are no controlled sources, the circuit will be stable, which means that $Rec_i < 0$ and the first term corresponding to the initial conditions will decay down to zero as $t \to \infty$.

6.8 Solved problems

1. The density of copper is ρ Kg/m^3. The atomic number is A. A copper sphere of radius R is deprived of N electrons by a charging scheme. Determine the percentage decrease in the total number of electrons in the sphere.

Ans: Let m_p be the mass of a proton. Then, the mass of one copper atom is Am_p, neglecting the electron mass (The proton mass is 1840 times the electron mass). The mass of the copper sphere given is $4\pi R^3 \rho/3$. Hence, the number of copper atoms in this sphere is $4\pi R^3 \rho/3Am_p$. If N electrons are deprived, the percentage decrease is given by $3Am_p N/4\pi R^3 \times 100$.

2. If the current in a conductor is I, the cross sectional area of the conductor is S and the density of free electrons is n, determine the drift velocity of an electron.

Ans: $J = I/S$ is the current density. Thus, $nev = J$ gives $v = I/neS$ for the drift velocity. Here, e is the electronic charge.

3. State the relation connecting charge and current: Ans $I = \frac{dQ}{dt}$. Current is rate of flow of charge.

4. If the capacitance between two plates varies with time as $C(t)$. If a constant voltage V is applied between the plates, then determine an expression for the current. Ans:

$$i(t) = \frac{d}{dt}(C(t)V) = V\frac{dC(t)}{dt}$$

5. Determine the current in problem 4 if $C(t)$ is periodic with period T over the positive time axis and $C(t) = at$ for $0 < t < T/2$ and $C(t) = a(T - t)$ for $T/2 < t < T$.

Ans: $\frac{dC(t)}{dt} = a$ for $0 < t < T/2$ and $= -a$ for $T/2 < t < T$. Thus if we define

$$P(t) = a(u(t) - u(t - T/2)) - a(u(t - T/2) - u(t - T)) = au(t) - 2au(t - T/2) + au(t - T)$$

then

$$\frac{dC(t)}{dt} = \sum_{n=0}^{\infty} p(t - nT)$$

and $i(t) = V\frac{dC(t)}{dt}$. We advise you to sketch a graph of this.

6. Assume that in a circuit, the capacitance varies periodically with time and has a Fourier series

$$C(t) = \sum_{k=-\infty}^{\infty} C_k exp(jk\omega t)$$

The voltage across the capacitor also varies periodically with time and has Fourier series

$$V(t) = \sum_{k=-\infty}^{\infty} V_k exp(jk\omega t)$$

Of course since both of these are real quantities, we must have $C_{-k} = \bar{C}_k, V_{-k} = \bar{V}_k$. Determine the Fourier series for the current through the capacitor.

Ans: $i = \frac{d}{dt}(CV) = V\frac{dC}{dt} + C\frac{dV}{dt}$ can be used to obtain the current. However, it is easier to first obtain the Fourier series for the charge $Q(t)$ directly, i.e.,

$$Q(t) = C(t)V(t) = \sum_{k,m=-\infty}^{\infty} C_k V_m exp(j(k+m)\omega t) = \sum_{n=-\infty}^{\infty} Q_n exp(jn\omega t)$$

where

$$Q_n = \sum_{k=-\infty}^{\infty} C_k V_{n-k}$$

Then,

$$i(t) = \sum_{n=-\infty}^{\infty} in\omega C_n exp(in\omega t)$$

Solved Problems, Study Projects, Suggestions and Remarks

7. Determine an expression for the magnetic energy stored in an inductance in terms of the flux linkage through it.

Ans: $v = L\frac{di}{dt}$. $w_L = \int_0^t vidt = \int Lidi = Li^2/2$. The flux linkage $\psi = Li$, so $w_L = \psi^2/2L$.

8. Draw the circuit diagram for the following controlled source: The input impedance is infinite. The output current $i_2 = -gv_1$ where v_1 equals the input voltage. Write down the two port equations for such a source.

Ans: Infinite input impedance implies zero input current. So the input terminals are open circuited. Thus, $i_1 = 0, i_2 = -gv_1$. This can be cast in the Y parameter form as

$$\begin{pmatrix} i_1 \\ i_2 \end{pmatrix} = \begin{pmatrix} 0 & 0 \\ -g & 0 \end{pmatrix} \begin{pmatrix} v_1 \\ v_2 \end{pmatrix}$$

The input terminal consists of two wires, the bottom one going to the ground and the top one open ended. The output terminal consists of a current source gv_1 going from the ground to the top terminal. The voltage v_1 is marked between the top and bottom grounded wires of the input terminal.

9. Write down the two port equations for the following controlled source. The input terminal wires are shorted. The voltage across the output terminals is ri_1.

Ans: $v_1 = 0, v_2 = ri_1$. In terms of Z parameters

$$\begin{pmatrix} v_1 \\ v_2 \end{pmatrix} = \begin{pmatrix} 0 & 0 \\ r & 0 \end{pmatrix} \begin{pmatrix} i_1 \\ i_2 \end{pmatrix}$$

10. Consider a circuit consisting of the following topology. The circuit has four nodes n_1, n_2, n_3 and n_4 the last of which is grounded. An inductor L joins n_1 and n_2, a capacitor C joins n_2 and ground. An inductor L joins n_2 and n_3 and finally a resistor R joins n_3 and ground. A voltage source v_1 is applied between n_1 and ground. Determine the voltage v_2 across the resistor.

Ans: Let i_1 be the current through the inductor L joining n_1 and n_2 and i_2 the current through the inductor joining n_2 and n_3. Then, the KVL equations for the two meshes are

$$I_1(s)sL + (I_1(s) - I_2(s))/sC = V_1(s)$$

$$I_2(s)(sL + R) + (I_2(s) - I_1(s))/sC = 0$$

These equations can be cast in matrix form as

$$\begin{pmatrix} sL + 1/sC & -1/sC \\ -1/sC & sL + R + 1/sC \end{pmatrix} \begin{pmatrix} I_1(s) \\ I_2(s) \end{pmatrix} = \begin{pmatrix} V_1(s) \\ V_2(s) \end{pmatrix}$$

and solved. We leave the details to the student.

11. Consider a circuit with the following topology. One end of a switch K is connected to the node c. The other end can toggle between nodes a and b. Node c is connected to the series combination of a resistor R and an inductor L which is connected to the ground. Node a is connected to the positive terminal of a dc voltage source V, the negative end of which is grounded. Node b is connected to a capacitor, the other end of which is grounded. Initially steady state is reached with the switch K in position a. At $t = 0$, the position of the switch is changed to position b. Let $i(t)$ denote the current through the inductor. Draw the circuit and solve for $i(t)$.

Ans: When the switch is in position a and steady state is reached, the capacitor disconnected and the inductor is short circuited. The inductor current in steady state is therefore V/R. The voltage across the capacitor is zero. Now, when the switch is moved to position b, the capacitor, resistor and inductor are in one loop and the differential equation for the current is

$$L\frac{di}{dt} + iR + \frac{1}{C}\int_0^t i\,dt = 0$$

The initial condition for this integro-differential equation is $i(0) = V/R$ and we get on Laplace transforming,

$$L(sI(s) - V/R) + I(s)R + I(s)/sC = 0$$

This gives

$$I(s) = \frac{VL}{R}/(sL + R + 1/sC) = \frac{sVLC}{R}/(s^2 LC + RCs + 1)$$

The inverse Laplace transform of this yields the current waveform $i(t)$.

12. A network consists of the series connection of a dc voltage source V with a resistance R and an inductor L and a switch K. The switch K is closed at $t = 0$ with zero current in the inductor. Find the values of $i, di/dt.d^2i/dt^2$ at $t = 0+$.

Ans. The differential equation for the circuit is

$$V = Ri + Ldi/dt + \frac{1}{C}\int_0^t idt$$

Setting $t = 0+$ and noting that $i(0+) = 0$ gives

$$\frac{di}{dt}(0+) = V/L$$

Differentiate the above equation to get

$$Rdi/dt + Ld^2i/dt^2 + i/C = 0$$

Set $t = 0+$ in this equation to get

$$d^2i(0+)/dt^2 = -(R/L)di(0+)/dt = -RV/L^2$$

13. Consider a network having the following topology. The circuit has four nodes numbered n_1, n_2, n_3, n_4, the last of which is grounded. A switch K connects nodes n_1 and n_2. In parallel with this switch, is a capacitor C between the same nodes n_1 and n_2. An inductor L connects nodes n_2 and n_3. A resistor R connects node n_3 to the ground node n_4. The network is in steady state with the switch K closed. At $t = 0$, the switch is opened. Determine the voltage across the switch v_K and dv_K/dt at $t = 0+$.

Ans: When the switch is closed, the circuit consists of the series connection of the voltage source, the resistor and the inductor. When steady state is reached, the current is V/R since the inductor is short circuited in steady state. The voltage across the switch and the capacitor is zero at steady state. When the switch is opened, the inductor current and the capacitor voltage cannot change instantaneously. So, at $t = 0+$, the capacitor voltage is zero and the inductor current is V/R. Note that when the switch is opened, the circuit consists of the series connection of the voltage source, the capacitor, the resistor and the inductor. Let $i(t)$ denote the current through the circuit. Applying KVL gives

$$Ldi/dt + Ri + \frac{1}{C}\int_0^t idt = 0$$

The voltage $v_K = \frac{1}{C}\int_0^t i\,dt$. At $t = 0+$, this evaluates to zero. $dv_K/dt = i/C$ which at time $t = 0+$ evaluates to $i(0+)/C = V/RC$. We can also solve the above integro-differential equation for the current at any time t. This is done by taking the Laplace transform to get

$$L(sI(s) - i(0+)) + RI(s) + I(s)/sC = 0$$

or using $i(0+) = V/R$, we get

$$I(s) = \frac{\frac{LV}{R}}{sL + R + 1/sC}$$

Inverse transforming this function gives the current. We leave the details to the student.

14. Basic equations of two port network. Let (v_1, i_1) denote the input voltage and current into a two port network and (v_2, i_2) the output voltage and current. The convention is usually chosen so that the currents i_1 and i_2 enter into the port. Assuming that the network consists of memoryless elements (no capacitors or inductors), one can choose two independent variables out of the four (v_1, i_1, v_2, i_2) and express the remaining two in terms of these variables. For example, if (v_1, v_2) are chosen as the independent variables, then the two port equations are specified as $i_1 = f(v_1, v_2), i_2 = g(v_1, v_2)$. For example in the Ebers-Moll model for a *pnp* transistor, v_{EB}, v_{CB} are chosen as the independent variables and i_E, i_C are expressed as

$$i_E = I_{E0}(exp(v_{EB}/V_T) - 1) - \alpha_E I_{C0}(exp(v_{CB}/V_T) - 1)$$

$$i_C = I_{C0}(exp(v_{CB}/V_T) - 1) - \alpha_C I_{E0}(exp(v_{EB}/V_T) - 1)$$

If the network has memory but is linear and time invariant, then we work with the Laplace transforms of the currents and voltages, i.e., $V_i(s), I_i(s), i = 1, 2$. The two port Z parameters are defined by

$$V_1(s) = Z_{11}(s)I_1(s) + Z_{12}(s)I_2(s)$$

$$V_2(s) = Z_{21}(s)I_1(s) + Z_{22}(s)I_2(s)$$

or in matrix notation,

$$\begin{pmatrix} V_1(s) \\ V_2(s) \end{pmatrix} = \begin{pmatrix} Z_{11}(s) & Z_{12}(s) \\ Z_{21}(s) & Z_{22})(s) \end{pmatrix} \begin{pmatrix} I_1(s) \\ I_2(s) \end{pmatrix}$$

If $z_{ij}(t)$ denotes the inverse Laplace transform of $Z_{ij}(s)$, i.e., $Z_{ij}(s) = \int_0^\infty z_{ij}(t)exp(-st)dt$, then the above equations are equivalent to the following time domain equations(Causality is being assumed):

$$v_a(t) = \int_0^t z_{a1}(\tau)i_1(t - \tau)d\tau + \int_0^t z_{a2}(\tau)i_2(t - \tau)d\tau, a = 1, 2$$

The inverse of the Z matrix is the Y matrix:

$$Y(s) = \begin{pmatrix} Y_{11}(s) & Y_{12}(s) \\ Y_{21}(s) & Y_{22}(s) \end{pmatrix}$$

$$= Z(s)^{-1} = \begin{pmatrix} Z_{11}(s) & Z_{12}(s) \\ Z_{21}(s) & Z_{22}(s) \end{pmatrix}^{-1}$$

so that

$$Y_{11}(s) = \frac{Z_{22}(s)}{Z_{11}(s)Z_{22}(s) - Z_{12}(s)Z_{21}(s)}$$

$$Y_{22}(s) = \frac{Z_{11}(s)}{Z_{11}(s)Z_{22}(s) - Z_{12}(s)Z_{21}(s)}$$

$$Y_{12}(s) = -\frac{Z_{12}(s)}{Z_{11}(s)Z_{22}(s) - Z_{12}(s)Z_{21}(s)}$$

$$Y_{21}(s) = -\frac{Z_{21}(s)}{Z_{11}(s)Z_{22}(s) - Z_{12}(s)Z_{21}(s)}$$

We can equivalently choose (I_1, V_2) as our independent variables and (V_1, I_2) as the dependent variables. Then,

$$V_2 = Z_{21}I_1 + Z_{22}I_2$$

is rearranged as

$$I_2 = -\frac{Z_{21}}{Z_{22}}I_1 + \frac{V_2}{Z_{22}}$$

and

$$V_1 = Z_{11}I_1 + Z_{12}I_2$$

can be expressed as

$$V_1 = Z_{11}I_1 + Z_{12}\left(\frac{V_2}{Z_{22}} - \frac{Z_{21}}{Z_{22}}I_1\right)$$

$$= \left(Z_{11} - \frac{Z_{12}Z_{21}}{Z_{22}}\right)I_1 + \frac{Z_{12}}{Z_{22}}V_2$$

or in matrix notation,

$$\begin{pmatrix} I_2 \\ V_2 \end{pmatrix} = \begin{pmatrix} -\frac{Z_{21}}{Z_{22}} & \frac{1}{Z_{22}} \\ Z_{11} - \frac{Z_{12}Z_{21}}{Z_{22}} & \frac{Z_{12}}{Z_{22}} \end{pmatrix} \begin{pmatrix} I_1 \\ V_2 \end{pmatrix}$$

This matrix is called the H-parameter matrix. If instead, we choose (V_1, I_2) as our independent variables and (I_1, V_2) as the dependent variables, we get the G parameter matrix. We leave it as an exercise to work out the details.

Study project 1

A basic course in circuit theory can be covered in one seminar. The important aspects to be discussed include graph theory, the current voltage equations for a resistor, capacitor, inductor, formulation of the KCL and KVL equations, the graph theoretic formulation should explain how the KCL equations are written down using the incidence and f-cutset matrix and how the KVL equations are written down using the f-loop and mesh matrix. The definition of a graph as a set of nodes or vertices numbered $1, 2, ..., v$ connected to each other by edges. The notion of a planar graph should follow this. How one counts the number of loops or meshes f in a planar graph. If v is the number of vertices, e the number of edges and f the number of loops, then Euler's relation $v - e + f = 1$ can be derived by an induction argument along the following lines. Clearly this equation is true when the graph has just one edge. For then it has two vertices and for such a graph, $v = 2, e = 1, f = 0$ which satisfy the equation $v - e + f = 1$. We want to prove that the equation holds for a planar graph having v vertices e edges and f loops when $e > 1$. Delete one edge. Then the resulting graph will have $f - 1$ loops, $e - 1$ edges and v vertices. By the induction hypothesis, $v - (e - 1) + (f - 1) = 1$, that is $v - e + f = 1$ and this proves that the relation holds for the given graph, i.e., Euler's identity follows by induction. The statement of the KCL equation is that the algebraic sum of the currents leaving any node equals zero. In other words, if i is the node number and edges $e_{k_1}, ..., e_{k_r}$ are leaving this node, i.e. they are directed away from the ith node and i_{k_j} is the current along e_{k_j}, then $\sum_{j=1}^{r} i_{k_j} = 0$. In this context, the notion of a directed graph must be introduced. In a directed graph, each edge is assigned a direction. For example, if edges e_1, e_2, e_3, e_4 are the only edges that meet at node 1 and e_1, e_3 are directed away from this node while e_2 and e_4 are directed towards this node, then the KCL equation for node 1 will read $i_1 - i_2 + i_3 - i_4 = 0$. More generally, if at node i, $e_{k1}, ..., e_{kr}$ are the only edges that meet, and edges $e_{ka}, a \in E$ are directed away from this node while edges $e_{kb}, b \in E^c$ are directed towards the node, where E is a subset of $\{1, 2, ..., r\}$, then the KCL reads

$$\sum_{a \in E} i_{ka} - \sum_{k \in E^c} i_{kb} = 0$$

Given a circuit, we can draw a directed graph that represents the topology of the circuit and write down its incidence matrix. The matrix consists of v rows numbered $1, 2, ..., v$ where v is the number of vertices or nodes and e columns numbered $1, 2, ..., e$, where e is the number of edges. If the kth edge is directed from node i to node j, then the $(i, k)^{th}$ entry of the matrix is one and the $(j, k)^{th}$ entry of the matrix is minus one. All the other entries in the kth column are zero. This means that each column of the incidence matrix has exactly one and one minus one. Thus, the sum of the rows of the incidence matrix is zero. If we define an edge current vector $\mathbf{I}_e = [i_1, ..., i_e]^T$ such that i_k is the current flowing along the kth edge along the direction of the edge, then it is easy to see that $\mathbf{AI}_e = 0$ where $\mathbf{A}$ is the incidence matrix. In fact, take the ith equation here. The ith row of $\mathbf{A}$ has one at the positions $(i, k_r), r = 1, 2, ..., p$ and minus one at the positions $(i, m_r), r = 1, 2, ..., q$ and zeros at the other positions where edges $k_1, ..., k_p$ are incident on node i and directed away from it while edges $m_1, ..., m_q$ are incident on node i and directed towards it. Thus, the ith equation of the system $\mathbf{AI}_e = 0$ reads $\sum_{r=1}^{p} i_{k_r} - \sum_{r=1}^{q} i_{m_r} = 0$ which is precisely the KCL for node i.

Study project 2

Consider a pn junction diode. The p side has holes of concentration N_A where N_A is the concentration of the acceptor impurities and the n side has electrons of concentration N_D where N_D is the concentration of donor impurities. Each donor atom releases a free electron that contributes to the sea of electrons on the n side while each acceptor atom absorbs an electron from the medium that results in a hole contributing to the sea of holes on the p side. The holes from the p side diffuse across the barrier into the sea of electrons on the n side while the electrons from the n side diffuse across the barrier into the sea of holes on the p side. As a result, on the p side, we have a negative charge density of $-eN_A$ $coulombs/m^3$ while on the n side, we have a positive charge density of $+eN_D$ $coulombs/m^3$. These is called the space charge. Assume that the width of the space charge region on the p side equals d_p while the width of the space charge region on the n side equals d_n. Let V be the potential in the space charge region. Thus the charge density can be written as $\rho(x) = -eN_A, -d_p < x < 0, \rho(x) = eN_D, 0 < x < d_n$ and the boundary condition is that the electric field vanishes at $x = -d_p, d_n$ and of course charge neutrality requires that $d_p N_A = d_n N_D$. By solving Poisson's equation $V''(x) = -\rho(x)/\epsilon$ with the boundary condition, determine the shape of the potential and the electric field in the space charge region.

Study project 3

Perturbation theoretic analysis of a diode resistor capacitance circuit with an arbitrary voltage input. The diode D is connected in series to the parallel combination of a resistor R and a capacitance C. An input voltage $V_i(t)$ is applied to the input of this circuit. Let $V_o(t)$ be the output voltage taken across the capacitor. Then the current through the capacitor equals $i_C = C\frac{dV_o}{dt}$, the current through the resistor equals $i_R = V_o/R$ and the current through the diode equals $i_D = f(V_i - V_o)$ where $f(x) = I_o(exp(x/V_T) - 1)$ is the diode voltage current characteristic. The current through the diode equals the sum of the currents through the resistor and the capacitor: $i_D = i_C + i_R$. This gives

$$f(V_i - V_o) = C\frac{dV_o}{dt} + \frac{V_o}{R}$$

We solve this equation using perturbation theory for nonlinear differential equations.

$$f(x) = I_o x/V_T + \psi(x)$$

where

$$\psi(x) = I_o(exp(x/V_T) - 1 - x/V_T) = I_0 \sum_{n=2}^{\infty} (x/V_T)^n/n!$$

is the nonlinear component of the diode voltage-current characteristics. The above differential equation can be written as

$$I_o(V_i - V_o)/V_T + \psi(V_i - V_o) = C\frac{dV_o}{dt} + \frac{V_o}{R}$$

Write

$$V_o = V_{o,0} + V_{o,1} + V_{o,2} + ...$$

where $V_{o,n}$ is of nth order smallness. The nonlinear function ψ is assumed to be of the first order of smallness. To keep track of the degree of smallness of each term, we introduce a parameter ϵ

so that the above series for V_o becomes

$$V_o = \sum_{n=0}^{\infty} \epsilon^n V_{o,n}$$

and the differential equation itself reads

$$I_o(V_i - V_o)/V_T + \epsilon\psi(V_i - V_o) = C\frac{dV_o}{dt} + \frac{V_o}{R}$$

The zeroth order terms give

$$I_oV_i/V_T = V_{o,0}(1/R + I_o/V_T) + C\frac{dV_{o,0}}{dt}$$

This equation represents a capacitor C in parallel with a resistance $R' = 1/(1/R+I_o/V_T)$ connected to a current source giving current I_oV_i/V_T. This circuit can equivalently be realized by connecting the input voltage source V_i to a resistance $r = V_T/I_o$ whose other end is connected to the parallel combination of the resistance R and capacitance C. This follows in view of the fact that the above equation can equivalently be expressed as

$$I_o(V_i - V_{o,0})/V_T = V_{o,0}/R + C\frac{dV_{o,0}}{dt}$$

The first order terms give

$$(1/R + I_o/V_T)V_{o,1} + C\frac{dV_{o,1}}{dt} = \psi(V_i - V_{o,0})$$

The circuit for this output can be constructed in the following way. Take the input V_i and the output $V_{o,0}$ of the zeroth order section. These two inputs are fed to a nonlinear element with two inputs that produces an output current $\psi(x)$ where x is the difference of the inputs. This current is fed into the parallel combination of a resistance $R' = 1/(1/R + I_o/V_T)$ and a capacitance C. The voltage across the capacitance measured equals $V_{o,1}$.

Study project 4

This project deals with the analysis of basic circuits and equipment used in the experiments of a basic electronics laboratory course in the second year syllabus for the undergraduate curriculum of electronics engineering. The basic experiments studied are (1) Operation of the cathode ray oscilloscope, (2) Design of differentiator and integrator using RC circuits, (3) Design of clipping and clamping circuits using diodes, resistors, capacitors and dc voltage sources, (4) Design of half-wave and full-wave rectification circuits and evaluation of the ripple factor and output rms voltage (5) Design of BJT transistor amplifier and its analysis.

(1) The operation of the CRO can be explained in terms of the presence of two horizontal plates and two vertical plates. A ramp voltage is applied to the horizontal plates causing a

transverse electric field of intensity increasing proportional to time to develop between the plates. The electron beam is thus deflected from the left to the right plate and back with the deflection increasing proportional to time. Now, if a voltage waveform is applied to the vertical plates, we get a scaled replica of the waveform on the fluorescent screen. The x deflection of the beam can be expressed as $x(t) = at$ for $0 < t < T$, where T is the period of the time base and the y deflection as $y(t) = bv(t)$ where $v(t)$ is the voltage waveform applied to the vertical plates. Thus, the graph of the wave form on the screen is $y = bv(x/a)$ which is a time and amplitude scaled version of the waveform $v(t)$. If $v(t)$ is periodic with period τ, then in order to obtain a stable waveform on the screen, one requires $T = n\tau$ where n is a positive integer. Explain this. Apart from using the CRO to plot waveforms, it can be used to constructs graphs of one waveform versus another. This is useful for example in problems involving determination of the input-output characteristic of a device in which we need a graph of the output versus the input signal. For example, if the input is $x(t) = A.sin(\omega t)$ and the output is $y(t) = B.sin(\omega t + \phi)$, then applying $x(t)$ to the x plates and $y(t)$ to the y plates and operating the CRO in the xy mode yields an ellipse. When $x = 0$, we have $t = n\pi/\omega$ and then $y = B.(-1)^n sin\phi$. Thus by noting the intercept of the graph on the y axis, the phase difference ϕ between the input and output waveforms can be determined. The use of the voltage and time sensitivity knob is to be explained.

Chapter 7

Optics

7.1 Monochromatic light sources and Fourier decomposition

Light can be used in place of electricity to transmit messages from one point to another. The primary advantage is speed. All the processes of digital and analog communication can be translated to the optical domain. The signal sources in an optical communication system are lasers which emit monochromatic light, i.e., light of a single frequency. Actually, no source can generate monochromatic light. A Fourier decomposition of the light signal coming out of a laser will have the form

$$f(t) = \int_{\omega_0 - \Delta}^{\omega_0 + \Delta} F(\omega) exp(i\omega t) d\omega$$

i.e., localized over a small band centred around the primary frequency ω_0. The energy spectrum $|F(\omega)|^2$ is therefore zero for $|\omega - \omega_0| > \Delta$. Usually, one models the spectrum of laser light as having the Gaussian shape, i.e. $F(\omega) = exp(i\phi(\omega)).exp(-(\omega - \omega_0)^2/4\sigma^2)$. This means that the light contains all the frequencies but the power in the higher frequencies is negligible for frequencies more than ten times the square root of the dispersion.

7.2 Cylindrical optical fibers and the cylindrical wave equation

Optical fibres can be viewed as cylindrical waveguides made of glass which could have a refractive index that varies with the distance from the fibre axis. The basic equations that govern the propagation of light through such a fibre are Maxwell's equations expressed in the cylindrical coordinate system. The wave equation in the cylindrical system is obtained by expressing the Laplacian in cylindrical coordinates. The wave equation that describes the propagation of light inside a cylindrical fibre is

$$\frac{1}{\rho}\frac{\partial}{\partial\rho}\rho\frac{\partial\psi}{\partial\rho} + \frac{1}{\rho^2}\frac{\partial^2\psi}{\partial\phi^2} + \frac{\partial^2\psi}{\partial z^2} - \mu\epsilon\frac{\partial^2\psi}{\partial t^2} = 0$$

For a definite frequency ω, this equation is replaced by the Helmholtz equation in cylindrical coordinates:

$$\frac{1}{\rho}\frac{\partial}{\partial\rho}\rho\frac{\partial\psi}{\partial\rho} + \frac{1}{\rho^2}\frac{\partial^2\psi}{\partial\phi^2} + \frac{\partial^2\psi}{\partial z^2} + \omega^2\mu\epsilon\psi = 0$$

In the first expression, ψ is a function of ρ, ϕ, z, t while in the second, it is just a function of ρ, ϕ, z.

7.3   Geometric optics and the Eikonal equation

The fundamental equation of geometric optics is the eikonal equation which is obtained as follows. In the wave equation, we set $\psi(\mathbf{r}, t) = A(\mathbf{r}, t)exp(iS(\mathbf{r}, t))$ where A is the amplitude and S the phase. First we calculate the Laplacian of ψ:

$$\nabla\psi = (\nabla A + iA\nabla S)exp(iS)$$

$$\nabla^2\psi = (\nabla^2 A + 2i\nabla A.\nabla S + iA\nabla^2 S - A|\nabla S|^2)exp(iS)$$

$$\frac{\partial\psi}{\partial t} = (\frac{\partial A}{\partial t} + iA\frac{\partial S}{\partial t})exp(iS)$$

$$\frac{\partial^2\psi}{\partial t^2} = (\frac{\partial^2 A}{\partial t^2} + 2i\frac{\partial A}{\partial t}\frac{\partial S}{\partial t} + iA\frac{\partial^2 S}{\partial t^2} - A(\frac{\partial S}{\partial t})^2)exp(iS)$$

Plugging these into the wave equation gives

$$\nabla^2 A + 2i\nabla A.\nabla S + iA\nabla^2 S - A|\nabla S|^2 - \frac{1}{c^2}(\frac{\partial^2 A}{\partial t^2} + 2i\frac{\partial A}{\partial t}\frac{\partial S}{\partial t} + iA\frac{\partial^2 S}{\partial t^2} - A(\frac{\partial S}{\partial t})^2) = 0$$

Now equate the real and imaginary parts to get

$$\nabla^2 A - A|\nabla S|^2 - \frac{1}{c^2}\frac{\partial^2 A}{\partial t^2} + \frac{1}{c^2}A(\frac{\partial S}{\partial t})^2 = 0$$

$$\nabla A.\nabla S + A\nabla^2 S - \frac{2}{c^2}\frac{\partial A}{\partial t}\frac{\partial S}{\partial t} - \frac{1}{c^2}A\frac{\partial^2 S}{\partial t^2} = 0$$

These can be regarded as the basic equations for optics. $|\nabla S|$ is of the order of the reciprocal of the wavelength, i.e., the wave-vector. If the wavelength is much smaller than the characteristic dimensions of the system over which light propagation takes place, then $\nabla^2 A$ can be neglected as compared to $A|\nabla S|^2$. The quantity $\frac{\partial S}{\partial t}$ is of the order of the frequency of the light. $\frac{\partial^2 A}{\partial t^2}$ is of the order of c^2/d^2 where d is the characteristic dimension of the system. This is much smaller than the frequency square, i.e. ω^2 when the wave length is much smaller than the characteristic dimension of the system. This is precisely the condition of geometric optics. Under this condition, the first equation approximates to

$$|\nabla S|^2 - \frac{1}{c^2}(\frac{\partial S}{\partial t})^2 = 0$$

and this is the fundamental equation of geometric optics. For waves of a definite frequency ω, we have $S(\mathbf{r}, t) = S_0(\mathbf{r}) - \omega t$ and this equation reduces to the "eikonal equation":

$$|\nabla S|^2 = k^2$$

where $k = \omega/c$ is the wavenumber. All the properties of light in the geometric optica approximation can be derived from this equation. (Landau and Lifshitz).

7.4 Total internal reflection

Consider a ray of light propagating inside a medium having refractive index n_1. Let the refractive index of the medium outside be n_2. Let θ_1 denote the angle of incidence of the incident beam with the normal to the interface and θ_2 the angle of refraction. Then, by Snell's law, $n_1 sin\theta_1 = n_2 sin\theta_2$. Suppose $n_1 > n_2$, then when $\theta_1 > sin^{-1}(n_2/n_1)$, we get $sin\theta_2 > 1$ which means that the refracted beam does not exist. What happens in such a situation is that the incident beam suffers internal reflection in the medium n_1. This is called the phenomenon of total internal reflection and forms the basis of all optical fibers. The beam inside the fibre gets reflected successively at the walls of the fibre owing to its angle being greater than the critical angle given above.

7.5 Study Project

Study project 1

Deals with equations governing wave propagation in a cylindrical fibre. Assume that the fibre has permitivity ϵ and is of radius R. Let the z axis be along the fibre central axis. The z dependence of the fields is $exp(-\gamma z)$. This means that we can write the complex electric and magnetic field phasors as

$$\mathbf{E}(\rho, \phi, z) = (E_\rho(\rho, \phi)\hat{\rho} + E_\phi(\rho, \phi)\hat{\phi} + E_z(\rho, \phi)\hat{z})exp(-\gamma z)$$

$$\mathbf{H}(\rho, \phi, z) = (H_\rho(\rho, \phi)\hat{\rho} + H_\phi(\rho, \phi)\hat{\phi} + H_z(\rho, \phi)\hat{z})exp(-\gamma z)$$

The frequency ω of the fields is assumed to be fixed. The frequency is fixed because the source of light is a monochromatic laser. The Maxwell curl equations in cylindrical coordinates have to be written down to express the $\hat{\rho}$ and $\hat{\phi}$ components of the electric fields in terms of the $\hat{z}$ components. The details of these equations are available

Study project 2

Write down the Maxwell equations in cylindrical coordinates assuming that the permeability μ is constant but the permitivity ϵ is a function of ρ, the radial distance from the z axis,i.e., $\epsilon(\rho)$. We are assuming a definite frequency ω, so that the fundamental equations are

$$\nabla \times \mathbf{E} = -i\omega \mathbf{B}$$

$$\nabla \times \mathbf{B} = i\omega\mu\epsilon(\rho)\mathbf{E}$$

The other two Maxwell equations $\nabla.(\epsilon\mathbf{E}) = 0$ and $\nabla.\mathbf{B} = 0$ can be derived from these two. The first step is to express the curl operation in the cylindrical system:

$$\nabla \times \mathbf{F} = det \begin{pmatrix} \hat{\rho}/\rho & \hat{\phi} & \hat{z}/\rho \\ \frac{\partial}{\partial\rho} & \frac{\partial}{\partial\phi} & \frac{\partial}{\partial z} \\ F_\rho & \rho F_\phi & F_z \end{pmatrix}$$

When expanded, this reads

$$\nabla \times \mathbf{F} = \frac{\hat{\rho}}{\rho}\left(\frac{\partial F_z}{\partial\phi} - \frac{\partial \rho F_\phi}{\partial z}\right) + \hat{\phi}\left(\frac{\partial F_\rho}{\partial z} - \frac{\partial F_z}{\partial\rho}\right) + \frac{\hat{z}}{\rho}\left(\frac{\partial}{\partial\rho}(\rho F_\phi) - \frac{\partial F_\rho}{\partial\phi}\right)$$

Chapter 8

Solid State Physics

This branch of science deals with properties of insulators or dielectrics, conductors, semiconductors and superconductors derivable from quantum mechanical models. The starting point for the analysis of such materials is the crystal which consists of atoms located at the sites of a periodic lattice with electrons moving about in the potential generated by the nuclei of the atoms.

8.1 Quantum Motion in a periodic potential : The Bloch wave functions

A major part of solid state physics therefore deals with the solution of the Schrodinger equation in a periodic potential. In the simplest case, we consider a single electron moving about in a periodic potential $V(x)$ in one dimension. The period of the potential is L, i.e. $V(x+L) = V(x)$. There is a theorem due to F.Bloch which states that the solutions to the corresponding stationary Schrodinger equation for such a periodic potential must have the form $\psi_k(x) = u_k(x)exp(ikx)$, where $u_k(x+L) = u(x)$. The proof is a consequence of the following argument. If $\psi(x)$ is a solution of the stationary Schrodinger equation corresponding to the periodic potential $V(x)$ and energy eigenvalue E, then so is $\psi(x+L)$ as can be seen from the equation

$$\psi''(x+L) + \frac{8\pi^2 m}{h^2}(E - V(x))\psi(x+L) = \psi''(x+L) + \frac{8\pi^2 m}{h^2}(E - V(x+L))\psi(x+L) = 0$$

Thus, $\psi(x+L) = C\psi(x)$ where C is a complex constant of unit magnitude. We consider periodic boundary conditions for this, i.e. assume that there are N atoms in the crystal which means that $\psi(x) = \psi(x+NL)$. This leads to the equation $C^N = 1$ or $C = exp(2\pi i s/N)$ where s is some integer in $\{0, 1, ..., N-1\}$. Now define $u(x) = exp(-2\pi i s x/L)\psi(x)$. Then,

$$u(x+L) = exp(-2\pi i s(x+L)/L)\psi(x+L) = exp(-2\pi i(x+L)/L)C\psi(x) = exp(-2\pi i s x/L)\psi(x) = u(x)$$

implying that $u(x)$ is a periodic function of x with period L proving the theorem. The argument can be extended to three dimensions. Substituting for the Bloch wave function $\psi(x) = u_k(x)exp(ikx)$ into the Schrodinger equation, we get

$$\frac{d^2}{dx^2}(u_k(x)exp(ikx)) + \frac{8\pi^2 m}{h^2}(E - V(x))u_k(x)exp(ikx) = 0$$

which gives

$$u_k''(x) - k^2 u_k(x) + 2iku_k'(x) + \frac{8\pi^2 m}{h^2}(E - V(x))u_k(x) = 0$$

or

$$u_k''(x) + 2iku_k'(x) + ((\frac{8\pi^2 mE}{h^2} - k^2)u_k(x) - \frac{8\pi^2 m}{h^2}V(x)u_k(x)) = 0$$

Remark: If $u_k(x)$ is a periodic function of x with period L, then $\psi(x) = exp(ikx)u_k(x)$ is also an eigenfunction of the translation operator T_L with eigenvalue $exp(ikL)$. This condition is to be imposed for a crystal since the Hamiltonian of an electron moving in a periodic potential commutes with translations by the length L.

8.2   A model for electrons moving in a crystal lattice, notion of effective mass

Assume that the lattice sites are numbered as $i, i = 0, \pm 1, \pm 2, ...$ An electron can make a transition from the ith site to the $i + 1^{'th}$ site or to the $i - 1^{th}$ site. Or it may not make a transition at all. Thus, the only nonvanishing matrix elements of the Hamiltonian are $< i|H|i + 1 >= a, < i - 1|H|i + 1 >= a, < i|H|i >= b$. Symmetry is being assumed, i.e. the probability amplitude for a left transition equals the probability amplitude for a right transition. The Schrodinger equation reads

$$\frac{ih}{2\pi}\frac{dc_n(t)}{dt} = a(c_{n+1}(t) + ac_{n-1}(t)) + bc_n(t)$$

Here, $c_i(t)$ is the amplitude for the electron being located at the lattice site i at time t. We want to look at a solution corresponding to the propagation of waves in the crystal, so we try a solution of the form

$$c_k(t) = exp(-i(\omega t - Knd))$$

where d is the distance between two neighbouring lattice sites. ω determines the frequency of oscillation and K the wave number, i.e. 2π times the reciprocal of the wavelength. Plugging this into the differential equation gives

$$h\omega/2\pi = a(exp(iKd) + exp(-iKd)) + b = 2a.cos(Kd) + b$$

Since $E = h\omega/2\pi$ is the energy according the quantum postulates, we obtain the following dispersion relation between energy and wave number: $E(K) = 2a.cos(Kd)+b$. For small spacings, we can approximate the cosine function as $cos(Kd) \approx 1 - K^2 d^2/2$. This gives $E(K) \approx 2a(1 - K^2 d^2/2) + b$. The constant term represents the zero point energy and can be ignored. The effective dispersion relation is therefore $E(K) \approx -ad^2 K^2$ and hence a must be negative. Equating this to $h^2 K^2/8\pi^2 m_{eff}$, we obtain the following formula for the effective mass for the electron moving in the crystal lattice: $m_{eff} = -h^2/8\pi^2 a d^2$.

8.3 Motion in a periodic potential, Fourier domain description

We denote a typical element of the reciprocal lattice vector by the symbol $\mathbf{G}$. Then, for an electron moving in the periodic potential of the lattice, the wave function can be expanded as a Fourier series

$$\psi(\mathbf{r}) = \sum_{\mathbf{K}} C(\mathbf{K}) exp(i\mathbf{K}.\mathbf{r})$$

where the sum is over all $\mathbf{K}$ having the form $\mathbf{k} + \mathbf{G}$ where $\mathbf{k}$ is a fixed vector (appearing in the Bloch wave function) and $\mathbf{G}$ varies over all the reciprocal lattice vectors. The periodic potential $V(\mathbf{r})$ of the crystal can also be expanded as a Fourier series:

$$V(\mathbf{r}) = \sum_{\mathbf{G}} V(\mathbf{G}) exp(i\mathbf{G}.\mathbf{r})$$

Plugging the wave function and the potential into the Schrodinger equation

$$-\frac{h^2}{8\pi^2 m}\nabla^2\psi + V\psi = E\psi$$

results in the following algebraic equation for the coefficients $C(\mathbf{K})$:

$$\frac{h^2 K^2}{8\pi^2 m}\sum_{\mathbf{K}} C(\mathbf{K}) exp(i\mathbf{K}.\mathbf{r}) + \sum V(\mathbf{G}) C(\mathbf{K}) exp(i(\mathbf{K}+\mathbf{G}).\mathbf{r})) = E\sum_{\mathbf{K}} C(\mathbf{K}) exp(i\mathbf{K}.\mathbf{r})$$

The second term here can be rearranged to give

$$\frac{h^2 K^2}{8\pi^2 m}\sum_{\mathbf{K}} C(\mathbf{K}) exp(i\mathbf{K}.\mathbf{r}) + \sum_{\mathbf{G},\mathbf{K}} V(\mathbf{G}) C(\mathbf{K}-\mathbf{G}) exp(i\mathbf{K}.\mathbf{r}) = E\sum_{\mathbf{K}} C(\mathbf{K}) exp(i\mathbf{K}.\mathbf{r})$$

Equating coefficients of $exp(i\mathbf{K}.\mathbf{r})$ on both sides gives

$$\frac{h^2 K^2}{8\pi^2 m} C(\mathbf{K}) + \sum_{\mathbf{G}} V(\mathbf{G}) C(\mathbf{K}-\mathbf{G}) = E C(\mathbf{K})$$

This is an algebraic equation for the coefficients $\{C(\mathbf{K})\}$ that determine the wave function. The energy bands of a solid are computed from this equation.

8.4   Reciprocal lattice

Let $\mathbf{a}, \mathbf{b}, \mathbf{c}$ be the basic translation vectors of a crystal lattice, i.e. the symmetry properties of the crystal remain invariant under translations by $k\mathbf{a} + m\mathbf{b} + n\mathbf{c}$ where k, m, n are integers. We define the vectors

$$\mathbf{A} = 2\pi \frac{\mathbf{b} \times \mathbf{c}}{\mathbf{a}.\mathbf{b} \times \mathbf{c}}$$

$$\mathbf{B} = 2\pi \frac{\mathbf{c} \times \mathbf{a}}{\mathbf{a}.\mathbf{b} \times \mathbf{c}}$$

$$\mathbf{C} = 2\pi \frac{\mathbf{a} \times \mathbf{b}}{\mathbf{a}.\mathbf{b} \times \mathbf{c}}$$

These are known as the basic vectors of the reciprocal lattice. They satisfy the following orthogonality relations with respective to the basis vectors of the lattice:

$$\mathbf{a}.\mathbf{A} = 2\pi, \mathbf{a}.\mathbf{B} = \mathbf{a}.\mathbf{C} = 0$$

$$\mathbf{b}.\mathbf{A} = 0, \mathbf{b}.\mathbf{B} = 2\pi, \mathbf{b}.\mathbf{C} = 0$$

$$\mathbf{c}.\mathbf{A} = \mathbf{c}.\mathbf{B} = 0, \mathbf{c}.\mathbf{C} = 2\pi$$

It can easily be seen from these properties, that any function $f(\mathbf{r})$ defined on the space of the lattice that has period with periods $\mathbf{a}, \mathbf{b}, \mathbf{c}$ can be expanded as a Fourier series

$$f(\mathbf{r}) = \sum_{k,m,n=-\infty}^{\infty} c(k, m, n) exp(i\mathbf{r}.(k\mathbf{A} + m\mathbf{B} + n\mathbf{C}))$$

This follows from the 2π periodicity of the complex exponential function and the following relations deduced from the above orthogonality relations:

$$(\mathbf{r} + p\mathbf{a} + q\mathbf{b} + t\mathbf{c}).(k\mathbf{A} + m\mathbf{B} + n\mathbf{C}) = \mathbf{r}.(k\mathbf{A} + m\mathbf{B} + n\mathbf{C}) + 2\pi(pk + qm + tn)$$

8.5   Diffraction

Consider a plane wave with wave vector $\mathbf{k}$ incident upon a crystal. The phasor associated with such a wave is $exp(i\mathbf{k}.\mathbf{r})$. At a given time, this factor determines the relative phase difference between the wave field amplitude at different points of space. We assume that this wave is incident upon two points in the crystal separated by the vector $\mathbf{r}$. The path difference between the two rays of the wave at these two respective points is given by $\mathbf{k}.\mathbf{r}$. Let $\mathbf{k}'$ denote the wave vector of the scattered wave field. The path difference between the two scattered rays coming from these two points is $-\mathbf{k'}.\mathbf{r}$. The overall path difference for the two rays during the incidence and scattering process is therefore the sum of these two path differences, i.e. $(\mathbf{k} - \mathbf{k'}).\mathbf{r}$. We choose our first point as the origin of the coordinate system so that the second point $\mathbf{r}$ is an arbitrary point in the crystal. The complex amplitude of the scattered wave is proportional to $exp(i(\mathbf{k} - \mathbf{k'}).\mathbf{r})$. The amplitude is also proportional to the density $n(\mathbf{r})$ of electrons in the crystal at the point $\mathbf{r}$, for higher the density, more the number of scattering centers. From superposition, it then follows that the total amplitude of the scattered wave field from the entire crystal is proportional to the Fourier integral

$$A = \int_{crystal} n(\mathbf{r})exp(i(\mathbf{k} - \mathbf{k'}).\mathbf{r})dV(\mathbf{r})$$

Now, for a crystal $n(\mathbf{r})$ is periodic with periods defined by the translation vectors. Assuming that $\mathbf{G}$ stands for an integer linear combination of the reciprocal lattice vectors, $n(\mathbf{r})$ can be expanded as a Fourier series

$$n(\mathbf{r}) = \sum_{\mathbf{G}} n(\mathbf{G})exp(i\mathbf{G}.\mathbf{r})$$

This gives the total amplitude of the scattered beam as

$$A = \sum_{\mathbf{G}} n(\mathbf{G}) \int dV exp(i(\mathbf{k} - \mathbf{k'} + \mathbf{G}).\mathbf{r}))$$

When $\mathbf{k} - \mathbf{k'} = -\mathbf{G}_0$ for some linear combination $\mathbf{G}_0$ of the reciprocal lattice vectors, then that particular term in the integral contributes $n(\mathbf{G}_0)$. The other terms are oscillatory and integrate to zero. Thus, the above integral evaluates to $A = n(\mathbf{G}_0)V$. When $\mathbf{k} - \mathbf{k'}$ is not a linear combination of the reciprocal lattice vectors, the integral consists of purely oscillatory terms and is negligible. In elastic scattering the energy of the photon is conserved and since by Planck's law, the photon energy is proportional to the magnitude of the wave number, the condition for scattering is that the following relationship hold between the incident wave vectors $\mathbf{k}$: $|(\mathbf{k} + \mathbf{G}_0)| = |\mathbf{k}|$, where $\mathbf{G}_0$ is some linear combination of the reciprocal lattice vectors. This condition is the same as $|\mathbf{G}_0|^2 + 2\mathbf{k}.\mathbf{G}_0 = 0$. Equivalently, replacing $\mathbf{G}_0$ by $-\mathbf{G}_0$, we get the condition as $2\mathbf{k}.\mathbf{G}_0 = |\mathbf{G}_0|^2$.

8.6 Applications of group theory to the analysis of crystal structure

Most of the crystals exhibit certain symmetry properties. For example a cubic crystal with the same kind of atom at its eight vertices and a different atom at its centre exhibits symmetry with respect to rotation by ninety degrees about an axis passing through the centers of any two opposite faces. It also exhibits symmetry with respect to reflection about a vertical plane or a horizontal plane passing through its centre. Symmetry properties of crystals are best described through symmetry groups and their representations. A group is simply a collection G of objects with a binary operation called composition and an inverse operation. The binary operation of composition is associative. If a, b are two elements of the group, then their composition is denoted by $a.b$. The group must be closed under this operation, i.e. $a.b \in G$ for all $a, b \in G$. By associativity, we mean that $a.(b, c) = (a.b).c$ for all $a, b, c \in G$. A better notation for composition is $\phi(a, b)$ as it conveys the meaning more clearly. The group need not be commutative, i.e., one need not have $b.a = a.b$. There is a unique identity element $e \in G$ with the property that $a.e = e.a = a$ for all $a \in G$. For each $a \in G$, there is a unique inverse element $a^{-1} \in G$ with the property that $a.a^{-1} = a^{-1}.a = e$. A typical example of a group is the following. Consider three points a, b, c which appear as vertices of an equilateral triangle. We can permute these elements and we know that there are exactly $6 = 3!$ permutations. These permutations can be written down as

$$e = \begin{pmatrix} 1 & 2 & 3 \\ 1 & 2 & 3 \end{pmatrix}$$

$$\sigma_1 = \begin{pmatrix} 1 & 2 & 3 \\ 1 & 3 & 2 \end{pmatrix}$$

$$\sigma_2 = \begin{pmatrix} 1 & 2 & 3 \\ 2 & 1 & 3 \end{pmatrix}$$

$$\sigma_3 = \begin{pmatrix} 1 & 2 & 3 \\ 2 & 3 & 1 \end{pmatrix}$$

$$\sigma_4 = \begin{pmatrix} 1 & 2 & 3 \\ 3 & 1 & 2 \end{pmatrix}$$

$$\sigma_5 = \begin{pmatrix} 1 & 2 & 3 \\ 3 & 2 & 1 \end{pmatrix}$$

In general a set having n objects can be arranged in $n!$ ways. Thus, the set of permutations of n objects is $n!$ in number. This group of permutations is denoted by S_n and is called the symmetric group of degree n. The group property follows from the fact that applying one permutation of n objects followed by another permutation results in a third permutation of the n objects. There are certain permutations which have special properties. These are the cyclic permutations. A cyclic permutation of objects numbered $(1, 2, ..., n)$ has the form $(i_1, i_2, ..., i_k)$ where $i_1, ..., i_k$ are distinct numbers in the set $(1, 2, ..., n)$ and this permutation sends i_1 to i_2, i_2 to i_3,..., i_{n-1} to i_n and finally i_n to i_1. The remaining objects are left unchanged by the permutation. The basic theorem concerning cyclic permutations is that every permutation can be expressed as a product of cyclic permutations. We can consider the subgroup of permutations of the vertices of an equilateral triangle obtained by rotations by angles of $2\pi/3$ and $4\pi/3$ and reflections about the axes obtained by joining each vertex of the triangle to the midpoint of the opposite side. These permutations are $e, (1, 2, 3), (1, 3, 2), (12), (13), (23)$. There are six in number and hence all the permutations are covered by rotations and reflections. This means that we have obtained a faithful representation of the permutation group. To see how these permutations are represented by matrices, we place 1 at $(sin(\pi/3), -cos(\pi/3))$, 2 at the point $(0, 1)$ and 3 at the point $(-sin(\pi/3), -cos(\pi/3))$. Then rotation about the z axis by $2\pi/3$ anticlockwise corresponds to the cyclic permutation (123) that sends 1 to 1, 2 to 3 and 3 to 1. This matrix is given by

$$\begin{pmatrix} cos(2\pi/3) & -sin(2\pi/3) \\ sin(2\pi/3) & cos(2\pi/3) \end{pmatrix}$$

Rotation about the z axis by $4\pi/3$ anticlockwise corresponds to rotation clockwise by $2\pi/3$ and has a matrix given by

$$\begin{pmatrix} cos(2\pi/3) & sin(2\pi/3) \\ -sin(2\pi/3) & cos(2\pi/3) \end{pmatrix}$$

Reflection about the y axis corresponds to the permutation (13) and has matrix given by

$$\begin{pmatrix} -1 & 0 \\ 0 & 1 \end{pmatrix}$$

The line joining vertex 1 to the midpoint of the line joining 2 and 3 passes through the origin and makes an angle $5\pi/6$ with the x axis and finally the line joining vertex 3 to the midpoint of the line joining vertices 1 and 2 passes through the origin and makes an angle $\pi/6$ with the x axis. The expression for the matrix of reflection about a line passing through the origin and making an angle θ with the x axis is given by

$$\begin{pmatrix} cos(2\theta) & sin(2\theta) \\ sin(2\theta) & -cos(2\theta) \end{pmatrix}$$

In this way, we obtain a faithful matrix representation for the group S_3 based on symmetries of an equilateral triangle. We now consider the symmetry group of a square. Label its vertices as $1, 2, 3, 4$ counterclockwise. The symmetry group is a subgroup of all permutations S_4 of four objects. S_4 consists of $4! = 24$ elements while the elements of the symmetry group are rotation by $\pi/2$, rotation by π, rotation by $3\pi/2$, no rotation at all, two reflections respectively about lines joining midpoints of opposite sides, two reflections respectively about the two diagonals. There are in all 8 elements of the symmetry group. Thus, in the case of a square, not all the permutations appear in the symmetry group. Consider now the group of symmetries of an n sided polygon. Suppose n is even. Then under reflections about any line joining the midpoints of a pair of opposite sides of the n given, the structure of the polygon is preserved. There are $n/2$ such reflections. Again reflection about any line joining a pair of opposite vertices preserves the polygon structure. There are $n/2$ such reflections. Finally rotations by angles $2k\pi/n$ with $k = 0, 1, ..., n-1$ preserves the polygonal structure. There are therefore in all $n/2 + n/2 + n = 2n$ elements in the group that preserves the polygonal structure of an n sided polygon having an even number of sides. Now consider the case when n is odd. In this case (take the example of a regular pentagon, $n = 5$). In this case, we do not have pairs of opposite sides, nor do we have pairs of opposite vertices. However, there are n pairs of elements consisting of a vertex and an edge opposite to each other. We can in other words draw a line that passes through a vertex and the middle point of an opposite side. n such lines can be drawn and the structure of the polygon is preserved under reflections about such lines. The structure of the polygon is also preserved under rotations by angles $2k\pi/n, k = 0, 1, ..., n - 1$. In all we have n reflections and n rotations that preserve the structure of a regular polygon having n sides with n odd. The total number of such transformations is $n+n = 2n$. Thus, for any n sided regular polygon, the group of symmetries has $2n$ elements. This group is denoted by D_n. Note that D_n has a subgroup C_n consisting of rotations through angles $2k\pi/n, k = 0, 1, ..., n - 1$. C_n is an n element Abelian subgroup of the $2n$ element group D_n. A group G can act on a set M. This notion is fundamental to our understanding of the crystal structure. For example, in the situation just considered, the group D_n acts on the set consisting of the vertices of the n sided regular polygon preserving it. A crystal may possess translational symmetry which means that under a certain group of translations, the structure of the crystal is preserved, which means that each atom in the transformed crystal now occupies an atomic positions of the original crystal. A group of translations has the form $\mathbf{x} \to \mathbf{x}+n\mathbf{a}+m\mathbf{b}+k\mathbf{c}$ where n, m, k are integers and $\mathbf{a}, \mathbf{b}, \mathbf{c}$ are vectors. Letting $T_\mathbf{u}$ denote the translation operator by the vector $\mathbf{u}$, one finds that the general translation operator is given by $T_\mathbf{a}^n T_\mathbf{b}^m T_\mathbf{c}^k$. This is an Abelian or a commutative group. One can also regard translations as acting upon functions. Suppose we are given a crystal with translational symmetry defined by the vectors $\mathbf{a}, \mathbf{b}, \mathbf{c}$. We can define function $f(\mathbf{x})$ on the crystal lattice. For example, f can be the potential field generated by the nuclei of the crystals. In this case, f is periodic with period vectors $\mathbf{a}, \mathbf{b}, \mathbf{c}$. We can regard the translation operators as acting on such functions. For example $T_\mathbf{a} f(\mathbf{x}) = f(\mathbf{x} - \mathbf{a})$, with similar expressions for $T_\mathbf{b} f(\mathbf{x})$ and $T_\mathbf{c} f(\mathbf{x})$. In general, we have the rule

$$T_{n\mathbf{a}+m\mathbf{b}+k\mathbf{c}} f(\mathbf{x}) = f(\mathbf{x} - n\mathbf{a} - m\mathbf{b} - k\mathbf{c})$$

It is easily verified that the group property is satisfied by these operators. In fact, we can verify that

$$T_{\mathbf{u}+\mathbf{v}} = T_\mathbf{u} T_\mathbf{v}$$

The fundamental result in the theory of group actions for a finite group is that if G is a finite group acting on a set M, then for any $m \in M$, we can define the orbit under G as $G.m = \{g.m : g \in G\}$. We also define the stabilizer of m as consisting of all elements $g \in G$ that fix m, i.e. $G_m = \{g \in G : g.m = m\}$. Then, our result states that $\mu(G)/\mu(G_m) = \mu(G.m)$, where $\mu(E)$ denotes the number of elements in the set E. This is proved by defining a map $\phi : G.m \to G/G_m$, where G/G_m is the set of all cosets of G_m, by the rule $g.m \to g.G_m$. This map is well defined since $g.m = g'.m$ implies $g'^{-1}g \in G_m$ which in turn implies $g'G_m = gG_m$. The map ϕ is also one-one since $g.G_m = g'.G_m$ implies $g'^{-1}g \in G_m$ which in turn implies $g'.m = g.m$. Since ϕ is one-one onto, it follows that $\mu(G.m) = \mu(G/G_m)$. But since G_m is a subgroup of G, it must follow from Lagrange's theorem that $\mu(G/G_m) = \mu(G)/\mu(G_m)$ and this completes the proof.

Remark: Groups play a fundamental role in solid state physics. The groups here are usually translation and rotation groups. Another group however plays a fundamental role in the theory of relativity. This is the Lorentz group, i.e. the group of linear transformations of $\mathbb{R}^4$ that preserves the metric $x_0^2 - \sum_{i=1}^3$. The vector in $\mathbb{R}^4$ here is $[x_0, x_1, x_2, x_3]^T$. The Lorentz transformation acting on $\mathbb{R}^4$ can be very conveniently described using matrices of the group $SL(2, \mathbb{C})$. Note that $SL(n, \mathbb{C})$ is the group of $n \times n$ complex matrices having determinant unity. We map the point $(x_0, x_1, x_2, x_3) \in \mathbb{R}^4$ to the 2×2 Hermitian matrix

$$\phi(\mathbf{x}) = \begin{pmatrix} x_0 + x_3 & x_1 - ix_2 \\ x_1 + ix_2 & x_0 - x_3 \end{pmatrix}$$

By Hermitian, we mean that the conjugate transpose of $\phi(\mathbf{x})$ equals itself. We then note that

$$det\phi(\mathbf{x}) = x_0^2 - x_1^2 - x_2^2 - x_3^2$$

Let $A \in SL(2, \mathbb{C})$. Then, we can consider the matrix $A\phi(\mathbf{x})A^*$. This matrix is clearly Hermitian and its determinant equals $det\phi(\mathbf{x})$. Hence, it is of the form $\phi(\mathbf{y})$ for some $\mathbf{y} = (y_0, y_1, y_2, y_3) \in \mathbb{R}^4$. It is clear that the transformation that carries the four vector $\mathbf{x}$ to the four vector $\mathbf{y}$ is linear. This transformation is given by

$$\phi(\mathbf{y}) = A\phi(\mathbf{x})A^*$$

or equivalently,

$$\mathbf{y} = \phi^{-1}(A\phi(\mathbf{x})A^*)$$

This transformation is therefore a Lorentz transformation. By varying the matrix A over $SL(2, \mathbb{C})$, we can generate a large class of Lorentz transformations. An example is provided by taking $A = \begin{pmatrix} r & 0 \\ 0 & r^{-1} \end{pmatrix}$ where $r > 0$. We find that

$$\phi(\mathbf{y}) = \begin{pmatrix} r(x_0 + x_3) & x_1 - ix_2 \\ x_1 + ix_2 & r(x_0 - x_3) \end{pmatrix}$$

This gives

$$y_0 + y_3 = r(x_0 + x_3) \qquad y_1 + iy_2 = x_1 - ix_2 \qquad y_0 - y_3 = r^{-1}(x_0 - x_3) \qquad y_1 - iy_2 = x_1 - ix_2$$

or

$$y_1 = x_1, y_2 = x_2, y_0 = (r + r^{-1})x_0 + (r - r^{-1})x_3, y_3 = (r - r^{-1})x_0 + (r + r^{-1})x_3$$

It is easily seen that this defines a Lorentz boost along the x_3 axis with speed v defined by

$$r + r^{-1} = (1 - v^2/c^2)^{-1/2}$$

A basic property of homomorphisms between representations. Let $T(g)$ be a representation of a group G in the space V and let $S(g)$ be a representation of the same group in the space W. Let $\pi_1, ..., \pi_k$ be all the irreps of G. We can then decompose

$$V = \bigoplus_{j=1}^{p_1+...+p_k} V_j \qquad W = \bigoplus_{j=1}^{q_1+...+q_k} W_j$$

where the restriction of T to the spaces $V_j, j = p_1 + ... + p_{r-1} + 1, ..., p_1 + ... + p_r$ equals π_r and likewise the restriction of S to $W_j, j = q_1 + ... + q_{r-1} + 1, ..., q_1 + ... + q_r$ equals π_r. Here, r runs from 1 to k. Now, let $A : V \to W$ intertwine T and S. Then, for any $x = x_1 + ... + x_{p_1+...+p_k} \in V$ with $x_j \in V_j$, we can write

$$Ax = \sum_{j=1}^{q_1+...+q_k} \sum_{m=1}^{p_1+...+p_k} A_{jm} x_m$$

with $A_{jm} x_m \in W_j$. Now,

$$S(g)Ax = \sum_{j,m} S(g) A_{jm} x_m = \sum_{r=1}^{k} \sum_{j=q_1+...+q_{r-1}+1}^{q_1+...+q_r} \sum_{m=1}^{p} \pi_r(g) A_{jm} x_m$$

where $p = p_1 + ... + p_k$. Likewise,

$$T(g)x = \sum_{r=1}^{k} \sum_{m=p_1+...+p_{r-1}}^{p_1+...+p_r} \pi_r(g) x_m$$

so that

$$AT(g)x = \sum_{j=1}^{q} \sum_{m=p_1+...+p_{r-1}}^{p_1+...+p_r} A_{jm} \pi_r(g) x_m$$

where $q = q_1 + ... + q_k$. Suppose A intertwines the two representations T, S, i.e., $AT(g) = S(g)A$. Then, we get by equating the two above expressions,

$$A_{jm} \pi_r(g) = \pi_s(g) A_{jm}$$

where $j = q_1 + ... + q_{s-1} + 1, ..., q_1 + ... + q_s, m = p_1 + ... + p_{r-1} + 1, ..., p_1 + ... + p_r$ with $1 \le r, s \le k$. Schur's lemma then gives $A_{jm} = 0$ if $s \ne r$ and $A_{jm} = c(r, j, m)\mathbf{I}$ if $r = s$. This means that the set of all intertwining operators forms a $\sum_{i=1}^{k} p_i q_i$ dimensional vector space. This space is denoted by $Hom_G(V, W)$ where it is understood that the G action on V is defined by the representation T while the G action on W is defined by the representation S.

8.7 Lattice vibrations and phonons

Consider a one dimensional crystal lattice with atomic planes located at $x = na, n = 0, \pm 1, \pm 2, ...$ Let $u_n(t)$ denote the displacement from equilibrium of the plane at $x = na$. We can visualize the plane at $x = na$ as being bound to the plane at $x = ma$ by a spring having force constant $C(|n - m|)$, i.e., the force constant depends only upon the separation between the planes. The equations of motion of the planes are

$$M\frac{d^2 u_n}{dt^2} = \sum_{m=-\infty}^{\infty} C(|n - m|)(u_m - u_n) = \sum_{p=-\infty}^{\infty} C(|p|)(u_{n+p} - u_n)$$

Here, M denotes the total mass of all the atoms in a given plane. We are interested in vibrations corresponding to a definite frequency ω. Then, $\frac{d^2 u_n}{dt^2} = -\omega^2 u_n$ and we get the following eigen-equation for the normal modes and the possible set of frequencies of vibration:

$$-M\omega^2 u_n = \sum_p C(|p|)(u_{n+p} - u_n)$$

We look for plane wave solutions to this eigen-equation, i.e., solutions having the following spatial dependence:

$$u_n = u.exp(ikna)$$

Substituting this into the eigen-equation gives us

$$-M\omega^2 = \sum C(|p|)(exp(ikpa) - 1) = 2\sum_{p=1}^{\infty} C(p)(cos(kpa) - 1)$$

or equivalently,

$$\omega^2 = \frac{2}{M}\sum_{p=1}^{\infty} C(p)(1 - cos(kpa))$$

This equation defines the general dispersion relation for vibrational waves in a crystal. The effective range of k is $-\pi/a \leq k < \pi/a$. This is the first Brillouin zone. The group velocity can be calculated from this equation. We can more generally consider a one dimensional crystal with two kinds of atoms. Let the displacement of the nth plane of atoms of type 1 be u_n and the displacement of the nth plane of atoms of 2 be v_n. The atomic planes of the two types are placed in an alternating fashion. Then, the force on the nth plane of type 1 atoms receives contribution from the nth plane of type 2 atoms as well as the $n - 1^{th}$ plane of type 1 atoms. We are assuming that only adjacent planes of atoms interact. This means that the dynamics of the type one planes can be expressed as

$$M_1 \frac{d^2 u_n}{dt^2} = K(v_n - u_n - (u_n - v_{n-1}) = K(v_n + v_{n-1} - 2u_n)$$

and likewise the dynamics of the type two planes can be written as

$$M_2 \frac{d^2 v_n}{dt^2} = K(u_{n+1} - v_n - (v_n - u_n)) = K(u_{n+1} + u_n - 2v_n)$$

Looking at plane waves of frequency ω and wave number k with a denoting the spacing between two adjacent planes of type one and type two atomic planes, we have

$$u_n(t) = u.exp(i(\omega t - kna)) \qquad v_n(t) = v.exp(i(\omega t - kna))$$

Substituting both of these into the above pair of differential equations gives

$$-M_1\omega^2 u = K(v(1 + exp(-ika)) - 2u) \qquad -M_2\omega^2 v = K(u(1 + exp(ika)) - 2v)$$

This gives us the following dispersion relation

$$det \begin{pmatrix} M_1\omega^2 - 2K & K(1+exp(-ika)) \\ K(1+exp(ika)) & M_2\omega^2 - 2K \end{pmatrix} = 0$$

The same idea can be generalized to a crystal having N different kinds of atomic planes. Each value of k yields a definite vibration characteristic of the crystal called a phonon. A phonon can thus be regarded as a sound particle because sound propagates as longitudinal waves in a solid. The general vibration pattern of the solid is a superposition of phonons of different $k's$. What we've done to a one dimensional vibration can be extended to three dimensional vibrations.

8.8   Einstein-Debye theory of specific heat of a solid

The energy levels of a quantum mechanical harmonic oscillator of frequency ω are given by $E_n = (n + 1/2)h\omega/2\pi, n = 0, 1, 2, ...,$. The number of oscillators in the energy state E_n, according to Boltzmann, is proportional to $exp(-E_n/kT)$. The average thermal energy of an oscillator of frequency ω is thus given by

$$U = \sum_{n=0}^{\infty} E_n exp(-E_n/kT) / \sum_{n=0}^{\infty} exp(-E_n/kT)$$

$$= -\frac{\partial}{\partial \beta} log \sum exp(-\beta E_n)$$

Now,

$$\sum exp(-\beta E_n) = exp(-\beta h\omega/4\pi) \sum_{n=0}^{\infty} exp(-\beta n h\omega/2\pi)$$

$$= exp(-\beta h\omega/4\pi)/(1 - exp(-\beta h\omega/2\pi))$$

and

$$-\frac{\partial}{\partial \beta} \sum exp(-\beta E_n) = -\frac{\partial}{\partial \beta} exp(-\beta h\omega/4\pi)/(1 - exp(-\beta h\omega/2\pi))$$

$$= (h\omega/4\pi)exp(-\beta h\omega/4\pi)/(1 - exp(-\beta h\omega/2\pi))$$

$$+ (h\omega/2\pi)exp(-\beta 3 h\omega/4\pi)/(1 - exp(-\beta h\omega/2\pi))^2$$

$$= (h\omega/2\pi)(\frac{1}{2}exp(\beta h\omega/4\pi)/(exp(h\omega/2\pi) - 1) + exp(h\omega/4\pi)/(exp(\beta h\omega/2\pi) - 1)^2)$$

$$= (h\omega/4\pi)((exp(\beta h\omega/2\pi) + 1)exp(\beta h\omega/4\pi)/(exp(\beta h\omega/2\pi) - 1)^2)$$

Dividing this by $\sum exp(-\beta E_n)$ gives

$$U = (h\omega/4\pi)(exp(\beta h\omega/2\pi) + 1)/(exp(\beta h\omega/2\pi) - 1) = (h\omega/4\pi))(1 + 2/(exp(\beta h\omega/2\pi) - 1))$$

With neglect of a temperature independent constant, (the zero point energy of the oscillator), we have

$$U(\omega, T) = (h\omega/2\pi)(exp(\beta h\omega/2\pi) - 1)^{-1}$$

Let $D(\omega)d\omega$ denote the number of oscillators (modes) in the frequency interval $[\omega, \omega + d\omega]$ per unit volume. Then, the average energy density of the solid is given by

$$<U>(T) = \int_0^\infty D(\omega)U(\omega, T)d\omega$$

The specific heat of the solid is found by differentiating this expression with respect to T. If the solid is a cube of length L, then we associate one mode with each triplet (K_x, K_y, K_z) corresponding to a phonon wave. The periodic boundary conditions applied to such a phonon wave imply that K_x, K_y, K_z must be integer multiples of $2\pi/L$. Thus, one phonon or mode corresponds to a volume $(2\pi/L)^3$ in **K**-space. The total number of modes having wavenumber magnitude lesser than K is thus

$$N = (4\pi K^3/3)/(2\pi/L)^3 = VK^3/6\pi^2$$

If v denotes the velocity of sound in the solid, then $\omega = Kv$ (Actually, the dispersion relation being non-linear, v will generally be K-dependent but for low frequencies, a linear relation can be assumed. This is same as the long wavelength limit). This gives

$$D(\omega) = \frac{dN}{d\omega} = \frac{d}{d\omega}V\omega^3/6\pi^2v^3 = V\omega^2/2\pi^2v^3$$

8.9 Free electron gas and Fermi-Dirac Statistics

In a metal, the valence electrons are the most loosely bound. Nevertheless, the concentration of such electrons is around the nuclei. However as a first order approximation, one can regard the electrons as moving freely and the metal can be approximated as an electron gas. The simplest model for a free electron is obtained from the Schrodinger equation for a free particle in a box, i.e.

$$\psi''(x) + \frac{8\pi^2 m}{h^2}E\psi(x) = 0$$

with the boundary conditions $\psi(0) = \psi(L) = 0$. The solutions are

$$\psi_n(x) = (\frac{2}{L})^{1/2}sin(n\pi x/L), n = 1, 2, ...,$$

with energy eigenvalues

$$E_n = n^2h^2/8mL^2, n = 1, 2, ...,$$

Corresponding to each quantum number n, we can have two electrons having opposite spin in accord with the Pauli exclusion principle. If n_F denotes the maximum quantum number occupied by the electrons in the metal, then the Fermi energy is the corresponding energy, i.e. $E_F = n_F^2 h^2/8mL^2$. Clearly $N = 2n_F$ is the number of electrons, since each quantum number can have two electrons. Here, we are assuming that all the quantum numbers $n \leq n_F$ are filled with two electrons. Thus, the Fermi energy of a metal is given by $N^2 h^2/32mL^2$. The probability that an energy level E will be occupied by an electron in an electron gas (Fermi gas) at the temperature T is given by the Fermi-Dirac distribution:

$$f(E) = \frac{1}{exp((E - \mu)/kT) + 1}$$

This formula is obtained by maximizing the entropy subject to the total energy constraint. When $T = 0$, we have $f(E) = 0$ for $E > \mu$ and $f(E) = 1$ for $E < \mu$. Note that $f(\mu) = 1/2$. This means that all energy levels $E < \mu$ are fully occupied and all energy levels $E > \mu$ are not occupied at all. That is, at zero temperature, the Fermi energy is μ. The wave functions for a free electron moving in a three dimensional box of length L are

$$\psi_n(\mathbf{r}) = A.sin(n_x\pi x/L)sin(n_y\pi y/L)sin(n_z\pi z/L)$$

where the quantum numbers (n_x, n_y, n_z) assume values $1, 2, ...$ This solution is obtained by solving the three dimensional Schrodinger equation with vanishing boundary conditions at $x = 0, L, y = 0, L, z = 0, L$. If however the metal is infinite, then we impose periodic boundary conditions on the wave function, i.e. $\psi(x + L, y, z) = \psi(x, y, z) = \psi(x, y + L, z) = \psi(x, y, z + L)$. This gives us the following wave functions

$$\psi_{\mathbf{k}}(\mathbf{r}) = exp(i\mathbf{k}.\mathbf{r})$$

where

$$k_x, k_y, k_z = 2m\pi/L, m = 0, \pm 1, \pm 2, ...$$

The energy of a level having quantum numbers (k_x, k_y, k_z) is given by

$$E(\mathbf{k}) = \frac{h^2}{8mL^2}(k_x^2 + k_y^2 + k_z^2)$$

Corresponding to each quantum number triplet $\mathbf{k}$, there are two electrons. Thus, the total number of electrons having energy lesser than E_F the Fermi energy is approximately given by

$$N = 2(L/2\pi)^3 \int_{h^2(k_x^2+k_y^2+k_z^2)/8\pi^2 m \leq E_F} dk_x dk_y dk_z$$

$$= (L^3/4\pi^3) \int_{0 < k < (8\pi^2 mE_F/h^2)^{1/2}} 4\pi k^2 dk$$

N is the total number of electrons. This evaluates to

$$(V/4\pi^3)4\pi/3(8\pi^2 mE_F/h^2)^{3/2} = (V/\pi/3)(8mE_F)^{3/2}/h^3$$

From this formula, we can obtain an expression for Fermi level as a function of the number of free electrons N. The number of orbitals per unit volume that have energy less than E is as seen from the above argument, proportional to $E^{3/2}$. This means that the number of orbitals in the energy range $[E, E + dE]$ per unit volume is given by $D(E)dE = A.\sqrt{E}dE$, where A is a function of the effective mass of the electron and Planck's constant. We now look at the problem of determining the heat capacity of the free electron gas. At temperature zero kelvin, all the electrons have energy lesser than the Fermi energy E_F as follows from the Fermi-Dirac distribution according to which the probability that an orbital of energy E will be occupied by an electron at the temperature T is given by $f(E) = \frac{1}{exp((E-E_F)/kT)+1}$. When $T = 0$, this formula gives $f(E) = 1$ when $E < E_F$ and $f(E) = 0$ when $E > E_F$. Thus, the average electronic energy per unit volume at zero Kelvin is given by $\int_0^{E_F} ED(E)dE$ and the average electronic energy at T degree Kelvin is given by $\int_0^\infty ED(E)f(E)dE$. The increase in energy of the electron gas on heating it from zero Kelvin to T Kelvin is the difference of these two, i.e.,

$$U = \int_0^\infty ED(E)f(E)dE - \int_0^{E_F} ED(E)dE$$

The total average number of electrons per unit volume is $N = \int_0^\infty D(E)f(E)dE$. Thus,

$$E_F N = E_F \int_0^\infty D(E)f(E)dE$$

The Fermi level E_F as well as the number of electrons N are independent of temperature. Thus, differentiating the two above equations gives

$$C = \frac{\partial U}{\partial T} = \int_0^\infty ED(E)\frac{\partial f(E)}{\partial T}dE$$

$$0 = \frac{\partial}{\partial T}(E_F N) = E_F \int D(E)\frac{\partial f(E)}{\partial T}dE$$

Subtracting the second from the first equation gives us

$$C = \int_0^\infty (E - E_F)D(E)\frac{\partial f(E)}{\partial T}dE$$

Now,

$$\frac{\partial f}{\partial T} = \frac{\partial}{\partial T}(exp((E-E_F)/kT) + 1)^{-1} = \frac{(E - E_F)exp((E-E_F)/kT)/kT^2}{(exp((E-E_F)/kT) + 1)^2}$$

From this formula, it follows that if $E \neq E_F$, then $\frac{\partial f}{\partial T} \to 0$ as $T \to 0$. In fact, suppose $E << E_F$. Then at low temperatures, $exp((E-E_F)/kT)$ is nearly zero and hence $\frac{\partial f}{\partial T}$ is negligible. On the other hand, suppose $E >> E_F$. Then at low temperatures, $exp((E-E_F)/kT)/(1 + exp((E-$

$E_F)/kT))^2$ is nearly zero. Thus, only in the vicinity of E_F does $\frac{\partial f}{\partial T}$ assume significant values at low temperatures, i.e., when $T << E_F/k$. This means that we have the following approximation:

$$C \approx D(E_F) \int_0^\infty (E - E_F)\frac{\partial f}{\partial T}dE$$

$$= D(E_F) \int_0^\infty ((E - E_F)^2/kT^2)exp((E - E_F)/kT)dE/(1 + exp((E - E_F)/kT))$$

This formula gives us the approximate dependence upon temperature of the specific heat of the electron gas at low temperatures.

8.10 Solved Problems, Study Projects, Suggestions and Remarks

Study project 1

A seminar on the basic matrix groups and their properties will be interesting from the standpoint of the quantum theory and crystallography. The groups to be covered include the rotation group $O(3)$, the proper rotation group $SO(3)$, the matrix of a reflection about a plane, the unitary group in n dimensions $U(n)$ and the special unitary group in n dimensions $SU(n)$. The generators of these groups can also be mentioned in the course of the seminar. If for example we take our group as $O(n)$, the orthogonal group in n dimensions, then every $A \in O(n)$ can be expressed as $A = exp(H)$ where $AA^T = I$ implies $H + H^T = 0$ or equivalently that H is a skew symmetric real matrix. Note that we are talking of $O(n, \mathbb{R})$ which means that our orthogonal matrix A has real entries. Thus, the generators of the group $O(n)$ comprise skew symmetric real matrices. Likewise if the orthogonal group consists of complex matrices, i.e. our group is $O(n, \mathbb{C})$, then the generators are skew symmetric complex matrices. These facts are usually cast by saying that the vector space of all skew symmetric real matrices forms the Lie algebra of the Lie group of all real orthogonal matrices and likewise the vector space of all skew symmetric complex matrices forms the Lie algebra of all complex orthogonal matrices. The dimension of the Lie algebra of $O(n, \mathbb{R})$ is also easily computed. Any skew symmetric real matrix H of size $n \times n$ requires $n(n-1)/2$ independent parameters, corresponding to the entries along the first,second,..., n^{th} subdiagonal. The entries along the first, second,..., n^{th} superdiagonal are merely obtained by changing the signs of the corresponding transposed elements on the subdiagonal. A skew symmetric matrix has zero entries along its diagonal. Similarly, the generators of the group $SU(n)$ are $n \times n$ complex matrices A satisfying $A + A^* = 0$ and $Tr(A) = 0$, i.e., $n \times n$ skew Hermitian matrices with vanishing trace. More advanced references on representation theory may be consulted by the interested student to complete this project.

Chapter 9

Digital Signal Processing

The basic issues of digital signal processing are representation of discrete time signals wherein different techniques for signal synthesis are discussed, linear and nonlinear systems in discrete time, time invariant systems, techniques for determining the response of a discrete time system to an input signal, spectral analysis of signals wherein we are interested in decomposing a signal into different frequency components which amounts to representing signals as a superposition of sinusoids, the application of spectral analysis to the determination of the response of a linear time invariant system to an input signal, the transmission of random processes through systems i.e. computation of the moments of the output and cross moments of input and output, different structures for implementing linear and nonlinear systems using basic elements like delays, coefficient and signal multipliers, design of filters of various kinds, i.e., lowpass, highpass and bandpass filters which are frequency selective filters that amplify a certain band of frequencies of the input and attenuate the complementary band. Such filters are used in applications like noise removal from signals or deblurring of images when the filter is two dimensional, representation of bandpass signals using the Hilbert transform and finally finite register effects wherein we are interested in how quantization errors propagate through a filter. By quantization errors, we mean the following. Finite memory registers are used to store the signal samples and filter coefficients. When two such samples or a coefficient and a sample is multiplied, we need to round off or truncate the resultant to accommodate it in a register of the same size. This rounding off and truncation creates an error that propagates to the output and appears as noise in the output. Different realizations of the same system lead to different noise power spectra at the output. The idea is to choose that realization that leads to the minimum noise power at the output. We shall discuss these issues in this chapter.

9.1   Classification of signals

Signals are classified based on their properties. The typical classifications include discrete time versus continuous time, energy versus power signals, periodic versus aperiodic signals, finite duration versus infinite duration signals, random versus nonrandom signals, one dimensional versus multidimensional signals, single channel versus multichannel signals. A continuous time signal is specified by specifying the signal value as a function of a parameter that ranges over a continuous range, like $x(t)$ with t varying over the real line $(-\infty, \infty)$ or $x(t)$ with t varying over the positive half real line $[0, \infty)$ or over any continuous interval (a, b). For a discrete time signal, the time parameter assumes discrete values, i.e. ranges over a countable set like integers $\mathbb{Z}$, or positive integers $\mathbb{Z}_+$. The samples of such a signal can be represented as $x(n)$. Usually, when t is used for the independent parameter, it is assumed that the signal is continuous time and when n is used, it is assumed that the signal is discrete time. A continuous time signal $x(t)$ can be converted into a discrete time signal by the process of sampling on a discrete set, like $t = n\Delta, n = 0, \pm1, \pm2, ...$. Here, Δ is called the sampling time interval. Conversely, a discrete time signal can be converted into a continuous time signal by the process of holding or interpolation. For example given the discrete time signal $x(n)$, we define $x(t)$ to be $x[n]$ when $n\Delta \leq t < (n+1)\Delta$ where n is an integer. This representation can be expressed in the following form: We define the pulse $p(t)$ to equal 1 when $0 \leq t < \Delta$ and zero otherwise. Then the held signal is given by

$$x(t) = \sum_{n=-\infty}^{\infty} x(n)p(t - n\Delta)$$

Such a representation occurs in pulse amplitude modulation systems in digital communication. A periodic signal $x(t)$ in continuous time satisfies the identity $x(t + T) = x(t)$ for all $t \in \mathbb{R}$, where T is some positive real number. The smallest positive number T for which this identity is valid is called the period of the signal. For example, the signal $x(t) = \sum_{n=1}^{N} A_n sin(n\omega_0 t + \phi_n)$ is periodic with period $T = 2\pi/\omega_0$. If ω_1/ω_2 is an irrational number, the signal $A.sin(\omega_1 t) + B.sin(\omega_2 t)$ is aperiodic when $A, B \neq 0$. When there is no positive real number T for which the identity $x(t + T) = x(t)$ is valid for all $t \in \mathbb{R}$, the signal is called aperiodic. The energy of a signal $x(t)$ in continuous time is defined as the integral of its modulus square: $E_x = \int_{-\infty}^{\infty} |x(t)|^2 dt$. If the signal is real valued, then $E_x = \int_{-\infty}^{\infty} x(t)^2 dt$, while if the signal is complex valued, we can write $x(t) = x_1(t) + ix_2(t)$ where $x_1(t) = Rex(t)$ and $x_2(t) = Rex_2(t)$ are real valued signals. Then,

$$E_x = \int_{-\infty}^{\infty} (x_1(t)^2 + x_2(t)^2) dt = E_{x_1} + E_{x_2}$$

which in words reads : The energy of a complex valued signal equals the sum of the energies in its real and imaginary parts. A signal having finite energy is called an energy signal. For example, the energy of the signal $x(t) = exp(-\sigma t)u(t)$ with $\sigma > 0$ is given by $\frac{1}{2\sigma}$ and hence this is an energy signal. More generally, the energy of the signal $x(t) = \sum_{k=1}^{p} A_k exp(-(\sigma_k + j\omega_k)t)$ with $\sigma_k > 0$ for all k is finite. In fact

$$E_x = \sum_{k,m=1}^{p} \frac{A_k \bar{A}_m}{\sigma_k + \sigma_m + j(\omega_k - \omega_m)}$$

The energy of a signal $x(n), n \in \mathbb{Z}$ in discrete time is defined as $E_x = \sum_{n=-\infty}^{\infty} |x(n)|^2$. For example, if $x(n) = \sum_{k=1}^{p} A_k z_k^n u(n)$ with $|z_k| < 1$, we have

$$E_x = \sum_{k,m=1}^{p} \frac{A_k \bar{A}_m}{1 - z_k \bar{z}_m} < \infty$$

The power of a continuous time signal $x(t)$ is defined to be the limit $P_x = lim_{T \to \infty} \frac{1}{2T} \int_{-T}^{T} |x(t)|^2 dt$ and the power of a discrete time signal $x(n)$ is defined to be $P_x = lim_{N \to \infty} \frac{1}{2N+1} \sum_{n=-N}^{N} |x(n)|^2$. Signals having finite power are called power signals. It is easily seen that finite energy signals have zero power and finite positive power signals have infinite energy. For example, the power of the harmonic signal $x(t) = \sum_{k=1}^{p} A_k * exp(j(\omega_k t + \phi_k))$ with the $\omega_k's$ all distinct is given by

$$P_x = \sum_{k=1}^{p} |A_k|^2$$

and likewise in discrete time, the power of the harmonic signal $x(n) = \sum_{k=1}^{p} A_k exp(j(\omega_k n + \phi_k))$ is given by

$$P_x = \sum_{k=1}^{p} |A_k|^2$$

It is being assumed in the latter case that the $\omega_k's$ are distinct numbers in $[0, 2\pi)$.

9.2 Linear systems

Let S_x, S_y be linear spaces of signals, i.e. S_x consists of signals $\{x(n)\}$ with signal addition defined sample by sample in the sense that $\{x(n)\} + \{y(n)\} = \{x(n) + y(n)\}$ and scalar multiplication also defined sample by sample in the sense that $c\{x(n)\} = \{c.x(n)\}$. Likewise S_y also consists of signals with addition and scalar multiplication. S_x and S_y are closed under the operations of signal addition and scalar multiplication. The scalars are chosen from the same field for both S_x and S_y. For example, if the field is $\mathbb{R}$, then the signal samples $x(n)$ are real while if the field is $\mathbb{F}$, then the signal samples $x(n)$ all belong to the field $\mathbb{F}$. In digital communication theory, the field $\mathbb{F}$ is a finite field like $\{0, 1, ..., p-1\}$ with addition and multiplication defined modulo p where p is a prime number. If $T : S_x \to S_y$ is a mapping that preserves addition and scalar multiplication, then T is said to be a linear system with input signal space S_x and output signal space S_y. By preserving addition and scalar multiplication, we mean that

$$T(c.x + y) = cT(x) + T(y)$$

for all $c \in \mathbb{F}$ and $x, y \in S_x$. A more common notation is the following:

$$T\{cx(n) + y(n)\} = cT\{x(n)\} + T\{y(n)\}$$

for all scalars c and all signals $\{x(n)\}, \{y(n)\} \in S_x$. A typical example of a linear signal space is the space of bounded complex valued signals, i.e.

$$S = \{\{x(n)\} : sup_n|x(n)| < \infty\}$$

Linearity of the signal space follows from the inequality

$$sup_n|cx(n) + y(n)| \leq |c|sup_n|x(n)| + sup_n|y(n)|$$

An example of a linear system defined from $S \rightarrow S$ is $y = T(x)$ with

$$y(n) = \sum_{k=-\infty}^{\infty} h(n, k)x(k)$$

where $sup_n \sum_k |h(n, k)| < \infty$. It is easily verified that T maps bounded signals to bounded signals and hence is a well defined map from $S \rightarrow S$. That T is linear follows from the identity

$$\sum_k h(n, k)(cx_1(k) + x_2(k)) = c \sum_k h(n, k)x_1(k) + \sum_k h(n, k)x_2(k)$$

The corresponding example in continuous time is the following. Let

$$S = \{\{x(t)\} : sup_t|x(t)| < \infty\}$$

This is the space of bounded signals in continuous time. $T : S \rightarrow S$ is defined by the rule

$$y(t) = \int_{-\infty}^{\infty} h(t, \tau)x(\tau)d\tau$$

where $sup_t \int_{-\infty}^{\infty} |h(t, \tau)|d\tau < \infty$. Then, T is a well defined linear system. We leave it as an exercise to prove that T is a well defined linear map from $S \rightarrow S$. An example of a linear system is provided by electrostatics. The input is the charge density field $\{\rho(\mathbf{r})\}$ and the output is the electric field $\mathbf{E}(\mathbf{r})$. The output is related to the input by a convolution equation

$$\mathbf{E}(\mathbf{r}) = \frac{1}{4\pi\epsilon} \int \frac{\rho(\mathbf{r}')(\mathbf{r} - \mathbf{r}')}{|\mathbf{r} - \mathbf{r}'|^3} d^3\mathbf{r}'$$

This is a linear relation. If $\mathbf{E}_1(\mathbf{r})$ is the field produced by $\rho_1(\mathbf{r})$ and $\mathbf{E}_2(\mathbf{r})$ is the field produced by $\rho_2(\mathbf{r})$, then the field produced by $c\rho_1(\mathbf{r})+\rho_2(\mathbf{r})$ equals $c\mathbf{E}_1(\mathbf{r})+\mathbf{E}_2(\mathbf{r})$. Without even using the above solution to the equations of electrostatics, it can be seen easily that the partial differential equations of electrostatics define a linear system. We start with the Poisson equation for electrostatics.

Assume that charge is distributed within a volume v in accord with the density $\rho(\mathbf{r})$. Let s denote the boundary of v. On the surface s, we assume that the potential is zero. Then, the potential $V(\mathbf{r})$ inside the region v is given by the solution to Poisson's equation

$$\nabla^2 V(\mathbf{r}) = -\rho(\mathbf{r})/\epsilon, \mathbf{r} \in v$$

with the boundary condition $V(\mathbf{r}) = 0$ for $\mathbf{r} \in s$. From the uniqueness theorem, we know that the solution for V is unique. Likewise, let V' denote the solution to the same boundary value problem with the charge density as ρ'. Then it is easily seen that $cV + V'$ is the solution to the same boundary value problem with charge density as $c\rho + \rho'$:

$$\nabla^2(cV + V') = c\nabla^2 V + \nabla^2 V' = c(-\rho/\epsilon) - \rho'/\epsilon = -(c\rho + \rho')/\epsilon$$

and $cV(\mathbf{r}) + V'(\mathbf{r}) = 0$ for $\mathbf{r} \in s$ because, $V(\mathbf{r}) = V'(\mathbf{r}) = 0$ for $\mathbf{r} \in s$. Thus, $cV + V'$ is the potential when the charge density is $c\rho + \rho'$. The electric field corresponding to ρ is $\mathbf{E} = -\nabla V$ and that corresponding to ρ' is $-\nabla V'$. The electric field corresponding to $c\rho + \rho'$ is then $-\nabla(cV + V') = -c\nabla V - \nabla V' = c\mathbf{E} + \mathbf{E}'$ and this proves linearity of the problem. Note that we have assumed zero potential on the boundary. If this were not the case, then linearity would not follow.

9.3 Time Invariance

Let $x : \mathbb{R} \to \mathbb{R}$ be a continuous time signal defined on the entire real line. Its shifted version is the signal $S_\tau(x)$ where $S_\tau(x)(t) = x(t - \tau)$. τ is any real number, positive or negative. Let S denote the space of input signals and S' the space of output signals. We assume S and S' to be invariant under time shifts, i.e., if $x \in S$, then $S_\tau(x) \in S$ and likewise, if $y \in S'$, then $S_\tau(y) \in S'$. A typical example is the space of all bounded signals defined on the real line. If $x(t)$ is a bounded signal, then it is clear that $x(t - \tau)$ is also a bounded signal, no matter how τ is chosen. Another example is the space of energy signals. If $x(t)$ is an energy signal, i.e., $\int |x(t)|^2 dt < \infty$, then $\int |x(t - \tau)|^2 dt < \infty$. In fact, the energy of $x(t - \tau)$ coincides with the energy of $x(t)$. More generally, the L^p space of signals, i.e., signals for which $\int |x(t)|^p dt < \infty$ is invariant under time shifts as follows from the obvious identity

$$\int |x(t - \tau)|^p dt = \int |x(t)|^p dt$$

Here $p > 0$ is arbitrary. Another example of a space of signals invariant under time shifts is the space of all solutions to an autonomous differential equation whose parameters do not vary with time and which vanish at $t = -\infty$, i.e., all solutions to the differential equation

$$\frac{d^n x(t)}{dt^n} = F(x(t), \frac{dx(t)}{dt}, ..., \frac{d^{n-1}x(t)}{dt^{n-1}}), -\infty < t < \infty$$

Given two signal spaces S, S' invariant under time shifts, i.e., for which $S_\tau(S) \subset S$ and $S_\tau(S') \subset S'$ for all $\tau \in \mathbb{R}$ and a system $T : S \to S'$, we say that the system T is time or shift invariant if $TS_\tau = S_\tau T$ for all $\tau \in \mathbb{R}$. This means that if $y = T(x)$ is the response to the input x, then $S_\tau(y)$ is the response to $S_\tau(x)$. Or equivalently, if $y(t)$ is the response to $x(t)$, then $y(t - \tau)$ is the response to $x(t - \tau)$, i.e., $T\{x(t - \tau)\} = y(t - \tau)$. This idea can be generalized to include multidimensional systems. For example consider the signal space S of all k dimensional signals $\{x(t_1, ..., t_k) : t_1, ..., t_k \in \mathbb{R}\}$ with the signal assuming real or complex values. if the signal assumes real values, we denote the corresponding signal space by $F(\mathbb{R}^k, \mathbb{R})$ meaning all real valued functions on $\mathbb{R}^k$. This space is shift invariant, the shift operators being defined by

$$S_{\tau_1, ..., \tau_k}(x)(t_1, ..., t_k) = x(t_1 - \tau_1, ..., t_k - \tau_k)$$

The multidimensional signal $x(t_1, ..., t_k)$ can also be denoted by $x(\mathbf{t})$ with the understanding that the symbol $\mathbf{t}$ stands for the ordered k-tuple $(t_1, ..., t_k)$. Letting $\tau = (\tau_1, ..., \tau_k)$, the shift operator $S_{\tau_1, ..., \tau_k}$ is denoted by S_τ. If T is a system defined on S, then T will be called shift invariant if $TS_\tau = S_\tau T$ for all shifts $\tau \in \mathbb{R}^k$. This means that if $y(t_1, ..., t_k)$ is the response of the system to $x(t_1, ..., t_k)$, then the response of the system to $x(t_1 - \tau_1, ..., t_k - \tau_k)$ equals $y(t_1 - \tau_1, ..., t_k - \tau_k)$. The discrete time scenarios are similar. For x defined on $\mathbb{Z}^k$, i.e. $x : \mathbb{Z}^k \to \mathbb{R}$, the shift operators are defined as $S_{m_1, ..., m_k}$ with

$$S_{m_1, ..., m_k}(x)(n_1, ..., n_k) = x(n_1 - m_1, ..., n_k - m_k)$$

This is to be read as the operator $S_{m_1, ..., m_k}$ acts on the signal $x(n_1, ..., n_k)$ to produce the signal $x(n_1 - m_1, ..., n_k - m_k)$. More generally, we can consider a set X and signals defined as mappings from this set to another set Y, i.e., $x : X \to Y$. Let G be a group of transformations acting on X. In the case of $\mathbb{R}^k$, $\mathbb{R}^k$ is the Abelian group acting on the set $\mathbb{R}^k$ according to the rule $(t_1, ..., t_k) \to (t_1 + \tau_1, ..., t_k + \tau_k)$. By saying that G acts on X, we mean that $e.x = x$ and $g_2.(g_1.x) = (g_2.g_1).x$ for all $x \in X$ and all $g_1, g_2 \in G$. Then we can define a G shift operator S_g acting on the signal $f(x)$ according to the rule $S_g(f)(x) = f(g^{-1}.x)$. If S is the signal space $F(X, Y)$ (All mappings from X to Y) and S' is the signal space $F(X', Y')$ (All mappings from X' to Y'), and suppose G acts on X' too, then we can define a system $T : F(X, Y) \to F(X', Y')$ to be G-shift invariant if $TS_g = S_g T$ for all $g \in G$, i.e., the response to the signal $f(g^{-1}x)$ equals the shifted response $f'(g^{-1}x)$ where $f' = T(f)$. An example of a shift invariant system in continuous time is the system defined by the differential equation

$$\frac{d^n y(t)}{dt^n} = F(y(t), \frac{dy(t)}{dt}, ..., \frac{d^{n-1}y(t)}{dt^{n-1}}, x(t), \frac{dx(t)}{dt}, ..., \frac{d^{m-1}x(t)}{dt^{m-1}})$$

where $x(t)$ is the input and $y(t)$ is the output. The system is assumed to be in operation from time $t = -\infty$. Time invariance follows from the fact that if $y(t)$ satisfies the above differential equation with $x(t)$ as the input, then $y(t - \tau)$ also satisfies the same differential equation with $x(t - \tau)$ as the input. The initial condition is that y and all its derivatives vanish as $t \to -\infty$.

However, if F happens to depend upon time explicitly, then the system fails to be time invariant, for example, the system defined by the differential equation

$$\frac{dy(t)}{dt} = F(y(t), x(t), t)$$

is not time invariant. If $x(t), y(t)$ is a pair satisfying this equation, and if $y_1(t)$ is the response to the input $x_1(t) = x(t - \tau)$, then

$$\frac{dy_1(t)}{dt} = F(y_1(t), x(t - \tau), t)$$

However since

$$\frac{dy(t - \tau)}{dt} = F(y(t - \tau), x(t - \tau), t - \tau)$$

it follows that y_1 and $S_\tau(y)$ do not satisfy the same differential equation and hence cannot be equal. y_1 coincides with $S_\tau(y)$ only when F does not depend upon time explicitly, i.e., when $F(y, x, t) = F(y, x)$. A system that is both linear and time invariant is called a linear time invariant system, or simply an LTI system. An example of such a system is that defined by the differential equation

$$D^n y(t) + a_1 D^{n-1} y(t) + \ldots + a_{n-1} D y(t) + a_n y(t) = b_0 D^m x(t) + b_1 D^{m-1} x(t) + \ldots$$

$$+ b_{m-1} D x(t) + b_m x(t)$$

where $a_1, \ldots, a_n, b_0, \ldots, b_m$ are constants. The system is assumed to be in operation from $t = -\infty$ with zero initial conditions. It is clear that if (x_1, y_1) and (x_2, y_2) are two input-output pairs that satisfy this differential equation, then so is the pair $(c_1 x_1 + c_2 x_2, c_1 y_1 + c_2 y_2)$ for all constants c_1, c_2. Thus linearity holds good. Linearity will hold good even if the $a'_j s$ and $b'_j s$ are functions of time. Time invariance follows from the fact that if $(x(t), y(t))$ is a pair that satisfies the above differential equation, then $(x(t - \tau), y(t - \tau))$ is also a pair that satisfies this equation with the same zero initial condition at $t = -\infty$. However time invariance will fail to hold if the $a'_j s$ and $b'_j s$ are functions of time. For then, $(x(t - \tau), y(t - \tau))$ will be a pair not for the same differential equation but for the differential equation obtained by replacing the functions $a_j(t)$ and $b_j(t)$ respectively by $a_j(t - \tau)$ and $b_j(t - \tau)$.

9.4 Convolution

Consider a linear time invariant system in discrete time. Let T denote the system operator. Any signal $x(n)$ can be represented as a superposition of delayed versions of unit impulse signals, i.e., $x(n) = \sum_{k=-\infty}^{\infty} x(k)\delta_k(n)$ where $\delta_k(n) = \delta(n - k)$. We can write $x(n) = lim_{N \to \infty} x_N(n)$ where $x_N(n) = \sum_{k=-N}^{N} x(k)\delta_k(n)$. Linearity of T implies $T(x_N) = \sum_{k=-N}^{N} x(k)T(\delta_k)$. Write $h_k = T(\delta_k)$.

By time invariance of T, $h_k = S_k(h_0)$, where S_k is the operator corresponding to delay by k samples. In terms of time samples, $h_k(n) = h_0(n-k)$. Thus, $T(x_N)(n) = \sum_{k=-N}^{N} h_0(n-k)x(k)$. If we assume that T is a smooth operator, then $lim_{N \to \infty} T(x_N) = T(lim_{N \to \infty} x_N) = T(x) = y$. Then,

$$y(n) = lim_{N \to \infty} \sum_{k=-N}^{N} h_0(n-k)x(k) = \sum_{k=-\infty}^{\infty} h_0(n-k)x(k)$$

The right side is a convolution between the sequences $\{h_0(n)\}$ and $\{x(n)\}$. We say that the output $\{y(n)\}$ is obtained by convolving the impulse response $h_0(n)$ of the system with the input signal $x(n)$.

9.5   Solution to linear constant coefficient difference equations

Consider a system defined by the differential equation

$$p(D)y(t) = (D^n + a_1 D^{n-1} + ... + a_n)y(t) = x(t)$$

In order to obtain the general solution to this equation, we first of all factorize the polynomial operator appearing on the left:

$$p(D)y(t) = (D^n + a_1 D^{n-1} + ... + a_{n-1}D + a_n) = \Pi_{i=1}^{k}(D - c_i)^{m_i}$$

with $m_i \geq 1$ $c_1, ..., c_k$ are the distinct roots of the polynomial. Now consider the equation

$$p(D)h(t) = \delta(t)$$

Its solution for $t > 0$ is of the form

$$h(t) = \sum_{i=1}^{k} \sum_{r=0}^{m_i-1} A_{i,r} exp(c_i t) t^r$$

This can be verified by showing that

$$(D - c)^m (exp(ct)t^r) = 0, r = 01, 2, ..., m - 1$$

The coefficients $A_{i,r}$ are determined from the conditions

$$\int_{0-}^{0+} p(D)h(t)dt = 1, \quad \int_{0-}^{0+} dt \int_{-\infty < \tau_1 < ... < \tau_p < t} p(D)h(\tau_1)d\tau_1...d\tau_p = 0, p = 1, 2, ..., n - 1$$

along with the causality assumption $h(t) = 0, t < 0$. These give us the equations

$$Dh(0+) = ... = D^{n-2}h(0+) = 0, D^{n-1}h(0+) = 1$$

using which the coefficients $A_{i,r}$ are obtained.. The general solution to the differential equation is then given by

$$y(t) = \int_0^\infty h(\tau)x(t-\tau)d\tau + \sum_{i=1}^{k} \sum_{r=0}^{m_i-1} B_{i,r}exp(c_i t)t^r$$

The first integral on the right is called the forced response. It is zero if the input is zero. The second term is called the natural response. It is independent of the input and depends only on the initial conditions in the system. For example, if the system is a linear circuit, then the initial conditions are the initial capacitor voltages and the inductor currents.

9.6 Sampling of analog signals and the sampling theorem

Let $x(t)$ be a continuous time signal. Its sampled version is the discrete time signal $x[n] = x(nT), n = 0, \pm 1, \pm 2, ...$ T is called the sampling time interval and $F = 1/T$ is called the sampling rate. Consider the impulse sampled train

$$x_\delta(t) = \sum_{n=-\infty}^{\infty} x(nT)\delta(t - nT)$$

Here, $\delta(t)$ is the Dirac delta signal, also called the unit impulse in continuous time. It is defined by the equation $\int_{-\infty}^{\infty} f(t)\delta(t)dt = f(0)$ for all continuous signals $f(t)$. Using the identity $x(t)\delta(t-\tau) = x(\tau)\delta(t - \tau)$, we can write

$$x_\delta(t) = x(t) \sum_{n=-\infty}^{\infty} \delta(t - nT)$$

Now take the Fourier transform on both sides of this equation to get

$$X_\delta(f) = X(f) * \sum_{n=-\infty}^{\infty} exp(-i2\pi nfT)$$

The Fourier transform of a signal $z(t)$ is defined by the equation

$$Z(f) = \int_{-\infty}^{\infty} z(t)exp(-i2\pi ft)dt$$

We now use the fact that

$$\sum_{n=-\infty}^{\infty} exp(-i2\pi nfT) = \frac{1}{T}\sum_{n} \delta(f - n/T) = F\sum_{n} \delta(f - nF)$$

Thus,

$$X_\delta(f) = F \sum_n X(f - nF)$$

If $X(f) = 0$ for $|f| > B$ and F is chosen greater than $2B$, then $X_\delta(f)$ consists of nonoverlapping displaced replicas of $X(f)$ scaled in magnitude by F. By passing $x_\delta(t)$ through a low pass filter having characteristics $H(f) = 1$ for $|f| \leq B$ and $H(f) = 0$ for $|f| > B$, the signal $x(t)$ can be recovered from $x_\delta(t)$. The inverse transform of $H(f)$ is $h(t) = \frac{sin(2\pi Bt)}{2\pi Bt}$ and the Shannon reconstruction formula results:

$$x(t) = \sum_{n=-\infty}^{\infty} x(n/2B) \frac{sin(\pi(2Bt - n))}{\pi(2Bt - n)}$$

9.7　Low pass, High pass and Band pass filters in continuous time

Let $H(f)$ be large in magnitude for all f in a band B and small in magnitude for all f outside the band B. Let $h(t)$ be the inverse Fourier transform of $H(f)$, i.e., $h(t) = \int_{-\infty}^{\infty} H(f)exp(i2\pi ft)df$. If $x(t)$ is input to an LTI system having impulse response $h(t)$, the output is $y(t) = \int h(\tau)x(t-\tau)d\tau$, which in the frequency domain reads $Y(f) = H(f)X(f)$. $y(t)$ can also be expressed as the inverse Fourier transform of $Y(f)$, i.e.,

$$y(t) = \int_{-\infty}^{\infty} H(f)X(f)exp(i2\pi ft)df$$

$$= \int_B H(f)X(f)exp(i2\pi ft)df + \int_{B^c} H(f)X(f)exp(i2\pi ft)df$$

This formula shows that the frequency components of the input in the band B are amplified and those outside the band B are attenuated. If B is of the form $[-a, a]$, the filter is lowpass, if B is of the form $(-\infty, -a) \cup (a, \infty)$, the filter is highpass while if B is of the form $(-b, -a) \cup (a, b)$, the filter is bandpass. An example of a lowpass filter is provided by the differential equation

$$\frac{dy(t)}{dt} + y(t)/\tau = x(t)$$

The frequency domain equation corresponding to this differential equation is

$$Y(f) = \frac{\tau X(f)}{i2\pi f\tau + 1}$$

Thus, $H(f) = \frac{\tau}{i2\pi f\tau + 1}$ and $|H(f)| = \frac{\tau}{\sqrt{1+(2\pi f\tau)^2}}$. This is a symmetric function of f and decreases as $|f|$ increases to ∞. These ideas can be carried over to discrete time, where the Fourier transform of a sequence $x(n)$ is defined as $X(\omega) = \sum_{n=-\infty}^{\infty} x(n)exp(-i\omega n)$ and the frequency range is $[-\pi, \pi)$. The inverse Fourier transform is $x(n) = \frac{1}{2\pi}\int_{-\pi}^{\pi} X(\omega)exp(i\omega n)d\omega$ and from the convolution theorem, we deduce that if $y(n) = \sum_k h(k)x(n-k)$, then

$$y(n) = \frac{1}{2\pi}\int_{-\pi}^{\pi} H(\omega)X(\omega)exp(i\omega n)d\omega$$

Depending on the range of frequencies where $H(\omega)$ is large in magnitude and the range where it is small in magnitude, one can define lowpass, bandpass and highpass filters.

9.8 Fourier series and transforms in continuous and discrete time

If $x(t)$ is periodic with period T, i.e., $x(t+T) = x(t)$ for all $t \in \mathbb{R}$, then we have the expansion

$$x(t) = \sum_k c_k exp(i\omega kt), \omega = 2\pi/T$$

Using the orthogonality equation

$$\int_0^T exp(i\omega kt)dt = T\delta(k)$$

we deduce the following formula for the computation of the coefficients c_k:

$$c_k = \frac{1}{T}\int_0^T x(t)exp(-i\omega kt)dt$$

The above expansion is called the Fourier series expansion of the signal $x(t)$. The mapping that carries a T periodic signal to its Fourier series coefficient sequence (c_k) is linear and energy preserving, i.e.,

$$\frac{1}{T}\int_0^T |x(t)|^2 dt = \sum_k |c_k|^2$$

It also satisfies the modulation property that the coefficients of $x(t)exp(im\omega t)$ are c_{k-m} and the time shift property that the coefficients of $x(t-\tau)$ are $c_k exp(-ik\omega\tau)$. Finally, if $x(t)$ and $y(t)$ are T periodic, then their cross correlation is given by

$$\frac{1}{T}\int_0^T x(t)\bar{y}(t)dt = \sum_k c_k \bar{d}_k$$

where c_k are the coefficients of $x(t)$ and d_k those of $y(t)$. These can be verified easily. Fourier series can be used to obtain the steady state response of a linear time invariant system to a periodic input. Let $x(t)$ be a T periodic input and $h(t)$ the impulse response of a linear time invariant system. The output is given by $y(t) = \int_{-\infty}^{\infty} h(\tau)x(t-\tau)d\tau$. Plugging into this expression the Fourier series expansion of $x(t)$, i.e., $x(t) = \sum_k c_k exp(ik\omega t)$ gives

$$y(t) = \sum_k H(k\omega)c_k exp(ik\omega t)$$

where

$$H(\omega) = \int_{-\infty}^{\infty} h(\tau)exp(-i\omega\tau)d\tau$$

is the frequency response of the system. Suppose the system is defined by a differential equation with constant coefficients

$$(D^n + a_1 D^{n-1} + ... + a_n)y(t) = (b_0 D^m + b_1 D^{m-1} + ... + b_m)x(t)$$

Then, we can determine the frequency response by setting $x(t) = exp(i\omega t), y(t) = H(\omega)exp(i\omega t)$ in this differential equation to get

$$H(\omega)Q(\omega) = P(\omega)$$

where

$$P(\omega) = b_0(i\omega)^m + b_1(i\omega)^{m-1} + ... + b_m, Q(\omega) = (i\omega)^n + a_1(i\omega)^{n-1} + ... + a_n$$

Thus

$$H(\omega) = \frac{P(\omega)}{Q(\omega)}$$

and if $x(t) = \sum c_k exp(ik\omega t)$, then

$$y(t) = \sum_k \frac{P(k\omega)}{Q(k\omega)} c_k exp(ik\omega t)$$

Now consider the discrete time case. Let $x(n)$ be N-periodic, i.e., $x(n+N) = x(n)\forall n \in \mathbb{Z}$. The discrete time signals $e_k(n) = exp(i2\pi ikn/N), k = 0, 1, ..., N-1$ are all linearly independent and have period N. Since any arbitrary N periodic signal is completely determined by its samples at times $n = 0, 1, ..., N-1$, it follows that $x(n)$ can be expressed as a linear combination of the $e_k(n)'s$, that is

$$x(n) = \sum_{k=0}^{N-1} c_k e_k(n)$$

Using the orthogonality relationship

$$\sum_{n=0}^{N-1} e_k(n)\bar{e}_j(n) = N\delta[k-j], k, j = 0, 1, ..., N-1$$

we derive the inverse of the above formula:

$$c_k = \frac{1}{N} \sum_{n=0}^{N-1} x(n)\bar{e}_k(n)$$

These are called the discrete Fourier series relations. $c_0, c_1, ..., c_{N-1}$ are called the discrete Fourier series coefficients of the N periodic signal $x(n)$. The mapping that carries an N-periodic signal $\{x(n)\}$ to its Discrete Fourier series coefficients $\{c_0, c_1, ..., c_{N-1}\}$ is linear. It satisfies the delay property, i.e., if the DFT coefficients of the delayed signal $x(n-m)$ are $c_k exp(-i2\pi km/N)$. It also satisfies the modulation property, that is the DFT coefficients of the modulated signal $x(n)exp(i2\pi k_0 n/N)$ are c_{k-k_0}. In other words, modulation causes a shift in the spectrum. The final and most important property is the energy equation:

$$\frac{1}{N} \sum_{n=0}^{N-1} |x(n)|^2 = \sum_{k=0}^{N-1} |c_k|^2$$

A slight generalization of this formula is that involving computation of the correlation between two N-periodic signals $x(n)$ and $y(n)$. If $\{c_k\}$ are the DFT coefficients of $x(n)$ and $\{d_k\}$ the DFT coefficients of $y(n)$, then their cross correlation is given by the formula

$$\frac{1}{N} \sum_{n=0}^{N-1} x(n)\bar{y}(n) = \sum_{k=0}^{N-1} c_k \bar{d}_k$$

9.9 Fast Fourier transform

Let $x(n), n = 0, 1, 2, ..., N-1$ be a signal of duration N. Equivalently, one may assume the signal to be N-periodic. Its discrete Fourier transform is defined by

$$X(k) = \sum_{n=0}^{N-1} x(n)exp(-i2\pi kn/N), k = 0, 1, ..., N-1$$

Equivalently, we may define $X(k)$ for all integers k and then $X(k)$ becomes an N-periodic sequence. Given two N periodic signals $h(n)$ and $x(n)$, their cyclic convolution is defined as

$$y(n) = h(n) \otimes x(n) = \sum_{k=0}^{N-1} h(k)x(n-k)$$

It is easy to see that $y(n)$ is also an N-periodic sequence and that its DFT is given by

$$Y(k) = H(k)X(k)$$

Thus, one way to compute the cyclic convolution between $h(n)$ and $x(n)$ is to compute their DFT's, form their product and compute the inverse DFT of this product. The inverse DFT is easily shown to be given by

$$y(n) = \frac{1}{N} \sum_{k=0}^{N-1} Y(k) exp(i2\pi kn/N)$$

In applications, one can evaluate the usual linear convolution of two finite duration sequences using the DFT. If $x(n)$ has duration N and $h(n)$ has duration M, then the ordinary convolution $h(n)*x(n)$ has duration $N+M-1$. Thus, by padding the signal with zeroes, one may assume that $x(n)$ has duration $N+M-1$ and $h(n)$ also has duration $N+M-1$. We compute the $N+M-1$ point DFT's of these two sequences, form their product and then compute the inverse $N+M-1$ point DFT of this product to obtain the ordinary linear convolution of the two sequences. In effect, linear convolution and filtering can be achieved well provided that we have available with us a fast algorithm for computing the DFT. If $x(n)$ has duration N, then computing its N point DFT involves performing N^2 multiplications. Computing each sample $X(k)$ involves doing N multiplications and since there are N samples of the DFT, we require N^2 complex multiplications. We now discuss two methods by which the DFT can be computed using of the order $Nlog(N)$ multiplications. Assume first that N is even. Then,

$$X(k) = \sum_{n=0}^{N-1} x(n) exp(-i2\pi kn/N)$$

$$= \sum_{m=0}^{N/2-1} x(2m) exp(-i2\pi k2m/N) + \sum_{m=0}^{N/2-1} x(2m+1) exp(-i2\pi k(2m+1)/N)$$

$$= \sum_{m=0}^{N/2-1} x(2m) exp(-i2\pi km/(N/2)) +$$

$$exp(-2\pi ik/N) \sum_{m=0}^{N/2-1} x(2m+1) exp(-i2\pi km/(N/2))$$

Let $G(k)$ denote the $N/2$ point DFT of the sequence $\{x(2m) : m = 0, 1, ..., N/2 - 1\}$ and $H(k)$ the $N/2$ point DFT of the sequence $\{x(2m+1) : m = 0, 1, ..., N/2 - 1\}$. Then we can write the above formula as

$$X(k) = G(k) + exp(-i2\pi k/N)H(k), k = 0, 1, ..., N - 1$$

Since $G(k)$ and $H(k)$ are $N/2$ point DFT's, we have $G(k+N/2) = G(k), H(k+N/2) = H(k)$ and hence the above identity is equivalent to

$$X(k) = G(k) + exp(-i2\pi kn/N)H(k), k = 0, 1, ..., N/2 - 1$$

$$X(k + N/2) = G(k) - exp(-i2\pi kn/N)H(k), k = 0, 1, ..., N/2 - 1$$

Let $F(N)$ denote the number of complex multiplications needed to compute an N-point DFT. Then, the above formulas imply

$$F(N) = F(N/2) + F(N/2) + N/2 = 2F(N/2) + N/2$$

$F(N/2)$ multiplications are needed to determine the sequence $\{G(k)\}$, $F(N/2)$ multiplications are needed for $\{H(k)\}$ and $N/2$ multiplications are needed for forming the products $exp(-i2\pi kn/N)H(k)$, $0, 1, ..., N/2 - 1$. If $N = 2^r$, then we can by this process, reduce the computation of the N point DFT to the computation of 2 point DFT's. Writing $K(r) = F(2^r)$, the above iteration boils down to

$$K(r) = 2K(r-1) + 2^{r-1} = 2(2K(r-2) + 2^{r-2}) + 2^{r-1} = 2^2 K(r-2) + 2^r$$

$$= 2^2(2K(r-3) + 2^{r-3}) + 2 2^{r-1} = 2^3 K(r-3) + 3.2^{r-1}$$

$$= ... = 2^{r-1}K(1) + (r-1)2^{r-1} = r2^{r-1} = rN/2 = \frac{N}{2}log_2(N)$$

since $K(1) = 1$ (To compute a two point DFT only one complex multiplication is needed). The second method is as follows:

$$X(2k) = \sum_{n=0}^{N-1} x(n)exp(-i2\pi kn/(N/2)),$$

$$X(2k + 1) = \sum_{n=0}^{N-1} x(n)exp(-i2\pi n/N)exp(-i2\pi kn/(N/2))$$

Write $N/2 = M$ and using M periodicity of the function $exp(-i2\pi kn/M)$, we get

$$X(2k) = \sum_{n=0}^{M-1} (x(n) + x(n+M))exp(-i2\pi kn/M)$$

$$X(2k + 1) = \sum_{n=0}^{M-1} (x(n) - x(n+M))exp(-i2\pi n/N)exp(-i2\pi kn/M)$$

Now define the M point sequence $g(n) = x(n) + x(n + M), n = 0, 1, ..., M - 1$ and the M point sequence $h(n) = (x(n) - x(n + M))exp(-i2\pi n/N), n = 0, 1, ..., M - 1$. Then $X(2k), k = 0, 1, ..., M - 1$ is the M point DFT of $g(n)$ and $X(2k + 1), k = 0, 1, ..., M - 1$ is the M-point DFT of $h(n)$. The computation of an N point DFT has thus been reduced to the computation of two M point DFT's. The number of multiplications needed to obtain $h(n)$ is $N/2 = M$. Thus, we get the iteration

$$F(N) = 2F(N/2) + N/2$$

which results in the same complexity $\frac{N}{2}log_2(N)$. The former method of obtaining the DFT is called decimation in time and the latter method is decimation in frequency for obvious reasons.

9.10   Random signals in discrete time

A random process in discrete time is simply a sequence of random variables defined on a fixed probability space $(\Omega, \mathcal{F}, P)$. The process is thus a map $X : \mathbb{Z} \times \Omega \to \mathbb{R}$ that is measurable. We can denote the samples of the process by $X(n)$ or by $X(n, \omega)$. The process can equivalently be specified by the joint distribution of its samples:

$$F(x_1, ..., x_k; n_1, ..., n_k) = P(X(n_i) \leq x_i, i = 1, 2, ..., k)$$

$$= P(\{\omega : X(n_i, \omega) \leq x_i, i = 1, 2, ..., k\})$$

$$= P(\bigcap_{i=1}^{k} X(n_i)^{-1}((-\infty, x_i]))$$

The statistics of the process can equivalently be specified by specifying the average values of functions of the samples. For example, if $g : \mathbb{R}^k \to \mathbb{R}$ is measurable, then the average value of $g(X(n_1), ..., x(n_k))$ is given by the integral

$$\mathbb{E}g(X(n_1), ..., X(n_k)) = \int g(X(n_1, \omega), ..., X(n_k, \omega))P(d\omega)$$

The construction of this integral is discussed in standard texts on probability and measure, see for example (An Introduction to Probability and Measure, by K. R. Parthasarathy, published by Hindustan Book Agency). Given a random process $X(n), n \in \mathbb{Z}$, we can look at its moments

$$M_{k+1}(\tau_1, ..., \tau_k; n) = \mathbb{E}(X(n)X(n + \tau_1)...X(n + \tau_k))$$

If $M_2(\tau; n)$ is independent of n, then the process $\{X(n)\}$ is said to be wide sense stationary. Moreover, if $M_{k+1}(\tau_1, ...\tau_k; n)$ is independent of n, the process is said to be $k+1^{th}$ order stationary. This corresponds to the moments depending only upon the time differences between the samples and not on where the set of times has been chosen. If the joint distribution of the samples $\{X(n), X(n + \tau_1), ..., X(n + \tau_k)\}$ is independent of n for all $k \geq 1$ and $\tau_1, ..., \tau_k$, then the process is called strict sense stationary or simply stationary. This corresponds to the statistics of the process being independent to time shifts, i.e., the statistics of the process $\{X(n + r) : n \in \mathbb{Z}\}$ coincides with that of the process $\{X(n) : n \in \mathbb{Z}\}$ for all integers r. Given a wide sense stationary random process $X(n)$ with autocorrelation function $R_X(\tau) = \mathbb{E}X(n)X(n + \tau)$, its power spectral density is defined by the formula

$$S_X(\omega) = \sum_{n=-\infty}^{\infty} R_X(n)exp(-i\omega n)$$

It can be shown that $S_X(\omega)$ equals the limit of $\frac{1}{2N+1}\mathbb{E}|\sum_{n=-N}^{N} X(n)exp(-i\omega n)|^2$ as $N \to \infty$, i.e., $S_X(\omega)$ is indeed the true power spectral density of the process. The total power in the process is

$$P_X = R_X(0) = \frac{1}{2\pi} \int_{-\pi}^{\pi} S_X(\omega)d\omega$$

One can also define the cross power spectral density between two processes $X(n)$ and $Y(n)$. First of all, we need the notion of joint wide sense stationarity of the process. The two processes are said to be jointly wide sense stationary if the correlations $\mathbb{E}(X(n)X(n+\tau)), \mathbb{E}((X(n)Y(n+\tau))$ and $\mathbb{E}(Y(n)Y(n+\tau))$ are independent of the time index n. In other words, the autocorrelation matrix of the vector process $\begin{pmatrix} X(n) \\ Y(n) \end{pmatrix}$ should be independent of the time origin. In general, this matrix has the form

$$\mathbb{E}\left\{ \begin{pmatrix} X(n) \\ Y(n) \end{pmatrix} (X(n+\tau), Y(n+\tau)) \right\} = \begin{pmatrix} \mathbb{E}(X(n)X(n+\tau)) & \mathbb{E}(X(n)Y(n+\tau)) \\ \mathbb{E}(Y(n)X(n+\tau)) & \mathbb{E}(Y(n)Y(n+\tau)) \end{pmatrix}$$

If this is independent of n, then its Fourier transform is the spectral density matrix of the vector process. The diagonal entries of this Fourier transform give respectively the spectral densities of the processes $X(n)$ and $Y(n)$, while the off diagonal entries give the cross spectral density of the two processes: The off diagonal entries are

$$S_{XY}(\omega) = \sum_{n=-\infty}^{\infty} R_{XY}(n)exp(-i\omega n)$$

and

$$S_{YX}(\omega) = \bar{S}_{XY}(\omega)$$

Sometimes, it is preferable to define $R_{XY}(\tau)$ as $\mathbb{E}(X(n)Y(n-\tau))$ rather than $\mathbb{E}(X(n)Y(n+\tau))$. If the processes are complex, we define

$$R_{XY}(\tau) = \mathbb{E}(X(n)\bar{Y}(n-\tau))$$

and $S_{XY}(\omega)$ as its Fourier transform. Suppose $X(n)$ and $Y(n)$ are jointly wide sense stationary complex processes. Suppose, $X(n)$ is passed through a linear time invariant system having impulse response $h(n)$ and $Y(n)$ is passed through a linear time invariant system having impulse response $g(n)$. The respective outputs are then given by

$$U(n) = \sum_k h(k)X(n-k), V(n) = \sum_k g(k)Y(n-k)$$

and their cross correlation is given by

$$R_{UV}(\tau) = \mathbb{E}(U(n)\bar{V}(n-\tau)) = \sum_{k,m} h(k)\bar{g}(m)\mathbb{E}(X(n-k)\bar{Y}(n-m-\tau))$$

$$= \sum_{k,m} h(k)\bar{g}(m)R_{XY}(m+\tau-k)$$

$$= h(\tau) * R_{XY}(\tau) * \bar{g}(\tau)$$

The cross spectral density between the processes $U(n)$ and $V(n)$ is the Fourier transform of this cross correlation function:

$$S_{UV}(\omega) = H(\omega)S_{XY}(\omega)\bar{G}(\omega)$$

In particular,

$$S_{UU}(\omega) = S_{XX}(\omega)|H(\omega)|^2$$

9.11 Quantization noise in a digital filter

Consider a digital filter defined by a linear constant coefficient difference equation

$$y(n) + a_1y(n-1) + ... + a_py(n-p) = b_0x(n) + b_1x(n-1) + ... + b_qx(n-q)$$

Suppose that this filter is implemented using delays and multipliers. The multiplying coefficients a_k, b_j are stored as b bit binary numbers in registers and the samples of the input $x(n)$ and output $y(n)$ are also stored as b bit binary numbers in registers. When two b bit numbers are multiplied, their product will exceed b bits and they will have to be rounded or truncated back to b bits. This rounding/truncation will introduce an error. The error is called quantization error since it results from quantizing the signal samples and multipliers using 2^b level quantizers. The quantization error is usually modeled as a uniformly distributed random variable over the range $[-\Delta/2, \Delta/2]$ or $[0, \Delta]$ where $\Delta = 2^b$ depending on whether rounding or truncation was used. Thus, the quantized filter has the model

$$y(n) + \sum_{k=1}^{p} a_ky(n-k) + \epsilon_k(n) = \sum_{k=0}^{q} b_kx(n-k) + \epsilon'_k(n)$$

Assuming rounding, the $\epsilon_k(n)$ and $\epsilon'_k(n)'s$ are modeled as independent random variables uniformly distributed over $[-\Delta/2, \Delta/2]$. Let $\epsilon(n) = \sum_{k=1}^{p} \epsilon_k(n)$ and $\epsilon(n)' = \sum_{k=0}^{q} \epsilon'_k(n)$. $\epsilon(n)$ is an iid sequence having zero mean and variance $p\sigma^2$ where $\sigma^2 = \int_{-\Delta/2}^{\Delta/2} x^2 dx/\Delta = \Delta^2/12$. Likewise, $\epsilon'(n)$ is an iid sequence with mean zero and variance $q\sigma^2$. Let $h(n)$ denote the impulse response of the system defined by the difference equation

$$y(n) + \sum_{k=1}^{p} a_ky(n-k) = x(n)$$

Then it is easy to see that when the input is zero to the system, the output is simply due to quantization noise and is given by

$$y_q(n) = h(n) * (\epsilon'(n) - \epsilon(n))$$

Since $\epsilon'(n) - \epsilon(n)$ is an iid sequence having zero mean and variance $(p+q)\sigma^2$, it follows that $y_q(n)$ has zero mean and variance

$$\sigma_{y,q}^2 = (p+q)\sigma^2 \sum h(n)^2$$

Suppose that the input $x(n)$ is an iid sequence uniformly distributed over the interval $[-A, A]$. Then, the output of the system due to this input is given by $y(n) = g(n) * x(n)$ where $g(n)$ is the impulse response of the system, i.e.,

$$g(n) = b_0 h(n) + b_1 h(n-1) + \ldots + b_q h(n-q)$$

This output is bounded by

$$|y(n)| = |\sum_k g(k)x(n-k)| \leq A \sum |g(k)|$$

For no overflow to occur, the right side should not exceed unity. This can happen only when $A \leq (\sum |g(k)|)^{-1}$. Taking A as $(\sum |g(k)|)^{-1}$, it follows that the variance of the output is given by

$$\sigma_y^2 = \frac{A^2}{3} \sum g(k)^2 = \frac{\sum g(k)^2}{3(\sum |g(k)|)^2}$$

The maximum signal to quantization noise ratio achievable is then given by the ratio $\frac{\sigma_y^2}{\sigma_{y,q}^2}$. These formulae can be specialized to first and second order filters.

9.12 Least squares method of filter design

We are given an unknown system and we wish to model it as an FIR filter. To do so, we give input $x(n)$ to this filter and measure the output $y(n)$. If $\{h(0), h(1), \ldots, h(p)\}$ is the impulse response of the approximating FIR filter, then we must have approximately

$$y(n) \approx \hat{y}(n) = \sum_{k=0}^{p} h(k)x(n-k)$$

We measure the output samples at times $n = p, p+1, \ldots, N$ and choose $h(0), \ldots, h(p)$ so that $\sum_{n=p}^{N}(y(n) - \hat{y}(n))^2$ is a minimum, that is, $h(k), k = 0, 1, \ldots, p$ satisfy the optimal normal equations

$$\frac{\partial}{\partial h(k)} \sum_{n=p}^{N}(y(n) - \hat{y}(n))^2 = 0, \, k = 0, 1, \ldots, p$$

These result in

$$\sum_{n=p}^{N}(y(n) - \hat{y}(n))x(n-k) = 0, \, k = 0, 1, \ldots, p$$

or

$$h(0) \sum_{n=p}^{N} x(n)x(n-k) + h(1) \sum_{n=p}^{N} x(n-1)x(n-k) + \ldots$$

$$+ h(p) \sum_{n=p}^{N} x(n-p)x(n-k) + \sum_{n=p}^{N} y(n)x(n-k) = 0, k = 0, 1, \ldots, p$$

The solution to these equations yields the optimal approximating FIR filter. This is called the least squares method of filter design. Sometimes we use the same term in another context. We are given a transfer function $G(\omega), \omega \in [-\pi, \pi)$ for a discrete time filter and we wish to construct an FIR filter having a transfer function close to this prescribed one. If $h(0), \ldots, h(p)$ are the FIR taps, then the least squares criterion involves selecting these so that $\int_{-\pi}^{\pi} |G(\omega) - H(\omega)|^2 d\omega$ is a minimum. In view of the Parseval identity, this amounts to minimizing $\sum_{n=-\infty}^{\infty} |g(n) - h(n)|^2$ and since $h(n) = 0$ for $n < 0$ and $n > p$, this is minimized when $h(n) = g(n), n = 0, 1, \ldots, p$, i.e., when the FIR filter impulse response is a truncated version of the given impulse response. One usually chooses a window function $w(n), n = 0, 1, \ldots, p$ that tapers gradually towards the ends and uses $g(n)w(n)$ as the designed impulse response. The window function is used in order to prevent a rapid decay of the impulse response to zero which causes the frequency spread to be large thereby being unsuitable if the filter to be designed is lowpass.

9.13　Introduction to adaptive filter theory

Suppose the parameters of a system vary slowly with time. Then it is pointless to talk about estimating these parameters using a block processing scheme, i.e., collecting all the input and output data and constructing estimates of the parameters from these. One should rather use a real time adaptive processing scheme which involves a continuous updating of the parameters based on recently collected input-output data. The updating would generally involve using the error involved in the construction of the output using previously obtained parameters. The simplest such adaptive algorithm is the least mean square algorithm which can be described as follows. We try to approximate the system by an FIR filter with slowly time varying coefficients $h(0, n), h(1, n), \ldots, h(p, n)$. Let $\{x(n)\}$ denote the input to this system and $\{y(n)\}$ the true output. The output constructed using the filter parameters at time n is given by

$$\hat{y}(n) = h(0, n)x(n) + h(1, n)x(n-1) + \ldots + h(p, n)x(n-p)$$

If $x(n)$ and $y(n)$ were jointly wide sense stationary, then the estimation error mean square value $\mathbb{E}(y(n) - h(0)x(n) - h(1)x(n-1) - \ldots - h(p)x(n-p))^2$ is independent of n and we can obtain estimates of the coefficients based on the correlations $R_{xx}(k), R_{xy}(k)$ by minimizing this mean squared error. These estimates will be independent of the time index n. The actual values of the correlations can be estimated using time averages. These will be good estimates provided the signals are jointly ergodic, for which stationarity is a necessary condition. Then for example, the correlation $R_{xy}(m)$ can be estimated as $\frac{1}{N}\sum_{n=0}^{N-1} x(n+m)y(n)$. If however the true parameters of the system vary with time, then $x(n)$ and $y(n)$ will no longer be jointly stationary and the mean square error will not carry any value. One can however modify this method slightly and thereby derive an adaptive algorithm for estimating the slowly time varying coefficients. The idea is to consider the gradient method of minimizing the mean square error. This involves using the following updating scheme:

$$h(k, n+1) = h(k, n) - \mu \frac{\partial}{\partial h(k, n)} \mathbb{E}(y(n) - h(0, n)x(n) - h(1, n)x(n-1) - \dots$$

$$-h(p, n)x(n-p))^2, k = 0, 1, ..., p$$

If the processes are jointly stationary, then it is easy to see that under certain restrictions on the range of the constant μ, the coefficients $h(k, n), k = 0, 1, ..., p$ will converge to the solution to the optimal normal equations

$$\frac{\partial}{\partial h(k)} \mathbb{E}(y(n) - h(0)x(n) - h(1)x(n-1) - \dots - h(p)x(n-p))^2 = 0, k = 0, 1, ..., p$$

When the processes are nonstationary, then we remove the expectation sign and obtain the least mean square algorithm

$$h(k, n+1) = h(k, n) + 2\mu e(n)x(n-k), k = 0, 1, ..., p$$

where

$$e(n) = y(n) - h(0, n)x(n) - h(1, n)x(n-1) - \dots - h(p, n)x(n-p)$$

is the estimation error based on the filter estimates at the epoch n. This is a stochastic algorithm and the important thing to recognize here is that the estimation error at time n is fed back into the algorithm to update the filter parameters. The convergence of the coefficients is not perfect here. The asymptotic values of the coefficients fluctuate randomly and expressions for the asymptotic mean square error can be obtained, but we are not interested in those expressions in this introduction. The true gradient of $\mathbb{E}e(n)^2$ with respect to $h(k)$ is $-2\mathbb{E}(e(n)x(n-k))$ and our algorithm involves removing the expectation sign in this expression, i.e., approximating the gradient by $-2e(n)x(n-k)$. A better approximation would involve averaging this product over a block of L data samples, thereby resulting in the block least mean square algorithm:

$$h(k, m+1) = h(k, m)$$

$$+(2\mu/L)\mu \sum_{n=mL}^{(m+1)L-1} (y(n) - h(0, m)x(n) - h(1, m)x(n-1) - \dots$$

$$-h(p, m)x(n-p))x(n-k), k = 0, 1, ..., p$$

9.14   Study Projects

Study project 1

Consider a filter, i.e. a linear time invariant system having impulse response $h(t) = \frac{sin(at)}{\pi t}$. Study the effect of this filter on various kinds of input signals, specifically on the signals (a) $x(t) = \frac{sin(bt)}{\pi t}$, (b) $x(t) = \sum_{k=1}^{\infty} c(k)sin(kbt)$, (c) $x(t) = \frac{sin(ct)^2}{(\pi t)^2}$.

Study project 2

A seminar presentation on basic techniques linear signal theory can be given. The issues discussed herein can include the following. Notion of a linear transformation on a vector space and how this idea can be used to define what a linear system acting on a signal space is. First the idea of a vector space should be given as a collection of objects V and a field $\mathbb{F}$ with two binary operations, one vector addition $+ : V \times V \rightarrow V$ and two scalar multiplication $. : \mathbb{F} \times V \rightarrow V$. The vector addition operation satisfies the commutativity and associativity axiom and the scalar multiplication satisfies the distributivity axiom, that is $x + y = y + x, x + (y + z) = (x + y) + z$ for all vectors $x, y, z \in V$. There is zero vector 0 defined by the rule $x + 0 = 0 + x = x$ and for each vector $x \in V$, we have a vector $-x$ satisfying $x + (-x) = 0$. Scalar multiplication satisfies $(c_1.c_2).x = c_1.(c_2.x)$ where c_1, c_2 belong to the field $\mathbb{F}$ and x belongs to the vector space V. By distributivity of scalar multiplication over vector addition, we mean that $c.(x + y) = c.x + c.y$ and by distributivity of scalar addition over scalar multiplication with a vector, we mean that $(c_1 + c_2).x = c_1.x + c_2.x$ where $c_1, c_2 \in \mathbb{F}$ and $x \in V$. Examples of vector spaces can be included here, the typical ones being $\mathbb{R}^n, \mathbb{C}^n$, All one sided sequences of real numbers, i.e. elements of the kind $(x_0, x_1, ...)$ with $x_j \in \mathbb{R}$ and with addition of two sequences being defined componentwise, i.e.

$$(x_0, x_1, ...) + (y_0, y_1, ...) = (x_0 + y_0, x_1 + y_1, ...)$$

and scalar multiplication also defined componentwise:

$$c(x_0, x_1, ...) = (cx_0, cx_1, ...)$$

Similarly for two sided real sequences. Instead of assuming real sequences, one can assume complex sequences and more generally sequences with entries taken from any field. These are all infinite dimensional vector spaces and are very important for signal analysis. The notion of a linear transformation or operator acting on a vector space follows next. For finite dimensional spaces, these operators are expressed using matrices, i.e. $\mathbf{x} \rightarrow \mathbf{Ax}$. In fact we can in this equation assume that the input vector space is $\mathbb{R}^n$ and the output vector space is $\mathbb{R}^m$ so that $\mathbf{A}$ becomes an $m \times n$ matrix with real entries. The more general idea of a linear transformation on a vector space can be introduced here as a mapping $T : V \rightarrow W$ from the input vector space V into the output vector space W preserving linear combinations, i.e. $T(cx + c'x') = cT(x) + c'T(x')$ for all $x, x' \in V, c, c' \in \mathbb{F}$. The idea of how an ordered basis in a vector space can be used to define the coordinates of a vector should be mentioned here. That is, we choose vectors $\{e_1, e_2, ..., \}$ in the vector space V such that any $x \in V$ can be expressed as a finite linear combination of these vectors: $x = \sum_{i=1}^{N} c_i.e_{\cdot i}$ for some sufficiently large positive integer N. These vectors are to be chosen as linearly independent, i.e. no finite linear combination of these vectors vanishes unless all the coefficients of the combination are zero. In other words, the set of the $e_i's$ is a linearly independent spanning set. For finite dimensional spaces, this set can be chosen to be finite. For infinite dimensional spaces, the set is infinite and is called a Hamel basis. Likewise one chooses a basis $\{f_1, f_2, ...,\}$ for the output vector space W. If $T : V \to W$ is a linear transformation from the input space V into the output space W, then $T(e_i)$ can be expressed as a finite linear combination of the $f_i's$, that is

$$T(e_i) = \sum_{j=1}^{N} a_{ji} f_j, i = 1, 2, ...$$

where for each i, at most a finite number of the $a'_{ji}s$ are non zero. Then the matrix $\mathbf{A} = ((a_{ij}))$ is called the matrix of the linear transformation T relative to the bases $B = \{e_1, e_2, ...\}$ and $B' = \{f_1, f_2, ...\}$ for the two spaces V, W. The notation used for this is

$$\mathbf{A} = [T]_{B,B'}$$

Linear systems in discrete time can be cast in this form. If $(x_n)_{n=0}^{\infty}$ is the input sequence the output of a linear system has the form $y_n = \sum_{m=0}^{\infty} h(n, m)x_m, n \geq 0$ and $\mathbf{H} = ((h(n, m)))_{0 \leq n,m < \infty}$ is the matrix of the linear system relative to the standard basis. The standard basis consists of the unit impulse and its delayed versions, i.e. $\delta(n - k), k = 0, 1, ...$ The Input signal is expressible as a linear combination of these basic signals in an elementary fashion:

$$x_n = \sum_{k=0}^{\infty} x_k \delta(n - k)$$

In abstract terms, denoting the signal $\delta(n - k)$ by δ_k, we have

$$x = \sum_{k=0}^{\infty} x_k \delta_k$$

and the action of the linear system on this input is given by

$$y = T(x) = \sum_{k=0}^{\infty} x_k T(\delta_k)$$

Writing

$$T(\delta_k) = \sum_{m=0}^{\infty} h(m, k)\delta_m$$

one finds that

$$y = \sum_{k=0}^{\infty} x_k \sum_{m=0}^{\infty} h(m, k)\delta_m$$

from which we deduce that y has samples given by $y(m) = \sum_{k=0}^{\infty} h(m, k)x_k$. Thus coordinate representations of vectors and linear transformations on vector spaces enable one to conceptualize the notion of signals and linear systems. The coordinate representation of vectors involves expressing $x \in V$ as $x = \sum_{k=1}^{\infty} x_k e_k$ where the $e'_k s$ form a Hamel basis for the vector space. In the example of signals, we are considering the vector space to consist of all one sided real sequences $(x_0, x_1, x_2, ...)$ having at most a finite number of nonzero $x'_i s$ and to represent the signal as $x = \sum_{k=0}^{\infty} x_k \delta_k$ so that x_k is the kth signal sample. The sum has at most a finite number of nonzero terms. The sequence (x_k) is called the coordinate representation of the signal x relative to the basis B.

Study project 3

Norm on a vector space is an important concept. This can be discussed in a seminar. This is to be introduced first via the abstract definition, i.e. properties and then followed up with examples and applications. The discussion can begin with the statement that norm defines something like the size of a vector. If V is a vector space and we have a map $\| \cdot \|: V \to [0, \infty)$ satisfying the properties (a) $\| x \| \geq 0$, (b) $\| x \| = 0$ iff $x = 0$, (c) $\| c.x \| = |c| \| x \|$ for all scalars c and all vectors x, (d) $\| x + y \| \leq \| x \| + \| y \|$ for all vectors x, y, then this map is called a norm on the linear space. The most characteristic example is the Euclidean vector norm on $\mathbb{R}^n$ and $\mathbb{C}^n$. If $x = (x_1, ..., x_n) \in \mathbb{C}^n$, then its Euclidean norm is given by

$$\| x \| = (\sum_{i=1}^{n} |x_i|^2)^{1/2}$$

All properties except the triangle inequality are obvious. The Triangle inequality is proved using the Cauchy inequality:

$$|\sum x_i y_i|^2 \leq \sum |x_i|^2 \sum |y_i|^2$$

which is in turn proved by considering

$$0 \leq \sum |x_i + zy_i|^2 = \sum |x_i|^2 + |z|^2 \sum |y_i|^2 + 2Re(\bar{z} \sum x_i \bar{y}_i)$$

$$= \sum |y_i|^2 (|z + (\sum x_i \bar{y}_i) / \sum |y_i|^2) + \sum |x_i|^2 - |\sum x_i \bar{y}_i|^2 / \sum |y_i|^2$$

When z is chosen to make the first term here zero, the result is the Cauchy inequality. Another important example of the norm on $\mathbb{C}^n$ is the L^1 norm:

$$\| x \|_1 = \sum |x_i|$$

The other important norm is the L^p norm. If $\mathbf{x} = (x_1, ..., x_N)$ is a vector in $\mathbb{C}^N$, its L^p norm is given by

$$\| \mathbf{x} \|_p = (\sum |x_i|^p)^{1/p}$$

This defines a norm provided that $p \geq 1$. The triangle inequality alone here requires a proof. It is proved using the Holder inequality according to which if $p, q > 1$ and $1/p + 1/q = 1$, then for any two complex vectors $(x_1, ..., x_N)$ and $(y_1, ..., y_N)$, one has

$$\left| \sum x_i \bar{y}_i \right| \leq \left(\sum |x_i|^p \right)^{1/p} \left(\sum |y_i|^q \right)^{1/q}$$

The same inequality is also valid for infinite sequences. The proof is based on the inequality

$$a^p/p + b^q/q \geq ab$$

for a, b positive. The student while presenting his seminar should present the proof of this elementary inequality first. Plug into this $a = |x_i|/ \parallel x \parallel_p, b = |y_i|/ \parallel y \parallel_q$ to get

$$|x_i|^p/p \parallel x \parallel_p^p + |y_i|^q/q \parallel x \parallel_q^q \geq |x_i y_i|/ \parallel x \parallel_p \parallel y \parallel_q$$

Summing over i then results in Holder's inequality. The inequality $a^p/p + b^q/q \geq ab$ with a replaced by $a^{1/p}$ and b by $b^{1/q}$ reads as

$$a/p + b/q \geq a^{1/p} b^{1/q}$$

which states that the arithmetic mean of a pair of numbers cannot be lesser than the geometric mean. The derivation of the triangle inequality from Holder's inequality is a little tricky affair and this can be presented in a seminar. For dealing with signals, one needs to introduce norms on function spaces. This can be presented as the concluding part of the seminar. For example let $\mathbb{C}[a, b]$ denote the space of continuous functions defined on the closed interval $[a, b]$. The L^p norm of the function x in this space is defined by

$$\parallel x \parallel_p = \left(\int_a^b |x(t)|^p dt \right)^{1/p}$$

This notion can be generalized to multidimensional signal spaces. For example on the space $C([a, b]^k)$ consisting of all k dimensional continuous signals $x(t_1, ..., t_k) a \leq t_1, ..., t_k \leq b$, we can introduce the L^p norm as

$$\parallel x \parallel_p = \left(\int_a^b ... \int_a^b |x(t_1, ..., t_k)|^p dt_1 ... dt_k \right)^{1/p}$$

Norms on sequence spaces are also important. For example, on the space of one sided sequences $(x_n)_{n=0}^\infty$ with $\sum_{n=0}^\infty |x_n|^p < \infty$ where $p \geq 1$, one can introduce the l^p-norm of the sequence $(x_n)_{n=0}^\infty$ as

$$\parallel x \parallel_p = \left(\sum_{n=0}^\infty |x_n|^p \right)^{1/p}$$

Using Holder's inequality for infinite sequences, one can once again deduce the triangle inequality. The L^p measure spaces can also be introduced while discussing norms. This involves introducing the measure space $(\Omega, \mathcal{F}, \mu)$ where $\mathcal{F}$ is a σ field of Ω subsets and the notion of measurability and integrability of functions, i.e. definition of $\int_\Omega f(x)d\mu(x)$ for an appropriate class of functions f. The $L^p - norm$ of f is then defined as

$$\| f \|_p = (\int_\Omega |f(x)|^p d\mu(x))^{1/p}$$

These issues are perhaps better suited for a seminar on basic theorems of measure theory.

Study project 4

An interesting seminar can be delivered on the basic aspects of Fourier series for continuous time periodic signals. The seminar should start with discussing about both the real and complex Fourier series representations, how the coefficients are computed from the signal and then introduce the basic properties of the Fourier series. An experiment can then be designed on verifying the properties of the Fourier series via computer experiments. The discussion can begin by generating computer plots of finite trigonometric sums of the kind

$$x(t) = a_0 \sum_{n=1}^{N} a_n cos(n\omega t) + b_n sin(n\omega t)$$

Simulation of this signal over one time period $T = 2\pi/\omega$ is carried out by dividing the time interval $[0, T]$ into N equal parts where $K \approx 1000$. The sequence $x(k\Delta), k = 0, 1, ..., 999$ is then plotted on a graph. Next, one introduces infinite sums of the kind

$$x(t) = a_0 + \sum_{n=1}^{\infty} a_n cos(n\omega t) + b_n sin(n\omega t)$$

and the Fourier integrals that tell us how to compute the coefficients $\{a_n, b_n\}$ in terms of the signal $\{x(t)\}$:

$$a_n = \frac{2}{T} \int_0^T x(t)cos(n\omega t)dt \qquad b_n = \frac{2}{T} \int_0^T x(t)sin(n\omega t)dt, n \geq 1$$

$$a_0 = \frac{1}{T} \int_0^T x(t)dt$$

The numerical evaluation of these Fourier integrals should be illustrated via computer programmes. For example to evaluate a_n, one would use the Riemann sum

$$\hat{a}_n = \frac{2}{T} \sum_{k=0}^{K-1} x(k\Delta)cos(nk\omega\Delta)\Delta$$

A comparison of a_n and $\hat{a}_n$ should be made for increasing values of K and the reconstruction of the signal from the partial trigonometric sums using the approximate value $\hat{a}_n$ of a_n should be discussed.

Study project 5

By means of examples, deduce the criteria on the locations of the poles and zeroes of the transfer function of a linear time invariant system for its stability and causality. The transfer function may be assumed to have a rational structure. The seminar can begin with the definition of the stability of a system as that which carries every bounded input to a bounded output. The definitions of boundedness should be given here. Now if $\{h(t)\}$ is the impulse response of a linear time invariant system, then it must be shown that a necessary and sufficient condition for the system to be stable is that $\{h(t)\}$ be integrable, i.e. $\int_{-\infty}^{\infty} |h(t)|dt < \infty$. The next step is to show that every rational transfer function can be expressed using partial fractions in the form

$$H(s) = P(s) + \sum_{k=1}^{d} \sum_{m=1}^{r_k} \frac{A_{k,m}}{(s-s_k)^m}$$

where $P(s)$ is a polynomial. Then one deduces by inverse transforming this that if the system is causal in addition, one must choose the region of convergence of the Laplace transform to be to the right of all the poles, i.e., $Re(s) > Re(s_k), k = 1, 2, ..., d$. Letting $P(s) = a_0 + a_1 s + ... + a_p s^p$, gives on inverse transforming the above function subject to the causality constraint,

$$h(t) = a_0\delta(t) + a_1\delta'(t) + ... + a_p\delta^{(p)}(t) + \sum_{k=1}^{d} \sum_{m=1}^{r_k} A_{k,m} t^{m-1} exp(s_k t)u(t)/(m-1)!$$

For this transform to be integrable, we require that $Re(s_k) < 0$ for all k, i.e. all the poles of the system fall to the left of the $j\omega$ axis and simultaneously, $a_1 = ... = a_p = 0$, i.e. the system should have no zero at $s = \infty$, or equivalently, the numerator polynomial of $H(s)$ have a degree not exceeding that of the denominator. Elaboration of the above points should be carried out in the seminar.

Study project 6

For dealing with linear signal theory, one of the projects has already emphasized the need of the idea of a vector space over a field. The definition of a field in terms of closure under operations of addition and multiplication should be dealt with along with the distributive property. Examples of fields like $\mathbb{R}, \mathbb{C}$ that one comes across in statistical signal processing and the finite fields that one comes across in coding theory should be covered in the seminar. The notion of a vector or a linear space as a set with the binary operations of addition and scalar multiplication should be introduced in the course of the seminar. Vector addition as a mapping $(x, y) \in V \times V \to x + y \in V$ and scalar multiplication as a mapping $(c, x) \in \mathbb{F} \times V \to c.x \in V$ should be dealt with in the seminar. Examples taken from $\mathbb{F}^n$ where each vector $x \in \mathbb{F}^n$ has the form

$$x = \begin{pmatrix} x_1 \\ x_2 \\ ... \\ x_n \end{pmatrix}$$

with $x_i \in \mathbb{F}$. The prototype of addition is provided by

$$x + y = \begin{pmatrix} x_1 + y_1 \\ x_2 + y_2 \\ ... \\ x_n + y_n \end{pmatrix}$$

where $x = [x_1, ..., x_n]^T, y = [y_1, ..., y_n]^T$. How this rule for vector addition in $\mathbb{F}^n$ satisfies commutativity and associativity $(x + y) = (y + x), x + (y + z) = (x + y) + z$ should be explained. More generally, the idea of a vector defined by a class of real valued or complex valued functions on a set should be explained. This idea is fundamental in signal theory. If X is a set, then a vector is a function $f : X \to \mathbb{R}$. If X is a finite set, say $\{x_1, ..., x_n\}$, then the vector can be represented in column form as

$$f = \begin{pmatrix} f(x_1) \\ f(x_2) \\ ... \\ f(x_n) \end{pmatrix}$$

The sum of two functions f and g defined on the set X into $\mathbb{R}$ is computed pointwise, i.e. $(f + g)(x) = f(x) + g(x), x \in X$. This can be regarded as vector addition. When the set X is finite, this scheme corresponds to the usual addition of vectors in $\mathbb{F}^n$:

$$\begin{pmatrix} f(x_1) \\ f(x_2) \\ ... \\ f(x_n) \end{pmatrix} + \begin{pmatrix} g(x_1) \\ g(x_2) \\ ... \\ g(x_n) \end{pmatrix}$$

$$= \begin{pmatrix} f(x_1) + g(x_1) \\ f(x_2) + g(x_2) \\ ... \\ f(x_n) + g(x_n) \end{pmatrix}$$

The space of all functions $f : X \to \mathbb{R}$ is denoted as $\mathbb{R}^X$ and we thus have proved that for finite sets X, $dim(\mathbb{R}^X) = \mu(X)$, with $\mu(E)$ denoting the number of elements in the set E. Elementary examples of vector spaces like $\mathbb{R}^2$ and $\mathbb{R}^3$ can be introduced. $\mathbb{R}^2$ as consisting of all ordered pairs (a, b) with $a, b \in \mathbb{R}$. Then, one notes that any such vector can be expressed as $a(1, 0) + b(0, 1)$. The vectors $\{(1, 0), (0, 1)\}$ span the whole of $\mathbb{R}^2$ and are linearly independent. The notion of a spanning set of vectors should be introduced before talking about linear independence. If S is a set of vectors in the vector space V, then S is called a spanning set if every $x \in V$ can be expressed as $x = \sum_{i=1}^{n} c_i x_i$ for some finite subset $\{x_1, ..., x_n\}$ of vectors in S. For example, the vectors $(1, 0), (1, 1), (0, 2)$ span $\mathbb{R}^2$, but from this set, we can extract a smaller set $(1, 0), (0, 2)$ that also spans $\mathbb{R}^2$. In general, we have the following rule: If a set S of vectors spans the vectors space V and if some vector $x_0 \in S$ can be expressed as a finite linear combination of some other vectors of S, S with x_0 deleted will also span V. Continuing in this fashion, we can extract a minimal spanning set and such a set is called a basis. Thus a linear independent spanning set is called a basis. A set S of vectors is said to be linearly independent provided the following holds: If $\{v_1, ..., v_m\}$ is any finite set of distinct vectors in S then the only linear combination $c_1 v_1 + ... + c_m v_m$ of these vectors that vanishes is when all the $c_i's$ are zero. For example, consider the vectors $v_1 = (1,0), v_2 = (0,1), v_3 = (1,1)$ in $\mathbb{R}^2$. We have $v_1 + v_2 - v_3 = 0$ implying that a nontrivial linear combination of these vectors vanishes and hence this set of three vectors is linearly dependent. If S is a basis for V consisting of a finite number of vectors, then we can order S as $S = \{v_1, ..., v_n\}$. Any vector $v \in V$ can then be expressed uniquely as a linear combination of these vectors, i.e. $v = c_1 v_1 + ... + c_n v_n$. That any vector can be expressed as such a linear combination follows from the spanning property of S. That such a representation is unique follows from the linear independence of S. Specifically, if $\sum c_i v_i = \sum c_i' v_i$ implies $\sum (c_i - c_i') v_i = 0$ and this implies $c_i - c_i' = 0$ for all i proving uniqueness of the representation. We say that the vector $[c_1, ..., c_n]^T \in \mathbb{R}^n$ equals $[v]_S$, the coordinate representation of v relative to the basis S. For any basis B of a vector space V having dimension n, the mapping $x \in V \to [x]_B \in \mathbb{R}^n$ is linear and is in fact an isomorphism.

Study project 7

Discuss the notion of region of convergence for the Laplace transforms. The following material can be included. If $x(t)$ is a signal, the two sided Laplace transform is defined by

$$X(s) = \int_{-\infty}^{\infty} x(t) exp(-st) dt$$

The one sided Laplace transform is the two sided transform of $x(t)u(t)$, i.e.

$$X_{os}(s) = \int_{0}^{\infty} x(t) exp(-st) dt$$

Taking s to be the complex number $s = \sigma + j\omega$, we find that

$$|X(s)| \leq \int_{-\infty}^{\infty} |x(t)| exp(-\sigma t) dt$$

This equation implies that the set of complex numbers s for which the transform converges is decided completely by its real part $Re(s)$.

Study project 8

A short seminar introducing the basic concepts of digital signal processing would be useful for an audience consisting of students who have done a basic course in signals and systems and wish to develop computer programmes for filter implementation, spectral analysis and other important signal processing problems. The following aspects of digital signal processing may be discussed:

The seminar can begin by introducing the notion of a discrete time signal obtained by sampling a continuous time signal $x(t)$: $x[n] = x(nT)$. How the sampling rate is chosen may be explained with reference to Shannon's sampling theorem: If $\{x(t)\}$ is a continuous time signal with Fourier transform $X(j\Omega) = \int_{-\infty}^{\infty} x(t)exp(-j\Omega t)dt$, then $\{x(t)\}$ is said to be bandlimited to $[-\sigma, +\sigma]$ radians per second if $X(j\Omega) = 0$ for $|\Omega| > \sigma$. Such a signal can be reconstructed from its samples taken T seconds apart when $T < \pi/\sigma$. The proof of this theorem is to be illustrated. The following one would do. Let $x_\delta(t) = \sum_n x(nT)\delta(t - nT)$, the impulse sampled train of the signal. Then one notes that $x_\delta(t) = x(t) \sum_n \delta(t - nT)$ and hence its Fourier transform equals

$$X_\delta(j\Omega) = \frac{1}{2\pi}X(j\Omega) * \sum_n exp(-j\Omega nT)$$

$$= \frac{1}{T}X(j\Omega) * \sum_n \delta(\Omega - 2n\pi/T) = \frac{1}{T}\sum_n X(j(\Omega - 2\pi n/T))$$

It follows that when $T < \pi/\sigma$ we have for $|\Omega| < \sigma$, the equation

$$X_\delta(j\Omega) = \frac{1}{T}X(j\Omega)$$

This is the same as saying that if $H(j\Omega)$ is a lowpass filter that equals unity for $|\Omega| < \sigma$ and zero for $|\Omega| > \sigma$, then

$$X(j\Omega) = T.H(j\Omega)X_\delta(j\Omega)$$

from which one deduces the Shannon reconstruction formula

$$x(t) = T \sum_n x(nT)h(t - nT)$$

The two dimensional sampling theorem can then be presented. Instead, one may directly state the multidimensional version of the Shannon theorem: Let $x(t_1, ..., t_k)$ be a k dimensional signal with multivariate Fourier transform

$$X(j\Omega_1, ..., j\Omega_k) = \int_{-\infty}^{\infty} ... \int_{-\infty}^{\infty} x(t_1, ..., t_k)exp(-j(\Omega_1 t_1 + ... + \Omega_k t_k))dt_1...dt_k$$

Assume that $X(j\Omega_1, ..., j\Omega_k) = 0$ when $|\Omega_r| > \sigma_r$ for some $r = 1, 2, ..., k$. Then taking $T_r < \pi/\sigma_r, r = 1, 2, ..., k$, one defines the multidimensional impulse train

$$x_\delta(t_1, ..., t_k) = \sum_{n_1,...,n_k=-\infty}^{\infty} x(n_1 T_1, ..., n_k T_k)\delta(t_1 - n_1 T_1)...\delta(t_k - n_k T_k)$$

We note that this train can be expressed as

$$x(t_1, ..., t_k). \sum_{n_1,...,n_k=-\infty}^{\infty} \delta(t_1 - n_1 T_1)...\delta(t_k - n_k T_k)$$

which on multivariate Fourier transforming gives

$$X_\delta(j\Omega_1, ..., j\Omega_k) = \frac{1}{T_1...T_k}X(j\Omega_1, ..., j\Omega_k) * \sum_{n_1,...,n_k} \delta(\Omega_1 - 2n_1\pi/T_1)...\delta(\Omega_k - 2n_k\pi/T_k)$$

$$= \frac{1}{T_1...T_k} \sum_{n_1,...,n_k} X(j(\Omega_1 - 2n_1\pi/T_1), ..., j(\Omega_k - 2n_k\pi/T_k))$$

so that if $H(j\Omega_1, ..., j\Omega_k)$ is a lowpass filter that equals unity when $|\Omega_r| < \sigma_r, r = 1, 2, ..., k$ and zero when any one of the inequalities $|\Omega_r| > \sigma_r$ holds then

$$x(t_1, ..., t_k) = T_1...T_k \sum_{n_1,...,n_k} x(n_1T_1, ..., n_kT_k)h(t_1 - n_1T_1, ..., t_k - n_kT_k)$$

It is easy to see that the function h is separable.

Study project 9

A short seminar on the discrete time Fourier transform can be delivered from the standpoint of processing of discrete time signals. If $\{x(t), t \in \mathbb{R}\}$ is an analog signal which is bandlimited to $[-\sigma, \sigma)$ radians per second, then it is known from the Shannon sampling theorem that the signal can be recovered from its samples taken at the rate of $1/T$ Hertz where $T < \pi/\sigma$. Thus, one can replace the analog signal by its sampled sequence $\{x(nT)\}$ which in turn can be replaced by its discrete time Fourier transform

$$X_d(exp(j\omega)) = \sum_{n=-\infty}^{\infty} x(nT)exp(-jn\omega)$$

This first thing to note is that the discrete time Fourier transform, also called the DTFT of the signal has period 2π. This follows from the 2π periodicity of the function $exp(-jn\omega)$ when n is any integer. The second thing to note is that the DTFT is defined provided that the signal is summable, i.e. $\sum_n |x(nT)| < \infty$. This follows from the triangle inequality

$$|\sum_n x(nT)exp(-jn\omega)| \le \sum_n |x(nT)|$$

Finiteness of the right side then implies finiteness of the left side. The third observation to be presented is that for an arbitrary continuous time signal, what is the relationship between the continuous time Fourier transform and the discrete time Fourier transform. This relationship can be obtained by the use of the inverse of the continuous time Fourier transform

$$x(t) = \frac{1}{2\pi} \int_{-\infty}^{\infty} X(j\Omega)exp(j\Omega t)d\Omega$$

where

$$X(j\Omega) = \int_{-\infty}^{\infty} x(t)exp(-j\Omega t)dt$$

is the continuous time Fourier transform of the original analog signal $\{x(t)\}$. Thus,

$$X_d(exp(j\omega)) = \sum_{n=-\infty}^{\infty} \frac{1}{2\pi} \int_{-\infty}^{\infty} X(j\Omega)exp(j\Omega nT)exp(-j\omega n)d\Omega$$

$$= \frac{1}{2\pi} \int_{-\infty}^{\infty} X(j\Omega)d\Omega \sum_{n=-\infty}^{\infty} exp(jn(\Omega T - \omega))$$

$$= \frac{1}{T} \int_{-\infty}^{\infty} X(j\Omega)d\Omega \sum_{n} \delta(\Omega - \omega/T - 2n\pi/T)$$

$$= \frac{1}{T} \sum_{n} X(j(\omega + 2n\pi)/T)$$

This is the exact relation between the continuous time Fourier transform and the discrete time Fourier transform valid even in the case when the signal is not bandlimited or when it is bandlimited but T is not selected to meet the Nyquist criterion. In particular, when $X(j\Omega) = 0$ for $|\Omega| > \pi/T$, one has for $|\omega| < \pi$ the equation

$$X_d(exp(j\omega)) = \frac{1}{T}X(j\omega/T)$$

This equation states that the discrete time Fourier transform for bandlimited signals with the sampling rate meeting the Nyquist criterion is simply a scaled version of the continuous time Fourier transform. The talk can then discuss the inversion formula for the DTFT, namely

$$x(n) = \frac{1}{2\pi} \int_{-\pi}^{\pi} X(exp(j\omega))exp(j\omega n)d\omega$$

The application of the DTFT to the solution of linear difference equations with constant coefficients can then be discussed. Firstly what must be discussed is how in practice does such a difference equation arise while attempting to numerically solve a linear differential equation in continuous time using the process of discretization. The example of the LCR circuit can be given here, the governing equation being

$$Lq''(t) + Rq'(t) + q(t)/C = v(t)$$

which on discretization, leads to the difference equation

$$\frac{L}{\Delta^2}(q[n] - 2q[n-1] + q[n-2]) + \frac{R}{\Delta}(q[n] - q[n-1]) + q[n]/C = v[n]$$

This can be cast in the form

$$a_0 q[n] + a_1 q[n-1] + a_2 q[n-2] = v[n]$$

and Fourier analyzing gives

$$\frac{Q(exp(j\omega))}{V(exp(j\omega))} = (a_0 + a_1 exp(-j\omega) + a_2 exp(-j2\omega))^{-1} = H(exp(j\omega))$$

say. The convolution theorem for the discrete time Fourier transform can be introduced at this stage explaining how this leads to the following solution to the above difference equation

$$q[n] = \sum_{k=0}^{n} h[k] v[n-k]$$

Study project 10

A short seminar on identification of linear time invariant systems using correlation methods would be of interest to an audience wishing to pursue research projects in statistical signal processing. The seminar can begin with the definition of a linear time invariant system in continuous time giving the prototype example of a first order linear system defined by the differential equation

$$\frac{dy(t)}{dt} + ay(t) = x(t)$$

Suppose one wishes to identify the parameter a of this system. We excite it with a white Gaussian random input $x(t)$. By white, we mean that $\mathbb{E}(x(t_1)x(t_2)) = R_{xx}(t_1 - t_2) = \sigma_x^2 \delta(t_1 - t_2)$. The output of this system is given by the equation

$$y(t) = \int_0^t h(\tau) x(t - \tau) d\tau, t \geq 0$$

where $h(t) = exp(-at)$. One can start with the computation of the cross correlation between the output and the input as

$$R_{yx}(t_1, t_2) = \mathbb{E}(y(t_1)x(t_2)) = \int_0^{t_1} h(\tau) \sigma_x^2 \delta(t_1 - \tau - t_2) d\tau$$

For $t_1 \geq t_2$, this evaluates to $\sigma_x^2 h(t_1 - t_2)$ and for $t_1 < t_2$ it is zero. This formula shows that if one can estimate the cross correlations between the output and the input, then the impulse response function $h(.)$ of the system may be determined from which a is obtained. The cross correlation can be estimated as a time average provided that joint ergodicity holds:

$$R_{yx}(\tau) \approx \frac{1}{T} \int_0^T y(t + \tau) x(t) dt$$

In this context, one may introduce the notion of ergodicity in the mean and ergodicity in the correlation. The notions of stationarity and wide sense stationarity should also be introduced while doing so. A random signal $\{x(t)\}$ is first of all sampled at times $\{t_1, ..., t_k\}$ and the joint probability density of the sampled variables $\{x(t_1), ..., x(t_k)\}$ is constructed. The joint distribution is given by

$$F_{t_1,...,t_k}(x_1, ..., x_k) = P\{x(t_1) \leq x_1, ..., x(t_k) \leq x_k\}$$

More precisely, the random process is specified by specifying the waveform $x(t, \omega), t \in \mathbb{R}$ for each chance outcome ω. The set of outcomes ω assumes values in a sample space Ω in accordance with a probability measure $P\{.\}$. The mean of a function $g(x(t_1), ..., x(t_k))$ of the variables is defined as

$$\mu(g, t_1, ..., t_k) = \mathbb{E}g(x(t_1), ..., x(t_k)) = \int_\Omega g(x(t_1, \omega), ..., x(t_k, \omega))P\{d\omega\}$$

The joint density of $\{x(t_1), ..., x(t_k)\}$ when it exists is defined as

$$f_{t_1,...,t_k}(x_1, ..., x_k) = \frac{\partial^k F_{t_1,...,t_k}(x_1, ..., x_k)}{\partial x_1...\partial x_k}$$

We can use the joint density to compute the average of the function g of the variables $x(t_j), j = 1, 2, ..., k$ as

$$\mu(g, t_1, ..., t_k) = \int_{\mathbb{R}^k} g(x_1, ..., x_k)f_{t_1,...,t_k}(x_1, ..., x_k)dx_1...dx_k$$

In terms of the joint distribution function, this is sometimes expressed as a Riemann-Stieltjes integral

$$\mu(g, t_1, ..., t_k) = \int_{\mathbb{R}^k} g(x_1, ..., x_k)dF_{t_1,...,t_k}(x_1, ..., x_k)$$

The process is said to be stationary if the joint distribution of the variables $\{x(t_1), ..., x(t_k)\}$ is the same as that of the variables $\{x(t_1 + \tau), ..., x(t_k + \tau)\}$ for all values of $k, t_1, ..., t_k, \tau$. This means that the statistics of the process remains invariant under time shifts. A consequence of stationarity is that

$$\mathbb{E}g(x(t_1), ..., x(t_k)) = \mathbb{E}g(x(t_1 + \tau), ..., x(t_k + \tau))$$

for all $k, t_1, ..., t_k, \tau$. The process $\{x(t)\}$ is ergodic in the mean when

$$\mathbb{E}x(t) = lim_{T \to \infty} \frac{1}{T} \int_{-T/2}^{T/2} x(t)dt$$

A necessary condition for this is clearly that the mean of $x(t)$ be independent of the time index t. The process $\{x(t)\}$ is correlation ergodic when

$$\mathbb{E}(x(t)x(t + \tau)) = lim_{T \to \infty} \frac{1}{T} \int_0^T x(t)x(t + \tau)dt$$

and a necessary condition is that the correlation of the process defined by the left hand side be independent of the time origin t. If however, we do not have available with us the input process, then we must estimate the parameter a using only the output process. Computing the output autocorrelation gives

$$\mathbb{E}(y(t)y(t+\tau)) = \int_0^t \int_0^{t+\tau} h(\alpha)h(\beta)\mathbb{E}(x(t-\alpha)x(t+\tau-\beta))d\alpha.d\beta$$

$$= \int_0^t \int_0^{t+\tau} h(\alpha)h(\beta)\sigma_x^2\delta(\tau-\beta+\alpha)d\alpha.d\beta$$

Noting that $\alpha+\tau$ varies from τ upto $t+\tau$ as α varies from 0 to t, it is clear that β is bound to cross $\tau+\alpha$ as the former varies from 0 upto $t+\tau$. This means that the above integral evaluates to

$$R_{yy}(t,t+\tau) = \sigma_x^2 \int_0^t h(\alpha)h(\alpha+\tau)d\alpha$$

the limiting form of which is given by

$$\bar{R}_{yy}(\tau) = lim_{t\to\infty} R_{yy}(t,t+\tau) = \sigma_x^2 \int_0^\infty h(\alpha)h(\alpha+\tau)d\alpha$$

This equation can be used to identify the parameter a. It can be used more generally to identify the impulse response of a linear time invariant system by exciting the system with a white random input and computing the time averaged autocorrelation function of the output. The spectral relationship is obtained by Fourier transforming the above equation and making use of the Wiener-Khintchine theorem that the power spectral density is the Fourier transform of the autocorrelation function:

$$S_{yy}(\omega) = \int_{-\infty}^\infty \bar{R}_{yy}(\tau)exp(-i\omega\tau)d\tau = \sigma_x^2|H(\omega)|^2$$

This equation shows that the power spectral density can be used to identify only the magnitude of the transfer function but not the phase. When the system is known to be of the minimum phase type, we can factorize the spectral density as $S_{yy}(\omega) = H(\omega)H(-\omega)$, where $H(\omega)$ is of the form

$$H(\omega) = \frac{\Pi_{i=1}^k(1-z_i exp(-j\omega)}{\Pi_{i=1}^m(1-p_i exp(-j\omega))}$$

with the $z_i's$ and $p_i's$ appearing either as reals or in complex conjugate pairs and $|z_i|, |p_i| < 1$. In other words, $H(\omega)$ is the minimum phase factor of S_{yy}. The frequency response of the system is then proportional to $H(\omega)$. The example of identifying the system parameters of a second order system can then be given along the following lines. The input-output equation is specified as

$$\frac{d^2y(t)}{dt^2} - (a+b)\frac{dy(t)}{dt} + aby(t) = x(t)$$

The system parameters are a, b. The transfer function in the Laplace domain is given by

$$\frac{Y(s)}{X(s)} = H(s) = \frac{1}{s^2 - (a+b)s + ab} = \frac{1}{(s-a)(s-b)} = \frac{1}{a-b}\left(\frac{1}{s-a} - \frac{1}{s-b}\right)$$

The impulse response of the system is given by

$$h(t) = \frac{1}{a-b}(exp(at) - exp(bt))u(t)$$

If the system is excited with white noise having zero mean and unit variance, the steady state autocorrelation function of the output is given by

$$R_{yy}(\tau) = \int_0^\infty h(t)h(t+\tau)dt$$

and by estimating this using a time average for two distinct values of τ, the parameters a, b may be estimated.

Study project 11

A seminar dealing with the Fourier series of elementary periodic signals like the square wave, both symmetric, antisymmetric and asymmetric can be delivered. For example if $x(t)$ is a square wave of period T, defined by $x(t) = 1$ for $0 < t < T/2$ and $x(t) = -1$ for $T/2 < t < T$, then $x(t)$ is symmetric about the origin. Its Fourier coefficients are computed as

$$c_n = \frac{1}{T}\int_0^{T/2} exp(-jn\omega_0 t)dt - \frac{1}{T}\int_{T/2}^T exp(-jn\omega_0 t)dt$$

with $\omega_0 = 2\pi/T$ as the fundamental frequency of the wave. The first term on the right side evaluates to

$$\frac{1}{jn\omega_0 T}(1 - (-1)^n)$$

when $n \neq 0$ and to $1/2$ when $n = 0$. Likewise, the second term evaluates to

$$((-1)^n - 1)/jn\omega_0 T$$

when $n \neq 0$ and $1/2$ when $n = 0$. This gives

$$c_0 = 0, c_n = (1 - (-1)^n)/jn\pi, n \neq 0$$

This gives the Fourier series of the square wave as

$$x(t) = \sum_{n=-\infty}^{\infty} \frac{2}{j(2n+1)\pi} exp(j(2n+1)\pi t/T)$$

$$= \sum_{n=0}^{\infty} \frac{4}{2n+1} sin((2n+1)\pi t/T)$$

The derivative of the square wave over the period $[0, T)$ is easily seen to be given by

$$\frac{dx(t)}{dt} = y(t) = 2\delta(t) - 2\delta(t - T/2), 0 \leq t < T$$

Thus, over the entire time interval,

$$y(t) = 2 \sum_{n=-\infty}^{\infty} \delta(t - nT) - \delta(t - (n + 1/2)T)$$

Taking the derivative of the Fourier series gives us

$$y(t) = \frac{4\pi}{T} \sum_{n=0}^{\infty} cos((2n+1)\pi t/T)$$

This equation can also be verified directly using the fact that the Fourier series of the δ train

$$\xi(t) = \sum_{n=-\infty}^{\infty} \delta(t - nT)$$

is given by

$$\xi(t) = \frac{1}{T} \sum_{n=-\infty}^{\infty} exp(j2\pi nt/T)$$

The differentiation property can be used to construct the Fourier series of more complicated signals given the Fourier series of their derivatives or integrals which are simpler waveforms.

Study project 12

A seminar on the stability of linear time invariant systems, stabilizability of open loop systems via feedback and plots of the magnitude and phase response can be delivered to an audience comprising control theorists. The seminar can also include the Nyquist plot and the root locus method. The seminar can begin with the basic definition of stability that a bounded input to the system should produce a bounded output. Then, if $\{h(t)\}$ is the impulse response of the linear time invariant system, the inequality

$$|\int_{-\infty}^{\infty} h(\tau)x(t - \tau)d\tau| \leq (\int_{-\infty}^{\infty} |h(\tau)|d\tau)sup|x(t)|$$

is derived from which the condition of integrability of the impulse response as a sufficient condition for the stability of the system is derived. If $G(s)$ is the Laplace domain transfer function of a linear time invariant system having the partial fraction expansion

$$G(s) = \sum_{k=1}^{p} \sum_{r=1}^{m_k} \frac{A_{k,r}}{(s - s_k)^r}$$

the impulse response of this system is given by

$$g(t) = \sum_{k,r} A_{k,r} t^{r-1} exp(s_k t) u(t)$$

assuming causality of the system. Integrability of this response follows when and only when $Re(s_k) < 0$ for all k. Thus, the system is stable when and only when all its poles fall in the left half s-plane. If the system is unstable, it may be stabilized using feedback. This involves feeding the output of the system through another system having transfer function $H(s)$ and subtracting the output of this auxiliary system from the input. Thus the effective input to the original system $G(s)$ is given in the Laplace domain by $X(s) - H(s)Y(s)$ and we have the equation $G(s)(X(s) - H(s)Y(s)) = Y(s)$ which gives the overall transfer function as

$$\frac{Y(s)}{X(s)} = \frac{G(s)}{1 + G(s)H(s)}$$

When the open loop system can have an arbitrary gain K, the stability of the closed loop system is governed by the roots of the function $1 + KG(s)H(s)$. The root locus technique involves plotting the roots of this function K varies from $-\infty$ to ∞. If Z denotes the number of zeroes of $1 + KG(s)H(s)$ that fall in the right half s-plane and P the number of its poles that fall in the right half s-plane, then one selects a contour consisting of the $j\omega$ axis in conjunction with a semicircle of infinite radius with diameter as the $j\omega$ axis and the contour encircling the entire right half s-plane. As we traverse this contour once clockwise, the number of times the complex number $1 + KG(s)H(s)$ encircles the origin clockwise equals $Z - P$. For stability of the closed loop system, one requires $Z = 0$. In other words, the function must encircle the origin P times counterclockwise. It is sufficient if s is assumed to run along the $j\omega$ axis alone from $\omega = -\infty$ to $\omega = +\infty$, the reason being that as s runs over the infinite semicircle, $G(s)H(s) = 0$ assuming that this open loop function has more poles than zeroes which will almost always be the case. This means that as s runs over the semicircular portion of the contour, the function $1 + KG(s)H(s)$ will stay fixed at unity and will not contribute to encirclement. These details can be presented succinctly in a seminar taking a variety of examples. The last part of the seminar can discuss magnitude and phase plots. If

$$G(s) = \frac{\Pi_{k=1}^{z}(s - \sigma_k - j\omega_k)}{\Pi_{k=1}^{p}(s - \sigma_k' - j\omega_k')}$$

then the magnitude of the frequency response is given by

$$|G(j\omega)| = \frac{\Pi_{k=1}^{z}\sqrt{\sigma_k^2 + (\omega - \omega_k)^2}}{\Pi_{k=1}^{p}\sqrt{\sigma_k'^2 + (\omega - \omega_k')^2}}$$

and the phase or argument function is given by

$$Arg(G(j\omega)) = -\sum_{k=1}^{z} tan^{-1}(\omega - \omega_k)/\sigma_k) + \sum_{k=1}^{p} tan^{-1}(\omega - \omega_k')/\sigma_k')$$

However it is more convenient to use the fact that the system transfer function $G(s)$ corresponds to a real system and hence its complex poles and zeroes occur in complex conjugate pairs. Thus,

$$G(s) = \frac{\Pi_{k=1}^{z_1}(s - \alpha_k)\Pi_{k=1}^{z_2}((s - \sigma_k)^2 + \omega_k^2)}{\Pi_{k=1}^{p_1}(s - \alpha_k')\Pi_{k=1}^{p_2}((s - \sigma_k')^2 + \omega_k'^2)}$$

where α_k, α_k' are real. We find for the magnitude of this transfer function

$$|G(j\omega)| = \frac{\Pi_{k=1}^{z_1}\sqrt{\omega^2 + \alpha_k^2}\Pi_{k=1}^{z_2}(\sqrt{\sigma_k^2 + (\omega - \omega_k)^2}\sqrt{\sigma_k^2 + (\omega + \omega_k)^2})}{\Pi_{k=1}^{p_1}\sqrt{\omega^2 + \alpha_k'^2}\Pi_{k=1}^{p_2}(\sqrt{\sigma_k'^2 + (\omega - \omega_k')^2}\sqrt{\sigma_k'^2 + (\omega + \omega_k')^2})}$$

and the phase or argument function is given by

$$Arg(G(j\omega)) = \sum_{k=1}^{z_1}\pi + tan^{-1}(\omega/\alpha_k) - \sum_{k=1}^{z_2}tan^{-1}((\omega + \omega_k)/\sigma_k) + tan^{-1}((\omega - \omega_k)/\sigma_k)$$

$$- \sum_{k=1}^{p_1}\pi + tan^{-1}(\omega/\alpha_k') + \sum_{k=1}^{p_2}tan^{-1}((\omega + \omega_k')/\sigma_k') + tan^{-1}((\omega - \omega_k')/\sigma_k')$$

The importance of the log magnitude plot should also be stressed here. The examples considered should include lowpass, highpass and bandpass filters.

Study project 13

A seminar on basic theory of broadcasting can be given. Several problems can be treated in this regard. One such problem involves partitioning each channel and broadcasting sections of each movie in each of these partitions. Fourier analysis of the resulting channel signals plays an important role here. The following item contains some interesting discussion material. N Movie Signals $\{s_1(t), 0 \le t \le T\}, ..., \{s_N(t), 0 \le t \le T\}$ to be broadcasted. Divide each movie into M parts. First part of ith first movie is $s_i(t), 0 \le t \le \Delta$, second part of the ith movie is $s_i(\Delta + t), 0 \le t \le \Delta$ and finally M^{th} part of the ith movie is $s_i((M - 1)\Delta + t), 0 \le t \le \Delta$, where $M\Delta = T$. Transmit the first part of the first movie over the first channel in the duration $0 \le t \le \Delta$. Then transmit the first part of the second movie over the first channel in the duration $\Delta + \tau_2 < t < 2\Delta + \tau_2$. Then transmit the first part of the third movie over the first channel in the duration $2\Delta + \tau_2 + \tau_3 < t < 3\Delta + \tau_2 + \tau_3$ and so on. In general, we transmit the first part of the kth movie over the first channel in the duration $(k-1)\Delta + \tau_2 + ... + \tau_k < t < k\Delta + \tau_2 + ... + \tau_k$. The wave form corresponding to the transmission of the first part of the kth movie over the first channel is given by $s_k(t - (k - 1)\Delta + \tau_2 + ... + \tau_k)(u(t - (k - 1)\Delta - \tau_2 - ... - \tau_k) - u(t - k\Delta - \tau_2 - ... - \tau_k)$, where $u(t)$ equals the unit step signal. Thus, we can express the overall signal transmitted over the first channel in the form

$$x_1(t) = s_1(t)(u(t) - u(t - \Delta))$$

$$+ \sum_{k=2}^{N} s_k(t - (k-1)\Delta - \tau_{12} + ... - \tau_{1k})(u(t - (k-1)\Delta - \tau_{12} - ... - \tau_{1k}) - u(t - k\Delta - \tau_{12} - ... - \tau_{1k})$$

We've used the notation τ_{1k} in place of τ_k to signify the fact that we are looking at signals being transmitted over the first channel. In general, one transmits the rth section of the first movie over the rth channel, then the rth section of the second movie over the same r^{th} channel after a duration of τ_{22} seconds and so on. The signal corresponding to the kth section being transmitted over the rth channel is given by

$$s_1(t + (r-1)\Delta), 0 \leq t \leq \Delta$$

if $k = 1$ and

$$s_k(t + (r-1)\Delta - (k-1)\Delta - \tau_{r2} - ... - \tau_{rk})$$

$$*(u(t - (k-1)\Delta - \tau_{r2} - ... - \tau_{rk}) - u(t - k\Delta - \tau_{r2} - ... - \tau_{rk}))$$

and the overall signal being transmitted over the rth channel is then given by

$$x_r(t) = s_1(t + (r-1)\Delta)(u(t) - u(t - \Delta))$$

$$+ \sum_{k=2}^{N} s_k(t + (r-1)\Delta - (k-1)\Delta - \tau_{r2} - ... - \tau_{rk})$$

$$*(u(t - (k-1)\Delta - \tau_{r2} - ... - \tau_{rk}) - u(t - k\Delta - \tau_{r2} - ... - \tau_{rk}))$$

We want to look at the phase of the Fourier transform of the signal being transmitted over the rth channel as a function of the jitter time variables $\tau_{r2}, ..., \tau_{rN}$. We use the fact that the Fourier transform of $u(t - a) - u(t - b)$ is given by

$$\frac{(exp(-j\omega a) - exp(-j\omega b))}{j\omega} = 2exp(-j\omega(a+b)/2)sin(\omega(b-a)/2)/\omega$$

Application of the multiplication theorem for the Fourier transform then gives us

$$X_1(\omega) = \int_{-\infty}^{\infty} x_1(t)exp(-j\omega t) =$$

$$= \frac{1}{\pi}S_1(\omega) * (exp(-j\omega(a+b)/2)sin(\omega(b-a)/2)/\omega)$$

$$+ \frac{1}{\pi}\sum_{k=2}^{N}[S_k(\omega)exp(-j\omega((k-1)\Delta - \tau_{12} - ... - \tau_{1k}))]$$

$$*[exp(-j\omega((k-1/2)\Delta - \tau_{12} - ... - \tau_{1k}))sin(\omega\Delta/2)/\omega]$$

The phase angle as a function of the $\tau'_{1j}s$ can be extracted from this expression. On similar lines, the Fourier transform of the signal being transmitted over the rth channel is computed.

Study project 14

A seminar can be given on state variable formulation of the input output equations for a system defined by ordinary differential equations. For example consider the second order linear differential equation

$$\frac{d^2x(t)}{dt^2} + a_1(t)\frac{dx(t)}{dt} + a_2(t)x(t) = u(t)$$

with $\{u(t)\}$ as the input and $\{x(t)\}$ as the output. We introduce the state variables $x_1(t) = x(t)$, $x_2(t) = \frac{dx(t)}{dt}$ so that

$$\frac{dx_1(t)}{dt} = x_2(t), \frac{dx_2(t)}{dt} = -a_1(t)x_2(t) - a_2(t)x_1(t) + u(t)$$

These are expressed in the standard matrix form as

$$\frac{d}{dt}\begin{pmatrix} x_1(t) \\ x_2(t) \end{pmatrix} = \begin{pmatrix} 0 & 1 \\ -a_2(t) & -a_1(t) \end{pmatrix}\begin{pmatrix} x_1(t) \\ x_2(t) \end{pmatrix} + u(t)\begin{pmatrix} 0 \\ 1 \end{pmatrix}$$

The solution of the general linear state variable differential equations

$$\frac{d\mathbf{x}(t)}{dt} = \mathbf{F}(t)\mathbf{x}(t) + \mathbf{G}(t)\mathbf{u}(t)$$

is obtained in terms of the state transition matrix function. This can also be discussed in the seminar. The state transition matrix function $\Phi(t,s)$ satisfies

$$\frac{\partial\Phi(t,s)}{\partial t} = \mathbf{F}(t)\Phi(t,s), t > s > 0$$

with initial condition $\Phi(s,s) = \mathbf{I}$. The solution to the equations is verified to be given by the formula

$$\mathbf{x}(t) = \Phi(t,s)\mathbf{x}(s) + \int_s^t \Phi(t,\tau)\mathbf{G}(\tau)\mathbf{u}(\tau)d\tau$$

How in practice the function $\Phi(t,s)$ is determined by a perturbation series may also be considered for discussion in the seminar.

Study project 15

Projection operators play a fundamental role in many applied problems. These include approximation theory, quantum mechanics and signal processing. A separate seminar introducing projections and their striking properties and methods of implementation would be of help to an audience interested in any of these fields. The seminar can begin by introducing the notion of a direct sum decomposition of a vector space. If V is a vector space and W_1, W_2 are subspaces of V, we write $V = W_1 \oplus W_2$ if every $x \in V$ is expressible uniquely as $x = x_1 + x_2$ with $x_i \in W_i, i = 1, 2,$. More generally, if $W_1, ..., W_k$ are subspaces of V we write $V = W_1 \oplus ... \oplus W_k$ if every $x \in V$ is expressible as $x = x_1 + ... + x_k$ with $x_i \in W_i$. It is an easily proved fact that this is equivalent to saying that every vector $x \in V$ is expressible as $x_1 + ... + x_k$ with $x_i \in W_i, i = 1, 2, ..., k$ and that for each $j = 1, 2, ..., k$,

$$W_j \bigcap (W_1 + ... + W_{j-1} + W_{j+1} + ... + W_k) = \{0\}$$

The following notation has been introduced If $U_1, ..., U_r$ are subspaces of a vector space V, then $U_1 + ... + U_r$ stands for the space of all vectors of the form $x_1 + ... + x_r$ with $x_i \in U_i, i = 1, 2, ..., r$. Now for $x \in V$, define $P_j x = x_j, j = 1, 2, ..., r$ where $x_1, ..., x_r$ are uniquely determined by the equation $x = x_1 + ... + x_r$ with $x_j \in W_j, j = 1, 2, ..., r$. Then, $P_j : V \to W_j$ is a linear operator satisfying the condition $P_j^2 = P_j$. It is clear from the decomposition of a general $x \in V$ that $\mathbf{I} = P_1 + ... + P_r$ and $P_j P_k = 0$ when $j \neq k$. The notion of diagonability of an operator can be formulated in terms of projections by noting that if a linear operator $T \in L(V)$ is diagonal in a basis B then we consider the distinct eigenvalues $c_1, ..., c_m$ of T and the corresponding eigenspaces $W_j = \mathcal{N}(T - c_j I), j = 1, 2, ..., r$. Diagonability of T is then equivalent to the direct sum decomposition $V = W_1 \oplus ... \oplus W_r$. We then have the following decomposition of the operator $T : T = \sum_{j=1}^{r} c_j P_j$ where P_j is the canonical projection onto W_j. The notion of a projections on an inner product space leads to the idea of an orthogonal projection which is fundamental in estimation theory and quantum mechanics. These ideas can be explained in detail. Expressing events in quantum mechanics using orthogonal projections decomposition of observables (Hermitian operators) using projections can be discussed in the seminar.

Study project 16

A seminar on the notion of transformations of the independent variable in signals will be of use to the beginner in signal theory. What is the purpose of this transformation and what are its applications should be discussed in the seminar. If $x(t)$ is a signal and $t \to \phi(t)$ is a transformation function, then the signal obtained after transformation of the independent variable is $x(\phi(t))$. One can take for instance $\phi(t) = at + b$ and interpret the signal shape $x(at + b)$ after this transformation has been effected. When $a > 0$ and b is negative, the transformed signal is obtained by translating the original signal to the right by $|b|$ units. when b is positive, the transformed signal is obtained by translating the original signal to the left by b units. When $b = 0$ and $a > 1$, the transformed signal $x(at)$ is obtained by compressing the signal in time by a factor of a. When $0 < a < 1$, the transformed signal $x(at)$ is obtained by expanding the signal in time by a factor of $\frac{1}{a}$. Similarly, when $a < -1$ and $b = 0$, the transformed signal $x(at)$ is obtained by compressing the signal in time and time reversing the resultant. When $-1 < a < 0$, the transformed signal is obtained by time reversing the signal and expanding the signal in time. When b is non zero, the transformed signal is obtained by applying time scaling and time translation and if $a < 0$, in addition a reversal in time.

Study project 17

A seminar on the notion of even and odd signals and the decomposition of a signal in terms of these components would be useful from the stand point of signal representations. Given a signal, one asks the question how to represent it in terms of signals having definite symmetry properties. Signal that possess particular symmetry properties can be processed more easily than the original signal and hence such representations prove to be useful. Signals having symmetry properties can also be analyzed mathematically more easily and hence it is important to represent a given signal in terms of more elementary signals. The following points can be discussed in the seminar: If $x(t)$ is a signal we can represent it as

$$x(t) = x_e(t) + x_o(t)$$

where

$$x_e(t) = \frac{x(t) + x(-t)}{2} \qquad x_o(t) = \frac{x(t) - x(-t)}{2}$$

x_e is an even signal in the sense that $x_e(-t) = x_e(t)$ and $x_o(t)$ is an odd signal in the sense that $x_o(-t) = -x_o(t)$. The graph of an even signal is symmetric with respect to reflections about the y axis while the graph of an odd signal is antisymmetric with respect to such a reflection. If $y(t)$ is another such signal, we have

$$y(t) = y_e(t) + y_o(t)$$

and we can evaluate the correlation between the two signals $\{x(t)\}$ and $\{y(t)\}$ easily:

$$\rho(x, y) = \int_{-\infty}^{\infty} x(t)y(t)dt = \int_{-\infty}^{\infty} (x_e(t) + x_o(t))(y_e(t) + y_o(t))dt$$

$$= \int_{-\infty}^{\infty} x_e(t)y_e(t)dt + \int_{-\infty}^{\infty} x_e(t)y_o(t)dt + \int_{-\infty}^{\infty} x_o(t)y_e(t)dt + \int_{-\infty}^{\infty} x_o(t)y_o(t)dt$$

$$= \int_{-\infty}^{\infty} x_e(t)y_e(t)dt + \int_{-\infty}^{\infty} x_o(t)y_o(t)$$

$$= 2\int_{0}^{\infty} x_e(t)y_e(t)dt + 2\int_{0}^{\infty} x_o(t)y_o(t)dt$$

the reason being that $x_e(t)y_o(t)$ and $x_o(t)y_e(t)$ are odd signals and hence their integrals over the entire real line evaluate to zero. Taking $y(t) = x(t)$, gives for the signal energy

$$E_x = \int_{-\infty}^{\infty} x(t)^2 dt = 2\int_{0}^{\infty} x_e(t)^2 dt + 2\int_{0}^{\infty} x_o(t)^2 dt$$

This formula indicates that the correlation between two signals can be determined by summing the correlations between their even and odd parts. No interference terms i.e. correlations between even and odd parts appear in the expression. Likewise, the signal energy equals the sum of the energies in the even and odd parts of the signal. No cross terms like the correlation between the even and odd parts appear in the final expression. This formula generalizes to multiple dimensions. Consider for example a two dimensional signal $x(t, s)$. We can define two operators R_1 and R_2. R_1 reflects the first variable while R_2 reflects the second variable: $R_1 x(t,s) = x(-t,s), R_2 x(t,s) = x(t,-s)$. Then, $R_2 R_1 x(t,s) = x(-t,-s)$. The even part of the signal is $x_e(t,s) = (1 + R_2 R_1)x(t,s)/2 = (x(t,s) + x(-t,-s))/2$. Similarly, the odd part is $x_o(t,s) = (1 - R_2 R_1)x(t,s)/2 = (x(t,s) - x(-t,-s))/2$. Clearly, $R_2 R_1 x_e = x_e, R_2 R_1 x_o = -x_o$. We can also consider the partially symmetrized and antisymmetrized signals $x_{e1} = (1 + R_1)x/2, x_{e2} = (1 + R_2)x/2, x_{o1} = (1 - R_1)x/2, x_{o2} = (1 - R_2)x/2$. It is clear that $x_{e1}(t,s) = (x(t,s) + x(-t,s))/2, x_{e2}(t,s) = (x(t,s) + x(t,-s))/2, x_{o1}(t,s) = (x(t,s) - x(-t,s))/2, x_{o2}(t,s) = (x(t,s) - x(t,-s))/2$. We see that R_1 and R_2 commute, i.e., $R_2 R_1 = R_1 R_2$ and $(1 + R_2)(1 + R_1)/4 = (1 + R_2 + R_1 + R_2 R_1)/4$. This operator when acting on x produces the signal $x_{e12} = (x(t,s) + x(t,-s) + x(-t,s) + x(-t,-s))/4$ which is symmetric with respect to reversal of any one variable or both. This can be termed as the completely symmetrized version of the signal $x(t,s)$. This fact can also be seen via the following formula $R_1^2 = 1, R_2^2 = 1$ implies $R_1(1 + R_2)(1 + R_1) = (1 + R_2)(R_1 + R_1^2) = (1 + R_2)(1 + R_1)$, with an analogous formula for the second operator acting on the product. The operator formulas

$$1 = (1 + R_1)/2 + (1 - R_1)/2 = (1 + R_2)/2 + (1 - R_2)/2$$

$$1 = (1 + R_1)(1 + R_2)/4 + (1 + R_1)(1 - R_2)/4 + (1 - R_1)(1 + R_2)/4 + (1 - R_1)(1 - R_2)/4$$

yield the signal representations

$$x = x_{e1} + x_{o1} = x_{e2} + x_{o2}$$

and

$$x = x_{e12} + x_{e1o2} + x_{o1e2} + x_{o12}$$

where x_{e12} is even with respect to reversal of either index, x_{e1o2} is even with respect to reversal of the first index and odd with respect to reversal of the second index. x_{o1e2} is odd with respect to reversal of the first index and even with respect to reversal of the second index and finally, x_{o12} is odd with respect to reversal of either one index. Of course, we also have the formula

$$1 = (1 + R_2 R_1)/2 + (1 - R_2 R_1)/2$$

which yields

$$x = x_e + x_o$$

We leave it as an exercise to derive formulas for the signal energy in terms of the component energies and the correlations of the components appropriately setting equal to zero those correlations that involve integrals over the real line with respect to a variable when the integrand is odd with respect to that particular variable.

Study project 18

A seminar on the various kinds of singularity functions that one comes across while constructing signal and system representations would be of use to the student interested in doing research in distribution theory and its applications to signal analysis. The basic idea of a distribution function is based upon integration by parts. Let ϕ be a well behaved test function. Suppose we want to define the derivative of a non-differentiable function f. Consider a linear map that carries the function f to the function $\phi(f) = \int_{-\infty}^{\infty} \phi(t)f(t)dt$. Integration by parts shows that

$$\phi(f') = \int_{-\infty}^{\infty} \phi(t)f'(t)dt = - \int_{-\infty}^{\infty} \phi'(t)f(t)dt = -\phi'(f)$$

This formula can be used to define the derivative. What we need is a sufficiently large class of test functions ϕ so that knowledge of all the numbers $\int \phi f dt$ as ϕ varies over the class almost certainly defines the function f. One can consider a linear map that carries the function f to its value $f(0)$ at $t = 0$. We want to define a function $\delta(t)$ that performs this operation, i.e.

$$\int_{-\infty}^{\infty} \delta(t)f(t)dt = f(0)$$

However, it is easy to see that $\delta(t)$ cannot be a well defined function, if this identity is to hold for all f continuous at $t = 0$. It is easy to see that $\delta(t)$ must have all its energy concentrated at the origin and hence must be infinite at the origin and zero elsewhere. Taking $f = 1$, we find that $\int_{-\infty}^{\infty} \delta(t)dt = 1$, i.e., the area under the δ function must be unity. Motivated by this fact, we define a pulse $\delta_\Delta(t)$ that equals $\frac{1}{\Delta}$ when $|t| \leq \Delta$ and zero when $|t| > \Delta$. This is a rectangular pulse of width Δ and height $1/\Delta$ symmetric about the origin. The area under this pulse is unity and we can intuitively conceive of the limit of this pulse as $\Delta \to 0$. The pulse's height becomes infinite and its width becomes zero with the area under it remaining unity. This limit is called the Dirac δ pulse. It is then easily deduced that

$$\int_{-\infty}^{\infty} f(\tau)\delta(t - \tau)d\tau = f(t)$$

a formula that in the language of signal processing means that the linear time invariant system having impulse response $\delta(t)$ is the identity system, i.e., takes as input a signal and outputs the same signal. The Dirac δ pulse can be used to define a linear time invariant system via its impulse response as follows. If $x(t)$ is a signal, then $x(t) = \int x(\tau)\delta(t-\tau)d\tau$. If T is the operator of a linear time invariant system, then $T(\delta(t)) = h(t)$ is called its impulse response and by time invariance, $T(\delta(t - \tau)) = h(t - \tau)$. Thus,

$$y(t) = T(x(t)) = T\left(\int x(\tau)\delta(t - \tau)d\tau\right) = \int_{-\infty}^{\infty} x(\tau)T(\delta(t - \tau))d\tau = \int_{-\infty}^{\infty} x(\tau)h(t - \tau)d\tau$$

Thus, any linear time invariant system is completely characterized by its impulse response function. For nonlinear systems, this theorem is invalid. Consider for example the second order Volterra system defined by the input-output equation

$$y(t) = \int h(\tau)x(t - \tau)d\tau + \int g(\tau, s)x(t - \tau)x(t - s)d\tau ds$$

Suppose we input $x(t) = \delta(t - t_0)$ to this system. The output is given by

$$y(t) = h(t - t_0) + g(t - t_0, t - t_0)$$

It is clear that no matter how t_0 is chosen, we cannot recover both the kernels $h(\tau)$ and $g(t, \tau)$ from the output. However, if we input a sum of two Dirac impulses $x(t) = A\delta(t - t_0) + B\delta(t - s_0)$, then by varying t_0 and s_0 and A, B, we can recover both the linear and quadratic kernels: The response is given by

$$y(t) = Ah(t - t_0) + Bh(t - s_0) + A^2 g(t - t_0, t - t_0) + B^2 g(t - s_0, t - s_0) + 2ABh(t - t_0, s - s_0)$$

assuming without loss of generality that $h(t, s) = h(s, t)$. The recovery equations can be derived from this formula.

Study project 19

A seminar giving examples for the computation of the convolution between different pairs of signals both in discrete time and in continuous time would be of interest to the beginner in signal processing. The examples covered can include (a) $x(n) = a^n u(n - n_a)$, $y(n) = b^n u(n - n_b)$, (b) $x(n) = u(n - n_a) - u(n - n_b)$, $y(n) = a^n u(n)$, (c) $x(t) = exp(-at)u(t), y(t) = exp(-|t|)$ and many similar examples. The computations can be illustrated by graphical arguments as well as using analytic arguments. The general formula for the convolution between piecewise exponential signals can also be given. A piecewise exponential signal has the form $x(n) = \sum_{k=1}^{p} A_k a_k^n (u(n - n_k) - u(n - m_k))$. Convolution examples for two dimensional signals would also enable the image processing student to get his concepts clear. For example, given $x(n, m) = 1$ over the rectangle $n_1 \leq n \leq n_2, m_1 \leq m \leq m_2$ and $x(n, m) = 0$ for other values of n, m and similarly another signal $h(n, m) = 1$ over the rectangle $n_1' \leq n \leq n_2', m_1' \leq m \leq m_2'$, a general expression for the convolution $y(n, m) = \sum_{k,l} h(k, l)x(n - k, m - l)$ can be derived using analytic arguments. Application of the Fourier transform to the determination of the convolution should also be illustrated. We shall illustrate with just a single computation, namely that discussed in (c). The Fourier transform of $exp(-at)u(t)$ equals $\frac{1}{j\omega + a}$ while that of $exp(-|t|)$ equals

$$\int_{-\infty}^{\infty} exp(-|t|)exp(-j\omega t)dt$$

$$= \int_{0}^{\infty} exp(-(j\omega + 1)t)dt + \int_{-\infty}^{0} exp(-(j\omega - 1)t)dt = \frac{1}{j\omega + 1} - \frac{1}{j\omega - 1} = \frac{2}{\omega^2 + 1}$$

Thus, the Fourier transform of the convolution of $exp(-at)u(t)$ and $exp(-|t|)$ is given by the product

$$\frac{2}{(j\omega + a)(1 + \omega^2)}$$

and inversion of this function can be accomplished by partial fraction expansion.

Study project 20

The computation of the impulse response of an LTI system defined by a difference equation or by a differential equation is of great importance in digital signal processing. General methods have been developed in this direction. The computation of the impulse response of a system described by the difference equation

$$y(n) + \sum_{k=1}^{p} a(k)y(n-k) = x(n)$$

proceeds by setting $x(n) = \delta(n)$ and solving the equation for $n \geq p$ assuming $y(n)$ for $n \geq 0$ to have the form z^n. The initial conditions are then deduced by setting $n = 0, 1, ..., p-1$. We do not go into the details.

Study project 21

Explain in the seminar how you would implement a first order system defined by the input output equation

$$\frac{dy(t)}{dt} = ay(t) + x(t)$$

using a differentiator and a multiplier. Also explain why use of the differentiator is impractical and hence one has to use an amplitude limiter to get over the difficulty. How would your input-output equation change on using a limiter ? To answer this question, we place two points on the paper. The input $x(t)$ is fed into the first point and the output $y(t)$ comes out of the other point. The output $y(t)$ is also fed into a differentiator producing the signal $z(t) = \frac{dy(t)}{dt}$. The input and the signal $z(t)$ are fed into a differencer producing the error signal $e(t) = x(t) - \frac{dy(t)}{dt}$. This error signal is fed into a multiplier by the factor $\frac{1}{a}$ yielding the signal $\frac{1}{a}(x(t) - \frac{dy(t)}{dt}))$. This signal is connected to the output point $y(t)$. The block diagram is easily drawn. Note that we've implemented the equation in the form

$$y(t) = \frac{1}{a}(x(t) - \frac{dy(t)}{dt})$$

When however $\frac{dy(t)}{dt}$ becomes very large in magnitude, a practical system based on such a differentiator element can break down. Thus, one would be forced to use an amplitude limiter following the differentiator. The amplitude limiter is a memoryless device having the input-output relation $\phi(x) = x$ when $|x| \leq A$ and $\phi(x) = A$ when $x \geq A$ and $\phi(x) = -A$ when $x \leq -A$. We can write this relation as $\phi(x) = x\chi_{|x|\leq A}(x) + A\chi_{|x|>A}(x)$. After incorporating the memoryless device, the differential equation dictating the system dynamics assumes the form

$$\phi(\frac{dy}{dt}) = ay + x$$

which is equivalent to saying that $y = \frac{A-x}{a}$ when $\frac{dy}{dt} \geq A$, $y = \frac{-A-x}{a}$ when $\frac{dy}{dt} \leq -A$ and $\frac{dy}{dt} = ay+x$ when $|\frac{dy}{dt}| \leq A$.

Solved problem 22

Explain how you could use an integrator to implement the first order system of first order system What are the advantages of using this in place of the differentiator ?

Ans: The clear advantage of using an integrator in place of the differentiator is that no spikes can occur. Integration is a smoothening operation and hence spikes will in fact disappear on using it. Consider the system

$$\frac{dy(t)}{dt} + ay(t) = x(t)$$

Integration from $t = -\infty$ to t transforms it into the system

$$y(t) = -a \int_{-\infty}^{t} y(\tau)d\tau + \int_{-\infty}^{t} x(\tau)d\tau$$

Denoting the integral operator by I, this can equivalently be expressed as

$$y = -aI(y) + I(x)$$

To draw a block diagram, we once again place two points. To the first point, we input x and from the second point, we take out y. y is then input into an integrator producing the signal $z = I(y)$. This signal is then passed into a multiplier by the factor a giving the signal $u = aI(y)$. The input signal x is also fed into an integrator producing the signal $w = I(x)$. The two signals u and w are then passed into a differencer producing $w - u = I(x) - aI(y)$. This signal is then fed into the output point y. The resulting block diagram produces a solution to the above integral equation.

Study project 23

Explain what you understand by the impulse response of a linear system and how it can be used to derive the response of a linear system to an arbitrary input in the form of an integral equation ?

Ans: An arbitrary input $\{x(t)\}$ can be expressed as a superposition of delayed versions of Dirac impulses, i.e. $x(t) = \int_{-\infty}^{\infty} x(\tau)\delta(t - \tau)d\tau$. This integral can be approximated by a discrete sum

$$x(t) \approx \sum_{k=-\infty}^{\infty} x(k\Delta)\Delta\delta_\Delta(t - k\Delta)$$

where $\delta_\Delta(t) = \frac{1}{\Delta}$ for $|t| \leq \Delta/2$ and $\delta_\Delta(t) = 0$ for $|t| > \Delta/2$. Note that $\delta_\Delta(t)$ is an approximation to the Dirac delta pulse that in the limit $\Delta \to 0$ becomes the Dirac delta signal. In fact, we have for any continuous function $f(t)$,

$$\int_{-\infty}^{\infty} f(t)\delta_\Delta(t)dt = \frac{1}{\Delta} \int_{-\Delta/2}^{\Delta/2} f(t)dt$$

which in the limit $\Delta \to 0$ converges to $f(0)$. The above discrete sum thus gives us a staircase approximation to the signal $x(t)$. As Δ shrinks to zero, this staircase approximation becomes finer and finer and in the limit converges to a smooth approximation to the signal.

Study project 24

Examples showing the computation of the response of a linear system to an input built out of a superposition of component signals when the response to each component is known. That is we are given p signals $x_1(t), ..., x_p(t)$ and the response $y_1(t), ..., y_p(t)$ to each of these signals for a system T:

$$y_k(t) = T\{x_k(t)\}, k = 1, 2, ..., p$$

We want to compute the response to the signal $x(t) = \sum_{k=1}^{p} c_k x_k(t)$ when the constants $\{c_k\}$ are specified. Linearity implies that the response will be $y(t) = \sum_{k=1}^{p} c_k y_k(t)$. An example is the following: Response to $exp(j\omega_1 t)$ is $exp(j\omega_2 t)$ and the response to $exp(-j\omega_1 t)$ is $exp(-j\omega_2 t)$. The response to $cos(\omega_1 t + \phi)$ can be determined by expressing this signal as a linear combination of the given ones:

$$cos(\omega_1 t + \phi) = \frac{1}{2} exp(j\phi) exp(j\omega_1 t) + \frac{1}{2} exp(-j\phi) exp(-j\omega_1 t)$$

Thus, the response of T to this signal is given by

$$T\{cos(\omega_1 t + \phi)\} = \frac{1}{2} exp(j\phi) T\{exp(j\omega_1 t)\} + \frac{1}{2} exp(-j\phi) T\{exp(-j\omega_1 t)\}$$

$$= \frac{1}{2} exp(j\phi) exp(j\omega_2 t) + \frac{1}{2} exp(-j\phi) exp(-j\omega_2 t) = cos(\omega_2 t + \phi)$$

Study project 25

Examples showing how the properties of linearity and time invariance for a system can be verified for specified situations. Like, the system $y(n) = x(-n)$ is linear but not time invariant while the system $y(n) = x^2(n-1)$ is nonlinear but time invariant. former is not causal while the latter is.

Study project 26

Analyze the first order lowpass filter built using a resistor and a capacitor. If $x(t)$ is the input voltage and $y(t)$ the output voltage, then the relation between x and y is defined by the differential equation

$$RC\frac{dy}{dt} + y(t) = x(t)$$

If $x(t) = exp(j\omega t)$, then $y(t) = A.exp(j\omega t)$ in steady state, where A can be determined by plugging into the differential equation:

$$j\omega RCA + A = 1 \qquad A = \frac{1}{j\omega RC + 1}$$

Thus,

$$y(t) = \frac{1}{1 + j\omega RC}exp(j\omega t) = \frac{1}{\sqrt{(omegaRC)^2 + 1}}exp(j(omegat - tan^{-1}(\omega RC)))$$

High frequencies are attenuated by this lowpass filter in comparison to the low frequencies. The phase shift of the output relative to the input is also frequency dependent. For $|\omega RC| \ll 1$, we have $A = 1$, which means that low frequency components are left nearly untouched by the filter. Higher frequencies are attenuated: for $|\omega RC| \gg 1$, $A \approx \frac{1}{j\omega RC}$ which means that at high frequencies, the filter acts as an integrator. The question posed is that can one do a similar analysis for a second order filter ?

Study project 27

A seminar on the second order analog filter would start by introducing the transfer function

$$H(s) = \frac{1}{s^2 + as + b}$$

or the frequency domain transfer function

$$H(j\omega) = \frac{1}{-\omega^2 + ja\omega + b}$$

The magnitude squared transfer function is

$$|H(j\omega)|^2 = \frac{1}{(b - \omega^2)^2 + a^2\omega^2}$$

The maximum is attained at the frequency ω when

$$\frac{d}{d\omega}((b - \omega^2)^2 + a^2\omega^2) = 0$$

or equivalently when

$$-4\omega(b - \omega^2) + 2a^2\omega = 0$$

or

$$\omega = \pm\sqrt{b - a^2/2}$$

We have three ranges of the frequency in which the filter exhibits different characteristics. The first range is $0 \le \omega \ll min(b/a, \sqrt{b})$. Then, $H(j\omega) \approx \frac{1}{b}$. The filter in the very low frequency range (i.e., near dc) simply passes through the signal after scaling its amplitude by $1/b$. Then, we have the rage $a >> \omega >> b/a$. In this range, $H(j\omega) \approx \frac{1}{ja\omega}$, which corresponds to an integrator. In this intermediate frequency range, the filter behaves like an integrator. Finally, we have the range $\omega >> a, \sqrt{b}$. In this range, $H(j\omega) \approx \frac{1}{(j\omega)^2}$ which corresponds to a double integrator. In the very high frequency range, the filter behaves as a double integrator.

Reference: Problem suggested by Dr.S.S.Gupta

Study project 28

A seminar on how the bandwidth of a lowpass filter can be defined via the rise time of the step response. The seminar can start by discussing the ideal lowpass filter which has a frequency response $H(\omega) = 1$ when $|\omega| \leq \sigma$, and $H(\omega) = 0$ when $|\omega| > \sigma$. The bandwidth of this filter is ideally σ/π Hertz. To see how the rise time corresponding to a step input is reciprocally related to the bandwidth, we first compute the impulse response of this system as

$$h(t) = \frac{1}{2\pi} \int_{-\sigma}^{\sigma} exp(j\omega t)d\omega = \frac{sin(\sigma t)}{\pi t}$$

The step response is given by

$$s(t) = \int_{-\infty}^{t} h(\tau)d\tau = \frac{1}{2} + \int_{0}^{t} \frac{sin(\sigma \tau)}{\pi \tau}d\tau$$

Clearly $s(\infty) = 1$. The rise time can be defined as the time t at which $s(t)$ reaches ninety percent of this maximum value. Denoting this time by t_r, it is clear that t_r satisfies the equation

$$\int_{0}^{t_r} \frac{sin(\sigma \tau)}{\pi \tau}d\tau = 0.9/2$$

or

$$\int_{0}^{\sigma t_r} \frac{sin(x)}{x}dx = 0.9\pi/2$$

If ξ denotes the solution to the equation

$$\int_{0}^{\xi} \frac{sin(x)}{x}dx = 0.9\pi/2$$

then the rise time is given by $t_r = \xi/\sigma$. Since ξ is a fixed constant, it follows that the rise time is reciprocally related to the bandwidth.

Study project 29

The equations of Einstein's general theory of relativity can be brought into the curriculum of signals and systems. This can be introduced in a separate seminar. The equations first of all introduce the Ricci tensor which is obtained by contracting the Riemann-Christoffel curvature tensor. The Ricci tensor is non-linearly related to the metric tensor and its first and second order partial derivatives. We can write this nonlinear relation as $R_{ij} = F(g_{ij}, g_{ij,k}, g_{ij,kl})$. The Ricci tensor contains information about the curvature of space time. The Einstein equations of gravitation state that $R_{ij} - \frac{1}{2}Rg_{ij} = -kT_{ij}$ where $T_{ij} = (\rho + p/c^2)v_i v_j + pg_{ij}$ with p as the pressure field, ρ as the density field and $\{v_i\}$ as the covariant four velocity field of the matter distribution. If charges and currents are present in addition, they generate an electromagnetic field which defines an electromagnetic stress tensor field. The matter density, velocity, pressure, charge density and current density can be regarded as sources that drive the gravitational field causing space-time to be curved. Thus Einstein's general theory of relativity is a problem in non-linear systems theory and it should be tackled using discretization techniques developed by engineers. The computation of the Ricci tensor from the metric tensor is itself a tough problem. A program can be developed for this purpose. The Riemann Christoffel curvature tensor is obtained by parallely displacing a vector around a small closed loop. An elementary computation based on the notion of parallel displacement shows that it is given by the formula

$$R^a_{bcd} = \Gamma^i_{bc}\Gamma^a_{id} - \Gamma^i_{bd}\Gamma^a_{ic} + \frac{\partial \Gamma^a_{bc}}{\partial x^d} - \frac{\partial \Gamma^a_{bd}}{\partial x^c}$$

If X^i is a vector and its displaced around a small loop in the $x^1 - x^2$ plane parallely, then its change can be shown to be given by the formula

$$\delta X^i = R^j_{p12}X^p \delta r \delta s$$

where δr and δs are respectively the lengths of the sides of the loop along the x^1 and x^2 directions. The basic property of the curvature tensor is that if it vanishes, then the parallel transport of a vector along any path is independent of the path. In other words, the vector returns back to its starting value on parallel displacement once around a closed loop when the curvature tensor vanishes. The Christoffel symbols are defined by the formulae

$$\Gamma^a_{bc} = \frac{1}{2}g^{ak}(g_{kb,c} + g_{kc,b} - g_{bc,k})$$

where $g_{ij,m}$ is the shorthand notation for the partial derivative of the component g_{ij} of the metric tensor with respect to the variable x^m, i.e. $g_{ij,m} = \frac{\partial g_{ij}}{\partial x^m}$. The covariant curvature tensor in terms of the metric tensor is given by

$$R_{abcd} = g_{ib}R^i_{acd} = \frac{1}{2}(g_{bc,ad} - g_{bd,ac} + g_{ad,bc} - g_{ac,bd}) + \Gamma^j_{ad}\Gamma_{bjc} - \Gamma^j_{ac}\Gamma_{bjd}$$

The expression has been separated into two parts, the first part being a linear function of the second order partial derivatives of the metric tensor and the second term being a quadratic function of the Christoffel symbols which in turn involve only first order partial derivatives of the metric tensor. The Ricci tensor is defined by

$$R_{ab} = R^i_{abi} = g^{ij}R_{ajbi} = \Gamma^j_{ab}\Gamma^i_{ji} - \Gamma^j_{ai}\Gamma^i_{jb} + \Gamma^i_{ab,i} - \Gamma^i_{ai,b}$$

The Einstein tensor is defined by

$$G^{ab} = R^{ab} - \frac{1}{2}Rg^{ab}$$

where $R^{ab} = g^{ai}g^{bj}R_{ij}$ is obtained by raising both the indices of the covariant Ricci tensor. The basic identity concerning the Einstein tensor is that its divergence vanishes, i.e. $G^{ab}_{;b} = 0$. Einstein's field equation in vacuum states that $R_{ab} = 0$ which is equivalent to $G_{ab} = 0$. We want to solve for the metric tensor corresponding a static distribution of matter inside a finite region of space. For this, we need to use iterative techniques. The metric corresponding to a static matter distribution can be brought to the general form

$$d\tau^2 = g_{00}(\mathbf{r})dt^2 - \sum_{r,s=1}^{3} g_{rs}(\mathbf{r})dx^r dx^s$$

One way to design an iterative or perturbative technique for solving the Einstein field equations is to regard the presence of matter as a weak perturbation to empty space by introducing a parameter ϵ. Thus, the right side of the Einstein field equations has the form $-\epsilon.KT_{ij}$. Correspondingly, we expand the metric tensor in powers of ϵ. Taking our coordinates as $x^1 = x, x_2 = y, x_3 = z, x^4 = ict$, the metric of flat space-time is $g_{ij}^{(0)} = \delta_{ij}$ This gives the invariant distance measure as

$$g_{ij}dx^i dx^j = dx^2 + dy^2 + dz^2 - c^2 dt^2$$

We assume that the actual metric in the presence of matter is given by

$$g_{ij} = \delta_{ij} + \epsilon.g_{ij}^{(1)} + \epsilon^2.g_{ij}^{(2)} + ...$$

We now develop an expansion for the Ricci tensor in powers of the same parameter, i.e.

$$R_{ij} == R_{ij}^{(0)} + \epsilon.R_{ij}^{(1)} + \epsilon^2.R_{ij}^{(2)} + ...$$

We shall for the sake of illustration, evaluate $R_{ij}^{(0)}, R_{ij}^{(1)}$ and $R_{ij}^{(2)}$.

$$\Gamma^j_{ab}\Gamma^i_{ji} = g^{jk}g^{ir}\Gamma_{kab}\Gamma_{rji}$$

Now, Γ_{abc} involves only derivatives of the metric tensor and is therefore of order ϵ. Thus, to approximate the above expression upto $O(\epsilon^2)$, we need retain only the zeroth order term in g^{jk}, i.e. δ_{jk}. This gives

$$\Gamma^j_{ab}\Gamma^i_{ji} = \epsilon^2 \delta_{jk}\delta_{ir}\Gamma^{(1)}_{kab}\Gamma^{(1)}_{rji} + O(\epsilon^3) = \epsilon^2\Gamma^{(1)}_{jab}\Gamma^{(1)}_{iji} + O(\epsilon^3)$$

where

$$\Gamma^{(1)}_{jab} = \frac{1}{2}(g^{(1)}_{ja,b} + g^{(1)}_{jb,a} - g^{(1)}_{ab,j})$$

The full expression is

$$\Gamma^j_{ab}\Gamma^i_{ji} = \frac{\epsilon^2}{4}(g^{(1)}_{ja,b} + g^{(1)}_{jb,a} - g^{(1)}_{ab,j})g^{(1)}_{ii,j} + O(\epsilon^3)$$

$$\Gamma^j_{ai}\Gamma^i_{jb} = \Gamma_{jai}\Gamma_{ijb} + O(\epsilon^3) = \frac{\epsilon^2}{4}(g^{(1)}_{ja,i} + g^{(1)}_{ji,a} - g^{(1)}_{ia,j})(g^{(1)}_{ij,b} + g^{(1)}_{ib,j} - g^{(1)}_{bj,i}) + O(\epsilon^3)$$

Study project 30

A basic project about implementation of electromagnetic problems on the 8085 microprocessor can be carried out. This will be a first step in programming the TMS card in assembly language for the implementation of electromagnetic problems. The 8085 has instructions for addition of eight bit binary numbers. Using this we can formulate programs for multiplication of binary numbers. For example, if x_1x_0 and y_1y_0 are two bit binary numbers that represent the decimal numbers $x_0 + 2x_1$ and $y_0 + 2y_1$ respectively, their product has the decimal representation $x_0y_0 + 2(x_0y_1 + x_1y_0) + 2^2x_1y_1$. If s is the modulo two sum of x_0y_1 and x_1y_0 and c is the carry, then this product is the same as $x_0y_0 + 2s + 2^2(x_1y_1 + c)$. again, if s' is the modulo two sum of x_1y_1 and c and c' the carry, then the product is the same as $x_0y_0 + 2s + 2^2s' + 2^3c'$ which has the four bit binary representation $c's's(x_0y_0)$. Thus multiplication can be implemented using the logical operations of exclusive or and. In all computations for electromagnetic problems, addition and multiplications are required. For example, we want to write an 8085 programme to compute the divergence of a vector field. The first step is to store the vector field in the registers of the microprocessor. Assume that the vector field is two dimensional, say $\mathbf{A}(\mathbf{x}, \mathbf{y}) = A_x(x, y)\hat{x} + A_y(x, y)\hat{y}$. Its divergence is defined as

$$\nabla.\mathbf{A} = \frac{\partial A_x}{\partial x} + \frac{\partial A_y}{\partial y}$$

We want to calculate and store the divergence in the region $[0, L] \times [0, L]$. The first step is to divide this square into square pixels of side length Δ and evaluate the divergence at the corresponding grid vertices $(k\Delta, m\Delta), 0 \leq k, m \leq N-1$ where $N\Delta = L$. The first step is to input the vector field values $A_x[k, m], A_y[k, m]$ at these pixel sites. The values of A_x and A_y will be stored as eight bit numbers followed by eight bit fractions. Each value will therefore require two memory locations. Any such number will have the representation $\sum_{i=0}^{7} 2^i x_i + \sum_{i=1}^{8} y_i/2^i$. The binary representation of this number is $x_7x_6...x_0\Delta y_1y_2...y_8$. We input the 2.2^{N^2} values of the vector field into memory locations from the input port. The input port may be taken as the keyboard of the computer. The command to be used is *Mov*. Alternately, we can directly input the data into the memory location by using the MVI command. A timing cycle of this can be drawn and presented during the project presentation. After doing this, the divergence can be evaluated approximately using the finite difference scheme. For example, if $A_x(k\Delta, m\Delta)$ is stored in the memory location 2000, $A_x((k+1)\Delta, m\Delta)$ in the memory location 2001, $A_y(k\Delta, (m+1)\Delta)$ in the memory location 2002, $A_y(k\Delta, m\Delta)$ in the location 2003, then we transfer the contents of the memory location 2001 to the accumulator and use the SUB command to subtract the contents of the memory location 2000 from this. In this way, we are able evaluate the approximate value of the partial derivative of A_x with respect to x. Likewise, we evaluate the partial derivative of A_y with respect to the variable y.

Study project 31

The solutions to the differential equation $(D - s)^2 y(t) = 0$ are $exp(st)$ and $t.exp(st)$. This We want to verify this numerically using the 8085 microprocessor. This is done by inputting samples of the two signals sampled at the times $k\Delta, k = 0, 1, 2, ...$ via a simple program that keeps incrementing a counter and the contents of the HL register pair. The input is fed sequentially to the A register (the accumulator) which is transferred to the memory location M pointed to by the address contained in the HL pair. Thus, the input sequence is transferred to the memory. We transfer in this fashion the two sampled sequences $exp(sk\Delta), k\Delta.exp(sk\Delta), k = 0, 1, 2, ...$ and evaluate the signal $(D - s)^2 y$ by a finite difference scheme, i.e. $(D - s)^2 y(t)$ at the point $k\Delta$. This is evaluated as $\epsilon[k] = (y[k+2] - 2y[k+1] + y[k])/\Delta^2 + s^2 y[k] - 2s(y[k+1] - y[k])/\Delta)$. If the values of this sequence do not exceed a threshold, then we agree that the identity has been satisfied.

Study project 32

We want to check via elementary programmes on the 8085 microprocessor whether or not the order in which the two linear time invariant systems are cascaded commutes or not. In other words, we want to compute the convolution of two sequences in different orders and check whether they yield the same result or not. If $x[n], n = 0, 1, ..., N - 1$ is the first sequence and $y[n], n = 0, 1, 2, ..., N-1$ is the second sequence, then their convolution $x*y[n]$ is defined as $\sum_{k=0}^{N-1} x[k]y[n-k]$ and their convolution in the opposite order $y * x[n]$ is defined as $\sum_{k=0}^{N-1} y[k]x[n - k]$. We know that both these sums evaluate to the same sequence. Assume that the sequence $h1(n), n = 0, 1, ..., M-1$ has been stored in a sequence of memory locations and the sequence $h2(n), n = 0, 1, ..., M - 1$ has been stored in another sequence of memory locations. The input sequence $x(n), n = 0, 1, ..., N-1$ is input serially to the accumulator. We want to write a simple program to compute the convolution of this input first with $h1$ sequence and then with $h2$ sequence and then first with $h2$ sequence and then with $h1$ sequence and tally the two outputs. The first step is therefore to write a program to read the input serially into the memory locations. This is done using the HL register pair by incrementing it successively so that it points to a sequence of memory locations into which the data from the accumulator is loaded. Then we must write a program to compute the convolution of the a sequence $h(n)$ stored in the memory with the input sequence stored in the memory. This requires multiplications and additions. The routine for multiplications has already been discussed. For example, the convolution $y(n) = x(n) * h(n)$ at the point $n = 1$ is evaluated as

$$y(1) = \sum_{m=0}^{N-1} x(1 - m)h(m) = x(1)h(0) + x(0)h(1)$$

The value at $n = 2$ is evaluated as

$$y(2) = x(2)h(0) + x(1)h(1) + x(0)h(2)$$

To compute $y(1)$, we transfer data $x(0), x(1)$ from the input to the accumulator A via the "IN" command. The first IN command will transfer $x(0)$ from the input port to the accumulator. We then use the "MOV M,A" command to transfer the accumulator data to the memory having address given in the HL register pair. Let us assume that this address is a. The next command is to increment the address of the HL pair so that it contains the address $a + 1$. The next IN A command obtained after a jump will transfer the next data $x(1)$ from the input to the A register, i.e. the accumulator. The next $MOV M, A$ command which is obtained after implementing a JMP operation to the original will transfer the data $x(1)$ in the A register to the memory having address $a+1$. Similarly, we transfer the filter data $h(0), h(1)$ the memory having addresses $b, b+1$. Then a multiplication subroutine taking inputs from the memory locations $a + 1, b$ gives $x(1)h(0)$ and likewise the same subroutine taking inputs from the memory locations $a, b+1$ gives $x(0)h(1)$. Finally the "ADD" command is used to sum up these two results to obtain $y(1)$.

Study project 33

An 8085 microprocessor programme can be written to compute the electrostatic force between two point charges q_1 and q_2 placed at positions $\mathbf{r}_1$ and $\mathbf{r}_2$ in space. The coordinates (x_1, y_1, z_1) of $\mathbf{r}_1$ and the coordinates (x_2, y_2, z_2) of $\mathbf{r}_2$ are imputed to the memory. We also input the values of the charges q_1 and q_2 in binary form to the memory and using the operations of multiplication and division, we evaluate the three components of the force. Division is required here, namely, division by the distance squared between the two charges. How to implement the division operation for numbers input in binary format ? This problem requires consideration.

Study project 34

A simple 8085 programme to evaluate the mean and variance of a random variable X assuming values $x_1, ..., x_N$ with probabilities $p_1, ..., p_N$ is of use to the engineer. The mean value is defined by $\mu_X = \mathbb{E}(X) = \sum_{i=1}^{N} p_i x_i$ and its variance by the formula $V(X) = \sum_{i=1}^{N} p_i (x_i - \mu_X)^2$. The first step is to read the values $x_1, ..., x_N$ of the data and the corresponding probabilities $p_1, ..., p_N$ into memory from the input. This is done using the INA command followed by the $MOV M, A$ command with the address M being stored in the HL register pair. The address stored in the HL pair is then incremented and another INA command transfers the next input to the A register and the $MOV M, A$ command transfers the data in A to the next memory location. In this way, the data $x_1, ..., x_N$ is transferred to successive memory locations. Likewise, we can transfer the probability data $p_1, ..., p_N$ to another successive set of memory locations. Then the multiplication and addition commands are used to compute the mean of the random variable. A similar method can be used to compute the variance of the random variable. In this case, we would load the computed mean in one of the registers say B and subtract it from each of the data $\{x_i\}$ that are stored in consecutive memory locations. This is done by loading each of the x_i into the accumulator and using the "SUB B" command to subtract the contents of the register B from the accumulator. Then the multiplication routine is used to square the contents of the accumulator and these squared values are written into consecutive memory locations. The weighted mean of these squared values is then evaluated in a similar fashion.

Study project 35

A seminar on basic properties of the derivative, how it is defined and examples involving its computation would be useful for a student beginning to learn the subject of signals and systems. Systems that take a signal as input and output its derivative are important in signal processing. So the rigorous definition of the derivative and its properties should be presented during the course. The following aspects can be discussed as a first step.

Successive differentiation: If $f(t)$ is a function of one real variable, its derivative at the point t is defined to be the following limit when its exists

$$f'(t) = \frac{df(t)}{dt} = lim_{s \to t} \frac{f(s) - f(t)}{s - t}$$

This may equivalently be expressed as

$$f'(t) = lim_{\epsilon \to 0} \frac{f(t + \epsilon) - f(t)}{\epsilon}$$

A necessary condition for the derivative of f at the point t to exist is that f be continuous. However, there do exist functions f continuous at a point t but not differentiable at t. For example, $f(t) = |t|$ is continuous everywhere but not differentiable at $t = 0$. The basic properties of the differential of a function f are the following: (2) If f and g are both differentiable at t, then so is $cf + g$ where c is an arbitrary scalar. Further $(cf + g)'(t) = cf'(t) + g'(t)$. This is verified by using the standard property of limit, namely that the limit of a finite sum of several functions equals the sum of the individual limits provided that each component function has the corresponding limit existing. In other words, if $f_1, ..., f_k$ are functions with $lim_{s \to t} f_k(s) = c_k$, then

$$lim_{s \to t} \sum_{j=1}^{k} f_j(s) = \sum_{j=1}^{k} lim_{s \to t} f_j(s) = \sum_{j=1}^{k} c_j$$

The proof of the linearity of the derivative follows immediately from this:

$$lim_{s \to t} \frac{f(s) + g(s) - f(t) - g(t)}{s - t}$$

$$= lim_{s \to t} \frac{f(s) - f(t)}{s - t} + lim_{s \to t} \frac{g(s) - g(t)}{s - t}$$

$$= f'(t) + g'(t)$$

Given a function f we say define its nth order derivative at the point t recursively:

$$D^n f(t) = D(D^{n-1} f(t))$$

where D stands for the derivative operator $\frac{d}{dt}$. A necessary condition for $D^n f(t)$ to exist is that $D^{n-1} f(s)$ must be defined in a small neighborhood of t and should be continuous at the point t.

Examples: (1) Let $f(t) = t^n$ where n is an integer. Then,

$$D(t^n) = lim_{\epsilon \to 0} \frac{(t+\epsilon)^n - t^n}{\epsilon}$$

$$= lim_{\epsilon \to 0} \frac{\sum_{k=0}^{n} \binom{n}{k} t^k \epsilon^{n-k} - t^n}{\epsilon}$$

$$= lim_{\epsilon \to 0} \frac{\sum_{k=0}^{n-1} \binom{n}{k} t^k \epsilon^{n-k}}{\epsilon}$$

$$= lim_{\epsilon \to 0} nt^{n-1} + O(\epsilon) = nt^{n-1}$$

Successive application of this formula yields

$$D^m t^n = n(n-1)...(n-m+1)t^{n-m}$$

if $m \leq n$ and zero if $m > n$.

(2) Let $f(t) = sin(at)$. We want to compute $D^n f(t)$. The first thing to evaluate is $Df(t) = f'(t)$. This is done by noting that

$$\frac{f(t+\epsilon) - f(t)}{\epsilon} = \frac{sin(at)cos(a\epsilon) + cos(at)sin(a\epsilon) - sin(at)}{\epsilon}$$

We now make use of Euler's formula and the definition of e:

$$cos(a\epsilon) = \frac{1}{2}(exp(ia\epsilon) + exp(-ia\epsilon)) = \frac{1}{2}(1 + ia\epsilon + 1 - ia\epsilon + O(\epsilon^2)) = 1 + O(\epsilon^2)$$

and

$$sin(a\epsilon) = \frac{1}{2i}(exp(ia\epsilon) - exp(-ia\epsilon)) = \frac{1}{2i}(1 + ia\epsilon - 1 + ia\epsilon + O(\epsilon^2)) = a\epsilon + O(\epsilon^2)$$

Using this, we get

$$\frac{f(t+\epsilon) - f(t)}{\epsilon} = sin(at)\frac{cos(a\epsilon) - 1}{\epsilon} + cos(at)\frac{sin(a\epsilon)}{\epsilon}$$

$$= sin(at)O(\epsilon^2)/\epsilon + cos(at)(a\epsilon + O(\epsilon^2))/\epsilon$$

The limit of this quantity as $\epsilon \to 0$ is clearly $a.cos(at)$. By saying that a given function $F(\epsilon)$ of ϵ is $O(\epsilon^2)$, we mean that $F(\epsilon) \leq C.\epsilon^2$ for some finite positive number C and all sufficiently small ϵ. The basic property of such a function is that if δ is any nonnegative real number smaller than 2, then $F(\epsilon)/\epsilon^\delta$ will converge to zero when ϵ converges to zero. We've thus shown that

$$D(sin(at)) = a.cos(at)$$

Similarly, one can show that $D(cos(at)) = -a.sin(at)$. In fact, if $f(t)$ is any function differentiable at $t - s$, then

$$\frac{d}{dt} f(t-s) = f'(t-s) = \frac{d}{d(t-s)} f(t-s)$$

In the proof of the derivative of $sin(at)$, we've made use of Euler's identity $exp(ix) = cos(x) + isin(x)$ and hence $cos(x) = \frac{exp(ix) + exp(-ix)}{2}$, $sin(x) = \frac{exp(ix) - exp(-ix)}{2i}$, where $exp(x)$ is the function defined by the series

$$exp(x) = 1 + x + \frac{x^2}{2!} + \frac{x^3}{3!} + \dots = \sum_{n=0}^{\infty} \frac{x^n}{n!}$$

From the formulas for the derivative of cos and sin functions, it is easy to see that

$$D^{2n} cos(t) = (-1)^n sin(t) \qquad D^{2n+1} cos(t) = (-1)^n cos(t)$$

$$D^{2n} sin(t) = (-1)^n sin(t) \qquad D^{2n+1} sin(t) = (-1)^n cos(t)$$

Leibnitz' theorem: This theorem states that if f and g are n times differentiable at t, then

$$D^n(fg) = \sum_{k=0}^{n} \binom{n}{k} (D^k f)(D^{n-k} g)$$

One way to see how this formula is proves is to note that when D acts on the product of two functions, it is equivalent to applying the operator $D_f + D_g$ where D_f acts only on the function f while D_g acts only on the function g. This is called the product rule:

$$D(fg) = fDg + gDf = D_g(fg) + D_f(fg) = (D_f + D_g)(fg)$$

It is easy to see that the operators D_f and D_g commute, i.e. it does not matter in which order we apply them to the product. Thus the binomial theorem can be applied to $D^n = (D_f + D_g)^n$. This results in

$$D^n(fg) = (D_f + D_g)^n(fg) = \sum_{r=0}^{n} \binom{n}{r} D_f^r D_g^{n-r}(fg) = \sum_{r=0}^{n} \binom{n}{r} (D^r f)(D^{n-r} g)$$

Study project 36

An 8085 microprocessor programme for evaluating the flux of the electric displacement field $\mathbf{D(r)}$ over a cube having faces $x = 0, a, y = 0, b, z = 0, c$ can be considered. Two different computations can be done based upon Gauss' divergence theorem. The first is to evaluate it directly as a surface integral:

$$\int_S \mathbf{D}.\mathbf{n}dS = \int_0^b \int_0^c (D_x(a,y,z) - D_x(0,y,z))dydz + \int (D_y(x,b,z) - D_y(x,0,z))dxdz$$

$$+ \int_0^a \int_0^b (D_z(x,y,c) - D_z(x,y,0))dxdy$$

These double integrals are evaluated numerically via the discretization process. Specifically, dividing each face of the cube into $N \times N$ little squares, we obtain the following approximate formula for the flux:

$$\int_S \mathbf{D}.\mathbf{n}dS \approx \sum_{r,s=0}^{N-1} (D_x(a,rb/N,sc/N) - D_x(0,rb/N,sc/N))bc/N^2$$

$$+ \sum_{r,s=0}^{N-1} (D_y(ra/N,b,rc/N) - D_y(ra/N,0,rc/N))ac/N^2$$

$$+ \sum_{r,s=0}^{N-1} (D_z(ra/N,sb/N,c) - D_z(ra/N,sb/N,0))ab/N^2$$

These sums are implemented via assembly language programs. The basic program required is the computation of the sum of N numbers specified in 8 bit binary format. The numbers

$$D_x(0,rb/N,sc/N), D_x(a,rb/N,sc/N), D_y(ra/N,b,rc/N),$$

$$D_y(ra/N,0,rc/N), D_z(ra/N,sb/N,0)$$

are stored in different memory locations having addresses increasing from 1 to $6N^2$ and the processing is done using the subtract command, a routine for multiplying two numbers and a program for summing numbers in a sequence of memory locations. If data is stored in memory locations beginning from address a and ending at address b, then to sum up these numbers, the routine used can be the following: We load the HL register pair with the address a so that at the beginning of the loop the pointer points at the memory location M. We load the data from M to the accumulator. Then increase the address of the HL pair by one

Study project 37

Computing the volume of a region in orthogonal curvilinear coordinates by means of the Lame's coefficients using the 8085 microprocessor. The Lame coefficients are

$$H_i = \sqrt{(\frac{\partial x}{\partial q_i})^2 + (\frac{\partial y}{\partial q_i})^2 + (\frac{\partial z}{\partial q_i})^2}, i = 1,2,3$$

The volume is given by the expression

$$\int H_1 H_2 H_3 dq_1 dq_2 dq_3$$

with the limits for integration chosen appropriately. The microprocessor must be programmed to compute the partial derivatives $\frac{\partial x}{\partial q_i}$, $\frac{\partial y}{\partial q_i}$ and $\frac{\partial z}{\partial q_i}$ by means of discretization schemes and the integral replaced by a Riemann sum. In the special case of cylindrical and spherical polar coordinates, the volume elements are respectively $dV = \rho d\rho d\phi dz$ and $dV = r^2 sin(\theta) dr d\theta d\phi$.

Study project 38

The idea of a linear functional on a vector space and examples of such functionals taken from electromagnetic theory could be dealt with in a seminar. If V is a vector space over a field $\mathbb{F}$, then a linear functional is a linear map $f : V \to \mathbb{F}$, i.e., $f(cx + y) = cf(x) + f(y)$. If the vector space V is finite dimensional with basis $B = \{e_1, ..., e_n\}$, and $x_1, ..., x_n$ are the coordinates of a vector x relative to this basis, i.e., $x = \sum_{i=1}^{n} x_i.e._i$, then we can define the coordinate functionals $f_i, i = 1, 2, ..., n$ by the rule $f_i(x) = x_i$. This is equivalent to saying that $f_i(e_j) = \delta_{ij}$. $\{f_1, ..., f_n\}$ forms a basis for the dual space V^* of all linear functionals on V. A typical example of a linear functional important in applications is the following: Consider the vector space V of all functions $f : X \to \mathbb{R}$ where X is a set with a measure μ. Then, given a set of fixed functions $\phi_1, ..., \phi_n$ defined on X, we can define n linear functionals

$$F_i(f) = \int_X f(x)\phi_i(x)d\mu(x), i = 1, 2, ..., n$$

For example, in electromagnetic theory, $\rho(\mathbf{r})$ is the charge density in space and the potential $V(\mathbf{r}_0)$ at a given point $\mathbf{r}_0$ is a linear functional of the charge density:

$$V(\mathbf{r}_0) = \frac{1}{4\pi\epsilon} \int \frac{\rho(\mathbf{r})}{|\mathbf{r}_0 - \mathbf{r}|} d^3\mathbf{r}$$

In fact, the map that takes the charge density field $\{\rho(\mathbf{r})\}$ to the potential field $\{V(\mathbf{r})\}$ is linear and when we evaluate the potential at a given point $\mathbf{r}_0$ we get a linear functional, i.e., a linear mapping from a function or a field to a scalar or a number. Many important linear functionals arise in this way, i.e. we have a linear transformation from a vector space V_1 of functions to another vector space V_2 of functions and when we evaluate the output of the transformation at a given point, we obtain a linear functional. Thus, If $f(x)$ denotes the input function and $g(x)$ denotes the output function, a linear map relating the output to the input is $g(x) = \int K(x, y)f(y)d\mu(y)$. The map $f \to g$ is a linear transformation. The map $f \to g(x_0)$ with x_0 a fixed point, defines a linear functional. Another typical example of a linear functional is the map that takes the electric field to the potential difference between two points along a curve. Let Γ be a fixed curve in space connecting two points $\mathbf{a}$ and $\mathbf{b}$. Then, if $\mathbf{E}(\mathbf{r})$ is the electric field, the potential difference between the two points is given by $V_{ab} = \int_\Gamma \mathbf{E}.d\mathbf{r}$. The map $\{\mathbf{E}(\mathbf{r})\} \to V_{ab}$ defines a linear functional. An example of a functional that is non-linear is the map that takes as input the electric field in space and outputs the total energy of the field. This map is

$$\mathbf{E} \to \int \frac{\epsilon}{2}|\mathbf{E}|^2 d^3\mathbf{r}$$

Study project 39

Differential equations with time varying coefficients can be studied as a part of a project dealing with simulation of continuous time systems on a digital computer. The idea of this project is to develop the solutions in the form of Taylor series with the coefficients also expressed in the form of Taylor series. The idea of a power series is elementary. Consider for example the nth order linear differential equation with constant coefficients

$$\frac{d^n y}{dt^n} + a_1 \frac{d^{n-1} y}{dt^{n-1}} + ... + a_{n-1} \frac{dy}{dt} + a_n y = 0$$

where $a_1, ..., a_n$ are constants. We substitute $y(t) = exp(st)$ into this equation and derive the following polynomial equation satisfied by the parameter s

$$s^n + a_1 s^{n-1} + ... + a_{n-1}s + a_n = 0$$

If this polynomial has distinct roots $s_1, ..., s_n$, then the general solution is given by

$$y(t) = \sum_{\alpha=1}^{n} A_\alpha exp(s_\alpha t)$$

where the constants $\{A_\alpha : \alpha = 1, 2, ..., n\}$ can be chosen arbitrarily. This solution can be expressed as a power series

$$y(t) = \sum_{\alpha=1}^{n} A_\alpha \sum_{m=0}^{\infty} (s_\alpha t)^m / m! = \sum_{m=0}^{\infty} y_m t^m$$

with

$$y_m = \sum_{\alpha=1}^{n} A_\alpha s_\alpha^m / m!$$

We can obtain this series expansion by directly substituting the series into the differential equation. This gives us

$$\sum_{m=n}^{\infty} y_m m(m-1)...(m-n+1)t^{m-n} + a_1 \sum_{m=n-1}^{\infty} y_m m(m-1)...(m-n+2)t^{m-n+1}$$

$$+ ... + a_n \sum_{m=0}^{\infty} y_m t^m = 0$$

Equating coefficients of t^m on both sides gives us

$$(m+n)(m+n-1)...(m+1)y_{m+n} + a_1(m+n-1)(m+n-2)...(m+1)y_{m+n-1} + ... + a_n y_m = 0$$

which is equivalent to

$$y_{m+n} + a_1 y_{m+n-1}/(m+n) + a_2 y_{m+n-2}/(m+n)(m+n-1) + ...$$

$$+a_n y_m/(m+n)(m+n-1)...(m+1) = 0$$

We substitute $y_m = s_\alpha^m/m!$ into this equation and obtain

$$s_\alpha^{m+n}/(m+n)! + a_1 s_\alpha^{m+n-1}/(m+n)! + ... + a_n s_\alpha^m/(m+n)! = 0$$

which is the same as

$$s_\alpha^m + a_1 s_\alpha^{m-1} + ... + a_n = 0$$

an identity. Thus, the general linear combination $y_m = \sum A_\alpha s_\alpha^m/m!$ also satisfies the above difference equation. When the coefficients of the differential equation vary with time and they are analytic, then also the power series technique can be used. For example suppose the input output equation is described by the differential equation

$$\frac{dy(t)}{dt} + a(t)y(t) = x(t)$$

where $a(t) = \sum_{n=0}^{\infty} a_n t^n$ and $x(t) = \sum_{n=0}^{\infty} x_n t^n$, then we look for a solution having the form

$$y(t) = \sum_{n=0}^{\infty} y_n t^n$$

Plugging this into the differential equation gives us

$$\sum_{n=1}^{\infty} n y_n t^{n-1} + \left(\sum_{n=0}^{\infty} a_n t^n\right)\left(\sum_{n=0}^{\infty} y_n t^n\right) = \sum_{n=0}^{\infty} x_n t^n$$

Equating coefficients of equal powers of t on both sides gives us the following recursive scheme for computing the coefficients of the output:

$$(n+1)y_{n+1} + \sum_{k=0}^{n} a_k y_{n-k} = x_n$$

This procedure can be extended to higher order linear differential equations. In this context, the method of moments is also useful. To illustrate this, we consider the above first order linear differential equation with time varying coefficient $a(t)$. Consider a set of linearly independent functions $\phi_k(t), k = 1, 2, ...$ We assume that the input and the output can be expanded as linear combinations of these functions, i.e.

$$x(t) = \sum_{n=1}^{\infty} x_n \phi_n(t) \qquad y(t) = \sum_{n=1}^{\infty} y_n \phi_n(t)$$

Plugging these into the differential equation gives us

$$\sum_n y_n \phi'_n(t) + a(t) \sum_n y_n \phi_n(t) = \sum_n x_n \phi_n(t)$$

We deduce from this equation linear equations for the coefficients $\{y_n\}$ by multiplying by $\phi_m(t)$ and integrating over $[0, T]$. These equations are

$$\sum_n c_{mn} y_n = \sum_n d_{mn} x_n$$

where

$$c_{mn} = \int_0^T \phi_m(t)\phi'_n(t)dt + \int_0^T \phi_m(t)\phi_n(t)dt$$

$$d_{mn} = \int_0^T \phi_m(t)\phi_n(t)dt$$

Study project 40

Transmission of a signal through a linear periodically time varying filter. The impulse response of the filter is $h(t, \tau)$. If it is to be periodically time varying, we must have $h(t + T, \tau + T) = h(t, \tau)$ for all $t, \tau \in \mathbb{R}$. Let $x(t)$ be any input to this system. The output is given by $y(t) = \int_{-\infty}^{\infty} h(t, \tau)x(\tau)d\tau$. Suppose we apply the input $x_1(t) = x(t - T)$, i.e. the original input delayed by T seconds, T being the period of the system. Then the output is given by

$$y_1(t) = \int h(t, \tau)x(\tau - T)d\tau = \int h(t, \tau + T)x(\tau)d\tau = \int h(t - T, \tau)x(\tau)d\tau = y(t - T)$$

By successive application of this result, it follows that when the input $x_n(t) = x(t - nT)$ is applied to the system, the corresponding output is given by $y_n(t) = y(t - nT)$. Delaying the input to a periodic system by any integral multiple of the period results in the output being delayed by the same amount of time. However if the input is delayed by an amount that is not an integral multiple of the system period, the output will not suffer the corresponding delay. This is owing to the fact that for a periodically time varying system having impulse response $h(t, \tau)$, in general, we do not have $h(t + \alpha, \tau + \alpha) = h(t, \tau)$ when α is not an integral multiple for T. This happens only for linear time invariant systems. Let us compute the Fourier transform of the output of a periodically time varying system.

$$Y(\omega) = \int exp(-j\omega t)dt \int h(t,\tau)x(\tau)d\tau = \int H(\omega,\tau)x(\tau)d\tau$$

where

$$H(\omega,\tau) = \int h(t,\tau)exp(-j\omega t)dt$$

Now, express $x(t)$ in terms of its Fourier transform:

$$x(t) = \frac{1}{2\pi} \int X(\omega)exp(j\omega t)d\omega$$

Then,

$$Y(\omega) = \frac{1}{2\pi} \int H(\omega,\tau)X(\theta)exp(j\theta\tau)d\tau d\theta$$

Define

$$G(\omega,\theta) = \int H(\omega,\tau)exp(j\theta\tau)d\tau = \int h(t,\tau)exp(j(\theta\tau - \omega t))dtd\tau$$

Then,

$$Y(\omega) = \int H(\omega,\theta)X(\theta)d\theta/2\pi$$

This formula is valid for any linear system, not necessarily periodically time varying. Now, let us see what the periodically time varying condition translates to in the frequency domain.

$$h(t,\tau) = (2\pi)^{-2} \int G(\omega,\theta)exp(j(\theta\tau - \omega t))d\omega d\theta$$

$$= h(t+T,\tau+T) = (2\pi)^{-2} \int G(\omega,\theta)exp(j(\theta\tau - \omega t))exp(j(\theta - \omega)T)d\theta d\omega$$

Equating coefficients of $exp(j(\theta\tau - \omega t))$ on both sides results in the identity

$$G(\omega,\theta) = G(\omega,\theta))exp(j(\theta - \omega)T)$$

for all $\omega, \theta \in \mathbb{R}$. This means that whenever $G(\omega,\theta) \neq 0$, we must have $\omega - \theta = 2\pi n/T$ for some integer n. This is equivalent to saying that the only nonzero values G occur at the frequency pairs $(2\pi n/T + \theta, \theta)$ for some integer n. So we can set

$$G(\omega,\theta) = \sum_{n=-\infty}^{\infty} G_n(\omega)\delta(\omega - \theta - 2n\pi/T)$$

Fourier inverting this expression with respect to the θ variables gives us

$$H(\omega, \tau) = \frac{1}{2\pi} \int G(\omega, \theta) exp(-j\theta\tau) d\theta = \frac{1}{2\pi} \sum_{n=-\infty}^{\infty} G_n(\omega) exp(j(\omega - 2n\pi/T)\tau)$$

This formula can be seen via another route also. The function $\psi(t) = h(t + x, t)$ for each fixed x is periodic in the variable t with a period of T. So it can be developed into a Fourier series

$$\psi(x) = h(t + x, t) = \sum_{n=-\infty}^{\infty} c_n(x) exp(-i2\pi nt/T)$$

It follows that

$$h(t, \tau) = \sum_{n=-\infty}^{\infty} c_n(t - \tau) exp(-i2\pi n\tau/T)$$

Fourier transforming with respect to the first variable gives

$$H(\omega, \tau) = \sum_{n=-\infty}^{\infty} C_n(\omega) exp(i(\omega - 2\pi n/T)\tau)$$

and $C_n(\omega)$ can be identified with $G_n(\omega)$.

Study project 41

In a course on signals and systems, reduction formulas which are recursive equations for computing the indefinite integrals $\int (sinx)^n dx, \int (tanx)^n dx, \int (secx)^n dx, \int x^n sinxdx, \int x^n cos(x)dx,$ $\int x^n exp(ax)dx, \int (logx)^n dx, \int (sinx)^n (cosx)^m$ are taught. The proof of these theorems can be dealt with in a seminar. It constitutes an integral part of the course on signals and systems. The proofs of these theorems are based on integration by parts. For evaluating $\int (sinx)^n dx$, choose the $f(x) = (sinx)^{n-1}$ and $g(x) = sinx$ and apply the integration by parts rule

$$\int fgdx = f \int gdx - \int f'(\int gdx)dx$$

$$I_n = \int (sinx)^n dx = -(sinx)^{n-1} cos(x)|_0^x + \int (n-1)(cosx)(sinx)^{n-2} cosxdx$$

$$= -(sinx)^{n-1}(cosx) + (n-1) \int (sinx)^{n-2}(cosx)^2 dx$$

$$= -(sinx)^{n-1}(cosx) + (n-1) \int (sinx)^{n-2}(1 - (sinx)^2)dx$$

$$= -(sinx)^{n-1} cosx + (n-1)(I_{n-2} - I_n)$$

or

$$nI_n = -(sinx)^{n-1}cosx + (n-1)I_{n-2}$$

or

$$I_n = -\frac{(sinx)^{n-1}cosx}{n} + \frac{n-1}{n}I_{n-2}$$

For evaluating $\int (tanx)^n dx$, choose $f = (tanx)^{n-2}, g = (secx)^2$ and write $\int (tanx)^n dx = \int (tanx)^{n-2}(secx)^2 dx - \int (tanx)^{n-2}dx$.

$$I_n = \int (tanx)^n dx = \int (tanx)^{n-2}(tanx)^2 dx = \int (tanx)^{n-2}(tanx)^2 dx$$

$$= \int (tanx)^{n-2}((secx)^2 - 1)dx = \int (tanx)^{n-2}(secx)^2 dx - I_{n-2}$$

$$(tanx)^{n-2}tanxdx - \int (secx)^2(n-2)(tanx)^{n-3}tanxdx - I_{n-2}$$

$$= (tanx)^{n-1} - (n-2)\int (tanx)^{n-2}((tanx)^2 + 1) - I_{n-2}$$

$$= (tanx)^{n-1} - (n-2)I_n - (n-2)I_{n-2} - I_{n-2}$$

or

$$(n-1)I_n = (tanx)^{n-1} - (n-1)I_{n-2}$$

or

$$I_n = \frac{(tanx)^{n-1}}{n-1} - I_{n-2}$$

$$\int (secx)^n dx = \int (secx)^{n-2}(secx)^2 dx$$

$$= (secx)^{n-2}(tanx) - \int (n-2)(secx)^{n-3}secxtanx.tanxdx$$

$$= (secx)^{n-2}(tanx) - (n-2)\int (secx)^{n-2}(tanx)^2 dx$$

$$= (secx)^{n-2}(tanx) - (n-2)\int (secx)^{n-2}((secx)^2 - 1)dx$$

$$= (secx)^{n-2}(tanx) - (n-2)\int (secx)^n dx + (n-2)\int (secx)^{n-2}dx$$

Thus,

$$\int (secx)^n dx = \frac{(secx)^{n-2}tanx}{n-1} + \frac{n-2}{n-1}\int (secx)^{n-2}dx$$

$$\int x^n sin(x)dx = -x^n cosx + n\int x^{n-1}(cosx)dx$$

$$\int x^n(\cos x)dx = x^n \sin x - n\int x^{n-1}(\sin x)dx$$

Defining $I_n = \int x^n \sin x\, dx$ and $J_n = \int x^n \cos x\, dx$, it follows that these two recursions can be put in matrix form:

$$\begin{pmatrix} I_n \\ J_n \end{pmatrix} = \begin{pmatrix} -x^n \cos x \\ x^n \sin x \end{pmatrix}$$

$$+ \begin{pmatrix} 0 & n \\ -n & 0 \end{pmatrix}\begin{pmatrix} I_{n-1} \\ J_{n-1} \end{pmatrix}$$

Study project 42

Give an account of how wavelets can be applied to image compression and synthesis, to problems of identifying the structure of a mechanical system and to problems of sonar. The account should be more descriptive than mathematical.

This project will be primarily aimed at applying wavelet analysis techniques to problems of mechanics and image processing. Wavelets have already been extensively used in image compression applications. Wavelets for image processing constitute an orthonormal basis for $L^2(\mathbb{R} \times \mathbb{R})$ obtained by translating and dilating a given mother wavelet function, i.e. orthonormal bases having the form $2^{(n+m)/2}\psi(2^n x - k, 2^m y - r), n, k, m, r \in \mathbb{Z}$. This structure is particularly suitable for approximating an image once the resolution level is specified and leads to image compression algorithms. The first part of the project will be to study the compression algorithms already developed by researchers and to develop TMS packages for this, i.e. assembly language programmed cards for compression. The study will also include theoretical analysis of the nature of the multiresolution subspaces $\{V_j : j \in \mathbb{Z}\}$ generated by the wavelets. The researcher in question will also be delivering seminars on this work. The second phase of the work will aim at applying wavelet analysis methods to problems of mechanics. The prototype example will be that of a spinning top whose motion is described by the three Eulerian angles ϕ, θ, ψ. Using the idea of rotation matrices $R(\phi, \theta, \psi)$, one evaluates the angular velocity tensor of the top relative to the principal axes fixed to the top. The kinetic energy of the top is then expressed in terms of the Eulerian angles and the principal moments of inertia I_1, I_2, I_3. This enables one to construct the Lagrangian function for the top $L(\phi, \theta, \psi, \phi', \theta', \psi')$. The Lagrange equations of motion yield coupled ordinary nonlinear differential equations for the tops motion and one obtains the solution in terms of quadratures. These integrals cannot be explicitly evaluated in closed form and hence one looks qualitatively at the motion leading to three kinds of motion, nutation, precession and spin. By approximately solving the equations of motion using numerical techniques, our aim will be to perform a wavelet analysis of these components and from the coefficients, try to infer something about the top's structure, i.e. moments of inertia about the three axes. We shall also look at other mechanical problems like the simple pendulum, the double pendulum and the anharmonic oscillator. The last one corresponds to the motion of a particle attached to a spring nearly obeying Hooke's law but for a small nonlinear perturbation:

$$\frac{d^2x}{dt^2} + \omega^2(1 + \epsilon.x^3)x = 0$$

Periodic solutions to this equation can be analyzed using Fourier series. We shall apply wavelet analysis to this problem when the oscillator is subject to nonstationary noise generated externally and try to use the wavelet coefficients to infer something about the parameters ω and ϵ. We shall also look at wavelet analysis of this oscillator when the natural frequency ω is a slowly varying function of time. Wavelet analysis is particularly appropriate for this problem since the underlying system is time varying leading to nonstationary responses even when the input is stationary Gaussian noise. The following lines indicate our tentative approach: Suppose $x_i(t, \theta_1, ..., \theta_p), t \geq 0, i = 1, 2, ..., p$ are the motions of a mechanical system whose parameters $\theta_1, ..., \theta_p$ are unknown. We start with an orthonormal wavelet basis $\{\psi_{n,k} : n, k \in \mathbb{Z}\}$ for the signal space $L^2(\mathbb{R})$ and compute the wavelet coefficients for the mechanical signals $x_i(t)$ relative to this basis:

$$c(n, k, i) = < x_i, \psi_{n,k} > = \int_{-\infty}^{\infty} x_i(t, \theta_1, ..., \theta_p)\psi_{n,k}(t)dt$$

These coefficients, as the above formula indicates, will be functions of the parameters $\{\theta_j\}$, i.e. $c(n, k, i, \theta_1, ..., \theta_p)$. To compress the motion, we store only the coefficients for $n \leq N$ and try to synthesize the motion from these compressed coefficients. From the synthesized motion, we try to infer what the mechanical system would look like. For example, if we know that the motion corresponds to that of a top, then the unknown parameters are I_1, I_2, L the moments of inertial corresponding to the principal axes and the distance of the centre of gravity of the top from the point of contact of the top with the ground. The wavelet coefficients $c(n, k, I_1, I_2, L), n \leq N$ can be used to estimate these parameters and synthesize the shape of the top. This application has reference to control systems rather than to signal processing. We shall also be looking at signal processing applications like processing the signals coming from the turbines of a submarine (sonar applications). A submarine underwater has turbines rotating at different speeds. Let $\omega_1, ..., \omega_p$ denote the angular frequencies of these turbines. The water waves generated by such a submarine will generally be resolvable as a fourier series having the general form

$$x(t) = \sum_{n_1, ..., n_p = -\infty}^{\infty} c(n_1, ..., n_p)exp(i(n_1\omega_1 + ... + n_p\omega_p)t)$$

We shall try to collect real submarine wave data from the navy and apply wavelet analysis to this signal. The signal $x(t)$ is then compressed by storing only the wavelet coefficents upto only a given resolution and from these coefficients synthesize the original submarine signal. From the synthesized submarine signal, we get information about the angular frequencies $\omega_i, i = 1, 2, ..., p$ of the turbine rotation which would enable one to reconstruct the structure of the submarine.

Study project 43

General aspects of linear coding theory can be presented as a separate seminar intended to introduce the audience to more fundamental aspects like error correcting codes. The notion of a binary code as a subspace $\mathcal{C}$ in $\mathbb{F}_q^n$ where $\mathbb{F}_q$ is a finite field consisting of q elements. If the dimension of the code is q^k, then we call it an (n,k) code. Such a code can always be constructed using a generator matrix, i.e. a matrix $\mathbf{G}$ over the field $\mathbb{F}_q$ of size $k \times n$ having linearly independent rows $\mathbf{g}_1^T, \mathbf{g}_2^T, ..., \mathbf{g}_k^T$. Specifically, if $\mathbf{g}_i^T = [g_{i1}, g_{i2}, ..., g_{in}]$ where $g_{ij} \in \mathbb{F}_q$, then the generator matrix can be expressed as

$$\mathbf{G} = \begin{pmatrix} g_{11} & g_{12} & \cdots & g_{1n} \\ g_{21} & g_{22} & \cdots & g_{2n} \\ \cdots & \cdots & \cdots & \cdots \\ g_{k1} & g_{k2} & \cdots & g_{kn} \end{pmatrix}$$

Any codeword $\mathbf{g} \in \mathcal{C}$ can be represented as $\mathbf{g} = \mathbf{x}^T \mathbf{G}$, for some $\mathbf{x} \in \mathbb{F}_q^k$. Note, that this equation implies that $\mathbf{g}^T = \mathbf{G}^T \mathbf{x}$, which is the same as saying that $\mathcal{C}$ is simply the transpose of the range space of $\mathbf{G}^T$. The finite field $\mathbb{F}_4$ consisting of the four elements $\{0, 1, \omega, \bar{\omega}\}$ is very important in applications. Its addition table can be described using any standard reference book on coding theory. The notion of a parity check matrix should be introduced here as an $n - k \times n$ matrix $\mathbf{H}$ that satisfies the property $\mathbf{H}\mathbf{g}^T = 0$ for all $\mathbf{g} \in \mathcal{C}$. The rows of the parity check matrix are assumed to be linearly independent. Equivalently, the parity check matrix satisfies the property $\mathbf{H}\mathbf{G}^T = \mathbf{0}$ since every $\mathbf{g} \in \mathcal{C}$ is of the form $\mathbf{x}\mathbf{G}$ and hence $\mathbf{g}^T = \mathbf{G}^T \mathbf{x}^T$. We have $rank(\mathbf{G}) = k$, so that by the rank-nullity theorem, $nullity(\mathbf{G}) = n - k$. Likewise, $rank(\mathbf{H}) = n - k$. The parity check matrix has been defined in such a way that the range of $\mathbf{G}^T$ which is simply $\mathcal{C}$ transposed equals the nullspace of $\mathbf{H}$. This means that $\mathbf{H}$ annihilates every codeword in $\mathcal{C}$. This can be used to check whether or not a codeword belongs to $\mathcal{C}$. The notion of the Hamming distance between two codewords should also be presented. If $\mathbf{g}$ and $\mathbf{g}'$ are two elements of $\mathbb{F}_q^n$, then $d(\mathbf{g}, \mathbf{g}')$ equals the number of spots where the entries of $\mathbf{g}$ and $\mathbf{g}'$ differ.

Study project 44

Choose n as a suitable positive integer and choose a basis $\mathbf{B}_1 = \{\mathbf{e}_1, ..., \mathbf{e}_n\}$ for $\mathbb{R}^n$. Choose an arbitrary vector $\mathbf{x} \in \mathbb{R}^n$ and solve the equations $\mathbf{x} = \sum_{i=1}^n a_i \mathbf{e}_i$ for the coefficients $\{a_i\}$ and hence obtain the coordinate representation of $\mathbf{x}$ relative to this basis. Do this problem by considering appropriate numerical examples. Now choose another basis $\mathbf{B}_2 = \{\mathbf{f}_1, ..., \mathbf{f}_n\}$ for $\mathbb{R}^n$ and construct the matrix that transforms the former basis to the latter, i.e., solve the linear equations $\mathbf{f}_i = \sum_{j=1}^n a_{ji} \mathbf{e}_j, i = 1, 2, ..., n$ for the $a_{ji}, 1 \leq i, j \leq n$. Do this by taking appropriate numerical examples.

Study project 45

Choose a convenient linearly independent set $\{\phi_1(t), ..., \phi_n(t)\}$ of signals in $L^2[0,T]$ and construct an operator $K : L^2[0,T] \to L^2[0,T]$ defined by $Kx(t) = \int_0^T K(t,\tau)x(\tau)d\tau$ so that K equals the orthogonal projection of $L^2[0,T]$ onto the space W spanned by the signals $\phi_1, ..., \phi_n$. In order to do this, show that

$$P_w x(t) = \sum_{j=1}^{n} c_j \phi_j(t)$$

where the $c'_j s$ are determined from the solution of the system of linear equations

$$< x - P_W x, \phi_j >= 0, j = 1, 2, ..., n$$

Study project 46

Seminar on the basic equations of the discrete Fourier transform. If $\{x(n) : n = 0, 1, ..., N-1\}$ is a random signal, we define its Fourier series coefficients $\{a_k : k = 0, 1, ..., N-1\}$ by means of the linear equations

$$x(n) = \sum_{k=0}^{N-1} a_k exp(j2\pi kn/N), n = 0, 1, ..., N-1$$

This transform can be cast in a matrix form by introducing the vectors

$$\mathbf{x} = [x(0), x(1), ..., x(N-1)]^T$$

$$\mathbf{a} = [a_0, a_1, ..., a_{N-1}]^T$$

and the matrix

$$\mathbf{W} = ((w_{nk}))_{0 \leq n, k \leq N-1} \qquad w_{nk} = exp(j2\pi kn/N)$$

Then, the above equation is the same as $x(n) = \sum_{k=0}^{N-1} w_{nk} a_k$ which in matrix notation, translates to

$$\mathbf{x} = \mathbf{W a}$$

Inversion gives $\mathbf{a} = \mathbf{W}^{-1}\mathbf{x}$. The identity

$$\sum_{n=0}^{N-1} \bar{w}_{nk} w_{nm} = \sum_{n=0}^{N-1} exp(j2\pi n(m-k)/N) = N\delta[m-k]$$

when $0 \leq k, m \leq N-1$ when cast in matrix notation, reads $\mathbf{W}^*\mathbf{W} = N.\mathbf{I}$ which is equivalent to saying that the matrix $\frac{1}{\sqrt{N}}\mathbf{W}$ is unitary. The inverse equation that tells us how to compute the Fourier coefficients from the signal can be cast in the matrix from using the above identity: $\mathbf{a} = \mathbf{W}^{-1}\mathbf{x} = \frac{1}{N}\mathbf{W}^*\mathbf{x}$ or in terms of components,

$$a_k = \frac{1}{N} \sum_{n=0}^{N-1} exp(-j2\pi kn/N)x(n)$$

We now consider a random signal consisting of N samples. Let $\mathbf{R_x} = \mathbb{E}(\mathbf{xx^*})$ be its correlation matrix. Then, the correlation matrix of its Fourier coefficients is given by the formula $\mathbf{R_a} = \frac{1}{N^2}\mathbf{W^*R_xW}$ whose inverse reads $\mathbf{R_x} = \mathbf{WaW}$

Study project 47

This project involves the study of some of the important results in ring and field theory which play a fundamental role in the construction of error correcting codes.

1. Let F, K, L be three fields such that $F \subset K \subset L$. Then, L is a vector space over K and also a vector space over F. Also, K is a vector space over F. To see that L is a vector space over K, we only need to note that if $c \in K$ and $x \in L$, then $c.x \in L$. This defines scalar multiplication over L with scalars chosen from K. Also, $x, y \in L$ implies $x + y \in L$ since L is a field. Distributivity of scalar multiplication is obvious. Also addition in L is commutative and associative and the other properties of a vector space are easily verified. Suppose that L has dimension n over K and K has dimension m over F. Then, we claim that L has dimension nm over F. This can be seen from the following argument. Let $\{x_1, ..., x_m\}$ be a basis for K viewed as a vector space over F and let $\{y_1, ..., y_n\}$ be a basis for L viewed as a vector space over K. Let $y \in L$. Then, $y = c_1 y_1 + ... + c_n y_n$ for some $c_1, ..., c_n \in K$. Also $c_i = a_{i1}x_1 + ... + c_{im}x_m$ for $a_{ij} \in F$. Thus, $y = \sum c_k y_k = \sum_{i,j} a_{ij}x_j y_k$. Now, $x_j y_k \in L$. Hence $\{x_j y_k\}$ spans L viewed as a vector space over F. It remains to prove that the set $\{x_j y_k\}$ is linearly independent. For $b_{jk} \in F$, $\sum_{j,k} b_{jk}x_j y_k = 0$ implies $\sum(\sum_j b_{jk}x_j)y_k = 0$ and since $b_{jk}x_j \in K$ and $\{y_k\}$ forms a linearly independent set in L over K, it follows that $\sum_j b_{jk}x_j = 0$ for all k. Since $\{x_j\}$ forms a linearly independent set in K over F, it follows that $b_{jk} = 0$ for all k, j. This proves the claim made: $dim_F K.dim_K L = dim_F L$.

2. Galois theory: Let $\mathbb{K}'$ be a field and K a subfield of K. $G(K', K)$ is the group of automorphisms of K' that leaves every element of K fixed. If K' is an algebraically closed field, then K' is called the Galois extension of K. Note that $K = \{c \in K' : s(c) = c \forall s \in G(K', K)\}$. For example, $G(\mathbb{C}, \mathbb{R})$ consists of two automorphisms of $\mathbb{C}$: s_1, s_2 where $s_1(a + ib) = a + ib$ is the identity and $s_2(a + ib) = a - ib$ is conjugation.

3. Let R be a commutative ring with unit element whose only ideals are (0) and R. Then, R is a field. To prove this theorem, we must show that if $a \in R$ is nonzero, then a $b \in R$ can be determined such that $ab = 1$. Consider the set Ra. It is easy to see that $Ra + Ra = Ra$ and for any $r \in R$, $rRa = Ra$, i.e. Ra is a left ideal in R.. By assumption, $Ra = (0)$ or $Ra = R$. Now, $a \in Ra$, so $Ra = R$. In particular, a $b \in R$ can be determined such that $ba = 1$.

4. An ideal M in a ring R is said to be a maximal ideal of R if U is any ideal in R containing M implies $U = M$ or $U = R$. For example, if R is the ring of integers and p is a prime, then all integer multiples of p constitute a maximal ideal. For if U is any ideal in R, then let a be the generator of U. a is an integer and can be taken as positive. $(p) \subset U$ implies that p is divisible

by a. This is possible only if $a = 1$ or $a = p$. In the former case, $U = R$ and in the latter case, $U = (p)$. Maximal ideals thus are generalizations of the notion of prime numbers. The role played by prime numbers in the ring of integers is played by maximal ideals in a ring.

5. Let R, R' be rings and $\phi : R \to R'$ a surjective homomorphism, i.e., $\phi(r+r') = \phi(r)+\phi(r')$ and $\phi(rr') = \phi(r)\phi(r')$ for all $r, r' \in R$. In other words, ϕ is a group homomorphism from R to R' both as Abelian additive and Abelian multiplicative groups. Let $U = ker\phi$, i.e.,

$$U = \{r \in R : \phi(r) = 0\}$$

Then, R' and R/U are isomorphic. Moreover, there is a one-one correspondence between the set of ideals of R' and the set of ideals of R containing U. This correspondence is achieved by associating with each ideal W' of R', the ideal $W = \phi^{-1}(W')$ of R. Then, R/W is isomorphic to R'/W'. The first thing to do is to show how the operations on R/U are defined. The elements of R/U are $r+U$ with $r \in R$. In the language of group theory, $r+U$ is a coset of the subgroup U of R regarded as an Abelian group under addition. The sum of two such cosets is defined by the rule $(r + U) + (s + U) = (r + s) + U$. This is well defined since if $r + U = r' + U$ and $s + U = s' + U$, then $r - r', s - s' \in U$ and this implies $r + s - r' - s' \in U$ so that $r + s + U = r' + s' + U$. Likewise, the product of two elements $r + U$ and $s + U$ in R/U is defined as $(r + U)(s + U) = rs + U$. This operation is well defined since $r + U = r' + U$ and $s + U = s' + U$ imply $r = r' + u_1, s = s' + u_2$ for some $u_1, u_2 \in U$ and hence $rs = r's'+r'u_2+s'u_1+u_1u_2 \in rs+U$ and hence $rs+U = r's'+U$. In this, we've used the fact that for any $r \in R$, we have $rU \subset U$ since if $u \in U$, then $\phi(ru) = \phi(r)\phi(u) = 0$. Thus, R/U is a ring. We now want to construct a ring isomorphism between R/U and R'. To this end, define $\psi(r + U) = \phi(r)$. We want to show that ψ is an isomorphism between R/U and R'. First we must show that ψ is well defined. $r + U = r' + U$ implies $r - r' \in U$ and this implies $\phi(r - r') = 0$ or $\phi(r) = \phi(r')$ and hence, $\psi(r + U) = \psi(r' + U)$. Now, to show that ψ is one-one. $\phi(r) = \phi(r')$ implies $\phi(r - r') = 0$ implies $r - r' \in U$ implies $r + U = r' + U$ proving ψ is one-one. ψ is surjective since ϕ is surjective. This shows that R/U and R' are isomorphic. Let I' denote the set of all ideals in R' and I_U the set of all ideals in R containing U. By saying that I' is an ideal of R', we mean that $I' + I' \subset I', r'I' \subset I'$ for all $r' \in R'$. Define $T : I' \to I_U$ by $T(W') = \phi^{-1}(W')$. We claim that T is a bijection. First note that if W' is an ideal in R', then $\phi^{-1}(W')$ is an ideal in R containing U. To prove that $\phi^{-1}(W')$ is an ideal we must first show that $\phi^{-1}(W') + \phi^{-1}(W') \subset \phi^{-1}(W')$. Indeed, $r, s \in \phi^{-1}(W')$ implies $\phi(r), \phi(s) \in W'$ implies $\phi(r + s) = \phi(r) + \phi(s) \in W'$ since W' is an ideal and this implies $r + s \in \phi^{-1}(W')$. We must next prove that $r.\phi^{-1}(W') \subset \phi^{-1}(W')$. Indeed $s \in r.\phi^{-1}(W')$ implies $s = rw$ where $w \in R, \phi(w) \in W'$. This implies $\phi(s) = \phi(r)\phi(w) \in W'$ since W' is an ideal. Thus, $s \in \phi^{-1}(W')$ completing the proof that W' is an ideal. Next, we must show that if W' is an ideal in R', then $\phi^{-1}(W')$ contains $U = ker\phi$. Indeed, let $u \in U$. Then, $\phi(u) = 0 \in W'$ and hence, $u \in \phi^{-1}(W')$ proving the claim. Next we must show that $W' \to \phi^{-1}(W')$ is a one-one correspondence between ideals of R' and ideals of R containing U. Indeed, let $W'_1 \neq W'_2$ be two ideals in R'. We want to show that $\phi^{-1}(W'_1) \neq \phi^{-1})(W'_2)$. Suppose, $\phi^{-1}(W'_1) = \phi^{-1}(W'_2)$. Choose any $w' \in W'_1$. Then, since ϕ is onto, there exist $w \in R$ such that $\phi(w) = w'$. In particular, $w \in \phi^{-1}(W'_1)$. By hypothesis,

$w \in \phi^{-1}(W_2')$,i.e., $\phi(w) \in W_2'$, or equivalently, $w' \in W_2'$. Thus, $W_1' \subset W_2'$. By a symmetrical argument, $W_2' \subset W_1'$ a contradiction. Thus, T is one. Finally, we must show that T is onto. Indeed, let W be any ideal of R containing U. Let $W' = \phi(W)$. Then, W' is an ideal in R' since $w_1, w_2 \in W$ implies $\phi(w_1) + \phi(w_2) = \phi(w_1 + w_2) \in \phi(W) = W'$ and $r \in R, w \in W$ implies $rw \in W$ implies $\phi(r)\phi(w) = \phi(rw) \in W$, i.e., $\phi(r)\phi(W) \subset \phi(W)$. But as r ranges over R, $\phi(r)$ ranges over R' since ϕ is surjective. Thus, $r'\phi(W) \subset \phi(W)$, or $r'W' \subset W'$, proving that W' is an ideal in R'. We claim that $\phi^{-1}(W') = W$. Indeed, let $w \in \phi^{-1}(W')$. Then $\phi(w) \in W'$. Since $\phi(W) = W'$, we can find a $w_1 \in W$ such that $\phi(w_1) = \phi(w)$, i.e., $\phi(w - w_1) = 0$, i.e., $w - w_1 \in U$. But by hypothesis $U \subset W$. Hence $w - w_1 \in W$ and hence $w \in w_1 + W \subset W$ since W is an ideal. Thus, $\phi^{-1}(W') \subset W$. conversely, if $w \in W$, then $\phi(w) \in W'$ and hence $w \in \phi^{-1}(W')$, i.e., $W \subset \phi^{-1}(W')$, proving that $\phi^{-1}(W') = W$. Thus, T is surjective. Finally, we want to show that the ring R/W is isomorphic to the ring R'/W', where $W = \phi^{-1})(W')$ with W' any ideal in R'. Corresponding to each element $r + W$ of R/W, we define the element $\phi(r) + W'$ of R'/W'. First note that this correspondence is well defined for $r_1 + W = r_2 + W$ implies $r_1 - r_2 \in W$ implies $\phi(r_1) - \phi(r_2) = \phi(r_1 - r_2) \in \phi(W) \subset W'$ implies $\phi(r_1) + W' = \phi(r_2) + W'$. Next note that

$$r_1 + W + r_2 + W = r_1 + r_2 + W \to \phi(r_1 + r_2) + W = \phi(r_1) + \phi(r_2) + W = \phi(r_1) + W + \phi(r_2) + W$$

so the correspondence preserves ring addition. Finally,

$$(r_1 + W)(r_2 + W) = r_1 r_2 + W \to \phi(r_1 r_2) + W = \phi(r_1)\phi(r_2) + W = (\phi(r_1) + W)(\phi(r_2) + W)$$

so the correspondence also preserves ring multiplication. Finally, the correspondence is surjective since $\phi : R \to R'$ is surjective.

6. Let R be a commutative ring with unit element and let M be an ideal of R. Then, M is a maximal ideal of R iff R/M is a field. First assume that M is an ideal of R such that R/M is a field. Then, the only ideals of R/M are (0) and R/M. Now, R/M is the homomorphic image of R as a ring. The homomorphism is $r \to r + M$. M equals the kernel of this homomorphism. Thus by the previous discussion, there is a one-one correspondence between the ideals of R/M and the ideals of R containing M. The ideal M of R is sent to the ideal (0) of R/M i.e., M and the ideal R of R is sent to the ideal R/M. There is no ideal of R between (0) and M since this correspondence is one and the only ideals of R/M are (0) and R/M. In other words, M is a maximal ideal of R. Now assume that M is a maximal ideal of R. In view of the same correspondence, the only ideals of R/M are (0) and R/M. Finally, R/M is commutative with a unit element and hence R/M is a field.

7. If H, K are finite subgroups of a group G of orders $o(H)$ and $o(K)$ respectively, then $o(HK) = \frac{o(H)o(K)}{o(H \cap K)}$. To show this, we construct a map $\psi : H \times K \to HK$ by $\psi(h, k) = hk$. Let g be any element in HK. Then, $g = hk$ for some $h \in H, k \in K$. Now $\psi(h', k') = g$ iff $h'k' = g = hk$ iff $h^{-1}h' = kk'^{-1}$. Equivalently, $\psi(h', k') = g$ iff $h' = hx, k' = x^{-1}k$ for some $x \in H \cap K$. This means that the number of pairs $(h', k') \in H \times K$ for which $h'k' = g$ equals the number of elements

in $H \cap K$ and this shows that in the listing of the elements of HK as images of the ordered pair (h, k), each element occurs $o(H \cap K)$ times. Thus, $o(HK) = \frac{o(H)o(K)}{o(H \cap K)}$.

8. If G is a finite group and A, B are subgroups of G, then the number of elements in the double coset AxB equals $\frac{o(A)o(B)}{o(A \cap xBx^{-1})}$. To see this, we define a map $\psi : A \times B \to AxB$ by $\psi(a, b) = axb$. Let $g \in AxB$. Then, $g = axb$ for some $a \in A, b \in B$. For $a' \in A, b' \in B$, we have $\psi(a', b') = g$ iff $a'xb' = axb$ iff $a^{-1}a' = xbb'^{-1}x^{-1}$, or equivalently, iff there is an element $y \in A \cap xBx^{-1}$ such that $a' = ay, b' = (b^{-1}x^{-1}yx)^{-1} = x^{-1}y^{-1}xb$. This means that in the listing of the elements of AxB as $\psi(a, b)$, each element occurs $o(A \cap xBx^{-1})$ times. Thus, $o(AxB) = \frac{o(A)o(B)}{o(A \cap xBx^{-1})}$.

Study project 48

The decomposition of a linear transformation on a finite dimensional vector space into more elementary factors plays a fundamental role in the study of linear systems. One of the most fundamental results is the primary decomposition theorem which is stated below. A seminar can be presented stating the result and proving it.

Primary decomposition theorem: Let T be a linear operator on a finite dimensional vector space V over the field F. Let p be the minimal polynomial of T and let $p = p_1^{r_1}...p_k^{r_k}$ where the $p_i's$ are distinct irreducible polynomials over F and the $r_i's$ are positive integers. Let $W_i = \mathcal{N}(p_i^{r_i}(T))$. Then, $V = W_1 \oplus ... \oplus W_k$, $W_i's$ are invariant subspaces and if T_i is the restriction of T to W_i, then the minimal polynomial of T_i is $p_i^{r_i}$. The polynomials $f_i = p/p_i^{r_i}, i = 1, 2, ..., k$ have gcd 1. Hence, polynomials $g_1, ..., g_k$ can be determined so that $\sum g_i f_i = 1$. The proof of the primary decomposition theorem involves showing that the operators $E_i = g_i(T)f_i(T), i = 1, 2, ..., k$ constitute a complete family of projectors and that the range of E_i coincides with the null space of $p_i(T)^{r_i}$.

Solved problem 1

Let $x(t) = \sum_{n=-\infty}^{\infty} p(t - nT)$ where $p(t) = 1$ when $0 \le t \le T/2$ and $p(t) = 0$ for other t. Clearly, $x(t)$ is a square wave having period T and duty cycle $1/2$. Let τ be a random variable uniformly distributed over $[0, T]$. We want to determine the autocorrelation function of $x(t - \tau) = \sum_{n=-\infty}^{\infty} p(t - nT - \tau)$. The autocorrelation is given by

$$R(t, s) = \mathbb{E}(x(t - \tau)x(s - \tau)) = \sum_{n,m=-\infty}^{\infty} \mathbb{E}(p(t - nT - \tau)p(s - mT - \tau))$$

Now

$$\mathbb{E}(p(t - nT - \tau)p(s - mT - \tau)) = \frac{1}{T} \int_0^T p(t - nT - \tau_p(s - mT - \tau)d\tau$$

The integrand is unity when $0 \le t - nT - \tau, s - mT - \tau \le T/2$ and zero otherwise. We shall evaluate

$$\sum_{n,m=-\infty}^{\infty} p(t - nT - \tau)p(s - mT - \tau)$$

The sum need extend over only all those integers n, m for which $0 \le t - nT - \tau \le T/2$ and $0 \le s - mT - \tau \le T/2$, i.e. over all n, m for which $(t - \tau)/T - 1/2 \le n \le (t - \tau)/T$ and $(s - \tau)/T - 1/2 \le m \le (s - \tau)/T$. It is clear that for each $t, s \in \mathbb{R}$, there exist integers k, r such that $kT \le t < (k + 1)T$ and $rT \le s < (r + 1)T$. Write $t = kT + \delta, s = rT + \mu$, where $0 \le \delta, \mu < T$. Then, the range of n, m is given by $k + (\delta - \tau)/T - 1/2 \le n \le k + (\delta - \tau)/T$ and $r + (\mu - \tau)/T - 1/2 \le m \le r + (\mu - \tau)/T$. Since δ, μ, τ vary over $[0, T)$, it is clear that $(\delta - \tau)/T$ and $(\mu - \tau)/T$ vary over $(-T, T)$. Consider first $(\delta - \tau)/T$ falling in the range -1 to $-1/2$. It is clear that in this range n can assume only the value $k - 1$. When $(\delta - \tau)/T$ falls in the range $-1/2$ to 0, no integer value of n will satisfy the equation. When $(\delta - \tau)/T$ falls in the range 0 to $1/2$, n can assume only the value k. When $(\delta - \tau)/T$ falls in the range $1/2$ to 1, n no integral value of n will satisfy the stated inequality. A similar remark goes for m. Taking into account these considerations, the function

$$\psi(t, s, \tau) = \psi(kT + \delta, rT + \mu, \tau) = \sum_{n,m=-\infty}^{\infty} p(t - nT - \tau)p(s - mT - \tau)$$

assumes the value 1 when $(\delta - \tau)/T$ falls in the range $(-1, -1/2)$ or in the range $(0, 1/2)$ and $(\mu - \tau)/T$ falls in the range $(-1, -1/2)$ or in the range $(0, 1/2)$. For all other values of t, s, τ this function equals zero. The autocorrelation of the given process is then given by the expression

$$R(t, s) = R(kT + \delta, rT + \mu) = \frac{1}{T} \int_0^T \psi(kT + \delta, rT + \mu, \tau)d\tau$$

$$= \frac{1}{T} \int_E d\tau$$

where E denotes the union of the following four disjoint sets

$$E_1 = \{\tau : 0 < \tau < T, -1 < (\delta - \tau)/T < -1/2, -1 < (\mu - \tau)/T < -1/2\}$$

$$E_2 = \{\tau : 0 < \tau < T, -1 < (\delta - \tau)/T < -1/2, 0 < (\mu - \tau)/T < 1/2\}$$

$$E_3 = \{\tau : 0 < \tau < T, 0 < (\delta - \tau)/T < 1/2, -1 < (\mu - \tau)/T < -1/2\}$$

$$E_4 = \{\tau : 0 < \tau < T, 0 < (\delta - \tau)/T < 1/2, 0 < (\mu - \tau)/T < 1/2\}$$

We note that E_1 is the same as $max(0, \delta + T/2, \mu + T/2) < \tau < min(T, T + \delta, T + \mu)$. E_2 is the same as $max(0, \delta + T/2, \mu - T/2) < \tau < min(T, \delta + T, \mu)$. E_3 is the same as $max(0, \delta - T/2, \mu + T/2) < \tau < min(T, \delta, \mu + T)$ and finally, E_4 is the same as $max(0, \delta - T/2, \mu - T/2) < \tau < min(T, \delta, \mu)$. Since δ, μ vary over $[0, T)$, it follows that

$$E_1 = \{max(\delta + T/2, \mu + T/2) < \tau < T\}$$

$$E_2 = \{\delta + T/2 < \tau < \mu\}$$
$$E_3 = \{\mu + T/2 < \tau < \delta\}$$
$$E_4 = \{max(0, \delta - T/2, \mu - T/2) < \tau < min(\delta, \mu)\}$$

Thus,

$$\int_{E_1} d\tau = min(T/2 - \delta, T/2 - \mu)$$

provided $\delta, \mu < T/2$. The integral equals zero otherwise.

$$\int_{E_2} d\tau = \mu - \delta - T/2$$

provided $\mu - \delta > T/2$. The integral equals zero otherwise.

$$\int_{E_3} d\tau = \delta - \mu - T/2$$

provided $\delta - \mu > T/2$. The integral equals zero otherwise. Finally,

$$\int_{E_4} d\tau = \mu - \delta + T/2$$

provided $\mu < \delta$, $\delta > T/2$ and $\mu - \delta + T/2 > 0$. If $\mu < \delta$ and $\delta < T/2$, get

$$\int_{E_4} d\tau = \mu$$

We can without loss of generality assume that $\delta > \mu$. Then consider the following case first:

$$0 < \delta < T/2$$

This case thus automatically reduces to the case $0 < \mu < \delta < T/2$. This automatically implies that $0 < \delta - \mu < T/2$. The integrals over E_2 and E_3 do not contribute. We get by summing the above contributions,

$$\int_E d\tau = \sum_{i=1}^{4} \int_{E_i} d\tau = T/2 - \delta + \mu$$

Now consider the case $0 < \mu < T/2 < \delta$ with $\delta - \mu < T/2$. Again, we get on summing the four integrals,

$$\int_E d\tau = T/2 + \mu - \delta$$

Next consider the case $0 < \mu < T/2 < \delta$ with $\delta - \mu > T/2$. Then,

$$\int_E d\tau = \delta - \mu - T/2$$

Applying symmetry to the variables t, s, this gives us the following final result

$$R(t, s) = \frac{1}{T}(T/2 - |\delta - \mu|)$$

if $|\delta - \mu| < T/2$. This completes the form of the correlation function once we note that $R(t, s)$ is a periodic function of t and s with period T.

Solved problem 2

Same as the previous problem but now we want to determine the time averaged autocorrelation function and verify whether or not the signal is ergodic. Clearly the time averaged autocorrelation function of $x(t - \tau)$ is the same as that of $x(t)$:

$$R_x(s) = lim_{N \to \infty} \frac{1}{2NT} \int_{-NT}^{NT} x(t+s)x(t)dt = lim_{N \to \infty} \frac{1}{2NT} \int_{-NT}^{NT} x(t-\tau+s)x(t-\tau)dt$$

Now,

$$\int_{-NT}^{NT} x(t+s)x(t)dt = \int_{-NT}^{NT} \sum_{n=-\infty}^{\infty} p(t+s-nT) \sum_{m=-\infty}^{\infty} p(t-mT)dt$$

We can assume that $s \geq 0$ without loss of generality since $R_x(-s) = R_x(s)$. It easily follows that

$$R_x(s) = lim_{N \to \infty} \frac{1}{2NT} \sum_{n,m=-\infty}^{\infty} \int_{-NT}^{NT} p(t+s-nT)p(t-mT)dt$$

The integrand equals unity when $0 \leq t+s-nT \leq T/2, 0 \leq t-mT \leq T/2$ and zero otherwise. For a given t in the range $[-N, N]$ and $s > 0$, the range over which n, m must vary for the summand in the expression $\sum_{n,m} p(t+s-nT)p(t-mT)$ to be non zero is thus given by $(t+s)/T - 1/2 \leq n \leq (t+s)/T$ and $t/T - 1/2 \leq m \leq t/T$. IT is clear that if k is any integer and $k \leq t/T < (k+1/2)$, then $k - 1/2 \leq t/T - 1/2 < k$. These two together imply $t/T - 1/2 < k \leq t/T$ and hence there is exactly one integer $m = k$ that satisfies the inequality for m. On the other hand if $k + 1/2 < t/T < k + 1$, then $k < t/T - 1/2 < t/T < k + 1$ and there is no integer m that falls in the interval $[t/T - 1/2, t/T]$. Thus as t/T ranges over the interval $[-N, N]$, the set of points t in each interval $t/T \in [k, k+1/2), k = -N, -N+1, ..., N-1$, we are able to find an integer $m = k$ satisfying the inequality. Now the autocorrelation $R(s)$ is a periodic function of s with a period of T. Consider $s \in [0, T/2)$. Assume that $t/T \in [k, k+1/2)$. We want to determine an integer n such that $(t+s)/T - 1/2 < n < (t+s)/T$. It is clear that if $(t+s)/T - 1/2 > k$ and $(t+s)/T < k+1$, then we cannot find any such integer n caught between the two bounds. This means that when $k + 1 - t/T > s/T > k + 1/2 - t/T$ we cannot find any such integer. When however $(t+s)/T - 1/2 < k$ we can find exactly one integer. This amounts to saying that when $k < t/T < k + 1/2 - s/T$ we can find one integer. In the computation of the time averaged autocorrelation, we have taking into account the above considerations,

$$R(s) = lim_{N \to \infty} R_N(s)$$

where

$$R_N(s) = \frac{1}{2NT} \int_{-NT}^{NT} \sum_{n,m=-\infty}^{\infty} p(t+s-nT)p(t-mT)dt$$

$$= \frac{1}{2NT} \sum_{k=-N}^{N-1} \int_{kT}^{(k+1/2)T} \sum_{n,m} p(t+s-nT)p(t-mT)dt$$

$$= \frac{1}{2NT} \sum_{k=-N}^{N-1} \int_{kT}^{(k+1/2)T-s} dt$$

$$= \frac{1}{2NT} 2N(T/2 - s) = 1/2 - s/T$$

This argument shows that $R(s) = 1/2 - |s|/T$ for $|s| \leq T/2$ and that $R(s)$ has period T. This formula for the time averaged autocorrelation coincides with the ensembled average autocorrelation proving that the square wave signal with random time delay is ergodic.

Study project 49

Read up about transforming a square or rectangular matrix into the row reduced echelon form by row operations and how this can be used to solve linear equations. Specifically, assume that the rows of A are $R_1, ..., R_n$ where each R_i is an $1 \times m$ vector. We want to obtain a basis for the nullspace of A, i.e., the set of all $m \times 1$ vectors X for which $AX = 0$. To do this we define the following operations: (1) Multiplication of a row R_i of A by a non-zero scalar c, (2) Interchange of two rows R_i and R_j, (3) replacing the ith row R_i of A by $R_i + cR_j$ where $j \neq i$ and c is any scalar. It is easily seen that each of this operations has the effect of transforming A to EA where E is a non-singular matrix. If we apply a sequence of these operations to A with the matrices corresponding to this sequence being $E_1, ..., E_k$, then this is equivalent to replacing A by $E_k E_{k-1}...E_1 A$ and $E_k E_{k-1}...E_1$ is a non-singular matrix. Thus, X is a solution of $AX = 0$ iff X is a solution of $E_k E_{k-1}...E_1 AX = 0$. It is easily seen that any A can be brought to the row reduced echelon form by a sequence of such operations. A matrix B is said to be in row reduced echelon form if (1) The nonzero rows of B occur at the start, i.e. the first r rows of B are nonzero and the last $n - r$ rows of B are zero, (2) If the first nonzero entry in the first row of B occurs at the k_1^{th} position, the first nonzero entry in the second row of B occurs in the k_2^{th} column and the first nonzero entry in the rth row of B occurs in the k_r^{th} column, then $k_1 > k_2 > ... > k_r$ and the entries $(1, j)$ with $j \notin \{k_2, ..., k_r\}$ are zero, the entries $(2, j)$ with $j \notin \{k_3, ..., k_r\}$ are zero and so on upto the entries $(r-1, j)$ with $j \neq k_r$ is zero. After transforming A to its row reduced echelon form B, the solution to $AX = 0$ which is equivalent to $BX = 0$ is obtained in a straightforward manner. This system is equivalent to

$$b_{i,k_i} x_{k_i} + \sum_{j \in J} b_{ij} x_j = 0, i = 1, 2, ..., r$$

where J is the complement of the set $\{k_1, ..., k_r\}$. The general solution is obtained by assigning arbitrary values to $x_j, j \in J$ and expressing x_{k_i} in terms of these using the above equations. Suppose we now want to investigate whether the system of linear equations $AX = b$ has a solution. We apply a sequence of row operations to the matrix A as well as correspondingly to b and this gives us $BX = c$ where $B = E_k E_{k-1}...E_1 A$ and $c = E_k E_{k-1}...E_1 b$. The row operations are selected so that B appears in row reduced echelon form. Assume that A has rank r and is of size $m \times n$. Then, the first r rows of B are nonzero and the last $m - r$ rows of B are zero. If even one of the last $m - r$ entries of c is non-zero, then the system will have no solution. Assume then that all the last $m - r$ entries of c are zero. Also assume that the first nonzero entry in the ith row of B appears in the k_i^{th} column with $k_1 > k_2 >, , , > k_r$. Then, with J denoting the complement of the set $\{k_1, ..., k_r\}$, we see that the general solution to the system is given by assigning arbitrary values to x_i for $i \in J$ and setting

$$x_{k_i} = (c_i - \sum_{j \in J} b_{i,j} x_j)/b_{i,k_i}, i = 1, 2, ..., r$$

In fact, the general solution to the system $AX = b$ is given by $X_0 + X_1$, where X_0 is any particular solution to $AX = b$ and X_1 is an arbitrary vector in the nullspace of A. Comparing this with the above formula, we can conclude that the vector having $c_i/b_{i,k_i}$ as its ith entry and zeroes as the other entries is a particular solution to $AX = b$ and the vector having arbitrary $x_i's$ for $i \in J$ and $x_{k_i} = -\sum_{i \in J} b_{i,j} x_j/b_{i,k_i}$ is a general vector in $\mathcal{N}(A)$.

Study project 50

If $x(t_1, ..., t_k)$ is a k dimensional signal, its Fourier transform is defined by the equation

$$X(\omega_1, ..., \omega_k) = \int_{-\infty}^{\infty} ... \int_{-\infty}^{\infty} x(t_1, ..., t_k) exp(-j(\omega_1 t_1 + ... + \omega_k t_k)) dt_1 ... dt_k$$

This equation can be cast in vectorial form

$$X(\omega) = \int x(\mathbf{t}) exp(-i\omega^T \mathbf{t}) d\mathbf{t}$$

The inverse Fourier relation is given by the equation

$$x(\mathbf{t}) = (2\pi)^{-k} \int X(\omega) exp(i\omega^T \mathbf{t}) d\omega$$

This follows by virtue of the following identity for the multidimensional Dirac δ function

$$\delta(\mathbf{t}) = \delta(t_1)...\delta(t_k) = (2\pi)^{-k} \int exp(i\omega^T \mathbf{t}) d\omega$$

$$= (2\pi)^{-k} \Pi_{\alpha=1}^{k} \int_{-\infty}^{\infty} exp(i\omega_\alpha t_\alpha) d\omega_\alpha$$

Study project 51

Analysis of an RC circuit excited by an exponential wave over applied over a finite interval. Let $x(t) = exp(-ct)(u(t-a) - u(t-b))$ be the input voltage to an RC circuit and let $y(t)$ be the output voltage taken across the capacitor. It satisfies the differential equation

$$C\frac{dy}{dt} + y/R = x$$

Let us determine first the impulse response $h(t)$ of this circuit. It satisfies the differential equation

$$RC\frac{dh(t)}{dt} + h(t) = R\delta(t)$$

For $t > 0$, the right side equals zero. Thus, $h(t) = exp(-t/RC)h(0)u(t)$. Integrating the above equation from $t = 0-$ to $t = 0+$ and using $h(0-) = 0$, we get $RCh(0+) = R$ or $h(0+) = 1/C$. Thus, $h(t) = \frac{1}{C}exp(-t/RC)u(t)$. The response to the given input assuming $0 < a < b$ is given by

$$y(t) = \int_0^\infty h(\tau)x(t-\tau)d\tau = \frac{1}{C}\int_{\tau>0,a<t-\tau<b} exp(-\tau/RC)exp(-c(t-\tau))d\tau$$

$$= \frac{1}{C}\int_{max(0,t-b)<\tau<t-a} exp(-ct)exp(\tau(c-1/RC))d\tau$$

The range of t for which the output is non-zero is given by $t > a$. For all t in this range, we have $max(0, t-b) = 0$ when $a < t < b$ and $max(0, t-b) = t-b$ when $t > b$. Thus, for $a < t < b$, we have

$$y(t) = \frac{1}{C}exp(-ct)(exp((t-a)(c-1/RC)) - 1)/(c-1/RC)$$

and for $t > b$, we have

$$y(t) = \frac{1}{C}exp(-ct)(exp(-(t-a)(c-1/RC)) - exp(-(t-b)(c-1/RC)))/(c-1/RC)$$

This signal can be generated using the 8085 microprocessor as follows. Assume T to be large number compared to all the three numbers $a, b, 1/c$. We want to generate samples of the input signal $\{x(t)\}$ over the interval $[0, T]$ and then to store these samples in the memory location. Assume that samples are being taken at the rate of $1/\Delta$ and let $N = T/\Delta, p = a/\Delta, q = b/\Delta, c\Delta = k$. Then for $n < p$ and for $n > q$ we have $x(n) = 0$ while for $p \leq n \leq q$, we have $x(n\Delta) = exp(-kn)$. Assuming that Δ is sufficiently small, storing the input can be considered as storing its samples taken at the rate of $1/\Delta$, i.e. $\{x(n\Delta) : n = 0, 1, ..., N-1\}$. Storing the input amounts to computing the exponential function. This is done via approximation, i.e. $exp(x)$ is replaced by its truncated expression $1 + x + x^2/2! + ... + x^M/M!$ with M sufficiently large. Likewise, samples of the impulse response $h(n\Delta), n = 0, 1, ..., N-1$ are also stored and the convolution integral is replaced by the convolution sum based on the discrete sum approximation of the Riemann integral:

$$\int_0^T h(t)dt \approx \sum_{k=0}^{N-1} h(k\Delta)\Delta, T = N\Delta$$

Study project 52

Elementary problems on periodic signals and their Fourier series: One example is

If $x[n]$ is a periodic signal with period N, then the Fourier series coefficients of the signal $g(t) = \sum_k x(k)\delta(t - kT)$ are periodic with period N. This fact can be seen by first noting that

$$g(t + NT) = \sum_k x(k)\delta(t - (k - N)T) = \sum_k x(k + N)\delta(t - kT) = \sum_k x(k)\delta(t - kT) = g(t)$$

so that g has period NT. Its Fourier series coefficients are obtained as

$$c_m = \frac{1}{NT} \int_{0-}^{NT-} g(t)exp(-i2\pi mt/NT)dt = \frac{1}{NT} \sum_{k=0}^{N-1} x(k)exp(-i2\pi mk/N)$$

from which it easily follows that $c_{m+N} = c_m$. Note that if $g(t)$ is transmitted through an LTI system having impulse response $h(t)$, the output will be

$$\psi(t) = g(t) * h(t) = \sum_k x(k)h(t - kT)$$

This signal can also be shown to be periodic with period NT. In fact

$$\psi(t + NT) = \sum_k x(k)h(t - (k - N)T) = \sum_k x(k + N)h(t - kT)$$

$$= \sum_k x(k)h(t - kT) = \psi(t)$$

We evaluate its Fourier series coefficients as $d_m = H(j2\pi m/NT)c_m$. The signal $\psi(t)$ is a PAM wave, i.e. a pulse amplitude modulated wave. The discrete time sequence $\{x(k)\}$ modulates the continuous time pulse $\{h(t)\}$ to produce the PAM wave. The Fourier transform of $\psi(t)$ is given by

$$\hat{\psi}(\omega) = \int_{-\infty}^{\infty} \psi(t)exp(-i\omega t)dt = \sum_k x(k)exp(-i\omega kT)H(\omega)$$

Now

$$\sum_k x(k)exp(-i\omega kT) = \sum_{m=-\infty}^{\infty} \sum_{r=0}^{N-1} x(Nm + r)exp(-i\omega(Nm + r)T)$$

$$= \sum_{r=0}^{N-1} x(r) exp(-i\omega rT) \sum_{m=-\infty}^{\infty} exp(-i\omega NmT)$$

$$= \sum_{r=0}^{N-1} x(r) \sum_{m=-\infty}^{\infty} \delta(\omega NT/2\pi - m)$$

$$= \sum_{r=0}^{N-1} x(r) exp(-i\omega rT) \frac{2\pi}{NT} \sum_{m} \delta(\omega - 2\pi m/NT)$$

$$= \frac{2\pi}{NT} \sum_{m}(\sum_{r=0}^{N-1} x(r) exp(-i2\pi mr/N)) \sum \delta(\omega - 2\pi m/NT)$$

The PAM wave can be simulated easily on the 8085 microprocessor.

Study project 53

An interesting project for the mathematically inclined student would involve the study of various kinds of metrics and their application to engineering problems. The metrics to be considered are (1) The L^p metrics on the space of signals defined on an interval I: $d_p(x, y) = (\int_I |x(t) - y(t)|^p dt)^{1/p}$. When $p \to \infty$, this becomes the sup metric: $d_\infty(x, y) = sup\{|x(t) - y(t)| : t \in I\}$. The corresponding metrics on the space of sequences is $d_p(x, y) = (\sum_n |x(n) - y(n)|^p)^{1/p}$ and $d_\infty(x, y) = sup\{|x(n) - y(n)| : n \in \mathbb{Z}\}$. Integral inequalities based on these metrics are used in control theoretic problems. The study should include proofs of completeness and the Holder and Minkowski inequalities.

Study project 54

Programme the 8085 microprocessor to store a two dimensional signal $x(n, m)$, $0 \le n, m \le N - 1$ (i.e., image) and do processing on this. The kinds of processing involve (1) computation of the discrete two dimensional Fourier transform of the image (2) Computation of the expansion coefficients relative to an orthonormal basis, (3) Forming a single vector out of the image data, (4) Operating on the sequence by a two dimensional filter both in the time domain and in the frequency domain. The filter convolution operation is assumed to be cyclic (5) Performing a smoothening operation, by iteratively replacing each pixel value with an average of the neighbouring pixel values (6) Computing the autocorrelation function and the discrete spectral density of the image. The operations should use only the multiplication routine based on shifting and modulo two addition operations.

Study project 55

Representing a two dimensional filter having the rational structure using a state variable model. An example is the filter $H(Z_1, Z_2) = b(0, 0) + b(1, 0)Z_1^{-1} + b(0, 1)Z_2^{-1} + b(1, 1)Z_1^{-1}Z_2^{-1}$.

If $x(n,m)$ is the input and $y(n,m)$ is the output, then the model for the image is given by the equation

$$y(n,m) = b(0,0)x(n,m) + b(1,0)x(n-1,m) + b(0,1)x(n,m-1) + b(1,1)x(n-1,m-1)$$

We introduce the state variables $x_1(n,m) = x(n-1,m), x_2(n,m) = x(n,m-1), x_3(n,m) = x(n-1,m-1)$. Equivalently, in the z domain,

$$X_1(Z_1, Z_2) = Z_1^{-1}X(Z_1, Z_2), X_2(Z_1, Z_2)$$

$$= Z_2^{-1}X(Z_1, Z_2), X_3(Z_1, Z_2) = Z_1^{-1}Z_2^{-1}X(Z_1, Z_2)$$

Then,

$$Z_1X_1 = X, Z_2X_2 = X, Z_1X_3 = X_2, Z_2X_3 = X_1, Y$$

$$= b(0,0)X + b(1,0)X_1 + b(0,1)X_2 + b(1,1)X_3$$

These can be regarded as the defining state equations.

Study project 56

Design a two dimensional lowpass filter and write simple 8085 programmes to operate the filter on a two dimensional sinusoidal signal $x(n,m) = cos(\omega_1 n + \omega_2 m)$. The filter frequency response is given by $H(\omega_1, \omega_2) = 1$ for $\omega_1^2 + \omega_2^2 \leq R^2$ and $H(\omega_1, \omega_2) = 0$ for $\omega_1^2 + \omega_2^2 > R^2$. To evaluate the impulse response of this lowpass filter, we need

$$h(n,m) = (2\pi)^{-2} \int_{-\pi}^{\pi} \int_{-\pi}^{\pi} H(\omega_1, \omega_2)exp(j(\omega_1 n + \omega_2 m))d\omega_1 d\omega_2$$

$$= (2\pi)^{-2} \int_{\omega_1^2 + \omega_2^2 \leq R^2} exp(j(\omega_1 n + \omega_2 m))d\omega_1 d\omega_2$$

The integration is carried out using polar frequency variables $r = \sqrt{\omega_1^2 + \omega_2^2}$ and $\phi = tan^{-1}(\omega_2/\omega_1)$. The integral equals

$$h(n,m) = (2\pi)^{-2} \int_0^T \int_0^{2\pi} exp(jr\sqrt{n^2 + m^2}cos\phi)rdrd\phi$$

$$= (2\pi)^{-1} \int_0^R J_0(r\sqrt{n^2 + m^2})rdr$$

Study project 57

The basic equations of a switched capacitor network can be presented in the form of state variable equations. If $\mathbf{u}(t)$ denotes the input to such a circuit and the switching frequency equals $1/T$, then during the nth clock cycle, the state equations are

$$\frac{d\mathbf{X}(t)}{dt} = \mathbf{A}_n\mathbf{X}(t) + \mathbf{B}_n\mathbf{u}(t), nT \le t < (n+1)T$$

with the initial state at the start of the nth switching cycle being given as a linear transformation of the state at the end of the $(n-1)^{th}$ switching cycle, i.e. $\mathbf{X}(nT+) = \mathbf{F}_n\mathbf{X}(nT-)$. The matrices $\mathbf{A}_n$ and $\mathbf{B}_n$ vary with the switching cycle number, since in each cycle, certain capacitors are switched on and certain others are switched off. Analysis of such state equations constitutes the theory of switched capacitor networks. Suppose that we wish to determined the sequence of states at the beginning of each cycle, i.e. $\mathbf{X}(nT+)$. solving the above state equations gives us

$$\mathbf{X}((n+1)T-) = exp(\mathbf{A}_nT)\mathbf{X}(nT+) + \int_{nT}^{(n+1)T)} exp(\mathbf{A}_n(t-\tau))\mathbf{B}_n\mathbf{u}(\tau)d\tau$$

and hence

$$\mathbf{X}((n+1)T+) = \mathbf{F}_{n+1}\mathbf{X}((n+1)T-)$$

$$= \mathbf{F}_{n+1}exp(\mathbf{A}_nT)\mathbf{X}(nT+) + \int_{nT}^{(n+1)T} \mathbf{F}_{n+1}exp(\mathbf{A}_n(t-\tau))\mathbf{B}_n\mathbf{u}(\tau)d\tau$$

Setting $\mathbf{X}_n = \mathbf{X}(nT+)$ and

$$\mathbf{g}_{n+1} = \int_{nT}^{(n+1)T} \mathbf{F}_{n+1}exp(\mathbf{A}_n(t-\tau))\mathbf{B}_n\mathbf{u}(\tau)d\tau$$

the above recursion can be expressed in the form

$$\mathbf{X}_{n+1} = \mathbf{F}_{n+1}\mathbf{X}_n + \mathbf{g}_{n+1}$$

which is a sequence of discrete time state variable equations. Noise analysis of a switched capacitor circuit defined by the state equations of the previous problem can also be carried out assuming $\mathbf{u}(t)$ to be white noise with a covariance kernel $\mathbb{E}(\mathbf{u}(t)\mathbf{u}(s)^T) = \mathbf{Q}\delta(t-s)$. Basically, we need to determine the statistics of the state $\mathbf{X}_n$ at the beginning of each cycle. Noting that the whiteness of the noise implies that the vector $\mathbf{g}_{n+1}$ is uncorrelated with $\mathbf{X}_n$, we obtain the following recursion equation for the covariance matrices:

$$\mathbb{E}(\mathbf{X}_{n+1}\mathbf{X}_{n+1}^T) = \mathbf{F}_{n+1}\mathbb{E}(\mathbf{X}_n\mathbf{X}_n^T)\mathbf{F}_{n+1}^T + \mathbb{E}(\mathbf{g}_{n+1}\mathbf{g}_{n+1}^T)$$

where

$$\mathbb{E}(\mathbf{g}_{n+1}\mathbf{g}_{n+1}^T) = \mathbf{F}_{n+1}\int_{nT}^{(n+1)T} exp(\mathbf{A}_n\tau))\mathbf{B}_n\mathbf{Q}\mathbf{B}_n^T exp(\mathbf{A}_n\tau))d\tau$$

Solving this recursion determines the covariance $\mathbf{R}_n$ of $\mathbf{X}_n$. In order to compute the correlations between the state at the start of two intervals $m < n$ we need to do more work.

Study project 58

Proof of the singular value decomposition. Let A be any matrix and let $|A| = (A^*A)^{1/2}$. Define a map $U : R(|A|) \to R(A)$ by $U|A|x = Ax$. $|A|x = 0$ implies $Ax = 0$ and hence this map is a well defined linear map. It is an isometry, i.e., preserves inner products since $< |A|x, |A|y >= y^*|A|^2x = y^*A^*Ax =< Ax, Ay >$. The dimension of $R(|A|)^\perp = N(|A|)$ is $n - r$ where n is the number of columns of A and $r = rank(A)$. Also the dimension of $R(A)^\perp$ is $n - r$. So, we can choose an onb for the former space and map it to an onb for the latter space. This is used to extend the domain of U to $\mathbb{C}^n$. This means that U is a well defined unitary operator on $\mathbb{C}^n$ such that $U|A| = A$. If $|A| = WDW^*$ is the spectral resolution of $|A|$, then we get $A = UWDW^* = V\!DW^*$ where $V = UW$ is a unitary operator. This is precisely the singular value decomposition.

Study project 59

Coordinate proof of the singular value decomposition. Let A be a matrix of size $m \times n$ AA^* has size $m \times m$. Let r be the rank of A. Then, $r \leq m$ and we can find an onb for $\mathbb{C}^m$ $\{v_1, ..., v_m\}$ such that $AA^*v_i = \lambda_i v_i, i = 1, 2, ..., m$ where $\lambda_i > 0$ for $i = 1, 2, ..., r$ and $\lambda_i = 0$ for $i = r+1, ..., m$. Now, $< A^*v_i, A^*v_j >= v_j^*AA^*v_i = \lambda_i v_j^*v_i = \lambda_i \delta_{ij}$. Thus, $u_i = A^*v_i/\sqrt{\lambda_i}, i = 1, 2, ..., r$ forms an orthonormal set in $\mathbb{C}^n$ and $Au_i = AA^*v_i/\sqrt{\lambda_i} = \sqrt{\lambda_i}v_i, i = 1, 2, ..., r$. Extend the orthonormal set $\{u_1, ..., u_r\}$ to an onb $\{u_1, ..., u_r, u_{r+1}, ..., u_n\}$ for $\mathbb{C}^n$. It is clear that $u_1, ..., u_r$ forms a basis for $R(A^*)$. Thus, $u_{r+1}, ..., u_n$ is a basis for $N(A)$, in particular, $Au_j = 0, j = r + 1, ..., n$. Let $U = [u_1, ..., u_n]$. U is an $n \times n$ unitary matrix and

$$AU = [Au_1, ..., Au_r, Au_{r+1}, ..., Au_n] = [\sqrt{\lambda_1}v_1, ..., \sqrt{\lambda_r}v_r, 0, ..., 0] = V\Lambda$$

where

$$\Lambda = \begin{pmatrix} \Lambda_1 & 0 \\ 0 & 0 \end{pmatrix}$$

is an $n \times m$ matrix with $\Lambda_1 = diag[\lambda_1, ..., \lambda_r]$. This is the singular value decomposition.

Study project 60

Shannon's sampling theorem is fundamental in the design of digital processors for analog signals. Let $\{x(t) : t \in \mathbb{R}\}$ be an analog signal with Fourier transform $X(f) = \int_{-\infty}^{\infty} x(t)exp(-j2\pi ft)dt$. Assume that $X(f) = 0$ for $|f| > B$. Then for $f \in [-B, B]$, we have

$$X(f) = \sum_{n=-\infty}^{\infty} c_n exp(-j\pi nf/B)$$

with

$$c_n = \frac{1}{2B} \int_{-B}^{B} X(f) exp(j\pi n f / B) df = \frac{1}{2B} x(n/2B)$$

from which we obtain

$$x(t) = \sum_{n=-\infty}^{\infty} c_n \int_{-B}^{B} exp(j2\pi f(t - n/2B)) df = \sum_{n=-\infty}^{\infty} c_n \frac{sin(2\pi B(t - n/2B))}{\pi(t - n/2B)}$$

$$= \sum_{n=-\infty}^{\infty} x(n/2B) \frac{sin(\pi(2Bt - n))}{\pi(2Bt - n)}$$

This is also called the Shannon reconstruction formula.

Study project 61

Reconstruction formula for periodic bandlimited signals. Suppose $\{x(t)\}$ is periodic with period T and bandlimited so that there is an integer N such that $x(t)$ has no frequency component outside the band $[-2\pi N/T, 2\pi N/T]$. Then, the Fourier series of $x(t)$ has the form

$$x(t) = \sum_{n=-N}^{N} c_n exp(j2\pi nt/T)$$

Taking $t = kT/2N + 1, k = -N, -N + 1, ..., N$ gives us

$$x(kT/2N + 1) = \sum_{n=-N}^{N} c_n exp(j2\pi kn/2N + 1), |k| \leq N$$

so that

$$c_n = \frac{1}{2N + 1} \sum_{k=-N}^{N} x(kT/2N + 1) exp(-j2\pi kn/2N + 1)$$

In other words the samples of the signal and its Fourier coefficients form a DFT pair. The reconstruction formula for such signals is

$$x(t) = \sum_{n=-N}^{N} exp(j2\pi nt/T) \frac{1}{2N + 1} \sum_{k=-N}^{N} x(kT/2N + 1) exp(-j2\pi kn/2N + 1)$$

$$= \frac{1}{2N + 1} \sum_{k=-N}^{N} x(kT/2N + 1) \sum_{n=-N}^{N} exp(j2\pi n(t/T - k/2N + 1))$$

$$= \sum_{k=-N}^{N} x(kT/2N + 1) P_k(t)$$

where

$$P_k(t) = \frac{1}{2N+1} \sum_{n=-N}^{N} exp(j2\pi n(t/T - k/2N + 1))$$

Study project 62

Direct product of groups

1.   Let $G_1, G_2, ..., G_n$ be n groups. Their direct product $G_1 \times ... \times G_n$ is defined to be the set of all ordered n-tuples $(g_1, ..., g_n)$ with $g_i \in G_i$ and the product of two elements defined componentwise, i.e.,

$$(g_1, g_2, ..., g_n) \cdot (g_1', g_2', ..., g_n') = (g_1 g_1', g_2 g_2', ..., g_n g_n')$$

More specifically, if $(g, g') \in G \times G \rightarrow \phi_i(g, g')$ is the composition on G_i, then the composition ϕ on $G = G_1 \times ... \times G_n$ is defined by the rule

$$\phi((g_1, ..., g_n), (g_1', ..., g_n')) = (\phi_1(g_1, g_1'), \phi_2(g_2, g_2'), ..., \phi_n(g_n, g_n'))$$

Under this composition, G becomes a group with the identity element as $e = (e_1, e_2, ..., e_n)$ where e_i is the identity element of G_i. For example, $\mathbb{Z}$ under addition is an Abelian group and its direct product with itself r times $\mathbb{Z}^r$ is also an Abelian group whose elements are $(n_1, n_2, ..., n_r)$ with $n_i \in \mathbb{Z}$. An element $\mathbf{n} \in \mathbb{Z}^r$ is thus an r tuple of integers with addition defined component wise in the usual vectorial fashion. Another example is obtained by taking $G_i = \{0, 1, ..., N_i - 1\}$ with addition defined modulo N_i where N_i is a positive integer. Then, $G_1 \times ... \times G_n$ consists of n-vectors $(m_1, m_2, ..., m_n)$ with $0 \le m_i \le N_i - 1$ and with composition defined as

$$(m_1, m_2, ..., m_n) + (m_1', m_2', ..., m_n') = (m_1 + m_1' mod N_1, m_2 + m_2' mod N_2, ..., m_n + m_n' mod N_n)$$

We can also look at the following situation. Let $V_i, i = 1, 2, ..., n$ be n vector spaces over perhaps different fields. Then, each vector space under vector addition is an Abelian group and we define $V = V_1 \times ... \times V_n$ as the set of all elements having the form $(x_1, x_2, ..., x_n)$ with $x_i \in V_i$. Then V is an Abelian group under usual componentwise addition.

2.   Let $G_1, ..., G_n$ be n groups. Define G to be their direct product. Now for each $g_i \in G_i$, define the element $\tilde{g}_i = (e_1, ..., e_{i-1}, g_i, e_{i+1}, ..., e_n) \in G$. Let $\tilde{G}_i$ consist of all the $\tilde{g}_i's$ as g_i varies over G_i with i fixed. Then, $\tilde{G}_i$ is clearly a normal subgroup of G. Normality follows from the identity

$$(g_1, ..., g_n)(e_1, ..., e_{i-1}, h_i, e_{i+1}, ..., e_n) = (e_1, ..., e_{i-1}, g_i h_i g_i^{-1}, e_{i+1}, ..., e_n)$$

for all $g_j \in G_j, j = 1, 2, ..., n, h_i \in G_i$. It is also easy to see that every element $g \in G$ has a unique representation of the form $g = \tilde{g}_1 \tilde{g}_1 ... \tilde{g}_n$ with $\tilde{g}_i \in \tilde{G}_i$. The equation $G = G_1 \times ... \times G_n$ is sometimes read as G is the external direct product of the groups $G_1, ..., G_n$ while the equation

$G = \tilde{G}_1...\tilde{G}_n$ with the unique representation is read as G is the internal direct product of the subgroups $\tilde{G}_1, ..., \tilde{G}_n$. Suppose that $N_1, ..., N_n$ are normal subgroups of G and every $g \in G$ is uniquely expressible as $g = n_1...n_n$, i.e., G is the internal direct product of the groups $N_1, ..., N_n$. Then we claim that G is isomorphic to the group $N_1 \times ... \times N_n$. In fact, one can define a map $\psi : N_1 \times ... \times N_n \to G$ by $\psi(n_1, ..., n_n) = n_1...n_n$ and prove that this is an isomorphism. First step to prove is that $n_i n_j = n_j n_i$ if $n_i \in N_i, n_j \in N_j$ and $i \neq j$. Now, $n_i n_j n_i^{-1} n_j^{-1}$ is an element of N_j as well as of N_i by virtue of normality of N_i, N_j. However suppose $g \in N_i \cap N_j$ and $g \neq e$. Then, g has two distinct representations of the form $n_1...n_n$, one with $n_i = g, n_j = e_j, j \neq i$ and the other with $n_j = g, n_i = e_i, i \neq j$. This is a contradiction to our assumption. Hence, $n_i n_j n_i^{-1} n_j^{-1} = e$, so that $n_i n_j = n_j n_i$ proving the claim. The homomorphism property of ψ follows easily from this:

$$\psi(n_1, ..., n_n)\psi(n'_1, ..., n'_n) = n_1...n_n n'_1...n'_n = n_1 n'_1 n_2 n'_2...n_n n'_n = \psi(n_1 n'_1, ..., n_n n'_n)$$

$$= \psi((n_1, ..., n_n)(n'_1, ..., n'_n))$$

where we've used the result that n'_1 commutes with $n_2, ..., n_n$, n'_2 commutes with $n_3, ..., n_n$ and so on.

Study project 63

1. If G has no nontrivial subgroups, then G must be finite of prime order. For suppose G is not finite. Then, choose any $a \in G, a \neq e$. Let (a) denote the subgroup of G generated by a, i.e., (a) consists of all elements of the form a^n with n ranging over the integers. By hypothesis, $G = (a)$. We consider two cases. Since G is assumed to be infinite, for every integer $n \neq 0$, $a^n \neq e$. Now let $m \neq 0, 1$. Consider the subgroup (a^m) generated by a^m. By hypothesis $(a^m) = G = (a)$. Thus, $a = a^{rm}$ for some $r \neq 0$ and hence $a^{rm-1} = e$ which means $rm = 1$, i.e., $r = m = 1$, a contradiction. So G must be finite. Now suppose $o(G) = pq$ where p, q are positive integers differing from 1. This means that $o(G)$ is not of prime order. Now choose any $a \in G, a \neq e$. Then, $(a) = G$ by hypothesis. Thus, a has order pq. Now let $b = a^p$. Then, b has order q. By hypothesis however $(b) = G$ which means that G has q elements, a contradiction. Hence, $o(G)$ must be finite of prime order and in fact cyclic and we are done.

2. Cauchy's theorem for Abelian groups: Suppose G is a finite Abelian group and p divides $o(G)$ where p is a prime number. Then there is an element $a \neq e \in G$ such that $a^p = e$. Suppose that the theorem holds for all Abelian groups having fewer elements than G. The theorem is trivially true for groups having a single element, namely the identity. If G has no subgroups H other than $(e), G$, then by the above problem, G must be cyclic of prime order. This prime must be p and every $a \neq e$ must have order p since $(a) = G$ for all $a \neq e$. In particular, we've shown that G has $p - 1$ elements a such that $a^p = a^{o(G)} = e$. Suppose G has a nontrivial subgroup N, i.e. $N \neq (e), G$. If p divides $o(N)$, by the induction hypothesis, since N is Abelian and $o(N) < o(G)$, there is an element $b \in N, n \neq e$ such that $b^p = e$. Since $e \neq b \in G$, we are through. So assume that p does not divide $o(N)$. Since G is Abelian, N is a normal subgroup of G, so G/N is a group. Moreover $o(G/N) = o(G)/o(N)$ (Lagrange's theorem on the number of cosets of a subgroup) and

since p does not divide $o(N)$ while p divides $o(G)$, it follows that p divides $o(G)/o(N) < o(G)$. Since G is Abelian, G/N is Abelian. Thus, by the induction hypothesis, there is an $X \in G/N$ such that $X^p = N, X \neq N$ (note that N is the identity element of G/N). Write $X = bN$ with $b \in G$. Then, $b \notin N, b^p \in N$. Thus the order of b^p divides $o(N)$ (Lagrange's theorem), and hence $(b^p)^{o(N)} = e$ or $b^{po(N)} = e$. Let $c = b^{o(N)}$. Then, $c^p = e$. We claim that $c \neq e$. If $c = e$, then $b^{o(N)} = e$ and so, $(bN)^{o(N)} = N$. Also, $(bN)^p = N$ and p does not divide $o(N)$. Since $b \notin N$, and hence $bN \neq N$ and the equation $(bN)^p = N$ implies that the order of bN divides p. But p is a prime and hence the order of bN must be equal to p. since $(bN)^{o(N)} = N$, it follows that p divides $o(N)$, a contradiction. Hence $b \in N$, again a contradiction. Thus, $c \neq e$ and c is the desired element.

Study project 64

Another proof of Sylow's theorem

We shall show that if p is a prime that divides $o(G)$, then G has a p-Sylow subgroup, i.e. if p^m divides G but p^{m+1} does not divide G, then a subgroup of G having order p^m is called a p-Sylow subgroup. Suppose that the theorem is true for all groups having order lesser than $o(G)$.

Study project 65

This project deals with general topology. Its importance in signal theory stems from the fact that notions of continuity of signals and convergence can be extended to signals defined over arbitrary sets. The first step to extend the idea of a signal is to regard it as a mapping from an arbitrary set to another, i.e., $f : X \to Y$ where X and Y are arbitrary sets. Using this idea, we can define the notion of an image as a two dimensional signal. However, in order to be able to perform meaningful operations on the signals, we must impose some additional structure on the sets. One such structure makes the set into a vector space. Another structure makes it into a group. Another structure makes it into a ring, another structure makes it into an algebra. Using these structures, we can add signals, multiply signals by scalars, i.e., amplify them or attenuate them, change the signal phase, evaluate the distance between signals (for doing this, we must impose the metric space structure on the set of signals), In order to integrate signals, we must impose the structure of a measure space on the sets, i.e. introduce the idea of measurable subsets. Perhaps the most important and basic structure that one imposes on a set is that of a topology for it enables one to define notions such as continuity and convergence. The structure of a vector space and a topology can enable one to define infinite sums of signals. Let X be a set. A topology on X is a collection τ of subsets of X such that (1) the empty set ϕ and the entire set X are both elements of τ, τ is closed under arbitrary unions and finite intersections, i.e., if $A_\alpha, \alpha \in \Gamma$ is a family of subsets of X with $A_\alpha \in \tau$ for all $\alpha \in \Gamma$, then $\bigcup \{A_\alpha : \alpha \in \Gamma\} \in \tau$ and if $A_1, ..., A_n \in \tau$ for some finite positive integer n, then $\bigcap_{i=1}^{n} A_i \in \tau$. In other words, τ is closed under arbitrary unions and finite intersections. The elements of τ are called open sets. Thus, any union of open sets is an open set and any finite intersection of open sets is again an open set. For example, consider

$\mathbb{R}$, the real line. We define an open interval to be a subset of $\mathbb{R}$ having one of the following forms $\phi, (a, \infty), (-\infty, a), \mathbb{R}, (a, b)$ with $-\infty < a < b < \infty$. A subset O of $\mathbb{R}$ is said to be an open set if it is a union of open intervals. It is easily seen that O is an open set iff either O is empty, or else, for any $x \in O$, there are numbers a, b such that $a < x < b$ and $(a, b) \in O$. More generally, we can consider a metric space, i.e., a set X with a metric, i.e., a mapping $d : X \times X \to [0, \infty)$ satisfying (1) $d(x, y) \geq 0$, (2) $d(x, y) = 0$ iff $x = y$ and (3) $d(x, y) + d(y, z) \leq d(x, z)$ for all $x, y, z \in X$. Then, an open ball is defined to be a set having the form $B(x, \delta) = \{y \in X, d(y, x) < \delta\}$ for some $x \in X$ and $\delta > 0$. $B(x, \delta)$ is the open ball with centre x and radius δ. A closed ball has the form $B[x, \delta] = \{y \in X : d(y, x) \leq \delta\}$. This is the closed ball with centre x and radius δ. An open ball in a metric space is a generalization of the notion of an open interval in $\mathbb{R}$. An open set in a metric space X is defined to be an arbitrary union of open balls. it is clear that given any non-empty open set $O \subset X$, for each $x \in O$, we can determine a $\delta > 0$ such that $B(x, \delta) \subset O$. For each $x \in O$, let $\delta_x > 0$ be such that $B(x, \delta_x) \subset O$. Then, $O = \bigcup_{x \in O} B(x, \delta_x)$. A basis for a topological space (X, τ) is a collection $\mathcal{B}$ of open sets, i.e., sets in τ satisfying the following conditions: Given any open set U and a point $x \in U$, there is a $B \in \mathcal{B}$ such that $x \in B \subset U$. For example, suppose $X = \mathbb{R}$ with the topology generated by the open intervals. A base for the topology consists of the open intervals in $\mathbb{R}$. This can be seen from the following argument. Let U be any open set and let $x \in U$ be arbitrary. Then, there exists an $\epsilon > 0$ such that $(x - \epsilon, x + \epsilon) \subset U$. Since further $x \in (x - \epsilon, x + \epsilon)$, the first condition for the base is satisfied. The second one is obvious: If (a, b) and (c, d) are any two open intervals, then their intersection is one of the following: $\phi, (a, b), (c, d), (a, d), (c, b)$ all of which are open intervals. Likewise in a metric space, a base is formed by the open balls. The finite intersection property of open balls is a little tricky to demonstrate. Let $B(x, \delta)$ and $B(y, \epsilon)$ be any two open balls. If $x = y$, then the intersection of these two open balls is once again an open ball and so there is nothing to prove. Suppose $x \neq y$. We want to show that $B(x, \delta) \cap B(y, \epsilon)$ is an open set. If this set is empty there is nothing to prove. So suppose that this set is non-empty. Let z be any element in this intersection. Then, $d(z, x) < \delta, d(z, y) < \epsilon$. Suppose $z = x$. Then $d(u, z) < \alpha$ implies $d(u, x) < \alpha < \delta$ and $d(u, y) \leq d(u, x) + d(x, y) = d(u, x) + d(z, y) < \alpha + \epsilon - \alpha = \epsilon$ provided we select $\alpha = min(\epsilon - d(z, y), \delta)$. Thus, $B(z, \alpha) \subset B(x, \delta) \cap B(y, \epsilon)$ and we are through. A similar result holds when $z = y$. So assume that $z \neq x, y$. Then, $d(z, x) < \delta, d(z, y) < \epsilon$. $d(u, z) < \alpha$ implies $d(u, x) \leq d(u, z) + d(z, x) < \alpha + \delta - \alpha = \delta$ and $d(u, y) \leq d(u, z) + d(z, y) < \alpha + \epsilon - \alpha = \epsilon$ where $\alpha = min(\delta - d(z, x), \epsilon - d(z, y))$. This shows that $B(z, \alpha) \subset B(x, \delta) \cap B(y, \epsilon)$ proving the assertion.

Study project 66

This project deals with more advanced concepts in general topology. We cover the following issues

1. Closed sets and closure: A subset C of a topological space X is called a closed set if $X - C$ is open. An arbitrary intersection of closed sets is closed and so also is any finite union of closed sets. These facts follow from De-Morgan's identities: If $C_\alpha, \alpha \in \Gamma$ is a family of closed sets,

then

$$X - \left(\bigcap C_\alpha\right) = \bigcup (X - C_\alpha)$$

is a union of open sets and is therefore open. Its complement $\bigcap C_\alpha$ is therefore a closed set. Likewise, suppose $C_n, n = 1, 2, ..., N$ is a finite collection of closed sets. Then,

$$X - \bigcup_{n=1}^{N} C_n = \bigcap_{n=1}^{N} (X - C_n)$$

is a finite intersection of open sets and is therefore open. Its complement $\bigcup_{n=1}^{N} C_n$ is therefore closed. Since X, ϕ are open sets, their complements ϕ, X are closed sets.

2. Basis and subbasis: Let (X, τ) be a topological space. A collection $\mathcal{B}$ of subsets of X is a basis if $\mathcal{B}$ is a subset of τ, if every finite intersection of elements from $\mathcal{B}$ is an element of $\mathcal{B}$ and if every open set is a union of elements of $\mathcal{B}$. For example, let X and Y be topological spaces and $\mathcal{B}$ the collection of all subsets of $X \times Y$ having the form $U \times V$ with U open in X and V open in Y. Then $\mathcal{B}$ is closed under finite intersections and we can define the product topology on $X \times Y$ for which every open set is a union of elements of $\mathcal{B}$. $\mathcal{S}$ is a subbasis for the topology on X if every open set in X is a union of finite intersections of elements of $\mathcal{S}$. For the product topology on $X \times Y$, a subbasis consists of elements having the form $X \times V$ and $U \times Y$ with V open in Y and U open in X.

3. Continuity of maps: If (X, τ_X) and (Y, τ_Y) are topological spaces, then a map $f : X \to Y$ is continuous if $f^{-1}(O) \in \tau_X$ for all $O \in Y$. In case X, Y are metric spaces, then continuity is equivalent to the $\epsilon - \delta$ definition, i.e. $f^{-1}(B(f(x), \epsilon))$ contains the ball $B(x, \delta)$ for some $\delta > 0$.

4. Homeomorphic topological spaces: If X, Y are topological spaces, they are said to be homeomorphic if there exists a bijection $f : X \to Y$ such that $O \subset Y$ is open iff $f^{-1}(O) \subset X$ is open. Homeomorphic spaces are equivalent as far as topological properties are concerned, for if Y is homeomorphic to X, then the elements of Y are obtained by renaming the elements of X and likewise the open subsets of Y are obtained by renaming the open subsets of X.

Study project 67

Axiom of choice: A is an infinite set iff there exists a one map $f : \mathbb{Z}_+ \to A$ iff there is a bijection $f : A \to B$ where B is some proper subset of A. Suppose that A is infinite. Choose an infinite sequence $B = \{a_n : n = 1, 2, ..., \}$ of distinct elements of A. Define $f(x) = x$ if $x \notin B$ and $f(a_n) = a_{n+1}, n = 1, 2, ...$ Then, f is a bijection from A onto $A - \{a_1\}$. There are various ways of defining sets. The first is by direct listing of the elements of the set. The other is to consider a bigger set A and define a subset B of A by specifying the properties that the elements of B are to satisfy. The third is to specify the subset as unions and intersections of certain more elementarily definable subsets of the bigger set. The other is by using the cartesian product. A function $f : \mathbb{Z}_+ \to A$ can be defined by specifying a subset of $\mathbb{Z}_+ \times A$ whose elements are $(n, f(n)), n = 1, 2, ...$ There is another method of constructing sets that does not fall in the above list of methods. This method is called the axiom of choice. Let $\mathcal{R}$ be a collection of disjoint non-empty sets. The axiom of choice states that we can construct a set having exactly one element from each of the members of $\mathcal{R}$. This is equivalent to defining a function $f : \mathcal{R} \to \bigcup \mathcal{R}$ such that $f(R) \in R$ for each $R \in \mathcal{R}$. The axiom of choice states that such a function exists even if $\mathcal{R}$ is a collection of not necessarily disjoint subsets of a set A. From the axiom of choice, it follows easily that if $\mathcal{R}$ is a collection of disjoint subsets, then there is a set C such that for each $R \in \mathcal{R}$, $C \cap R$ contains exactly one element.

Axiom of choice states that given a collection $\mathcal{R}$ of disjoint non-empty sets, there exists a set C having exactly one element in common with each element of $\mathcal{R}$. From this axiom, we can show that given a collection $\mathcal{B}$ of non-empty sets, there is a function $c : \mathcal{B} \to \bigcup \{B : B \in \mathcal{B}\}$ such that $c(B)$ is an element of B. For given any $B \in \mathcal{B}$, define B' as $B' = \{(B, x) : x \in B\}$. B' is a subset of $\mathcal{B} \times \bigcup \{B : B \in \mathcal{B}\}$. Since B contains at least one element x, the set B' contains at least one element (B, x), so it is non-empty. It is easy to see that if B_1, B_2 are two different sets in $\mathcal{B}$, then B_1' and B_2' are disjoint. For suppose (B, x) is in B_1' and also in B_2'. Then since the first entry of every element of B_1' is B_1 while the first entry of every element of B_2' is B_2, it must follow that $B = B_1 = B_2$, a contradiction. Consider $\mathcal{C} = \{B' : B \in \mathcal{B}\}$. The elements of $\mathcal{C}$ are disjoint and non-empty. Disjointness was just proved. Clearly, $\mathcal{C}$ is a subset of $\mathcal{B} \times \bigcup \{B : B \in \mathcal{B}\}$. By the axiom of choice, there is a set c having exactly one element in common with each element of $\mathcal{C}$. First note that c is a subset of $\mathcal{B} \times \bigcup \{B : B \in \mathcal{B}\}$. Secondly, c contains exactly one element from each set B', i.e., for each $B \in \mathcal{B}$, c contains exactly one ordered pair (B, x) whose first coordinate is B. Thus, c can be viewed as a function from $\mathcal{B}$ to $\bigcup \{B : B \in \mathcal{B}\}$. Finally, since $(B, x) \in c$ implies $x \in B$, it follows that $c(B) \in B$ and we are through.

The importance of the axiom of choice in signal analysis: Axiom of choice states that given an arbitrary family of sets, we can define a set containing exactly one element from each of the sets in the given family. This apparently trivial statement cannot be proved and hence it acquires the status of an axiom. Given a countable sequence of non-empty sets $X_n, n = 1, 2, ...$, we can be recursion select one element from each member, i.e., choose $x_n \in X_n$ and define a sequence or a signal $(x_n, n = 1, 2, ...)$. In this manner, we can construct many kinds of discrete time signals. Note that this permits construction of discrete signals, each sample of which falls in a different space. To do so for continuous time signals, the axiom of choice is required. Given a family of sets $\{X_t : t \in \mathbb{R}\}$ indexed by the real parameter t, we want to define a continuous time signal $\{x_t : t \in \mathbb{R}\}$ whose t^{th} sample x_t belongs to the set X_t, i.e., $x_t \in X_t$ for all $t \in \mathbb{R}$. To do this the axiom of choice is required. More generally, given any set Γ and a family of sets $X_\alpha, \alpha \in \Gamma$ indexed by Γ, we want to define a signal taking Γ as the domain, i.e., $x : \Gamma \to \mathbb{R}$ in such a way that $x(\alpha) \in X_\alpha$. To do this, we require the axiom of choice. In conventional signal theory, we are used to defining signals whose amplitude varies over a given single set. The axiom of choice in the above setting however permits the definition of signals whose amplitude at each time point belongs to a different set. This axiom thus significantly extends the scope of signal theory.

Study project 68

Deals with basic problems related to topological groups. One of the important problems is outlined below.

1. Let G be a Lie group and H a closed Lie subgroup. One important problem involves defining a natural analytic structure on the G-space G/H consisting of all cosets gH with g ranging over G. This structure must be imposed so that G/H becomes an analytic manifold on which the group G acts analytically. The action of G on a topological space M is defined by the mapping $G \times M \to M$ by the rule $(g, x) \to g.x$ such that $e.x = x, g_2.(g_1.x) = (g_2.g_1).x$. This map should be continuous. It then follows that for each $g \in G$, the map $x \to t_g(x) = g.x$ is a homeomorphism from M into M. These operations make M into a G-space. G acts transitively on M if there is just a single orbit, i.e., for any one $x \in M$ (and hence for all $x \in M$), we have $G.x = \{g.x, g \in G\} = M$. This condition is equivalent to requiring that for any $x, y \in M$, there is a $g \in G$ such that $y = g.x$. M then becomes a transitive G-space. The stability subgroup G_{x_0} of $x_0 \in M$ is the set of $g's$ that fix x_0, i.e.,

$$G_{x_0} = \{g \in G : g.x_0 = x_0\}$$

Let x be any other element in M which is assumed to be a transitive G-space. Then, choose an $h \in G$ such that $x = hx_0$. Clearly, $g \in G_x$ iff $gx = x$ iff $ghx_0 = hx_0$ iff $h^{-1}ghx_0 = x_0$ iff $h^{-1}gh \in G_{x_0}$. Hence, $G_x = hG_{x_0}h^{-1}$. Thus the stability subgroups of a transitive G space are conjugate to each other. Let G be a locally compact second countable Hausdorff group. By locally compact, we mean that the topology on G is such that given any $g \in G$, there exists an open set $N \subset G$ such that $g \in N$ and $Cl(N)$ is compact. By second countable, we mean that as a topological space, G has a countable base. Let G_0 be a closed subgroup of G and set $N = G/G_0$, the space of all cosets gG_0 of G_0. Let $\beta : G \to N$ be defined by $\beta(a) = aG_0$. Call a set $V \subset N$ as open iff $\beta^{-1}(V)$ is open in G. Then, N becomes a topological space. This topology on N is called the quotient topology. It is known that N is Hausdorff, locally compact and second countable as a topological space. The map $(g, \beta(a)) \to \beta(ga) = g\beta(a)$ is well defined and continuous from $G \times N \to N$. $\beta(1) = G_0$ and a belongs to the stability group of $\beta(1)$ in G/G_0 iff $aG_0 = G_0$ iff $a \in G_0$.

Unsolved problems

1. Consider the vectors $\mathbf{x}_1 = [0, 0, 2, 4]$, $\mathbf{x}_2 = [1, 2, 5, 6]$, $\mathbf{x}_3 = [1, 0, 3, 4]$, $\mathbf{x}_4 = [1, 2, 4, 6]$. Are these vectors linearly independent ? If not, find a basis for the space spanned by these vectors. Also find an orthonormal basis for this space by using the Gram-Schmidt orthonormalization procedure.

2. Let V denote the space of all polynomials of degree lesser than 3 with complex coefficients. Show that V is a vector space. What is the dimension of this space ? Let t_1, t_2, t_3 be three distinct

complex numbers and define for each polynomial $p \in V$, $L_i(p) = p(t_i)$, $i = 1, 2, 3$. Show that L_i is a linear functional. Show that $\{L_1, L_2, L_3\}$ are linearly independent vectors in V^*. Determine the dual of this set.

Study project 69

This project deals with transmitting a signal whose bit rate is R through modems that can accommodate rates $R' < R$. One can use several such modems for the transmission. One method is to use multiplexing along the following lines. Suppose the signal samples are numbered as $n = 0, \pm1, \pm2, ...$ We take the samples $mp + j$ with $m = 0, \pm1, \pm2, ...$ for a fixed $j = 0, 1, ..., p - 1$ and transmit these samples through the jth modem. If R is the original rate in samples per second, then the rate at which samples are being transmitted through the jth modem equals R/p. We select p large enough so that $R/p < R'$ and then by using p modems, the message can be transmitted over the p modems. At the receiver end, we combine the outputs of each modem by interleaving and pass the resultant through a low pass filter to reconstruct the transmitted message.

Reduction of data rate by decimation: The first step is to understand how the sampling of the original analog signal is performed and how the analog signal is reconstructed from the signal samples. Let $x(t)$ be a signal bandlimited to $[-B, B]$ Hertz and let it be sampled at the rate of $F = 1/T$. We shall evaluate the discrete time Fourier transform of the sampled signal. Set $x_d(n) = x(nT)$, $n \in \mathbb{Z}$. Then,

$$X_d(exp(j\omega)) = \sum_{n=-\infty}^{\infty} x_d(n)exp(-j\omega n) = \sum_{n=-\infty}^{\infty} x(nT)exp(-j\omega n)$$

$$= \sum_{n=-\infty}^{\infty} \frac{1}{2\pi} \int X(j\Omega)exp(j(\Omega T - \omega)n)d\Omega$$

$$= \int_{-\infty}^{\infty} X(j\Omega) \sum_{n} \delta((\Omega T - \omega) - 2\pi n)$$

$$= \frac{1}{T} \sum_{n} X(j(\omega/T + 2\pi n/T))$$

where $X(j\Omega)$ denotes the continuous time Fourier transform of $x(t)$, i.e.,

$$X(j\Omega) = \int_{-\infty}^{\infty} x(t)exp(-j\Omega t)dt$$

Now, $X(j\Omega)$ is zero for $|\Omega| > 2\pi B$ and hence, if $|\omega| < \pi$ and $\pi/T > 2\pi B$, i.e., $T < 1/2B$, we have from the above formula

$$X_d(exp(j\omega)) = \frac{1}{T}X(j\omega/T)$$

This formula shows that for bandlimited signals and the sampling rate satisfying the Nyquist criterion, the digital spectrum of the sampled signal is a scaled replica of the spectrum of the original analog signal. In case the Nyquist criterion $T < 1/2B$ is not met, there will be aliasing in the digital spectrum and we will not obtain a scaled replica of the analog spectrum. Suppose that $T < 1/2B$. Then the above formula shows that $X_d(exp(j\omega))$ is restricted to the band $[-2\pi BT, 2\pi BT]$ with $2BT < 1$. In other words, the entire digital frequency band $[-\pi, \pi]$ is not covered. Suppose to be specific that $2BT = 1/D$ where $D > 1$ is an integer. This means that the digital spectrum is concentrated over the band $[-\pi/D, \pi/D]$. Then, the data rate can be reduced by decimation. In other words, complete recovery of the data from the decimated version $x(nD), n \in \mathbb{Z}$ is possible. The reconstruction formula can be found in any standard textbook or can be deduced from the following identity:

$$DTFT\{x(nD)\} = \sum_{n=-\infty}^{\infty} x(nD)exp(-j\omega n)$$

$$= \sum_{n} exp(-j\omega n)\frac{1}{2\pi} \int_{-\pi}^{\pi} X(exp(j\omega'))exp(j\omega' nD)d\omega'$$

$$= \frac{1}{2\pi} \int_{-\pi}^{\pi} d\omega' X(exp(j\omega'))d\omega' \sum_{n} exp(j(\omega'D - \omega)n)$$

$$= \int_{-\pi}^{\pi} X(exp(j\omega'))d\omega' \sum_{m} \delta(\omega'D - \omega - 2\pi m)$$

The integral above can be replaced by an integral over any interval of length 2π giving

$$DTFT\{x(nD)\} = \frac{1}{D} \sum_{m=0}^{D-1} X(exp(j(\omega + 2\pi m)/D))$$

This is the general formula for the Fourier transform of the decimated sequence. If $R/D < R'$, then using just a single modem the original data can be transmitted through. Thus, if the original sampled sequence was an oversampled version of the signal whose rate exceeded the Nyquist rate by a factor of D, with $D > R/R'$, then just a single modem would suffice in the data transmission.

Reduction of data rate by using fewer quantizers. Suppose that the samples of a signal vary over the range $[-A, A]$. Let $p(x)$ denote the probability density of a sample. We place quantizer levels at the levels $l_1, ..., l_{N-1}$ where $-A < l_1 < l_2 < ... < l_{N-1} < A$ and select numbers $\xi_1 \in (-A, l_1), \xi_2 \in (l_1, l_2), ..., \xi_{N-1} \in (l_{N-2}, l_{N-1}), \xi_N \in (l_{N-1}, A)$. We then use the quantizer function

$$Q(x) = \sum_{i=1}^{N} \xi_i \chi_{(l_{i-1}, l_i)}(x)$$

where $l_0 = -A, l_N = A$ and $\chi_E(x)$ denotes the indicator function of the set E with $\chi_E(x) = 1$ when $x \in E$ and $\chi_E(x) = 0$ when $x \notin E$. If X is a signal sample and $Q(X)$ the quantized sample, the mean square distortion is given by

$$\sigma_Q^2(X) = \mathbb{E}(Q(X) - X)^2 = \sum_{i=1}^{N} \int_{l_{i-1}}^{l_i} (x - \xi_i)^2 p(x) dx$$

The basic problem here is to minimize $\sigma_Q^2(X)$ with respect to the numbers $l_1, ..., l_{N-1}, \xi_1, ..., \xi_N$. Normally one is given the prescribed tolerance ϵ for the mean squared distortion error and we choose the minimum value of N for which $\sigma_Q^2(X)_{min} < \epsilon$. If the original signal arrived at a rate of R_s samples per second and we have available with us a modem that can transmit at the rate of R_b bits per second, then for successful transmission with the condition of prescribed tolerance for the mean squared error, we must have $R_s.log_2(N) < R_b$. If not, we decrease the number N of quantizer levels so that this inequality is met thereby increasing the mean squared distortion error beyond the prescribed tolerance. On the other hand, suppose we have with us p modems that can transmit at the rate of R_b bits per second each, we do the following. Select the smallest value of N so that $R_s log_2(N)/p < R_b$. If this N meets the prescribed tolerance for distortion,then fine otherwise decrease N till the above inequality is met and make a compromise on the tolerance. The other option is to increase the number of modems and N to meet both the transmission rate for each modem and the mean square distortion tolerance.

Reduction of data rate by LPC coding: Given a data sequence, $X(n)$ obtained after sampling the analog signal, we consider the following parametric model for its generation

$$X(n) = -\sum_{k=1}^{p} a(k)X(n-k) + W(n)$$

where $W(n)$ is a noise sequence. In case that the samples are highly correlated, we can choose the coefficients $\{a(k)\}$ so that $\hat{X}(n) = -\sum_{k=1}^{p} a(k)X(n-k)$ is a reliable estimate of $X(n)$. This is achieved by minimizing the sum of error square $E(a(1), ..., a(p)) = \sum_{n=p}^{N}(X(n) - \hat{X}(n))^2$ via the optimal equations

$$\frac{\partial}{\partial a(k)} E(a(1), ..., a(p)) = 0, k = 1, 2, ..., p$$

These lead to equations for the $a(k)'s$. If the data correlations are high, then $W(n)$ has a much smaller dynamic range than $X(n)$ and hence fewer quantizer levels are required to encode $W(n)$ than those required to encode the original sequence $X(n)$. Compression is achieved by representing the process $X(n)$ using the process $W(n)$ and the parameters $a(k), k = 1, 2, ..., p$.

Study project 70

This project deals with some digital signal processing techniques applied to problems of analog electronics and electromagnetic engineering. There are several problems of this nature but we outline only a few of them here to give a flavour of the kinds of techniques involved.

1. Resistor-capacitor circuit. Resistance R is connected in series with a capacitor C to a voltage source $x(t)$. Let $y(t)$ be the voltage across the capacitor. Then $y(t)$ satisfies the differential equation

$$C\frac{dy}{dt} + (y - x)/R = 0$$

or

$$RC\frac{dy}{dt} + y = x$$

The time constant $\tau = RC$ of the circuit is $\tau = RC$. In terms of this parameter, the differential equation is given by

$$\frac{dy}{dt} + y/\tau = x/\tau$$

the solution to which is

$$y(t) = exp(-t/\tau)y(0) + \int_0^t exp(-(t - s)/\tau)x(s)ds$$

If the circuit is in operation from $t = -\infty$ and $y(-\infty) = 0$, then the solution is expressible as

$$y(t) = \int_{-\infty}^t exp(-(t - s)/\tau)x(s)ds = \int_0^\infty exp(-s/\tau)x(t - s)ds$$

This equation can be expressed as

$$y(t) = \int_{-\infty}^\infty h(s)x(t - s)ds$$

where $h(t) = exp(-t/\tau)u(t)$ is the impulse response of the circuit. To simulate it, we discretize the signal $x(t)$ giving the discrete time signal $x_n = x(nT), n \in \mathbb{Z}$. The derivative $\frac{dx}{dt}$ is replaced by the scaled finite difference $\frac{x_n - x_{n-1}}{T}$ and the differential equation for the circuit goes over to the following difference equation

$$\frac{y_n - y_{n-1}}{T} + y_{n-1}/\tau = x_n/\tau$$

or

$$y_n = (1 - T/\tau)y_{n-1} + Tx_n/\tau$$

If no input is applied, i.e., $x_n = 0$ for all values of n, then the solution is given by $y_n = (1-T/\tau)^n y_0$ and this decays down to zero as $n \to \infty$ provided $0 < 1-T/\tau < 1$, i.e., $0 < T < \tau$. Let $v_n = Tx_n/\tau$ and $a = 1 - T/\tau$. The above difference equation can be expressed as

$$y_n = ay_{n-1} + v_n$$

To solve this equation, we divide both sides by a^n and sum from $n = 1$ to $n = k$ to get

$$y_k/a^k - y_0 = \sum_{n=1}^k v_n/a^n$$

or

$$y_k = a^k y_0 + \sum_{n=1}^{k} a^{k-n} v_n$$

If the process is in operation from $n = -\infty$, then with zero initial conditions, the solution can be expressed as

$$y_k = \sum_{n=-\infty}^{k} a^{k-n} v_n = \frac{T}{\tau} \sum_{n=-\infty}^{k} (1 - T/\tau)^{k-n} x_n = \frac{T}{\tau} \sum_{n=0}^{\infty} (1 - T/\tau)^n x_{k-n}$$

Thus, the impulse response of the discretized system is given by

$$h_n = \frac{T}{\tau}(1 - T/\tau)^n u(n)$$

2. Diode-Resistance circuit. This circuit consists of a diode in series with a resistor connected in series with a voltage source $x(t)$. The current-voltage characteristics of the diode is given by $i_D(t) = I_0(exp(v_D(t)/V_T) - 1)$. Thus, if $y(t)$ is the current through the circuit, the voltage across the diode is given by $v_D(t) = x(t) - y(t)R$ and the current $y(t)$ through the circuit satisfies the algebraic equation

$$y(t) = I_0(exp((x(t) - y(t)R)/V_T) - 1)$$

We write the diode voltage current characteristic as the sum of a linear characteristic and a nonlinear perturbation:

$$i_D = I_0 v_D/V_T + \epsilon\psi(v_D)$$

where

$$\epsilon\psi(v_D) = I_0 \sum_{n=2}^{\infty} \frac{1}{n!}(v_D/V_T)^n$$

Then, y satisfies the equation

$$y = I_0(x - yR)/V_T + \epsilon.\psi(x - yR)$$

or

$$(1 + I_0R/V_T)y = I_0 x/V_T + \epsilon.\psi(x - yR)$$

or

$$y = \frac{I_0}{V_T + I_0R}x + \epsilon\frac{V_T}{V_T + I_0R}\psi(x - yR)$$

One can obtain in principle a perturbation series for y in powers of the parameter ϵ from this equation. The computations are of a purely algebraic nature and are left as exercises. An alternative iterative method is to obtain the sequence $y_n, n = 0, 1, 2, ...$ successively as solutions to the iteration

$$y_{n+1} = I_0(exp((x - y_nR)/V_T) - 1), n = 0, 1, 2, ...$$

where y_0 is an initial guess solution. y_0 can be taken as the solution to the linearized equation

$$y_0 = I_0(x - y_0 R)/V_T$$

or

$$y_0(1 + RI_0/V_T) = I_0 x/V_T$$

or

$$y_0 = \frac{I_0}{V_T + RI_0}x$$

A proof of the convergence of the iteration is required here. Thus the diode resistor circuit can be viewed as a non-linear difference equation.

3. Simulation of plane wave starting from the Maxwell equations, an application of the Fourier transform. Maxwell equations in free space are

$$\nabla.E = 0, \nabla \times E = -\mu\frac{\partial H}{\partial t}, \nabla \times H = \epsilon\frac{\partial E}{\partial t}, \nabla.H = 0$$

From these, we deduce that E, H satisfy the three dimensional wave equation

$$\nabla^2\psi - \frac{1}{c^2}\frac{\partial^2\psi}{\partial t^2} = 0$$

The solution is obtained by the Fourier transform

$$\hat{\psi}(t, \mathbf{k}) = \int \psi(t, \mathbf{x})exp(-i\mathbf{k}.\mathbf{x})d^3\mathbf{x}$$

with the inverse transform

$$\psi(t, \mathbf{x}) = (2\pi)^{-3}\int \hat{\psi}(t, \mathbf{k})exp(i\mathbf{k}.\mathbf{x})d^3\mathbf{k}$$

Plugging this into the wave equation gives

$$\frac{\partial^2\hat{\psi}(t, \mathbf{k})}{\partial t^2} = -k^2 c^2\hat{\psi}(t, \mathbf{k})$$

with solution

$$\hat{\psi}(t, \mathbf{k}) = A(\mathbf{k})exp(ikct) + B(\mathbf{k})exp(-ikct)$$

Letting $\omega(\mathbf{k}) = kc$ and superposing the solutions gives

$$\psi(t, \mathbf{x}) = \int (A(\mathbf{k})exp(i(\omega(\mathbf{k})t - \mathbf{k}.\mathbf{x})) + B(\mathbf{k})exp(-i(\omega(\mathbf{k})t + \mathbf{k}.\mathbf{x})))d^3\mathbf{k}$$

We can also look at the solution by discretization. Consider for simplicity the one dimensional wave equation. Its discretized version reads

$$\frac{1}{\Delta^2}(\psi[m,n] - 2\psi[m-1,n] + \psi[m-2,n]) - \frac{1}{c^2\tau^2}(\psi[m,n] - 2\psi[m,n-1] + \psi[m,n-2]) = 0$$

To solve this equation, we introduce the discrete space Fourier transform of the wave amplitude:

$$\hat{\psi}[\omega,n] = \sum_{m=-\infty}^{\infty} \psi[m,n]exp(-i\omega m)$$

Plugging this into the wave equation gives

$$\hat{\psi}[\omega,n](1 - 2exp(-j\omega) + exp(-j2\omega)) = \frac{\Delta^2}{c^2\tau^2}(\hat{\psi}[\omega,n] - 2\hat{\psi}[\omega,n-1] + \hat{\psi}[\omega,n-2]) = 0$$

This is a second order difference equation for the function $\hat{\psi}[\omega,n]$ and can be solved manually and Fourier inverted to get a discretized version of the plane wave solution.

MATLAB exercises in signal theory

1. Learn to install MATLAB on the Pentium machine.

2. Learn how to store a matrix of numbers and how to access them. Example: N and M are integers to be inputed by the user. The numbers a_{ij} of an $N \times M$ matrix are also input by the user.

3. Learn how to test whether an equality or equality is true or not.

4. Learn to write simple programs using the "For Loop" for recursive computation like the solution of a simple difference equation $y(n) = ay(n-1) + x(n) + bx(n-1), n \geq 2$, where $\{x(n)\}$ is an array of numbers imputed by the user.

5. Learn the use of "Plot" command to plot (a) A vector of numbers, (b) several vectors on the same plot. Learn also the use of the "subplot" command in which two or four windows are created in the plot with each one having a figure in it.

6. Learn the following simple functions : How to compute a^b where a and b are real numbers, $log_a(X)$ where a and X are positive real numbers, multiplication of two matrices, inversion of a matrix.

7. Learn to write a simple program using the FOR loop for computing the cross correlation between two arrays $\{x(n) : n = 0, 1, ..., N\}$ and $\{y(n) : n = 0, 1, ..., N\}$. The cross correlation is defined by $R_{xy}(k) = \frac{1}{N-k+1}\sum_{n=0}^{N-k} x(n)y(n+k), k = 0, 1, ..., N$.

8. Learn about generation of random numbers (both uniform and Gaussian).

9. If a zero mean Gaussian sequence $\{x(n)\}$ of unit variance is passed through the filter $H(z) = \frac{1}{1-az^{-1}}$, the output sequence $\{y(n)\}$ will satisfy the difference equation $y(n) - ay(n-1) = x(n)$ and its spectral density will be $|H(e^{j\omega})|^2$. Experimentally estimate the spectrum using $|X(k)|^2$,where $X(k) = \sum_{n=0}^{N-1} x(n)exp(-j2\pi kn/N)$. Make a plot of this and compare with the true spectrum.

10. Learn the use of FFT command.

11. Look at the temperature data in the daily newspaper $\{T(n)\}$ and design a linear predictor as follows: Suppose that $T(.)$ is observed on days 1,2,..., N. Then, the predicted value on day n is $\hat{T}(n) = a(1)T(n-1) + a(2)T(n-2) + ... + a(p)T(n-p)$, where p is something like 10 and $a(1), ..., a(p)$ are chosen to minimize $\sum_{n=p+1}^{N}(T(n) - \hat{T}(n))^2$. From the parameters $a(1), ..., a(p)$ obtained in this way, predict the temperature on day $N+1$ and check that it nearly agrees with the true value.

12. Do problem (i) for rainfall data and share prices in the stock market based on newspaper information.

13. Given an $N \times M$ matrix with $N > M$ write a program to calculate the vector $\mathbf{X} \in \mathbb{R}^M$ such that $\| \mathbf{Y} - \mathbf{AX} \|$ is a minimum. Here, $\mathbf{Y} \in \mathbb{R}^N$ is a fixed vector and for any vector $\mathbf{Z} \in \mathbb{R}^N$, we define $\| \mathbf{Z} \|^2 = \sum_{i=1}^{N} Z_i^2$.

14. Write a simple MATLAB programme to solve the series LCR circuit for a given input voltage by the process of discretization, i.e., replacement of $dX(t)/dt$ by a finite difference $(X(nT) - X((n-1)T))/T$. The step size T should be chosen at least one twentieth of the smaller of the time constants involved RC and $2\pi\sqrt{LC}$.

15. Write a simple MATLAB programme to plot the output of an OP-AMP working as (a) an integrator, (b) a differentiator when the OP AMP gain is chosen as $\frac{A_v}{1+jf/f_c}$ and the amplifier has finite input and finite output impedance. Choose $A_v, f_c, Z_{in}, Z_{out}$ from the standard manual available in the electronics laboratory.

16. The graph of a network is specified by its incidence matrix. Think about how you would write a programme to select a tree from this graph and write down the cutset matrix for it.

17. Based on the Ebers's Moll equation, write down a MATLAB programme to plot the response of a common emitter amplifier to an input sinusoid. The values of the non-linear model

should be chosen from the textbook and the design should be the one you have done in the electronics laboratory. Differentials are to be replaced by finite differences.

18. To store a matrix of $N \times M$ size do the following:

for n=1: N

for m=1: M

a(n,m)= the given function (eg $sin(n.m), 1/(n+m), n^2/m^2$)

end;

end;

Suppose the functional form of $a(n, m)$ is not given and entries have to be made manually. Then suppose $N = 4, M = 3$. Then, simply say

A=[3,3,4;4,2,1;2,2,1;4,5,5];

If you want to input a matrix of random numbers uniformly distributed over $[0,1]$, then use rand command:

A=rand(4,3); This outputs a 4×3 matrix whose entries are uniformly distributed random numbers.

19. Suppose you want sum numbers $1, 2, ..., N$ where N is to be input from the command window.

sum=0;

for n=1: N

sum =sum+n;

end;

Typing sum after the command prompt will output the required sum.

20. Suppose you are given an array x of N numbers. To access the ith number. Then simply type $x(i)$ followed by enter. i runs from $1, 2, ...N$. To write a simple program to calculate the first N Fibonacci numbers $1, 1, 2, 3, 5, 8, 13, ...$ The sequence is defined by $x(1), x(2), x(3)...,,$ where $x(1) = 1 = x(2)$ and $x(n+2) = x(n+1) + x(n), n \geq 1$. To implement using for loop

x(1)=1;x(2)=2;

for n=1: N

x(n+2)=x(n+1)+x(n);

end;

Typing x followed by enter lists the entire array x. Remember that the value of N has to be input from the command window.

21. Suppose you are given a simple function like $sin(\omega t)$. You wish to calculate its integral from $t = a$ upto $t = b$. The numbers a, b, ω are input by the user from the command prompt. The idea is to discretize the interval $[a, b]$ into N steps and set $\Delta = (b - a)/N$. The integral is approximately calculated as $\sum_{n=0}^{N-1} sin(\omega.(a + \Delta.n))\Delta$.

$sum = 0;$

for $n = 0 : N - 1$

$sum = sum + sin(\omega.(a + \Delta.n))\Delta$

end;

$I = sum;$

Try the same for $cos(\omega t), t^2, exp(t), exp(-t^2/2)$ etc.

22. To generate the output of an IIR filter defined by the transfer function $H(z) = \frac{B(z)}{A(z)}$ where $B(z) = b(1) + b(2)z^{-1} + ... + b(p+1)z^{-p}$ and $A(z) = 1 + a(1)z^{-1} + ... + a(p)z^{-p}$, you have to solve the difference equation $y(n) = (b(1)x(n) + b(2)x(n-1) + ... + b(p+1)x(n-p)) - (a(1)y(n-1) + ... + a(p)y(n-p))$.

for n=p+1: N

sum1=0;

for k=1: p

sum1=sum1+a(k)*y(n-k);

end;

sum2=0;

for k=0: p

sum2=sum2+b(k+1)*x(n-k);

end;

y(n)=sum2-sum1;

end;

In the command window, you'll have to input $y(1), ..., y(p)$ to initiate the recursion.

23. $Y = A * X$ where A is an $n \times m$ matrix and X an $m \times 1$ vector, outputs an $n \times 1$ vector Y obtained by multiplying A into X. Write a program to do this multiplication without using the inbuilt command.

A and X are imputed from the command window.

for i=1: n

```
sum=0;

for j=1: m

sum=sum+a(i,j)*X(j);

end;

Y(i)=sum;

end;
```

24. Write a program to multiply and $n \times m$ matrix A with an $m \times p$ matrix B.

```
for i=1: n

for j=1: p

sum=0;

for k=1: m

sum=sum+a(i,k)*b(k,j);

end;

c(i,j)=sum;

end;

end;
```

25. Write a program to calculate approximately the length of a curve defined by the equation $y = f(x)$ between $x = a$ and $x = b$. The values of f are stored at the points $a + k\Delta, k = 0, 1, ..., N$ where $\Delta = (b-a)/N$. We set $f(k) = f(a+k\Delta)$. The length of the segment joining $(a+k\Delta, f(k))$ and $(a + (k+1)\Delta, f(k+1))$ is given by $\sqrt{\Delta^2 + (f(k+1) - f(k))^2}$ and hence the total length of the curve is approximately given by $\sum_{k=0}^{N-1} \sqrt{\Delta^2 + (f(k+1) - f(k))^2}$.

```
sum=0;

for k=0: N-1

sum=sum+√(Δ² + (f(k+1) − f(k))²)

end;

l=sum;
```

26. Write a simple program using the for loop to calculate the sequence of successive sums of an array of real numbers. Thus, if $A = [a(1), ..., a(n)]$ is the array, your output should be the array $B = [b(1), ..., b(n)]$ where $b(i) = \sum_{j=1}^{i} a(j), i = 1, 2, ..., n$.

27. Write a program to multiply to $n \times n$ matrices using the for loop. Compare your result with

the value of the product obtained using the "star" command built in MATLAB. Your program can progress along the following lines. If $A = ((a(i,j)))_{1 \leq i,j \leq n}$ and $B = ((b(i,j)))_{1 \leq i,j \leq n}$, then $C = AB$ has entries

$$c(i,j) = \sum_{k=1}^{n} a(i,k)b(k,j)$$

Your program should incorporate three for loops. One for makes i vary from 1 to n, another for makes j vary from 1 to n and the third for calculates the above sum.

28. Write a simple program to plot the magnitude and phase of the transfer function of a series RLC circuit taking $R = 1, L = 1, C = 1$ in MKS units. The input is a voltage source and the output is the current through the circuit. Now input a complex sinusoid $exp(i\omega t)$ to this circuit and by numerically solving the differential equation, plot the amplitude of the output as the input frequency ω varies. Check that the ratio of output and input amplitudes coincides with the magnitude transfer function plot.

29. A diode having the i-v characteristics $i = 0.1(exp(v/0.5) - 1)$ is connected in series with a resistor of 1 Ohm value and a dc voltage source V. Write a program to plot the current through the circuit as a function of the input voltage V. If i is the current through the circuit, then we have

$$i = 0.1(exp((V - iR)/0.5) - 1)$$

Your program must consist of a iterative scheme to solve this non-linear algebraic equation. For example the Newton Raphson scheme, or a scheme of the form

$$i_{n+1} = 0.1(exp((V - i_n R)/0.5) - 1)$$

30. The one dimensional wave equation is given by

$$\frac{\partial^2 \psi(t,x)}{\partial t^2} - \frac{\partial^2 \psi(t,x)}{\partial x^2} = 0$$

Its solution is a superposition of the waves

$$\psi(t,x) = sin(\omega(t \pm x))$$

Write a MATLAB program to plot these waves for various values of t as a function of x. The numerical solution to this wave equation in the region $[0, L]$ is obtained by discretizing this with respect to time and position. This leads to a difference equation. Write a program to solve this difference equation numerically.

31. A random variable U is said to be uniform if its probability density is given by $p_U(u) = 1$ for $0 \leq u \leq 1$ and $p_U(u) = 0$ for $u < 0$ and $u > 1$. If U and V are two uniform random variables independent of each other, the density of $U+V$ is $\int_{-\infty}^{\infty} p_U(u-x)p_V(x)dx$ where $p_V = p_U$. Evaluate this convolution by hand and also numerically by discretizing the derivative. Make a plot of both the functions and compare them for different discretization step sizes.

32. Consider a circuit with the following topology. The circuit consists of four nodes numbered $1, 2, 3, 4$. 4 is the ground. A voltage source *sint* is connected between node 1 and 4. A resistor 1 Ohm is connected between node 1 and node 2. A capacitor 1 Farad is connected between node 2 and node 3. A resistor 1 Ohm is connected between node 1 and 4 and if i is the current flowing through this resistor, a controlled current source $0.1i$ is connected from 3 to 4. Write down the circuit differential equations for this circuit and solve these numerically using MATLAB. To discretize the circuit equations, we've got first of all to select an appropriate sampling step size T and if $f(t)$ is an analog signal, we replace $\frac{df(t)}{dt}$ by the finite difference $\frac{f((n+1)T)-f(nT)}{T}$. This circuit is a model for a transistor amplifier.

33. The one dimensional Schrodinger equation is given by

$$\psi''(x) + (E - V(x))\psi(x) = 0$$

We want to solve this differential equation in the interval $[0, 1]$ taking $V(x) = x^2$. The boundary conditions are $\psi(0) = \psi(1) = 0$. Determine the possible values of E by discretizing the above equation and casting it as a matrix eigenvalue problem.

34. Let $\mathbf{A}$ be an $n \times p$ matrix with $p < n$. Let $\mathbf{b}$ be an $n \times 1$ vector. Write a program to determine the $p \times 1$ vector $\mathbf{x}$ for which $\| \mathbf{b} - \mathbf{Ax} \|$ is a minimum. Prepare a table taking N choices of the vector $\mathbf{b}$ and plotting the ratio of the minimum error square $\| \mathbf{b} - \mathbf{Ax} \|^2 / \| \mathbf{b} \|^2$. In this model, $\mathbf{b}$ can be regarded as the unknown signal, $\mathbf{Ax}$ as the best estimate of the signal based on the columns of the data matrix $\mathbf{A}$ and $\| \mathbf{b} - \mathbf{Ax} \|^2$ as the estimation error variance. The above ratio can be regarded as the noise to signal ratio.

35. Write a program to evaluate the roots of a third degree polynomial using the Newton-Raphson scheme. If $f(x) = 0$ is to be solved and x_0 is an approximation to a solution of this equation and $x_0 + \delta x$ is the next order approximation, then $f(x_0 + \delta x) = f(x_0) + \delta x f'(x_0) = 0$ gives $\delta x = -f(x_0)/f'(x_0)$ so that the next order approximation is given by $x_0 - f(x_0)/f'(x_0)$. This is the Newton Raphson scheme. The method works well only when the initial guess is good.

36. Let $\{x_0, x_1, ..., x_{n-1}\}$ and $\{y_0, y_1, ..., y_{n-1}\}$ be two sequences. Their convolution is defined by

$$z_m = \sum_{k=0}^{n-1} x_k y_{m-k}$$

The range of m for which z_m is non-zero is $0, 1, ..., 2n-2$ and in the above sum $m-k$ falls outside the range $0, 1, ..., n-1$, then y_{m-k} is set equal to zero. Write a MATLAB program to evaluate this sum and plot it. Compare with the Conv command built inside MATLAB.

37. If $\psi(x)$ is the wave function of a quantum mechanical particle, then the normalization condition $\int_{-\infty}^{\infty} |\psi(x)|^2 dx = 1$ must hold. Using the discretization procedure for evaluating the integrals, evaluate the normalization factor for the wave function $exp(-|x|)sin(\alpha x)$ as a function of α and make a comparison with the theoretically evaluated value. Repeat this for the wave function corresponding to a particle in a box. The wave function has the shape $sin(n\pi x/L), 0 \leq x \leq L$.

38. If $(x_n, n = 0, 1, 2, ..., N-1)$ is a sequence, its autocorrelation function is defined by the formula

$$R_{xx}(k) = \sum_{n=0}^{N-1} x_n x_{n+k}$$

Write a program to compute this function and plot it for various lags k.

39. If $(x_n, n = 0, 1, ..., N-1)$ is a sequence, its discrete Fourier transform is defined by the rule

$$X_k = \sum_{n=0}^{N-1} x_n exp(-i2\pi kn/N), k = 0, 1, ..., N-1$$

Write a program to compute this quantity using a FOR loop and plot it. Do it for the sequences (a) $x_n = 1, n = 0, 1, ..., N-1$, (b) $x_n = 1, n = 0, 1, ..., N/2-1$, $x_n = 0, n = N/2, N/2+1, ..., N-1$, (c) $x_n = sin(2\pi k_0 n/N), n = 0, 1, ..., N-1$, (d) $x_n = \sum_{i=1}^{p} A_i sin(2\pi k_i n/N + \phi_i)$, where $k_0, k_1, ..., k_p$ are some integers in the range $\{0, 1, ..., N-1\}$.

40. If $(x(n, m), n, m = 0, 1, ..., N-1)$ is a two dimensional sequence, its two dimensional DFT is defined by the rule

$$X(k, r) = \sum_{n,m=0}^{N-1} x(n, m) exp(-j2\pi(kn + rm)/N), k, r = 0, 1, ..., N-1$$

Write a program to calculate this and plot it when the data $x(n, m)$ is obtained by discretizing an image.

Study project 71

Transmission of ascii characters using harmonic processes: Suppose we have N^2 symbols to be transmitted. We select a set $E = \{\omega_1, ..., \omega_p\}$ of p frequencies that the communications channel can transmit with less attenuation. We then prepare a look up table with rows indexed by the frequencies $\omega_1, ..., \omega_p$ and the columns also indexed by the same frequencies. Each symbol is indexed by its row number and column number and hence by a pair of frequencies $(\omega_\alpha, \omega_\beta)$. The corresponding signal transmitted over the channel is given by

$$x(t) = A_1 cos(\omega_\alpha t + \phi_\alpha) + A_2 cos(\omega_\beta t + \phi_\beta)$$

where ϕ_α, ϕ_β are random phases. The channel also adds noise to this signal. If there were no noise, the $x(t)$ would satisfy the fourth order ordinary differential equation

$$D^4 x(t) + a_1 D^3 x(t) + a_2 D^2 x(t) + a_3 D x(t) + a_4 x(t) = 0$$

where $D = \frac{d}{dt}$ and $exp(\pm i \omega_\alpha t)$ and $exp(\pm i \omega_\beta t)$ are roots of the biquadratic equation

$$s^4 + a_1 s^3 + a_2 s^2 + a_3 s + a_4 = 0$$

or equivalently,

$$s^4 + a_1 s^3 + a_2 s^2 + a_3 s + a_4 = (s - exp(i\omega_\alpha))(s - exp(-i\omega_\alpha))(s - exp(i\omega_\beta))(s - exp(-i\omega_\beta))$$

$$= (s^2 - 2cos(\omega_\alpha)s + 1)(s^2 - 2cos(\omega_\beta)s + 1) = 0$$

T determine the pair of frequencies at the receiver end from received signal, we simply compute the peaks of the periodogram, i.e., of the function

$$P_T(\omega) = \int_{-T/2}^{T/2} x(t) exp(-i\omega t) dt$$

If no noise were present, this would equal

$$A_1 \frac{sin((\omega - \omega_\alpha)T/2)}{(\omega - \omega_\alpha)} + A_1 \frac{sin((\omega + \omega_\alpha)T/2)}{(\omega + \omega_\alpha)}$$

$$+ A_2 \frac{sin((\omega - \omega_\beta)T/2)}{(\omega - \omega_\beta)} + A_2 \frac{sin((\omega + \omega_\beta)T/2)}{(\omega + \omega_\beta)}$$

The two frequencies can be resolved provided $|\omega_\alpha - \omega_\beta| > 4\pi/T$. Note that $4\pi/T$ is the width of sinc pulse $sin(\omega T/2)/\omega$, for the zeroes of this function closest to the origin are at $\pm 2\pi/T$. The main lobe of the pulse falls in between these first zeroes. The periodogram has poor resolution if the data length is short or if the data is noise-corrupted. An alternate method having a higher resolution is the parametric modeling which involved minimizing the quantity

$$\int_{-T/2}^{T/2} (D^4 x(t) + a_1 D^3 x(t) + a_2 D^2 x(t) + a_3 D x(t) + a_4)^2 dt$$

with respect to the parameters a_1, a_2, a_3, a_4. The optimal normal equations obtained by setting the partial derivatives of this quantity with respect to these parameters equal to zero are

$$\sum_{i=1}^{4} a_i \int_{-T/2}^{T/2} D^{4-i}x(t)D^j x(t)dt + \int_{-T/2}^{T/2} D^4 x(t)D^j x(t)dt = 0, j = 0, 1, 2, 3$$

Using the inner product notation, these equations can be expressed as

$$\sum_{i=1}^{4} a_i < D^{4-i}x, D^j x > + < D^4 x, D^j x >= 0, j = 0, 1, 2, 3$$

Study project 72

Ascoli's theorem:

If X and Y are topological spaces and $\mathcal{F}$ is a family of functions from X into Y, then this family is called an equicontinuous family of functions if for each $x \in X$ and each open nhood V of $f(x)$ in Y, there is an open nhood U of x in X such that $f(U) \subset V$ for all $f \in \mathcal{F}$. If (Y, d) is a metric space then the statement that the family $\mathcal{F}$ is equicontinuous boils down to the statement that for each $\epsilon > 0$ and each $x \in X$, there is an open nhood U of x such that $y \in U$ implies $d(f(y), f(x)) < \epsilon$ for all $f \in \mathcal{F}$.

Preliminary result: Let X be a compact space. Let (Y, d) be a compact metric space and $\mathcal{F}$ a subset of $C(X, Y)$, the space of all continuous maps from X into Y. Then, $\mathcal{F}$ is equicontinuous iff $\mathcal{F}$ is totally bounded with respect to the sup metric.

Suppose $\mathcal{F}$ is totally bounded in the sup metric. Let $\epsilon > 0$ be given. Choose positive numbers ϵ_1, ϵ_2 so that $2\epsilon_1 + \epsilon_2 \leq \epsilon$. Cover $\mathcal{F}$ by finitely many open balls $B(f_j, \epsilon_1), j = 1, 2, ..., n$. Since each f_i is continuous, we can choose a nhood U of x_0 such that $x \in U$ implies $d(f_i(x), f_i(x_0)) < \epsilon_2, i = 1, 2, ..., n$. Let f be any member of $\mathcal{F}$. Then, f belongs to $B(f_i, \epsilon_1)$ for some i. Thus, $d(f(x), f_i(x)) < \epsilon_1$ for all $x \in X$. Then, for $x \in U$,

$$d(f(x), f(x_0)) \leq d(f(x), f_i(x)) + d(f_i(x), f_i(x_0)) + d(f_i(x_0), f(x_0)) \leq \epsilon_1 + \epsilon_2 + \epsilon_1$$

$$= 2\epsilon_1 + \epsilon_2 \leq \epsilon$$

proving equicontinuity of $\mathcal{F}$. Conversely, assume that $\mathcal{F}$ is equicontinuous. Let $\epsilon > 0$ be given. Choose ϵ_1, ϵ_2 so that $2\epsilon_1 + \epsilon_2 \leq \epsilon$. For each $x \in X$, choose a nhood U_x of x such that $y \in U_x$ implies $d(f(y), f(x)) < \epsilon_1$ for all $f \in \mathcal{F}$. This is possible by virtue of assumed equicontinuity of $\mathcal{F}$. Then $\{U_x : x \in X\}$ is an open cover of X and since X is compact, there is a finite subcover $\{U_{x_i} : i = 1, 2, ..., n\}$. For simplicity set $U_i = U_{x_i}$. Then, $x_i \in U_i$ and $x \in U_i$ implies $d(f(x), f(x_i)) < \epsilon_1$ for all $f \in \mathcal{F}$. Cover Y by finitely many open sets $V_1, ..., V_m$ having diameter smaller than ϵ_2. Let J be the collection of all functions $\alpha : \{1, 2, ..., k\} \to \{1, 2, ..., m\}$. There are in all m^k such functions. Given $\alpha \in J$, if there exists a function $f \in \mathcal{F}$ such that $f(x_i) \in V_{\alpha(i)}$ for each $i = 1, 2, ..., k$, choose one such function and label it f_α. The collection $\{f_\alpha\}$ is indexed by a subset J' of J and is therefore finite. We claim that the open balls $B(f_\alpha, \epsilon)$ with α running over J' cover $\mathcal{F}$. Let $f \in \mathcal{F}$. For each $i = 1, 2, ..., k$, is clear that $f(x_i) \in V_{r_i}$, for some integers $r_1, ..., r_k$ in $\{1, 2, ..., m\}$. Hence if α denotes the function $\alpha(i) = r_i, i = 1, 2, ..., k$, we have $f(x_i) = V_{\alpha(i)}$ and $\alpha \in J'$. Let $x \in X$. Then, $d(f(x), f(x_i)) < \epsilon_1, d(f(x_i), f_\alpha(x_i)) < \epsilon_2, d(f_\alpha(x_i), f_\alpha(x)) < \epsilon_1$ and adding these three inequalities results in $d(f(x), f_\alpha(x)) < \epsilon$. This holds for all $x \in X$ and hence we conclude total boundedness of $\mathcal{F}$ relative to the sup metric.

Statement of Ascoli's theorem: Let X be a compact space and $C(X, \mathbb{R}^n)$ the space of continuous maps from X into $\mathbb{R}^n$. Let ρ denote the sup metric on $C(X, \mathbb{R}^n)$, i.e., $\rho(f, g) = sup\{d(f(x), g(x)) : x \in \mathbb{R}^n\}$. A subset $\mathcal{F}$ of $C(X, \mathbb{R}^n)$ is compact iff it is closed, bounded and equicontinuous.

Study project 73

Imbedding theorems play a fundamental role in analysis. How to imbed one space into a larger space having a more comfortable structure is a fundamental idea. The Ideas centre around the fundamental theorems of Tychonoff and Urysohn. However to start with we simply aim at proving how a regular topological space can be imbedded into a metric space formed using the real numbers. A consequence of this will be the imbedding of a separable metric space into a compact metric space. This theorem also plays a fundamental role in probability theory where one wishes to define the notion of a metric on the space of measures.

An imbedding theorem: A topological space X is said to be completely regular if one point sets are closed in X and if for each $x \in X$ and each closed subset $A \subset X$ such that $x \notin A$, there is a continuous function $f : X \to [0, 1]$ such that $f(x_0) = 1$ and $f(A) = 0$.

Regularity of a topological space: A space X is regular if one point sets are closed and if for each $x \in X$ and a closed set B not containing x, there exist open sets U, V such that $x \in U$ and $B \subset V$. Obviously a regular space is Hausdorff.

Normality of a topological space: A space X is said to be normal if one point sets are closed and if for each pair A, B of disjoint closed subsets of X, there exist disjoint open sets U, V such that $A \subset U, B \subset V$. Obviously a normal space is regular.

The Urysohn lemma: Let X be a normal space and A, B disjoint closed subsets of X. Let $[a, b]$ be a closed interval in $\mathbb{R}$. Then there is a continuous map $f : X \to [a, b]$ such that $f(A) = \{a\}, f(B) = \{b\}$

Complete regularity of a space: A space X is completely regular if one point sets are closed in X and if for each $x \in X$ and each closed set A not containing x, there is a continuous function $f : X \to [0,1]$ such that $f(x) = 1$ and $f(A) = \{0\}$.

Suppose X is a normal space. Then, one point sets are closed. Further suppose x is a point in X and A is a closed subset of X not containing x. Then by the Urysohn lemma there is a continuous function $f : X \to [0,1]$ such that $f(A) = 0$, $f(x) = 1$ and this proves that every normal space is completely regular.

Let X be a regular space with a countable basis. We claim that U can be expressed as a countable union of closed sets. To see this, first choose $x \in U$. Then, by regularity of X, there exist disjoint open sets V_x, W_x such that $x \in V_x$ and $U^c \subset W_x$. Now, $x \in V_x \subset W_x^c \subset U$. Let $B_n, n = 1, 2, ...$ be a countable basis for X. Then, there exists a basis element $B_{n(x)}$ such that $x \in B_{n(x)} \subset V_x \subset W_x^c$. Since W_x^c is closed, it follows that $Cl(B_{n(x)} \subset W_x^c \subset U$. Thus, U is the union of all $Cl(B_{n(x)}$ with x varying over U. But the collection $B_{n(x)}, x \in U$ is a subset of $\{B_1, B_2, ...\}$ and is therefore countable. It follows that U is a countable union of closed sets. We write $U = \cup_{n=1}^{\infty} C_n$ where C_n is closed.

Let X be a regular space with a countable basis. We want to show that X is normal. Let A, B be disjoint closed subsets of X. Each $x \in A$ has a nhood U not intersecting B by the definition of regularity. Using regularity, chose a nhood V of x such that $Cl(V) \subset U$. That this is possible follows from the following argument. U^c is a closed set disjoint from x. Hence, there is a nhood V of x and a nhood W of U^c such that V and W are disjoint. It follows that $V \subset W^c \subset U$. Since W^c is closed, $Cl(V) \subset W^c$ and hence, $Cl(V) \subset U$. Finally, choose a basis element $B_{n(x)}$ from the given countable basis $\mathcal{B} = \{B_1, B_2, ...\}$ such that $x \in B_{n(x)} \subset V$. As x varies over A, $B_{n(x)}$ varies over $\mathcal{B}$. This collection of basis elements $B_{n(x)}, x \in A$ is countable since $\mathcal{B}$ is countable. Denote this collection by $\{U_n : n = 1, 2, ..., \}$. Then $\{U_n\}$ is a countable open covering for A. In the above construction, U was disjoint from B and since $Cl(V)$ was contained in U and $B_{n(x)}$ was contained in V, it follows that $Cl(B_{n(x)})$ is disjoint from B. Thus for all n, $Cl(U_n) \cap B = \phi$. Similarly choose a countable collection $\{V_n\}$ of open sets covering B such that $Cl(V_n) \cap A = \phi$ for all n. The sets $\cup U_n$ and $\cup V_n$ are open sets containing A and B respectively but they need not be disjoint. Define $U_n' = U_n - \cup_{i=1}^n Cl(V_i)$ and $V_n' = V_n - \cup_{i=1}^n Cl(U_i)$. Since the finite union of closed sets is closed and since the complement of a closed set is open, both U_n' and V_n' are open. We claim that $A \subset \cup U_n'$ and $B \subset \cup V_n'$. Indeed, suppose $x \in A$. Then $x \in U_n$ for some n and since $A \cap Cl(V_k) = \phi$ for all k, it follows that $x \notin \cup_{i=1}^n Cl(V_i)$. Thus, $x \in U_n'$. Likewise, $B = \subset \cup V_n'$. Further, $\cup U_n'$ and $\cup V_n'$ are disjoint for suppose x belongs to both. Then $x \in U_n'$ for some n and $x \in V_m'$ for some m. Suppose $m \geq n$. Then $x \cap V_m' = V_m - \cup_{k=1}^m Cl(U_k)$ implies $x \notin U_k, k = 1, 2, ..., m$ and in particular $x \notin U_n$. Thus, $x \notin U_n'$, a contradiction. This proves that A and B can be separated using open sets and hence X is normal.

Let X be a regular space with a countable basis. Let U be any non-empty open subset of

X. Then, we've seen that $U = \cup C_n$ where C_n is closed. We've also seen that X is normal. Now, U^c and C_n are disjoint closed sets for each n. The Urysohn lemma thus implies the existence of a continuous function $f_n : X \to [0, a_n]$ such that $f_n(C_n) = a_n$ and $f_n(U^c) = \{0\}$. We may take a_n to be a sequence such that $0 < a_n \le 1$ and a_n decreases to zero as $n \to \infty$. We regard f_n as a function $f_n : X \to [0, 1]$. Then, f_n is continuous and assumes values in $[0, a_n]$. Define f on X by $f(x) = \sup\{f_n(x) : n \ge 1\}$. We claim that f is continuous. Let $0 \le a \le 1$. Then,

$$f^{-1}([0, a]) = \{x : f(x) \le a\} = \{x : f_n(x) \le a \; for \; all \; n\} = \cap_{n=1}^{\infty} f_n^{-1}([0, a])$$

this being an intersection of closed sets is closed. Now assume that $0 < a \le 1$. Then,

$$\{x : f(x) \ge a\} = \cap_{r=1}^{\infty}\{x : f_n(x) \ge a - 1/r \; for \; some \; n\}$$

To prove that this is a closed set, it suffices therefore to prove that for each $a \in (0, 1]$,

$$\cup_{n=1}^{\infty} f_n^{-1}([a, 1])$$

is closed. Let N be such that $a_N \ge a > a_{N+1}$. Then the above equals

$$\cup_{n=1}^{N} f_n^{-1}([a, 1])$$

which is closed by continuity of f_n and the fact that a finite union of closed sets is closed. We've thus proved continuity of f. Let $x \in U$. Then, $x \in C_n$ for some n and hence $f_n(x) = a_n > 0$. Thus, $f(x) \ge f_n(x) > 0$. This proves that f is positive on U and zero on U^c and that f is a continuous function.

Let X be a regular space with a countable basis $\{B_n, n = 1, 2, ...\}$. According to the above discussion, we can choose a function f_n for each n such that $f_n(x) > 0$ for all $x \in B_n$ and $f_n(x) = 0$ for $x \notin B_n$. Let U be any open set and $x \in U$. Then, there is a basis element B_n such that $x \in B_n \subset U$ and clearly $f_n(x) > 0$ and $f_n(U^c) = \{0\}$. In other words we have a sequence $\{f_n\}$ of continuous functions such that given any open set U and an $x \in U$, there is an integer n such that $f_n(x) > 0$ but $f_n(U^c) = \{0\}$. Given these functions f_n, take on $\mathbb{R}^{\omega}$ the product topology and define a map $F : X \to \mathbb{R}^{\omega}$ by the rule $F(x) = (f_1(x), f_2(x), ...)$. We claim that this is an imbedding. F is continuous because $\mathbb{R}^{\omega}$ has the product topology. Indeed, a general basis element is of the form $B = \Pi E_i$ where $E_i = \mathbb{R}$ for all $i \in \omega - J$ with J a finite subset of ω. and for $i \in J$, E_i is an open subset of $\mathbb{R}$. Then

$$F^{-1}(B) = \{x : f_i(x) \in E_i \forall i \in \omega\} = \{x : f_i(x) \in E_i \forall i \in J\} = \bigcap_{i \in J} f_i^{-1}(E_i)$$

This is an open subset of X, since continuity of f_i implies that $f_i^{-1}(E_i)$ is an open subset of X and a finite intersection of open sets is again open. F is injective for the following reason. Let $x \ne y$. By regularity, we can find an open set U such that $x \in U$ and $y \notin U$. Now, there is an index n

such that $f_n(x) > 0$ and $f_n(U^c) = 0$. The latter implies $f_n(y) = 0$. Hence, $F(x) \neq F(y)$. Finally, we prove that F is a homeomorphism of X onto its image $Z = F(X) \subset \mathbb{R}^\omega$. We need only prove that for each open set U in X, $F(U)$ is open in Z. Let $z_0 \in F(U)$. We shall find an open set W of Z, such that $z_0 \in W \subset F(U)$. Let $x_0 \in U$ be the unique point such that $F(x_0) = z_0$. Choose an index N for which $f_N(x_0) > 0$ and $f_N(X - U) = \{0\}$. Let $V = \pi_N^{-1}((0, \infty))$ and $W = V \cap Z$. Since π_N is continuous ($\pi_N : \mathbb{R}^\omega \to \mathbb{R}$ is defined by $\pi_N((x_1, x_2, ...)) = x_N$. It is continuous by the definition of product topology), it follows that W is open in Z. (If Y is a topological space and S is a subspace of Y, then a subset E of Y is said to be open in Y if $E = Y \cap O$ where O is open in Y. The subsets of Y that are open in Y define a topology on Y called the subspace topology). Now observe that $z_0 \in W$ since $\pi_N(z_0) = \pi_N(F(x_0)) = f_N(x_0) > 0$. Second, $W \subset F(U)$ for if $z \in W$, then $z = F(x)$ for some $x \in X$ and $\pi_N(z) > 0$. Since $\pi_N(z) = \pi_N(F(x)) = f_N(x)$ and f_N is zero outside U, it must follow that $x \in U$. Thus, $z = F(x) \in F(U)$. This proves that $F(U)$ is open. Thus, F is an imbedding of X in $\mathbb{R}^\omega$. Thus we've proved that every regular space with a countable basis is metrizable.

Let X be a separable metric space, i.e. a metric space having a countable dense subset. Then, X is a regular space with a countable basis. In fact every metric space is regular. This can be seen via the following argument. Let B be any closed subset of X and $x \in X, x \notin B$. Then, $d(x, B) = inf\{d(x, y) : y \in B\} > 0$. This follows from the fact that if $d(x, B) = 0$, then a sequence $x_n \in B$ can be found such that $d(x, x_n) \to 0$. This would imply that $x \in Cl(B)$. But B being closed, $Cl(B) = B$ and this contradicts the hypothesis that $x \notin B$. Let $d(x, B) = \epsilon$ and define $U = \{y : d(y, B) < \epsilon/2\}$. Since the function $y \to d(y, B)$ is continuous, in fact $|d(y, B) - d(z, B)| \leq d(y, z)$, it follows that U is open. Obviously $B \subset U$. If $z \in U$, then $d(z, B) < \epsilon/2$. Hence, there exists a $y \in B$ such that $d(z, y) < \epsilon/2$. Since $d(x, B) = \epsilon$ and $y \in B$, it follows that $d(x, y) \geq \epsilon$. Thus, $d(z, x) \geq d(x, y) - d(z, y) > \epsilon - \epsilon/2 = \epsilon/2$. This argument shows that $B(x, \epsilon/2)$ and U are disjoint, i.e., B and x can be separated using open sets, proving regularity of X. That X has a countable basis follows from the following argument. Let $D = \{x_1, x_2, ...\}$ be a countable dense subset of X. For each $n = 1, 2, ...$ consider the collection $S_n = \{B(x_k, 1/n) : k = 1, 2, ...\}$. Then, $S = \bigcup S_n$ is a countable collection of open subsets of X and is easily seen to form a base for the topology on X. Indeed, let U be any non-empty open subset of X and let $x \in U$. Then, there is a $\delta > 0$ such that $B(x, \delta) \subset U$. Let n be such that $1/n < \delta/2$. Since D is dense in X, there exists a k such that $d(x_k, x) < 1/n$. Now, $y \in B(x_k, 1/n)$ implies $d(x_k, y) < 1/n$ implies $d(y, x) \leq d(x_k, y) + d(x_k, x) < \delta/2 + \delta/2 = \delta$ implies $y \in B(x, \delta)$ implies $y \in U$. In other words, $x \in B(x_k, 1/n) \subset U$ proving that S forms a basis for X.

Chapter 10

Probability

10.1 Probability spaces

A probability space consists of the triplet $(\Omega, \mathcal{F}, P)$ where Ω is a set whose elements are called the elementary outcomes of the experiment, Ω itself is called the sample space, $\mathcal{F}$ is a collection of subsets of Ω called the events of the experiment and is closed under countable unions and complements, P is a mapping $P : \mathcal{F} \to [0, 1]$ called the probability measure and satisfies $P(\Omega) = 1$ and countable additivity, i.e., if $E_n, n = 1, 2, ...$, is a sequence of disjoint events, then $P(\cup E_n) = \sum P(E_n)$. When two events E, F are disjoint, i.e., $E \cap F$ is empty, then the occurrence of E rules out the occurrence of F, i.e., both E and F cannot occur simultaneously. For any event E, $P(E)$ is called the probability of occurrence of E. As an example, when a coin is tossed, the sample space consists of two points: $\Omega = \{H, T\}$. $\mathcal{F} = \{\phi, \Omega, \{H\}, \{T\}\}$. We assign probability p to $\{H\}$ and probability $q = 1 - p$ to $\{H\}$. Then, the probability of the different events are

$$P(\{H\}) = p, P(\{T\}) = q, P(\phi) = 0, P(\Omega) = 1$$

Countable additivity is easily verified. In fact, when the class $\mathcal{F}$ of events is finite, countable additivity is equivalent to finite additivity. As another example, consider the throw of a die. Let $p_1, ..., p_6$ be the respective probabilities of $1, 2, ..., 6$. The sample space for this experiment is $\Omega = \{1, 2, ..., 6\}$. If we are interested in all the outcomes occurring, then our class of events $\mathcal{F}$ consists of all the 32 subsets of Ω. If $\{i_1, ..., i_k\}$ is any such subset, then the probability assigned to this set is $P(\{i_1, ..., i_k\}) = \sum_{j=1}^{6} p_{i_j}$. It is easily verified that P is finitely additive and hence also countably additive. If however, we are interested only in whether an even number shows up or an odd number, then our class of events can be taken as $\{\phi, \Omega, \{1, 3, 5\}, \{2, 4, 6\}\}$ with $P(\phi) = 0, P(\Omega) = 1, P(\{1, 3, 5\}) = p_1 + p_3 + p_5$ and $P(\{2, 4, 6\}) = p_2 + p_4 + p_6$. A class of subsets of a set closed under countable union and complementation is also called a σ field or a σ algebra. A class of subsets closed only under finite unions and complementation is called a field or an algebra. Thus, every σ field is a field but not every field is a σ field. A σ field is also closed under countable intersections. This can be proved using De-Morgan's laws:

$$\cap E_n = (\cup E_n^c)^c$$

Now use closure of $\mathcal{F}$ under complementation and countable union. Some of the properties of the probability measure are (1) If $E_n, n = 1, 2, ..., N$ is a finite collection of events, then

$$P(\cup_{n=1}^N E_n) = \sum P(E_n) - \sum_{n<m} P(E_n E_m) + \sum_{n<m<k} P(E_n E_m E_k) + ... + (-1)^{N-1} P(E_1 ... E_N)$$

This can be proved by induction. In particular, for two events E, F,

$$P(E \cup F) = P(E) + P(F) - P(E \cap F)$$

(2) If $E \subset F$ with both E, F events, then the identity $F = E \cup (FE^c)$ expresses F as a disjoint union of two events. Thus, $P(F) = P(E) + P(FE^c)$ from which one deduces that $P(F) \geq P(E)$. (3) Let $E_n, n = 1, 2, ...,$ be a sequence of events such that $E_n \subset E_{n+1}$ for all n. Then $P(E_n) \uparrow P(\cup E_n)$ as $n \uparrow \infty$. This can be proved using countable additivity of P: Let $F_1 = E_1, F_n = E_n - E_{n-1}, n > 1$. Then, $\{F_n\}$ is a sequence of disjoint events and $E_n = F_1 \cup ... \cup F_n, n \geq 1$. Thus, $\cup E_n = \cup F_n$ and

$$P(\cup E_n) = P(\cup F_n) = \sum P(F_n) = \sum P(E_n) - P(E_{n-1}) = lim P(E_n)$$

Likewise, if $E_n, n = 1, 2, ...$ is a sequence of events with $E_{n+1} \subset E_n, n \geq 1$, then $P(E_n) \downarrow P(\cap E_n)$. This follows by taking complements and applying the previous result. (4) If E, F are two events, we define their symmetric difference $E \Delta F = EF^c \cup FE^c$. This set consists of all points of E not in F and all points of F not in E. Let $\mathbb{R}$ denote the real line and $\mathcal{I}$ the class of all intervals in $\mathbb{R}$. Let $\mathbb{B}(\mathbb{R})$ be the σ field generated by all the intervals $\mathcal{I}$. In other words, $\mathbb{B}(\mathbb{R})$ equals the intersection of all σ fields on $\mathbb{R}$ containing $\mathcal{I}$ (Show that an arbitrary intersection of σ fields of subsets of a set Ω is once again a σ-field). Then, we say that a map $X : \Omega \to \mathbb{R}$ is a random variable if $X^{-1}(B) \in \mathcal{F}$ for every $B \in \mathbb{B}(\mathbb{R})$. It can be shown that this condition is equivalent to requiring that $X^{-1}(I) \in \mathcal{F}$ for all intervals $I \in \mathcal{I}$. $X(\omega)$ is the value assumed by the random variable X when the outcome of the experiment is ω.

10.2　Distribution and density functions

The probability distribution of the random variable X is the function $F_X(x) = P(X^{-1}((-\infty, x]))$. This is the same as $P(\{\omega : X(\omega) \leq x\})$ or using a shorthand notation, $P\{X \leq x\}$. Given a finite collection $X_1, ..., X_n$ of random variables on $(\Omega, \mathcal{F}, P)$, their joint distribution is defined as

$$F(x_1, ..., x_n) = P\{X_i \leq x_i, i = 1, 2, ..., n\} = P(\bigcap_{i=1}^n X_i^{-1}((-\infty, x_i]))$$

This is defined for all $x_1, ..., x_n \in \mathbb{R}$. The following properties easily follow from the σ additivity of P, the fact that $P(A) \leq P(B)$ when $A \subset B$ and $P(\Omega) = 1, P(\phi) = 0$: (1) F is non-decreasing in each argument, i.e., $x_i < y_i$ for some i implies $F(x_1, ..., x_i, ..., x_n) \leq F(x_1, ..., y_i, ..., x_n)$, (2) F is right-continuous in each of its arguments, i.e., $F(x_1, ..., x_i+\epsilon, ..., x_n)$ converges to $F(x_1, ..., x_i, ..., x_n)$ when ϵ decreases to zero. Moreover, $F(x_1, ..., x_n)$ converges as $x_i \to \infty$ to the distribution of $(X_1, .X_{i-1}, X_{i+1}, ..., X_n)$ evaluated at $x_1, ..., x_{i-1}, x_{i+1}, ..., x_n$. Finally, as $x_i \to -\infty$, $F(x_1, ..., x_n) \to 0$. We suggest the reader to prove these properties using properties of the probability measure. Note that by $X^{-1}(B)$, we mean the inverse image of the set B under the mapping X. Specifically, if $f : S_1 \to S_2$ is any mapping from the set S_1 into the set S_2, then for any subset $E \subset S_2$, $f^{-1}(E) = \{s \in S_1 : f(s) \in S_2\}$. We emphasize that f^{-1} does not stand for the inverse map. The inverse map may or may not exist, but the inverse image is always defined for any map. The basic properties of inverse image are, (1) if $\{E_\alpha : \alpha \in \Gamma\}$ is any family of subsets S_2, then

$$f^{-1}(\cup\{E_\alpha : \alpha \in \Gamma\}) = \cup\{f^{-1}(E_\alpha) : \alpha \in \Gamma\}$$

Also if $E \subset F$, then $f^{-1}(E) \subset f^{-1}(F)$. We suggest the reader to prove these statements. If $F(x)$ is the distribution function of a random variable X, then the density of X is defined by $f(x) = F'(x)$ provided the derivative exists. When it does not exist, we often use generalized functions to define the derivative. The distribution can be obtained from the density using the formula $F(x) = \int_{-\infty}^{x} f(\xi)d\xi$. The density is always non-negative at each point. If B is a subset of the real line, then

$$P(X \in B) = P\{\omega : X(\omega) \in B\} = P(X^{-1}(B)) = \int_B f(x)dx = \int_B dF(x)$$

equals the probability that the random variable X will assume values in the set B. For example, consider a random variable X uniformly distributed over the interval $[a, b]$. This means that the density of the random variable is given by $f(x) = 1/(b-a)$ for $x \in [a, b]$ and $f(x) = 0$ for $x < a$ and $x > b$. The total probability is one: $\int_{-\infty}^{\infty} f(x)dx = 1$. The distribution of the random variable is given by $F(x) = 0$ if $x < a$, $F(x) = (x-a)/(b-a)$ if $a \leq x < b$ and $F(x) = 1$ if $x > b$. Note that F is not left continuous but is right continuous. As another example, assume that X assumes values only in the finite set $\{a_1, a_2, ..., a_N\}$ where $a_1 < a_2 < ... < a_N$. Let p_i denote the probability that X assumes the value a_i: $P(X = a_i) = p_i$. Then, in terms of the Dirac-Delta function, the density of X is given by $f(x) = \sum_{i=1}^{N} p_i\delta(x - a_i)$. This follows from the fact that if B is any Borel subset of $\mathbb{R}$ containing the points $a_{i_1}, ..., a_{i_k}$ but not containing the other $a'_j s$, then

$$\int_B f(x)dx = \int_B \sum_{i=1}^{N} p_i\delta(x - a_i)dx = \sum_{j=1}^{k} p_{i_j}$$

One of the important densities is the normal density. It is defined by

$$f(x) = (2\pi\sigma^2)^{-1/2}exp(-(x - \mu)^2/2\sigma^2)$$

where σ^2 is a positive real number and μ is any real number. Another important density is the exponential density

$$f(x) = \lambda.exp(-\lambda x)u(x)$$

where $\lambda > 0$. One can also define a joint density for a finite collection of random variables. Let $X_1, ..., X_n$ be a finite collection of random variables having joint distribution F, i.e.,

$$F(x_1, ..., x_n) = P(X_1 \le x_1, ..., X_n \le x_n) = P(\bigcap_{i=1}^{n} X_i^{-1}((-\infty, x_i]))$$

If the following n fold derivative exist:

$$f(x_1, ..., x_n) = \frac{\partial^n F(x_1, ..., x_n)}{\partial x_1 ... \partial x_n}$$

then f is called the joint density of $(X_1, ..., X_n)$. Even when it does not exist, we can define it in terms of generalized functions. We can express the probability that the n-tuple $(X_1, ..., X_n)$ of random variables will assume values in a Borel set $B \in \mathbb{B}(\mathbb{R}^n)$. $\mathbb{B}(\mathbb{R}^n)$ is defined as the σ field generated by sets of the form $I_1 \times ... \times I_n$ where $I_1, ..., I_n$ are intervals in $\mathbb{R}$. It can also be shown that $\mathbb{B}(\mathbb{R}^n)$ is the σ field generated by sets of the form $B_1 \times ... \times B_n$ where $B_1, ..., B_n \in \mathbb{B}(\mathbb{R})$. The reader is referred to the textbook "An introduction to probability and measure" by K.R.Parthasarathy for the details.

10.3 Mean and variance of random variables

If X is a random variable having F as its distribution function and f as its density function, then the mean or average of the random variable X is defined as

$$\mathbb{E}X = \mu_X = \int_{-\infty}^{\infty} xf(x)dx = \int_{-\infty}^{\infty} xdF(x)$$

The integral may not exist. One can in fact define the mean entirely in terms of the probability measure P without even having to introduce the notion of a density function. This involves first of all introduction of the notion of a simple random variable having the general form

$$X(\omega) = \sum_{i=1}^{N} c_i \chi_{E_i}(\omega)$$

where $E_1, ..., E_N$ are disjoint events. A simple random variable, in other words, is a random variable that assumes only a finite number of distinct values. Note that some authors also refer to random variables as measurable functions on the probability space. By measurability, one means that the inverse image of Borel sets are events. The integral of the simple random variable X defined above with respect to the probability measure P is defined as

$$\int X dP = \int X(\omega) dP(\omega) = \int X(\omega) P\{d\omega\} = \sum_{i=1}^{N} c_i P(E_i)$$

Now if X is any non-negative random variable(not necessarily simple), then there is a theorem that states that $X = lim X_n$, where $X_1, X_2, \ldots$ are simple non-negative random variables with $X_1 \le X_2 \le \ldots$. Then, the integrals $\int X_n dP$ form a nondecreasing sequence. We define $\int X dP$ to be the limit of this sequence as $n \to \infty$. One can show that the limit is independent of the sequence X_n chosen. This defines the integral for an arbitrary non-negative random variable. Now let X be any random variable. We can write $X = X_+ - X_-$ where X_+ and X_- are both non-negative random variables. We define $\int X dP = \int X_+ dP - \int X_- dP$ provided both the integrals are finite. If one of them is infinite and the other finite, then $\int X dP$ is either set equal to $+\infty$ or equal to $-\infty$ depending on whether $\int X_+ dP$ is infinite or $\int X_- dP$ is infinite. It can be shown that when X has a finite mean, then $\int X dP = \int x dF(x)$. This identity is usually stated in the form of the change of variables theorem:

$$\int X(\omega) dP(\omega) = \int x dP X^{-1}(x)$$

More generally, if g is any function of a real variable such that g is measurable, i.e., $g^{-1}(B) \in \mathbb{B}(\mathbb{R})$ for all $B \in \mathbb{B}(\mathbb{R})$, and X is a random variable on a probability space then $g(X)$ defined by $g(X)(\omega) = g(X(\omega))$ is also a random variable, since if B is any Borel set, $g(X)^{-1}(B) = X^{-1}g^{-1}(B)$ is an event. If $g(X)$ has finite mean, then $\int g(X) dP = \mathbb{E}g(X) = \int g(x) dF(x)$, where F is the distribution function of X. In particular, the mean square value of a random variable X is defined by $\mathbb{E}X^2 = \int x^2 dF(x)$ and the variance by $V(X) = \mathbb{E}(X - \mathbb{E}X)^2$. The mean operator is a linear operator in the sense that if X, Y are random variables defined on the same probability space $(\Omega, \mathcal{F}, P)$ and c_1, c_2 are constants, then $\mathbb{E}(c_1 X + c_2 Y) = c_1 \mathbb{E}X + c_2 \mathbb{E}Y$. Apart from being linear, the mean operator (also called the expectation operator and the average operator) is non-negative, in the sense that if X is a non-negative random variable, then $\mathbb{E}X \ge 0$. The other important property is that the expectation of a constant random variable is a constant, i.e., $\mathbb{E}c = c$ which can be deduced from linearity and the property $\mathbb{E}(1) = 1$. From the linearity of the expectation operator, we can deduce an alternate formula for the variance:

$$V(X) = \mathbb{E}(X - \mathbb{E}X)^2 = \mathbb{E}(X^2 + (\mathbb{E}X)^2 - 2X\mathbb{E}X) = \mathbb{E}X^+(\mathbb{E}X)^2 - 2(\mathbb{E}X)^2 = \mathbb{E}X^2 - (\mathbb{E}X)^2$$

Perhaps the most important result related to the variance of a random variable is the Chebyshev inequality which roughly states that if the variance is small, then the probability of the random variable deviating much from the mean must be small. The precise inequality is

$$P(|X - \mathbb{E}X| > \epsilon) \le \frac{V(X)}{\epsilon^2}$$

We shall prove it assuming a density. Let $\mu = \mathbb{E}X, \sigma^2 = V(X)$. Then,

$$V(X) = \int_{-\infty}^{\infty} (x - \mu)^2 f(x)dx \geq \int_{|x-\mu|>\epsilon} (x - \mu)^2 f(x)dx \geq \int_{|x-\mu|>\epsilon} \epsilon^2 f(x)dx$$

$$= \epsilon^2 P(|X - \mu| > \epsilon)$$

yielding the desired inequality.

10.4   Characteristic function and moments

Let $X_1, ..., X_n$ be random variables on a probability space. Given any measurable function $g : \mathbb{R}^n \to \mathbb{R}$, we can define the random variable $g(X_1, ..., X_n)$ by the rule $g(X_1, ..., X_n)(\omega) = g(X_1(\omega), ..., X_n(\omega))$. We can then define the mean value of this random variable. In particular, we can take $g(x_1, ..., x_n) = cos(t_1x_1 + ... + t_nx_n)$ and $g(x_1, ..., x_n) = sin(t_1x_1 + ... + t_nx_n)$ with $t_1, ..., t_n$ as arbitrary real numbers. One can prove that both of these are measurable by expanding them as Taylor series and using the fact that measurability is preserved under addition, multiplication and formation of limits. We define the characteristic function of the variables as

$$\psi(t_1, ..., t_n) = \mathbb{E}exp(i(t_1X_1 + ... + t_nX_n)) = \mathbb{E}cos(t_1X_1 + ... + t_nX_n) + i\mathbb{E}sin(t_1X_1 + ... + t_nX_n)$$

Note that both the expectations are defined since they are bounded by unity. It is easily verified by differentiating under the expectation sign that

$$\frac{\partial^{r_1+...+r_n}}{\partial t_1^{r_1}....\partial t_n^{r_n}}\Big|_{\mathbf{t}=\mathbf{0}} = i^{r_1+...+r_n}\mathbb{E}(X_1^{r_1}...X_n^{r_n})$$

Equivalently, Taylor expansion of the characteristic function yields

$$\psi(t_1, ..., t_n) = \sum_{r_1,...,r_n=0}^{\infty} i^{r_1+...+r_n} \frac{t_1^{r_1}...t_n^{r_n}}{r_1!...r_n!}\mathbb{E}(X_1^{r_1}...X_n^{r_n})$$

10.5   Independence of random variables

A finite collection $X_1, ..., X_n$ of random variables is said to be an independent set if for every collection $B_1, ..., B_n \in \mathbb{B}(\mathbb{R})$, the events $X_i^{-1}(B_i), i = 1, 2, ..., r$ are independent events, i.e., if

$$P(\bigcap_{i=1}^{n} X_i^{-1}(B_i)) = \Pi_{i=1}^{n} P(X_i^{-1}(B_i))$$

This is equivalent to saying that the joint distribution factors into a product:

$$F(x_1, ..., x_n) = \Pi_{i=1}^{n} F_i(x_i)$$

If such a factorization takes place, then automatically, F_i becomes the marginal distribution of X_i. As an example, if $X_1, ..., X_n$ are jointly Gaussian random variables so that the vector $\mathbf{X} = [X_1, ..., X_n]^T$ has mean $\mathbb{E}\mathbf{X} = \mathbf{m}$ and covariance $\mathbf{A} = Cov(\mathbf{X}) = \mathbb{E}(\mathbf{X} - \mathbf{m})(\mathbf{X} - \mathbf{m})^T)$, then the joint density of these variables is given by

$$f(\mathbf{x}) = (2\pi)^{-n/2} |\mathbf{A}|^{-1/2} exp(-\frac{1}{2}(\mathbf{x} - \mathbf{m})^T \mathbf{A}^{-1}(\mathbf{x} - \mathbf{m}))$$

and its characteristic function can be evaluated as

$$\psi(\mathbf{t}) = \int_{\mathbb{R}^n} f(\mathbf{x}) exp(i\mathbf{t}^T \mathbf{x}) d\mathbf{x} = exp(i\mathbf{m}^T \mathbf{t} - \mathbf{t}^T \mathbf{A}\mathbf{t}/2)$$

By differentiating this expression, we can show that

$$\mathbb{E}X_j = -i \frac{\partial \psi(\mathbf{t})}{\partial t_j}\Big|_{\mathbf{t}=\mathbf{0}} = m_j$$

$$\mathbb{E}(X_j X_k) = -\frac{\partial^2 \psi(\mathbf{t})}{\partial t_j \partial t_k} = a_{jk} + m_j m_k$$

If $X_1, ..., X_n$ are independent random variables, then their joint density factors into a product of marginal densities. Thus, their joint characteristic function is given by

$$\psi(t_1, ..., t_n) = \mathbb{E}exp(i(t_1 X_1 + ... + t_n X_n)) = \int f(x_1, ..., x_n) exp(i(t_1 x_1 + ... + t_n x_n)) dx_1 ... dx_n$$

$$= \int f_1(x_1) ... f_n(x_n) exp(i(t_1 x_1 + ... + t_n x_n)) dx_1 ... dx_n$$

$$= \Pi_{j=1}^{n} \int f_j(x) exp(it_j x) dx = \Pi_{j=1}^{n} \psi_j(t_j)$$

where $\psi_j(t) = \mathbb{E}exp(itX_j)$ is the characteristic function of X_j.

10.6   The notion of a random process

A random process in discrete time is simply a sequence $\{X_n : n \in \mathbb{Z}\}$ of random variables defined on the same probability space $(\Omega, \mathcal{F}, P)$. To be more specific, we perform the experiment, note the outcome $\omega \in \Omega$ and generate the sequence $X_n(\omega), n \in \mathbb{Z}$. Thus, a random process is simply an ensemble of sequences, one sequence corresponding to each outcome. We can do the same in continuous time. A random process here is simply an ensemble of functions $\{X_t(\omega) : t \in \mathbb{R}\}$, one function corresponding to one elementary outcome $\omega \in \Omega$. We can represent such random process on a graph by taking the abscissa as the time axis and the ordinate as the process amplitude axis. On this graph, corresponding to each ω, we plot the function $t \to X_t(\omega)$ and label the corresponding function by the symbol ω that indicates which outcome of the experiment does the given function correspond to. A typical example of a random process that one comes across in statistical mechanics is the following. Consider a system of gas particles whose positions and momenta at time t are represented by the vector $\mathbf{x}(t)$. This vector will have $6N$ components when N is the number of particles. The evolution of this vector will be governed by Hamilton's equations of motion and the initial coordinates and momenta. Thus, we can write $\mathbf{x}(t, \xi)$ as the state of the system at time t, where the vector $\xi \in \mathbb{R}^{6N}$ represents the initial state, i.e. $\mathbf{x}(0, \xi) = \xi$. The initial configuration is not known and hence assumed to be random in accordance with the Gibbs distribution $dF(\xi)$. A function of the state of the gas like energy, entropy, enthalpy, pressure, temperature can be represented as $\phi(\mathbf{x})$ where ϕ is a mapping from $\mathbb{R}^{6N} \to \mathbb{R}$. This quantity as a function of time constitutes a random process. Each initial state ξ defines an elementary outcome of the experiment and the corresponding observed process is given by $\phi(\mathbf{x}(t, \xi)), t \geq 0$. This is a classic example of a random process that one comes across in statistical physics.

10.7   Examples of random processes

Consider a sequence of coin tosses. If at the nth toss, a Head shows up, we set $X_n = 1$ while if at the nth toss a Tail shows up, we set $X_n = -1$. Assuming that on each toss, the probability of a Head is p, we get a sequence $\{X_n : n = 1, 2, ..., \}$ of identically distributed independent Bernoulli random variables with $P(X_n = 1) = p, P(X_n = -1) = q = 1 - p$. This is an example of a discrete time stochastic process. The joint probability density of this process is given by the formula

$$f(x_1, ..., x_k; n_1, ..., n_k) = \Pi_{i=1}^{k} p\delta(x_i - 1) + q\delta(x_i + 1)$$

To see how this random process can be put in the framework of a stochastic process defined as a family of random variables on a probability space, we consider our sample space to be the space Ω of all sequences of $1's$ and $-1's$. It can be shown that one can define a σ field $\mathcal{F}$ and a probability measure P on this space such that the sets $\{x_1\} \times ... \times \{x_n\} \times \{-1, 1\} \times \{-1, 1\} \times ...$ are all events when $x_i \in \{-1, 1\}$ and the probability of this event equals $p(x_1)...p(x_n)$ where $p(1) = p, p(-1) = q$. Then, each random variable X_n can be regarded as the nth coordinate of the outcome $(x_1, x_2, ...)$ and hence a function on the sample space. An example of a random process derivable from this coin tossing process is the random walk process defined as $S_0 = 0, S_n = X_1 + ... + X_n, n \geq 1$. This construction can be expressed iteratively as $S_{n+1} = S_n + X_{n+1}$. At time n, if the particle is located at S_n, then at time $n + 1$, it moves one step right to the position $S_n + 1$ with probability p or one step left to the position $n - 1$ with probability $q = 1 - p$. $\{S_n\}$ can thus be viewed as a drunken man's walk in one dimension, in which motion takes place in steps of one unit and time also flows in units of one second, the man at each stage taking a step to the right with probability p or a step to the left with probability $q = 1 - p$. Suppose we want to calculate $P(S_n = r)$, i.e. at time n, the man is located at r units from the origin. Note that if r is positive, it means that the man is located to the right of the origin while if r is negative, then the man is located to the left of the origin. To arrive at r, he must take m steps to the right and $n - m$ steps to the left where $2m - n = r$, i.e., $m = (n + r)/2$. This can happen only when n and r have the same parity, i.e. either both are even or both are odd. The probability of this happening is given by the binomial expression

$$P(S_n - r) = \binom{n}{(n+r)/2} p^{(n+r)/2} q^{(n-r)/2}$$

The random walk process is encountered typically in statistical mechanics when one is interested in analyzing the motion of a pollen particle in a liquid. The molecules of the liquid are in random motion at any finite temperature. Each molecular kick on the pollen grain can cause the grain either to get deflected to the left or to the right. As a result, the pollen executes Brownian motion which is a limit of the random walk process.

Another example of a random process that one typically comes across in speech processing applications is the autoregressive process. This can best be illustrated by considering an $AR(1)$ process. We are given a speech segment $\{X_n\}$. Speech has highly correlated neighbouring samples, so we can think of predicting X_n based on X_{n-1} as $\hat{X}_n = aX_{n-1}$. a is determined by minimizing the mean square error $\mathbb{E}(X_n - aX_{n-1})^2$ with respect to a. $\frac{d}{da}\mathbb{E}(X_n - aX_{n-1})^2 = 0$ gives $\mathbb{E}((X_n - aX_{n-1})X_{n-1}) = 0$. This gives $a = \frac{R(1)}{R(0)}$ where $R(m) = \mathbb{E}(X_n X_{n-m})$ is the autocorrelation function of the speech process. The orthogonality relation implies that the prediction error $e_n = X_n - aX_{n-1}$ is orthogonal to X_{n-1}. The dynamic range of the error e_n is much smaller than the dynamic range of the original speech sample X_n when the correlations are large. Thus, fewer quantization levels are required to store the error sequence e_n than the original speech sequence X_n. By storing the sequence e_n and the prediction parameter a, we can reconstruct the process by the algorithm $X_n = aX_{n-1} + e_n$. This motivates us to define an $AR(1)$ process by the recursion $X_n = aX_{n-1} + W_n$, where $\{W_n\}$ is an iid sequence of $N(0, \sigma^2)$ random variables. Assuming that the process has been running from time $n = -\infty$, we note that X_n is the output of a linear time invariant system with transfer function $H(z) = \frac{1}{1 - az^{-1}}$ and input W_n. The corresponding impulse response of the filter is $h(n) = a^n u(n)$ and hence the output can be expressed as $X_n = \sum_{k=0}^{\infty} a^k W_{n-k}$. Using the fact that $\mathbb{E}(W_n W_m) = \sigma^2 \delta(n - m)$, one obtains the formula

$$R_{XX}(m) = \sum_{k,r=0}^{\infty} a^k a^r \mathbb{E}(W_{n-k} W_{n+m-r})$$

$$= \sigma^2 \sum_{k,r=0}^{\infty} a^{k+r} \delta(m - r + k) = \sigma^2 \sum a^{2k+m} = \sigma^2 \frac{a^m}{1 - a^2}$$

provided $m \geq 0$. For arbitrary m, the expression for the autocorrelation is obtained by replacing m by $|m|$. Such an $AR(1)$ process also occurs in statistical mechanics as the following example illustrates. Let V_n denote the velocity of the pollen particle moving in the liquid after the nth molecular impact. A random force F_{n+1} due to the molecular kick changes the momentum by an amount τF_{n+1} where τ is the duration of the impact. Thus, $V_{n+1} - V_n = \tau F_{n+1}$. The mass of the pollen grain has been scaled to unity. Thus, $V_{n+1} = V_n + \tau F_{n+1}$ however, if viscous effects are taken into account, then the velocity should also be scaled by a factor a. This gives $V_{n+1} = aV_n + a\tau F_{n+1}$, which is an $AR(1)$ process. A third example of a random process is a sinewave with initial phase random and uniformly distributed over $[0, 2\pi)$. The process is given by $x(t) = A.sin(\omega t + \phi)$ where ϕ has the uniform distribution $d\phi/2\pi$ over 2π. The density of $x(t)$ for a given t can be obtained by the formula for transformation of random variables: $x(t)$ assumes values in $[-A, A]$ and for a given value of $x(t)$ say x, ϕ can assume two values in $[0, 2\pi)$. These two values are $\phi_1 = sin^{-1}(x/A) - \omega t mod 2\pi$ and $\phi_2 = \pi - sin^{-1}(x/A) - \omega t mod 2\pi$. We find $\frac{dx}{d\phi} = A.cos(\omega t + \phi)$ and hence $|dx/d\phi| = \sqrt{A^2 - x^2}$. Thus, the density of $x(t)$ is given by $\frac{1}{\pi\sqrt{A^2-x^2}}$. We can also ask what is the joint density of $x(t_1)$ and $x(t_2)$ where t_1, t_2 are two distinct times. For this we shall first evaluate the conditional density of $x(t_2)$ given $x(t_1)$. If $x(t_1) = x_1$ is fixed, ϕ can assume two values $sin^{-1}(x_1/A) - \omega t_1 mod 2\pi$ and $\pi - sin^{-1}(x_1/A) - \omega t_1 mod 2\pi$. From symmetry, the probability with which it assumes any one of these values equals $1/2$. When ϕ assumes the first value, $x(t_2) = A.sin(\omega(t_2 - t_1) + sin^{-1}(x_1/A))$ and when ϕ assumes the second value, $x(t_2) = A.sin(\omega(t_2 - t_1) + \pi - sin^{-1}(x_1/A)) = A.sin(sin^{-1}(x_1/A) - \omega(t_2 - t_1))$. Denoting these values by $\psi(x_1)$ and $\chi(x_1)$, we get the conditional density of $x(t_2)$ given $x(t_1)$ as $\frac{1}{2}(\delta(x_2 - \psi(x_1)) + \delta(x_2 - \chi(x_1)))$. This multiplied by the density of $x(t_1)$ gives us the joint density of $(x(t_1), x(t_2))$.

10.8   Study Projects

Study Project 1

Read up about the axiomatic foundations of probability theory and analyze some simple examples of probability distributions, i.e. in which applications they arise and how one uses these distributions to compute important statistics. Your survey should include

1. Binomial distribution

$$p_r = \binom{n}{r} p^r (1-p)^{n-r}, r = 0, 1, ..., n$$

How it occurs in problems involving sampling with replacement, or in the computation of the probability distribution of the number of times an event occurs in n repetitive independent performances of an experiment. The computation of its mean value

$$\mu = \sum_{r=0}^{n} r p_r = np$$

How the binomial random variable can be expressed as the sum of n independent Bernoulli random variables assuming value one with probability p and zero with probability q, i.e.

$$X = X_1 + ... + X_n$$

By using the result that if $X_1, ..., X_n$ are discrete random variables, then

$$\mathbb{E}(X_1 + ... + X_n) = \mathbb{E}(X_1) + ... + \mathbb{E}(X_n)$$

and the representation of the binomial random variable in terms of Bernoulli variables, derive the formula for the expectation of the binomial random variable. Your presentation should include a proof of the general formula which can proceed along the following lines: If $X_1, ..., X_n$ assume values in the set $E_1 \times E_2 \times ... \times E_n$ where $E_i = \{x_{i1}, ..., x_{i,m_i}\}$ with probabilities

$$P\{X_1 = x_{1,r_1}, ..., X_n = x_{n,r_n}\} = p(r_1, ..., r_n)$$

then

$$\mathbb{E}(X_1 + ... + X_n) = \sum_{r_1,...,r_n} (r_1 + ... + r_n) p(r_1, ..., r_n)$$

$$= \sum r_1 p_1(r_1) + \sum r_2 p_2(r_2) + ... + \sum r_n p_n(r_n)$$

where

$$p_j(r_j) = \sum_{r_1,...,r_{j-1},r_{j+1},...,r_n} p(r_1, ..., r_n)$$

equals the marginal probability distribution of X_j. Your discussion can also include the corresponding result for continuous random variables, namely if $f(x_1, ..., x_n)$ denotes the joint probability density of the variables $(X_1, ..., X_n)$, then the marginal density of X_i is obtained by integrating f over all the $x'_j s$ excluding x_i, i.e.

$$f_i(x_i) = \int_{\mathbb{R}^{n-1}} f(x_1, ..., x_n) dx_1 ... dx_{i-1} dx_{i+1} ... dx_n$$

and hence

$$\mathbb{E}(X_1 + ... + X_n) = \int (x_1 + ... + x_n) f(x_1, ..., x_n) dx_1 ... dx_n$$

$$= \int x_1 f_1(x_1) dx_1 + \int x_2 f_2(x_2) dx_2 + ... + \int x_n f_n(x_n) dx_n = \mathbb{E}(X_1) + ... + \mathbb{E}(X_n)$$

Illustrate the computation of marginal distributions from joint distributions for discrete random variables by means of an example. The example can include the following: Two urns containing respectively a_1 white, b_1 black and a_2 white, b_2 black balls are present. The first urn is selected with a probability of p and the second urn with a probability of $q = 1 - p$. From the selected urn, two balls are drawn successively without replacement. Determine the joint probability distribution of the two balls and the marginal probability distributions. Ans:

$$p(b, w) = p\frac{b_1}{b_1 + a_1}\frac{a_1}{b_1 + a_1 - 1} + q\frac{b_2}{b_2 + a_2}\frac{a_2}{b_2 + a_2 - 1}$$

with similar expressions for $p(b, b), p(w, b), p(w, w)$. From these, evaluate $p_1(b)$, $p_1(w)$, $p_2(b)$, $p_2(w)$ as

$$p_1(b) = p(b, w) + p(b, b) \qquad p_1(w) = p(w, b) + p(w, w)$$
$$p_2(b) = p_2(b, b) + p_2(w, w) \qquad p_2(w) = p(w, b) + p(w, w)$$

2. The geometric distribution: If one tosses a coin successively until a Head is obtained, then count the number of tosses required. Assuming that the coin need not be biased, if p denotes the Heads probability and q the Tails probability, then the probability that n trials are needed for a Head to occur is $p(n) = q^{n-1}p, n = 1, 2, ...$ Illustrate the computation of the mean and variance of this distribution and in general higher moments

$$\mu_m = \sum_{n=1}^{\infty} n^m q^{n-1} p$$

3. For discrete random variables, another important tool exists which facilitates the computation of various statistics like moments. This tool is the moment generating function. If X assumes values $\{0, \pm 1, \pm 2, ...\}$ with probabilities $P(X = n) = p(n)$, then

$$\Gamma_X(z) - \mathbb{E}(z^X) = \sum_{n=-\infty}^{\infty} p(n)z^n$$

The mean for example is computed by differentiation

$$\frac{d}{dz}\mathbb{E}(z^X)|_{z=1} = \mathbb{E}(Xz^{X-1})|_{z=1} = \mathbb{E}(X)$$

Equivalently,

$$\frac{d}{dz}\sum_n p(n)z^n|_{z=1} = \sum_n np(n)z^{n-1}|_{z=1} = \sum_n np(n) = \mathbb{E}(X)$$

Again the second moment is obtained using

$$\frac{d^2}{dz^2}\Gamma_X(z)|_{z=1} = \frac{d^2}{dz^2}\mathbb{E}(z^X)|_{z=1} = \mathbb{E}(X(X-1)z^{X-2})|_{z=1} = \mathbb{E}(X(X-1))$$

so that

$$\mathbb{E}(X^2) = \Gamma''(1) + \Gamma'(1)$$

Likewise the third moment is computed using

$$\Gamma^{(3)}(1) = \mathbb{E}(X(X-1)(X-2)) = \mathbb{E}(X^3) - 3\mathbb{E}(X^2) + 2\mathbb{E}(X)$$

giving

$$\mathbb{E}(X^3) = \Gamma^{(3)}(1) + 3(\Gamma''(1) + \Gamma'(1)) - 2\Gamma'(1) = \Gamma^{(3)}(1) + 3\Gamma^{(2)}(1) + \Gamma'(1)$$

Study project 2

(a) A survey of the maximum likelihood and maximum aposteriori techniques for estimating parameters. Coverage includes : If X is a random variable depending upon a parameter θ, then θ is estimated by the maximum likelihood method by maximizing $p(X|\theta)$, the probability of occurrence of X given the parameter θ. This means that θ is estimated as that value which leads to the most probable value of X. If however, θ is itself a random parameter like the observation X, then one is given the joint probability density $p(X,\theta)$ and the engineer chooses to maximize this with respect to θ. Since $p(X,\theta) = p(\theta|X)p(X)$, this maximization is equivalent to maximizing $p(\theta|X)$, which means that we are interested in determining the most probable value of θ for the given observation X. An example of the maximum likelihood method is the Gaussian model $X = \theta.U + V$, where U, V are Gaussian random variables independent of each other and having means zero and variances σ_U^2, σ_V^2. From the observation X, we want to estimate $\theta, \sigma_U, \sigma_V$. Observe that the probability density of X given $\theta, \sigma_U, \sigma_V$ is given by

$$p(X|\theta, \sigma_U, \sigma_V) = (2\pi(\theta^2\sigma_U^2 + \sigma_V^2))^{-1/2}.exp\left(-\frac{X^2}{2(\theta^2\sigma_U^2 + \sigma_V^2)}\right)$$

and the maximum likelihood estimators of these parameters are determined by maximizing the logarithm of this which is equivalent to minimizing

$$\frac{X^2}{(\theta^2\sigma_U^2 + \sigma_V^2)} + log(\theta^2\sigma_U^2 + \sigma_V^2)$$

Maximization of $p(\theta|X)$ for estimating the parameter θ is called the maximum aposteriori technique. Aposteriori stands for the fact that prior to making the observation X, θ has the density $p(\theta)$ and its maximization will lead to an apriori estimate, i.e. the estimate of the parameter prior to making any observation. However, after making the observation X, θ has the density $p(\theta|X)$ and its maximization will lead to the estimate of the parameter after (aposteriori) making the observation.

(b) Other criteria based on minimizing an appropriate distance measure. Suppose θ is a non-random parameter and X is the observation. Let $\hat{\theta}(X)$ denote the estimate of the parameter based on the observation. Let $d(\theta, \hat{\theta}(X))$ denote a distance measure between the parameter and its estimate. Then, an appropriate criterion is to minimize $\mathbb{E}(d(\theta, \hat{\theta}(X))$ with respect to the parameter θ to derive its estimate. For example, if $d(\theta, \hat{\theta}) = (\theta - \hat{\theta})^2$, then we can easily show that $\hat{\theta}(X) = \mathbb{E}(\theta|X)$ is the conditional expectation of the parameter given the observation X. This can be proved via a variational scheme. Other distance measures like the L^p measure can also be adopted. This amounts to minimizing $\mathbb{E}|\hat{\theta} - \theta|^p$. Here, $p \geq 1$. However for p other than 2, the resulting optimizing equations are not linear.

Study project 3

Identification of minimum phase systems based on power spectral analysis. If $\{h(n)\}$ is the impulse response of a linear time invariant system with frequency response $H(\omega) = \sum_{n=-\infty}^{\infty} h(n)^*$ $exp(-j\omega n)$ and if we input to this system a stationary random process $\{x(n)\}$ having autocorrelation function $R_{xx}(k) = \mathbb{E}(x(n)x(n-k))$ and power spectral density $S_{xx}(\omega) = \sum_{k=-\infty}^{\infty} R_{xx}(k)$ $exp(-j\omega k)$, then it is shown via a straightforward computation (this computation should be illustrated by the student in a seminar either using transparencies or on the blackboard) that the autocorrelation function of the output $y(n) = \sum_{k=-\infty}^{\infty} h(k)x(n-k)$ is given by

$$R_{yy}(m) = h(m) * h(-m) * R_{xx}(m) = \sum_{\alpha,\beta=-\infty}^{\infty} h(\alpha)h(\beta)R_{xx}(m - \alpha + \beta)$$

and the corresponding power spectral density of the output is given by

$$S_{yy}(\omega) = \sum_{m=-\infty}^{\infty} R_{yy}(m)exp(-j\omega m) = |H(\omega)|^2 S_{xx}(\omega)$$

This means that the magnitude of the transfer function can be estimated by measurement of the input and output power spectral densities according to the formula

$$|H(\omega)| = \sqrt{\frac{S_{yy}(\omega)}{S_{xx}(\omega)}}$$

If it is known that the system is minimum phase, then the transfer function can be completely determined from knowledge of the magnitude of the transfer function. The idea in brief should be mentioned. This involves spectral factorization, i.e. factorization of the following function of a complex variable

$$F(z) = \frac{S_{yy}(z)}{S_{xx}(z)}$$

into its minimum and maximum phase factors and retain only the minimum phase factor. The problem of identifying an FIR system can also be presented here. The discussion should involve the following idea. Let $\{h(n) : n = 0, 1, ..., p\}$ be an FIR filter and let $x(n)$ be a white Gaussian input to this system having zero mean and unit variance, i.e. $\mathbb{E}(x(n)) = 0, \mathbb{E}(x(n)x(n-m)) = \delta(m)$. Then, the autocorrelation function of the output is given by $R_{yy}(m) = \sum_k h(k)h(k+m)$. One can attempt therefore to identify the coefficients $\{h(m)\}$ of the FIR filter by minimizing the quantity $\sum_m (\hat{R}_{yy}(m) - \sum_k h(k)h(k+m))^2$ with respect to the $h(k)'s$. Here, $\hat{R}_{yy}(m)$ is a finite data estimate of the autocorrelation function of the output. It may for example be taken to be $\hat{R}_{yy}(m) = \frac{1}{N} \sum_{n=0}^{N-1} y(n)y(n+m)$

Study project 4

Give simple examples of the computation of the probability density for functions of random variables. The examples can include (1) $Z = X.Y$ where X, Y are independent random variables having densities concentrated over $(0, \infty)$, (2) $Z = X/Y$ with again X, Y being independent random variables with densities concentrated over $(0, \infty)$, (3) $Z = X_1....X_n$ with $X_1, ..., X_n$ being independent random variables with densities concentrated over $(0, \infty)$. Also evaluate the corresponding densities when the component random variables are not independent and their joint densities are specified. In the first case, if f_X denotes the density of X and f_Y the density of Y, then

$$P(Z \leq z) = P(XY \leq z) = \int_0^\infty P(Y \leq z/x) f_X(x) dx = \int_0^\infty F_Y(z/x) f_X(x) dx$$

and differentiating this with respect to z gives

$$f_Z(z) = \int_0^\infty \frac{1}{x} f_Y(z/x) f_X(x) dx$$

In case, X, Y are not independent but their joint density $f(x, y)$ is specified, we have

$$P(Z \leq z) = \int_0^\infty P(Y \leq z/x | X = x) f_X(x) dx = \int_0^\infty F_{Y|X}(z/x | x) f_X(x) dx$$

and on differentiation, we find

$$f_Z(z) = \int_0^\infty \frac{1}{x} f_{Y|X}(z/x | x) f_X(x) dx = \int_0^\infty \frac{1}{x} f(z/x, x) dx$$

This can be confirmed by a computer simulation. First simulate N samples of the pair of random variables (X, Y) using a Monte Carlo scheme. For each sample, compute $Z = XY$ and obtain the empirical distribution of Z via a histogram plot. Compare with the theoretically predicted density of Z. The second case is treated similarly:

$$P(Z \leq z) = P(X/Y \leq z) = \int_0^\infty P(X \leq yz | Y = y) f_Y(y) dy =$$

$$\int_0^\infty F_{X|Y}(yz | y) f_Y(y) dy$$

and on differentiation, we find that

$$f_Z(z) = \int_0^\infty y f_{X|Y}(yz|y) f_Y(y) dy = \int_0^\infty y f(yz, y) dy$$

Finally the case $Z = X_1 ... X_n$. Assume that the $X_i's$ are independent with density $f_i(x)$ concentrated over $(0, \infty)$. Then,

$$P(Z \le z) =$$

$$\int P(X_n \le z/x_1 x_2 ... x_{n-1} | X_1 = x_1, ..., X_{n-1} = x_{n-1}) g(x_1, ..., x_{n-1}) dx_1 ... dx_{n-1}$$

where g is the joint density of $X_1, ..., X_{n-1}$. This is the same as

$$\int F_{X_n|X_1,...,X_{n-1}}(z/x_1 ... x_{n-1} | x_1, ..., x_{n-1}) g(x_1, ..., x_{n-1}) dx_1 ... dx_{n-1}$$

Differentiating this with respect to z gives us

$$f_Z(z) = \int \frac{1}{x_1 ... x_{n-1}} f_{X_n|X_1 ... X_{n-1}}(z/x_1 ... x_{n-1} | x_1, ..., x_{n-1}) g(x_1, ..., x_{n-1}) dx_1 ... dx_{n-1}$$

$$= \int \frac{1}{x_1 ... x_{n-1}} f(z/x_1 ... x_{n-1}, x_1, ..., x_{n-1}) dx_1 ... dx_{n-1}$$

The problem of simulating samples of a collection of random variables $(X_1, ..., X_n)$ having a prescribed joint distribution needs to be looked into carefully, For example if (X, Y) has a joint distribution, then the random variables $F_{X|Y}(X|Y)$ and $F_Y(Y)$ are independent and uniformly distributed over $[0, 1]$. For

$$P(F_{X|Y}(X|Y) \le u, F_Y(Y) \le v) = P(X \le F_{X|Y}^{-1}(u|Y), Y \le F_Y^{-1}(v))$$

$$- \int_{-\infty}^{F_Y^{-1}(v)} P(X \le F_{X|Y}^{-1}(u|Y')|Y' = y) dF_Y(y) = u \int_{-\infty}^{F_Y^{-1}(v)} dF_Y(y) = uv$$

proving the claim. Thus, to simulate samples of the pair (X, Y), we first simulate N samples of the pair (U, V) and then compute $Y = F_Y^{-1}(V), X = F_{X|Y}^{-1}(U|Y)$.

Study project 5

A seminar on the characteristic function and its characteristic properties and applications to problems of signal analysis can be given. The seminar should commence with the fact that the characteristic function is defined as the Fourier transform of the probability density function. If X is a random variable having distribution F and density f, its characteristic function is defined as

$$\Phi(\omega) = \mathbb{E} exp(i\omega X) = \int_{-\infty}^\infty exp(i\omega x) dF(x) = \int_{-\infty}^\infty exp(i\omega x) f(x) dx$$

The basic inequality

$$|\Phi(\omega)| = |\mathbb{E}exp(i\omega X)| \le \mathbb{E}|exp(i\omega X)| = 1$$

should be mentioned to start with. Then, for any complex scalars $c_1, ..., c_n$ and real numbers $t_1, ..., t_n$

$$\sum_{i,j=1}^{n} c_i\bar{c}_j\Phi(t_i - t_j) = |\mathbb{E}\sum_j c_j exp(it_j X)|^2 \ge 0$$

should be mentioned. This is called the positive definite property of the characteristic function. Bochner's theorem should be mentioned, namely that if $\Phi(\omega)$ is any positive definite function that evaluates to unity at the origin, then Φ is the characteristic function of a random variable, i.e. the Fourier transform of a probability density function. After that the talk should mention how the moments of a random variable are computed using the characteristic function:

$$\mathbb{E}(X^n) = i^{-n}\frac{d^n}{d\omega^n}\Phi(\omega)|_{\omega=0}$$

The Taylor expansion

$$\Phi(\omega) = \sum_{n=0}^{\infty} \frac{(i\omega)^n}{n!}\mathbb{E}(X^n)$$

should be mentioned. Finally, the computation of the characteristic function of a number of random variables should be illustrated. For example, if X is binomial with parameters n, p,

$$\Phi(\omega) = \sum_{r=0}^{n} exp(ir\omega)\binom{n}{r}p^r q^{n-r} = (p.exp(i\omega) + q)^n$$

If X is Bernoulli assuming values $1, 0$ with probabilities p, q, then

$$\Phi(\omega) = p.exp(i\omega) + q$$

The multiplication theorem for characteristic functions should be mentioned: If $X_1, ..., X_n$ are independent random variables having characteristic functions $\Phi_1(\omega), ..., \Phi_n(\omega)$, then the characteristic function of the sum $X_1 + ... + X_n$ is given by

$$\Phi(\omega) = \mathbb{E}exp(i\omega(X_1 + ... + X_n)) = \Pi_{k=1}^{n}\mathbb{E}exp(i\omega X_k) = \Pi_{k=1}^{n}\Phi_k(\omega)$$

This fact should also be proved using the convolution theorem for Fourier transforms: If X_i has density $f_i(x), i = 1, 2, ..., n$ and the $X_i's$ are independent, then the density of the sum $X_1 + ... + X_n$ can be computed as follows

$$P(X_1 + ... + X_n \le x) = \int P(X_n \le x - x_1 - ... - x_{n-1})f_1(x_1)...f_{n-1}(x_{n-1})dx_1...dx_{n-1}$$

$$= \int F_n(x - x_1 - ... - x_{n-1})f_1(x_1)...f_{n-1}(x_{n-1})dx_1...dx_{n-1}$$

and hence on differentiation, the density of $X_1 + ... + X_n$ turns out to be

$$f(x) = \int f_n(x - x_1 - ... - x_{n-1}) f_1(x_1)...f_{n-1}(x_{n-1}) dx_1...dx_{n-1} = f_1 * ... * f_n(x)$$

with star denoting convolution. The convolution theorem states that the Fourier transform of the convolution of functions equals the product of the Fourier transforms of each of the functions. This gives

$$\Phi(\omega) = \int_{-\infty}^{\infty} f(x) exp(i\omega x) dx = \Pi_{i=1}^{n} \int_{-\infty}^{\infty} f_i(x) exp(i\omega x) dx$$

$$= \Pi_{i=1}^{n} \Phi_i(\omega)$$

This theorem should be illustrated using the Bernoulli and Binomial distributions. If $X_1, ..., X_n$ are independent Bernoulli random variables each assuming values 1 or 0 with probabilities p, q, then $X_1 + ... + X_n$ is Binomial with parameters n, p. The multiplication theorem therefore gives

$$\Phi(\omega) = (p.exp(i\omega) + q)^n$$

Unsolved problem 1

Write down the Fokker-Planck equation for the motion of a particle in a central potential when there is a random force perturbing the motion that can be modeled using white noise. Explain how by linearization, one can obtain a solution to the original stochastic differential equation.

Unsolved problem 2 with hint

Consider the diffusion process $\mathbf{x}(t)$ satisfying the following stochastic differential equation

$$dx_i(t) = f_i(\mathbf{x}(t), t) dt + \sum_{j=1}^{n} g_{ij}(\mathbf{x}(t), t) dB_j(t)$$

Using the Ito rule $dB_i dB_j = dt \delta_{ij}$,derive the following equation for the differential of a function $\phi(\mathbf{x}(t), t)$:

$$d\phi(\mathbf{x}(t), t) = \sum_i \frac{\partial \phi(\mathbf{x}(t), t)}{\partial x_i} dx_i(t) + \frac{1}{2} \sum_{i,j=1}^{n} \frac{\partial^2 \phi(\mathbf{x}(t), t)}{\partial x_i \partial x_j} dx_i(t) dx_j(t)$$

$$= \frac{\partial \phi(\mathbf{x}(t), t)}{\partial t} dt + \sum_i \frac{\partial \phi(\mathbf{x}(t), t)}{\partial x_i} (f_i(\mathbf{x}(t), t) dt + \sum_j g_{ij}(\mathbf{x}(t), t) dB_j(t))$$

$$+ \frac{1}{2} \sum_{i,j} \frac{\partial^2 \phi(\mathbf{x}(t), t)}{\partial x_i \partial x_j} g_{ik}(\mathbf{x}(t)) g_{jk}(\mathbf{x}(t), t) dt$$

Show that this equation can be cast in vectorial form

$$d\phi(\mathbf{x}(t), t) = \frac{\partial \phi}{\partial t}dt + (\nabla\phi)^T \mathbf{f}dt + (\nabla\phi)^T \mathbf{G}d\mathbf{B}(t) + \frac{1}{2}Tr(\mathbf{G}\mathbf{G}^T\nabla\nabla^T\phi)dt$$

Note that the original stochastic differential equation can be put in the following vector form

$$d\mathbf{x}(t) = \mathbf{f}(\mathbf{x}(t), t)dt + \mathbf{G}(\mathbf{x}(t), t)d\mathbf{B}(t)$$

Assuming that $\phi(\mathbf{x})$ does not depend explicitly on time, deduce that

$$\frac{d}{dt}\mathbb{E}\phi(\mathbf{x}(t)) = \mathbb{E}\mathbf{f}^T\nabla\phi + \frac{1}{2}Tr(\mathbb{E}(\mathbf{G}\mathbf{G}^T\nabla\nabla^T\phi))$$

Show that this implies

$$\frac{d}{dt}\int \phi(\mathbf{x}p(t, \mathbf{x})d\mathbf{x} = \int \mathbf{f}(\mathbf{x}, t)^T\nabla\phi(\mathbf{x})p(t, \mathbf{x})d\mathbf{x} + \frac{1}{2}Tr\int p(t, \mathbf{x})\mathbf{G}\mathbf{G}^T(\mathbf{x}, t)\nabla\nabla^T\phi(\mathbf{x})d\mathbf{x}$$

where $p(t, \mathbf{x})$ equals the probability density of $\mathbf{x}(t)$. Using integration by parts with vanishing boundary conditions, deduce that

$$\int \phi(\mathbf{x})(\frac{\partial p}{\partial t} + \nabla^T(fp) - \frac{1}{2}Tr(\nabla\nabla^T(\mathbf{G}\mathbf{G}^T p)))d\mathbf{x} = 0$$

and hence the validity of the Fokker-Planck equation

$$\frac{\partial p(t, \mathbf{x})}{\partial t} + \nabla^T(f(\mathbf{x}, t)p(t, \mathbf{x})) - \frac{1}{2}Tr(\nabla\nabla^T(\mathbf{G}\mathbf{G}^T(\mathbf{x}, t)p(t, \mathbf{x}))) = 0$$

Unsolved problem 3 with hint

Consider the process $\mathbf{x}(t)$ as the solution to the stochastic differential equation

$$d\mathbf{x}(t) = \mathbf{f}(\mathbf{x}(t), t)dt + \mathbf{G}(\mathbf{x}(t), t)d\mathbf{B}(t)$$

Define the conditional moment generating function of the process

$$M(\mathbf{s}, t, \mathbf{x}(t_0), t_0) = \mathbb{E}exp(\mathbf{s}^T(\mathbf{x}(t) - \mathbf{x}(t_0)))|\mathbf{x}(t_0))$$

Study project 6

The set theoretical preliminaries required for the formulation of problems of probability theory is an important topic for seminar discussion. This discussion can begin with what one understands by a set, i.e. how a set of elements is defined in terms of properties or by direct enumeration. Examples of both kinds of specification may be given here. For example, to specify the set of numbers from one to hundred, one may either directly write down the elements of the set as $\{1, 2, 3, ..., 100\}$ where the dots are filled with numbers or in a more compact notation as $\{n \in \mathbb{Z} : 1 \leq n \leq 100\}$. The latter scheme of specification is in terms of a defining property, i.e. all integers sharing the property of being greater than or equal to one and less than or equal to one hundred. Likewise the set of positive integer pairs that add up to six can either be specified by direct enumeration as $\{(1, 5), (2, 4), (3, 3), (4, 2), (5, 1)\}$ or in terms of the defining property $\{(n, m) : n \in \mathbb{Z}, n, m \geq 1, n + m = 6\}$. The seminar should then introduce the basic Boolean operations of set theory, i.e. union, intersection and complementation. The expressions for these operations should be given in the form of properties and illustrated by direct listing for finite sets. For example, if E, F are two sets, then

$$E \cup F = \{x : x \in E \, or \, x \in F\}$$

More generally, for any finite collection of sets $E_1, ..., E_n$, the equation

$$E_1 \cup E_2 \cup ... \cup E_n = \{x : x \in E_1 \, or \, x \in E_2 \, or ... or \, x \in E_n\}$$

This can be expressed in the more compact language

$$\cup_{i=1}^n E_i = \{x : x \in E_i, for \, some \, i = 1, 2, ..., n\}$$

One can extend this to arbitrary countable collections of set $E_1, E_2, ...$:

$$\cup_{i=1}^\infty E_i = \{x : x \in E_i \, for \, some \, i = 1, 2, ...\}$$

and more generally, even to an uncountable family of sets $E_\alpha, \alpha \in \Gamma$

$$\cup_{\alpha \in \Gamma} E_\alpha = \{x : x \in E_\alpha \, for \, some \, \alpha \in \Gamma\}$$

The discussion can illustrate with numerical examples such as (a) the union of $\{1, 2, 3\}$ and $\{2, 3, 4\}$ equals $\{1, 2, 3, 4\}$, the union of the intervals $(1, 3]$ and $(2, 4]$ equals $(1, 4]$, (c) the union of the countable collection of intervals $(1, 2 - 1/n], n = 2, 3, ...$, equals $(1, 2)$. The fact that the union of sets is independent of the order in which the sets are taken should be mentioned in this discussion. Likewise the intersection of a family of sets $E_\alpha, \alpha \in \Gamma$ is defined as

$$\cap_{\alpha \in \Gamma} E_\alpha = \{x : x \in E_\alpha \forall \alpha \in \Gamma\}$$

and the complement of a subset E of the set Ω is

$$E^c = \{x \in \Omega : x \notin E\}$$

The Demorgan's rules

$$(\cup E_\alpha)^c = \cap E_\alpha^c, \qquad (\cap E_\alpha)^c = \cup E_\alpha^c$$

$$E \cap \cup F_\alpha = \cup E \cap F_\alpha \qquad E \cup \cap F_\alpha = \cup E \cap F_\alpha$$

are to be derived via set theoretic arguments. The notions of union, intersection and complementation are to be illustrated using Venn diagrams. Finally, the notion of symmetric difference of sets is to be illustrated via operations and also diagrammatically: $A \Delta B = AB^c \cup BA^c$ with the statement that it consists of those elements that belong to A but not to B or those that belong to B but not to A.

Study project 7

The rigorous construction of the conditional expectation operator using measure theoretic arguments can be given to enable the engineer get familiar with rigorous mathematical arguments. The seminar can begin with the notion of a probability space $(\Omega, \mathcal{F}, P)$ and the Hilbert space $\mathcal{H} = L^2(P) = L^2(\Omega, \mathcal{F}, P)$ as consisting of all random variables X defined on this probability space with finite mean square value, i.e. $\mathbb{E}(X^2) = \int X^2 dP < \infty$. The proof of the Radon-Nikodym theorem is to be covered first of all according to which if μ is another measure on the measurable space $(\Omega, \mathcal{F})$ such that whenever $P(E) = 0$, then $\mu(E) = 0$, which is usually stated as μ is absolutely continuous with respect to P and the notation $\mu << P$ reserved for this, then there exists a non-negative measurable function f on the probability space such that for every $E \in \mathcal{F}$, one has $\mu(E) = \int_E f dP$. The function f is uniquely determined upto almost sure P equivalence, i.e. if g is another measurable function with $\mu(E) = \int_E g dP$ for all $E \in \mathcal{F}$, then $g = f a.s. P$, i.e. $P\{g \neq f\} = 0$. Here, the meaning of a property holding true for almost all $\omega \in \Omega$ with respect to a measure can be introduced. If $(\Omega, \mathcal{F})$ is a measure space and ν a σ finite measure on this space, then a measurable function f on this space is said to be equal to zero a.sν when $\nu\{E \cap \{f \neq 0\}\} = 0$ for every event E with $\nu(E) < \infty$. If ν is a finite measure, i.e. $\nu(\Omega) < \infty$, then $f = 0$ holds good a.s.P when $\nu(\{f \neq 0\}) = 0$. If ν is a probability measure, then this is equivalent to $\nu(\{f = 0\}) = 1$. The notion of a σ-finite measure is to be given here with the example of the Lebesgue measure on $(\mathbb{R}, \mathbb{B}(\mathbb{R}))$. How the Radon-Nikodym theorem can be exploited to construct the conditional expectation of an integrable random variable given a sub σ algebra is to be dealt with next along the following lines. Suppose $\mathcal{G}$ is a sub σ algebra of $\mathcal{F}$. This means that $\mathcal{G}$ is a σ algebra and every element of $\mathcal{G}$ is an element of $\mathcal{F}$. Then one can consider another probability space $(\Omega, \mathcal{G}, P)$ and the associated Hilbert space $\mathcal{H}' = L^2(\Omega, \mathcal{G}, P)$. $\mathcal{H}'$ consists of all square integrable random variables X that are measurable with respect to the sub σ algebra $\mathcal{G}$, i.e. $X^{-1}(B) \in \mathcal{G}$, for all $B \in \mathbb{B}(\mathbb{R})$. Note that $\mathcal{H}'$ is a Hilbert subspace of $\mathcal{H}$. Now take an $X \in \mathcal{H}$ and define the measures ν_+ and ν_- on $(\Omega, \mathcal{G})$ by

$$\nu_+(E) = \int_E X_+ dP \qquad \nu_-(E) = \int_E X_- dP, \qquad E \in \mathcal{G}$$

where $X = X_+ - X_-$ is the standard decomposition of X into its positive and negative parts. The precise notation for these derived variables is $X_+ = X I_{X \geq 0}, X_- = -X_- I_{X < 0}$. Then, both ν_+ and ν_- are measures on $(\Omega, \mathcal{G})$ that are absolutely continuous with respect to the probability measure P. This is seen via the following basic result of integration: If X is an integrable random variable and $P(E) = 0$, then $\int_E X dP = 0$. The theorem is proved first for simple random variables, then for non-negative random variables by representing these as monotone limits of simple random variables and finally for arbitrary integrable random variables by decomposing the variable into the difference of a non-negative variable and a negative variable. To be precise, ν_+ and ν_- are absolutely continuous with respect to the restriction of P to the sub σ algebra $\mathcal{G}$. Thus, by the Radon-Nikodym theorem, one can construct $\mathcal{G}$ measurable non-negative functions f_+, f_- such that

$$\nu_+(E) = \int_E f_+ dP \qquad \nu_-(E) = \int_E f_- dP, \qquad E \in \mathcal{G}$$

and thus

$$\int_E X dP = \nu_+(E) - \nu_-(E) = \int_E f dP, \qquad E \in \mathcal{G}$$

where $f = f_+ - f_-$. f is $\mathcal{G}$-measurable and is denoted by $\mathbb{E}(X|\mathcal{G})$ read as the conditional expectation of X given the sub σ-algebra $\mathcal{G}$. The seminar should stress that the conditional expectation operation is in fact an orthogonal projection of the Hilbert space $L^2(\Omega, \mathcal{F}, P)$ onto the subspace $L^2(\Omega, \mathcal{G}, P)$. and the properties of the orthogonal projection may be explained here.

Solved problem 8

Let $W_1, ..., W_n$ be independent random variables with characteristic functions $\psi_i(t), i = 1, 2, ..., n$. Let $X_i = W_1 + ... + W_i, i = 1, 2, ..., n$. Determine the joint characteristic function of $X_1, ..., X_n$. First observe that

$$t_1 X_1 + ... + t_n X_n = t_1 W_1 + t_2(W_1 + W_2) + ... + t_n(W_1 + ... + W_n)$$

$$= (t_1 + ... + t_n)W_1 + (t_2 + ... + t_n)W_2 + ... + t_n W_n$$

Thus,

$$\mathbb{E} exp(i(t_1 X_1 + ... + t_n X_n)) = \psi_1(t_1 + ... + t_n)\psi_2(t_2 + ... + t_n)...\psi_{n-1}(t_{n-1} + t_n)\psi_n(t_n)$$

Solved problem 9

Let $B(t)$ be standard Brownian motion and $L_1(x, t), L_2(x, t)$ two twice continuously differentiable functions. Evaluate using Ito's formula the stochastic differential $d(L_1(B(t), t)L_2(B(t), t))$.

Ans:

$$d(L_1 L_2) = (L_1 + dL_1)(L_2 + dL_2) - L_1 L_2 = L_1 dL_2 + L_2 dL_1 + dL_1 dL_2$$

where

$$dL_1 = \frac{\partial L_1}{\partial t} dt + \frac{\partial L_1}{\partial x} dB(t) + \frac{1}{2}\frac{\partial^2 L_1}{\partial x^2} dt$$

$$= dt(\frac{\partial L_1}{\partial t} + \frac{1}{2}\frac{\partial^2 L_1}{\partial x^2}) + \frac{\partial L_1}{\partial x}dB$$

Similarly for dL_2. Thus,

$$dL_1 dL_2 = \frac{\partial L_1}{\partial x}\frac{\partial L_2}{\partial x}dt$$

where Ito's formula has been used: $dB(t)^2 = dt, dB(t)dt = 0, dt.dt = 0$. Note that $dB(t)$ has order $\sqrt{dt}$ and hence, $dB(t)dt$ has order $dt^{3/2}$ and hence is negligible in comparison with dt. Similarly $dtdt = dt^2$ is negligible in comparison with dt. Note the notation used: By dL, we mean $dL(B(t), t)$.

Solved problem 10

Evaluate the differential of the function $X(t) = exp(aB(t) - a^2t/2)$.

Ans: Let $\xi(t) = aB(t) - a^2t/2$. Then, Ito's formula gives

$$dX(t) = exp(\xi(t))d\xi(t) + \frac{1}{2}exp(\xi(t))(d\xi(t))^2$$

Now,

$$d\xi(t) = adB(t) - a^2dt/2$$

$$(d\xi(t))^2 = a^2dt$$

since $dB(t)^2 = dt, dB(t)dt = 0, dt^2 = 0$. Thus,

$$dX(t) = exp(\xi(t))dB(t) = X(t)dB(t)$$

Study project 11

The problem of detecting signals in white Gaussian noise will constitute an interesting seminar cum mini-project. One such problem is when the signal being transmitted has a random parameter and averaging with respect to this parameter is to be carried out. The hypothesis testing problem is then

$$H_1 : x(t) = s(t,\theta) + w(t), t \in [0,T]$$

$$H_0 : x(t) = w(t), t \in [0,T]$$

This amounts to transmitting the signal $s(t,\theta)$ when hypothesis H_1 is true and no signal when H_0 is true. Here, $\{w(t)\}$ equals white Gaussian noise with mean zero and spectral density $N_0/2$. θ is a random parameter having density $p(\theta)$. The likelihood ratio for this problem is given by

$$\Lambda(x) = \frac{\int p(\theta)d\theta exp(\frac{1}{N_0}\int_0^T (x(t) - s(t,\theta))^2 dt)}{exp(\frac{1}{N_0}\int_0^T x(t)^2 dt)}$$

Assuming equal apriori probabilities, the test amounts to deciding H_1 if $\Lambda(x) \geq 1$ and H_0 if $\Lambda(x) < 1$. Noting that

$$\Lambda(x) = \int p(\theta)d\theta.exp(-\frac{2}{N_0}\int_0^T x(t)s(t,\theta)dt + \frac{1}{N_0}\int_0^T s^2(t,\theta)dt)$$

we can compute for different $p(\theta)'s$ and signal waveforms $s(t,\theta)'s$, the error probability and implement the receiver. The implementation of this receiver using Matlab would proceed along the following lines. We discretize the signal duration $[0,T]$ into N equal parts of length Δ each. We also discretize the parameter space $\theta \in [a,b]$ into M equal parts of length δ each. Then, the likelihood ratio for the test can be numerically computed based on the discrete observations $\{s(k\Delta) : k = 0, 1, ..., N-1\}$ as

$$\Lambda_d(x) = \sum_{m=0}^{M-1} p(k\delta)\delta.exp(\frac{\Delta}{N_0}\sum_{k=0}^{N-1} s(k\Delta,,m\delta)^2 - \frac{2\Delta}{N_0}\sum_{k=0}^{N-1} x(k\Delta)s(k\Delta,m\delta))$$

Study project 12

One seminar on the estimation of the amplitude of a signal using detection methods is of interest especially from the view point of signal processing problems for gravitational waves. In such problems, the signal shape is known but the amplitude strength is unknown and the signal is corrupted by white Gaussian noise. The measurement sequence is given by

$$x_j = \epsilon.s_j + n_j, j = 1, 2, ..., N$$

where ϵ is an unknown parameter and $\{n_1, ..., n_N\}$ are independent Gaussian random variables with mean zero and variance σ^2. The signal shape $\{s_1, ..., s_N\}$ is known. Another problem involves estimating the shape of the signal, i.e. the samples $\{s_1, ..., s_N\}$ when the probability density of the amplitude $p(\epsilon)$ is known. The first problem would estimate the amplitude parameter ϵ by maximizing the probability density of the observed signal samples $\{x_1, ..., x_N\}$ given the parameter ϵ, i.e. maximize

$$p(x_1, ..., x_N|\epsilon) = (2\pi\sigma^2)^{-N/2}exp(-\sum_{j=1}^{N}(x_j - \epsilon.s_j)^2/2\sigma^2))$$

Equivalently, ϵ is determined by minimizing the quantity

$$F(\epsilon) = \sum_{j=1}^{N}(x_j - \epsilon.s_j)^2$$

The optimal equations $F'(\epsilon) = 0$ give for the parameter estimate

$$\hat{\epsilon} = \frac{\sum x_j s_j}{\sum s_j^2}$$

Study project 13

A seminar can be given on basic definitions of cumulants and their characteristic properties that help one to design signal processing algorithms. The talk can begin with defining the joint moment generating function of n random variables $\{X_1, ..., X_n\}$:

$$\psi(t_1, ..., t_n) = \mathbb{E}\{exp(t_1 X_1 + ... + t_n X_n)\}$$

and deriving the Taylor expansion in terms of the moments of the collection

$$\psi(t_1, ..., t_n) = \sum_{m_1,...,m_n=0}^{\infty} \mathbb{E}(X_1^{m_1}...X_n^{m_n}) t_1^{m_1}...t_n^{m_n}/m_1!...m_n!$$

which is equivalent to the expressing the moments of the sequence in terms of the generating function as

$$\mathbb{E}(X_1^{m_1}...X_n^{m_n}) = \frac{\partial^{m_1+...+m_n}}{\partial t_1^{m_1}...\partial t_n^{m_n}} \psi(t_1, ..., t_n)|_{t_1=...=t_n=0}$$

Study project 14

The problem of estimating the amplitudes and frequencies of a harmonic signal corrupted by white Gaussian noise can be addressed. This is of importance in the problem of gravitational wave detection. The received signal has the form

$$x(t) = \sum_{k=1}^{p} \alpha_k cos(\omega_k t + \phi_k) + w(t), t \in [0, T]$$

One assumes that the noise $\{w(t)\}$ is white Gaussian and that the phases $\{\phi_k\}$ are independent and uniformly distributed over $[0, 2\pi)$. The likelihood function to be maximized is then given by

$$p(\mathbf{x}|\alpha_1, ..., \alpha_p, \omega_1, ..., \omega_p)$$

$$= (2\pi)^{-p} \int_{[0,2\pi)^p} exp(-\frac{1}{N_0} \int_0^T (x(t) - \sum \alpha_k cos(\omega_k t + \phi_k))^2 dt) d\phi_1...d\phi_p$$

where T, the duration of observation is much larger than $2\pi/\omega_k, k = 1, 2, ..., p$. This is approximately equivalent to maximizing

$$(2\pi)^{-p} \int_{[0,2\pi)^p} exp(\frac{2}{N_0} \sum_k \alpha_k \int_0^T x(t) cos(\omega_k t + \phi_k) dt - \frac{T}{2N_0} \sum_k \alpha_k^2) d\phi_1, ...d\phi_p$$

We define the random variables

$$L_{c,k} = \frac{2}{N_0} \int_0^T x(t)cos(\omega_k t)dt, L_{s,k} = \frac{2}{N_0} \int_0^T x(t)sin(\omega_k t)dt$$

Then the problem amounts to maximizing

$$\int_0^{2\pi} ... \int_0^{2\pi} d\phi_1...d\phi_p exp(\sum_k \alpha_k(L_{c,k}cos(\phi_k) - L_{s,k}sin(\phi_k)) - \frac{T}{2N_0}\alpha_k^2)$$

which in turn is equivalent to maximizing

$$\Pi_k exp(-\frac{T}{2N_0}\alpha_k^2) \int_0^{2\pi} exp(\alpha_k \sqrt{L_{c,k}^2 + L_{s,k}^2})cos(\phi))d\phi$$

Study project 15

A seminar on the notion of integral in measure theoretic probability theory can be delivered for the benefit of research workers in signal processing. Measure theory is seldom taught in courses on signal analysis, the reason being that engineers think it too abstract a topic and not relevant to making computations. However, it has been recently felt that a good grinding in measure theory would enable the signal processing engineer to understand the algorithms better and perhaps even come up with newer techniques. Of course abstract concepts in measure theory like Caratheodory's theorem on extension of measures apparently do not find too many practical applications but sometimes we have probabilities defined only on a field of events and to make computations, we need to extend this measure to the σ-field generated by the field. Proper identification of the field and the σ field in the given engineering problem is therefore important even in applications. The discussion of the integral can start with the idea of an outer measure given a field of subsets $\mathcal{F}$ on a sample space Ω along with a measure μ defined on the field. What one means by a measure on a field should also be explained in detail. If $E_n, n = 1, 2, ...$ are pairwise disjoint events in the field such that their union is also a member of the field, then μ to fulfill the requirement of being a measure on this field must satisfy the axiom of countable additivity: $\mu(\cup E_n) = \sum_n \mu(E_n)$. Note that unlike the definition of a measure on a σ field where the countable union of the events is automatically an element of the σ field, in this case, we require the condition that $\cup E_n$ belong to the field. A field is not generally closed under countable union. This fact can be stressed in the seminar. Then, for any $A \subset \Omega$, we define $\mu^*(A)$ to be the infimum of $\sum_n \mu(A_n)$ over all $A'_n s$ belonging to the field covering A, i.e. $A \subset \cup A_n$. The proof that μ^* when restricted to the σ field generated by the field is a measure is then to be discussed. Integration with respect to a measure is finally elaborated. The partition method can be used here ,i.e. for non-negative measurable functions f one defines for each finite partition $P = \{A_1, ..., A_n\}$ of the sample space

$$L(f, P) = \sum_i inf\{f(\omega) : \omega \in A_i\}\mu(A_i)$$

and $\int f\mu$ as the superemum of $L(f, P)$ over all partitions P.

Study project 16

A discussion on the multivariate normal distribution, highlighting how one arrives at the density function starting from the definition of a joint normal collection would be of interest to the student of signal processing. Many problems involve generating normal random processes with a given mean and autocorrelation function. The histogram is used to verify whether or not the generated random process has the prescribed density. So one needs to derive the expression for the density. If $X_1, ..., X_n$ are jointly normal random variables, this means that for every n-tuple $(t_1, ..., t_n)$ of real numbers, $\sum_{i=1}^{n} t_i X_i$ has the normal distribution. Hence

$$\mathbb{E}exp(\sum t_i X_i) = exp(\mathbb{E}\sum(t_i X_i) + \frac{1}{2}Var(\sum(t_i X_i)) =$$

$$exp(\sum t_i \mu_i + \frac{1}{2}\sum t_i t_j C_{ij})$$

where $C_{ij} = Cov(X_i, X_j)$. The joint characteristic function of $(X_1, ..., X_n)$ is obtained by replacing t_α by it_α:

$$\mathbb{E}exp(i\sum t_\alpha X_\alpha) = exp(i\sum_\alpha t_\alpha \mu_\alpha - \frac{1}{2}\sum_{\alpha\beta} t_\alpha t_\beta C_{\alpha\beta})$$

Inversion of this Fourier transform yields the joint density of the $X_j's$:

$$f\mathbf{x}) = (2\pi)^{-n} \int exp(i\mu^T \mathbf{t} - \frac{1}{2}\mathbf{t}^T \mathbf{C}\mathbf{t})exp(-i\mathbf{t}^T \mathbf{x})dt$$

$$= (2\pi)^{-n/2}det(\mathbf{C})^{-1/2}exp(-\frac{1}{2}(\mathbf{x} - \mu)^T \mathbf{C}^{-1}(\mathbf{x} - \mu))$$

Study project 17

A seminar can be given on the derivation of the Jensen inequality for conditional expectations and its applications to the proof of several inequalities like the conditional Holder's inequality. The idea can be based on the definition of a convex function $\phi(x_1, ..., x_n)$ of n variables. Assuming ϕ to be twice continuously differentiable, ϕ is said to be convex if the Hessian matrix $\mathbf{H}(\mathbf{x}) = ((\frac{\partial^2 \phi(\mathbf{x})}{\partial x_i \partial x_j}))_{1 \leq i,j \leq n}$ is positive definite. When this happens, we have from the Taylor expansion formula with remainder

$$\phi(\mathbf{t}) = \phi(\mathbf{x}) + (\mathbf{t} - \mathbf{x})^T \nabla\phi(\mathbf{x}) + \frac{1}{2}(\mathbf{t} - \mathbf{x})^T \mathbf{H}(\mathbf{x} + \theta(\mathbf{t} - \mathbf{x}))(\mathbf{t} - \mathbf{x})$$

where $\theta \in [0, 1]$, the inequality

$$\phi(\mathbf{t}) \geq \phi(\mathbf{x}) + (\mathbf{t} - \mathbf{x})^T \nabla\phi(\mathbf{x})$$

Replacing $\mathbf{t}$ by the measurable vector valued function $\mathbf{g}$ and $\mathbf{x}$ by $\mathbb{E}(g|\mathcal{B}_0)$, one gets

$$\phi(\mathbf{g}) \geq \phi(\mathbb{E}(\mathbf{g}|\mathcal{B}_0)) + (\mathbf{g} - \mathbb{E}(\mathbf{g}|\mathcal{B}_0))^T \nabla\phi(\mathbb{E}(\mathbf{g}|\mathcal{B}_0))$$

Taking the conditional expectation on both sides given $\mathcal{B}_0$ gives

$$\mathbb{E}(\phi(\mathbf{g})|\mathcal{B}_0) \geq \phi(\mathbb{E}(\mathbf{g}|\mathcal{B}_0))$$

which is the Jensen inequality. Its applications include for example the conditional Holder inequality. Consider the function $\phi(x,y) = -x^\alpha y^\beta$ in the region $x, y \geq 0$. The function is easily verified to be convex. We compute the Hessian matrix and show that its diagonal elements and determinant are positive. Thus, application of the conditional Jensen inequality leads to

$$\mathbb{E}(|f|^\alpha |g|^\beta|\mathcal{B}_0) \leq \mathbb{E}(|f||\mathcal{B}_0)^\alpha \mathbb{E}(|g||\mathcal{B}_0)^\beta$$

Taking $|f| = \psi_1^\alpha, |g| = \psi_2^\beta$ leads to the conditional Holder inequality. Another application is the following. Rather than stating the result, we derive it from first principles. Consider $p \geq 1$. The function t^p is convex in the region $t \geq 0$ as follows from the nonnegativeness of its second derivative $p(p-1)t^{p-2}$. Thus, conditional Jensen's inequality gives for non-negative measurable g,

$$\mathbb{E}(g^p|\mathcal{B}_0) \geq \mathbb{E}(g|\mathcal{B}_0)^p$$

Now assume that $p > p_1 \geq 1$. Then we can replace p by p/p_1 in the above equation to get

$$\mathbb{E}(g^{p/p_1}|\mathcal{B}_0) \geq \mathbb{E}(g|\mathcal{B}_0)^{p/p_1}$$

Replace g by g^{p_1} here to get

$$\mathbb{E}(g^p|\mathcal{B}_0)^{1/p} \geq \mathbb{E}(g^{p_1}|\mathcal{B}_0)^{1/p_1}$$

This is an important inequality in the theory of Banach spaces. Setting $p_1 = 1$ here gives us for any integrable g

$$\mathbb{E}(|g|^p|\mathcal{B}_0)^{1/p} \geq \mathbb{E}(|g||\mathcal{B}_0)$$

Raising both sides to the power p and taking expectations gives us

$$\mathbb{E}(|g|^p) \geq \mathbb{E}\mathbb{E}(|g||\mathcal{B}_0)^p$$

which can be cast in the form

$$\| g \|_p \geq \| \mathbb{E}(g|\mathcal{B}_0) \|_p$$

There are many more inequalities that one can derive from the Jensen inequality, but we do not discuss these here.

Study project 18

When $X_1, ..., X_n$ are uniformly distributed independent random variables over different intervals, the distribution of their sum plays an important role in applications. A seminar can be delivered on the computation of the moments of these random variables. For example if X_i is uniform over $[a_i, b_i]$, the Laplace transform of its density is given by

$$\mathbb{E}exp(-sX_i) = \frac{exp(-sa_i) - exp(-sb_i)}{\Delta_i s}, i = 1, 2, ..., n$$

where $\Delta_i = b_i - a_i$. The Laplace transform of the distribution of the sum $X_1 + ... + X_n$ is given by their product:

$$\Psi(s) = (\Delta_1...\Delta_n)^{-1}s^{-n}\Pi_{i=1}^{n}(exp(-sa_i) - exp(-sb_i))$$

The product can be evaluated as a sum of exponentials:

$$\Pi_{i=1}^{n}(exp(-sa_i) - exp(-sb_i)) = exp(-s(a_1 + ... + a_n)) - \sum_{i} exp(-s(a_i + \sum_{j \neq i} b_j))$$

$$+ \sum_{i<j} exp(-s(a_i + a_j + \sum_{k \notin \{i,j\}} b_k)) + ... + (-1)^{n-m}$$

$$* \sum_{i_1 < i_2 < ... < i_m} exp(-s(a_{i_1} + ... + a_{i_m} + \sum_{k \notin \{i_1,...,i_m\}} b_k)) + ...$$

$$= \sum_{m=0}^{n}(-1)^{n-m} \sum_{i_1 < i_2 < ..., < i_m} exp(-s(a_{i_1} + ... + a_{i_m} + \sum_{k \notin \{i_1,...,i_m\}} b_k))$$

This can be expressed in using a compact notation as

$$\sum_{i=1}^{r} exp(-s\alpha_i)(-1)^{m_i}$$

with the quantities $\alpha_1, ..., \alpha_r$ and the integers $m_1, ..., m_r$ chosen appropriately. The Laplace inverse of this quantity gives the density of the sum of the $X_i's$:

$$f(x) = \sum_{i=1}^{r}(-1)^{m_i} = \Delta^{-1} \sum_{i=1}^{r} \frac{(x - \alpha_i)^{n-1}}{(n - 1)!} u(x - \alpha_i)(-1)^{m_i}$$

with $\Delta = \Delta_1...\Delta_n$. It is clear that the distribution is concentrated on the interval $[\sum a_i, \sum b_i]$. The moments can be evaluated keeping this in mind.

Study project 19

A seminar dealing with the definition, properties and simulation of the Brownian motion process would be of interest to engineers involved in the analysis of noise removal from signals. The definition of the Brownian motion process as a stochastic process $\{B(t) : t \geq 0\}$ starting at the origin, i.e. $B(0) = 0$ and having independent normal increments with each increment being a zero mean normal random variable with a variance equal to the length of the interval can be supplied to begin with: For $0 \leq t_1 < t_2 < ... < t_k$,

$$P(B(t_1) \leq x_1, B(t_2) \leq x_2, ..., B(t_k) \leq x_k)$$

$$= (2\pi)^{-k/2} \frac{1}{\sqrt{t_1(t_2 - t_1)...(t_k - t_{k-1})}}$$

$$\times \int_{-\infty}^{x_1} ... \int_{-\infty}^{x_k} exp(-y_1^2/2t_1 - (y_2 - y_1)^2/2(t_2 - t_1)... - (y_k - y_{k-1})^2/2(t_k - t_{k-1}))dy_1...dy_k$$

Then the process is constructed by application of the Kolmogorov existence theorem. Einstein's derivation of the probability density of the process starting from first principles is to be discussed next. If $p(t, x)$ denotes the probability density of the position of the particle at time t and a kick on this particle which lasts for a duration of τ takes the particle from x to $x + \Delta$, where Δ has the density $\phi(\Delta)$, then one has the obvious Chapman-Kolmogorov equation

$$p(t + \tau, x) = \int_{-\infty}^{\infty} p(t, x - \Delta)\phi(\Delta)d\Delta$$

We assume that the displacement of the particle in one molecular kick has zero mean, i.e. $\int \Delta\phi(\Delta)d\Delta = 0$ and that its variance $\int \Delta^2\phi(\Delta)d\Delta = \sigma^2$. Then the Chapman equation approximates to

$$\frac{\partial p(t, x)}{\partial t} = D\frac{\partial^2 p(t, x)}{\partial x^2}, D = \sigma^2/\tau$$

This is the diffusion equation first derived by Einstein and its solution leads to the normal distribution with zero mean and variance $2Dt$. From measurements of the variance, the value of D can be estimated from which in turn, the molecular size determined.

Study project 20

The state variable equations with white noise input can be described. Basically, the equations are determined by the stochastic differential system

$$d\mathbf{X}(t) = \mathbf{F}(t)\mathbf{X}(t)dt + \mathbf{G}(t)d\mathbf{B}(t)$$

where

$$\mathbf{B}(t) = [B_1(t), ..., B_n(t)]^T$$

is a vector of independent standard Brownian motion processes. This means that the processes $\{B_k(t), t \geq 0\}, k = 1, 2, ..., n$ are independent standard Gaussian processes with $\mathbb{E}(B_k(t)) = 0, \mathbb{E}(B_k(t)B_k(s)) = min(t, s)$. An alternate way to describe the vector Brownian motion process is that if $f_1(t), ..., f_n(t)$ are non-random functions, then for any finite interval $[a, b]$ of the real line,

$$\mathbb{E}exp(\sum_{k=1}^{n} \int_a^b f_k(t)dB_k(t)) = exp(\frac{1}{2}\sum_{k=1}^{n} \int_a^b f_k(t)^2 dt)$$

Equivalently, the random variables $X_k = \int_a^b f_k(t)dB_k(t), k = 1, 2, ..., n$ are independent Gaussian random variables with mean zero and variances $\sigma_k^2 = \int_a^b f_k(t)^2 dt$. With $\Phi(t, s)$ as the state transition matrix of the system, we have the following solution:

$$\mathbf{X}(t) = \Phi(t, 0)\mathbf{X}(0) + \int_0^t \Phi(t, s)\mathbf{G}(s)d\mathbf{B}(s)$$

The correlation is given by

$$R(t, t') = \mathbb{E}(\mathbf{X}(t)\mathbf{X}(t')^T) = \Phi(t, 0)R(0, 0)\Phi(t', 0)^T + \int_0^{t'} \Phi(t, s)\mathbf{G}(s)\mathbf{G}(s)^T\Phi(t', s)^T ds, t > t'$$

which on differentiation with respect to t gives

$$\frac{\partial R(t, t')}{\partial t} = \mathbf{F}(t)\mathbf{R}(t, t')$$

This equation could also have been deduced directly from the original differential equation by noting that $\frac{d\mathbf{B}(t)}{dt}$ is independent of $\mathbf{X}(t')$ for $t' < t$:

$$\mathbb{E}(\frac{d\mathbf{X}(t)}{dt}\mathbf{X}(t')^T) = \frac{\partial R(t, t')}{\partial t} = \mathbb{E}\mathbf{F}(t)\mathbf{X}(t)\mathbf{X}(t')^T + \mathbb{E}(\mathbf{G}(t)\frac{d\mathbf{B}(t)}{dt}\mathbf{X}(t')^T)$$

$$= \mathbf{F}(t)R(t, t'), t' < t$$

We also find that

$$R(t, t) = \mathbb{E}(\mathbf{X}(t)\mathbf{X}(t)^T) = \Phi(t, 0)R(0, 0)\Phi(t, 0)^T + \int_0^t \Phi(t, s)\mathbf{G}(s)\mathbf{G}(s)^T\Phi(t, s)^T ds$$

and the differential equation satisfied by this autocorrelation function is given by

$$\frac{dR(t, t)}{dt} = \mathbf{F}(t)R(t, t) + R(t, t)\mathbf{F}(t)^T + \mathbf{G}(t)\mathbf{G}(t)^T$$

These and related issues can be discussed in the seminar. Then, the Ito calculus, according to which if $B(t)$ is standard Brownian motion, then $dB(t)^2 = dt$. The Ito rule for differentiating functions $df(B(t)) = f'(B(t))dB(t) + \frac{1}{2}f''(B(t))dt$, or equivalently, in integral form

$$f(B(t)) - f(B(s)) = \int_s^t f'(B(\tau))dB(\tau) + \frac{1}{2}\int_s^t f''(B(\tau))d\tau$$

the first integral on the right being the Ito integral. It should be mentioned how the Ito calculus is a generalization of the Newtonian calculus: $df(x) = f'(x)dx$. Of course, for a general independent increment process $x(t)$, contributions $dx(t)^n$ to the differential for $n \geq 3$ must also be accounted for. This gives the general formula

$$df(x(t)) = \sum_{n=1}^\infty f^{(n)}(x(t))dx(t)^n/n!$$

Study project 21

Nonlinear systems driven by white noise, analysis using the Ito calculus formulation. These include for example, systems of the kind

$$dx(t) = f(x(t))dt + \sigma.dB(t)$$

If $\phi(x)$ is any twice differentiable function, then application of the Ito calculus gives

$$d\phi(x(t)) = \phi'(x(t))dx(t) + \frac{1}{2}\phi''(x(t))dx(t)^2$$

$$= \phi'(x(t))f(x(t))dt + \phi'(x(t))\sigma.dB(t) + \frac{\sigma^2}{2}\phi''(x(t))dt$$

which gives on taking expectations

$$\frac{d}{dt}\mathbb{E}\phi(x(t)) = \mathbb{E}(\phi'(x(t))f(x(t))) + \frac{\sigma^2}{2}\mathbb{E}\phi''(x(t))$$

from which one deduces the Fokker-Planck equation for $p(t, x)$, the probability density of $x(t)$:

$$\frac{\partial p(t, x)}{\partial t} = -\frac{\partial}{\partial x}(f(x)p(t, x)) + \frac{\sigma^2}{2}\frac{\partial^2 p(t, x)}{\partial x^2}$$

Second order systems with white noise forcing as a perturbation. Consider the motion of a particle in the potential field $V(x)$ and subject to a white noise perturbation. The differential equation of motion is

$$m\frac{d^2 x(t)}{dt^2} = -V'(x(t)) + \sigma\frac{dB(t)}{dt}$$

which can be cast as a pair of stochastic differential equations. Taking $m = 1$ without any loss of generality,

$$dx(t) = v(t)dt \qquad dv(t) = -V'(x(t))dt + \sigma.dB(t)$$

The joint probability density of the position-velocity pair $p(t, x, v)$ satisfies the Fokker-Planck equation

$$\frac{\partial p(t, x, v)}{\partial t} = -v\frac{\partial p(t, x, v)}{\partial x} + V'(x)\frac{\partial p(t, x, v)}{\partial v} + \frac{\sigma^2}{2}\frac{\partial^2 p(t, x, v)}{\partial v^2}$$

Numerical evaluation of the density $p(t, x, v)$ by discretizing the Fokker-Planck equation is an interested digital signal processing problem. Suppose one starts with a density $p(0, x, v)$. For example, if the initial displacement is x_0 and the initial velocity is v_0, then $p(0, x, v) = \delta(x - x_0)\delta(v - v_0)$. Assume that position and velocity space have been discretized with the unit of discretization of position space being Δ_x and the unit of discretization of velocity space being Δ_v. The unit of discretization of time is τ, The discretized density reads $p[n, k, m] = p(k\tau, n\Delta_x, m\Delta_v)$. The discretized Fokker-Planck equation reads

$$\frac{p[n+1, k, m] - p[n, k, m]}{\tau} = -m\frac{\Delta_v}{\Delta_x}(p[n, k, m] - p[n, k-1, m])$$

$$+V'[k]\left(\frac{p[n,k,m] - p[n,k,m-1]}{\Delta_v}\right)$$

$$+\frac{\sigma^2}{2}\left(\frac{p[n,k,m] - 2p[n,k,m-1] + p[n,k,m-2]}{\Delta_v^2}\right)$$

For a given time index n, we can define a state vector having components $\{p[n,k,m]\}_{k,m}$ and then we end up with a linear state variable equation of the form

$$\mathbf{p}[n+1] = \mathbf{F}[n]\mathbf{p}[n]$$

Study project 22

The most important theorem in the theory of stochastic processes is the Kolmogorov existence theorem. A separate seminar can be delivered on this subject. The following points should be elaborated while discussing the proof.

1. Let T be any index set and let $\{X_t : t \in T\}$ be a stochastic process whose time index comes from this set and which is defined on a probability space $(\Omega, \mathcal{F}, P)$. Define the finite dimensional distributions

$$\mu_{t_1,\dots,t_k}(H) = P((X_{t_1}, \dots, X_{t_k}) \in H), H \in \mathbb{B}(\mathbb{R}^k), t_1, \dots, t_k \in T$$

Here, $\mathbb{B}(\mathbb{R}^k)$ is the Borel σ field on $\mathbb{R}^k$. It is generated by rectangles of the form $H_1 \times \dots \times H_k$ with $H_j \in \mathbb{B}(\mathbb{R})$.

2. Note the following consistency equations are satisfied. Firstly,

$$\mu_{t_1,\dots,t_k}(H_1 \times \dots \times H_k) = \mu_{t_{\pi 1},\dots,t_{\pi k}}(H_{\pi 1} \times \dots \times H_{\pi k})$$

for all permutations π of $\{1, 2, \dots, k\}$ and secondly

$$\mu_{t_1,\dots,t_k,t_{k+1}}(H_1 \times \dots \times H_k \times \mathbb{R}) = \mu_{t_1,\dots,t_k}(H_1 \times \dots \times H_k)$$

The first property can be described using a more compact notation by defining for each permutation π a mapping $\phi_\pi : \mathbb{R}^k \to \mathbb{R}^k$ by

$$\phi_\pi(x_1, \dots, x_k) = (x_{\pi^{-1}1}, \dots, x_{\pi^{-1}k})$$

Then,

$$\phi_\pi^{-1}(H_1 \times \dots \times H_k) = \{(x_1, \dots, x_k) : (x_{\pi^{-1}1}, \dots, x_{\pi^{-1}k}) \in H_1 \times \dots \times H_k\}$$

$$= \{(x_1, \dots, x_k) : (x_1, \dots, x_k) \in H_{\pi 1} \times \dots \times H_{\pi k}\} = H_{\pi 1} \times \dots \times H_{\pi k}$$

Then the first consistency property is the same as

$$\mu_{t_{\pi 1},\dots,t_{\pi k}}(\phi_\pi^{-1}(H_1 \times \dots \times H_k)) = \mu_{t_1,\dots,t_k}(H_1 \times \dots \times H_k)$$

or equivalently, for all rectangles H in $\mathbb{R}^k$,

$$\mu_{t_{\pi 1},\ldots,t_{\pi k}}(\phi_\pi^{-1}H) = \mu_{t_1,\ldots,t_k}(H)$$

Likewise, if one defines $\phi : \mathbb{R}^k \to \mathbb{R}^{k-1}$, then

$$\phi^{-1}(H_1 \times \ldots \times H_{k-1}) = H_1 \times \ldots \times H_{k-1} \times \mathbb{R}$$

and hence, the second property can be expressed as

$$\mu_{t_1,\ldots,t_k}\phi^{-1} = \mu_{t_1,\ldots,t_{k-1}}$$

3. Kolmogorov's theorem is the converse of this. It states that given a consistent family of finite dimensional distributions, one can construct a process defined on some probability space whose finite dimensional distributions agree with the given consistent family.

4. Both properties can be combined into a single one by defining $\psi : \mathbb{R}^m \to \mathbb{R}^k$ with $m \geq k$ as

$$\psi(x_1,\ldots,x_m) = (x_{\pi^{-1}1},\ldots,x_{\pi^{-1}k})$$

where π is a permutation of $\{1,2,\ldots,k\}$. Both the properties are then equivalent to the single one

$$\mu_{t_1,\ldots,t_m}\psi^{-1} = \mu_{t_{\pi 1},\ldots,t_{\pi k}}$$

Indeed, we note that

$$\psi^{-1}(H_{\pi 1} \times \ldots \times H_{\pi k}) = \{(x_1,\ldots,x_m) : (x_{\pi 1},\ldots,x_{\pi k}) \in H_{\pi 1} \times \ldots \times H_{\pi k}\}$$

$$= \{(x_1,\ldots,x_m) : (x_1,\ldots,x_k) \in H_1 \times \ldots \times H_k\} = H_1 \times \ldots \times H_k \times \mathbb{R} \times \ldots \times \mathbb{R}$$

so that

$$\mu_{t_1,\ldots,t_m}\psi^{-1}(H_{\pi 1} \times \ldots \times H_{\pi k}) = \mu_{t_1,\ldots,t_m}(H_1 \times \ldots \times H_k \times \mathbb{R} \times \ldots \times \mathbb{R})$$

$$= \mu_{t_1,\ldots,t_k}(H_1 \times \ldots \times H_k)$$

if one assumes the second consistency property. The converse is easy to see.

Study project 23

A seminar on applications of the theory of random fields to engineering problems makes an interesting seminar. Applications include fluid dynamics, quantum mechanics and electromagnetism. For example, consider the propagation of a wave inside a waveguide with the source excitation being a random field. The source excitation appears at one end of the guide. Assume that the wave propagates along the z direction so that the field phasors have the dependence

$\mathbf{E}(x,y)exp(-\gamma z)$, $\mathbf{H}(x,y)\ exp(-\gamma z)$, the frequency of excitation being ω, the Maxwell equations $\nabla \times \mathbf{E} = -i\omega\mu\mathbf{H}$ and $\nabla \times \mathbf{H} = (\sigma + i\omega\epsilon)\mathbf{E}$ read

$$\frac{\partial E_z}{\partial y} + \gamma E_y = -i\omega\mu H_x$$

$$-\gamma E_x - \frac{\partial E_z}{\partial x} = -i\omega\mu H_y$$

$$\frac{\partial E_y}{\partial x} - \frac{\partial E_x}{\partial y} = -i\omega\mu H_z$$

$$\frac{\partial H_z}{\partial y} + \gamma H_y = (\sigma + i\omega\epsilon)H_x$$

$$-\gamma E_x - \frac{\partial E_z}{\partial x} = (\sigma + i\omega\mu)H_y$$

$$\frac{\partial E_y}{\partial x} - \frac{\partial E_x}{\partial y} = (\sigma + i\omega\mu)H_z$$

The first two equations and the third and fourth equations can be used to solve for the field components E_x, E_y, H_x, H_y in terms of the derivatives of E_z, H_z with respect to the coordinates x, y. Thus, if the components E_z, H_z are known as functions of x, y, the entire electromagnetic field inside the wave-guide can be determined. The above equations imply that E_z, H_z satisfy the following versions of the wave equation

$$\frac{\partial^2 E_z}{\partial x^2} + \frac{\partial^2 E_z}{\partial y^2} + (\gamma^2 + \omega^2\mu\epsilon)E_z = 0$$

with a similar equation for H_z. We define $h^2 = \gamma^2 + \omega^2\mu\epsilon$. Under the boundary condition that E_z vanishes at the guide walls $x = 0, a, y = 0, b$, we obtain the following solution for the z component of the electric field:

$$E_z(x, y, z) = \sum_{m,n=0}^{\infty} A_{mn}sin(m\pi x/a)sin(n\pi y/b)exp(-\gamma_{mn}z)exp(i\omega t)$$

where

$$\gamma_{mn} = \sqrt{m^2\pi^2/a^2 + n^2\pi^2/b^2 - \omega^2\mu\epsilon}$$

The cutoff frequency for the $(m, n)^{th}$ mode is given by

$$\omega_c(m, n) = \frac{1}{\sqrt{\mu\epsilon}}\sqrt{m^2\pi^2/a^2 + n^2\pi^2/b^2}$$

Suppose that the excitation phasor at $z = 0$ (the source end) is given by $E_0(x, y)exp(i\omega t)$, then the coefficients A_{mn} are determined from the Fourier series

$$E_0(x, y) = \sum_{m,n=0}^{\infty} A_{mn}sin(m\pi x/a)sin(n\pi y/b), 0 < x < a, 0 < x < b$$

Standard computation based on orthogonality of the sinusoidal functions shows that

$$A_{mn} = \frac{4}{ab} \int_0^a \int_0^b E_0(x,y) sin(m\pi x/a) sin(n\pi y/b) dx dy$$

Thus, if we define the propagation kernel

$$K(x,y,\xi,\eta,z) = \frac{4}{ab} \sum_{m,n=0}^{\infty} sin(n\pi x/a) sin(n\pi\xi/a) sin(m\pi y/b) sin(m\pi\eta/b) exp(-\gamma_{mn}z)$$

we find that

$$E_z(x,y,z) = \int_0^a \int_0^b K(x,y,\xi,\eta,z) E_0(\xi,\eta) d\xi d\eta$$

When the excitation source $E_0(\xi,\eta)$ is random with given correlation function, it is an interesting problem to determine the correlations in the fields at two different points inside the waveguide.

Study project 24

A project on the computer simulation of the Brownian motion process and processes derived from this would be useful to the engineer wishing to specialize in theory and applications of stochastic processes in engineering systems. The simulation of the Brownian motion process involves first of all simulating a sequence of identically distributed independent random variables $X_1, X_2, ...$ with $P(X_j = 1) = P(X_j = -1) = 1/2$. The approximate Brownian process at time t is then defined by the formula

$$B(t) \approx \sqrt{\tau} S_{[t/\tau]}$$

where $S_n = X_1 + ... + X_n$. We see that, the random variable on the right side is for large t $(t >> \tau)$ approximately normal with a mean of zero and a variance of t. The increments are also independent. In the limit of $\tau \to 0$, the process on the right converges to the Brownian motion process. To simulate the process on the interval $[0,1]$, we choose a large integer N, say $N = 1000$ and define

$$B(t) \approx \frac{1}{\sqrt{N}}(X_1 + ... + X_n)$$

where $n = [Nt]$. This follows by taking $\tau = 1/N$ in the earlier expression. Note that as t varies over $[0,1]$, $n = [Nt]$ will vary over the set $\{0, 1, ..., N\}$. Next we look at the simulation of a diffusion process defined by the stochastic differential equation

$$dX(t) = \mu(X(t))dt + \sigma(X(t))dB(t)$$

The differential equation is replaced by the difference equation

$$\Delta X_n = \mu(X_n)\Delta + \sigma(X_n)\Delta B_n$$

where $\Delta X_n = X_{n+1} - X_n$ and $\Delta B_n = B_{n+1} - B_n$, with $B_n = B(n\Delta)$. It is clear that $\{\Delta B_n\}$ is a sequence of identically distributed independent normal random variables with zero mean and variance Δ: $\Delta B_n \approx N(0, \Delta)$. Thus, if $\xi_n, n = 1, 2, ...$, is a sequence of identically distributed independent standard normal random variables, the process $\{X_n\}$ can be simulated using the difference equation

$$X_{n+1} = X_n + \mu(X_n)\Delta + \sigma(X_n)\sqrt{\Delta}\xi_n$$

Study project 25

Techniques for obtaining the dynamic time varying spectrum of a vector valued process defined by means of a linear stochastic differential equation find applications in the theory of switched capacitor networks. The differential equations are of the standard kind used in Kalman filter theory:

$$dX_t = AX_t dt + GdB_t$$

The time varying spectral density is defined by the equation

$$S(\omega, t) = \mathbb{E}\left(\int_0^t X_s exp(-i\omega s)ds \int_0^t X_s^* exp(i\omega s)ds \right)$$

So, we need a differential equation for the process

$$Z_t = \int_0^t X_s exp(-i\omega s)ds \int_0^t X_s^* exp(i\omega s)ds$$

We evaluate its differential:

$$dZ_t = X_t exp(-i\omega t)dt \int_0^t X_s^* exp(i\omega s)ds + \int_0^t X_s exp(-i\omega s)ds X_t^* exp(i\omega t)dt$$

Taking expectations on both sides gives

$$d\mathbb{E}Z_t = dt \int_0^t \mathbb{E}(X_t X_s^*)exp(i\omega(s - t))ds + dt \int_0^t \mathbb{E}(X_s X_t^*)exp(i\omega(t - s))ds$$

It is clear that we need a formula for $\mathbb{E}(X_t X_s^*)$. Consider $Y_t = X_t X_s^*$ with $s \leq t$. Its differential with respect to t is computed as

$$d(X_t X_s^*) = (dX_t)X_s^* = (AX_t dt + GdB_t)X_s^*$$

Taking expectations on both sides gives us

$$d\mathbb{E}(X_t X_s^*) = A\mathbb{E}(X_t X_s^*)dt$$

For the initial condition, we need the value of $\mathbb{E}(X_t X_s)$ at $t = s$. Thus, we need to compute $dK_t = d\mathbb{E}(X_t X_t^*)$. Applying the Ito calculus, we find that

$$d(X_t X_t^*) = (dX_t)X_t^* + X_t(dX_t^*) + (dX_t)(dX_t)^*$$

Taking expectations on both sides gives us

$$\frac{d}{dt}\mathbb{E}(X_t X_t^*) = dK_t = AK_t + K_t A^* + GG^*$$

Using these formulas, the time varying spectrum of the state process is calculated.

Study project 26

The basic equations of non-linear filtering theory can be discussed in a separate seminar. The seminar can include simulation of the optimal filtering differential equations.

Study project 27

A seminar on the important matrix decomposition theorems that find importance in digital and statistical signal processing involves the discussion of (1) Bringing a matrix into upper triangular form by an appropriate choice of the basis, (2) Bringing a family of commuting matrices into upper triangular form by an appropriate choice of a basis, (3) Diagonability of a matrix and methods of choosing an ordered basis to bring a matrix to diagonal form. (4) Simultaneous diagonability of a family of commuting matrices and methods of choosing an ordered basis to bring the family into diagonal form. That any square matrix can be brought into upper triangular form can be proved by using properties of the minimal polynomial, the monic polynomial of least degree that annihilates the matrix. The t-conductor of a vector relative to a matrix and an invariant subspace is also introduced at this stage. If W is an invariant subspace for a linear operator T, then the T-conductor of a vector x is denoted by $S_T(x, W)$. It can either stand for the collection of all polynomials g for which $g(T)x \in W$, or since $S_T(x, W)$ is an ideal in the algebra of polynomials, it may equivalently stand for the generator of this ideal, i.e., the monic polynomial s_T of least degree for which $s_T(x) \in W$. If p denotes the minimal polynomial, then clearly p belongs to the ideal hence it is divisible by the monic generator s_T. Hence, the roots of s_T are precisely the eigenvalues of the operator T. Now, let $x \notin W$. Then s_T is the monic generator of $S_T(x, W)$, it follows that $s_T(T)(x) \in W$. Choose any one root of s_T, say c. Then, $s_T(t) = (t - c)h(t)$, where $deg h(t) = deg s_T(t) - 1$. Hence, by definition of monic generator, $y = h(T)x \notin W$. Clearly, $s_T(T)(x) = (T - c)y \in W$. This shows that there exists a vector $y \notin W$ such that $(T - c)y \in W$ for some eigenvalue c of T (the roots of the minimal polynomial are its eigenvalues). The principle behind proving the upper triangulability of any linear operator is this result. For proving simultaneous triangulability of a collection $\{T_1, ..., T_r\}$ of linearly independent operators, we first show that if W is a proper subspace invariant under this family, then there exists an $x \notin W$ and constants $c_1, ..., c_r$ such that $(T_j - c_j)x \in W$ for all $j = 1, 2, ..., r$. First of all one chooses a vector

$z_1 \notin W$ and a constant c_1 such that $(T_1 - c_1)z_1 \in W$. Then, one defines V_1 to be the subspace consisting of all such vectors, i.e., all vectors z with the property $(T_1 - c_1)z \in W$. V_1 is an invariant subspace for the family as follows from the fact that $(T_1 - c_1)Tz = T(T_1 - c_1)z \in W$ by virtue of commutativity of the family and invariance of W with respect to the family. Moreover, V_1 contains W properly. Thus, one can choose a vector z_2 belonging to V_1 but not belonging to W such that $(T_2 - c_2)z_2 \in W$. This z_2 will by construction also satisfy $(T_1 - c_1)z_2 \in W$. Continuing upto r in this way, we finally end up with a $z_r \notin W$ such that $(T_j - c_j)z_r \in W, j = 1, ..., r$ which is the basic result from which simultaneous triangulability can be deduced. One can think of exploring signal estimation problems in which this result will find applications. For example, suppose one wishes to solve using parallel processing the system of equations $T_j x_j = b_j, j = 1, 2, ..., r$ where $\{T_j\}$ is a commuting family of matrices. We can set $T_j = SR_jS^{-1}, j = 1, 2, ..., r$,where the S is a non-singular matrix and the $R_j's$ are upper triangular matrices. The system of linear equations is clearly equivalent to the system $R_j y_j = c_j, j = 1, 2, ..., r$ where $y_j = S^{-1}x_j, c_j = S^{-1}b_j$. This modified system can easily be solved recursively starting from the last element and moving upwards upto the first element. The advantage with this procedure is that the triangulating matrix S is common to all the $T_j's$, i.e. if b_j varies, we have to write a program of multiplying it by just a single matrix S^{-1} and for recovering x_j from y_j, once again, we have to write the program for multiplying y_j by just the single matrix S. The computational load for solving linear equations with commuting matrices is therefore very little.

Study project 28

A seminar on the computation of the statistics (i.e., mean, second and higher moments) of random variables derived from a random process $\{X(t)\}$ having prescribed joint distributions is important to the engineer. Consider $Y = \int_0^T X(t)dt$. Its mean is given by $\mathbb{E}(Y) = \int_0^T \mathbb{E}X(t)dt = \int_0^T m_x(t)dt$, where $m_x(t) = \mathbb{E}X(t)$, $\mathbb{E}Y^2 = \int_0^T \int_0^T R_{xx}(t, s)dtds$, where $R_{xx}(t, s) = \mathbb{E}(X(t)X(s))$ is the autocorrelation function of $\{X(t)\}$. In general, the kth order moment of Y is given by

$$\mu_k(Y) = \mathbb{E}(Y^k) = \int_{[0,T]^k} \mu_{X,k}(t_1, ..., t_k)dt_1...dt_k$$

where

$$\mu_{X,k}(t_1, ..., t_k) = \mathbb{E}(X(t_1)...X(t_k))$$

is the kth order moment of the process $\{X(t)\}$. We can also look at the functional $Y = \int_0^T (t, X(t))dt$. Its kth order moment is given by

$$\mu_k(Y) = \mathbb{E}(Y^k) = \int_{[0,T]^k} \mathbb{E}(\phi(t_1, X(t_1))...\phi(t_k, X(t_k)))dt_1...dt_k$$

$$= \int_{[0,T]^k} dt_1...dt_k \int_{\mathbb{R}^k} \phi(t_1, x_1)...\phi(t_k, x_k)f_X(t_1, ..., t_k, x_1, ..., x_k)dx_1...dx_k$$

where $f_X(x_1, ..., x_k, t_1, ..., t_k)$ equals the joint density of the random variables $\{X(t_1), ., X(t_k)\}$.

Study project 29

A seminar on gambling theory can be delivered. This is an important topic in applied probability and would enable the listener to formulate problems in this subject on his own. The set up is a sequence $\{X_n, n = 1, 2, ...,\}$ of independent Bernoulli random variables assuming values 1 and -1. When $X_n = 1$, the gambler gains a rupee on the nth bet and when $X_n = -1$, the gambler loses a rupee on the nth bet. One can adopt gambling policies. For example, the gambler bets on the nth trial iff $\sum_{k=1}^{n-1} X_k \geq 0$, this amounts to the number of successes before time n being more than the number of failures. Indeed, the gambler may feel that if there are more successes than failures before time n, then it is wiser to bet at time n while if there are more failures than successes before time n, then it is wiser to avoid betting at time n. More generally, one can consider a sequence $\{B_n : n = 1, 2, ...,\}$ of random variables with B_n assuming values $+1$ and -1 being measurable with respect to the σ field $\sigma(X_1, ..., X_{n-1})$. The gambler chooses to bet at the epoch n iff $B_n = 1$. Otherwise, he skips the turn. Now, one can ask, what is the probability that at time k or earlier, the gambler will place his n^{th} bet ? Let N_n denote the epoch at which the gambler places his n^{th} bet. Then, the event $\{N_n \leq k\}$ occurs iff $B_1 + ... + B_k \geq n$. This event is evidently $\sigma(X_1, ..., X_{k-1}\}$ measurable. The event $\{N_n = k\}$ is the same as $\{N_n \leq k\} - \{N_n \leq k - 1\}$ and this is clearly measurable with respect to the σ field $\sigma\{X_1, ..., X_{k-1}\}$. computation of the probability $P\{B_1 + ... + B_k \geq n\}$ is generally a hard task but an important problem in gambling theory. For the game to continue endlessly, the $B_n's$ must be chosen so that $P(B_n, i.o) = 1$. The gambler's gain on the nth bet is evidently $Y_n = X_{N_n}$. We want now to determine the event on which all the $Y_n's$ are defined. Note that Y_n is defined provided N_n is defined, i.e., provided that the gambler places his n^{th} bet. This happens provided that $B_1 + ... + B_k \geq n$ for some finite k. For this to be true no matter how large n is, one must have $B_1 + ... + ...$ is infinite, i.e. provided that $\{B_n = 1, i.o\}$. Thus, all the $Y_n's$ are defined only on the set $\{B_n = 1, i.o\}$ and hence for all the $Y_n's$ to be well defined random variables, one must have $P(B_n, i.o) = 1$.

Study project 30

A lecture on basic electroencephalography can be given based on triggering of signals by the Poisson process. The standard model of the brain can be explained in this lecture as consisting of isolated spots which generate the so called rhythms, i.e. signals of definite frequencies. External stimuli cause these rhythms to get triggered by a Poisson train. Thus, if $s(t) = \sum_{k=1}^{p} A_k cos(\omega_k t)$ is the total signal with the ω_k's being the rhythm frequencies, and $\{t_i : i = 1, 2, ...\}$ the arrival times of a Poisson train, then the signal measured by a sensor placed on the surface of the head is given by

$$x(t) = \sum_{i=1}^{N(t)} s(t - t_i) = \int_0^t s(t - \tau) dN(\tau)$$

One can consider evaluating the higher moments and spectra of this triggered signal. The of this process is particularly important since the signal $\{x(t)\}$ is non-Gaussian. In fact, one can derive an expression for the characteristic functional of the process. Let $g(t)$ be any deterministic signal.

Then,

$$\int_0^T g(t)x(t)dt = \int_0^T g(t)dt \int_0^t s(t-\tau)dN(\tau) = \int_0^T dN(\tau) \int_\tau^t g(t)s(t-\tau)dt$$

Determining the characteristic function of this random variable is therefore equivalent to determining the characteristic function of the variable $X = \int_0^T \phi(t)dN(t)$, where ϕ is an arbitrary non-random function. We approximate this integral by the discrete sum $\sum \phi(\alpha_k)\Delta N(\alpha_k)$ and use the fact that the Poisson increments are independent with $P(\Delta N(\alpha) = 1) = \lambda\Delta\alpha$ and $P(\Delta N(\alpha) = 0) = 1 - \lambda\Delta\alpha$. Thus,

$$\mathbb{E}exp(\sum_k \phi(\alpha_k)\Delta N(\alpha_k)) = \Pi_k\mathbb{E}exp(\phi(\alpha_k)\Delta N(\alpha_k))$$

$$= \Pi_k((exp(\phi(\alpha_k))\lambda\Delta\alpha_k + 1 - \lambda\Delta\alpha_k)$$

$$= \Pi_k exp(\lambda(exp(\phi(\alpha_k)) - 1)) = exp(\lambda\sum_k(exp(\phi(\alpha_k)) - 1))$$

Taking limits transforms the sum in the exponential into an integral. This gives

$$xp(\int_0^T \phi(\alpha)dN(\alpha)) = exp(\lambda\int_0^T (exp(\phi(t) - 1)dt)$$

from which the various moments and higher order spectra of the Poisson triggered process can be evaluated.

Study project 31

The general jump process as comprising of a superposition of Poisson processes with variable intensity can be discussed. It is fundamental in filtering theory applications. The general construction proceeds by introducing the process $P(t, E)$ as being a Poisson process with intensity $\lambda\mu(E)$ where μ is a measure (Dave Applebaum, Levy Processes and Stochastic Calculus). The process has the additional property that if $E_1, ..., E_k$ are disjoint events, then the processes $\{P(t, E_j), t \geq 0\}, j = 1, 2, ..., k$ are independent processes. The superposition of these processes is constructed as

$$X(t) = \int c(x)P(t, dx)$$

This process has independent increments and one can develop filtering theory for stochastic differential equations driven by this process. More generally, one can make the coefficient $c(x)$ time dependent, i.e. define the increment of $X(t)$ as

$$dX(t) = \int c(t, x)P(dt, dx)$$

with he integration being understood to be with respect to the variable x. If c is a simple function assuming values $c_1, ..., c_N$ on the disjoint sets $E_1, ..., E_N$, then

$$X(t) = \sum_{k=1}^{N} c_k P(t, E_k)$$

and it is easy to see that $dX(t)$ assumes the values $c_1, ..., c_k$ with probabilities $\lambda\mu(E_k)dt, k = 1, 2, ..., N$ respectively and the value zero with probability $1 - dt \sum_k \lambda\mu(E_k)$. Thus, the characteristic functional of the $X(t)$ process is given by

$$\mathbb{E}exp(\int_0^T \phi(t)dX(t)) = exp(\sum_k \lambda\mu(E_k) \int_0^T (exp(c_k\phi(t) - 1))dt)$$

We can generalize this formula to the situation when c is not simple:

$$\mathbb{E}exp(\int_0^T \phi(t)dX(t)) = exp(\int \lambda\mu(dx) \int_0^T (exp(c(x)\phi(t)) - 1)dt)$$

The Ito formula for this process follows by noting that at any time at most one of the Poisson components can have a nonzero increment with non-negligible probability. Thus,

$$df(X(t)) = \int (f(X(t) + c(x)) - f(X(t))P(dt, dx)$$

Study project 32

If X and Y are jointly Gaussian random variables with zero mean and covariance matrix $\mathbf{R} = \begin{pmatrix} \sigma^2 & \rho\sigma^2 \\ \rho\sigma^2 & \sigma^2 \end{pmatrix}$, we want to design an 8085 microprocessor programme to evaluate the conditional mean of X given Y and the conditional variance of X given Y. The joint density of X and Y is given by the formula

$$f(X, Y) = \frac{1}{2\pi\sigma^2\sqrt{1-\rho^2}}exp(-\frac{1}{2\sigma^2(1-\rho^2)}(X^2 + Y^2 - 2\rho XY))$$

Note that this expression is obtained from the general formula that if $\mathbf{R}$ is the autocorrelation matrix of a random vector $\mathbf{X}$, then the density of $\mathbf{X}$ is given by Fourier inverting the characteristic function

$$\mathbb{E}exp(i\mathbf{u}^T\mathbf{X}) = exp(-\frac{1}{2}\mathbf{u}^T\mathbf{R}\mathbf{u})$$

The general formula for the density is

$$f(\mathbf{X}) = \frac{1}{(2\pi)^{N/2}det\mathbf{R})}exp(-\frac{1}{2}\mathbf{X}^T\mathbf{R}^{-1}\mathbf{X})$$

and for the bivariate case, this evaluates to the above expression. We want to calculate and store the values $\phi[m] = \mathbb{E}(X|Y = m\Delta)$ where Δ can be taken as $\sigma/1000$. The formula

$$\mathbb{E}(X|Y) = \frac{\int_{-\infty}^{\infty} X f(X,Y) dX}{\int_{-\infty}^{\infty} f(X,Y) dX}$$

evaluates to ρY. To check this we require an 8085 microprocessor programme. The above quantity can be numerically approximated as

$$\mathbb{E}(X|Y = m\Delta) \approx \frac{\sum_{k=-\infty}^{\infty} k\Delta f(k\Delta, m\Delta)}{\sum_{k=-\infty}^{\infty} f(k\Delta, m\Delta)}$$

We read the values $f(k\Delta, m\Delta), k, m = -N, -N+1, ..., N$ from the input into the memory and evaluate the above sum.

Study project 33

Computing the percentile of a random variable is an important problem. If X is a random variable with distribution function $F(x)$ and u is a number in $[0, 1]$, then the u percentile of the random variable X is defined as x_u where $F(x_u) = u$. In words, x_u is that value for which the probability of X taking values less than or equal to equals u. Given a distribution function F, how does one numerically evaluate its u percentile for a given u ? The answer is by provided by discretization. We choose a large value A so that with negligible probability, the random variable X will fall outside the interval $[-A, A]$. We then discretize the interval $[-A, A]$ into steps of size Δ, i.e. $N\Delta = A$.

Study project 34

Hidden Markov model for speech compression: Speech is segmented into two records. The data sequence for the first record is $\{X_n^{(1)} : n = 0, 1, ..., N - 1\}$ and the data sequence for the second record is $\{X_n^{(2)} : n = 0, 1, ..., N-1\}$. We want to construct a hidden Markov model for the overall process which is the first sequence followed by the second sequence. The first sequence is modeled as an AR(1) process with parameter a and error variance of σ_1^2. The second sequence is likewise modeled as an AR(1) process with parameter b and error variance σ_2^2. The parameters a, σ_1, b, σ_2 are assumed to be random with (a, σ_1) having a probability density $\mu(a, \sigma_1)$ and the conditional probability density of (b, σ_2) given (a, σ_1) having a probability density $\mu(b, \sigma_2|a, \sigma_1)$. The AR model for the first sequence is given by $X_n^{(1)} = aX_{n-1}^{(1)} + W_n^{(1)}$ and the AR model for the second sequence is given by $X_n^{(2)} = bX_{n-1}^{(2)} + W_n^{(2)}$. Assuming the error to be Gaussian, we have obtain the following expression for the joint density of the process and the parameters:

$$\mu(X_n^{(1)}, X_n^{(2)}, n = 0, 1, ..., N - 1, a, \sigma_1, b, \sigma_2) =$$

$$exp(-\frac{1}{2\sigma_1^2} \sum_{n=1}^{N} (X_n^{(1)} - aX_{n-1}^{(1)})^2) exp(-\frac{1}{2\sigma_2^2} \sum_{n=1}^{N} (X_n^{(2)} - bX_{n-1}^{(2)})^2) \mu(a, \sigma_1) \mu(b, \sigma_2|a, \sigma_1)$$

We normally assume that the distribution of (a, σ_1) is normal with a mean vector of (α, β) and a diagonal covariance matrix.

Study project 35

Programming the 8085 microprocessor to evaluate the probability distribution of a function of a random variable X. If $Y = g(X)$ is a function of the random variable X, it density function is given by the expression $f_Y(y) = \sum_i \frac{f_X(x_i)}{|g'(x_i)|}$ where $\{x_i\}$ are solutions to the equation $g(x_i) = y$. We want to numerically evaluate this and verify this identity. For this, we simulate N samples of the random variable X, say $X_1, ..., X_N$ by applying the transformation F_X^{-1} to a sequence $U_1, ..., U_N$ of uniformly distributed random variables and follow up by application of the function g, i.e. we compute $Y_j = g(X_j) = g(F_X^{-1}(U_j)), j = 1, 2, ..., N$. Then, we empirically evaluate the distribution function of the $Y_j's$ by dividing the range of the $Y's$ into M points and evaluating the proportion of the $Y_j's$ that are lesser than or equal to each point in this partition. A routine for evaluating the function g on a given argument needs to be written on the 8085 processor. One convenient method is via interpolation, we choose p different points $u_1, ..., u_p$ and fit a polynomial curve of degree $p - 1$ to the p pairs of points $(u_j, g(u_j)), j = 1, 2, ..., p$. The polynomial $\phi(u)$ that approximates the function g can be implemented using the multiplication routine and the ADD command.

Study project 36

The random variable Z is given by $Z = X.cos\theta$, where X has the density $f_X(x)$ and θ is uniformly distributed over the interval $[-\pi, \pi)$. We want to evaluate the density of Z. Observe that the density of Z is the same as that of the process $Z(t) = X.cos(\omega t + \theta)$ at any given time t. Thus the density is same as that of a sinusoid with purely random phase and random amplitude at any given time t when the amplitude is statistically independent of the phase. One way to do this is to perform a Monte-Carlo simulation of the random variable Z, i.e. generate N independent samples of the pair (X, θ), say $(X_i, \theta_i), i = 1, 2, ..., N$ and evaluate $Z_i = X_i.cos(\theta_i), i = 1, 2, ..., N$ and then evaluate the empirical distribution of Z by the standard procedure. This must be compared with the theoretically predicted value. Assuming that X is positive, it follows that for a given X, $|Z| \leq X$ and there are two values of θ that produce a given Z, namely $\pm cos^{-1}(Z/X)$. We find that $f(Z|X) = \sum \frac{1}{2\pi |\frac{d\theta}{dZ}|}$ the sum being over both the values of θ that produce this value of Z. This gives

$$f(Z|X) = \frac{1}{\pi\sqrt{X^2 - Z^2}}, |Z| \leq X$$

The same result holds for negative values of X except that the range of Z is now $|Z| \leq -X$. We can combine these two equations and say that for all values of X positive or negative,

$$f(Z|X) = \frac{1}{\pi\sqrt{X^2 - Z^2}}, |Z| \leq |X|$$

Thus,

$$f(Z) = \int_{-\infty}^{\infty} f(Z|X)f_X(X)dX = \int_{|X|\geq|Z|} \frac{f_X(X)}{\pi\sqrt{X^2 - Z^2}}dX$$

Note: The process $Z(t) = X.cos(\omega t + \theta)$ is not ergodic and hence if one tries to estimate the distribution of Z by evaluating the proportion of time that the process $Z(t)$ has values lesser than a prescribed z, the wrong result will ensue. The ensemble average autocorrelation $Z(t)$ equals $R_{ZZ}(\tau) = \frac{1}{2}\mathbb{E}(X^2)cos(\omega\tau)$ while the time averaged autocorrelation of $Z(t)$ equals $\frac{1}{2}X^2 cos(\omega\tau)$. The former is nonrandom while the latter is in general random.

Study project 37

The probability density of the sum of squares of n independent standard normal random variables, i.e. the χ^2 distribution. If $X_1, ..., X_n$ are identically distributed independent $N(0, 1)$ random variables, then set $Z = X_1^2 + ... + X_n^2$. It moment generating function which coincides with the Laplace transform of its density function is given by

$$\mathbb{E}exp(-sZ) = (\mathbb{E}exp(-sX_1^2))^n$$

Now,

$$\mathbb{E}exp(-sX_1^2) = \int_{-\infty}^{\infty} exp(-sx^2)exp(-x^2/2)dx/\sqrt{2\pi} = \frac{1}{\sqrt{2\pi}} \int_{-\infty}^{\infty} exp(-(s+1/2)x^2)dx$$

$$= (2s+1)^{-1/2}$$

Thus,

$$\mathbb{E}exp(-sZ) = \int_{0}^{\infty} f_Z(z)exp(-sz)dz = (2s+1)^{-n/2}$$

Inverting this transform yields the density of Z:

$$f_Z(z) = \frac{2^{-n/2}z^{n/2-1}}{\Gamma(n/2)}exp(-z/2)$$

for $z \geq 0$ and zero for $z < 0$.

Study project 38

If $\{Z(t)\}$ is a Gaussian stochastic process with zero mean and autocorrelation $R_{ZZ}(t, s)$, we want to evaluate the moments of the random variable

$$X = \int_{0}^{T} \phi(Z(t), t)dt$$

where ϕ is a nonrandom function. This is of importance in the theory of random signal processing. In fact, if $\{Z(t)\}$ is any process with joint density function $f_Z(z_1, ..., z_n, t_1, ..., t_n)$ for the random variables $Z(t_1), ..., Z(t_n)$, then we find that

$$\mathbb{E}(X^n) = \int_{0}^{T} ... \int_{0}^{T} \mathbb{E}(\phi(Z(t_1), t_1)...\phi(Z(t_n), t_n))dt_1...dt_n$$

$$= \int_0^T ... \int_0^T \int_{\mathbb{R}^n} f_Z(z_1, ..., z_n, t_1, ..., t_n)\phi(z_1, t_1)...\phi(z_n, t_n)dz_1...dz_n$$

(Pugachev)

Study project 39

Basic fact about the normal distribution. X is a normally distributed random variable with zero mean and variance one denoted by $N(0, 1)$ if its density is given by $f(x) = \frac{1}{\sqrt{2\pi}}exp(-x^2/2)$. The first thing to verify is that its integral over $(-\infty, \infty)$ is unity. This is seen by noting that if $I = \int_{-\infty}^{\infty} exp(-x^2/2)dx$, then

$$I^2 = \int_{\mathbb{R}^2} exp(-(x^2 + y^2)/2)dxdy = 2\pi \int_0^{\infty} exp(-r^2/2)rdr = 2\pi$$

The mean $\int xf(x)dx = 0$ and the variance $\int x^2 f(x)dx = 1$. The first follows from the oddness of the function $xf(x)$ and the second by transforming the integral into a Γ integral.

Study project 40

If X is a standard normal random variable, then $Y = aX + b$ is a normal random variable with a mean of b and a variance of a^2. This is seen via linearity of the expectation operator:

$$\mathbb{E}(aX + b) = a\mathbb{E}(X) + b = a.0 + b = b$$

$$\mathbb{E}(aX + b)^2 = \mathbb{E}(a^2 X^2 + b^2 + 2abX) = a^2\mathbb{E}(X^2) + b^2 + 2ab\mathbb{E}(X) = a^2 + b^2$$

The density of $Y = aX + b$ is computed easily. Assuming $a > 0$, we have

$$P(Y \leq y) = P(X \leq (y - b)/a) = F_X((y - b)/a)$$

which gives on differentiation,

$$f_Y(y) = \frac{d}{dy}P(Y \leq y) = \frac{1}{a}F'_X((y - b)/a) = \frac{1}{a}f_X((y - b)/a)$$

$$= \frac{1}{a\sqrt{2\pi}}exp(-(y - b)^2/2a^2)$$

A random variable having this density function is said to be an $N(b, a^2)$ random variable.

Study project 41

In many applications in probability theory, one needs to compute the distribution of a function of several random variables or the joint distribution function of several functions of several random variables. The Jacobian formula plays here a fundamental role. The idea is based on the fact that if $y^1, ..., y^n$ are n functions of n variables $x^1, ..., x^n$, then the differential volume elements in the two spaces (x^i) and (y^i) are related via the determinant of the $n \times n$ matrix $((\frac{\partial y^i}{\partial x^j}))_{1 \le i,j \le n}$. Denoting this determinant by J, we have

$$dy^1...dy^n = J dx^1...dx^n$$

If $f(y^1, ..., y^n)$ is the joint density of the variables $(y^1, ..., y^n)$, then the probability that the vector $(y^1, ..., y^n)$ falls inside the region V is given by the integral of f over V:

$$p = P((y^1, ..., y^n) \in V) = \int_V f(y^1, ..., y^n) dy^1...dy^n$$

By adopting the above transformation of variables, this integral can equivalently be expressed as

$$\int_{T^{-1}(V)} f J dx^1...dx^n$$

where T is the transformation that carries $(x^1, ..., x^n)$ to $(y^1, ..., y^n)$.

Study project 42

Prediction: Given a process $\{X_n : n \in \mathbb{Z}\}$, we want to construct the best non-linear predictor of order r of this process based on $p+1$ samples. The predictor can have only second degree non-linearities:

$$\hat{X}_{n+r} = \sum_{k=0}^{p} h(k) X_{n-k} + \sum_{k,m=0}^{p} g(k,m) X_{n-k} X_{n-m}$$

The coefficients $\{h(k)\}$ and $\{g(k,m)\}$ are selected by minimizing the sum of squares of the differences between the two sides, i.e.

$$E(\{h(k)\}, \{g(k,m)\}) = \sum_{n=p}^{N} (X_{n+r} - \hat{X}_{n+r})^2$$

We leave it as an exercise to derive the normal equations to carry out this minimization.

Study project 43

Suppose that we are given a speech sequence. We segment it into K records with each record having N samples. Denote the samples of the ith record by $X_i(n), n = 0, 1, ..., N-1$ with $i = 1, 2, ..., K$. We model the data sequence in the ith record using a nonlinear difference equation having the form

$$X_i(n) = f_i(X_i(n-1), ..., X_i(n-p), \theta_{i1}, ..., \theta_{ir}) + W_i(n)$$

Here, the parameters $(\theta_{i1}, ..., \theta_{ir})$ are unknown and need to be estimated. The process $W_i(n)$ is a zero mean Gaussian sequence with a variance of σ_i^2. Assume that $X_i(0), ..., X_i(p-1)$ have the density $\psi_i(x(0), ..., x(p-1))$ which is a known function of p variables. Then conditioned on the values of the parameters $\theta_{i1}, ..., \theta_{ir}, \sigma_i^2$, we can write down the joint density of the entire process as

$$f(X_i(n), n = 0, 1, ..., N-1, i = 1, 2, ..., K | \theta_{i1}, ..., \theta_{ir}, \sigma_i^2, i = 1, 2, ..., K)$$

$$= (\Pi_{i=1}^{K} \psi_i(X_i(0), X_i(1), ..., X_i(p-1)))$$

$$\Pi_{i=1}^{K}(2\pi\sigma_i^2)^{-1/2} exp(-\frac{1}{2\sigma_i^2}(\sum_{n=p}^{N-1}(X_i(n) - f_i(X_i(n-1), ..., X_i(n-p)|\theta_{i1}, ..., \theta_{ir})^2)$$

To estimate the parameters $\theta_{i1}, ..., \theta_{ip}, \sigma_i^2$ given the observation sequence, we simply have to maximize the above function with respect to these parameters. It is easily seen that such a maximization can be achieved in a decoupled form, i.e. maximize individually the likelihood function for each record. This means that for each record number i, we must maximize the function

$$g_i(X_i(n), n = 0, 1, ..., N-1|\theta_{i1}, ..., \theta_{ip}, \sigma_i^2) =$$

$$(2\pi\sigma_i^2)^{-1/2} exp(-\frac{1}{2\sigma_i^2}$$

$$\times \sum_{n=p}^{N-1}(X_i(n) - f_i(X_i(n-1), ..., X_i(n-p)|\theta_{i1}, ..., \theta_{ir}))^2$$

with respect to the parameters $\theta_{i1}, ..., \theta_{ip}, \sigma_i^2$. This is equivalent to minimizing

$$L_i(\mathbf{X}_i|\sigma_i, \theta_{i1}, ..., \theta_{ir})$$

$$= \frac{1}{2\sigma_i^2} \sum_{n=p}^{N-1}(X_i(n) - f_i(X_i(n-1), ..., X_i(n-p)|\theta_{i1}, ..., \theta_{ir})^2 + log(\sigma_i)$$

with respect to the parameters of the ith record. The optimal equations are obtained by setting the derivatives of this log likelihood function equal to zero, i.e.,

$$\frac{\partial L_i(\mathbf{X}_i|\sigma_i, \theta_{i1}, ..., \theta_{ir})}{\partial \sigma_i} = 0$$

$$\frac{\partial L_i(\mathbf{X}_i|\sigma_i, \theta_{i1}, ..., \theta_{ir})}{\partial \theta_{ij}} = 0, j = 1, 2, ..., r$$

The first partial derivative gives

$$-\frac{1}{\sigma_i^3} \sum_{n=p}^{N-1}(X_i(n) - f_i(X_i(n-1), ..., X_i(n-p), \theta_{i1}, ..., \theta_{ir}))^2 + \frac{1}{\sigma_i} = 0$$

where $i = 1, 2, ..., K$. The second partial derivative gives

$$\sum_{n=p}^{N-1} \frac{\partial f_i(X_i(n-1), ..., X_i(n-p), \theta_{i1}, ..., \theta_{ir})}{\partial \theta_{ij}} \quad *$$

$$(X_i(n) - f_i(X_i(n-1), ..., X_i(n-p), \theta_{i1}, ..., \theta_{ir})) = 0$$

where $j = 1, 2, ..., r, i = 1, 2, ..., K$. These are nonlinear equations and can be solved only via iterative techniques. This system of equations can be expressed as

$$R_i(\sigma_i, \theta_{i1}, ..., \theta_{ir}) = 0, i = 1, 2, ..., Kr + K$$

Instead, we look directly at the problem of minimizing $L_i(\mathbf{X}|\sigma_i, \theta_{i1}, ..., \theta_{ir})$ by a gradient scheme:

$$\sigma_i(n+1) = \sigma_i(n) - \mu \frac{\partial L_i(\mathbf{X}|\sigma_i(n), \theta_{i1}(n), ..., \theta_{ir}(n))}{\partial \sigma_i}$$

$$\theta_{ij}(n+1) = \theta_{ij}(n) - \mu \frac{\partial L_i(\mathbf{X}|\sigma_i(n), \theta_{i1}(n), ..., \theta_{ir}(n))}{\partial \theta_{ij}}, j = 1, 2, ..., r$$

Another method is to apply the least mean square algorithm directly to the model

$$X_i(n) = f_i(X_i(n-1), ..., X_i(n-p), \theta_{i1}, ..., \theta_{ir}) + W_n$$

This leads to the following recursions for the parameters θ_{ij}:

$$\theta_{ij}(n+1) = \theta_{ij}(n) - \mu \frac{\partial}{\partial \theta_{ij}(n)}(X_i(n) - f_i(X_i(n-1), ..., X_i(n-p), \theta_{i1}(n), ..., \theta)_{ir}(n))^2$$

which expands to

$$\theta_{ij}(n+1) = \theta_{ij}(n) + 2\mu \frac{\partial f_i(X_i(n-1), ..., X_i(n-p), \theta_{i1}(n), ..., \theta_{ir}(n))}{\partial \theta_{ij}(n)}$$

$$.(X_i(n) - f_i(X_i(n-1), ..., X_i(n-p), \theta_{i1}(n), ..., \theta_{ir}(n))$$

We can look at the more general situation in which the parameters θ_{ij}, σ_i^2 are random in each record. The typical model that takes this into account is the Hidden Markov model. According to this model, we consider the parameter vector $\mathbf{P}_i = (\theta_{i1}, ..., \theta_{ir}, \sigma_i^2)$ in each record. The parameter vector $\mathbf{P}_1$ for the first record is assumed to have a density $\mu(\mathbf{P}_1)$ and the sequence $\{\mathbf{P}_1, ..., \mathbf{P}_m\}$ is assumed to form a Markov chain with transition probability density $\pi(\mathbf{P}_{i+1}|\mathbf{P}_i)$. Then, the joint density of the observation sequence and the parameters is given by

$$[\Pi_{i=1}^{K}\psi_i(X_i(0), ..., X_i(p-1))g_i(X_i(n), n = 0, 1, ..., N-1|\mathbf{P}_i)]$$

$$*\mu(\mathbf{P}_1)\pi(\mathbf{P}_2|\mathbf{P}_1)\pi(\mathbf{P}_3|\mathbf{P}_2)...\pi(\mathbf{P}_K|\mathbf{P}_{K-1})$$

Usually, the parameter vector for the first record as well as the transition densities from each record to the succeeding record depend upon a set of unknown parameters that need to be identified from the observation sequence. Thus writing $\mu(\mathbf{P}_1|\mathbf{q}_0)$ for the initial record parameter density and $\pi(\mathbf{P}_{i+1}|\mathbf{P}_i,\mathbf{q}_i)$ for the transition density from the ith to the $i+1^{th}$ record, we can for practical purposes, express the density of the observation sequence given these parameters as

$$g(\mathbf{X}|\mathbf{q}_0,\mathbf{q}_1,...,\mathbf{q}_{K-1})$$

$$= \int [\Pi_{i=1}^K g_i(X_i(n), n = 0, 1, ..., N-1)]$$

$$\cdot \mu(\mathbf{P}_1|\mathbf{q}_0)\pi(\mathbf{P}_2|\mathbf{P}_1,\mathbf{q}_1)\pi(\mathbf{P}_3|\mathbf{P}_2,\mathbf{q}_2)...\pi(\mathbf{P}_K|\mathbf{P}_{K-1},\mathbf{q}_{K-1})d\mathbf{P}_1...d\mathbf{P}_K$$

Here, $\mathbf{q}_0$ are the parameters on which the density of $\mathbf{P}_1$ depends and $\mathbf{q}_i$ are the parameters on which the transition density of $\mathbf{P}_i$ to $\mathbf{P}_{i+1}$ depends. Actually the required density is obtained by multiplying this by the product of $\psi_i(X_i(0), ..., X_i(p-1)), i = 1, 2, ..., K$, but since these functions are independent of the parameters $\mathbf{P}_i$, we have neglected their presence. This is the essence of using the phrase "for all practical purposes". An example illustrating this circle of ideas is as follows: We model the observation sequence in each record as an AR process with parameters $a_i(k), k = 1, 2, ..., p$. Thus, the model for the ith record is given by

$$X_i(n) = \sum_{k=1}^{p} a_i(k)X_i(n-k) + W_i(n)$$

where $W_i(n)$ is assumed to be a white Gaussian noise sequence with a mean of zero and a variance of σ_i^2. Assume that the initial parameter vector $\mathbf{a}_1$ is normal with a mean vector of μ_1 and covariance matrix Σ_1 and conditioned on $\mathbf{a}_i$, the density of $\mathbf{a}_{i+1}$ is normal with mean $\mathbf{a}_i$ and a covariance matrix Σ_{i+1}. Then, the joint density of the observation and the parameters can be expressed as

$$g(X_i(n), n = 0, 1, ..., N-1, i = 1, 2, ..., K, \mathbf{a}_1, ..., \mathbf{a}_K|\sigma_1, ..., \sigma_K, \Sigma_1, ..., \Sigma_K)$$

$$= [\Pi_{i=1}^K \psi_i(X_i(0), ...X_i(p-1))][\Pi_{i=1}^k \Pi (2\pi\sigma_i^2)^{-1/2}$$

$$*exp(-\frac{1}{2\sigma_i^2} \sum_{n=p}^{N-1} (X_i(n) - \sum_{k=1}^{p} a_i(k)X_i(n-k))^2)]$$

$$[exp(-\frac{1}{2}(\mathbf{a}_1 - \mu_1)^T \Sigma_1^{-1}(\mathbf{a}_1 - \mu_1))]\Pi_{i=1}^{K-1} exp(-\frac{1}{2}(\mathbf{a}_{i+1} - \mathbf{a}_i)^T \Sigma_{i+1}^{-1}(\mathbf{a}_{i+1} - \mathbf{a}_i))$$

The integral of this function over all $\mathbf{a}_1, ..., \mathbf{a}_K$ gives the density of the observation sequence as a function of the parameters $\sigma_1, ..., \sigma_K, \Sigma_1, ..., \Sigma_K$ of the Hidden Markov model.

Study project 44

The notion of vector valued Brownian motion with possible non-stationary taken account and construction of the stochastic integral with respect to such a process can be discussed in a separate seminar. Once again, such a discussion would be useful from the standpoint of engineers wishing to delve deeper into stochastic filtering aspects and apply these ideas to practical problems like mechanical systems perturbed by white or coloured Gaussian noise. Let $B_1(t), ..., B_n(t)$ be n standard independent Brownian motion processes. The Ito rule for these processes is $dB_i(t)dB_j(t) = \delta_{ij}dt$. If $f(x_1, ..., x_n)$ is a twice differentiable function of n variables, then the Ito rule gives

$$df(B_1(t), ..., B_n(t)) = \sum_{i=1}^{n} \frac{\partial f}{\partial x_i}(B_1(t), ..., B_n(t))dB_i(t)$$

$$+ \frac{1}{2}\sum_{i,j=1}^{n} \frac{\partial^2 f}{\partial x_i \partial x_j}(B_1(t), ..., B_n(t))dB_i(t)dB_j(t)$$

$$= \sum_{i=1}^{n} \frac{\partial f}{\partial x_i}(B_1(t), ..., B_n(t))dB_i(t) + \frac{1}{2}\sum_{i=1}^{n} \frac{\partial^2 f}{\partial x_i^2}(B_1(t), ..., B_n(t))dt$$

This differential rule is to be interpreted in terms of the following stochastic integral equation:

$$f(B_1(t), ..., B_n(t)) - f(B_1(s), ..., B_n(s)) = \int_s^t \sum_{i=1}^{n} \frac{\partial f}{\partial x_i}dB_i(\tau)$$

$$+ \int_s^t \sum_{i=1}^{n} \frac{\partial^2 f}{\partial x_i^2}(B_1(\tau), ..., B_n(\tau))d\tau$$

Study project 45

A seminar on the determination of the phase of an LTI system using bispectral analysis. It is known that the magnitude spectrum can be determined from a power spectral analysis of the output with a white noise input. Let $\{w(n)\}$ be white noise with zero mean, variance σ^2 and skewness γ. This means that

$$\mathbb{E}(w(n)w(m)) = \sigma^2 \delta(n - m), \mathbb{E}(w(n)w(m)w(k)) = \gamma \delta(n - m)\delta(n - k)$$

If $\{h(n)\}$ is the impulse response of an LTI system and $w(n)$ is input to this system, then the power spectrum of the output $y(n) = \sum_k h(k)w(n - k)$, defined as the Fourier transform of the output autocorrelation function, is given by

$$S_{yy}(\omega) = \sum_{k=-\infty}^{\infty} \mathbb{E}(y(n)y(n + k))exp(-j\omega k) = \sigma^2|H(\omega)|^2$$

where

$$H(\omega) = \sum_{k=-\infty}^{\infty} h(m)exp(-j\omega m)$$

is the frequency response or transfer function of the system. The bispectrum of the output, defined as the Fourier transform of the third moment sequence

$$B(\omega_1, \omega_2) = \sum_{k,m=-\infty}^{\infty} \mathbb{E}(y(n)y(n+k)y(n+m))exp(-j(\omega_1 k + \omega_2 m))$$

evaluates to

$$B(\omega_1, \omega_2) = \gamma.H(\omega_1)H(\omega_2)\bar{H}(\omega_1 + \omega_2)$$

The magnitude of the transfer function can be determined from the output power spectrum as

$$|H(\omega)| = \sqrt{S_{yy}(\omega)/\sigma^2}$$

The phase of the transfer satisfies the following equation (which is derived by taking arguments of the bispectrum equation)

$$\psi(\omega_1, \omega_2) = \phi(\omega_1) + \phi(\omega_2) - \phi(\omega_1 + \omega_2)$$

where $\psi(\omega_1, \omega_2) = Arg(B(\omega_1, \omega_2))$ and $\phi(\omega) = Arg(H(\omega))$. Differentiating the above equation with respect to ω_2 and setting $\omega_2 = 0$ gives us the following differential equation for the transfer function phase:

$$\frac{\partial \psi}{\partial \omega_2}(\omega_1, 0) = \phi'(0) - \phi'(\omega_1)$$

This integrates to give

$$\phi(\omega_1) = \phi(0) + \phi'(0)\omega_1 - \int_0^{\omega_1} \frac{\partial \psi}{\partial \omega_2}(u, 0)du$$

The of the transfer function at zero frequency is an arbitrary additive factor. The phase $\phi(\omega)$ can be determined upto an arbitrary linear phase factor, the slope of which is given by $\phi'(0)$.

Study project 46

Levy oscillation property of Brownian motion can be discussed in a seminar. The essential content of this property is contained in the formula $\int_0^T (dB(t))^2 = T$ in the mean square sense. This is equivalent to stating that the limit in the mean square sense of $\sum_{k=0}^{N-1}(B(t_{k+1}) - B(t_k))^2$ as the partition size $max(t_{k+1} - t_k)$ tends to zero equals T where $0 = t_0 < t_1 < ... < t_N = T$. The proof of this property is based upon the following argument: Call $B(t_{k+1}) - B(t_k)$ as δB_k. Then,

$$(\sum_{k=0}^{N-1}(\delta B_k)^2 - T)^2 = \sum_{k=0}^{N-1}(\delta B_k)^4 + 2\sum_{0 \leq k < j \leq N-1}(\delta B_k)^2(\delta B_j)^2 + T^2 - 2T\sum_{k=0}^{N-1}(\delta B_k)^2$$

For $k < j$, δB_k and δB_j are independent random variables and hence

$$\mathbb{E}((\delta B_k)^2(\delta B_j)^2) = \mathbb{E}(\delta B_k)^2\mathbb{E}(\delta B_j)^2 = \delta t_k \delta t_j$$

Also $\mathbb{E}(\delta B_k)^4 = 3(\delta t_k)^2$ since δB_k is a Gaussian random variable with mean zero and variance δt_k. Thus,

$$\mathbb{E}(\sum_{k=0}^{N-1}(\delta B_k)^2 - T)^2 = 3\sum_{k=0}^{N-1}(\delta t_k)^2 + 2\sum_{0\leq k<j\leq N-1}\delta t_k \delta t_j - 2T\sum_{k=0}^{N-1}\delta t_k + T^2$$

Clearly,

$$\sum_{k=0}^{N-1}(\delta t_k)^2 \leq max\delta t_k \sum \delta t_j = T.max\delta t_k \to 0$$

as the partition size converges to zero. Also,

$$2\sum_{0\leq k<j\leq N-1}\delta t_k \delta t_j = (\sum \delta t_k)^2 - \sum(\delta t_k)^2 \to T^2$$

Thus, the limit of the above quantity equals

$$2T^2 - 2T^2 = 0$$

which proves the Levy oscillation property. The construction of the Ito stochastic integral should also be discussed in the context of stochastic filtering theory. The essential point is the notion of an adapted process. Let $\{B(t), t \geq 0\}$ be standard Brownian motion, i.e. a separable Gaussian random process having continuous sample trajectories, i.e., $P(lim_{s\to t}B(s) = B(t) for all t \in [0,1]) = 1$ and $\mathcal{F}_t = \sigma\{B(s), s \leq t\}$ the σ field generated by the samples of the Brownian motion process upto time t and $\mathbb{E}B(t) = 0, \mathbb{E}B(t)B(s) = min(t, s)$. Note that $\{\mathcal{F}_t\}_{t\geq 0}$ forms an increasing family of sub-σ-fields of the entire probability space $(\Omega, \mathcal{F}, P)$ on which the process is defined. A process $\xi(t)$ is said to be adapted to the Brownian motion process, if for each t, the random variable $\xi(t)$ is measurable with respect to the σ field $\mathcal{F}_t$, i.e. writing $\xi(t)$ as $\xi(t, \omega)$, we require that $\{\omega : \xi(t, \omega) \in B\} \in \mathcal{F}_t$ for all $t \geq 0$ where B is any Borel subset of $\mathbb{R}$. For such a process, we want to construct the integral $X(t) = \int_0^t \xi(s)dB(s)$. The integral is constructed as the limit of the sequence $\sum_{k=0}^{N-1} \xi(t_k)(B(t_{k+1}) - B(t_k))$, where $P : 0 = t_0 < t_1 < ... < t_N = t$ is a partition of $[0, t]$. For engineers, it suffices to consider this limit in the mean square sense. The proof of the existence of the mean square limit is based on proving that this sequence of partial Riemann-Stieltjes sums is Cauchy. The adaptedness of the process $\xi(.)$ implies the following two properties: $\mathbb{E}\int_0^T \xi(s)dB(s) = \int_0^t \mathbb{E}(\xi(s)dB(s)) = 0$ and if ξ, η are two adapted processes, then

$$\mathbb{E}(\int_0^t \xi(s)dB(s) \int_0^t \eta(s)dB(s)) = \int_0^t \mathbb{E}(\xi(s)\eta(s))ds$$

The first property follows from the independence of $\xi(t_k)$ and δB_k. This implies

$$\mathbb{E}\sum \xi(t_k)\delta B_k = \sum \mathbb{E}\xi(t_k)\delta B_k = \sum \mathbb{E}\xi(t_k)\mathbb{E}\delta B_k = 0$$

$$\mathbb{E}(\sum \xi(t_k)\delta B_k).(\sum \eta(t_j)\delta B_j) = \sum_{k,j} \mathbb{E}(\xi(t_k)\eta(t_j)\delta B_k\delta B_j)$$

For $k = j$, we have

$$\mathbb{E}(\xi(t_k)\eta(t_j)\delta B_k\delta B_j) = \mathbb{E}(\xi(t_k)\eta(t_k)\delta B_k^2) = \mathbb{E}(\xi(t_k)\eta(t_k))\delta t_K$$

since the random variable $\xi(t_k)\eta(t_k)$ is independent of δB_k. For $k > j$, we shall prove that

$$\mathbb{E}(\xi(t_k)\eta(t_j)\delta B_k\delta B_j) = 0$$

and by symmetry this will also equal zero for $k < j$. For $k > j$, the random variable $\xi(t_k)\eta(t_j)\delta B_j$ is independent of δB_k. Thus the above expectation equals

$$\mathbb{E}(\xi(t_k)\eta(t_j)\delta B_j)\mathbb{E}\delta B_k = 0$$

Thus,

$$\mathbb{E}(\sum_k \xi(t_k)\delta B_k \sum_j \eta(t_j)\delta B_j) = \sum_k \mathbb{E}(\xi(t_k)\eta(t_k))\delta t_k$$

and by taking mean square limits, we infer that

$$\mathbb{E}(\int_0^t \xi(s)dB(s) \int_0^t \eta(s)dB(s)) = \int_0^t \mathbb{E}(\xi(s)\eta(s))ds$$

which is the desired property. This last property can also be stated as in the form of a Hilbert space isomorphism. If $\mathcal{H}$ denotes the space of all adapted processes ξ defined over the time interval $[0, T]$, with the property $\int_0^T \mathbb{E}\xi(s)^2 ds < \infty$, we can introduce an inner product on this space by the formula

$$< \xi, \eta >= \int_0^T \mathbb{E}(\xi(s)\eta(s))ds$$

Then, the map $\xi \to \int_0^T \xi(s)dB(s)$ is an isomorphism between the Hilbert spaces $\mathcal{H}$ and the Hilbert subspace of all random variables in $L^2(\Omega, \mathcal{F}, P)$ which are stochastic integrals of L^2 adapted processes ξ with respect to Brownian motion.

Study project 47

The solution to the stochastic differential equation

$$dX(t) = (a + bX(t))dt + (c + fX(t))dB(t)$$

where $B(.)$ is standard Brownian motion. This can be cast in the form

$$dX(t) = (adt + cdB(t)) + (bdt + fdB(t))X(t)$$

We try a solution of the form

$$X(t) = exp(\lambda t + \mu B(t))X(0) + \int_0^t exp(\lambda(t - \tau) + \mu(B(t) - B(\tau)))(ad\tau + cdB(\tau))$$

Observe that for such an $X(t)$,

$$dX(t) = ((\lambda + \mu^2/2)dt + \mu dB(t))exp(\lambda t + \mu B(t))X(0)$$

$$+(adt + cdB(t)) + ((\lambda + \mu^2/2)dt + \mu dB(t))\int_0^t exp(\lambda(t - \tau) + \mu(B(t) - B(\tau)))(ad\tau + cdB(\tau))$$

For $X(t)$ to satisfy the given stochastic differential equation, this must agree with

$$(adt + cdB(t)) + (bdt + fdB(t))X(t) =$$

$$adt + cdB(t) + (bdt + fdB(t))exp(\lambda t + \mu B(t))X(0)$$

$$+(bdt + fdB(t))\int_0^t exp(\lambda(t - \tau) + \mu(B(t) - B(\tau)))(ad\tau + cdB(\tau))$$

It follows that
$$\lambda + \mu^2/2 = a, \mu = f$$

Thus,

$$\lambda = f - f^2/2$$

Study project 48

The notion of a compound Poisson process as a prototype of a Markov process can be introduced in a separate seminar. Events occur at times $t_i, i = 1, 2, ...$, forming the arrival times of a Poisson process. At each occurrence, a random quantity X_i is added to the system with X_i having the distribution F. At time t, $N(t)$ occurrences have taken place with $N(t)$ having the Poisson distribution with mean λt. The total quantity added to the system at time t is a Markov process described by the formula $Z(t) = \sum_{i=1}^{N(t)} X_i$. Its distribution is given by

$$F_Z(t, x) = \sum_{n=0}^{\infty} exp(-\lambda t)\frac{(\lambda t)^n}{n!}F^{n*}(x)$$

where $F^{n*} = F * F * ... * F$ is the n fold convolution of F with itself. Expressing the distribution function as a set function defined on the Borel sets of $\mathbb{R}$, we can equivalently express this as

$$F_Z(t, E) = \sum_{n=0}^{\infty} exp(-\lambda t)\frac{(\lambda t)^n}{n!}F^{n*}(E)$$

$F_Z(t, E)$ satisfies the forward Kolmogorov or Fokker-Planck equation:

$$\frac{\partial F_Z(t, E)}{\partial t} = -\lambda F_Z(t, E) + \lambda \sum_{n=1}^{\infty} exp(-\lambda t)\frac{(\lambda t)^{n-1}}{(n-1)!}F^{n*}(E)$$

$$= -\lambda F_Z(t, E) + \lambda \int F_Z(t, E - z)dF(z)$$

Equivalently in terms of distribution functions,

$$\frac{\partial F_Z(t, x)}{\partial t} = -\lambda F_Z(t, x) + \lambda \int F_Z(t, x - z)dF(z)$$

Study project 49

This project deals with studying filtering of random processes through linear systems and the relationships between the correlation functions obtained after the filtering. Suppose $\{u(t)\}$ and $\{v(t)\}$ are two random processes with cross correlation function $< u(t)v(s) >= R_{uv}(t, s)$. Suppose $\{u(t)\}$ is passed through a linear system having impulse response $h(t, s)$ and $\{v(t)\}$ is passed through a linear system having impulse response $g(t, s)$. The respective outputs are then given by $x(t) = \int_{-\infty}^{\infty} h(t, s)u(s)ds$ and $y(t) = \int_{-\infty}^{\infty} g(t, s)v(s)ds$. The cross correlation between the outputs is given by

$$R_{xy}(t, s) =< \int h(t, \alpha)u(\alpha)d\alpha \int g(s, \beta)v(\beta)d\beta >$$

$$= \int h(t, \alpha)g(s, \beta) < u(\alpha)v(\beta) > d\alpha d\beta$$

$$= \int_{-\infty}^{\infty} \int_{-\infty}^{\infty} h(t, \alpha)g(s, \beta)R_{uv}(\alpha, \beta)d\alpha d\beta$$

This identity can be described in a more picturesque way. Let T denote the action of the first system $h(t, s)$ upon a signal, i.e. $Tf(t) = \int h(t, s)f(s)ds$ and let S denote the action of the second system $g(t, s)$ upon a signal, i.e., $Sf(t) = \int g(t, s)f(s)ds$. T and S can also act on functions of two variables, i.e., if $R(t, s)$ is a function of two variables like a correlation function, then

$$T_1\{R(t, s)\} = \int h(t, \alpha)R(\alpha, s)d\alpha$$

$$T_2\{R(t, s)\} = \int h(s, \alpha)R(t, \alpha)d\alpha$$

with analogous definitions for S_1 and S_2. Then,

$$x(t) = T\{u(t)\}, y(t) = S\{v(t)\}$$

$$R_{xy}(t,s) = < x(t)y(s) > = < T\{u(t)\}.S\{v(s)\} > = T_1 S_2 < u(t), v(s) > = T_1 S_2 R_{uv}(t,s)$$

In particular,

$$R_{xx}(t,s) = T_1 T_2 R_{uu}(t,s) \qquad R_{yy}(t,s) = S_1 S_2 R_{vv}(t,s)$$

Study project 50

If X, Y are two independent random variables, we wish to determine the density of $X.Y, X/Y$. This has the following application in statistical mechanics. Suppose X is the energy of the system and we add a random number Y of identical copies to this system. Then, the energy of the system becomes $X.Y$. Likewise, if we split the given system into a random number Y of systems having the same energy, the energy of each subsystem becomes X/Y. So determining the statistical properties of the system resulting after a number of copies have been added or after the system has been split into identical copies, requires computing the distributions of the variables $X.Y$ and X/Y. Assuming Y to take only positive values, we have

Study project 51

For the AR(1) process, the dynamics is given by the first order difference equation $x(n) = ax(n-1) + w(n)$ where $w(n)$ is an $N(0, \sigma^2)$ sequence with autocorrelation function $\sigma^2 \delta(k)$. Determine the autocorrelation sequence of $x(n)$. In general, for the $AR(p)$ process $x(n) + a_1 x(n-1) + ... + a_p x(n-p) = w(n)$ where $w(n)$ is an iid $N(0, \sigma^2)$ sequence, determine the autocorrelation sequence of $x(n)$ in terms of the roots of the polynomial $z^p + a_1 z^{p-1} + ... + a_{p-1} z + a_p$. Assume all the roots have magnitudes lesser than unity. For the same process, determine via computer simulations and ergodicity assumptions, the dependence of the autocorrelation function on the parameters $a_1, ..., a_p$.

Study project 52

Given a random variable X and another finite set of random variables $Y_1, Y_2, ..., Y_n$ having a joint density $f_{X,Y_1,...,Y_n}(x, y_1, ..., y_n)$, we want to construct the optimal mean square estimate of X based on the collection $(Y_1, ..., Y_n)$. The most general such estimate is a Borel function of the variables $(Y_1, ..., Y_n)$, i.e.,

$$\hat{X} = \phi(Y_1, ..., Y_n)$$

Adopting vectorial notation,

$$\hat{X} = \phi(\mathbf{Y})$$

By optimal mean square estimate, we mean that the function ϕ must be chosen so that the mean square error $\mathbb{E}(X - \hat{X})^2$ is a minimum. Let ϕ_0 denote the optimal Borel function, i.e.

$$\phi_0 = argmin_\phi \mathbb{E}(X - \phi(\mathbf{Y}))^2$$

If $\psi(\mathbf{Y})$ is any other function of the variables $\mathbf{Y}$, and ϵ any real number, then using the estimate $\phi_0(\mathbf{Y}) + \epsilon\psi(\mathbf{Y})$ will produce a larger mean square estimation error than using $\phi_0(\mathbf{Y})$, i.e.,

$$\mathbb{E}(X - \phi_0(\mathbf{Y})^2 \leq \mathbb{E}(X - \phi_0(\mathbf{Y}) - \epsilon\psi(\mathbf{Y}))^2$$

for all $\epsilon \in \mathbb{R}$. In other words, the function

$$F(\epsilon) = \mathbb{E}(X - \phi_0(\mathbf{Y}) - \epsilon\psi(\mathbf{Y}))^2$$

attains its minimum when $\epsilon = 0$. This condition can be expressed as $F'(0) = 0$. Now,

$$F'(\epsilon) = \frac{d}{d\epsilon}\mathbb{E}(X - \phi_0(\mathbf{Y}) - \epsilon\psi(\mathbf{Y}))^2$$

$$= -2\mathbb{E}((X - \phi_0(\mathbf{Y}) - \epsilon\psi(\mathbf{Y}))\psi(\mathbf{Y}))$$

Thus,

$$F'(0) = -2\mathbb{E}((X - \phi_0(\mathbf{Y}))\psi(\mathbf{Y}))$$

Setting this equal to zero, we obtain the following condition for $\phi_0(\mathbf{Y})$ to be the optimal mean squared estimate of X: For all Borel functions ψ, we must have

$$\mathbb{E}((X - \phi_0(\mathbf{Y}))\psi(\mathbf{Y})) = 0$$

This is equivalent to the condition

$$\mathbb{E}\psi(\mathbf{Y})\mathbb{E}(X - \phi_0(\mathbf{Y}|\mathbf{Y}) = 0$$

If this is to hold for all ψ, we must have

$$\mathbb{E}(X - \phi_0(\mathbf{Y}|\mathbf{Y}) = 0$$

or

$$\phi_0(\mathbf{Y}) = \mathbb{E}(X|\mathbf{Y})$$

In other words, the optimal mean square estimate of X based on the data $\mathbf{Y}$ equals the conditional expectation of X given $\mathbf{Y}$. For example, suppose

Study project 53

The notion of a confidence interval for constructing reliable estimates of parameters is fundamental in mathematical statistics. A typical example is the following. Suppose $X_1, X_2, ..., X_n$ are independent random variables with mean μ and variance σ^2. If n is large, then we can expect by the central limit theorem, $Z = \frac{\sum_{i=1}^{n} X_i - n\mu}{\sigma\sqrt{n}}$ to have nearly the standard normal distribution. Let $S_n = \sum_{i=1}^{n} X_i$. Then, if Φ denotes the standard normal distribution, we have approximately,

$$P(-c \leq \frac{S_n - n\mu}{\sigma\sqrt{n}} \leq c\} = \Phi(c) - \Phi(-c) = 2\Phi(c) - 1$$

Suppose that σ is known and we want to estimate μ with a given degree of accuracy. Let α be number close to unity and choose c so that $2\Phi(c) - 1 = \alpha$. Then we get from the above equation

$$P(-c\sigma\sqrt{n} \leq S_n - n\mu \leq c\sigma\sqrt{n}) = \alpha$$

or

$$P((S_n - c\sigma\sqrt{n})/n\mu \leq (S_n + c\sigma\sqrt{n})/n) = \alpha$$

If we have taken measurements $X_1, ..., X_n$, then the above equation asserts that with a confidence of 100α percent, the mean μ of X will fall in the interval $[(S_n - c\sigma\sqrt{n})/n, (S_n + c\sigma\sqrt{n})/n]$. The closer the value of α to unity, the larger will be the value of c and this would imply a larger confidence interval, i.e. more uncertainty in our estimate of the mean. Another example is as follows. Suppose that $X_1, ..., X_n$ are independent measurements of a Gaussian random variable with unknown mean μ and known variance σ^2. We want to construct a confidence interval for the next measurement X based on these n measurements. First observe that $\bar{X} = S_n/n$ is normal with mean μ and variance σ^2/n. The next measurement X is independent of S_n and hence of $\bar{X}$. The random variable $Y = \bar{X} - X$ is therefore normal with zero mean and variance $\sigma^2(1 + 1/n)$. Thus,

$$P(-c \leq (\bar{X} - X)/\sigma\sqrt{1 + 1/n} \leq c) = 2\Phi(c) - 1$$

Choose c so that $2\Phi(c) - 1 = \alpha$. Then, the above equation can be expressed as

$$P(\bar{X} - c\sigma\sqrt{1 + 1/n} \leq X \leq \bar{X} + c\sigma\sqrt{1 + 1/n}) = \alpha$$

that is, with a confidence of 100α, the new measurement X will fall in the interval defined as $[\bar{X} - c\sigma\sqrt{1 + 1/n}, \ \bar{X} + c\sigma\sqrt{1 + 1/n}]$. Note that the unknown mean μ has disappeared from the picture completely.

Study project 54

Let $X_1, ..., X_n$ be identically distributed independent exponential random variables with mean $1/\theta$. The density of X_i is $f(x, \theta) = \theta.exp(-\theta x)$. We want to determine the maximum likelihood estimate of θ based on $X_1, ..., X_n$. Observe that this amounts to maximizing the log likelihood function

$$L(\mathbf{X}, \theta) = \sum_{i=1}^{n} f(X_i, \theta) = \sum_{i=1}^{n} -\theta X_i + log(\theta) = -n\theta\bar{X} + nlog(\theta)$$

where $\bar{X} = \frac{1}{2}\sum_{i=1}^{n} X_i$. The likelihood equation

$$\frac{\partial L(\mathbf{X}, \theta)}{\partial\theta}\Big|_{\theta=\hat{\theta}} = 0$$

gives

$$-\bar{X} + 1/\hat{\theta} = 0$$

or

$$\hat{\theta} = \frac{1}{\bar{X}}$$

The maximum likelihood estimator of θ equals the reciprocal of the sample mean. The maximum likelihood estimator of $1/\theta$ equals $\bar{X}$. $(1/\theta) = 1/\hat{\theta} = \bar{X}$. Its mean is given by

$$\mathbb{E}(1/\hat{\theta}) = \mathbb{E}\bar{X} = 1/\theta$$

This shows that $1/\hat{\theta}$ is a an unbiased estimate of the mean.

Study project 55

Let N particles be distributed amongst n boxes numbered $1, 2, ..., n$ at random. The number of ways in which the ith box can contain r_i particles, $i = 1, 2, ..., n$ is given by the multinomial formula

$$W(r_1, ..., r_n) = \frac{N!}{r_1!...r_n!}$$

This can be seen in the following way. First select r_1 particles in $\binom{N}{r_1}$ ways. Put these in the first box. Out of the remaining $N - r_1$ particles, select r_2 particles in $\binom{N-r_1}{r_2}$ ways. Put these in the second box. Continue in this fashion until the last box. In general the r_i particles for the ith box are selected in $\binom{N-r_1-...-r_{i-1}}{r_i}$ ways. The total number of ways resulting from this process is the product

$$\binom{N}{r_1}\binom{N-r_1}{r_2}\binom{N-r_1-r_2}{r_3}...\binom{N-r_1-...-r_{n-1}}{r_n}$$

$$= \frac{N!}{r_1!...r_n!}$$

Study project 56

An experiment consists of throwing a die having n faces and observing which face occurs. The probability of the ith face occurring is $p_i, i = 1, 2, ..., k$. We now repeat the experiment N independent times and count the number of times that each face occurs in the sequence. A sequence in which the ith face occurs r_i times with $i = 1, 2, ..., n$ has probability $p_1^{r_1}p_2^{r_2}...p_n^{r_n}$. The total number of sequences having r_i occurrences of the ith face is $\frac{N!}{r_1!...r_n!}$. Thus, the probability that the ith face appears r_i times with $i = 1, 2, ..., n$ equals

$$p(r_1, ..., r_n) = \frac{N!}{r_1!...r_n!}p_1^{r_1}...p_n^{r_n}$$

This is the multinomial distribution with parameters $N, p_1, ..., p_n$ also denoted by $M(N, p_1, ..., p_n)$. The distribution is over all n tuples $(r_1, ..., r_n)$ of non-negative integers satisfying the constraint

$r_1 + r_2 + ... + r_n = N$. The probability generating function of this distribution is given by

$$P(z_1, ..., z_n) = \sum_{r_i \geq 0, \sum r_i = N} p(r_1, ..., r_n) z_1^{r_1} ... z_n^{r_n}$$

$$= (p_1 z_1 + ... + p_n z_n)^N$$

This generating function can be derived via an alternate route. Consider a collection of n random variables $(X_1, ..., X_n)$, or equivalently, a random vector $\mathbf{X} = [X_1, ..., X_n]^T$. When the experiment is performed if the ith face turns up, then $X_i = 1$ and $X_j = 0$ when $j \neq i$. Now consider N independent copies of this random vector $\mathbf{X}_1, ..., \mathbf{X}_N$. Let $\mathbf{X}_i = [X_1(i), ..., X_n(i)], i = 1, 2, ..., N$. Consider the sum of these N random vectors

$$\mathbf{S} = \sum_{i=1}^{N} \mathbf{X}_i = [\sum_{i=1}^{N} X_1(i), \sum_{i=1}^{N} X_2(i), ..., \sum_{i=1}^{N} X_n(i)]^T$$

$$= [S_1, S_2, ..., S_n]$$

Then, S_j gives the number of times that the jth face occurred in the N trials. The generating function of the random vector $\mathbf{S}$ is given by

$$P(z_1, ..., z_n) = \mathbb{E}(z_1^{S_1} ... z_n^{S_n}) = \mathbb{E}(z_1^{\sum X_1(i)} ... z_n^{\sum X_n(i)})$$

$$= \Pi_{i=1}^{N} \mathbb{E}(z_1^{X_1(i)} ... z_n^{X_n(i)})$$

Now, for any i

$$\mathbb{E}(z_1^{X_1(i)} ... z_n^{X_n(i)}) = p_1 z_1 + ... + p_n z_n$$

Thus,

$$P(z_1, ..., z_n) = (p_1 z_1 + ... + p_n z_n)^N$$

Study project 57

Ornstein-Uhlenbeck process. Diffusion process satisfying the following Fokker-Planck equation.

$$\frac{\partial u(t, x)}{\partial t} = -\rho x \frac{\partial u(t, x)}{\partial x} + \frac{1}{2} \frac{\partial^2 u(t, x)}{\partial x^2}$$

Make the transformation

$$x = y.exp(\rho t), \tau = (1 - exp(-2\rho t))/2\rho$$

The new function obtained is

$$u(t, x) = u(\frac{1}{2\rho} log(1 - 2\rho\tau), y.exp(\rho t)) = u(\frac{1}{2\rho} log(1 - 2\rho\tau), y(1 - 2\rho\tau)^{-1/2}) = w(\tau, y)$$

say. This equation is equivalent to the equation

$$u(t,x) = w((1 - exp(-2\rho t))/2\rho, x.exp(-\rho t))$$

Assume that w satisfies the diffusion equation, i.e.

$$\frac{\partial w}{\partial \tau} = \frac{1}{2}\frac{\partial^2 w}{\partial y^2}$$

We want to deduce from this that u will satisfy the Ornstein-Uhlenbeck Fokker-Planck equation described at the beginning. Observe that

$$\frac{\partial u}{\partial t} = exp(-2\rho t)w_\tau - x.\rho.exp(-\rho t)w_y$$

$$\frac{\partial u}{\partial x} = exp(-\rho t)w_y$$

$$\frac{\partial^2 u}{\partial x^2} = exp(-2\rho t)w_{yy}$$

Thus

$$\frac{\partial u}{\partial t} + \rho x \frac{\partial u}{\partial x} - \frac{1}{2}\frac{\partial^2 u}{\partial x^2}$$

$$= exp(-2\rho t)w_\tau - x\rho.exp(-\rho t)w_y + x\rho.exp(-\rho t)w_y - \frac{1}{2}exp(-\rho t)w_{yy}$$

$$= exp(-2\rho t)(w_\tau - \frac{1}{2}w_{yy}) = 0$$

proving that u satisfies the Fokker-Planck equation for the Ornstein-Uhlenbeck process.

Study project 58

Explain how you would determine $\mathbb{E}x(t)^2$ when $\{x(t)\}$ satisfies the differential equation

$$\frac{d^n x(t)}{dt^2} + a_1\frac{d^{n-1}x}{dt^{n-1}} + ... + a_{n-1}\frac{dx(t)}{dt} + a_n x = w(t)$$

with $w(t)$ being white noise with mean zero and autocorrelation $\mathbb{E}(w(t)w(s)) = \delta(t-s)$.

Ans: The transfer function of the system is given by the formula

$$H(s) = \frac{1}{s^n + a_1 s^{n-1} + ... + a_{n-1}s + a_n}$$

Assume that this has the following partial fraction expansion

$$H(s) = \sum_{i=1}^{d}\sum_{r=1}^{m_i-1}\frac{A_{i,r}}{s - s_i)^r}$$

The impulse response is then given by the formula

$$h(t) = \sum_{i=1}^{d} \sum_{r=1}^{m_i-1} A_{i,r} t^{r-1} exp(s_i t)/(r-1)!$$

for $t \geq 0$ and $h(t) = 0$ for $t < 0$. Assuming that the system has been in operation from $t = -\infty$, the solution for the process $\{x(t)\}$ is given by

$$x(t) = \int_0^\infty h(\tau)w(t-\tau)d\tau$$

Thus,

$$R_{xx}(\tau) = \mathbb{E}(x(t+\tau)x(t)) = \int_0^\infty h(t)h(t+\tau)ds$$

assuming that $\tau > 0$. In particular,

$$\mathbb{E}x(t)^2 = R_{xx}(0) = \int_0^\infty h(t)^2 dt = \int_0^\infty |h(t)|^2 dt$$

$$\sum_{i,j=1}^{d} \sum_{r=1}^{m_i-1} \sum_{s=1}^{m_j-1} A_{i,r}\bar{A}_{j,s} \int_0^\infty t^{r+s-2} exp((s_i + \bar{s}_j)t)dt/(r-1)!(s-1)!$$

$$= \sum_{i,j=1}^{d} \sum_{r=1}^{m_i-1} \sum_{s=1}^{m_j-1} A_{i,r}\bar{A}_{j,s} \frac{(r+s-2)!}{(r-1)!(s-1)!(s_i + \bar{s}_j)^{r+s-1}}$$

Study project 59

Explain how you would determine $\mathbb{E}x^2(t)$ in the preceding project when the system is in operation from $t = 0$. Assume zero initial conditions, i.e. $x(0) = x'(0) = ... = x^{n-1}(0) = 0$.

Ans: It is easily seen that $x(t) = \int_0^t h(\tau)w(t-\tau)d\tau$. Thus,

$$\mathbb{E}x(t)^2 = \int_0^t h(\tau)^2 d\tau$$

In the limit as $t \to \infty$, this formula for the variance converges to the formula of the preceding problem where we assumed that the system has been in operation from $t = \infty$.

Study project 60

Obtain the complete solution including the statistics to the Ornstein-Uhlenbeck stochastic differential equation

$$dV(t) = -\gamma V(t) + \sigma dB(t)$$

Ans:

$$V(t) = \int_0^t exp(-\gamma(t-\tau))\sigma dB(\tau)$$

This shows that $\{V(t)\}$ is a Gaussian process with mean zero and variance

$$\mathbb{E}V(t)^2 = \sigma^2 \int_0^t exp(-2\gamma\tau)d\tau = \sigma^2(1 - exp(-2\gamma t))/2\gamma$$

The probability density of $V(t)$ $p(t,v)$ satisfies the Fokker-Planck equation

$$\frac{\partial p(t,v)}{\partial t} = \gamma\frac{\partial}{\partial v}vp(t,v) + \frac{\sigma^2}{2}\frac{\partial^2}{\partial v^2}p(t,v)$$

By directly solving this partial differential equation, it can be shows that $V(t)$ is a Gaussian random variable with zero mean and variance $\sigma^2(1 - exp(-2\gamma t))/2\gamma$. Note that $\{V(t)\}$ is the velocity process of a pollen particle moving in liquid with no external potential and when the molecular kicks on the particle can be described by a white noise forcing term.

Solved problem 1: Describe the Cauchy distribution.

The Cauchy distribution: Consider a probability distribution with density $f(x)$ having the characteristic function

$$\psi(t) = \mathbb{E}exp(itX) = \int_{-\infty}^{\infty} f(x)exp(itx)dx = exp(-a|t|)$$

Fourier inversion gives

$$f(x) = \frac{1}{2\pi} \int_{-\infty}^{\infty} \psi(t)exp(-itx)dt = \frac{a}{\pi(a^2 + x^2)}$$

Denote this density by $f_a(x)$ and the associated characteristic function by $\psi_a(t)$. The equation $\psi_a(t)\psi_b(t) = \psi_{a+b}(t)$ shows that $f_a(x) * f_b(x) = f_{a+b}(x)$, i.e. the family of Cauchy densities $\{f_a(x)\}_{a>0}$ forms a one parameter convolution semigroup of densities. This means that if X_a, X_b are two independent Cauchy random variables with parameters a, b, then $X_a + X_b$ is a Cauchy random variable with parameter $a + b$, i.e. $X_a + X_b$ has the same distribution as X_{a+b}.

Study project 61

The backward and the forward equations for a diffusion process. Let $q_t(x,y)dy$ denote the transition probability density. One defines for a fixed $\phi(y)$, the function

$$u(t,x) = \int_{-\infty}^{\infty} q_t(x,y)\phi(y)dy$$

Assuming the regularity conditions (1)

$$\frac{1}{t}\int_{|x-y|>\delta} q_t(x,y)dy \to 0$$

as $t \to 0$, (2)

$$\frac{1}{t}\int_{|x-y|\leq\delta} (y-x)q_t(x,y)dy \to a(x)$$

as $t \to 0$ and (3)

$$\frac{1}{t}\int_{|x-y|\leq\delta} (y-x)^2 q_t(x,y)dy \to b(x)$$

as $t \to 0$, the backward equation for u is easily derived:

$$\frac{\partial u}{\partial t} = a(x)\frac{\partial u}{\partial x} + \frac{1}{2}b(x)\frac{\partial^2 u}{\partial x^2}$$

Using this equation and the method of adjoints of an operator, the forward equation can be derived. For a given function $\psi(x)$, define

$$v(s,y) = \int_{-\infty}^{\infty} \psi(x)q_s(x,y)dx$$

Then,

$$\int_{-\infty}^{\infty} v(s,y)u(t,y)dy = \int \psi(x)dxq_s(x,y)dyq_t(y,z)dz\phi(z)$$

$$= \int \psi(x)\phi(z)q_{t+s}(x,z)dxdz$$

in view of the Chapman-Kolmogorov equation

$$q_{t+s}(x,z) = \int q_s(x,y)dyq_t(y,z)$$

This equation implies that

$$\frac{\partial}{\partial s}\int v(s,y)u(t,y)dy = \frac{\partial}{\partial t}\int v(s,y)u(t,y)dy$$

which is equivalent to

$$\int \frac{\partial v(s,y)}{\partial s}u(t,y)dy = \int v(s,y)\frac{\partial u(t,y)}{\partial t}dy$$

$$= \int v(s,y)(a(y)\frac{\partial u(t,y)}{\partial y} + \frac{1}{2}b(y)\frac{\partial^2 u(t,y)}{\partial y^2})dy$$

$$= -\int u(t,y)\frac{\partial}{\partial y}(v(s,y)a(y))dy + \frac{1}{2}\int u(t,y)\frac{\partial^2}{\partial y^2}((v(s,y)b(y))dy$$

on integrating by parts with vanishing boundary conditions. We can conclude that v satisfies the forward Kolmogorov equation

$$\frac{\partial v(s,y)}{\partial s} = -\frac{\partial}{\partial y}(a(y)v(s,y)) + \frac{1}{2}\frac{\partial^2}{\partial y^2}(b(y)v(s,y))$$

Study project 62

Examine the approximate mean and variance propagation for a one dimensional stochastic differential equation. The equation satisfied by the state $x(t)$ is

$$dx(t) = f(x(t),t)dt + g(x(t),t)dB(t)$$

Thus,

$$d\mathbb{E}x(t) = \mathbb{E}dx(t) = \mathbb{E}f(x(t),t)dt$$

or

$$\frac{d}{dt}\mathbb{E}x(t) = \mathbb{E}f(x(t),t)$$

Study project 63

Elucidate properties of independent random variables and learn how to simulate such variables using the 8085 microprocessor. If (X,Y) is a pair of independent random variables assuming values (x_i,y_j) with probabilities having the form p_iq_j where $1 \le i,j \le N$, then the two are called independent random variables. The first objective is how to simulate samples of such a pair using Monte-Carlo schemes and verify that they are really independent variables. We divide the interval $[0,1]$ into N^2 disjoint subintervals having lengths

$$p_1q_1, p_1q_2, ..., p_1q_N, p_2q_1, p_2q_2, ..., p_2q_N, ..., p_Nq_1, ..., p_Nq_N$$

and generate a random number U that is uniformly distributed over $[0,1]$. If this number falls in the $(i-1)N + j^{th}$ subinterval, we assign X the value x_i and Y the value y_j. In this way, we generate a large number say K of samples of the pair (X,Y). We then compute empirically the average of the quantity $a(X)b(Y)$ where a,b are two functions. This is done as follows. Assume that the samples generated are $(X_n,Y_n), n = 1,2,...,K$ and evaluate $\frac{1}{K}\sum_{n=1}^{K} a(X_n)b(Y_n)$. If our sample is a faithful reproduction of the independent pair (X,Y), then this sample average will be nearly equal to $\sum_{i=1}^{N} p_ia(x_i)\sum_{j=1}^{N} q_jb(x_j)$.

Study project 64

Deals with the conditional mean of one vector given another when the two are jointly normal. Let $\mathbf{x}$ and $\mathbf{y}$ be jointly distributed zero mean normal random vectors. Consider $\mathbf{z} = \mathbf{y} - \mathbf{Ax}$, where the matrix $\mathbf{A}$ is chosen so that $\mathbf{z}$ is uncorrelated with $\mathbf{x}$, i.e., $\mathbb{E}\mathbf{zx}^T = \mathbf{0}$. This means that $\mathbf{A} = \mathbf{R}_{yx}\mathbf{R}_{xx}^{-1}$. Since $\mathbf{z}, \mathbf{x}$ are uncorrelated jointly normal random vectors, they are independent. Thus, $\mathbb{E}(\mathbf{z}|\mathbf{x}) = \mathbb{E}(\mathbf{z}) = \mathbf{0}$. This implies that

$$\mathbb{E}(\mathbf{y}|\mathbf{x}) = \mathbf{Ax}$$

Generalize this formula to obtain the conditional mean of $\mathbf{y}$ given $\mathbf{x}$ when the two variables have non-zero means.

Unsolved problems

1. Suppose X has the $N(\mu, \sigma^2)$ distribution. Evaluate the density of X^2 and X^3. Write down the expression for the characteristic function of X. Using this expression, write down the expression for the first three moments of X.

2. Let X have the density $f(x) = \lambda exp(-\lambda x)U(x)$. Calculate $\mathbb{E}X$ and $V(X)$. Evaluate the characteristic function of X and the conditional probability $P(X > t + s | X > t)$.

3. Assume that X is a random variable whose density is given by $f(x|\theta) = \theta.exp(-\theta x)U(x)$. Consider the following estimate for the parameter θ: $\hat{\theta}(X) = aX + bexp(-X)$. Determine the optimal values of the coefficients a, b so that $\mathbb{E}(\hat{\theta}(X) - \theta)^2$ is minimized. Compare the mean squared error with the mean squared error for the maximum likelihood estimate of θ.

4. Let $X(1, n), X(2, n), ..., X(p, n), n = 0, 1, ..., N - 1$ be p different signals of duration N and let $Y(n), n = 0, 1, ..., N - 1$ be another signal of duration N. Write down the equations of the optimal filter that takes the first p signals as input and outputs $\sum_{k=1}^{p} h(k)X(k, n)$ so that this output is the optimal mean square estimate of the process $Y(n)$. Evaluate the mean squared error explicitly when $X(k, n) = a_k^n, k = 1, 2, ..., p$ and $Y(n) = a^n$.

5. Let X be the base of a triangle and Y the altitude. Assuming that X, Y have the joint density $f(x, y)$, write down an integral expression for the density of the triangle's area. Also write down an integral expression for the angles of a right angle triangle having X as its base length and Y as its altitude.

6. Consider an FIR filter having three taps h_0, h_1, h_2 that takes as input a process $X(n)$ having autocorrelation function $a^{|n|}$ and outputs a process $Y(n)$. Write down an expression for the power spectral density of the output. The output of this filter is passed through a two tap

FIR filter g_0, g_1 to yield an optimal mean squared estimate of the input $X(n)$. Determine the coefficients g_0, g_1.

7. What do you understand by a positive definite matrix. Give an example of a 2×2 positive definite matrix. Show that if A is positive definite, there exists a Gaussian random vector X having A as its correlation matrix. What characteristics of the random vector X do the diagonal and off diagonal entries of A correspond to.

8. A vector valued signal is defined by the following linear model:

$$\begin{pmatrix} X_1 \\ X_2 \end{pmatrix} = \begin{pmatrix} a_{11} & a_{12} \\ a_{21} & a_{22} \end{pmatrix} \begin{pmatrix} s_1 \\ s_2 \end{pmatrix} + \begin{pmatrix} w_1 \\ w_2 \end{pmatrix}$$

where w_1, w_2 are independent random variables having mean zero and variance σ^2 and s_1, s_2 are independent random variables having mean zero and variances P_1, P_2 respectively. Further, $\mathbb{E}(s_i w_j) = 0$ for $i, j = 1, 2$. Write down an expression for the correlation matrix of the vector X and determine its eigenvalues. Write a short note on the importance of this problem in array signal processing.

Solved problem 2

Develop a theory of convergence of the parameters of a memoryless non-linear system with additive noise when the input is Gaussian and the least mean square technique is adopted for updating the parameters.

Ans: The system output is defined by the equation

$$y(n) = G(x(n), \theta) + v(n)$$

where the input process $\{x(n)\}$ is an iid $N(0, \sigma^2)$. $\{v(n)\}$ is additive Gaussian noise. θ is a parameter of the system to be estimated. The LMS algorithm for estimating θ proceeds via the recursion

$$\theta(n+1) = \theta(n) - \mu \frac{\partial}{\partial \theta}(y(n) - G(x(n), \theta(n)))^2$$

$$= \theta(n) + 2\mu \frac{\partial G(x(n), \theta(n))}{\partial \theta} \partial \theta (y(n) - G(x(n), \theta(n)))$$

As $n \to \infty$, it is hoped that $\theta(n)$ will converge to a value that has a small random fluctuation about the true parameter θ. Substituting for the actual value of $y(n)$ into the above recursion gives us

$$\theta(n+1) = \theta(n) + 2\mu \frac{\partial G(x(n), \theta(n))}{\partial \theta}(G(x(n), \theta) - G(x(n), \theta(n)) + v(n))$$

Make the approximation

$$G(x(n), \theta(n)) = G(x(n), \theta) + (\theta(n) - \theta)\frac{\partial G(x(n), \theta)}{\partial \theta}$$

Then retaining only first order terms in the error $\theta(n) - \theta$ gives us the following approximate recursion:

$$\theta(n+1) = \theta(n) + 2\mu\frac{\partial G(x(n),\theta)}{\partial\theta}(v(n) - \frac{\partial G(x(n),\theta)}{\partial\theta}(\theta(n) - \theta))$$

or writing

$$\epsilon(n) = \theta(n) - \theta$$

we get

$$\epsilon(n+1) = \epsilon(n) + 2\mu z(n)(v(n) - z(n)\epsilon(n))$$

where $z(n) = \frac{\partial G(x(n),\theta)}{\partial\theta}$. This recursion can be cast as

$$\epsilon(n+1) = (1 - 2\mu z(n)^2)\epsilon(n) + 2\mu z(n)v(n)$$

Note that this recursion implies that $\epsilon(n)$ is expressible as a function of $x(k), v(k), k \leq n-1$ and is therefore independent of $z(n), v(n)$. Taking mean values of the above equation gives

$$< \epsilon(n+1) >= (1 - 2\mu < z(n)^2 >) < \epsilon(n) >$$

since assumed independence of $x(n)$ and $v(n)$ implies independence of $z(n)$ and $v(n)$ and since $v(n)$ has zero mean, $< z(n)v(n) >=< z(n) >< v(n) >= 0$. Setting $\sigma_z^2 =< z(n)^2 >$ (Note that since $z(n)$ does not necessarily have zero mean, this is not the variance of $z(n)$), we get

$$< \epsilon(n+1) >= (1 - 2\mu\sigma_z^2) < \epsilon(n) >$$

If the adaptation constant μ is selected so that $0 < \mu < 1/\sigma_z^2$, then $0 < 1 - 2\mu\sigma_z^2 < 1$ and this implies convergence of the mean value of the parameter to the true value of the parameter, i.e.,

$$< \epsilon(n) >=< \theta(n) - \theta >=< \theta(n) > -\theta \to 0$$

Study project 65

Analysis of stochastic differential equations with coloured noise. Suppose x_t satisfies the differential equation

$$\frac{dx_t}{dt} = f(x_t) + w_t$$

where w_t is Gaussian noise with autocorrelation function $\sigma^2 exp(-a|t-s|)$. Then, w_t is an Ornstein-Uhlenbeck process and hence the stationary solution to the sde

$$dw_t = aw_t + b.dB_t$$

where B_t is Brownian motion. In fact, the stationary solution to this equation is

$$w_t = b\int_{-\infty}^{t} exp(-a(t-s))dB_s$$

and hence

$$< w_t w_{t+u} >= b^2 \int_{-\infty}^{t} exp(-a(2t + u - 2s))ds = \frac{b^2}{2a} exp(-au), u \geq 0$$

Taking $b = \sigma\sqrt{2a}$, the desired correlation for w_t is obtained. Now, (x_t, w_t) is the extended state vector satisfying the sde

$$d \begin{pmatrix} x_t \\ w_t \end{pmatrix} = \begin{pmatrix} f(x_t) \\ aw_t \end{pmatrix} dt + \begin{pmatrix} 0 \\ 1 \end{pmatrix} dB_t$$

Mean and variance propagation equations can be derived for this problem. Suppose however we try to directly develop the mean and variance propagation equations for the process x_t. Let $\mu_t =< x_t >$. Then,

$$\frac{d\mu_t}{dt} =< f(x_t) > + < w_t >=< f(x_t) >$$

Now,

$$f(x_t) \approx f(\mu_t + \delta x_t) = f(\mu_t) + \delta x_t f'(\mu_t) + (\delta x_t)^2 f''(\mu_t)/2$$

where $\delta x_t = x_t - \mu_t$. Letting $V_t =< (\delta x_t)^2 >$ gives us

$$< f(x_t) >\approx f(\mu_t) + V_t f''(\mu_t)/2$$

so that the approximate mean propagation equation is

$$\frac{d\mu_t)}{dt} = f(\mu_t) + V_t f''(\mu_t)/2$$

To obtain the variance propagation equation, we note that

$$d(\delta x_t)^2 = d(x_t - \mu_t)^2 = 2(x_t - \mu_t)(dx_t - d\mu_t)$$

giving

$$dV_t = 2 < (x_t - \mu_t)(dx_t - d\mu_t) >= 2 < (x_t - \mu_t)(f(x_t)dt + w_t dt - d\mu_t) >$$

$$= 2 < \delta x_t f(x_t) > dt + 2 < \delta x_t w_t > dt$$

Now,

$$< \delta x_t f(x_t) >\approx< \delta x_t (f(\mu_t) + f'(\mu_t)\delta x_t >= f'(\mu_t)V_t$$

giving

$$\frac{dV_t}{dt} = 2f'(\mu_t)V_t + 2 < \delta x_t w_t >$$

Since w_t is correlated noise, the term $< \delta x_t w_t >$ is nonzero and we cannot simplify it. This shows the necessity of modeling w_t using an sde and considering the dynamics of the extended state vector (x_t, w_t).

Study project 66

The connection between the Ito and Stratanovich stochastic differential equations. The Stratanovich integral $S \int_0^T f(B_t, t) dB_t$ is defined as the limit of the partial sums

$$\sum_{k=0}^{n-1} \frac{1}{2} (f(B_{t_k}, t_k) + f(B_{t_{k+1}}, t_{k+1}))(B_{t_{k+1}} - B_{t_k})$$

where $0 = t_0 < t_1 < ... < t_n = T$ as $max(t_{k+1} - t_k))$ tends to zero. This can be expressed using the following compact notation

$$\sum_{k=0}^{n-1} \frac{1}{2} (f(B_k, t_k) + f(B_k + \delta B_k, t_k + \delta t_k)) \delta B_k$$

Using the formula

$$f(B_k + \delta B_k, t_k + \delta t_k) = f(B_k, t_k) \frac{\partial f(B_k, t_k)}{\partial x} \delta B_k + \frac{1}{2} \frac{\partial^2 f(B_k, t_k)}{\partial x^2} \delta t_k + \frac{\partial f(B_k, t_k)}{\partial t} \delta t_k$$

and noting that $\delta B_k \delta t_k = 0, (\delta B_k)^2 = \delta t_k$ gives for the partial sum the value

$$\sum_{k=0}^{n-1} f(B_k, t_k) \delta B_k + \frac{1}{2} \sum_{k=0}^{n-1} \frac{\partial f(B_k, t_k)}{\partial x} \delta t_k$$

This gives us the following equation connecting the Stratanovich integral to the Ito integral:

$$S \int_0^T f(B_t, t) dB_t = I \int_0^T f(B_t, t) dB_t + \frac{1}{2} \int_0^T \frac{\partial f(B_t, t)}{\partial x} dt$$

The Stratanovich integral can be extended to more general diffusion processes. Suppose x_t is a process satisfying the Stratanovich sde

$$dx_t = f(x_t) dt + g(x_t) dB_t$$

This is to be interpreted as the integral equation

$$x_t - x_0 = \int_0^t f(x_s) ds + S \int_0^t g(x_s) dB_s$$

We define Ito integral $I \int_0^T h(x_t) dB_t$ as the limit of the partial sums $\sum_{k=0}^{n-1} h(x_{t_k}) \delta B_k$ and the Stratanovich integral $S \int_0^T h(x_t) dB_t$ as the limit of the partial sums $\sum_{k=0}^{n-1} \frac{1}{2} (h(x_{t_k}) + h(x_{t_{k+1}})) \delta B_k$ using an abbreviated notation

$$I \int_0^T h(x_t) dB_t = lim \sum_{k=0}^{n-1} h(x_k) \delta B_k$$

$$S \int_0^T h(x_t)dB_t = lim \sum_{k=0}^{n-1} \frac{1}{2}(h(x_k) + h(x_{k+1}))\delta B_k$$

We can write in terms of the Ito definition,

$$S \int_0^T h(x)dB = \int_0^T \frac{1}{2}(h(x) + h(x + dx))dB = \int \frac{1}{2}(h(x) + h(x) + h'(x)dx)dB$$

$$= \int_0^T h(x)dB + \frac{1}{2}\int h'(x)g(x)dt$$

since $dxdB = (f(x)dt + g(x)dB)dB = g(x)dt$. In particular,

$$S \int g(x)dB = \int_0^T g(x)dB + \frac{1}{2}\int_0^T g'(x)g(x)dt$$

and the Stratanovich sde is equivalent to the following Ito sde:

$$dx = (f(x) + g(x)g'(x)/2)dt + g(x)dB$$

More generally, consider the vector Stratanovich sde

$$d\mathbf{x}_t = f(t, \mathbf{x}_t)dt + \mathbf{G}(t, \mathbf{x}_t)d\mathbf{B}_t$$

which when expressed in terms of components, reads

$$dx_i(t) = f_i(t, \mathbf{x}(t))dt + \sum_{j=1}^{p} g_{ij}(t, \mathbf{x}(t))dB_j(t), i = 1, 2, ..., n$$

This is to be interpreted as the integral equation

$$x_i(t) - x_i(0) = \int_0^t f_i(s, \mathbf{x}(s))ds + \sum_{j=1}^{p} S \int_0^t g_{ij}(s, \mathbf{x}(s))dB_j(s)$$

where

$$S \int_0^t g_{ij}(s, \mathbf{x}(s))dB_j(s) = \sum_j \int_0^t \frac{1}{2}(g_{ij}(s, \mathbf{x}(s)) + g_{ij}(s + ds, \mathbf{x}(s) + d\mathbf{x}(s)))dB_j(s)$$

$$= \sum_j \int_0^t g_{ij}(s, \mathbf{x}(s))dB_j(s) + \frac{1}{2}\int_0^t \sum_{j,k} \frac{\partial g_{ij}(s, \mathbf{x}(s))}{\partial x_k}dx_k(s)dB_j(s)$$

$$= \sum_j \int_0^t g_{ij}(s,\mathbf{x}(s))dB_j(s) + \frac{1}{2}\sum_{j,k}\int \frac{\partial g_{ij}(s,\mathbf{x}(s))}{\partial x_k}(f_k(s,\mathbf{x}(s))ds$$

$$+ \sum_m g_{km}(s,\mathbf{x}(s))dB_m(s))dB_j(s)$$

$$= \sum_j \int_0^t g_{ij}(s,\mathbf{x}(s))dB_j(s) + \frac{1}{2}\sum_{k,j}\int_0^t g_{kj}(s,\mathbf{x}(s))\frac{\partial g_{ij}(s,\mathbf{x}(s))}{\partial x_k}dt$$

This means that the above Stratanovich sde is equivalent to the following Ito sde

$$dx_i(t) = (f_i(t,\mathbf{x}(t)) + \frac{1}{2}\sum_{k,j}g_{kj}(t,\mathbf{x}(t))\frac{\partial g_{ij}(t,\mathbf{x}(t))}{\partial x_k})dt + \sum_j g_{ij}(t,\mathbf{x}(t))dB_j(t)$$

Study project 67

Smoothing and Prediction. Problems of state estimation broadly fall in three classes: Prediction, filtering and smoothing. In prediction, data $z(s)$ is given upto time $\tau < t$ and we wish to estimate the state $\mathbf{x}(t)$. In filtering, we estimate $\mathbf{x}(t)$ given data upto time t and in smoothing, we are given data upto time $T > t$ and we construct an estimate of $\mathbf{x}(t)$. The message model is

$$d\mathbf{x}(t)/dt = \mathbf{F}(t)\mathbf{x}(t) + \mathbf{G}(t)\mathbf{w}(t)$$

$$\frac{d\mathbf{z}(t)}{dt} = \mathbf{H}(t)\mathbf{z}(t) + \mathbf{v}(t)$$

where $\mathbf{w}(t)$ is a white noise process and $\mathbf{v}(t)$ is another white noise process independent of the process $\mathbf{w}(t)$. For $t_1 < t_2$, the solution is given by

$$\mathbf{x}(t_2) = \Phi(t_2,t_1)\mathbf{x}(t_1) + \int_{t_1}^{t_2}\Phi(t_2,\tau)\mathbf{G}(\tau)\mathbf{w}(\tau)d\tau$$

where Φ is the state transition kernel given by the solution to the differential equation

$$\frac{\partial}{\partial t}\Phi(t,\tau) = \mathbf{F}(t)\Phi(t,\tau), t \geq \tau$$

with the initial condition $\Phi(\tau,\tau) = \mathbf{I}$. Let $\Phi(\tau,t) = \Phi(t,\tau)^{-1}$ with $\tau \leq t$. It is not hard to see that

$$\frac{\partial\Phi(\tau,t)}{\partial\tau} = \mathbf{F}(\tau)\Phi(\tau,t), \tau \leq t$$

with the final condition $\Phi(t,t) = \mathbf{I}$. In the special case when $\mathbf{F}(t) = \mathbf{F}$ is a constant matrix, we have $\Phi(t,\tau) = exp((t-\tau)\mathbf{F})$. Now,

$$\mathbf{0} = \frac{\partial}{\partial\tau}(\Phi(t,\tau)\Phi(\tau,t)) = \frac{\partial\Phi(t,\tau)}{\partial\tau}\Phi(\tau,t) + \Phi(t,\tau)\mathbf{F}(\tau)\Phi(\tau,t)$$

giving

$$\frac{\partial}{\partial \tau}\Phi(t,\tau) = -\Phi(t,\tau)\mathbf{F}(\tau)$$

A chaos expansion can be developed for $\Phi(t,\tau)$ based on the integral equation

$$\Phi(t,\tau) = \mathbf{I} + \int_{\tau}^{t} \mathbf{F}(s)\Phi(s,\tau)ds$$

This gives

$$\Phi(t,\tau)$$

$$= \mathbf{I} + \int_{\tau}^{t} \mathbf{F}(s)ds + \int_{\tau<s_2<...s_1<t} \mathbf{F}(s_1)\mathbf{F}(s_2)ds_2ds_1 + ...$$

$$+ \int_{\tau<s_n<...<s_1<t} \mathbf{F}(s_1)\mathbf{F}(s_2)...\mathbf{F}(s_n)ds_n...ds_1 + ...$$

The partial derivatives of this expansion with respect to the variables t,τ respectively give back the above differential equations. Let $\mathbf{Z}(t)$ stand for the collection of random vectors $\{\mathbf{z}(s) : s \leq t\}$ corresponding to the observations upto time t. Then

$$\mathbb{E}(\mathbf{x}(t_2)|\mathbf{Z}(t_1)) = \Phi(t_2,t_1)\mathbb{E}\mathbf{x}(t_1)|\mathbf{Z}(t_1))$$

since $\mathbb{E}(\mathbf{w}(\tau)|\mathbf{Z}(t_1)) = \mathbf{0}$ for $\tau > t_1$. This equation can be written as

$$\hat{\mathbf{x}}(t_2|t_1) = \Phi(t_2,t_1)\hat{\mathbf{x}}(t_1)$$

This is the fundamental equation for obtaining the prediction of the state in terms of the filtered estimate. Three kinds of prediction are important: One in which t_1 is fixed and $t_2 = t > t_1$ varies. Two in which t_2 is fixed and $t_1 < t$ varies and three in which $t_1 = t$ and $t_2 = t + T$ vary with the prediction lag T remaining fixed. In the first case, we obtain

$$\frac{\partial}{\partial t}\hat{\mathbf{x}}(t|t_1) = \frac{\partial}{\partial t}\Phi(t,t_1)\hat{\mathbf{x}}(t_1)$$

$$= \mathbf{F}(t)\Phi(t,t_1)\hat{\mathbf{x}}(t_1) = \mathbf{F}(t)\hat{\mathbf{x}}(t|t_1), t > t_1$$

This is the evolution of the predicted state estimate with time. In the second case we obtain

$$\frac{\partial}{\partial t}\hat{\mathbf{x}}(t_2|t) = \frac{\partial}{\partial t}\Phi(t_2,t)\hat{\mathbf{x}}(t_1)$$

$$= -\Phi(t_2,t)\mathbf{F}(t)\hat{\mathbf{x}}(t_1)$$

Finally,

$$\frac{\partial}{\partial t}\hat{\mathbf{x}}(t+T|t) = \frac{\partial}{\partial t}\Phi(t+T,t)\hat{\mathbf{x}}(t)$$

$$= (\mathbf{F}(t+T)\Phi(t+T,t)\hat{\mathbf{x}}(t) - \Phi(t+T,t)\mathbf{F}(t))\hat{\mathbf{x}}(t)$$

These differential equations can be implemented on the computer in discrete form.

Study project 68

Sequential decision theory: We have two hypotheses H_1, H_0 and a sequence of observations available to us. Let $z(k), k = 0, 1, ...,$ denote the sequence of observations and for simplicity of notation, we set $Z(k) = (z(0), ..., z(k)), k = 0, 1, ...$ The Likelihood ratio at stage k is defined by

$$L(k) = \frac{p(Z(k)|H_1)}{p(Z(k)|H_0)}$$

We define two threshold sequence $T_0(k) < T_1(k), k = 0, 1, ...$ If $L(k) > T_1(k)$ accept H_1, if $L(k) < T_0(k)$, accept H_0, while if $T_0(k) \leq L(k) \leq T_1(k)$, then make the $k+1^{th}$ observation. The thresholds $T_1(k), T_0(k)$ are computed using the apriori probabilities for H_1 and H_0 after the kth observation has been made. We denote these probabilities by $p(H_1, k)$ and $p(H_0, k)$. These apriori probabilities for the kth stage must coincide with the aposteriori probabilities for H_1, H_0 after the $k - 1^{th}$ observation. Thus,

$$p(H_1, k) = p(H_1|z(k-1)) = \frac{p(z(k-1)|H_1)p(H_1, k-1)}{p(z(k-1|H_1)p(H_1, k-1) + p(z(k-1)|H_0)p(H_0, k-1)}$$

since $p(H_1, k-1)$ is the apriori probability for H_1 before the $k - 1^{th}$ observation. Likewise,

$$p(H_0, k)p(H_0|z(k-1)) = \frac{p(z(k-1)|H_0)p(H_0, k-1)}{p(z(k-1)|H_1)p(H_1, k-1) + p(z(k-1)|H_0)p(H_0, k-1)}$$

Forming the ratio gives us

$$\frac{p(H_1, k)}{p(H_0, k)} = \frac{p(z(k-1)|H_1)}{p(z(k-1)|H_0)}\frac{p(H_1, k-1)}{p(H_0, k-1)}$$

Solving this recursion gives

$$\frac{p(H_1, k)}{p(H_0, k)} = \frac{p(z(k-1|H_1)p(z(k-2|H_1)...p(z(0)|H_1)}{p(z(k-1)|H_0)p(z(k-2|H_0)...p(z(0)|H_0)}\frac{p(H_1, 0)}{p(H_0, 0)}$$

Assuming independent sampling, we have

$$p(Z(k-1)|H_1) = p(z(k-1)|H_1)...p(z(0)|H_1)$$

$$p(Z(k-1)|H_0) = p(z(k-1)|H_0)...p(z(0)|H_0)$$

For convenience, we set $p(H_i) = p(H_i, 0)$, the apriori probability for H_i before the zeroth observation, i.e., the probability of H_i before any observation has been made. Then, the above equation can be written as

$$\frac{p(H_1, k)}{p(H_0, k)} = L(k-1)\frac{p(H_1)}{p(H_0)}$$

After the kth stage, let $\Omega(k)$ denote the decision region for H_1, i.e.,

$$\frac{p(Z(k)|H_1)}{p(Z(k)|H_0)} \geq T_1(k)$$

for all $Z(k) \in \Omega(k)$. Multiply both sides of this inequality by $p(Z(k)|H_0)$ and integrate with respect to $Z(k)$ over $\Omega(k)$ to get

$$\int_{\Omega(k)} p(Z(k)|H_1)dZ(k) \geq T_1(k) \int_{\Omega(k)} p(Z(k)|H_0)dZ(k)$$

Now, the integral on the left side equals $P_D(k)$ the probability of decision, i.e. deciding H_1 when H_1 is true after k observations. The integral on the right equals $P_F(k)$, the probability of false alarm, i.e., the probability of deciding H_1 when H_0 is true after k stages. Thus,

$$P_D(k) \geq T_1(k)P_F(k)$$

We can also write $P_D(k) \leq 1 - P_M(k)$ where $P_M(k)$ is the miss probability, i.e., the probability of deciding H_0 when actually H_1 is true after k stages. Thus,

$$T_1(k) \leq \frac{1 - P_M(k)}{P_F(k)}$$

Similarly, for all $Z(k) \in \Omega(k)'$, the decision region for H_0, we have

$$\frac{p(Z(k)|H_1)}{p(Z(k)|H_0)} \leq T_0(k)$$

from which we deduce that

$$\int_{\Omega(k)'} p(Z(k)|H_1)dZ(k) \leq T_0(k) \int_{\Omega(k)'} p(Z(k)|H_0)dZ(k)$$

Now, $\int_{\Omega(k)'} p(Z(k)|H_1)dZ(k)$ equals the probability of deciding H_0 when H_1 is actually true. This is $P_M(k)$. Likewise, $\int_{\Omega(k)'} p(Z(k)|H_0)dZ(k)$ is the probability of deciding H_0 when actually H_0 is true. This is clearly smaller than one minus the probability of deciding H_1 when H_0 is true. Thus,

$$P_M(k) \leq T_0(k)(1 - P_F(k))$$

or

$$T_0(k) \geq \frac{P_M(k)}{1 - P_F(k)}$$

Thus, the following inequalities are satisfied by the thresholds T_0, T_1:

$$\frac{P_M(k)}{1 - P_F(k)} \leq T_0(k) \leq T_1(k) \leq \frac{1 - P_M(k)}{P_F(k)}$$

Usually, we assume that $P_M(k)$ and $P_F(k)$ are independent of the stage number k and this amounts to selecting fixed thresholds T_1, T_0. This completes the discussion of the sequential decision problem.

Study project 69

Wiener Filter for vector processes: How to formulate the Wiener filter completely in the frequency domain with reference to vector valued processes. The message process $\mathbf{y}(t)$ is vector valued and is contaminated by additive noise $\mathbf{v}(t)$. The two processes are uncorrelated, i.e., $\mathbb{E}(\mathbf{y}(t)\mathbf{v}(t')^T) = \mathbf{0}$. Let

$$R_y(s) = \int_{-\infty}^{\infty} exp(-s\tau)\mathbb{E}(\mathbf{y}(t)\mathbf{y}(t-\tau)^T)d\tau$$

the spectral density of the $\mathbf{y}(t)$ process and

$$R_v(s) = \int_{-\infty}^{\infty} exp(-s\tau)\mathbb{E}(\mathbf{v}(t)\mathbf{v}(t-\tau))d\tau$$

the spectral density of the $\mathbf{v}(t)$ process. Let $\mathbf{W}(s)$ be the transfer function of the filter that takes as input $\mathbf{y}(t) + \mathbf{v}(t)$ and outputs the optimal estimate $\hat{\mathbf{y}}(t)$ of $\mathbf{y}(t)$. In the spectral domain, the transform of the error is given by

$$\tilde{\mathbf{Y}}(s) = \mathbf{Y}(s) - \hat{\mathbf{Y}}(s) = \mathbf{Y}(s) - \mathbf{W}(s)(\mathbf{Y}(s) + \mathbf{V}(s))$$

$$= (\mathbf{I} - \mathbf{W}(s))\mathbf{Y}(s) - \mathbf{W}(s)\mathbf{V}(s)$$

which when translated into the time domain, reads

$$\tilde{\mathbf{y}}(t) = \mathbf{y}(t) - \int_{-\infty}^{\infty} \mathbf{w}(t-\tau)(\mathbf{y}(\tau) + \mathbf{v}(\tau))d\tau$$

The spectral density of the error process is

$$\tilde{R}_y(s) = (\mathbf{I} - \mathbf{W}(s))R_Y(s)(\mathbf{I} - \mathbf{W}(-s))^T + \mathbf{W}(s)R_V(s)\mathbf{W}(-s)^T$$

The mean square error reads

$$MSE = Tr\{\frac{1}{2\pi} \int_{-\infty}^{\infty} R_y(j\omega)d\omega$$

$$= Tr\{\frac{1}{2\pi} \int (\mathbf{I} - \mathbf{W}(j\omega))R_Y(j\omega)(\mathbf{I} - \mathbf{W}(j\omega)^*)d\omega - \frac{1}{2\pi} \int_{-\infty}^{\infty} \mathbf{W}(j\omega)R_V(j\omega)\mathbf{W}(j\omega)^*d\omega\}$$

The problem is to select the matrix transfer function $\mathbf{W}(j\omega)$ such that this mean squared error is a minimum. Taking the variation of the above expression with respect to $W(j\omega)$ gives

$$\delta(MSE) = -\frac{1}{2\pi} \int \delta W(j\omega)R_Y(j\omega)(\mathbf{I} - \mathbf{W}(j\omega)^*)d\omega-$$

$$\frac{1}{2\pi}\int (\mathbf{I} - \mathbf{W}(j\omega))R_Y(j\omega)\delta W(j\omega)^*d\omega$$

$$+\int_{-\infty}^{\infty}\delta W(j\omega)R_V(j\omega)W(j\omega)^*d\omega + \int_{-\infty}^{\infty}W(j\omega)R_V(j\omega)\delta W(j\omega)^*d\omega = 0$$

Equating coefficients of $\delta W(j\omega)^*$ to zero gives

$$-(\mathbf{I} - \mathbf{W}(j\omega))R_Y(j\omega) + \mathbf{W}(j\omega)R_V(j\omega) = 0$$

so that

$$\mathbf{W}(j\omega) = R_Y(j\omega)(R_Y(j\omega) + R_V(j\omega))^{-1}$$

This is the complete solution to the non-causal Wiener filter.

Study project 70

Wiener Ito chaos expansions in the time and spectral domain as solutions to certain stochastic differential equations: Solution to a stochastic differential equation using the Wiener-Ito chaos expansion. Consider the sde

$$dx(t) = (a(t) + b(t)x(t))dt + (c(t) + g(t)x(t))dB(t)$$

where $a(t), b(t), c(t), g(t)$ are non-random functions of time and $B(t)$ is Brownian motion process. This equation can be interpreted in the white noise sense as

$$\frac{dx(t)}{dt} = (a(t) + b(t)x(t)) + (c(t) + g(t)x(t))w(t)$$

where $w(t) = \frac{dB(t)}{dt}$ is a white noise process. Its autocorrelation function is given by

$$< w(t)w(s) >= \delta(t - s)$$

It has the spectral representation

$$w(t) = \int_{-\infty}^{\infty} exp(i\omega t)dZ(\omega)$$

The statistics of the spectral measure $dZ(\omega)$ can be determined as follows:

$$< w(t)w(s) >= \delta(t - s) = \int exp(i(\omega t - \omega' s)) < dZ(\omega)dZ(\omega') >$$

or

$$\delta(\tau) = \int exp(i(\omega t - \omega'(t - \tau)) < dZ(\omega)dZ(\omega')^* >$$

$$= \int exp(i((\omega - \omega')t + \omega'\tau)) < dZ(\omega)dZ(\omega')^* >$$

It follows that

$$< dZ(\omega)dZ(\omega')^* >= d\omega.\delta_{\omega,\omega'}/2\pi$$

This orthogonality relation can be expressed as

$$< dZ(\omega)dZ(\omega')^* >= 0, \omega \neq \omega', \qquad < |dZ(\omega)|^2 >= d\omega/2\pi$$

Now, the differential equation can be expressed in the spectral domain after Fourier transforming as

$$i\omega X(\omega) = A(\omega) + \frac{1}{2\pi} \int B(\omega - \omega')X(\omega')d\omega'$$

$$+\frac{1}{2\pi} \int C(\omega - \omega')W(\omega')d\omega' + \frac{1}{4\pi^2} \int G(\omega - \omega' - \omega'')X(\omega')W(\omega'')d\omega'd\omega''$$

where

$$W(\omega) = 2\pi\frac{dZ(\omega)}{d\omega}$$

This suggests the following recursive scheme for determining the spectrum of the process $x(t)$:

$$X_{n+1}(\omega) = \frac{A(\omega)}{i\omega} + \frac{1}{2\pi i\omega} \int B(\omega - \omega')X_n(\omega')d\omega'$$

$$+\frac{1}{2\pi i\omega} \int C(\omega - \omega')W(\omega')d\omega' + \frac{1}{4\pi^2 i\omega} \int G(\omega - \omega' - \omega'')X_n(\omega')W(\omega'')d\omega'd\omega''$$

Another way to solve the above integral equation is to adopt a perturbation expansion. Introduce a parameter ϵ into the original differential equation writing it as

$$\frac{dx(t)}{dt} = (a(t) + b(t)x(t)) + (c(t) + \epsilon d(t)x(t))w(t)$$

Now let

$$x(t) = \sum_{n=0}^{\infty} \epsilon^n x_n(t)$$

Plugging this into the differential equation and equating coefficients of equal powers of ϵ results in the following sequence of differential equations:

$$\frac{dx_0(t)}{dt} = a(t) + b(t)x_0(t) + c(t)w(t)$$

$$\frac{dx_{n+1}(t)}{dt} = b(t)x_{n+1}(t) + d(t)x_n(t)w(t), n \geq 1$$

The solution to the first equation is

$$x_0(t) = \int_0^t \phi(t, s)(a(s) + c(s)w(s))ds, \qquad \phi(t, s) = exp(\int_s^t b(\tau)d\tau)$$

The solution to the second is

$$x_{n+1}(t) = \int_0^t \phi(t,s)d(s)x_n(s)w(s)ds$$

Thus,

$$x_1(t) = \int_0^t \phi(t,t_1)d(t_1)x_0(t_1)w(t_1)dt_1$$

$$= \int_{0<t_2<t_1<t} \phi(t,t_1)d(t_1)\phi(t_1,t_2)(a(t_2)+c(t_2)w(t_2))w(t_1)dt_2dt_1$$

In general,

$$x_n(t) = \int_{0<t_n<...<t_1<t} \phi(t,t_1)d(t_1)\phi(t_1,t_2)d(t_2)...\phi(t_{n-1},t_n)d(t_n)x_0(t_n)w(t_n)...w(t_1)dt_n...dt_1$$

This means that the solution to the sde can be expressed in the form of a chaos expansion:

$$x(t) = \psi(t) + \sum_{n=1}^{\infty} \int_{0<t_n<...<t_1<t} h_n(t,t_n,t_{n-1},...,t_1)w(t_n)...w(t_1)dt_n...dt_1$$

Now the chaos expansion can also be expressed in the frequency domain as follows. Consider

$$\int_{0<t_1<...<t_n<t} h(t_1,...,t_n)w(t_1)...w(t_n)dt_1...dt_n$$

$$= \int h(t_1,...,t_n)exp(i(\omega_1 t_1 + ... + \omega_n t_n))dt_1...dt_n dZ(\omega_1)...dZ(\omega_n)$$

$$= \int H(\omega_1,...,\omega_n)dZ(\omega_1)...dZ(\omega_n)$$

where

$$H(\omega_1,...,\omega_n) = \int h(t_1,...,t_n)exp(i(\omega_1 t_1 + ... + \omega_n t_n))dt_1...dt_n$$

Suppose we are given the process

$$\xi(t) = \int_{-\infty<t_1<...<t_n<t} h(t,t_1,...,t_n)w(t_1)...w(t_n)dt_1...dt_n$$

The Fourier transform of this process can be expressed as

$$\hat{\xi}(\omega) = \int H(\omega,\omega_1,...,\omega_n)dZ(\omega_1)...dZ(\omega_n)$$

where

$$H(\omega,\omega_1,...,\omega_n) = \int_{-\infty<t_1<...<t_n<t<\infty} h(t,t_1,...,t_n)exp(i(\omega_1 t_1 + ... + \omega_n t_n - \omega t))dt_1...dt_n dt$$

This means that the Fourier transform of the solution to the sde admits a spectral domain chaos expansion:

$$X(\omega) = \hat{\psi}(\omega) + \sum_{n=1}^{\infty} \int_{\mathbb{R}^n} H_n(\omega, \omega_1, ..., \omega_n) dZ(\omega_1)...dZ(\omega_n)$$

Study project 71

Wiener-Ito chaos expansions for the sde

$$d\mathbf{X}(t) = (\mathbf{a} + \mathbf{B}\mathbf{X}(t))dt + (\mathbf{c} + \epsilon\mathbf{D}\mathbf{X}(t))dB(t)$$

where $\mathbf{X}(t)$ is a $n \times 1$ vector valued diffusion process, $\mathbf{a}, \mathbf{c}$ are $n \times 1$ constant vectors and $\mathbf{B}, \mathbf{D}$ are $n \times n$ constant matrices. $B(t)$ is scalar Brownian motion. To solve this, we make the expansion

$$\mathbf{X}(t) = \sum_{n=0}^{\infty} \epsilon^n \mathbf{X}_n(t)$$

Plugging this into the sde and equating coefficients of equal powers of ϵ gives the sequence of equations

$$d\mathbf{X}_0(t) = (\mathbf{a} + \mathbf{B}\mathbf{X}_0(t))dt + \mathbf{c}dB(t) = \mathbf{B}\mathbf{X}_0(t) + (\mathbf{a}dt + \mathbf{c}dB(t))$$

$$d\mathbf{X}_{n+1}(t) = \mathbf{B}\mathbf{X}_{n+1}(t)dt + \mathbf{D}\mathbf{X}_n(t)dB(t), n = 0, 1, 2, ...,$$

The first equation has solution

$$\mathbf{X}_0(t) = \int_0^t exp((t - s)\mathbf{B})(\mathbf{a}ds + \mathbf{c}dB(s))$$

and the second

$$\mathbf{X}_{n+1}(t) = \int_0^t exp((t - s)\mathbf{B})\mathbf{D}\mathbf{X}_n(s)dB(s)$$

For example,

$$\mathbf{X}_1(t) = \int_{0<t_1<t} exp((t - t_1)\mathbf{B})\mathbf{D}\mathbf{X}_0(t_1)dB(t_1)$$

$$\mathbf{X}_2(t) = \int_{0<t_2<t_1<t} exp((t - t_1)\mathbf{B})\mathbf{D}exp((t_1 - t_2)\mathbf{B})\mathbf{D}\mathbf{X}_0(t_2)dB(t_2)dB(t_1)$$

and in general,

$$\mathbf{X}_n(t) = \int_{0<t_n<t_{n-1}<...<t_1<t} exp((t - t_1)\mathbf{B})\mathbf{D}exp((t_1 - t_2)\mathbf{B})\mathbf{D}$$

$$...exp((t_{n-1} - t_n)\mathbf{B})\mathbf{D}\mathbf{X}_0(t_n)dB(t_n)dB(t_{n-1})...dB(t_1)$$

This is a Wiener-Ito vector chaos expansion in the time domain. It can be generalized to include the case when $\mathbf{a}, \mathbf{B}, \mathbf{c}, \mathbf{D}$ depend upon time. Consider

$$d\mathbf{X}(t) = (\mathbf{a}(t) + \mathbf{B}(t)\mathbf{X}(t))dt + (\mathbf{c}(t) + \epsilon\mathbf{D}(t)\mathbf{X}(t))dB(t)$$

Make the expansion

$$\mathbf{X}(t) = \sum_{n=0}^{\infty} \epsilon^n \mathbf{X}_n(t)$$

Plugging this into the sde and equating coefficients of equal powers of ϵ gives

$$d\mathbf{X}_0(t) = (\mathbf{a}(t) + \mathbf{B}(t)\mathbf{X}_0(t))dt + \mathbf{c}(t)dB(t)$$

$$d\mathbf{X}_{n+1}(t) = \mathbf{B}(t)\mathbf{X}_{n+1}(t)dt + \mathbf{D}(t)\mathbf{X}_n(t)dB(t)$$

If $\Phi(t,\tau)$ is the matrix solution to

$$\frac{\partial\Phi(t,\tau)}{\partial t} = \mathbf{B}(t)\Phi(t,\tau), t \geq \tau$$

with initial condition $\Phi(\tau,\tau) = \mathbf{I}$, then

$$\mathbf{X}_0(t) = \int_0^t \Phi(t,s)(\mathbf{a}(s)ds + \mathbf{c}(s)dB(s))$$

$$\mathbf{X}_{n+1}(t) = \int_0^t \Phi(t,s)\mathbf{D}(s)\mathbf{X}_n(s)dB(s)$$

and hence

$$\mathbf{X}_n(t) = \int_{0<t_n<...<t_1<t} \Phi(t,t_1)\mathbf{D}(t_1)\Phi(t_1,t_2)\mathbf{D}(t_2)...$$
$$\Phi(t_{n-1},t_n)\mathbf{D}(t_n)\mathbf{X}_0(t_n)dB(t_n)dB(t_{n-1})...dB(t_1)$$

In the time invariant case, note that $\Phi(t,s) = exp((t-s)\mathbf{B})$.

Study project 72

Analysis of a linear sde using Fokker-Planck equation. Consider the sde

$$dx(t) = (a + bx(t))dt + (\mathbf{c}^T + x(t)\mathbf{d}^T)d\mathbf{B}(t)$$

where $x(t)$ is a scalar diffusion process, a, b are scalar constants, $\mathbf{c}$ is a $p \times 1$ constant column vector, $\mathbf{d}$ is another $p \times 1$ constant column vector and $\mathbf{B}(t)$ is $p \times 1$ vector Brownian motion. In terms of components,

$$dx(t) = (a + bx(t))dt + \sum_{j=1}^{p}(c_j + x(t)d_j)dB_j(t)$$

We want to determine the stationary density of $x(t)$. If $p(t,x)$ is the density of $x(t)$, then p satisfies the Fokker-Planck equation

$$\frac{\partial p(t,x)}{\partial t} = -\frac{\partial}{\partial x}((a+bx)p(t,x)) + \frac{1}{2}\frac{\partial^2}{\partial x^2}((\sum c_j + xd_j)^2 p(t,x))$$

Note that

$$\sum_j (c_j + xd_j)^2 = \parallel \mathbf{c} + x\mathbf{d} \parallel^2 = \parallel \mathbf{c} \parallel^2 + x^2 \parallel \mathbf{d} \parallel^2 + 2x\mathbf{c}^T\mathbf{d}$$

so the stationary density $p(x)$ satisfies the differential equation

$$\frac{d}{dx}((a+bx)p(x)) = \frac{1}{2}\frac{d^2}{dx^2}((\alpha + \beta x + \gamma x^2)p(x))$$

where

$$\alpha = \parallel \mathbf{c} \parallel^2, \beta = 2\mathbf{c}^T\mathbf{d}, \gamma = \parallel \mathbf{d} \parallel^2$$

Suppose $p(x)$ decays fast enough to zero as $|x| \to \infty$, i.e., so fast that $x^2 p(x) \to 0$, then integration of the above equilibrium equation gives

$$(a+bx)p(x) = \frac{1}{2}\frac{d}{dx}((\alpha + \beta x + \gamma x^2)p(x))$$

$$= \frac{1}{2}(\alpha + \beta x + \gamma x^2)p'(x) + (\beta + 2\gamma x)p(x)/2$$

or

$$p'(x) = \frac{2a - \beta + 2(b - \gamma)x}{\alpha + \beta x + \gamma x^2}p(x)$$

This differential equation can be integrated immediately.

Study Project 73

Stochastically perturbed two body problem. Consider the motion of a body of unit mass in the radial potential $V(r)$. The motion is assumed to be taking place in a plane. The equations of motion in polar coordinates are

$$r'' - r\phi'^2 = -V'(r) \qquad r\phi'' + 2r'\phi' = 0$$

These give two first integrals of motion, energy and angular momentum:

$$E = \frac{1}{2}(r'^2 + r^2\phi'^2) + V(r)$$

$$\beta = r^2\phi'$$

where prime denotes differentiation with respect to time. The fact that these are constants of the motion may be verified by direct differentiation

$$E' = r'r'' + 2rr'\phi'^2 + 2r^2\phi'\phi'' + V'(r)r' = r'(r'' - r\phi'^2 + V'(r)) + 2r\phi'(r\phi'' + 2r'\phi') = 0$$

$$\beta' = 2rr'\phi' + r^2\phi'' = r(r\phi'' + 2r'\phi') = 0$$

in view of the equations of motion. Dividing the energy integral by ϕ'^2 and using the angular momentum integral gives the differential equation for the orbit:

$$\frac{1}{2}((\frac{dr}{d\phi})^2 + r^2) + r^4 V(r)/\beta^2 = r^4 E/\beta^2$$

Letting $u = 1/r$ gives

$$\frac{1}{2}((\frac{du}{d\phi})^2 + u^2) + V(1/u)/\beta^2 = E/\beta^2$$

from which the orbit equation is obtained:

$$\int_{u_0}^{u} du(2E/\beta^2 - 2V(1/u)/\beta^2 - u^2)^{-1/2} = \phi - \phi_0$$

The trajectory as a function of time is obtained by combining the orbit equation with the angular momentum integral:

$$\beta(t - t_0) = \int_{\phi_0}^{\phi} r^2 d\phi = \int_{\phi_0}^{\phi} d\phi/u^2 = \int_{u_0}^{u} \frac{du}{u^2}\frac{d\phi}{du}$$

Now consider a perturbation to the two body problem by two white noise processes. This is an appropriate model for the perturbation caused by the gravitational force produced by dust distributed in the field of motion. Let $B_r(t)$ and $B_\phi(t)$ be two independent Brownian motion processes. Then, the perturbed equations of motion are

$$r'' - r\phi'^2 = \sigma_r B_r' - V'(r) \qquad r\phi'' + 2r'\phi' = \sigma_\phi B_\phi'$$

σ_r and σ_ϕ can be taken as constants. However, they can also be made functions of r and ϕ if one wishes to take into account the variation in the density of dust from point to point. These equations can be cast in the Ito Stochastic differential form by introducing the radial and angular velocities: $r' = v_r, \phi' = \omega$. Then,

$$dr = v_r dt, \qquad d\phi = \omega dt, \qquad dv_r = (r\omega^2 - V'(r))dt + \sigma_r dB_r \qquad d\omega = -2v_r\omega dt/r + \frac{\sigma_\phi}{r}dB_\phi$$

The Fokker-Planck equation for $p(t, r, v_r, \omega)$ can be written down:

$$\frac{\partial p}{\partial t} = -v_r\frac{\partial p}{\partial r} - \omega\frac{\partial p}{\partial \phi} - (r\omega^2 - V'(r))\frac{\partial p}{\partial v_r} + \frac{2v_r}{r}\frac{\partial(\omega p)}{\partial \omega} + \frac{\sigma_r^2}{2}\frac{\partial^2 p}{\partial v_r^2} + \frac{\sigma_\phi^2}{2r^2}\frac{\partial^2 p}{\partial \omega^2}$$

One approach to the stochastically perturbed two body problem is to perform a perturbation theoretic analysis of the Fokker-Planck equation or to look at eigenfunction expansions for the density. Another approach to perform a perturbation theoretic analysis directly of the original Ito stochastic differential equation regarding the Brownian terms as first order perturbations. This is done by introducing a parameter ϵ into the Brownian terms:

$$r'' - r\phi'^2 = -V'(r) + \epsilon\sigma_r B_r', \qquad r\phi'' + 2r'\phi' = \epsilon\sigma_\phi B_\phi'$$

Writing

$$r(t) = r(t, \epsilon) = r_0(t) + \epsilon r_1(t) + O(\epsilon^2), \qquad \phi(t) = \phi(t, \epsilon) = \phi_0(t) + \epsilon\phi_1(t) + O(\epsilon^2)$$

we find on equating coefficients of ϵ^0 that

$$r_0'' - r_0\phi_0'^2 = -V'(r_0) \qquad r_0\phi_0'' + 2r_0'\phi_0' = 0$$

which means that $t \to (r_0(t), \phi_0(t))$ satisfies the unperturbed two body equations of motion. Assume that we've solved for the unperturbed trajectory. Equating coefficients of ϵ^1 gives

$$r_1'' - 2r_0\phi_0'\phi_1' - r_1\phi_0'^2 = -V''(r_0)r_1 + \sigma_r B_r'$$

$$r_0\phi_1'' + r_1\phi_0'' + 2r_0'\phi_1' + 2r_1'\phi_0' = \sigma_\phi B_\phi'$$

These follow on noticing that the left sides of the above two equations are respectively the coefficients of ϵ in $(r_0'' + \epsilon r_1'') - (r_0 + \epsilon r_1)(\phi_0' + \epsilon\phi_1')^2$ and $(r_0 + \epsilon r_1)(\phi_0'' + \epsilon\phi_1'') + 2(r_0' + \epsilon r_1')(\phi_0' + \epsilon\phi_1')$. Defining the variables $v_1 = r_1', \omega_1 = \phi_1'$ enables us to cast the above two differential equations in the Ito sde format:

$$dr_1 = v_1 dt, d\phi_1 = \omega_1 dt, dv_1 = 2r_0\phi_0'\omega_1 dt + r_1\phi_0'^2 dt - V''(r_0)r_1 dt + \sigma_r dB_r$$

$$d\omega_1 = -\frac{\phi_0''}{r_0}r_1 dt - \frac{2r_0'}{r_0}\omega_1 dt - \frac{2\phi_0'}{r_0}v_1 dt + \frac{\sigma_\phi}{r_0}dB_\phi$$

These can be expressed in matrix format as linear time varying state variable equations:

$$d\begin{pmatrix} r_1 \\ \phi_1 \\ v_1 \\ \omega_1 \end{pmatrix} = \begin{pmatrix} 0 & 0 & 1 & 0 \\ 0 & 0 & 0 & 1 \\ -V''(r_0) + \phi_0'^2 & 0 & 0 & 2r_0\phi_0' \\ -\frac{\phi_0''}{r_0} & 0 & -\frac{2\phi_0'}{r_0} & -\frac{2r_0'}{r_0} \end{pmatrix} \begin{pmatrix} r_1 \\ \phi_1 \\ v_1 \\ \omega_1 \end{pmatrix} + \begin{pmatrix} 0 & 0 \\ 0 & 0 \\ \sigma_r & 0 \\ 0 & \frac{\sigma_\phi}{r_0} \end{pmatrix} d\begin{pmatrix} B_r \\ B_\phi \end{pmatrix}$$

The linearized equations will be completely solved once we determine the state transition matrix for the system.

Study project 74

Motion of a body in a rotating frame of reference with the angular velocity of rotation having a small random component. The equations of motion are

$$\frac{d^2\mathbf{r}}{dt^2} + 2\omega \times \frac{d\mathbf{r}}{dt} + \frac{d\omega}{dt} \times \mathbf{r} + \omega \times (\omega \times \mathbf{r}) = \mathbf{0}$$

Here, the derivatives of position are taken with respect to the rotating frame of reference. These equations are derived by noting that the rate of change of a vector $\mathbf{P}$ relative to the static frame $\frac{d\mathbf{P}}{dt}$ and relative to the rotating frame $\frac{\partial\mathbf{P}}{\partial t}$ are related via the equation

$$\frac{d\mathbf{P}}{dt} = \frac{\partial\mathbf{P}}{\partial t} + \omega \times \mathbf{P}$$

Taking another derivative gives

$$\frac{d^2\mathbf{P}}{dt^2} = \frac{d}{dt}\frac{\partial\mathbf{P}}{\partial t} + \frac{d}{dt}\omega \times \mathbf{P}$$

$$= \frac{\partial^2\mathbf{P}}{\partial t^2} + \omega \times \frac{\partial\mathbf{P}}{\partial t} + \frac{d\omega}{dt} \times \mathbf{P} + \omega \times \frac{d\mathbf{P}}{dt}$$

$$= \frac{\partial^2\mathbf{P}}{\partial t^2} + \omega \times \frac{\partial\mathbf{P}}{\partial t} + \frac{d\omega}{dt} \times \mathbf{P} + \omega \times (\frac{\partial\mathbf{P}}{\partial t} + \omega \times \mathbf{P})$$

If the rotation of the primed frame is about the z axis, then we get two equations for the x and y components of the motion:

$$x'' - 2\omega y' - \omega' y - \omega^2 x = 0$$

$$y'' + 2\omega x' + \omega' x - \omega^2 y = 0$$

To take into account random fluctuations in the rotation, we replace ω by $\omega(t) + w(t)$, where $\omega(t)$ is the nonrandom component of the rotation and $w(t)$ is the random component. Assuming w to be small, we neglect w^2. Then, we get

$$x'' - 2(\omega + w)y' - (\omega' + w')y - (\omega^2 + 2\omega w)x = 0$$

$$y'' + 2(\omega + w)x' + (\omega' + w')x - (\omega^2 + 2\omega w)y = 0$$

Assume that w is an Ornstein-Uhlenbeck process, i.e., it satisfies the sde

$$dw = -\gamma w dt + \sigma dB$$

where B is standard Brownian motion. Then, our state vector is (x, y, v_x, v_y, w) and the sde satisfied by this state vector is

$$dx = v_x dt, dy = v_y dt$$

$$dv_x = 2(\omega + w)v_y dt + (\omega' dt - \gamma w dt + \sigma dB)y + (\omega^2 + 2\omega w)x dt$$

$$dv_y = -2(\omega + w)v_x dt - (\omega' - \gamma w dt + \sigma dB)x + (\omega^2 + 2\omega w)y dt$$

This system of sde's contains bilinear coupling between the Brownian noise or white noise and the state vector in that terms having the form xdB, ydB appear in the forcing. Write down the Fokker-Planck equation for this system and analyze its solutions. First cast this equation in vectorial form:

$$d\begin{pmatrix} x \\ y \\ v_x \\ v_y \\ w \end{pmatrix} = \begin{pmatrix} v_x \\ v_y \\ 2(\omega + w)v_y + \omega'y - \gamma wy + \omega^2 x + 2\omega w x \\ -2(\omega + w)v_x - \omega'x + \gamma wx + \omega^2 y + 2\omega w y \\ -\gamma w \end{pmatrix} dt + \begin{pmatrix} 0 \\ 0 \\ \sigma y \\ -\sigma x \\ \sigma \end{pmatrix} dB$$

Study project 75

We suggest a scheme by which one can construct a queuing model for the issue of books in a library. Let us first assume that we are not bothered about which subject books are involved. Let there be n students in the library and let us assume that there are N counters for issuing books. Let the arrival rate of the students at the counter be λ. This arrival rate can be estimated by noting the times at which the students arrive at the counters. If $0 = t_0 < t_1 < t_2 < ... < t_n$ denotes such a sample of arrival times, then the interarrival times are $X_i = t_i - t_{i-1}, i = 1, 2, ..., n-1$ and the average arrival rate can be estimated by assuming $X_i's$ to be exponentially distributed with mean $1/\lambda$. Then, the joint density of $\{X_1, ..., X_n\}$ is given by

$$f(x_1, ..., x_n|\lambda) = \lambda^n exp(-\lambda(x_1, ..., x_n))$$

This has a maximum when

$$\frac{\partial f}{\partial \lambda} = 0$$

that is, when

$$n/\lambda - (x_1 + ... + x_n) = 0$$

so that the estimate is given by

$$\lambda^{-1} = (x_1 + ... + x_n)/n = t_n/n$$

This estimate depends only on the arrival time of the last student and may not be very reliable. Other schemes can be tried out. For example, in a time interval $[0, t]$, let N_t denote the number of arrivals. This is given by

$$N_t = max(k : t_k \leq t)$$

Then, λ can be estimated as

$$\lambda = \frac{1}{T} \int_0^T \frac{N_t}{t} dt$$

where T is the arrival time of the last student. Let us assume for the sake of generality that the interarrival times $X_i's$ have the distribution F_X and let the service time of each counter have the distribution G. The problem is to compute the statistics of the number of students in the queue (i.e., either being served at the counters or in the waiting line) at any given time t. If one assumes that F_X is the exponential distribution with mean $1/\lambda$ and G is exponential with mean $1/\mu$, then the number of students $X(t)$ in the queue at time t is a Markov process with transition probabilities

$$P(X(t+dt) = k+1|X(t) = k) = (n-k)\lambda dt, 0 \le k \le n$$

$$P(X(t+dt) = k-1|X(t) = k) = k\mu dt, 0 \le k \le N$$

$$P(X(t+dt) = k-1|X(t) = k) = N\mu dt, N < k \le n$$

$$P(X(t+dt) = k|X(t) = k) = 1 - P(X(t+dt) = k+1|X(t) = k) - P(X(t+dt) = k-1|X(t) = k)$$

Using these equations, the Chapman-Kolmogorov equations for the queue can be formulated. To apply it to the library problem, one also needs to estimate the service rate μ. This can be done by measuring the service times $Y_1, ..., Y_n$ for each student and forming the estimate

$$\mu^{-1} = (Y_1 + ... + Y_n)/n$$

Study project 76

The concept of progressive measurability of a random process is important in stochastic process theory. If $(\Omega, \mathcal{F}, P)$ is a measurable space and $\{\mathcal{F}_t\}_{t \ge 0}$ is a nondecreasing family of sub σ fields of the σ field $\mathcal{F}$, then a random process $X_t(\omega)$ defined on this space is said to be progressively measurable relative to this family of σ fields if $\{(t, \omega) : 0 \le t \le T, X_t(\omega) \in B\}$ is in $\mathcal{B}_{[0,t]} \times \mathcal{F}$ for every Borel subset B of the real line and for all nonnegative real T. Here, $\mathcal{B}_{[0,t]}$ is the smallest σ field containing all the intervals of $[0, t]$.

Let $(E, \mathcal{F})$ be a measurable space and let $\mathbb{R}$ be equipped with the Borel σ algebra. Let (X, D) be a separable metric space and let $\{\mathcal{F}_t\}_{t \ge 0}$ be an increasing family of sub-σ algebras of $\mathcal{F}$. Let $\theta : [0, \infty) \times E \to X$ be a map such that for each $q \in E$, the map $t \to \theta(t, q)$ is right continuous and for each $t \in \mathbb{R}$, the map $q \to \theta(t, q)$ is $\mathcal{F}_t$ measurable. Then, the map θ is progressively measurable, i.e., the map $(s, q) \to \theta(s, q)$ from $[0, t] \times E \to X$ is $\mathcal{B}_{[0,t]} \times \mathcal{F}_t$ measurable. Here, $\mathcal{B}_{[0,t]}$ is the σ algebra on $[0, t]$ generated by all Borel subsets of $[0, t]$, or equivalently by all intervals $I \subset [0, t]$.

Let $u_n = ([nu] + 1)/n$. For all $u \in [0, t]$, it is clear that $min(u_n, t) \ge u$ and $min(u_n, t) \to u$ as $n \to \infty$. Thus, by right continuity, $\theta_n(u, q) = \theta(min(u_n, t), q)$ will converge to $\theta(u, q)$ for all $u \in [0, t]$. To prove that the map $(u, q) \to \theta(u, q)$ from $([0, t] \times E, \mathcal{B}_{[0,t]} \times \mathcal{F})$ into (X, D) is measurable, it suffices therefore to show that the map $(u, q) \to \theta_n(u, q)$ from $([0, t] \times E, \mathcal{B}_{[0,t]} \times \mathcal{F})$ into (X, D) is measurable. Let B be any Borel set in (X, D). Then,

$$\{(u, q) \in [0, t] \times E : \theta_n(u, q) \in B\}$$

$$= \cup_{(k+1)/n \le t}\{(u, q) \in [0, t] \times E : k \le nu < k+1, \theta((k+1)/n, q) \in B\} \cup$$

$$\cup_{(k+1)/n > t}\{(u, q) \in [0, t] \times E : k \le nu < k+1, \theta(t, q) \in B\}$$

Further,

$$\cup_{(k+1)/n \le t}\{(u, q) \in [0, t] \times E : k \le nu < k+1, \theta((k+1)/n, q) \in B\}$$

$$\cup_{(k+1)/n \le t}[k/n, (k+1)/n) \times \{q : \theta((k+1)/n, q) \in B\}$$

By hypothesis, $\{q : \theta((k+1)/n, q) \in B\}$ is in $\mathcal{F}_{(k+1)/n}$ which is contained in $\mathcal{F}_t$ for $(k+1)/n \le t$. Thus,

$$\cup_{(k+1)/n \le t}[k/n, (k+1)/n) \times \{q : \theta((k+1)/n, q) \in B\}$$

is in $\mathcal{B}_{[0,t]} \times \mathcal{F}_t$. consider now the second term

$$\cup_{(k+1)/n > t}\{(u, q) \in [0, t] \times E : k \le nu < k+1, \theta(t, q) \in B\}$$

$$= \cup_{(k+1)/n > t}\{u \in [0, t], k/n \le u < (k+1)/n\} \times \{q : \theta(t, q) \in B\}$$

$$= ([0, t] \cap \cup_{(k+1)/n > t}[k/n, (k+1)/n)) \times \{q : \theta(t, q) \in B\}$$

$$= ([0, t] \cap \cup_{k > nt-1}[k/n, (k+1)/n)) \times \{q : \theta(t, q) \in B\}$$

$$= ([0, t] \cap \cup_{k \ge [nt]}[k/n, (k+1)/n)) \times \{q : \theta(t, q) \in B\}$$

$$= [[nt]/n, t] \times \{q : \theta(t, q) \in B\}$$

which is clearly in $\mathcal{B}_{[0,t]} \times \mathcal{F}_t$ and we are through.

Study project 77

A fundamental problem in probability theory involves the notion of convergence of a sequence of probability measures on the space. Weak convergence plays here an important role. The basic theorems about weak convergence deal with measures on separable metric spaces, i.e., Polish spaces. Convergence theory of probability measures finds fundamental applications in statistical mechanics, wherein one is interested in the convergence of non-equilibrium distributions to the equilibrium Gibbs distribution in accord with the second law of thermodynamics. The theorem of the following section can be taken as the starting point of all discussions in this direction.

(X, D) is a Polish space, i.e., X is a complete, separable metric space and D is the metric on X. $\mathcal{B} = \mathcal{B}_X$ is the Borel σ-field on X, i.e., the σ-field generated by the open subsets of X. $M(X)$ is the set of all probability measures on $(X, \mathcal{B}_X)$ and $C_b(X)$ the space of all bounded continuous functions on X. $M(X)$ is a subset of the dual space of $C_b(X)$. This follows from the fact that if P is a probability measure on $(X, \mathcal{B}_X)$, then the map $f \to \int_X f dP$ from $C_b(X)$ into $\mathbb{R}$ is a bounded linear functional. Linearity is obvious. Boundedness follows from the fact that $f \in C_b(X)$ has norm $\| f \| = sup|f(x)|$ that makes $C_b(X)$ into a Banach space and

$$\left| \int f dP \right| \le \| f \|$$

so that the linear functional $L_P \in C_b(X)^*$ defined by $L_P(f) = \int f dP$ has norm $\leq$ unity. Note that taking $f = 1$, we get $L_P(1) = 1$ and hence $\| L_P \| = 1$. The dual space $C_b(X)^*$ has the weak topology, i.e., the topology in which convergence of a sequence L_n to L takes place iff for each $f \in C_b$, $L_n(f) \to L(f)$. This topology is called the weak star topology. What is an open set of $C_b(X)^*$ in the weak star topology ? A subbasis for this topology consists of elements of the form

$$\{L \in C_b(X)^* : |L(f) - c| < \epsilon\}$$

where ϵ is any positive number, c is an arbitrary real number and $f \in C_b(X)$. Thus, if W is any open set in $C_b(X)^*$ relative to the weak star topology and $L_0 \in W$, then W contains a set of the form

$$\{L \in C_b(X)^* : |L(f_i) - L_0(f_i)| < \delta, i = 1, 2, ..., n\}$$

where $f_1, ..., f_n \in C_b(X)$. Suppose $L_n \to L_0$. Then it follows that $|L_n(f_i) - L_0(f_i)| \to 0$ for all $i = 1, 2, ..., n$ and hence L_n is eventually in W. The basic theorems regarding convergence of probability measures on a separable metric space is the equivalence of the following statements. Let $\mu_n, n = 1, 2, ...$ be a sequence in $M(X)$. Given $\mu \in M(X)$, the following are equivalent: (1) $\int f d\mu_n \to \int f d\mu$ for all $f \in C_b(X)$, (2) $\int f d\mu_n \to \int f d\mu$ for every bounded, uniformly continuous function f on (X, ρ) with ρ any equivalent metric, (3) $limsup \mu_n(C) \leq \mu(C)$ for any closed set C in X, (4) $liminf \mu_n(G) \geq \mu(G)$ for any open set G in X, (5) $lim \mu_n(B) = \mu(B)$ for any Borel set B whose boundary has zero measure. Some of the implications are obvious. The ideas around which the proof of the other implications is based are as follows. If B is any set whose boundary has zero μ measure and if we agree to the validity of (3) and (4), then

$$limsup \mu_n(B) \leq limsup \mu_n(Cl(B)) \leq \mu(Cl(B))$$

$$= \mu(int(B)) \leq liminf \mu_n(int(B)) \leq liminf \mu_n(B)$$

where $\mu(Cl(B)) = \mu(int(B))$ follows from the assumption that $\mu(\partial B) = 0$, i.e., the boundary of B has zero measure. The other fact used that if ρ is any equivalent metric and C any closed set, then we define

$$f_k(x) = (1 + \rho(x, C))^{-k}, k = 1, 2, ...$$

Then, f_k is bounded and uniformly continuous and $f_k(x)$ decreases to $\chi_C(x)$ as $k \to \infty$. Therefore if we agree to (2), then

$$\mu(C) = lim_k \int f_k d\mu = lim_k lim_n \int f_k d\mu_n$$

Now,

$$\int f_k d\mu_n \geq \mu_n(C)$$

since $f_k \geq \chi_C$. Thus,

$$\mu(C) \geq limsup \mu_n(C)$$

The last idea is based on the following. Let $f \in C_b(X)$ and $\epsilon > 0$. Choose N and $a_1, ..., a_{N-1}$ so that

$$- \| f \| - 1 = a_0 < a_1 < ... < a_N = \| f \| + 1$$

and $\mu\{f = a_i\} = 0$. This is possible because the set of atoms of μ can be atmost countably finite. For suppose we set $E_n = \{a : \mu(f = a) > 1/n\}$. Then E_n cannot have more than n elements since μ is a probability measure. Hence, $\cup E_n = \{a : \mu(f = a) > 0\}$ must be countable. We may also assume that $a_i - a_{i-1} < \epsilon$ for all i. Let $B_i = \{a_{i-1} \leq f < a_i\}$. The $B_i's$ are disjoint, their union is X. Also the boundary of B_i is $\{f = a_i\}$ which has zero μ-measure. Also,

$$\| f - \sum_{i=1}^{N} a_i \chi_{B_i} \| < \epsilon$$

since given any x, we have $x \in B_i$ for some i and then, $a_{i-1} \leq f(x) < a_i$ and $\sum a_j \chi_{B_j}(x) = a_i$ so that the difference of the two is smaller in magnitude than $a_i - a_{i-1}$ which is smaller than ϵ. If we assume (5), then it would follow that

$$limsup| \int f d\mu_n - \int f d\mu| \leq 2\epsilon + limsup| \int \sum a_i \chi_{B_i} d\mu_n - \sum a_i \chi_{B_i} d\mu|$$

$$\leq 2\epsilon + \sum a_i limsup|\mu_n(B_i) - \mu(B_i)| \leq 2\epsilon$$

letting ϵ decrease to zero yields (1).

Chapter 11

Statistical Physics

11.1 Microcanonical system

The energy of a system of N particles is given by $H(\mathbf{q}_1, ..., \mathbf{q}_N, \mathbf{p}_1, ..., \mathbf{p}_N)$, where $\mathbf{q}_i$ is the canonical coordinate of the ith particle and $\mathbf{p}_i$ is the canonical momentum. The energy of the system is assumed to fall in an infinitesimal interval $[E, E + dE]$. Then the microcanonical density of the distribution of the phase variables is a constant $1/A$ for all phase variables $\mathbf{x}$ with $E < H(\mathbf{x}) < E + dE$ and zero for $\mathbf{x}$ violating this inequality. Let $\theta(x)$ denote the unit step function, i.e., $\theta(x) = 1$ for $x > 0$ and $= 0$ for $x < 0$. Then, the density of the phase points can be expressed as

$$\rho(x) = \frac{1}{A}(\theta(E + dE - H(\mathbf{x}) - \theta(E - H(\mathbf{x}))$$

where

$$A = \int_{E<H(\mathbf{x})<E+dE} d\mathbf{x}$$

We are assuming that the system of particles is contained inside a definite volume V of space. Thus, the coordinates $\mathbf{q}_i$ of the particles in the above integral must be restricted to this range. In this way, the microcanonical density ρ is parameterized by the energy E, the volume V and the number of particles N of the system, i.e., we write $\rho(\mathbf{x}|E, V, N)$. The uncertainty principle of quantum mechanics states that a basic cell of phase space for one particle moving in one dimension must be taken as $dqdp \approx h$. Thus, the phase volume for N particles moving in three dimensions must be taken as h^{3N}. If the particles were distinguishable, then the probability of finding the system inside a basic cell Z_i of phase space satisfying the energy and volume constraint is given by h^{3N}/A. But when we assume indistinguishability, this probability must be multiplied by $N!$ and this leads to the probability of a basic cell as $p_i = h^{3N}N!/A$. The total number of basic cells satisfying the energy, volume and number constraint is the reciprocal of this, i.e. $\Omega(E, V, N) = A/h^{3N}N!$. The entropy of the microcanonical distribution is then given by $S(E, V, N) = k_B log(\Omega(E, V, N))$. As an example, suppose the system consists of a single free particle moving in a box of volume V. The energy of the system is $H(p_x, p_y, p_z) = (p_x^2 + p_y^2 + p_z^2)/2m$. Then

$$A = V \int_{E < H(\mathbf{p}) < E + dE} dp_x dp_y dp_z = V.dE.\frac{d}{dE} \int_{H(\mathbf{p}) < E} dp_x dp_y dp_z$$

$$= V.dE.\frac{d}{dE} \int_{p^2 < 2mE} 4\pi p^2 dp = V.dE.\frac{d}{dE}(4\pi(2mE)^{3/2}/3)$$

We leave it as an exercise to evaluate the entropy of the free particle. As another example which is a generalization of this one, consider N free particles moving in 3 dimensions. The energy of the system is given by $H(\mathbf{p}_1, ..., \mathbf{p}_N) = \sum_{i=1}^{N}(p_{xi}^2 + p_{yi}^2 + p_{zi}^2)/2m$. The normalization integral A is given by

$$A = V^N dE.\frac{d}{dE} \int_{\sum_{i=1}^{N}(p_{xi}^2+p_{yi}^2+p_{zi}^2) < 2mE} \Pi_{i=1}^{N} dp_{xi} dp_{yi} dp_{zi}$$

We denote the volume of the unit sphere in n dimensions as $c(n)$. This is a standard integral and can be evaluated easily. Since we are more concerned with the physical interpretation, we shall not describe its evaluation. Then,

$$A = V^N c(3N)\frac{d}{dE}(2mE)^{3N/2}$$

from which the entropy can be evaluated. The fundamental assumption in describing a system using the micro-canonical distribution is that its energy is a constant.

11.2   Entropy in quantum statistical mechanics

Divide a closed system into a large number of subsystems and consider any one of these. Let w_n be the probability that this subsystem is in the energy state E_n. In other words, w_n is the distribution function for this subsystem. It is assumed that the particles of the subsystem are enclosed inside a rigid box. This means that momentum and angular momentum are not constants of the motion and hence w_n is a function of energy alone, i.e., $w_n = w(E_n)$. The probability that the energy of this subsystem is in the range $[E, E + dE]$ is obtained by multiplying $w(E)$ with the number of quantum states in this interval. The number of quantum states of the subsystem having energy smaller than E is denoted by $\Gamma(E)$. Thus, the energy density function is given by $W(E)dE = w(E)\Gamma'(E)dE$. $W(E)$ has a sharp maximum at $\bar{E}$. Let ΔE denote the width of this maximum. Then, $W(\bar{E})\Delta E \approx 1$ and this gives for the number of quantum states $\Delta \Gamma$ in the energy interval ΔE about $\bar{E}$, $\Delta \Gamma = 1/w(\bar{E})$. In classical statistical mechanics, we take $\Delta \Gamma = \Delta p \Delta q/h^{3N}$ as seen above. The entropy of the subsystem is defined as the logarithm of the number of quantum states, i.e., $S = log(\Delta\Gamma)$. Now, we've seen that $logw(E_n)$ is an additive function of the energy. Thus, we can write $logw(E_n) = \alpha + \beta E_n$ and hence

$$S = log\Delta\Gamma = -logw(\bar{E}) = -\alpha - \beta\bar{E} = - <\alpha + \beta E_n> = - <logw(E_n)> = -\sum w_n log(w_n)$$

If ρ is the quantum density operator of the subsystem, then w_n are the eigenvalues of ρ and $-\sum w_n logw_n = -Tr(\rho log\rho)$. This is the correct expression for the entropy in quantum statistical mechanics. The entropy is an additive function. This can be seen from the assumption that the various parts of the original system do not interact. Let $\Delta\Gamma_j$ dempte the number of quantum states of the jth subsystem. This is also called the statistical weight of the jth subsystem. Then, $\Delta\Gamma = \Pi_j\Delta\Gamma_j$ gives the number of states for the entire system and its logarithm is the entropy of the entire system:

$$S = log\Delta\Gamma = \sum_j log\Delta\Gamma_j = \sum_j S_j$$

Another more direct way to see the reason why the entropy of a system must equal the sum of the entropies of its parts is to make use of the microcanonical distribution. Let E_0 be the total energy of the system. $d\Gamma_j(E_j)$ gives the number of microstates of the jth subsystem that fall in the energy range $[E_j, E_j + dE_j]$. The total number of microstates of the entire system that have energy lesser than E is given by the integral

$$\int_{\sum E_j < E} \Pi_j d\Gamma_j(E_j) = \int \theta(E - \sum E_j)\Pi_j\Gamma'_j(E_j)dE_j$$

The number of microstates of the entire system per unit energy range at E is then obtained by differentiating the above expression with respect to E. This number equals

$$\int \delta(E - \sum E_j)\Pi_j\Gamma'_j(E_j)dE_j$$

If the jth subsystem has energy E_j and the total energy of the system is $E_0 = \sum E_j$, then the number of microstates of the entire system per unit energy range is

$$dw = C.\delta(E - E_0)\Pi_j\Gamma'_j(E_j)dE_j$$

The derivative $\Gamma'_j(E_j)$ is replaced by $\Delta\Gamma_j/\Delta E_j$ and $\Delta\Gamma_j$ is replaced by $exp(S_j(E_j))$. This gives

$$dw = C.\delta(E - E_0)exp(\sum_j S_j(E_j))\Pi dE_j/\Delta E_j$$

This leads to the desired formula. This is to be regarded as heuristic derivation, not a mathematically rigorous one.

11.3 Systems in contact

Consider two systems A, B in thermal contact with each other. Let E be the total energy of the system, e the energy of system A, so that $E - e$ is the energy of system B. Let V the volume of system A and V_B the volume of system B. Let N_B be the number of particles in system B and N the number of particles in system A. Then, the number of microstates of the overall system in which the first system has energy e and the second $E - e$ is proportional to $\Omega(e, V, N)\Omega(E - e, V_B, N_B)$. Each microstate has the same probability. Thus, the probability density of the energy e of the first system is proportional to the above product, i.e.

$$p(e) = C.\Omega(e, V, N)\Omega(E - e, V_B, N_B)$$

The entropy of the first system is $S(e, V, N) = k_B log(\Omega(e, V, N))$ while the entropy of the second system is $S_B(E - e, V_B, N_B) = k_B log(\Omega(E - e, V_B, N_B)$. Thus,

$$p(e) = C.exp((S(e, V, N) + S_B(E - e, V_B, N_B))/k_B)$$

Now, assuming e to be small in comparison with E, we have

$$S_B(E - e, V_B, N_B) = S_B(E, V_B, N_B) - e\frac{\partial S_B(E, V_B, N_B)}{\partial E}$$

We define the temperature of system B T by $\frac{1}{T} = \frac{\partial S_B(E, V_B, N_B)}{\partial E}$. Then,

$$p(e) = C'.exp(-e/k_B T + S(e, V, N)/k_B) = C'.exp(-\beta e + S(e, V, N)/k_B)$$

The energy of each system particle is $y = e/N$ and the entropy per particle is $s(y) = S(e, V, N)/N$. It follows that the density of y has the form

$$p(y) = C'exp(-N(\beta y - s(y)/k_B)) = C'exp(-Ng(y))$$

Thus the energy per particle of the system satisfies the large deviation property. The most probable value of e satisfies the equation

$$\frac{d}{de}(-\beta e + S(e, V, N)/k_B) = 0$$

or

$$\frac{\partial S(e, V, N)}{\partial e} = 1/T$$

i.e., the most probable state of the system is that in which the system temperature coincides with the bath temperature. This is the condition of thermal equilibrium.

11.4 Statistical matrix

In quantum mechanics, the state of a system can be expressed as the superposition $\sum_n c_n \psi_n$ where $\psi_n(q)$ is the wave function corresponding to eigenvalue E_n. Then the distribution of the coordinates is given by

$$|\psi|^2 dq = \sum_{n,m} \bar{c}_n c_m \bar{\psi}_n \psi_m$$

For a mixed state, the constants c_n are taken as random variables distributed over some other probability space (the environment space) and an averaging over this space leads to the following distribution function

$$\rho(q,q)dq = \sum_{n,m} \bar{w}_{nm} \bar{\psi}_n(q) \psi_m(q) dq$$

where $w_{nm} =< \bar{c}_n c_m >$. The average value of an observable $X(q,q')$ expressed in the coordinate representation when no averaging over the environment space is performed is given by

$$< X >= \int \bar{\psi}(q) X(q,q') \psi(q) dq dq' = \int \bar{\psi}(q)(X\psi)(q) dq =< \psi|X|\psi >$$

This can be expressed as

$$< X >= \sum_{n,m} \bar{c}_n c_m < \psi_n|X|\psi_m >$$

After averaging is performed, this acquires the form

$$< X >= \sum w_{nm} < \psi_n|X|\psi_m >= \int \rho(q,q') X(q,q') dq dq' = Trace(\rho X)$$

where

$$\rho(q,q') = \sum w_{nm} \bar{\psi}_n(q) \psi_m(q')$$

is the density operator. Suppose ψ_p is the momentum wave function, i.e. $C.exp(i2\pi p.q/h)$. Then when the system is in the state ψ given above, the probability density of the momentum p is given by

$$| < \psi_p|\psi > |^2 = | < \psi_p| \sum c_n \psi_n > |^2 = \sum_{n,m} \bar{c}_n c_m < \psi_p|\psi_n >< \psi_p|\psi_m >^*$$

$$= \sum_{n,m} \bar{c}_n c_m \int \bar{\psi}_p(q) \psi_n(q) dq \int \psi_p(q') \bar{\psi}_m(q') dq'$$

After averaging over the environment is performed, this acquires the form

$$\sum_{n,m} w_{nm} \bar{\psi}_n(q) \psi_m(q') \bar{\psi}_p(q) \psi_p(q') dq dq' = \int \bar{\psi}_p(q) \rho(q,q') \psi_p(q') dq dq'$$

$$=< \psi_p|\rho|\psi_p >= Tr(\rho|\psi_p >< \psi_p|)$$

11.5   Liouville's theorem

Let $\rho(q, p, t)$ denote the density of phase points as a function of time. Each phase point moves along the trajectory dictated by Hamiltonian mechanics:

$$\frac{dq_i}{dt} = \frac{\partial H}{\partial p_i}, \frac{dp_i}{dt} = -\frac{\partial H}{\partial q_i}$$

The principle of conservation of phase points which is equivalent to conservation of the total number of particles can be formulated as a law analogous to the mass conservation law of fluid dynamics:

$$\frac{\partial \rho}{\partial t} + \sum_i \frac{\partial(\rho q_i')}{\partial q_i} + \frac{\partial(\rho p_i')}{\partial p_i} = 0$$

where $q_i' = dq_i/dt, p_i' = dp_i/dt$. In view of the Hamiltonian mechanics, this boils down to

$$\frac{\partial \rho}{\partial t} + \{\rho, H\} = 0$$

where $\{u, v\}$ is the Poisson bracket of two observables $u(q, p)$ and $v(q, p)$ defined by

$$\{u, v\} = \sum_i \frac{\partial u}{\partial q_i} \frac{\partial v}{\partial p_i} - \frac{\partial u}{\partial p_i} \frac{\partial v}{\partial q_i}$$

The steady state distribution $\rho(q, p)$ satisfies $\{\rho, H\} = 0$. Such a steady state distribution exists provided that the Hamiltonian does not depend explicitly on time, i.e., $H(q, p, t) = H(q, p)$. It can be seen that the phase density $\rho(q, p, t)$ is a constant along the phase trajectory of the system, i.e.,

$$\frac{d\rho}{dt} = \frac{\partial \rho}{\partial t} + \sum_i \frac{\partial \rho}{\partial q_i} q_i' + \frac{\partial \rho}{\partial p_i} p_i'$$

$$= \frac{\partial \rho}{\partial t} + \{\rho, H\} = 0$$

Thus, ρ must be a function of the integrals of the motion. ρ must be a multiplicative integral, i.e., for two independent systems $1, 2$ the overall density function $\rho_{12} = \rho_1 \rho_2$. Equivalently, $log\rho$ must be an additive function for independent systems. The additive integrals are energy, momentum and angular momentum. Thus, ρ has the general form

$$\rho = exp(A + \beta E(q, p) + \gamma.\mathbf{P}(q, p) + \delta.\mathbf{M}(q, p))$$

where $\mathbf{P}$ is the total momentum of the system and $\mathbf{M}$ is the total angular momentum of the system. E is the total energy.

11.6 Partition function and Boltzmann's statistics

Consider a system in contact with a heat bath. If the total energy is E and the system energy is e, then the density of e given V, N, V_B, N_B (system volume, number of system particles, bath volume, number of bath particles) is

$$\rho(e|V, T, N) = C.\Omega(e|V, N)\Omega_B(E - e|V_B, N_B)$$

where $\Omega(e|V, N)de$ is the number of microstates of the system having energy in the range $[e, e+de]$ and $\Omega_B(E-e|V_B, N_B)$ is the number of microstates of the bath having energy $E-e$. C is a constant of proportionality. Let $\mathbf{x}$ denote the phase variables of the system, i.e., the canonical coordinates and momenta of the different system particles. The density of $\mathbf{x}$ is given by the law of total probabilities:

$$\rho(\mathbf{x}|V, T, N) = \int \rho(\mathbf{x}|e, V, N)\rho(e|V, Y, N)de$$

When the phase variables of the system are $\mathbf{x}$, the system energy is $H(\mathbf{x})$ and for a given system energy in the range $e, e + \Delta E$, $H(\mathbf{x})$ varies over this range. Moreover, for all $\mathbf{x}$ such that $H(\mathbf{x}) \in [e, e+\Delta E]$, $\rho(\mathbf{x}|e, V, N)$ is a constant (the microcanonical distribution for the system phase variables). Thus, in the integration in the above expression, e ranges over $[H(\mathbf{x}) - \Delta E, H(\mathbf{x})]$ and $\rho(\mathbf{x}|e, V, N) = C'$ a constant. Thus,

$$\rho(e|V, Y, N) = CC'.\int_{H(\mathbf{x-\Delta E}}^{H(\mathbf{x})} \Omega(e|V, N)\Omega_B(E - e|V_B, N_B)de$$

Now,

$$\Omega_B(E - e|V_B, N_B) = exp(S_B(E - e|V_B, N_B)) = exp(S_B(E|V_B, N_B) - e/T)$$

by a Taylor expansion. Thus,

$$\rho(\mathbf{x}|V, T, N) = C'' \int_{H(\mathbf{x})-\Delta E}^{H(\mathbf{x})} exp(-e/T)de = C''\Delta E.exp(-H(\mathbf{x})/T)$$

In other words, we've proved that when a system is in contact with a heat bath and has attained thermal equilibrium with the bath at temperature T, the density of the phase variables of the system is proportional to $exp(-\beta H(\mathbf{x})$. This is called the Boltzmann distribution. The microcanonical distribution holds for an isolated system while the Boltzmann distribution, also called the canonical distribution holds for a system in equilibrium with a heat bath. The partition function is defined as the integral

$$Z(T, V, N) = \int exp(-\beta H(\mathbf{x}))d\mathbf{x}$$

The density of the phase variables of the system in the canonical distribution is thus

$$\rho(\mathbf{x}|V, T, N) = \frac{exp(-\beta H(\mathbf{x}))}{Z(T, V, N)}$$

11.7   Boltzmann's statistics from the maximum entropy principle

Suppose a system has possible energies $E_1, E_2, \ldots$ with degeneracies $g_1, g_2, \ldots$. We want to distribute N distinguishable particles amongst these energy levels. The number of ways in which we can distribute N distinguishable particles amongst these energy levels in such a manner that n_i particles occupy the energy state E_i is given by the multinomial formula

$$\Omega(\{n_i\}) = \frac{N!}{n_1! n_2! \ldots}$$

The entropy is simply the logarithm of this which after making the Stirling approximation and neglecting an additive constant is

$$log\Omega(\{n_i\}) = -\sum log(n_i!) = -\sum_i n_i log(n_i) - n_i$$

We want to maximize this (that is maximize the number of microstates corresponding to the distribution $\{n_i\}$ subject to the energy constraint $\sum n_i.e._i = E$ and the number constraint $\sum n_i = N$. The appropriate function after introducing Lagrange multipliers is given by

$$L(\{n_i\}) = \sum_i (n_i log(n_i) - n_i) + \lambda(\sum n_i - N) + \mu(\sum n_i.e._i - E)$$

Setting the partial derivatives equal to zero, i.e., $\frac{\partial L}{\partial n_i} = 0$ gives us

$$log(n_i) + \lambda + \mu E_i = 0$$

so that $n_i = C.exp(-\mu E_i)$. $1/\mu$ is interpreted as the temperature. This formula shows that the number of particles in the ith energy state is proportional to $exp(-E_i/T)$.

11.8   Pressure

Suppose a system is in contact with the environment and volume and energy can be exchanged between the two. The total number of microstates of the overall system is $\Omega(e, V, N)\Omega(e', V', N')$, where $e + e' = E$ is the total energy and $V + V' = V_0$ is the total volume. The entropy of the overall system plus environment is the logarithm of this, i.e.,

$$S = log\Omega(e, V, N) + log(\Omega(e', V', N') = S(e, V, N) + S(e', V', N')$$

Equilibrium is attained when the entropy is a maximum, i.e.

$$d(S(e, V, N) + S(e', V', N')) = 0$$

Since $de' = -de, dV' = -dV$, the condition for entropy maximum is

$$\left(\frac{\partial S}{\partial e} - \frac{\partial S'}{\partial e'}\right)de + \left(\frac{\partial S}{\partial V} - \frac{\partial S'}{\partial V'}\right)dV = 0$$

that is,

$$\frac{\partial S}{\partial e} = \frac{\partial S'}{\partial e'}$$

$$\frac{\partial S}{\partial V} = \frac{\partial S'}{\partial V'}$$

The pressure is defined by the thermodynamic formula, $T\frac{\partial S}{\partial V} = p$ where the temperature is given by $1/T = \frac{\partial S}{\partial e}$. Thus, the condition for equilibrium boils down to

$$T = T', p = p'$$

i.e. when a system is in thermodynamic equilibrium with the environment, the system and environment temperatures and pressures are equal. The condition for equality of temperatures is thermal equilibrium while the condition for equality of pressures is mechanical equilibrium. Thermodynamic equilibrium corresponds to both thermal and mechanical equilibrium.

11.9   The first thermodynamical equation and Free energy

We've seen that for a system in contact with a bath both the energy of the system and its volume can vary. If $S(e, V, N)$ denotes the entropy of the system as a function of its energy and volume and the number N of particles of the system is kept fixed, then we'd defined the temperature and pressure by the formulas $1/T = \frac{\partial S}{\partial e}, p/T = \frac{\partial S}{\partial V}$. This leads to the thermodynamical relation

$$TdS = dE + pdV$$

which coincides with our energy conservation equation when we note that $TdS = dQ$ is the heat added to the system under reversible conditions and $dW = pdV$ is the work done by the system on the bath. The free energy (also called Helmholtz free energy) is defined as $F = E - TS$. We find for its total differential $dF = dE - TdS - SdT = -pdV - SdT$ so that

$$\frac{\partial F}{\partial V} = -p, \frac{\partial F}{\partial T} = -S$$

Suppose that we start with the Gibbs density $exp(-\beta H(\mathbf{x}))$ for N particles. This will be the case when the system is in contact with a bath at constant temperature and the volume of the system if fixed, i.e. energy can be exchanged with the bath but neither volume nor particles can be exchanged. Assuming indistinguishability we can write the density of phase points as

$$\rho(\mathbf{x}|T, V, N) = exp(-\beta H(\mathbf{x}))/h^{3N}N!Z(T, V, N)$$

where the partition function

$$Z(T, V, N) = \frac{1}{h^{3N} N!} \int exp(-\beta H(\mathbf{x})) d\mathbf{x}$$

The Helmholtz free energy is defined by

$$exp(-\beta F(T, V, N)) = Z(T, V, N)$$

The density of phase points can be expressed in terms of this as

$$\rho(\mathbf{x}|T, V, N) = \frac{1}{h^{3N} N!} exp(\beta(F - H))$$

The average energy of the system is given by

$$E = <H> = exp(\beta F) \int H exp(-\beta H) d\mathbf{x} / h^{3N} N! = -\frac{exp(\beta F)}{h^{3N} N!} \frac{\partial}{\partial \beta} \int exp(-\beta H) d\mathbf{x}$$

$$= -exp(-\beta F) \frac{\partial}{\partial \beta} Z = -\frac{\partial}{\partial \beta} log(Z)$$

The energy fluctuation is given by

$$\sigma_H^2 = <H^2> - <H>^2 = \frac{1}{Z} \frac{\partial^2 Z}{\partial \beta^2} - (\frac{\partial Z}{\partial \beta})^2$$

$$= \frac{\partial}{\partial \beta} (\frac{1}{Z} \frac{\partial Z}{\partial \beta}) = \frac{\partial^2}{\partial \beta^2} log(Z)$$

The entropy of the Gibbs distribution is given by

$$S(T, V, N) = - <log\rho> = - <log(exp(-\beta H)/h^{3N} N! Z)>$$

$$= \beta <H> + 3N log(h) + log(N!) + log(Z)$$

Neglecting an additive constant and recalling that $\beta = 1/T$ we have

$$-T log(Z) = E - TS$$

Thus, $-T log(Z)$ equals the Helmholtz free energy of the system.

11.10   Distribution of energy and volume at constant temperature and pressure

Consider a system in contact with a bath. The temperature and pressure are assumed to be a constant. The volume of the system and the bath can vary in such a way that the total volume is a constant. The density of the energy and volume of the system are given by the product of the microcanonical densities of the system and the bath:

$$\rho(e, v | p, T, N) = C.\Omega(e, v, N)\Omega(E - e, V_0 - v, N_B) = C.\Omega(e, v, N)exp(S(E - e, V_0 - v, N_B))$$

A first order Taylor approximation gives

$$S(E - e, V_0 - v, N_B) = S(E, V_0, N_B) - e\frac{\partial S(E, V_0, N_B)}{\partial e} - v\frac{\partial S(E, V_0, N_B)}{\partial V}$$

$$= S(E, V_0, N_B) - e/T - pv/T$$

Thus,

$$\rho(e, v | p, T, N) = C'.\Omega(e, v, N).exp(-\beta(e + pv))$$

The joint density of $\mathbf{x}, V$ the phase variables and volume of the system is obtained by integrating the above over the range $e \in [H(\mathbf{x}), H(\mathbf{x}) + \Delta E]$ and $v \in [V, V + \Delta V]$ after multiplying by $\rho(\mathbf{x}|e, v, N)$ which is a constant over the above range. This gives

$$\rho(\mathbf{x}, V | p, T, N) = C''exp(-\beta H(\mathbf{x}) + pV))$$

The normalization constant C'' is given by

$$C''^{-1} = \int_{\mathbb{R}^{3N} \times [0,\infty)} exp(-\beta(H(\mathbf{x}) + pV))d\mathbf{x}dV$$

$$= \int_0^\infty Z(T, V, N)exp(-\beta pV)dV = Y(T, p, N)$$

say. Define $G(T, p, N)$ via the equation $Y(T, p, N) = exp(-\beta G(T, p, N))$. We shall now evaluate the entropy corresponding to the above joint density of the phase variables and volume.

$$S(T, p, N) = - < log(exp(-\beta(H + pV))/Y) >= \beta < H + pV > +logY = \beta(E + pV - G)$$

where for simplicity of notation, we've denoted the average volume $< V >$ by the same symbol V. This is a function of T, p, N. Since $\beta = 1/T$, this relation can be expressed as

$$G = E - TS + pV$$

and this corresponds to the well known definition of Gibbs free energy in thermodynamics. $E =< H >$ is the average system energy.

11.11 Systems with variable number of particles and thermodynamic potentials

A system is in contact with a bath and there can be exchange of energy, volume and particles between the system and the bath. The change in the entropy for small differential changes in energy, volume and number of particles can be expressed as

$$dS = dE/T + pdV/T - \mu dN/T$$

where $-\mu = T\frac{\partial S}{\partial N}$. Thus,

$$dE = TdS - pdV + \mu dN$$

The differential of the Helmholtz energy is

$$dF = d(E - TS) = dE - TdS - SdT = -pdV - SdT + \mu dN$$

and the differential of the Gibbs energy is

$$dG = d(E - TS + pV) = dE - TdS - SdT + pdV + Vdp = -SdT + Vdp + \mu dN$$

11.12 Computations for an ideal gas

For an ideal gas consisting of N particles enclosed inside a volume V, the total energy is pure kinetic: $H = \sum_{i=1}^{N}(p_{xi}^2 + p_{yi}^2 + p_{zi}^2)/2m$ where m is the mass of each gas particle. The partition function is evaluated as

$$Z(T, V, N) = V^N \int exp(-\beta H)\Pi_{i=1}^{N}dp_{xi}dp_{yi}dp_{zi} = V^N\left(\int_{-\infty}^{\infty} exp(-\beta p^2/2m)dp\right)^{3N}$$

$$= V^N(2\pi m/\beta)^{3/2N} = V^N(2\pi mkT)^{3N/2}$$

It should be noted however that the particles are indistinguishable and hence we've got to divide this expression by $N!$. Also, our phase volume element is not $dq_x dq_y\ dq_z dp_x dp_y dp_z$ but rather this divided by h^3. Thus, the correct expression for the partition function is

$$Z(T, V, N) = \frac{1}{N!}(\frac{V}{h})^N(2\pi mkT)^{3N/2}$$

Then,

$$Y(T, p, N) = \int_{0}^{\infty} Z(T, V, N)exp(-\beta pV)dV$$

$$= \frac{1}{N!h^N}(2\pi mkT)^{3N/2}\int_0^\infty V^N exp(-\beta pV)dV$$

$$= \frac{1}{h^N p^{N+1}}(2\pi mkT)^{3N/2}$$

The Gibbs free energy is given by

$$G(T, p, N) = -kTlog(Y(T, p, N))$$

$$= -\frac{3NkT}{2}log(2\pi mkT) + NkTlog(h) + (N+1)kTlog(p)$$

Then,

$$S = -\frac{\partial G}{\partial T}$$

We leave the job of evaluating this to the reader.

$$V = \frac{\partial G}{\partial p} = (N+1)/kTp$$

or

$$pV = (N+1)kT/p$$

or since N is a very large number (of the order 10^{23}), we get the equation of state as

$$pV = NkT$$

11.13   Grand canonical or macrocanonical ensemble

Consider now a system in contact with a bath when both energy and particles can be exchanged. Then assuming indistinguishability of the particles, the appropriate probability density of the phase variables and particle number is given by

$$\rho(x, N|T, V, \mu) = exp(-\beta(H(x) - \mu N))/h^{3N}N!Y$$

where

$$Y(T, V, \mu) = \sum_{N=0}^\infty \int exp(-\beta(H(x) - \mu N)dx/h^{3N}N!$$

Y is called the grand canonical partition function. The argument around which the derivation of the above density hinges once again begins with the microcanonical density which is $\Omega(e, V, N)\Omega(E - e, V_B, N_0 - N)$. Here, N, the number of system particles is a variable, e the system energy is also a variable. E, the total energy of the system and the bath is a constant, N_0, the total number of particles in the system and bath is also a constant. The volume of the system V is a constant. Now this density can be approximated as

$$\rho(e, N|T, V, \mu) = C.\Omega(e, V, N)exp(S(E - e, V_B, N_0 - N))$$

$$= C.\Omega(e, V, N)exp(S(E, V_B, N_0) - e/T - \mu N/T) = C'\Omega(e, V, N)exp(-\beta(e - \mu N))$$

where

$$1/T = \frac{\partial S(E, V_B, N_0)}{\partial E}, -\mu/T = \frac{\partial S(E, V_B, N_0)}{\partial N}$$

The density of the phase variables and number density is given by

$$\rho(\mathbf{x}, N|T, V, \mu) = \int \rho(\mathbf{x}|e, T, V, N)\rho(e, N|T, V, \mu)de$$

where $\rho(\mathbf{x}|e, T, V, N)$ is a constant when $\mathbf{x}$ is such that e falls in the range $[H(\mathbf{x}) - \Delta E, H(\mathbf{x})]$ and zero otherwise. Approximating the integral, we get

$$\rho(\mathbf{x}, N|T, V, \mu) = C''exp(-\beta(H(\mathbf{x}) - \mu N))$$

which is the desired grand canonical formula provided we divide this quantity by $N!$ corresponding to the indistinguishability of the particles. The formula for the entropy of the grand canonical distribution is given by

$$S(T, V, \mu) = - < log\rho >= \beta < H - \mu N > + < log(h^{3N} N!) > +logY$$

The second term is neglected since it is an infinite constant. $< log(h^{3N} N!) >= 3 < N > log(h) + < NlogN - N >$. h is nearly zero and hence $logh$ is nearly $-\infty$. $< NlogN >$ is infinite. $< N >$ is denoted by N and its is an additive constant, i.e., does not involve the chemical potential. Thus,

$$S = \beta(E - \mu N - K)$$

where

$$exp(\beta K(T, V, \mu)) = Y$$

This equation can also be expressed as

$$K = E - TS - \mu N$$

This is one of the thermodynamic potential, the generalization of the Helmholtz free energy to include the possibility of a variable number of particles in the system.

11.14 Large deviation property

Let $p(x)$ be a probability density function. Its moment generating function is given by $M(t) = \int_{-\infty}^{\infty} p(x)exp(tx)dx$. Its cumulant generating function equals $C(t) = log(M(t))$. $C(t)$ is a strictly convex function as follows from

$$C'(t) = M'(t)/M(t) = \int xp(x)exp(tx)dx/M(t)$$

$$C''(t) = \int x^2 p(x)exp(tx)dx/M(t) - (\int xp(x)exp(tx)dx)^2/M^2(t)$$

We define a new density function by

$$q(x) = p(x)exp(tx)/M(t)$$

and note that

$$C''(t) = \int x^2 q(x)dx - (\int xq(x)dx)^2$$

is the variance of the distribution corresponding to the density $q(x)$. Hence, $C''(t) > 0$. Note that t is assuming real values only. Let $f(t)$ be a convex function. Its Legendre transform is defined by $g(y) = inf_t(ty - f(t))$. The point t at which the function $ty - f(t)$ assumes it minimum satisfies $\frac{d}{dt}(ty - f(t)) = 0$, i.e., $y = f'(t)$. Let $t(y)$ denote the solution to this equation. Then, $g(y) = t(y)y - f(t(y))$. The differential of g is given by

$$dg(y) = ydt + tdy - f'(t)dt = (y - f'(t))dt + tdy = tdy$$

Let $X_1, X_2, ...$ be independent random variables having the density $p(x)$. The moment generating function of $Y_N = X_1 + ... + X_N/N$ is given by

$$M_y(t, N) = \mathbb{E}exp(tY_N) = (\int exp(tx/N)p(x)dx)^N$$

$$C_y(t, N) = log(M_y(t, N)) = N.log \int exp(tx/N)p(x)dx = NC_x(t/N)$$

Let $S(y)$ denote the Legendre transform of $C_x(t)$. By choosing an appropriate contour, the density of $p_N(y)$ of Y_N can be expressed as

$$p_N(y) = \int M_y(t, N)exp(-ty)dt = \int exp(N(C_x(t/N) - ty/N))dt$$

The quantity $C_x(t/N) - ty/N$ attains its maximum value of $-S(y)$ at a certain t_0 and hence we are intuitively led to believe that the dominant term in the integral is $exp(-NS(y))$. In other words, for large N, the density is proportional to $exp(-NS(y))$.

11.15   Relative entropy

Let $p(x)$ and $q(x)$ be two probability density functions. The relative entropy of p with respect to q is defined by

$$S[p|q] = -k \int_{-\infty}^{\infty} p(x)log(p(x)/q(x))dx$$

where k is a positive constant. Now, using the inequality $exp(z) \geq 1 + z$ with $z = log(q/p)$ we get $q/p \geq 1 + log(q/p)$ so that $q \geq p + p.log(q/p)$. Integrating both sides gives us

$$\int plog(q/p)dx \leq 0$$

or

$$\int plog(p/q)dx \geq 0$$

Thus,

$$S[p|q] \leq 0$$

Equality holds iff $p = q$ everywhere.

11.16   Legendre transform relations for the Helmholtz and Gibbs free energies

The starting point for all relations is the fundamental one concerning change in entropy:

$$TdS = dE + pdV - \mu dN$$

where E is the internal energy, p the pressure, V the volume, μ the chemical potential and N the number of particles. S and T are respectively the entropy and the temperature of the system. Regarding E as a function of S, V, N, we obtain from this the equations

$$\frac{\partial E}{\partial S} = T, \frac{\partial E}{\partial V} = -p, \frac{\partial E}{\partial N} = \mu$$

One may equivalently regard S as a function of E, V, N and then derive

$$\frac{\partial S}{\partial E} = 1/T$$

The free energy is $F = E - TS$ and its differential is obtained as

$$dF = dE - TdS - SdT = -pdV + \mu dN - SdT$$

from which we derive taking T, V, N as independent variables

$$\frac{\partial F}{\partial T} = -S, \frac{\partial F}{\partial V} = -p, \frac{\partial F}{\partial N} = \mu$$

The Gibbs free energy is defined by

$$G = E + pV - TS$$

and its differential is given by

$$dG = dE + pdV + Vdp - TdS - SdT = Vdp + \mu dN - SdT$$

so that taking T, p, N as our independent variables gives

$$\frac{\partial G}{\partial p} = V; \frac{\partial G}{\partial N} = \mu, \frac{\partial G}{\partial T} = -S$$

We can define another potential related to the free energy by

$$K = E - TS - \mu N$$

Its differential is given by

$$dK = dE - TdS - SdT - \mu dN - Nd\mu == -SdT - Nd\mu - pdV$$

so that taking T, V, μ as independent variables gives

$$\frac{\partial K}{\partial T} = -S, \frac{\partial K}{\partial \mu} = -N, \frac{\partial K}{\partial V} = -p$$

For a system in equilibrium with a bath, the probability density of the energy of the system is given by

$$p(E) = exp((S(E, V, N) + S_B(E_0 - E, V_B, N_B))/k)$$

where it is assumed that the system can exchange only energy with the bath not volume and particles. Then making a first order Taylor expansion

$$S_B(E_0 - y, V_B, N_B) = S_B(E_0, V_B, N_B) - E\frac{\partial S_B(E_0, V_B, N_B)}{\partial E} = -y/T + \alpha$$

where α is a constant, we get

$$p(E) = C.exp(-\beta E + S(E, V, N)/k)$$

where $\beta = 1/kT$. The energy per particle of the system is $y = E/N$ and its probability density is given by

$$p(y) = C.exp(-N(\beta y - s(y, V, N)/k))$$

where $s = S/N$ is the entropy per particle. The normalization constant is set equal to $\frac{1}{Z\Delta E}$, i.e.,

$$p(y) = \frac{1}{Z\Delta E}.exp(-N.g(y|T, V, N))$$

where

$$Z = \frac{1}{\Delta E} \int_0^\infty exp(-N.g(y|T, V, N))dy$$

and $g = \beta y - s/k$. The free energy is $F = -kT log Z = -\frac{1}{\beta} log Z$. This can be seen via the following argument. First the entropy of this density, i.e., entropy per particle is given by

$$< s > = -k \int p(y) log p(y) dy = -k < log p >$$

$$= k < Ng + log(Z\Delta E) > = Nk < g > + k log(Z\Delta E) = Nk < \beta y - s/k > + k log(Z\Delta E)$$

$$= k\beta < E > - < S > + k log(Z\Delta E)$$

since $< s >$ is negligible in comparison with $< S >$, it follows that

$$F = < E > - T < S > = -kT log Z$$

This equation may be expressed as

$$-\beta F = log(\frac{1}{\Delta E} \int_0^\infty dy.exp(-Ng)dy)$$

The free energy per particle is $f = F/N$. In the limit $N \to \infty$, what does this equal ?

$$\beta f_\infty = lim_{N\to\infty} \frac{\beta}{N} F = -lim_{N\to\infty} \frac{1}{N} log(\frac{1}{\Delta E} \int_0^\infty dy exp(-Ng)dy)$$

If g attains its minimum value of g_0 at y_0, the width of this minimum is $\Delta E/N$. Then, for large N,

$$\int_0^\infty exp(-Ng)dy \approx exp(-Ng_0)\Delta E/N$$

and

$$\beta f_\infty = -\frac{1}{N} log(exp(-Ng_0)\Delta E/N) = g_0 = inf(\beta y - s/k)$$

This leads us to the fundamental fact that the free energy per particle and the entropy of the microcanonical system are related by a Legendre transform.

Suppose that the system is in equilibrium with a bath and the equilibrium is such that both energy and volume can be exchanged. This means that the temperatures and pressures of the system and bath are equal and since particles cannot be exchanged between the system and the bath (we are assuming this), N for the system is fixed. Let x denote the phase variables of the system. Then, we've noted that the joint density of the phase variables and volume of the system has the form

$$\rho(x, V | T, P, N) = C.exp(-\beta(H(x) + pV))$$

We set

$$C = \frac{1}{h^{3N} N! \Delta V Y'}$$

where

$$Y'(T, p, N) = \frac{1}{h^{3N} N! \Delta V} \int exp(-\beta(H(x) + pV)) dx dV$$

The partition function was defined as

$$Z(T, V, N) = \frac{1}{h^{3N} N!} \int exp(-\beta H(x)) dx$$

Note that $x = (\mathbf{q}, \mathbf{p})$, with $\mathbf{q}, \mathbf{p}$ being the canonical coordinates and momenta of the system particles. Thus,

$$Y'(T, p, N) = \frac{1}{\Delta V} \int_0^\infty exp(-\beta pV) Z(T, V, N) dV$$

Let us evaluate the entropy of the system.

$$S = -k < log\rho >= k < \beta(H + pV) > + klog(h^{3N} N! \Delta V Y')$$

$$= < E > /T + p < V > /T + klog(h^{3N} N! \Delta V Y')$$

Thus the Gibbs free energy is given by

$$G(T, p, N) = E + p < V > -TS = -kTlog(h^{3N} N! \Delta V Y')$$

Since N is assumed to be a constant, we can take G as a function of T and p alone and then with neglect of a constant, we have

$$G(T, p) = -kTlog(Y') = -kTlog(\frac{1}{\Delta V} \int_0^\infty Z(T, V, N) exp(-\beta pV) dV$$

The Gibbs free energy per system particle is given by $g(T, p) = \frac{G(T,p)}{N}$ and we get

$$-\beta g(T, p) = \frac{1}{N} log(\frac{1}{\Delta V} \int_0^\infty Z(T, V, N) exp(-\beta pV) dV)$$

Now,

$$F = -kTlogZ$$

is the Helmholtz free energy and $f(T, V, N) = F/N$ is the Helmholtz free energy per system particle. Thus,

$$Z = exp(-N\beta f)$$

and we have

$$-\beta g = \frac{1}{N}log(\frac{1}{\Delta V}\int_0^\infty exp(-N\beta(f+pv))Ndv)$$

where $v = V/N$ is the volume per system particle. Since $log(N)/N$ is negligible, we can write this as

$$-\beta g = \frac{1}{N}log(\frac{1}{\Delta V}\int_0^\infty exp(-N(\beta f(T,v,N)+\beta pv))dv)$$

Suppose $f+pv$ attains its minimum at $v = v_0$ and that the width of this minimum is $\Delta v = \Delta V/N$. Then in the limit of very large N,

$$-\beta g = \frac{1}{N}log\frac{1}{\Delta V}\frac{\Delta V}{N}exp(-N(\beta f_0 + \beta pv_0))) = -\beta(f_0 + pv_0)$$

and

$$G = f_0 + pv_0 = inf_v(f+pv)$$

This means that the Gibbs free energy is related to the Helmholtz free energy by a Legendre transform.

11.17   Heat capacities

The heat dQ added to a system during an infinitesimal reversible transformation is given by $dQ = TdS$. Regarding the entropy S as a function of T, V, N, the heat added at constant volume per unit temperature change is given by

$$C_V = \frac{dQ}{dT}|_{V,N} = T\frac{\partial S(T,V,N)}{\partial T}$$

Now, the Helmholtz free energy is

$$F = E - TS$$

so that

$$dF = dE - TdS - SdT = -pdV - SdT$$

N remaining constant. Thus regarding F as a function of T, V, N, we have

$$\frac{\partial F(T,V,N)}{\partial T} = -S$$

and hence

$$C_V = -T\frac{\partial^2 F}{\partial T^2}$$

C_V is called the heat capacity at constant volume. The energy equation $TdS = dE + pdV$ shows that

$$T\frac{\partial S}{\partial T} = \frac{\partial E}{\partial T}$$

Again S and E are regarded as functions of T, V, N. Thus,

$$C_V = \frac{\partial E}{\partial T}$$

We can derive an important equation for the heat capacity in terms of the energy fluctuations by assuming the canonical distribution, i.e., a system obeying Boltzmann's statistics for the phase variables. This is valid for a system in equilibrium with a heat bath with which it can exchange energy but not volume or number of particles. Assuming that such a system has been quantized, we let $E_n, n = 1, 2, \ldots$ denote the energy levels of the system and $T = 1/k\beta$ the temperature. Then, the partition function of the system is given by

$$Z = \sum_{n=1}^{\infty} exp(-\beta E_n)$$

and the average energy of the system is

$$E = \sum E_n exp(-\beta E_n)/Z = -\frac{1}{Z}\frac{\partial Z}{\partial \beta} = -\frac{\partial}{\partial \beta} log(Z)$$

$$= kT^2 \frac{\partial Z}{\partial T}$$

Note that the energy levels E_n of the system are functions of V, N, the volume of the system as well as the number of particles in the system. In fact, writing down the Schrodinger equation for the system, we note that the number of particles appears as a parameter and the boundary conditions that the wave function vanishes at the walls of the system will force the volume of the system to also appear as a parameter in the wave functions and in the quantized energy levels. Now,

$$C_V = \frac{\partial E}{\partial T} = k\beta^2 \frac{\partial^2}{\partial \beta^2}(log Z)$$

$$= k\beta^2 \frac{\partial}{\partial \beta}(Z'(\beta)/Z(\beta)) = k\beta^2(Z(\beta)Z''(\beta) - Z'(\beta)^2)/Z(\beta)^2$$

$$= k\beta^2(Z''(\beta)/Z(\beta) - (Z'(\beta)/Z(\beta))^2) = k\beta^2(<H^2> - <H>^2) = \frac{1}{kT^2}Var(H)$$

where H is the energy of the system. Choosing as independent variables T, p, N, we define the specific heat at constant pressure to be

$$C_P = T\frac{\partial S(T, p, N)}{\partial T}$$

Now $G = E + pV - TS$ so that $dG = dE + pdV + Vdp - TdS - SdT = Vdp - SdT$ and hence if we regard G as a function of T, p, N (N is being taken as constant here, but the required modification when N varies can be made), then

$$\frac{\partial G}{\partial T} = -S$$

so that

$$C_P = -T\frac{\partial^2 G(T, p, N)}{\partial T^2}$$

The enthalpy is defined as $H = E + pV$ and its differential is given by $dH = dE + pdV + Vdp = TdS + Vdp$ so that

$$C_P = T\frac{\partial S}{\partial T}|_P = \frac{\partial H}{\partial T}|_P$$

Consider now a system that is in equilibrium with a bath and is such that both energy and volume can be exchanged. The condition for equilibrium means that the temperatures and pressures of the system and the bath are the same. Then, the joint density of the phase variables of the system particles and the system volume is given by

$$\rho(x, V | T, p, N) = \frac{1}{h^{3N} N! \Delta V Y'(T, p, N)} exp(-\beta(E(x) + pV))$$

We are using the notation $E(x)$ for the system energy in place of $H(x)$ in order not to confuse with the enthalpy H. The enthalpy of the system is

$$H =< E + pV >= \frac{1}{h^{3N} N! \Delta V Y'} \int (E(x) + pV) exp(-\beta(E(x) + pV)) dx dV$$

$$= -\frac{1}{h^{3N} N! \Delta V Y'} \frac{\partial}{\partial \beta} \int exp(-\beta(E(x) + pV)) dx dV$$

$$- -\frac{1}{Y'}\frac{\partial Y'}{\partial \beta} = -\frac{\partial}{\partial \beta} log(Y')$$

Thus,

$$C_P = \frac{\partial H}{\partial T} = k\beta^2 \frac{\partial^2}{\partial \beta^2} log(Y') = \frac{1}{kT^2} Var(E(x) + pV)$$

11.18   Equipartition theorem

Consider a system in equilibrium with a heat bath so that energy can be exchanged but volume and particles cannot be exchanged. Then, the system volume and temperature are constants and the phase variables of the system obeys the Gibbs distribution, i.e. the density of the phase variables is given by

$$\rho(x|T, V, N) = \rho(x) = \frac{1}{h^{3N} N! Z} exp(-\beta H(x))$$

where Z, the system partition function is given by

$$Z = \frac{1}{h^{3N} N!} \int exp(-\beta H(x)) dx$$

Let $f(x)$ be a function of the phase variables of the system, i.e., a system observable. Then,

$$< f(x) \frac{\partial H(x)}{\partial x_i} >= \frac{1}{h^{3N} N! Z} \int f(x) \frac{\partial H(x)}{\partial x_i} exp(-\beta H(x)) dx$$

$$= -\frac{1}{h^{3N} N! Z \beta} \int f(x) \frac{\partial}{\partial x_i} exp(-\beta H(x)) dx$$

$$= \frac{1}{h^{3N} N! Z \beta} \int \frac{\partial f(x)}{\partial x_i} exp(-\beta H(x)) dx$$

$$= kT < \frac{\partial f(x)}{\partial x_i} >$$

Taking $f(x) = x_j$ gives

$$< x_j \frac{\partial H(x)}{\partial x_i} >= \delta_{ij} kT$$

In particular, since $\sum_i p_i q_i' = 2T$ where T is the kinetic energy of the system, we get

$$< 2T >=< \sum p_i q_i' >=< \sum p_i \frac{\partial H}{\partial p_i} >= 3NkT$$

where N is the number of particles, so that $3N$ is the number of generalized coordinates and $3N$ is the number of generalized momenta. Thus the average kinetic energy of the system particles is given by

$$< T >= \frac{3NkT}{2}$$

This is Maxwell's equipartition theorem.

11.19   Virial expansion

Consider now a system in equilibrium with a bath such that energy and particles can be exchanged but volume cannot. The entropy of the system plus environment is given by $S(E, V, N) + S_B(E_0 - E, V_B, N_0 - N)$. Making a first order Taylor expansion of the second term gives

$$S_B(E_0 - E, V_B, N_0 - N) = S_B(E_0, V_B, N_0) - E\frac{\partial S_B(E_0, V_B, N_0)}{\partial E} - N\frac{\partial S_B(E_0, V_B, N_0)}{\partial N} N$$

Now, the chemical potential is defined by

$$\mu = -T\frac{\partial S_B(E_0, V_B, N_0)}{\partial N}$$

where the temperature is given by

$$\frac{1}{T}\frac{\partial S_B(E_0, V_B, N_0)}{\partial E}$$

The chemical potential and temperatures of the system and bath are equal at equilibrium. Thus the total entropy can be expressed as

$$S(E, V, N) + S_B(E_0, V_B, N_0) - E/T + \mu N/T$$

The energy per particle is $y = E/N$ and the joint probability density of the energy and number of particles in the system is proportional to $exp(totalentropy/k)$, that is

$$\rho(y, N|T, V, \mu) = C.exp(Ns(y, V, N)/k - Ny/kT + \mu N/kT) = C.exp(-N\beta(y - Ts))z^N$$

where $z = exp(\mu/kT)$ is the fugacity. The partition function Z for fixed number of system particles is defined by

$$Z(T, V, N) = \int_0^\infty exp(-N\beta(y - Ts))dy$$

The partition function with both energy and particle number variable, is defined by

$$Y(T, V, \mu) = \sum_{N=0}^\infty z^N \int_0^\infty exp(-N\beta(y - Ts))dy = \sum_{N=0}^\infty z^N Z(T, V, N)$$

It is easy to see that $logY = pV/kT$. To see this, we evaluate the entropy of this system. This system in which energy and particles can be exchanged between the system and the bath has been called the grand canonical ensemble or the macro-canonical ensemble. Noting that the normalization constant C is given by $\frac{1}{Y}$, it follows that the entropy per particle equals

$$< s > = -k < log\rho > = Nk\beta < y - Ts > -k < N > logz + klogY$$

$$= Nk\beta(< E > /N - T < S > /N) - k < N > logz + klogY$$

$$= < E > /T - < S > -k < N > logz + klogY$$

and since s is negligible compared with S, it follows that

$$K(T, V, \mu) = < E > -T < S > -\mu < N > = -kTlog(Y) = F - \mu < N >$$

We find that

$$dK = d < E > -Td < S > - < S > dT - \mu d < N > - < N > d\mu$$

$$= -pdV - <S> dT - <N> d\mu$$

in view of the energy equation

$$Td<S> = d<E> + pdV - \mu d<N>$$

so that

$$\frac{\partial K}{\partial V} = -p$$

If N_0, y_0 are the most probable values of N_0, y ,the number of particles and the energy per particle., then we have approximately,

$$Y = z^{N_0} exp(-N_0(y_0 - Ts_0))\Delta E/N_0$$

Thus,

$$K = -kTlogY = -kT(N_0 logz - N_0(y_0 - Ts_0)) - kTlog(\Delta E/N_0)$$

Now, $N_0 y_0 = E_0$ is the most probable value of the system energy and $S_0 = N_0 s_0$ is the most probable value of the system entropy. These are extensive quantities and hence are proportional to the volume. N_0 is also proportional to volume. It follows that K is also proportional to the volume and hence $\frac{\partial K}{\partial V} = K/V$ Thus,

$$K = -pV$$

or

$$logY = \beta pV = pV/kT$$

This is the equation of state of the gas. Written in full, this equation reads

$$pV/kT = log(\sum_{N=0}^{\infty} z^N Z(T,V,N))$$

The computation of Z involves knowledge of the Hamiltonian of the system as a function of the number of particles in the system. In general, the computation of the Nth order partition function $Z(T,V,N)$ is performed as follows.

A remark: We cannot assert the equation $K = -pV$ from the equation $\frac{\partial K}{\partial V} = -p$ by saying that the pressure is a constant since the system is in equilibrium with the bath, the reason being that the equilibrium is only with regard to exchange of energy and number of particles, not with regard to volume. The system has rigid walls which force the volume of the system to be a constant. That $logY$ is an extensive quantity, i.e., proportional to volume is more of a hypothesis regarding homogeneity of the system.

Consider the evaluation of $Z(T,V,N)$. Suppose that the interaction between the molecules of the gas is pairwise. Let $V(\mathbf{r} - \mathbf{r}')$ denote the potential energy of interaction between two molecules of the gas located at points $\mathbf{r}$ and $\mathbf{r}'$ respectively. The Hamiltonian of the system assumed to comprise of N molecules is given by

$$H(\mathbf{r}_1, ..., \mathbf{r}_N, \mathbf{p}_1, ..., \mathbf{p}_N) = \sum_{i=1}^{N} \frac{1}{2M}(p_{xi}^2 + p_{yi}^2 + p_{zi}^2) + \sum_{1 \leq i < j \leq N} V(\mathbf{r}_i - \mathbf{r}_j)$$

The partition function

$$Z(T, V, N) = \int exp(-\beta H) d^3\mathbf{r}_1, ..., d^3\mathbf{r}_N d^3\mathbf{p}_1 ... d^3\mathbf{p}_N$$

is to be evaluated over the region $V^N \times \mathbb{R}^{3N}$. The integral separates out into the product of an integral over the momenta and an integral over the coordinates. The momentum integral is evaluated by noting that

$$\int_{-\infty}^{\infty} exp(-\beta p^2/2M) dp = (2\pi M/\beta)^{1/2}$$

Thus if T denotes the kinetic energy, we get

$$\int exp(-\beta T) d^3\mathbf{p}_1 ... d^3\mathbf{p}_N = (2\pi M/\beta)^{3N/2}$$

The coordinate integral is evaluated by noting that

$$exp(-\beta V) = exp(-\beta \sum_{1 \leq i < j \leq N} V_{ij}) = \Pi_{1 \leq i < j \leq N} exp(-\beta V_{ij})$$

$$= \Pi_{1 \leq i < j \leq N}(1 + f_{ij})$$

where

$$f_{ij} = exp(-\beta V_{ij}) - 1, V_{ij} = V(\mathbf{r}_i - \mathbf{r}_j)$$

This gives

$$exp(-\beta V) = 1 + \sum_{i<j} f_{ij} + \sum_{i<j,k<l,(i,j)\neq(k,l)} f_{ij}f_{kl} + ... + \Pi_{i_1<j_1,...,i_k<j_k} f_{i_1 j_1} ... f_{i_k j_k} + ... + \Pi_{i<j} f_{ij}$$

If the potential energy of interaction is weak in comparison with the kinetic energy, we have $\beta V_{ij} << 1$ and hence $-1 << f_{ij} < 0$. Then only the first few terms need to be retained in the above expansion. We have

$$\int exp(-\beta V) d^3\mathbf{r}_1 ... d^3\mathbf{r}_N = V^N + \int \sum_{i<j} f_{ij} d^3\mathbf{r}_1 ..., d^3\mathbf{r}_N + ...$$

Retaining only the first term, this approximates to

$$V^N + (N(N-1)V^{N-2}/2) \int f_{12} d^3\mathbf{r}_1 d^3\mathbf{r}_2$$

$$= V^N + \frac{N(N-1)}{2} V^{N-2} \int_{V \times V} (exp(-\beta V(\mathbf{r}_1 - \mathbf{r}_2)) - 1) d^3\mathbf{r}_1 d^3\mathbf{r}_2 -$$

From this we can derive the approximate equation of state of the gas. Since the particles are indistinguishable, in the evaluation of the partition function $Z(T, V, N)$ we must divide by $N!$. This gives the following approximate expression for the partition function:

$$Z(T, V, N) \approx \frac{1}{h^{3N} N!} (2\pi M/\beta)^{3N/2} (V^N + \frac{N(N-1)}{2} V^{N-2} \int_{V \times V} f_{12} d^3\mathbf{r}_1 d^3\mathbf{r}_2)$$

Using this expression in Y, the approximate equation of state of the gas can be derived.

11.20   The Maxwell relations

Consider a system with a fixed number of particles. Then we can form several functions of state such as E, H, F, G. Their differentials are given by

$$dE = TdS - pdV, dF = d(E - TS) = -pdV - SdT$$

$$dH = d(E + pV) = TdS + Vdp, dG = d(E + pV - TS) = Vdp - SdT$$

from which we deduce

$$\frac{\partial E}{\partial S}\Big|_V = T, \frac{\partial E}{\partial V}\Big|_S = -p$$

$$\frac{\partial F}{\partial V}\Big|_T = -p, \frac{\partial F}{\partial T}\Big|_V = -S$$

$$\frac{\partial H}{\partial S}\Big|_p = T, \frac{\partial H}{\partial p}\Big|_S = V$$

$$\frac{\partial G}{\partial p}\Big|_T = V, \frac{\partial G}{\partial T}\Big|_p = -S$$

These in turn imply the following

$$\frac{\partial T}{\partial V}\Big|_S = -\frac{\partial p}{\partial S}\Big|_V$$

$$\frac{\partial p}{\partial T}\Big|_V = \frac{\partial S}{\partial V}\Big|_T$$

$$\frac{\partial T}{\partial p}\Big|_S = \frac{\partial V}{\partial S}\Big|_p$$

$$\frac{\partial V}{\partial T}\Big|_p = -\frac{\partial S}{\partial p}\Big|_T$$

In terms of Jacobians, these can be expressed as

$$\frac{\partial(T,S)}{\partial(V,S)} = \frac{\partial(p,V)}{\partial(V,S)}$$

$$\frac{\partial(p,V)}{\partial(T,V)} = \frac{\partial(T,S)}{\partial(T,V)}$$

$$\frac{\partial(T,S)}{\partial(p,S)} = \frac{\partial(p,V)}{\partial(p,S)}$$

$$\frac{\partial(p,V)}{\partial(T,p)} = \frac{\partial(T,S)}{\partial(T,p)}$$

These four equations are equivalent to the single equation

$$\frac{\partial(p,V)}{\partial(x,y)} = \frac{\partial(T,S)}{\partial(x,y)}$$

where x,y can be any two thermodynamic variables. This is in turn equivalent to

$$\frac{\partial(p,V)}{\partial(T,S)} = 1$$

11.21 Master equations in statistical mechanics

Consider a transition probability function $\rho(z,t+\tau|z',t)$. $P(E,t+\tau|z,t) = \int_E \rho(z,t+\tau|z',t)dz$ gives the probability that at time $t+\tau$ the state of the system will fall in the set E given that at time t the state was z'. The obvious condition $P(E,t|z,t) = 1$ if $z \in E$ and $= 0$ if $z \notin E$ implies the

$$lim_{\tau \to 0}\rho(z,t+\tau|z',t) = \delta(z - z')$$

For small transition times τ, we can make a Taylor expansion of the transition probability function and get

$$\rho(z,t+\tau|z',t) = \delta(z - z') + \tau(w(z,z',t) - a(z',t)\delta(z - z')) + O(\tau^2)$$

The probability that in the time interval $[t,t+\tau]$, the state will not make any transition from z' is obtained by integrating ρ with respect to z over an infinitesimally small neighborhood of z. This probability evaluates to $1 - \tau a(z',t)$ for small τ. The probability that in a small time interval $[t,t+\tau]$, the system will make a transition from z' is then given by $\tau a(z',t)$ and hence the jump probability per unit time from the state z' at time t is given by $a(z', t)$. Thus $a(z', t)$ has the physical interpretation of being the jump probability rate. Now,

$$1 = \int \rho(z, t + \tau | z', t) dz = 1 + \tau \left(\int w(z, z', t) dz - a(z', t) \right) + O(\tau^2)$$

Thus, the jump rate is also given by

$$a(z', t) = \int w(z, z', t) dz$$

We now write down the Chapman-Kolmogorov equations for the transition probability density:

$$\rho(z, t + \tau | z', t') = \int \rho(z, t + \tau | z'', t) \rho(z'', t | z', t') dz''$$

For small τ, the right side approximates to

$$\int ((1 - a(z'', t)\tau)\delta(z - z'') + \tau w(z, z'', t)) \rho(z'', t | z', t') dz''$$

$$= (1 - a(z, t)\tau)\rho(z, t | z', t') + \tau \int w(z, z'', t) \rho(z'', t | z', t') dz''$$

In the limit of $\tau \to 0$, we thus obtain the equation

$$\frac{\partial \rho(z, t | z', t')}{\partial t} = -a(z, t)\rho(z, t | z', t') + \int w(z, z'', t)\rho(z'', t | z', t') dz''$$

$$= \int w(z, z'', t)\rho(z'', t | z', t') dz'' - \int w(z'', z, t)\rho(z, t | z', t') dz''$$

This is called the master equation. It states that the probability of being in the state z can increase either via transitions from some other state z'' to the state z or can decrease via transitions from the state z to some other state z''.

11.22   Reversible state transformations, heat and work

Let $|n>$ be the quantum states of a system and p_n the probability that the system is in the state $|n>$. For the Gibbs canonical system corresponding to the physical situation in which the system is in equilibrium with a heat bath, we have $p_n = exp(-\beta E_n)/Z$, where E_n is the energy of the state $|n>$, i.e., if H is the Hamiltonian operator of the system, $H|n> = E_n|n>$ and Z is the partition function, i.e., $Z = \sum_n exp(-\beta E_n)$. The entropy is $S = -k \sum p_n log(p_n)$ and the energy of the system is given by $E = \sum p_n E_n$. Suppose now that the extrinsic system variables such as volume, temperature change by infinitesimal amounts, but the particle number is constant. The resulting change in the energy is given by $dE = \sum p_n dE_n + E_n dp_n$. The term $dQ = \sum E_n dp_n$ is called heat (absorbed by the system) and the term $dW = \sum p_n dE_n$ is called work (done on the system). The change dp_n in the probabilities produces a change dQ in the heat absorbed by the system and a corresponding change dS in the entropy given by

$$dS = -k \sum dp_n.log(p_n) + dp_n = -k \sum dp_n.log(p_n)$$

since $\sum dp_n = d \sum p_n = d(1) = 0$. Now,

$$logp_n = -\beta E_n - log(Z)$$

so that

$$dS = k\beta \sum E_n dp_n + klog(Z) \sum dp_n = k\beta \sum E_n dp_n = dQ/T$$

This is the fundamental relation connecting heat absorbed in an irreversible process with the change in entropy. The work done dW is due to the change in the energy eigenvalues with the probabilities remaining fixed. The eigenvalues are functions of the system volume and possibly other extensive parameters of the system. For a small volume change, we have

$$dW = \sum p_n dE_n = \sum p_n \frac{\partial E_n}{\partial V} dV = -pdV$$

with

$$p = -\sum p_n \frac{\partial E_n}{\partial V}$$

The overall change in the energy is given by

$$dE = dQ + dW$$

An infinitesimal transformation of the state of a system through non-equilibrium states is called irreversible. For such a transformation, the relation $dQ = TdS$ is replaced by

$$dS > dQ/T = \sum E_n dp_n/T$$

because the system always tends towards its equilibrium state of maximum entropy and hence entropy is created. We can write this equation as $dS_{irr} - dS_{rev} > 0$.

During reversible heating, no work is done and energy is added in the form of heat, i.e., $dE = dQ = TdS, dW = 0$. A process is adiabatic if no heat is exchanged during the whole process. An adiabatic reversible process is one in which the probabilities remain constant and only the energy eigenvalues change. Thus, $dQ = 0, dE = dW$. For an ideal gas, $E = 3NkT/2$ and $dE = 3NkdT/2$. But, $dE = dW = -pdV = -(NkT/V)dV$ and hence

$$\frac{3}{2}\frac{dT}{T} + \frac{dV}{V} = 0$$

which integrates to give $T^{3/2}V = Constt$ or $pV^{5/3} = Constt.$. The change in energy is given by $E_2 - E_1 = \frac{3Nk}{2}(T_2 - T_1)$ where $T_2^{3/2}V_2 = T_1^{3/2}V_1$. During reversible isothermal expansion of an ideal gas in contact with a heat bath, energy eigenvalues and probabilities both change in such a way that the equation $p_n = exp(-\beta E_n)/Z$ always holds. Since $E = 3NkT/2$ and T is a constant, we get $dE = 0$ and hence $dQ + dW = 0$ where

$$\Delta W = -\int_{V_1}^{V_2} pdV = -NkT \int_{V_1}^{V_2} dV/V = -NkTlog(V_2/V_1)$$

The heat supplied is

$$\Delta Q = \int_{S_1}^{S_2} TdS = T(S_2 - S_1) = -\Delta W = NkTlog(V_2/V_1)$$

In this case, the energy received by the gas from the bath in the form of heat is exactly balanced by the loss of energy due to work done by the gas.

11.23 Carnot cycle

Consider first a reversible isothermal expansion in which the gas expands from the volume V_1 to the volume V_2. The temperature during this expansion remains constant at T. The entropy increases from S_1 to S_2. The amount of heat added equals $\Delta Q = T(S_2 - S_1)$. We label the initial state of the system as one and the final state as two. From the state two, we consider an adiabatic expansion of the gas to the state 3, followed by isothermal compression of the gas to state four followed by an adiabatic compression back to state one. Since the entire process is reversible, we have the energy relation $dE = TdS - pdV$ valid throughout. The total change in the energy around the cycle must be zero since the gas returns to its original state, i.e.,

$$\int_{1-2-3-4-1} dE = \int_{1-2-3-4-1} TdS - \int_{1-2-3-4-1} pdV = 0$$

The work done by the gas around the complete cycle is given by

$$W = \int_{1-2-3-4-1} pdV$$

The heat absorbed by the gas during isothermal expansion from state one to state two is given by

$$Q_1 = \int_{1-2} TdS = T_1(S_2 - S_1)$$

and the heat rejected by the gas during isothermal compression from state 3 to state 4 is given by

$$Q_2 = -\int_{3-4} TdS = -T_2(S_4 - S_3) = T_2(S_3 - S_4) = T_2(S_2 - S_1)$$

Note that since the gas undergoes adiabatic change from state two to state 3, we have $S_2 = S_3$ and since it also undergoes adiabatic change from state 4 to state 1, we have $S_1 = S_4$. The total heat absorbed by the gas during the entire cyclic process is given by $Q_1 - Q_2$ and since the net change in the energy during the cycle is zero, this must also equal the work done by the gas, i.e., $W = Q_1 - Q_2$. The efficiency of the engine that undergoes this cyclic process is defined as the work done by the engine divided by the heat absorbed, i.e.,

$$\eta = \frac{W}{Q_1} = \frac{Q_1 - Q_2}{Q_1} = 1 - \frac{Q_2}{Q_1} = 1 - \frac{T_2}{T_1}$$

We now prove the fundamental theorem of Carnot, that any engine that operates between the temperatures T_1 and T_2 in a cyclic fashion must have efficiency lesser than η given by the above formula. To see the validity of this inequality, we consider a plot of the cycle on the T-S plane. Choose, the x axis as the S line and the y axis as the T line. Let γ be a cycle in this plane whose ordinate always falls in the range $[T_2, T_1]$. Then, we can bound this cycle by a rectangle having minimum ordinate T_1, maximum ordinate T_2, minimum abscissa S_1 and maximum abscissa S_2, where S_1 and S_2 are respectively the minimum and maximum abscissas of the original cycle γ. The rectangle represents a Carnot cycle operating between the temperatures T_1 and T_2. The rectangle has its side $S = S_1$ tangential to γ at the point $a = (S_1, T_a)$ and its side $S = S_2$ tangential to γ at the point $b = (S_2, T_b)$. Clearly, $T_2 < T_a, T_b < T_1$. The top portion of the cycle γ starts from (S_1, T_a) and goes till (S_2, T_b). Call this portion of γ γ_1. The bottom position of γ starts from (S_2, T_b) and goes till (S_1, T_a). Call this portion of γ γ_2. The curve γ_1 falls below the line $T = T_1$ and the curve γ_2 falls above the line $T = T_2$. During the process γ_1, the heat absorbed by the gas $\int_{\gamma_1} T dS$ is therefore always lesser than $T_1(S_2 - S_1)$ and during the process γ_2, the heat rejected $-\int_{\gamma_2} T dS$ is always greater than $T_2(S_2 - S_1)$. It follows that the work done during the cycle γ divided by the heat absorbed given by

$$\frac{\int_{\gamma_1} T dS + \int_{\gamma_2} T dS}{\int_{\gamma_1} T dS} = 1 - \frac{-\int_{\gamma_2} T dS}{\int_{\gamma_1} T dS} < 1 - T_2/T_1$$

Note that $\int_{\gamma_1} T dS$ and $-\int_{\gamma_2} T dS$ are both positive. This proves Carnot's fundamental theorem.

11.24   Study Projects

Study project 1

A seminar on the role of probability theory in statistical physics and statistical thermodynamics. The seminar should emphasize the fact that probability theory should be studied from the standpoint of applying it to problems of statistical mechanics. If $\epsilon_n, n = 1, 2, ...$, are the energy levels of a system, then the Gibbs postulate states that the probability of finding the system to be in the nth state at time t is proportional to $exp(-\beta\epsilon_n)$ where $\beta = 1/kT$ with T as the temperature of the system. The standard method is to introduce the partition function $Z(\beta) = \sum_{n=1}^{\infty} exp(-\beta\epsilon_n)$. Thus, $p_n(\beta) = exp(-\beta\epsilon_n)/Z(\beta)$ gives the probability of the system being in the nth state. The quantum Gibbs ensemble can also be introduced at this stage. If $\psi_n(q)$ is the wave function for the system when it is in the nth state, then the density operator of the system is given by $\rho(q,q') = \sum_{n=1}^{\infty} p_n\psi_n(q)\psi_n(q')$. $w(q)dq = \rho(q,q)dq$ gives the probability distribution of the coordinates of the quantum system. The average of any observable $X(q,q')$ in the Gibbs state equals $\int \rho(q,q')X(q',q)dqdq' = Tr(\rho X)$. Computation of ρ is the fundamental problem of quantum statistical mechanics.

Study project 2

The fundamental thermodynamic equation $dQ = dE + dW = dE + PdV$ where dQ is the infinitesimal amount of heat pumped into the system and dE is the increase in the internal energy of the system and $dW = PdV$ is the work done by the system in the form of pressure on its walls causing an expansion against the surrounding force can be taught in a course on probability theory. The idea is to introduce the thermodynamic functions of state as statistical averages of quantities defined on the state of the system. If the process is reversible, then $dQ = TdS$ where T is the temperature of the system and S is the entropy. Actually, if the energy of the system is expressed as a function of its generalized coordinates, generalized momenta and the coordinates of its walls, $E(\mathbf{q}, \mathbf{p}, \mathbf{r})$, then $\frac{\partial E}{\partial \mathbf{r}}.d\mathbf{r}$ is the change in the energy of the system when its walls undergo an expansion defined by the change $d\mathbf{r}$ of its coordinates. Noting that $pdS\mathbf{n} = -\frac{\partial E}{\partial \mathbf{r}}$ is the force exerted on an infinitesimal surface element dS whose normal is $\mathbf{n}$, we can express the change in the energy as $-pdS\mathbf{n}.d\mathbf{r}$ and since $dS\mathbf{n}.d\mathbf{r} = dV$ is the increment in the volume, this change in energy is $-pdV$ which results in the desired thermodynamic formula. When the energy levels $\{\epsilon_n\}$ of the system are specified, how does one compute the entropy of the system ? Observe that the energy levels $\{\epsilon_n\}$ are in fact functions of the volume of the container and hence the partition function and therefore all the thermodynamical functions depend upon the container volume. To get an expression for the entropy we use the equation $TdS = dE + PdV$ which results in $d(E - TS) = dE - TdS - SdT = -PdV - SdT$. The free energy is defined as $F = E - TS$. This is also one of the thermodynamic potentials. The above differential relation implies that $\frac{\partial F}{\partial T} = -S$. So the entropy can be computed once we know F. Boltzmann's equation for the entropy is $S = klog(1/P)$ where P is the probability of the state occurring. When the ensemble is a mixture determined by a probability distribution $\{p_n\}$, the entropy is obtained by averaging, i.e. $S = -k\sum_n p_n log(p_n)$. For the Gibbs ensemble,

$$S = -k\sum p_n log(exp(-\beta\epsilon_n)/Z) = k\beta\sum p_n\epsilon_n + klog(Z) = k\beta E + k.log(Z)$$

where $E = \sum p_n\epsilon_n$ is the average energy of the system, also called the internal energy. Now

$$Z'(\beta) = \frac{d}{d\beta}\sum exp(-\beta\epsilon_n) = -\sum \epsilon_n.exp(-\beta\epsilon_n) = -E.Z$$

Thus,

$$E = -\frac{d}{\beta}log(Z)$$

and this gives

$$S = -k\beta\frac{d}{d\beta}log(Z) + k.log(Z)$$

$$F = E - TS = E - (1/k\beta)S = -kT.log(Z)$$

This is the fundamental equation of statistical thermodynamics. The partition function Z can be used directly to determine the free energy of the system from which the other thermodynamical quantities are derived.

Study project 3

Look at the momentum equation for a plasma starting from the Boltzmann equation:

$$\frac{\partial f_\alpha}{\partial t} + \mathbf{v}.\nabla f_\alpha + \frac{\mathbf{F}}{m_\alpha}.\nabla_\mathbf{v} f_\alpha = \frac{\partial f}{\partial t}\Big|_{coll}$$

where the right side denotes the rate of change of the number density of particles in phase space due to binary collision. Multiply both sides by $m_\alpha\mathbf{v}$ and integrate over all velocities. Note that

$$\int m_\alpha\mathbf{v}\frac{\partial f_\alpha}{\partial t}d^3\mathbf{v} = \frac{\partial}{\partial t}\int m_\alpha\mathbf{v}f_\alpha d^3\mathbf{v}$$

$$= \frac{\partial}{\partial t}(n_\alpha m_\alpha\mathbf{u}_\alpha) = \frac{\partial}{\partial t}(\rho_{m\alpha}\mathbf{u}_\alpha)$$

This is the rate of change of momentum per unit volume at a given point in the fluid. Note that we are considering the fluid as consisting of several ions species numbered α, $\alpha = 1, 2, ..., N$. The collision term is given by

$$\frac{\partial f_\alpha}{\partial t}\Big|_{coll} = \sum_{\beta}\int \sigma(\mathbf{v}_\alpha\mathbf{v}_\beta|\mathbf{v}'_\alpha, \mathbf{v}'_\beta)(f_\alpha(\mathbf{r}, \mathbf{v}'_\alpha, t)f_\beta(\mathbf{r}, \mathbf{v}'_\beta, t) - f_\alpha(\mathbf{r}, \mathbf{v}_\alpha, t)f_\beta(\mathbf{r}, \mathbf{v}_\beta, t))d\Omega.d^3\mathbf{v}_\beta$$

The βth term in the sum represents the rate of increase of the number density of α type of particles due to their collision with the β type of particles. The second term is given by

$$\int m_\alpha\mathbf{v}\mathbf{v}.\nabla f_\alpha d^3\mathbf{v}$$

Its ith component equals

$$\sum_{j}\int m_\alpha v_i v_j\frac{\partial f_\alpha}{\partial x_j}d^3\mathbf{v}$$

$$= \sum_j \frac{\partial}{\partial x_j} \int m_\alpha v_i v_j f_\alpha d^3 \mathbf{v}$$

$$= \sum_j \frac{\partial}{\partial x_j} (m_\alpha n_\alpha < v_i v_j >)$$

$$= \sum_j \frac{\partial}{\partial x_j} (\rho_{m\alpha} < v_i v_j >_\alpha)$$

Now observe that

$$< v_i v_j >_\alpha = < (u_{\alpha i} + c_{\alpha i})(u_{\alpha j} + c_{\alpha j}) > = u_{\alpha i} u_{\alpha j} + < c_{\alpha i} c_{\alpha j} >$$

where $\mathbf{c}_\alpha = \mathbf{v}_\alpha - \mathbf{u}_\alpha$ with

$$\mathbf{u}_\alpha = < \mathbf{v} >_\alpha = \int \mathbf{v} f_\alpha d^3 \mathbf{v} / \int f_\alpha d^3 \mathbf{v}$$

Note that $n_\alpha(\mathbf{r}, t) = \int f_\alpha(\mathbf{r}, \mathbf{v}, t) d^3 \mathbf{v}$ is the number density of ions of the α species at the point $\mathbf{r}$ at time t. The third term is given by

$$\int \mathbf{v} \mathbf{F} . \nabla_\mathbf{v} . \nabla_\mathbf{v} f_\alpha d^3 \mathbf{v}$$

Its ith component equals

$$\int v_i \sum_j F_j \frac{\partial f_\alpha}{\partial v_j} f_\alpha d^3 \mathbf{v}$$

$$= - \int f_\alpha \sum_j \frac{\partial}{\partial v_j} (v_i F_j) d^3 \mathbf{v}$$

$$= -n_\alpha < F_j >$$

Finally, the fourth term is the collision term

$$\int m_\alpha v_i (\frac{\partial f}{\partial t})_{coll} d^3 \mathbf{v} = \frac{\partial}{\partial t} (\rho_{m\alpha} \mathbf{u}_\alpha)|_{coll} = A_{\alpha i}$$

with $A_{\alpha i}$ as the rate of increase of the ith component of the momentum due to collisions. Combining all the terms gives us the momentum equation

$$\frac{\partial}{\partial t} (\rho_{m\alpha} \mathbf{u}_\alpha) + \sum_j \frac{\partial}{\partial x_j} (\rho_{m\alpha} u_{\alpha i} u_{\alpha j}) + \sum_j \frac{\partial}{\partial x_j} p_{\alpha i j} - n_\alpha < F_j > = A_{\alpha i}$$

The mass conservation equation when collisions are taken into account reads

$$\frac{\partial \rho_{m\alpha}}{\partial t} + \sum_j \frac{\partial}{\partial x_j} (\rho_{m\alpha} u_j) = S_\alpha$$

with S_α as the rate of generation of matter per unit volume due to collisions. The momentum equation can be expressed as

$$u_{\alpha i}\left(\frac{\partial \rho_{m\alpha}}{\partial t} + \sum_j \frac{\partial}{\partial x_j}(\rho_{m\alpha}u_{\alpha j})\right)+$$

$$\rho_{m\alpha}\left(\frac{\partial u_{\alpha i}}{\partial t} + \sum_j u_{\alpha j}\frac{\partial u_{\alpha i}}{\partial x_j}\right)$$

$$+\sum_j \frac{\partial}{\partial x_j}p_{\alpha ij} - n_\alpha < F_{\alpha i} > -A_{\alpha i} = 0$$

Here,

$$p_{\alpha ij} = \rho_{m\alpha} < c_{\alpha i}c_{\alpha j} >$$

Substituting for the first bracketed equation from the equation of matter conservation, we get

$$\rho_{m\alpha}\left(\frac{\partial u_{\alpha i}}{\partial t} + \sum_j u_{\alpha j}\frac{\partial u_{\alpha i}}{\partial x_j}\right)$$

$$+u_{\alpha i}S_\alpha + \sum_j \frac{\partial}{\partial x_j}p_{\alpha ij} - n_\alpha < F_{\alpha i} > -A_{\alpha i} = 0$$

or adopting vector notation,

$$\rho_{m\alpha}\left(\frac{\partial \mathbf{u}_\alpha}{\partial t} + \mathbf{u}_\alpha.\nabla \mathbf{u}_\alpha\right) + \mathbf{u}_\alpha S_\alpha - \mathbf{A}_\alpha + \nabla.\mathbf{p} = 0$$

To be explicit,

$$S_\alpha = m_\alpha \sum_\beta \int \sigma(\mathbf{v}_\alpha, \mathbf{v}_\beta | \mathbf{v}'_\alpha, \mathbf{v}'_\beta)(f_\alpha(\mathbf{r}, \mathbf{v}'_\alpha, t)f_\beta(\mathbf{r}, \mathbf{v}'_\beta, t) - f_\alpha(\mathbf{r}, \mathbf{v}_\alpha, t)f_\beta(\mathbf{r}, \mathbf{v}_\beta, t))d^3\mathbf{v}_\beta d^3\mathbf{v}_\alpha d\Omega$$

and

$$A_\alpha = m_\alpha \sum_\beta \int \sigma(\mathbf{v}_\alpha, \mathbf{v}_\beta | \mathbf{v}'_\alpha, \mathbf{v}'_\beta)\mathbf{v}_\alpha(f_\alpha(\mathbf{r}, \mathbf{v}'_\alpha, t)f_\beta(\mathbf{r}, \mathbf{v}'_\beta, t) - f_\alpha(\mathbf{r}, \mathbf{v}_\alpha, t)f_\beta(\mathbf{r}, \mathbf{v}_\beta, t))d^3\mathbf{v}_\beta d^3\mathbf{v}_\alpha d\Omega$$

Study project 4

This project deals with looking at the Navier-Stokes fluid dynamical equations for a plasma when the pressure is given by a tensor and not a single scalar function. Let $\mathbf{B}$ be the magnetic field. The charged particle locally will gyrate around the magnetic field. Thus there will be an equipartition of energy along the two directions normal to $\mathbf{B}$. The pressure tensor then has the form

$$\mathbf{p} = \begin{pmatrix} p_\perp & 0 & 0 \\ 0 & p_\perp & 0 \\ 0 & 0 & p_\parallel \end{pmatrix}$$

relative to a local coordinate system in which the local z axis is along the magnetic field and the local x and y axes are normal to the magnetic field. This can be expressed in compact notation as

$$\mathbf{p} = p_\perp \mathbf{I} + (p_\parallel - p_\perp)\mathbf{B}\mathbf{B}^T/B^2$$

To see the validity of this equation, in our local coordinate system, $\mathbf{B} = (0,0,B)^T$ so that

$$\mathbf{B}\mathbf{B}^T/B^2 = \begin{pmatrix} 0 & 0 & 0 \\ 0 & 0 & 0 \\ 0 & 0 & 1 \end{pmatrix}$$

and hence in this coordinate system, the right side of the above equation equals

$$p_\perp \mathbf{I} + (p_\parallel - p_\perp) \begin{pmatrix} 0 & 0 & 0 \\ 0 & 0 & 0 \\ 0 & 0 & 1 \end{pmatrix}$$

$$= \begin{pmatrix} p_\perp & 0 & 0 \\ 0 & p_\perp & 0 \\ 0 & 0 & p_\parallel \end{pmatrix}$$

The equation of motion is

$$\rho\frac{D\mathbf{u}}{Dt} = \mathbf{J} \times \mathbf{B} - \nabla.\mathbf{p}$$

with

$$\nabla.\mathbf{p} = \nabla p_\perp + \nabla.((p_\parallel - p_\perp)\mathbf{B}\mathbf{B}^T/B^2)$$

The ith component of the second term on the right is given by

$$\sum_j \frac{\partial}{\partial x_j}((p_\parallel - p_\perp)B_i B_j/B^2)$$

$$= \sum_j B_j \frac{\partial}{\partial x_j}((p_\parallel - p_\perp)B_i/B^2)$$

Thus, the equation of motion is

$$\rho\frac{D\mathbf{u}}{Dt} = \mathbf{J} \times \mathbf{B} - \nabla p_\perp - \mathbf{B}.\nabla((p_\parallel - p_\perp)\mathbf{B}/B^2)$$

With neglect of the displacement current,

$$\mathbf{J} = \nabla \times \mathbf{B}/\mu$$

Thus,

$$\mathbf{J} \times \mathbf{B} = \frac{1}{\mu}(\mathbf{B}.\nabla\mathbf{B} - \frac{1}{2}\nabla B^2)$$

which means that the equation of motion can be expressed as

$$\rho\frac{D\mathbf{u}}{Dt} = \mathbf{B}.\nabla((1/\mu - (p_\parallel - p_\perp)/B^2)\mathbf{B}) - \nabla(p_\perp + B^2/2\mu)$$

Note that in this equation, the variables are $p_\parallel, p_\perp, \mathbf{B}, \mathbf{u}$. It is not clear at once what conditions on the plasma need to be imposed so that the pressure tensor assumes this form.

Study project 5

Consider a plasma consisting of several ions. The αth ion species has number density $n_\alpha(\mathbf{r}, t)$ and each charge in this species has value q_α Coulombs. each charge in the αth species has mass m_α killogrammes. The velocity of the αth species is $\mathbf{u}_\alpha(\mathbf{r}, t)$. Let $\phi(\mathbf{r}, t)$ be the potential generated by these charges. Assume that the magnetic field can be neglected due to small velocities of the charges. The governing equations for the plasma modeled using standard fluid dynamics are

$$\frac{\partial n_\alpha}{\partial t} + \nabla.(n_\alpha\mathbf{u}_\alpha) = 0$$

$$m_\alpha n_\alpha(\frac{\partial\mathbf{u}_\alpha}{\partial t} + \mathbf{u}_\alpha.\nabla\mathbf{u}_\alpha)$$

$$= -\nabla p_\alpha - q_\alpha\nabla\phi$$

$$\nabla^2\phi = -\frac{\sum q_\alpha n_\alpha}{\epsilon_0}$$

The equation of state determines the partial pressure $p_\alpha = n_\alpha KT$. We're assuming that the temperature T is a constant. The linearized equations read

$$\frac{\partial n_\alpha'}{\partial t} + n_{\alpha 0}\nabla.\mathbf{u}_\alpha = 0$$

$$n_{\alpha 0}m_\alpha\frac{\partial\mathbf{u}_\alpha}{\partial t} = -KT\nabla n_\alpha' - q_\alpha\nabla\phi$$

$$\nabla^2\phi = -\frac{\sum q_\alpha n_\alpha'}{\epsilon_0}$$

We want to investigate the dispersion relation for waves propagating in such a plasma. The waves are caused by a small perturbation to the equilibrium charge density of any one species at one point or equivalently by a small stir causing a small movement of the charges. The potential ϕ represents in fact a perturbation to the potential created by such a small disturbance in the equilibrium. We try plane wave solutions:

$$n'_\alpha(\mathbf{r}, t) = a_\alpha exp(i(\omega t - \mathbf{k}.\mathbf{r}))$$

$$\phi(\mathbf{r}, t) = b.exp(i(\omega t - \mathbf{k}.\mathbf{r}))$$

$$\mathbf{u}_\alpha(\mathbf{r}, t) = \mathbf{c}_\alpha.exp(i(\omega t - \mathbf{k}.\mathbf{r}))$$

Substituting these into the linearized equations gives us the following homogeneous linear equations in the variables $a_\alpha, b, \mathbf{c}_\alpha$:

$$\omega a_\alpha - n_{\alpha 0}\mathbf{k}.\mathbf{c}_\alpha = 0$$

$$n_{\alpha 0}m_\alpha\omega\mathbf{c}_\alpha = KT\mathbf{k}a_\alpha + q_\alpha\mathbf{k}b$$

$$-k^2 b + \sum q_\alpha a_\alpha/\epsilon_0 = 0$$

This is a homogeneous system of linear equations in the variables $a_\alpha, b, \mathbf{c}_\alpha, \alpha = 1, 2, ..., N$ The dispersion relation can be derived from these by setting the associated determinant equal to zero.

Study project 6

Kinetic transport equation of Boltzmann:

$$\frac{\partial f(\mathbf{r}, \mathbf{v}, t)}{\partial t} + \mathbf{v}.\nabla f(\mathbf{r}, \mathbf{v}, t) + \frac{\mathbf{F}}{m}.\nabla_\mathbf{v} f(\mathbf{r}, \mathbf{v}, t) = (\frac{\partial f(\mathbf{r}, \mathbf{v}, t)}{\partial t})_{coll}$$

where we are assuming that the external force $\mathbf{F}$ on each charged particle is of electromagnetic origin, i.e.,

$$\mathbf{F} = q(\mathbf{E}(\mathbf{r}, t) + \mathbf{v} \times \mathbf{B}(\mathbf{r}, t))$$

$$(\frac{\partial f}{\partial t})_{coll} = \int \sigma(\mathbf{v}, \mathbf{v}_1|\mathbf{v}', \mathbf{v}'_1)(f(\mathbf{r}, \mathbf{v}', t)f(\mathbf{r}, \mathbf{v}'_1, t) - f(\mathbf{r}, \mathbf{v}, t)f(\mathbf{r}, \mathbf{v}_1, t))d^3\mathbf{v}_1 d\Omega$$

Ω is the scattering angle corresponding to the direction of the final relative velocity vector $\mathbf{v}'_1 - \mathbf{v}$ relative to the initial relative velocity vector $\mathbf{v}_1 - \mathbf{v}$. This expression for the rate of change of the particle phase number density due to collisions can be expressed using a compact notation:

$$(\frac{\partial f}{\partial t})_{coll} = \int \sigma(\Omega)(f'f'_1 - ff_1)d\Omega d^3\mathbf{v}_1$$

Let f_0 be the equilibrium distribution. It is given by the expression

$$f_0(\mathbf{r}, \mathbf{v}) = n(\mathbf{r})(\frac{m}{2\pi kT(\mathbf{r})})^{3/2}exp(-\frac{m\mathbf{v}^2}{2kT(\mathbf{r}})$$

with $T(\mathbf{r})$ as the local temperature and $n(\mathbf{r}$ as the particle number density. This phase number density satisfies the steady state Boltzmann kinetic equation without forces and without the collision term, i.e.,

$$\mathbf{v}.\nabla f_0(\mathbf{r},\mathbf{v}) = 0$$

Now observe that

$$\nabla f_0 = (\frac{m}{2\pi kT})^{3/2}\nabla n$$

$$+(n\nabla T)(\frac{m}{2\pi k})^{3/2}(-3/2)T^{-5/2} - n(\frac{m}{2\pi kT})^{3/2}(mv^2/2kT^2)\nabla T).exp(-mv^2/2kT)$$

It is clear that $n.\nabla f_0$ cannot equal zero unless both n and T are constants. Thus, the only equilibrium distribution for the collisionless and forceless Boltzmann equation is the Maxwell velocity distribution function. Now let us look at the form of the equilibrium distribution for the Boltzmann equation without a collision term and in an external potential field dependent only upon the position. The partial differential equation is given by

$$\mathbf{v}.\nabla f - (\nabla U(\mathbf{r})/m).\nabla_{\mathbf{v}}f = 0$$

We claim that the equilibrium density is of the Gibbs type:

$$f(\mathbf{r},\mathbf{v}) = exp(-\beta(m\mathbf{v}^2/2 + U(\mathbf{r}))$$

For this density, we find that

$$\nabla f = -(\beta\nabla U)f$$
$$\nabla_{\mathbf{v}}f = -\beta m\mathbf{v}f$$

Thus,

$$\mathbf{v}.\nabla f - (\nabla U/m).\nabla_{\mathbf{v}}f = -\beta\mathbf{v}.(\nabla U)f + \beta\mathbf{v}.(\nabla U)f = 0$$

proving that f is indeed the equilibrium distribution. One problem that can be posed is to determine the equilibrium density for a particle in a static electromagnetic field, i.e. the solution to the partial differential equation

$$\mathbf{v}.\nabla f + e(\mathbf{E}(\mathbf{r}) + \mathbf{v} \times \mathbf{B}(\mathbf{r}).\nabla_{\mathbf{v}}f = 0$$

Note that this is a linear homogeneous equation and its solutions can be obtained numerically by discretization thereby transforming this into an eigenvalue equation.

The other problem is to obtain general solutions to the coupled Boltzmann and Maxwell equations of electrodynamics. Assume that the fluid consists of only a single charged species with each particle having charge e. If $f(\mathbf{r},\mathbf{v},t)$ is the distribution function, then the charge density is given by

$$\rho_e(\mathbf{r},t) = e\int f(\mathbf{r},\mathbf{v},t)d^3\mathbf{v}$$

and the current density is given by

$$\mathbf{J}_e(\mathbf{r}, t) = e \int \mathbf{v} f(\mathbf{r}, \mathbf{v}, t) d^3\mathbf{v}$$

The Boltzmann equation without the collision term is

$$\frac{\partial f}{\partial t} + \mathbf{v}.\nabla f + \frac{e}{m}(\mathbf{E} + \mathbf{v} \times \mathbf{B}).\nabla_\mathbf{v} f = 0$$

Given the electric and magnetic fields, we can solve this equation for f. The electric and magnetic fields are however determined from the Maxwell equations:

$$\nabla.\mathbf{E} = \frac{\rho_e}{\epsilon} = \frac{e}{\epsilon} \int f(\mathbf{r}, \mathbf{v}, t) d^3\mathbf{v}$$

$$\nabla \times \mathbf{E} = -\frac{\partial \mathbf{B}}{\partial t}$$

$$\nabla.\mathbf{B} = 0$$

$$\nabla \times \mathbf{B} = \mu e \int \mathbf{v} f(\mathbf{r}, \mathbf{v}, t) d^3\mathbf{v} + \epsilon \frac{\partial \mathbf{E}}{\partial t}$$

More generally, one can consider a charged fluid consisting of several ions numbered $\alpha, \alpha = 1, 2, ..., N$ with each particle of the αth species having charge e_α and the number density for the αth species being $f_\alpha(\mathbf{r}, \mathbf{v}, t)$. The collisionless Boltzmann equation is given by

$$\frac{\partial f_\alpha}{\partial t} + \mathbf{v}.\nabla f_\alpha + \frac{e_\alpha}{m}(\mathbf{E} + \mathbf{v} \times \mathbf{B}).f_\alpha = 0, \alpha = 1, 2, ..., N$$

This is coupled to the Maxwell equations driven by the charge density

$$\rho = \sum_{\alpha=1}^{N} e_\alpha \int f_\alpha d^3\mathbf{v}$$

$$\mathbf{J} = \sum_{\alpha=1}^{N} e_\alpha \int \mathbf{v} f_\alpha d^3\mathbf{v}$$

We can look at an iterative solution to this coupling of the Boltzmann and Maxwell equations. The general idea is first to assume an incident electromagnetic field $\mathbf{E}_0(\mathbf{r}, t), \mathbf{B}_0(\mathbf{r}, t)$ and to solve the Boltzmann equation for f_α. Call this solution as $f_\alpha^{(0)}$. In general, assume that we've got $E^{(n)}(\mathbf{r}, t), \mathbf{B}^{(n)}(\mathbf{r}, t)$, the nth iterate of the electric and magnetic fields. We plug this into the Boltzmann equation and solve for $f_\alpha^{(n)}$, i.e.,

$$\frac{\partial f_\alpha^{(n)}}{\partial t} + \mathbf{v}.\nabla f_\alpha^{(n)} + \frac{e_\alpha}{m}(\mathbf{E}^{(n)} + \mathbf{v} \times \mathbf{B}^{(n)}).f_\alpha^{(n)} = 0, \alpha = 1, 2, ..., N$$

Then, calculate using $f_\alpha^{(n)}$, the nth iterate for the charge and current densities, i.e.,

$$\rho^{(n)} = \sum_{\alpha=1}^{N} e_\alpha \int f_\alpha^{(n)} d^3\mathbf{v}$$

$$\mathbf{J}^{(n)} = \sum_{\alpha=1}^{N} e_\alpha \int \mathbf{v} f_\alpha^{(n)} d^3\mathbf{v}$$

Substitute this into the Maxwell equations and determine $\mathbf{E}^{(n+1)}, \mathbf{B}^{(n+1)}$.

Study project 7

Explain how one computes energy fluctuations in a thermodynamical system.

Ans: Let $H(q, p)$ denote the Hamiltonian of the system as a function of the phase coordinates. The partition function is defined as

$$Z(\beta) = \int exp(-\beta H(q, p)) dqdp$$

This is to be understood as an abbreviation for the formula

$$Z(\beta) = \int exp(-\beta H(q_1, ..., q_n, p_1, ..., p_n)) dq_1...dq_n dp_1...dp_n$$

where $2n$ is the number of degrees of freedom. Let $A = -\frac{1}{\beta} log(Z)$. Then, $Z = exp(-\beta A)$ so that

$$\int exp(\beta(A - H)) dqdp = 1$$

A is a function of β, V, or equivalently of the temperature and volume of the system. Differentiating the above expression with respect to β gives

$$0 = \int (A - H) exp(\beta(A - H)) dqdp + \beta \int \frac{\partial A}{\partial \beta} exp(\beta(A - H)) dqdp = 0$$

This simplifies to

$$0 = \int (A - H) exp(\beta(A - H)) dqdp + \beta \frac{\partial A}{\partial \beta} = 0$$

Which is the same as

$$A - U + \beta \frac{\partial A}{\partial \beta} = 0$$

where

$$U = <H> = \int H exp(\beta(A - H)) dqdp$$

is the average energy of the system. Noting that $\frac{\partial A}{\partial \beta} = -kT^2 \frac{\partial A}{\partial T}$, this formula gives us

$$A = U + T\frac{\partial A}{\partial T}$$

We now make use of the thermodynamical equations $TdS = dU + PdV, A = U - TS$ to get $dA = dU - TdS - SdT = -PdV - SdT$ so that $S = -\frac{\partial A}{\partial T}|_V$ and hence $A = U + T\frac{\partial A}{\partial T}$. Thus, A as defined by the above integral can be identified with the free energy of the system. Let $\int_{E<H(q,p)\leq E+dE} dqdp = \omega(E)dE$. $\omega(E)$ gives us the number of states per unit energy in the microcanonical ensemble. Using this function, we can write

$$1 = \int exp(\beta(A - H))dqdp = \int exp(\beta(A - E))\omega(E)dE$$

an expression that shows that the probability distribution of the energy is given by the formula $exp(-\beta E)\omega(E)dE$. Now, $S(E) = klog(\omega(E))$ and hence we can express the above equation as

$$1 = \int exp(\beta(A - E))exp(S(E)/k)dE$$

The distribution of the energy is determined by the formula $exp(-\beta E)exp(S(E)/k)dE$. The most probable value of the energy is that value $\bar{E}$ at which $S(E) - k\beta E = S(E) - E/T$ shows a maximum. Let $\bar{E}$ denote this value. Then

$$\frac{d}{dE}(S(E) - E/T)|_{E=\bar{E}} = 0$$

which is the same as

$$\frac{dS(E)}{dE}\Big|_{E=\bar{E}} = \frac{1}{T}$$

This formula combined with the thermodynamical equation $\frac{\partial S}{\partial U} = 1/T$ shows that $\bar{E}$ is to be identified with the average energy U of the system. The other condition for the function $S(E) - E/T$ to show a maximum at $E = \bar{E}$ is that $\frac{d^2}{dE^2}(S(E) - E/T)|_{E=\bar{E}} < 0$, that is $\frac{d^2 S(E)}{dE^2}|_{E=\bar{E}} < 0$. We now approximate the probability density of E by Taylor expanding the entropy function retaining only upto second degree terms in E:

$$S(E) - E/T \approx S(\bar{E}) - \bar{E}/T + (S'(\bar{E}) - 1/T) + S''(\bar{E})(E - \bar{E})^2 = S(\bar{E}) - \bar{E}/T + S''(\bar{E})(E - \bar{E})^2$$

This shows that the approximate energy distribution is proportional to $exp(S''(\bar{E})(E - \bar{E})^2/k)dE$ and hence the variance of the energy fluctuations is given by

$$< (H - U)^2 >=< (E - \bar{E})^2 >= -k/S''(U)$$

Now,

$$S'(U) = 1/T, S''(U) = -\frac{1}{T^2}\frac{\partial T}{\partial U} = -\frac{1}{T^2 C_V}$$

giving

$$< (H - U)^2 >= kT^2 C_V$$

where C_V equals the specific heat at constant volume of the gas.

Study project 8

Elementary properties of the conditional probability density, how it is defined and how it can be used to compute conditional expectations and its applications to problems of nonlinear estimation theory can be presented in a lecture. One can begin with a pair (X, Y) of random variables defined on a probability space $(\Omega, \mathcal{F}, P)$ and the definition of their joint distribution function $F(x, y) = P\{X \leq x, Y \leq y\}$. Then by means of a diagram, it may be shown that

$$P\{x < X \leq x + h, y < Y \leq y + k) = F(x + h, y + k) - F(x + h, y) - F(x, y + k) + F(x, y)$$

whose limiting case as $h, k \to 0$ leads to the expression

$$P\{X \in (x, x + dx], Y \in (y, y + dy]\} = f(x, y)dxdy$$

where

$$f(x, y) = \frac{\partial^2 F(x, y)}{\partial x \partial y}$$

is the joint density of the variables X, Y). To compute the probability that the vector (X, Y) falls in a borel subset E of $\mathbb{R}^2$, one simply has to integrate the density function over this region:

$$P((X, Y) \in E) = \int_E f(x, y)dxdy$$

Then the conditional density is defined by

$$f_{X|Y}(x|y)dx = P(X \in (x, x + dx]|Y = y) = P(X \in (x, x + dx]|Y \in (y, y + dy])$$

$$= \frac{P(X \in (x, x + dx], Y \in (y, y + dy])}{P(Y \in (y, y + dy])}$$

$$= \frac{f(x, y)dxdy}{f_Y(y)dy} = \frac{f(x, y)}{f_Y(y)}dx$$

where

$$f_Y(y) = \int_{-\infty}^{\infty} f(x, y)dx$$

is the density of Y. $\int_B f_{X|Y}(x|y)dx$ equals the probability that the variable X takes values in the Borel set B given that Y takes the value y. The conditional expectation of a function $g(X)$ of the random variable X given that $Y = y$ is defined by the integral

$$\mathbb{E}(X|Y = y) = \int_{-\infty}^{\infty} g(x) f_{X|Y}(x|y)dx$$

It satisfies the obvious property

$$\int_{-\infty}^{\infty} f_Y(y)\mathbb{E}(X|Y=y)dy = \mathbb{E}\mathbb{E}(X|Y) = \mathbb{E}(X)$$

These ideas can be used in statistical mechanics. Suppose that we are looking at a system S interacting with the environment S'. Let $\mathbf{X}$ denote the phase coordinates of the system and $\mathbf{Y}$ the phase coordinates of the environment. The total energy of the system and environment equals the energy of the system plus the energy of the environment plus the interaction energy between the system and the environment: $H(\mathbf{X}, \mathbf{Y}) = H_{sys}(\mathbf{X}) + H_{env}(\mathbf{Y}) + V(\mathbf{X}, \mathbf{Y})$. The Gibbs distribution determines the joint distribution of the system and environment phase variables:

$$f(\mathbf{x}, \mathbf{y}) = C.exp(-\beta H_{sys}(\mathbf{x})).exp(-\beta H_{env}(\mathbf{y}).exp(-\beta V(\mathbf{x}, \mathbf{y}))$$

If there were no interaction between the system and the environment, then this joint density of the system plus environment factorizes into the product of a function of the system variables alone and the environment variables alone:

$$f(\mathbf{x}, \mathbf{y}) = \psi(\mathbf{x})\phi(\mathbf{y})$$

where $\psi(\mathbf{x})$ is proportional to $exp(-\beta H_{sys}(\mathbf{x})$ and $\phi(\mathbf{y})$ is proportional to $exp(-\beta H_{env}(\mathbf{y})$. This means that in the absence of any interaction, the system variables are independent of the environment variables. Interaction removes this independence. Assuming the interaction energy to be small in comparison to the average molecular internal energy kT, we can approximate the joint density by the formula

$$f(\mathbf{x}, \mathbf{y}) \approx C.exp(-\beta H_{sys}(\mathbf{x}).exp(-\beta H_{env}(\mathbf{y})(1 - \beta V(\mathbf{x}, \mathbf{y}))$$

Suppose for example that the system consists of an oscillator having energy $H_{sys}(x, v) = \frac{m}{2}v^2 + \frac{kx^2}{2}$ and the environment consists of a free particle having energy $H_{env}(V) = \frac{M}{2}V^2$. The environment is coupled by a spring to the system leading to an interaction energy $V(x, Y) = \frac{K}{2}(x - Y)^2$. Here, x is the position of the system particle, v is its velocity, Y is the position of the environment particle and V is its velocity. Then, the joint density of the system and coordinate variables is given by

$$f(x, v, Y, V) = C.exp(-\beta mv^2/2 - \beta kx^2/2 - \beta MV^2/2 - \beta K(x - Y)^2/2)$$

This expression shows that the phase variables (x, v, Y, V) have a joint Gaussian distribution. When the environment is frozen at specific phase variables, one can determine the distribution of the system variables using the already discussed formula for the conditional density. Statistical quantities like internal energy, entropy, enthalpy, free energy, Gibbs free energy etc can be computed from the resulting joint density. Consider now the approximation to the joint density made for the general interaction model. The average system energy is given by

$$U_{sys} = \int H_{sys}(\mathbf{x})f(\mathbf{x}, \mathbf{y})d\mathbf{x}d\mathbf{y}$$

$$\approx C.\int exp(\beta H_{sys}(\mathbf{x})).exp(-\beta H_{env}(\mathbf{y}))H_{sys}(\mathbf{x})d\mathbf{x}d\mathbf{y}$$

$$-\beta.C\int exp(-\beta H_{sys}(\mathbf{x}).exp(-\beta H_{env}(\mathbf{y}).V(\mathbf{x},\mathbf{y})H_{sys}(\mathbf{x})d\mathbf{x}d\mathbf{y}$$

The partition function for the environment is given by

$$Z_{env}(\beta) = \int exp(-\beta.H_{env}(\mathbf{y}))d\mathbf{y}$$

and the partition function for the system is given by

$$Z_{sys}(\beta) = \int exp(-\beta.H_{sys}(\mathbf{x})d\mathbf{x}$$

The overall partition function for the system plus environment is given by

$$C^{-1} = \int exp(-\beta H_{sys}(\mathbf{x})).exp(-\beta H_{env}(\mathbf{y})exp(-\beta.V(\mathbf{x},\mathbf{y})d\mathbf{x}d\mathbf{y}$$

$$\approx \int exp(-\beta H_{sys}(\mathbf{x})exp(-\beta H_{env}(\mathbf{y})(1-\beta V(\mathbf{x},\mathbf{y}))d\mathbf{x}d\mathbf{y}$$

$$= Z_{sys}(\beta)Z_{env}(\beta) - \beta Z_{sys}(\beta)Z_{env}(\beta).<V>_0$$

where $<V>_0$ equals the average value of the interaction potential taken with respect to the distribution of the system plus environment assuming no interaction.

Study project 9

Consider a particle moving in one dimension with canonical position and momentum coordinates (q,p) at any instant of time. Let $U(q)$ denote the potential in which the particle is moving. The total energy of the particle is given by $E(q,p) = \frac{p^2}{2m} + U(q)$ and the Gibbs density is given by $f(q,p) = C.exp(-\beta E(q,p))$ where

$$C^{-1} = Z(\beta) = \int_{\mathbb{R}^2} exp(-\beta E(q,p))dqdp$$

The average position of the particle is given by the formula

$$<q> = \frac{\int_{-\infty}^{\infty} q.exp(-\beta U(q))dq}{\int_{-\infty}^{\infty} exp(-\beta U(q))dq}$$

Assume that the particle is constrained to move inside a box bounded by the walls $x = 0$ and $x = L$. The potential inside is $U(q)$. Assuming that the potential is much smaller in comparison to the mean kinetic energy $kT/2$ of the particle, we have the following approximation:

$$<q> \approx \frac{\int_0^L q(1-\beta U(q))dq}{\int_0^L (1-\beta U(q))dq}$$

Thus, we need evaluate only the integrals $I_1 = \int_0^q qU(q)dq$ and $I_2 = \int_0^L U(q)dq$. For let the minimum value of $U(q)$ be attained at $q = q_0$. We can then make a Taylor expansion

$$U(q) = U(q_0) + U'(q_0)(q - q_0) + U''(q_0)(q - q_0)^2/2 + \ldots + U^{(n)}(q_0)(q - q_0)^n/n! + \ldots$$

$$= \sum_{n=0}^{\infty} U^{(n)}(q_0)(q - q_0)^n/n!$$

Then,

$$\int_0^L U(q)dq = \sum_{n=0}^{\infty} \frac{U^{(n)}(q_0)}{(n+1)!}((L - q_0)^{n+1}/(n+1)! - (-q_0)^{n+1}/(n+1)!)$$

We also find that

$$\int_0^L qU(q)dq = \sum_{n=0}^{\infty} \frac{U^{(n)}(q_0)}{n!} \int_0^L q(q - q_0)^n dq$$

with

$$\int_0^L q(q - q_0)^n dq = \int (q - q_0)^{n+1} dq + q_0 \int_0^L (q - q_0)^n dq$$

$$= (q - q_0)^{n+2}/(n+2)! + q_0(q - q_0)^{n+1}/(n+1)!|_0^L$$

$$= ((L - q_0)^{n+2} - (-q_0)^{n+2})/(n+2)! + q_0((L - q_0)^{n+1} - (-q_0)^{n+1})/(n+1)!$$

This expression can be used to obtain the approximate value of $< q >$.

Study project 10

A course on probability theory can introduce the students to the computation of the average values of functions of the state. This will constitute an interesting exercise in integration with respect to probability densities. The entropy function has already been introduced. If $E_n, n = 1, 2, \ldots$, are the discrete energy levels obtained from the solution of the stationary Schrodinger equation $H(q, -\frac{ih}{\pi}\frac{d}{dq})\psi_n = E_n\psi_n$, then the probability of the system being in the state E_n is known to be given by the formula $p_n = exp(-\beta E_n)/Z(\beta)$ where $Z(\beta) = \sum_n exp(-\beta E_n)$. The entropy equals

$$S = -k \sum p_n log(p_n) = -k \sum_n p_n log(\frac{exp(-\beta E_n)}{Z}) = k\beta U + klog(Z)$$

where $U = \sum_n p_n E_n$ is the average energy. Note that

$$U = -\frac{d}{d\beta}log(Z)$$

Thus,

$$S = -k.\beta\frac{d}{d\beta}log(Z) + k.log(Z) = -kT\frac{d}{dT}log(Z)$$

The other thermodynamical quantities can be obtained easily from S using the equation $TdS = dU + PdV$. For example, the specific heat at constant volume is given by

$$C_V = T(\frac{\partial S}{\partial T})_V = \frac{\partial U}{\partial T}|_V + k\frac{\partial}{\partial T}log(Z)|_V$$

It is clear that C_V is a statistical quantity computed in the final analysis from the Gibbs distribution. Consider the example of a particle in a one dimensional box of length L. Exact solution of the Schrodinger equation leads to the energy levels $E_n = \frac{n^2\pi^2h^2}{2mL^2}, n = 1, 2, ...$ Gibbs distribution gives the following expression for the exact quantum mechanical partition function:

$$Z(\beta) = \sum_{n=1}^{\infty} exp(-\beta\pi^2h^2n^2/2mL^2)$$

This expression depends upon two parameters, one the mass m of the particle and two the length L of the box. To obtain the expression for the pressure exerted by the particle on the walls of the box, we need to use the thermodynamic equation $TdS = dU + PdV$. This gives $P = T\frac{\partial S}{\partial V}|_U$.

Study project 11

System has energy H depending upon a parameter θ. Let the phase variables of the system be denoted by the vector $\mathbf{X}$. The energy has the form $H(\mathbf{X}, \theta)$. The classical partition function is given by

$$Z(\beta, \theta) = \int exp(-\beta H(\mathbf{X}, \theta)d\mathbf{X}$$

Let $\phi(\mathbf{X})$ be any function of the state of the system. Its average value is given by

$$< \phi > (\beta, \theta) = \int \phi(\mathbf{X})exp(-\beta H(\mathbf{X}, \theta))d\mathbf{X}/Z(\beta, \theta)$$

and its mean square fluctuation is given by the expression

$$< \delta\phi^2 > = < (\phi - < \phi >)^2 > = < \phi^2 > - < \phi >^2$$

$$= \int (\phi(\mathbf{X}) - < \phi >)^2 exp(-\beta H(\mathbf{X}, \theta))d\mathbf{X}/Z(\beta)$$

From measurements of the average value of ϕ and its mean square fluctuation, we want to estimate the parameter θ and the temperature $1/k\beta$. For doing this, moment matching must be employed. More generally, one can compute the classical joint moment generating function of the phase variable vector $\mathbf{X}$ as

$$M_X(\mathbf{s}, \theta, \beta) = \mathbb{E}exp(\mathbf{s}^T\mathbf{X}) = \int exp(\mathbf{s}^T\mathbf{X})exp(-\beta H(\mathbf{X}, \theta))d\mathbf{X}/Z(\beta, \theta)$$

From this function, various moments of the phase vector $\mathbf{X}$ can be extracted by differentiation.

$$\frac{\partial^m}{\partial s_1^{r_1}...\partial s_n^{r_n}}M_X(\mathbf{s},\theta,\beta)|_{\mathbf{s}=\mathbf{0}}$$

$$=\int X_1^{r_1}...X_n^{r_n}exp(-\beta H(\mathbf{X},\theta))d\mathbf{X}/Z(\beta,\theta)$$

Matching of these moments with time averaged quantities can be done to extract the parameters θ,β. One can choose $\phi=H$. But this observable depends upon the parameter θ. The dependence on $<\phi>$ as a function of the parameter θ is investigated using a sensitivity analysis:

$$\frac{\partial}{\partial\theta}<\phi>=\frac{\partial}{\partial\theta}\frac{\int\phi(\mathbf{X})exp(-\beta H(\mathbf{X},\theta))d\mathbf{X}}{Z(\beta.,\theta)}$$

To evaluate this, we need the sensitivity of the partition function with respect to θ.

$$\frac{\partial}{\partial\theta}Z(\beta,\theta)=\frac{\partial}{\partial\theta}\int exp(-\beta H(\mathbf{X},\theta))d\mathbf{X}$$

$$=-\beta\int\frac{\partial H(\mathbf{X},\theta)}{\partial\theta}exp(-\beta H(\mathbf{X},\theta))d\mathbf{X}$$

$$-\beta.Z(\beta,\theta)<\frac{\partial H}{\partial\theta}>$$

Study project 12

A typical situation where the χ^2-distribution appears in statistical physics is the following. Consider a gas of N particles moving freely in space. If $\mathbf{v}_1,...,\mathbf{v}_N$ denote the velocity vectors of the particles, their total kinetic energy is given by the formula

$$T(\mathbf{v}_1,...,\mathbf{v}_N)=\frac{m}{2}\sum_{i=1}^{N}<\mathbf{v}_i,\mathbf{v}_i>$$

There is no applied external potential and neither is there any interaction potential between the particles. Thus, the Gibbs distribution directly gives the joint probability density of the velocity variables:

$$f(\mathbf{v}_1,...,\mathbf{v})_N)=C.exp(-\beta.T(\mathbf{v}_1,...,\mathbf{v}_N))=C.exp(-\frac{\beta m}{2}\sum_{i=1}^{N}<\mathbf{v}_i,\mathbf{v}_i>)$$

This is a joint Gaussian density in $3N$ variables with all the $3N$ components being statistically independent of each other. Suppose we want to compute the probability distribution of the total kinetic energy T. First observe that $\sqrt{\beta m}v_{xi}, \sqrt{\beta m}v_{yi}, \sqrt{\beta m}v_{zi}, i = 1, 2, ..., N$ are independent standard normal variates. The kinetic energy equals one half the sum of squares of these divided by β Thus, the kinetic energy has the distribution of $X/2\beta$ where X is a χ^2_N random variable.

Study project 13

Relativistic correction to the probability density of a free particle moving in one dimension. The kinetic energy is given by the relativistic formula: $T(v) = \frac{mc^2}{\sqrt{1-v^2/c^2}}$. For velocities small compared to the speed of light but not so small that the Newtonian formula can be used, we use the truncated binomial approximation: $T(v) = mc^2 + mv^2/2 + 3mv^4/4c^2$ and compute the Gibbs density: $f(v) = C.exp(-\beta mv^2/2).exp(-3\beta mv^4/4c^2)$. This is a relativistically perturbed Maxwellian density. We can further approximate this by

$$f(v) = C.exp(-\beta mv^2/2)(1 - 3\beta mv^4/4c^2)$$

where

$$C^{-1} = \int_{-\infty}^{\infty} exp(-\beta mv^2/2)(1 - 3\beta mv^4/4c^2)dv$$

The integral can be evaluated easily using Γ functions.

Study project 14

Consider a charged particle moving in one dimension along the positive real line $x \geq 0$. The electric field in this region is given by $\mathbf{E} = -E_0\hat{x}$. With q denoting the charge, the potential energy of the particle is given by $V(x) = qE_0x$ and the Gibbs density is given by $f(x) = C.exp(-\beta qE_0x)$ for $x \geq 0$. Letting $a = \beta qE_0$, we have $f(x) = a.exp(-ax), x \geq 0$. Thus the position of the particle follows the exponential law with mean $1/a$. This is how the exponential distribution occurs naturally in statistical mechanics when applied to electrostatic problems. Again, consider a particle moving in the positive quadrant of the two dimensional plane: $\{(x,y) : x, y \geq 0\}$. Assume that there is a constant electric field in this region given by $\mathbf{E} = -E_1\hat{x} - E_2\hat{y}$. The electrostatic potential is then given by $V(x,y) = E_1x + E_2y$ and if q denotes the charge on the particle, the potential energy of the particle is given by $qV(x,y) = qE_1x + qE_2y$. The Gibbs distribution for the position (x,y) is given by

$$f(x,y) = C.exp(-\beta qV(x,y)) = C.exp(-\beta qE_1x - \beta qE_2y), x, y \geq 0$$

This shows that the position coordinates (X,Y) of the particle are independently distributed exponential random variables with means of $1/\beta qE_1$ and $1/\beta qE_2$ respectively. Let $a = \beta qE_1, b = \beta qE_2$. Then the joint density of (X,Y) is given by $f(x,y) = ab.exp(-ax - by), x, y \geq 0$. Suppose we are interested in determining the distribution of the projection of the position of the particle along the direction (n_x, n_y). This projection is given by $Z = n_xX + n_yY$ and its moment generating function equals

$$\mathbb{E}exp(-sZ) = \mathbb{E}exp(-sn_xX)\mathbb{E}exp(-sn_yY) = \frac{1}{(sn_x + a)(sn_y + b)}$$

$$= \frac{1}{bn_x - an_y}\left(\frac{n_x}{sn_x + a} - \frac{n_y}{sn_y + b}\right)$$

Inverting the Laplace transform gives us the density of Z:

$$f_Z(z) = (bn_x - an_y)^{-1}(exp(-az/n_x) - exp(-bz/n_y)), z \geq 0$$

This problem illustrates how computation of the density of functions of random variables finds applications to problems of statistical mechanics.

Study project 15

Consider two independent systems not having any interaction with each other. Assume that each system is a harmonic oscillator and the characteristic frequencies are ω_1 and ω_2 respectively. The energy spectrum of the first oscillator is $E_n = (n + 1/2)h\omega/2\pi, n = 0, 1, 2, ...$, and the energy spectrum of the second oscillator is $E'_n = (n + 1/2)h\omega'/2\pi, n = 0, 1, 2, ...$ The partition function of the first oscillator is given by

$$Z(\beta) = \sum_{n=0}^{\infty} exp(-\beta(n + 1/2)h\omega/2\pi)$$

and that of the second oscillator is given by

$$Z(\beta)' = \sum_{n=0}^{\infty} exp(-\beta(n + 1/2)h\omega'/2\pi)$$

The probability distribution of the first oscillator is

$$p_n = exp(-\beta(n + 1/2)h\omega/2\pi)/Z(\beta), n = 0, 1, 2, ...$$

and that of the second oscillator is given by

$$q_n = exp(-\beta(n + 1/2)h\omega'/2\pi)/Z(\beta)$$

When both the oscillators are put together, the energy of the overall system assumes values $E_n + E'_m, n, m = 0, 1, 2, ...$ The partition function of the overall system is given by

$$Z_{tot}(\beta) = Z(\beta)Z(\beta)'$$

In the special case, when the angular frequencies of both the oscillators are equal, i.e., $\omega = \omega'$, the total energy assumes values $nh\omega/2\pi, n = 1, 2, 3, ...$ The probability that the total energy equals $(n + 1)h\omega/2\pi$ where n assumes values $0, 1, 2, ...$ is given by

$$p_{tot}(n) = \sum_{k=0}^{n} p_k p_{n-k} = \frac{1}{Z(\beta)^2} \sum_{k=0}^{n} exp(-\beta(k + 1/2)h\omega/2\pi)exp(-\beta(n - k + 1/2)h\omega/2\pi)$$

$$= \frac{1}{Z(\beta)^2}(n+1)exp(\beta(n+1)h\omega/2\pi)$$

This problem illustrates the application of the convolution theorem for discrete random variables to problems of statistical mechanics.

Study project 16

Consider a quantum mechanical harmonic oscillator moving with a random velocity. The total energy of the oscillator equals the sum of its kinetic energy of translation and the discrete energy level associated with its excited quantum state: $E(v,n) = \frac{m}{2}v^2 + (n+1/2)h\omega/2\pi$. Note that even in the quantum theory, the translational momentum of the oscillator can assume a continuum of values. The partition function of the oscillator is given by

$$Z(\beta) = \sum_{n=0}^{\infty} \int_{-\infty}^{\infty} exp(-\beta E(v,n))dv$$

$$= (\int_{-\infty}^{\infty} exp(-\beta mv^2/2)dv). \sum_{n=0}^{\infty} exp(-\beta(n+1/2)h\omega/2\pi)$$

Both the Gaussian integral and the geometric sum can be evaluated easily and the various thermodynamical quantities associated with the particle starting from the free energy and covering internal energy and entropy can be evaluated. The probability distribution of the energy can also be evaluated. The procedure for doing this, is to note that the variables v, n are independently distributed, with the former having the normal density $C.exp(-\beta mv^2/2)dv$ and the latter a discrete distribution $C'.exp(-\beta(n+1/2)h\omega/2\pi), n = 0, 1, 2, ...,$. Denote the former density by $f(v)dv$ and the latter probabilities by $p_n, n = 0, 1, 2, ...,$. The probability density of the translational kinetic energy $\epsilon = \frac{mv^2}{2}$ can be evaluated by a simple transformation of variable formula applied to the density $f(v)dv$. Denote this by $g(\epsilon)d\epsilon$. The density of the total energy is then given by

$$\sum_{n=0}^{\infty} p_n g(\epsilon - (n+1/2)h\omega/2\pi)$$

This formula illustrates the general principle that if X and Y are two independent random variables with X having the density $f(x)$ and Y assuming discrete values $a_n, n = 0, 1, 2, ...$ with respective probabilities $p_n, n = 0, 1, 2, ...,$, then the density of $Z = X+Y$ is given by the formula $\sum_{n=0}^{\infty} p_n f(z - a_n)$.

Study project 17

Explain the role of estimation theory in statistical physics. In equilibrium statistical physics, one deals with the Gibbs distribution, $f(\mathbf{x}) = exp(\beta H(\mathbf{x})$, with $H(\mathbf{x})$ as the energy of the system as a function of the phase coordinates $\mathbf{x} = (\mathbf{q}, \mathbf{p})$. The energy can involve a parameter θ or several such parameters. From measurements of the average values of certain observables like energy, entropy, pressure, free energy, we desire to estimate the values of these parameters. Herein estimation theory is involved. The temperature $T = 1/k\beta$ is also a parameter of the density function and needs to be estimated from average values of observables. First step is to discuss the basics of statistical hypothesis testing theory. The following can act as a starting point. Suppose $f(\mathbf{X}, \theta)d\mathbf{X}$ is the distribution of the observation $\mathbf{X}$ given the parameter θ. We want to test the hypothesis $H_0 : \theta = \theta_0$ against the hypothesi $H_1 : \theta \neq \theta_0$. To this end, we construct a statistic $q = g(\mathbf{X})$. Let $f(q, \theta)dq$ denote the distribution of q. The critical region R_c of the test is constructed. If $q \in R_c$, then we reject the hypothesis H_0, while if $q \notin R_c$, we then accept the hypothesis H_0. The construction of the test amounts to constructing both the statistic q and the critical region R_c. First, we construct q by the choice of an appropriate function g of the observations $\mathbf{X}$. Then we fix a false alarm probability α defined by the probability of rejecting H_0 when actually H_0 is true, i.e.

$$\alpha = P(q \in R_c | H_0) = \int_{R_c} f(q, \theta)dq$$

Subject to this constraint on R_c, we minimize the probability of miss i.e. of accepting H_0 when actually H_1 is true, i.e.

$$\beta(\theta_1) = P(q \in R_c | H_1) = \int_{R_c^c} f(q, \theta_1)dq$$

Here, we out hypothesis H_1 is $\theta = \theta_1$. Here, we need to specify the value of θ_1, while our objective is to test the hypothesis $H_1 : \theta \neq \theta_0$ against $H_0 : \theta = \theta_0$. To do this, we assume that $f(q, \theta)$ is concentrated on the right of $f(q, \theta_0)$ for $\theta > \theta_0$ and on the left of $f(q, \theta_0)$ for $\theta < \theta_0$. If $\theta > \theta_0$, the likely values of q are on the right of the graph of $f(q, \theta_0)$ while if $\theta < \theta_0$, the likely values of q are on its left. This motivates us to select our critical region R_c as $\{q : q < c_1 \, or \, q > c_2\}$ where c_1, c_2 are properly chosen constants. To maintain the constraint $P(q \in R_c | H_0) = \alpha$, c_1, c_2 are chosen so that

$$P(q < c_1 | H_0) = P(q > c_1 | H_0) = \alpha/2$$

If q_u is such that

$$\int_{-\infty}^{q_u} f(q, \theta)dq = u$$

then $c_1 = q_{\alpha/2}, c_2 = q_{1-\alpha/2}$ and the probability of miss is given by

$$\beta(\theta) = P(q \in R_c^c | H_1, \theta) = \int_{q_{\alpha/2}}^{q_{1-\alpha/2}} f(q, \theta)dq$$

This idea of testing hypothesis can be successfully applied to problems of statistical mechanics. Assume that the energy of the system depends upon a parameter θ and we want to test the hypothesis that $\theta = \theta_0$ against the hypothesis $\theta \neq \theta_0$. For example if our gas consists of a rotating fluid, then the parameter θ may be the angular velocity of rotation. Of if our gas consists of magnetic dipoles in a magnetic field, then the parameter θ can be the $\hat{z}$ component of the magnetic field and the strength of each molecular dipole. Or if our gas consists of particles in a harmonic potential, the parameters forming θ can be the value of the masses of the individual particles and the spring constants. If these parameters coincide with definite known values, then we can assert that the molecules of the gas belong to a certain element. In other words, the hypothesis testing problem boils down to testing whether or not the molecules of the gas are of a certain kind or not. We are not interested in what kind the molecules are when they are not of the given kind. This problems then amounts to testing the hypothesis $\theta = \theta_0$ against the hypothesis $\theta \neq \theta_0$. Using quantum mechanics, we can evaluate the energy levels $E_n(\theta)$ of the system by solving the Schrodinger equation $H(\theta)|\psi_n(\theta) >= E_n(\theta)|\psi_n(\theta) >$. The energy levels as well as the wave functions will depend upon the parameters θ. Then, we calculate the partition function corresponding to the temperature $T = 1/k\beta$:

$$Z(\beta) = \sum_{n=1}^{\infty} exp(-\beta E_n(\theta)) = Tr(exp(-\beta H))$$

Suppose that we want to test our hypothesis based on measurements of an observable X like position, momentum, angular momentum, energy. The first step is to evaluate the probability distribution of X given the parameter θ. If $|\phi_k >, k = 1, 2, ...$, denote the eigenfunctions of X and $x_k, k = 1, 2, ...$, denote the corresponding eigenvalues, i.e. $X|\psi_k >= x_k|x_k >$. Then, the probability that the X assumes the value x_k in the Gibbs state is given by

$$P(X = x_k|\theta) = \sum_{n=1}^{\infty} exp(-\beta E_n(\theta))| < \phi_k|\psi_n(\theta) > |^2$$

We can apply the hypothesis testing algorithm proposed above to this distribution function.

Study project 18

Boltzmann's equation for an electron gas in an electric field with the collision term replaced by a relaxation term is

$$\frac{\partial f}{\partial t} + \mathbf{v}.\nabla f - \frac{e}{m}\mathbf{E}.\nabla_{\mathbf{v}} f = \nu(f_0 - f)$$

where f_0 is the equilibrium distribution function. Assuming that steady state has been reached, $\frac{\partial f}{\partial t} = 0$ and f is a function of $\mathbf{r}$ and $\mathbf{v}$ only, i.e. $f = f(\mathbf{r}, \mathbf{v})$. This gives us the steady state Boltzmann equation for the electron gas:

$$\mathbf{v}.\nabla f - \frac{e}{m}\mathbf{E}.\nabla_{\mathbf{v}} f = \nu(f_0 - f)$$

Write $f = f_0 + f_1$ where f_1 is the perturbation to the particle density function caused by the weak electric field. We consider the gradients of f_1 to be of the second order of smallness. Substituting this into the steady state equation gives us the first order correction f_1 as

$$-\nu f_1 = \mathbf{v}.\nabla f_0 - \frac{e}{m}\mathbf{E}.\nabla_{\mathbf{v}} f_0$$

or

$$f_1 = \frac{e}{m\nu}\mathbf{E}.\nabla_{\mathbf{v}}f_0 - \frac{1}{\nu}\mathbf{v}.\nabla f_0$$

Assuming that the electron gas is adiabatic, we have the following equation connecting temperature with the particle number density: $n(\mathbf{r})T(\mathbf{r})^{-3/2} = constant$. Thus, for the unperturbed density, we use

$$f_0(\mathbf{r}, \mathbf{v}) = n(\mathbf{r})(\frac{m}{2\pi kT(\mathbf{r})})^{3/2}exp(-m\mathbf{v}^2/2kT(\mathbf{r})$$

Plugging this into the above equation, i.e. evaluating the spatial and velocity gradients, gives us the first order correction to the particle density function. First observe that the adiabatic equation implies that

$$f_0(\mathbf{r}, \mathbf{v}) = C.exp(-m\mathbf{v}^2/2kT)$$

Thus,

$$\nabla_{\mathbf{v}}f_0 = -\frac{m\mathbf{v}}{kT}f_0$$

$$\nabla f_0 = \frac{m\mathbf{v}^2}{2kT^2}\nabla T f_0$$

This gives

$$f_1 = \frac{e}{\nu kT}\mathbf{v}.\mathbf{E}f_0 - \frac{m\mathbf{v}^2}{2kT^2\nu}\nabla T f_0$$

Study project 19

The assumption that the velocities of particles in a gas follow a normal distribution when no potential is applied is one of the fundamental laws of statistical mechanics. It stems from the central limit theorem which qualitatively states that the sum of a very large number of independent random effects has an approximately normal distribution. The central limit theorem is one of the fundamental theorems of probability theory but it is finds equally fundamental applications to problems of statistical mechanics. Let $X_1, X_2, ..., X_n, ...$, be independent random variables with means $\mu_1, .., \mu_n, ...$, and variances $\sigma_1^2, \sigma_2^2, ..., \sigma_n^2, ...,...$ such that $\frac{\mu_1 + ... + \mu_n}{n} \to \mu$ and $\frac{\sigma_1^2 + ... + \sigma_n^2}{n} \to \sigma^2$ as $n \to \infty$. Then construct the sums $S_n = \frac{X_1 + ... + X_n}{,}n = 1, 2, ...,$. The central limit theorem states that the distribution of $(S_n - \sum_{i=1}^n \mu_i)/\sum \sigma_1^2 + ... + \sigma_n^2$ converges to the $N(0, 1)$ distribution. From this, it follows that S_n for large n is approximately distributed as $N(n\mu, n\sigma^2)$.

Study project 20

Multivariate normal distribution. This occurs in statistical mechanics in the context of the distribution of the phase coordinates of a system of coupled harmonic oscillators. Assume that masses $m_1, m_2, ..., m_N$ are connected to each other with springs such that the spring connecting mass m_i with mass m_j has the spring constant $k(i, j)$. The potential energy of the system is given by

$$U(\mathbf{r}_1, ..., \mathbf{r}_N) = \frac{1}{4} \sum_{i \neq j} k(i, j) |\mathbf{r}_i - \mathbf{r}_j|^2$$

$$= \frac{1}{4} \sum_{i \neq j} k(i, j)((x_i - x_j)^2 + (y_i - y_j)^2 + (z_i - z_j)^2)$$

The kinetic energy of the system is given by

$$T(\mathbf{v}_1, \mathbf{v}_2, ..., \mathbf{v}_N) = \frac{1}{2} \sum_i m_i |\mathbf{v}_i|^2 = \frac{1}{2} \sum_{i=1}^{N} m_i(v_{xi}^2 + v_{yi}^2 + v_{zi}^2)$$

The joint probability distribution of the phase coordinates $(\mathbf{r}_1, ..., \mathbf{r}_N, \mathbf{v}_1, ..., \mathbf{v}_N)$ is given by the Boltzmann density:

$$f(\mathbf{r}_1, ..., \mathbf{r}_N, \mathbf{v}_1, ... \mathbf{v}_N) = C.exp(-\beta(T(\mathbf{v}_1, ..., \mathbf{v}_N) + U(\mathbf{r}_1, ..., \mathbf{r}_N)))$$

This is a multivariate normal distribution in $6N$ variables. To evaluate the moments of the distribution, we need first of all evaluate the characteristic function of a multivariate distribution. Suppose $\mathbf{X} = [X_1, ..., X_n]^T$ is a Gaussian random vector. This means that if $\mathbf{c} = [c_1, ..., c_N]^T$ is any constant vector, then $\xi = \mathbf{c}^T \mathbf{X} = \sum_{i=1}^{N} c_i X_i$ has the normal distribution for a scalar variable. Its mean is $\mathbb{E}\xi = \mathbf{c}^T \mu_X$ where,

$$\mu_X = \mathbb{E}\mathbf{X} = [\mathbb{E}X_1, \mathbb{E}X_2, ..., \mathbb{E}X_n]^T$$

The variance of ξ is given by

$$V(\xi) = \mathbb{E}(\mathbf{c}^T \mathbf{X})^2 - (\mathbb{E}\mathbf{c}^T \mathbf{X})^2$$
$$= \mathbf{c}^T(\mathbb{E}(\mathbf{X}\mathbf{X}^T) - \mathbb{E}\mathbf{X}\mathbb{E}\mathbf{X}^T)\mathbf{c}$$
$$= \mathbf{c}^T \mathbf{C}_{XX} \mathbf{c}$$

where

$$\mathbf{C}_{XX} = \mathbb{E}\mathbf{X}\mathbf{X}^T - \mathbb{E}\mathbf{X}\mathbb{E}\mathbf{X}^T$$

is the covariance matrix of the random vector $\mathbf{X}$. This identity can equivalently be described using components:

$$V(\xi) = \mathbb{E}(\sum c_a X_a)^2 - (\mathbb{E}\sum c_a X_a)^2 = \sum c_a c_b \mathbb{E}(X_a X_b) - \sum c_a c_b \mathbb{E}X_a \mathbb{E}X_b$$

$$= \sum c_a c_b Cov(X_a, X_b)$$

where

$$Cov(X_a, X_b) = \mathbb{E}(X_a X_b) - \mathbb{E}X_a \mathbb{E}X_b = \mathbb{E}((X_a - \mathbb{X}_a)(X_b - \mathbb{X}_b))$$

Now, from the formula for the characteristic function of a single scalar Gaussian random variable, i.e.,

$$\mathbb{E}exp(i\xi) = exp(i\mathbb{E}\xi - \frac{1}{2}V(\xi))$$

we get the following expression for the joint characteristic function of the random vector $\mathbf{X}$:

$$\psi(\mathbf{t}) = \mathbb{E}exp(i\mathbf{t}^T\mathbf{X}) = exp(i\mathbf{t}^T\mu_x - \frac{1}{2}\mathbf{t}^T\mathbf{C}_{xx}\mathbf{t})$$

To evaluate the joint density of the random vector $\mathbf{X}$, we need the multivariate inverse Fourier transform of the characteristic function.

$$f(\mathbf{x}) = \frac{1}{(2\pi)^n} \int_{\mathbb{R}^n} \psi(\mathbf{t})exp(-i\mathbf{t}^T\mathbf{x})dt$$

$$= (2\pi)^{-n} \int_{\mathbb{R}^n} exp(-i\mathbf{t}^T(\mathbf{x}-\mu))exp(-\frac{1}{2}\mathbf{t}^T\mathbf{C}_{xx}\mathbf{t})dt$$

Now $\mathbf{C}_{xx}$ is a positive definite matrix. This can be seen via the formula

$$\mathbf{y}^T\mathbf{C}_{xx}\mathbf{y} = \mathbf{y}^T\mathbb{E}(\mathbf{X}-\mu)(\mathbf{X}-\mu)^T)\mathbf{y}$$

$$\mathbb{E}(\mathbf{X}-\mu)^T\mathbf{y})^2 \geq 0$$

Strict inequality will follow if we assume that the components of $\mathbf{X}-\mu$ are linearly independent. We now employ the spectral factorization of the matrix $\mathbf{C}_{xx}$. The spectral theorem states that every positive definite Hermitian matrix can be diagonalized with respect to an orthonormal basis, i.e. we can find vectors $\mathbf{v}_1, ..., \mathbf{v}_N$ and real numbers $d_1, ..., d_N$ such that $\mathbf{C}_{xx}\mathbf{v}_i = d_i\mathbf{v}_i, i = 1, 2, ..., N$. Setting $\mathbf{V} = [\mathbf{v}_1, \mathbf{v}_2, ..., \mathbf{v}_N]$, the orthonormality of the $\mathbf{v}_i's$ implies that $\mathbf{V}\mathbf{V}^T = \mathbf{V}^T\mathbf{V} = \mathbf{I}$. The defining eigen equations can be formulated as $\mathbf{C}_{xx} = \mathbf{V}\mathbf{D}\mathbf{V}^T$, or equivalently as $\mathbf{V}^T\mathbf{C}_{xx}\mathbf{V} = \mathbf{D}$. The integral that defines the density $f(\mathbf{x}$ can be expressed as

$$f(\mathbf{x}) = (2\pi)^{-n} \int_{\mathbb{R}^n} exp(-i\mathbf{t}^T(\mathbf{x}-\mu))exp(-\frac{1}{2}\mathbf{t}^T\mathbf{V}\mathbf{D}\mathbf{V}^T\mathbf{t})dt$$

We adopt now the following change of variables: $\mathbf{s} = \mathbf{V}^T\mathbf{t}$, or $\mathbf{t} = \mathbf{V}\mathbf{s}$. Then since $\mathbf{V}$ is an orthogonal matrix, its determinant equals ± 1. This follows by taking determinants on both sides of the equation $\mathbf{V}^T\mathbf{V} = \mathbf{I}$ to get $det\mathbf{V}^T.det\mathbf{V} = 1$ and since the determinant of the transpose of a matrix equals the determinant of the matrix, this gives $(det\mathbf{V})^2 = 1$, so that $det\mathbf{V} = \pm 1$. Thus the Jacobian of the transformation is unity and the integral becomes

$$f(\mathbf{x}) = (2\pi)^{-n} \int_{\mathbb{R}^n} exp(-i\mathbf{s}^T\mathbf{V}^T(\mathbf{x}-\mu))exp(-\frac{1}{2}\mathbf{s}^T\mathbf{D}\mathbf{s})ds$$

Define $\mathbf{V}^T(\mathbf{x}-\mu) = \mathbf{y}$. Then,

$$f(\mathbf{x}) = (2\pi)^{-n} \int_{\mathbb{R}^n} exp(-i\mathbf{s}^T\mathbf{y})exp(-\frac{1}{2}\mathbf{s}^T\mathbf{D}\mathbf{s})ds$$

$$= (2\pi)^{-n} \int \Pi_{a=1}^n exp(-is_a y_a) exp(-\frac{1}{2} d_a s_a^2) ds_a$$

$$= (2\pi)^{-n} \Pi_{a=1}^n \int_{-\infty}^{\infty} exp(-is_a y_a) exp(-\frac{1}{2} d_a s_a^2) ds_a$$

$$= (-2\pi)^{-n} (\Pi_{a=1}^n (2\pi/d_a)^{1/2}) exp(-y_a^2/2d_a)$$

$$= (2\pi)^{-n/2} (\Pi_{a=1}^n d_a)^{-1/2} exp(-\frac{1}{2} \sum_{a=1}^n y_a^2/d_a)$$

Now, $\Pi_{a=1}^n d_a = det\mathbf{C}_{xx}$ as follows by taking determinants on both sides of the equation $\mathbf{C}_{xx} = \mathbf{VDV}^T$:

$$det\mathbf{C}_{xx} = det\mathbf{V} det\mathbf{D} det\mathbf{V}^T = det\mathbf{D} = \Pi_{a=1}^n d_a$$

Also

$$\sum_{a=1}^n y_a^2/d_a = \mathbf{y}^T \mathbf{D}^{-1} \mathbf{y} = (\mathbf{x} - \mu)^T \mathbf{V} \mathbf{D}^{-1} \mathbf{V}^T (\mathbf{x} - \mu)$$

$$= (\mathbf{x} - \mu) \mathbf{C}_{xx}^{-1} (\mathbf{x}_\mu)$$

We finally arrive at the expression for the multivariate normal density:

$$f(\mathbf{x}) = (2\pi)^{-n/2} (det\mathbf{C}_{xx})^{-1/2} exp(-\frac{1}{2}(\mathbf{x} - \mu)^T \mathbf{C}_{xx}^{-1} (\mathbf{x} - \mu))$$

Study project 21

Collisionless Boltzmann equation

$$\frac{\partial f(\mathbf{r}, \mathbf{v}, t)}{\partial t} + \mathbf{v}.\nabla f(\mathbf{r}, \mathbf{v}, t) + \frac{\mathbf{F}}{m}.\nabla_\mathbf{v} f(\mathbf{r}, \mathbf{v}, t) = 0$$

Let $\chi(\mathbf{r}, \mathbf{v})$ be any physical quantity. We wish to obtain a differential equation dictating the evolution of its average value at each point $\mathbf{r}$.

$$n(\mathbf{r}, t) < \chi > (\mathbf{r}, t) = \int \chi(\mathbf{r}, \mathbf{v}) f(\mathbf{r}, \mathbf{v}, t) d\mathbf{v}$$

where

$$n(\mathbf{r}, t) = \int f(\mathbf{r}, \mathbf{v}, t) d\mathbf{v}$$

is the number density of particles at the point $\mathbf{r}$, at time t. Multiplying the Boltzmann equation by $\chi(\mathbf{r}, \mathbf{v})$ and integrating with respect to $\mathbf{v}$ gives us

$$\int \chi \frac{\partial f}{\partial t} d\mathbf{v} + \int \chi (\mathbf{v}.\nabla f) d\mathbf{v} + \int \chi \frac{\mathbf{F}}{m}.\nabla_\mathbf{v} f d\mathbf{v} = 0$$

Now,

$$\int \chi \frac{\partial f}{\partial t} d\mathbf{v} = \frac{\partial}{\partial t} \int \chi f d\mathbf{v} = \frac{\partial}{\partial t}(n < \chi >)$$

$$\int \chi \mathbf{v}.\nabla f d\mathbf{v} = \nabla. \int \chi \mathbf{v} f d\mathbf{v} - \int \mathbf{v}.\nabla \chi f d\mathbf{v}$$

$$= \nabla.(n < \mathbf{v}\chi >) - n < \mathbf{v}.\nabla \chi >$$

$$\int \chi \mathbf{F}/m.\nabla_{\mathbf{v}} f d\mathbf{v} = -\frac{1}{m} \int \nabla_{\mathbf{v}}.(\chi \mathbf{F}) f d\mathbf{v}$$

$$= -\frac{n}{m} < \nabla_{\mathbf{v}}.(\chi \mathbf{F})$$

We thus obtain the following equation for the average dynamics of χ:

$$\frac{\partial}{\partial t}(n < \chi >) + \nabla.(n < \mathbf{v}\chi >) - n < \mathbf{v}.\nabla \chi >= \frac{n}{m} < \nabla_{\mathbf{v}}.(\chi \mathbf{F}) >$$

Take $\chi = m$, a constant equal to the mass of the particle. Then, $n < \chi >= nm = \rho$, the particle mass density. $< \mathbf{v}\chi >=< m\mathbf{v} >= m\mathbf{u}$, where $\mathbf{u}$ is the average velocity field. This is the same as $n < \mathbf{v}\chi >= \rho\mathbf{u}$. $\nabla \chi = 0$ implies $< \mathbf{v}.\nabla \chi >= 0$. Finally, $< \nabla_{.\mathbf{v}}.(\chi \mathbf{F}) >= m < \nabla_{\mathbf{v}}.\mathbf{F} >= 0$ for electromagnetic forces in which the force on a given charged particle is given by $\mathbf{F} = e(\mathbf{E}+\mathbf{v} \times \mathbf{B})$. The averaged dynamics of χ thus gives

$$\frac{\partial \rho}{\partial t} + \nabla.(\rho\mathbf{u}) = 0$$

which is the matter conservation equation. Next take $\chi = mv_i$ the ith component of the momentum of each particle. Then, $n < \chi >= \rho u_i$, the averaged momentum per unit volume. $n < \mathbf{v}\chi >= nm < v_i\mathbf{v} >= \rho < v_i\mathbf{v} >$. Its jth component equals $\rho < v_i v_j >$ which is the averaged value of the momentum flux tensor. This can be expressed as

$$\rho < v_i v_j >= \rho < (v_i - u_i + u_i)(v_j - u_j + u_j) >$$

$$= \rho < (v_i - u_i)(v_j - u_j) > +\rho u_i u_j = P_{ij} + \rho u_i u_j$$

where $P_{ij} = \rho < (v_i - u_i)(v_j - u_j) >$ is the pressure tensor. when the velocities are uncorrelated, we have $P_{ij} = 0$ when $i \neq j$ and $P_{ii} = \rho < (v_i - u_i)^2 >$. If Maxwellian distribution is assumed about the mean value, then $\frac{m}{2} < (v_i - u_i)^2 >= kT/2$ and $P_{ii} = nkT/m$ which is the equation of state for an ideal gas if P_{11}, P_{22} and P_{33} are interpreted as the pressure of the plasma regarded as an ideal gas. Finally,

$$< \nabla_{\mathbf{v}}.(\chi \mathbf{F}) >= m < \nabla_{\mathbf{v}}.(v_i \mathbf{F}) >= m < F_i > +m < v_i \nabla_{.\mathbf{v}}.\mathbf{F} >= m < F_i >= mf_i$$

where f_i is the average external force on a particle. $\frac{n}{m}m_i = nf_i$ gives the averaged force per unit volume. Denote this by f_i'. Then, the averaged dynamics of χ is gives the equation

$$\frac{\partial}{\partial t}(\rho u_i) + \sum_{j=1}^{3} \frac{\partial P_{ij}}{\partial x_j} + \sum_{j=1}^{3} \frac{\partial}{\partial x_j}(\rho u_i u_j) = f_i'$$

Combining this with the mass conservation equation, this results in

$$\rho\left(\frac{\partial u_i}{\partial t} + \sum_{j=1}^{3} u_j \frac{\partial u_i}{\partial x_j} + \sum_{j=1}^{3} \frac{\partial P_{ij}}{\partial x_j}\right) = f_i'$$

which is the Navier-Stokes equation with the pressure replaced by the more general pressure tensor. The viscosity term has not appeared here since we have neglected the collision term in the Boltzmann equation. Taking into account collisions leads to a viscosity term in the Navier-Stokes equation. Finally, take $\chi = \frac{m}{2}|\mathbf{v} - \mathbf{u}(\mathbf{r}, t)|^2$. In this case, χ depends explicitly on time. So a small modification needs to be made in the analysis of the dynamics of averaged χ. The first term for such an observable reads

$$\int \chi \frac{\partial f}{\partial t} d^3\mathbf{v} = \frac{\partial}{\partial t}\int \chi f d^3\mathbf{v} - \int \frac{\partial \chi}{\partial t} f d^3\mathbf{v}$$

$$= \frac{\partial}{\partial t}(n<\chi>) - n<\frac{\partial \chi}{\partial t}>$$

Thus the equation for the averaged dynamics for a time varying observable gets modified to

$$\frac{\partial}{\partial t}(n<\chi>) - n<\frac{\partial \chi}{\partial t}> + \nabla.(n<\mathbf{v}\chi>) - n<\mathbf{v}.\nabla\chi> - \frac{n}{m}<\nabla_\mathbf{v}.(\chi\mathbf{F})> = 0$$

For the above chosen observable,

$$\frac{\partial \chi}{\partial t} = -\left(\frac{\partial \mathbf{u}}{\partial t}, m(\mathbf{v} - \mathbf{u})\right)$$

The average of this quantity is clearly zero since $<\mathbf{v}> = \mathbf{u}$. Also,

$$n<\chi> = \frac{nm}{2}<|\mathbf{v} - \mathbf{u}|^2> = 3nkT/2$$

where T is the temperature of the fluid defined by the kinetic relation in accordance with Maxwell's equipartition theorem:

$$\frac{3kT}{2} = \frac{m}{2}<|\mathbf{v} - \mathbf{u}|^2>$$

$$n<\mathbf{v}\chi> = \frac{nm}{2}<\mathbf{v}|\mathbf{v} - \mathbf{u}|^2> = \frac{\rho}{2}<(\mathbf{v} - \mathbf{u})|\mathbf{v} - \mathbf{u}|^2>$$

$$+\frac{\rho}{2}\mathbf{u}<|\mathbf{v} - \mathbf{u}|^2>$$

We define the heat flow vector

$$\mathbf{q} = \frac{m\rho}{2}<(\mathbf{v} - \mathbf{u})|\mathbf{v} - \mathbf{u}|^2>$$

Then,

$$n < \mathbf{v}\chi > = \mathbf{q}/m + \frac{3}{2}nkT\mathbf{u}$$

where the temperature T is defined in accordance with Maxwell's equipartition theorem:

$$\frac{3}{2}kT = \frac{m}{2} < |\mathbf{v} - \mathbf{u}|^2 >$$

$$< \mathbf{v}.\nabla\chi > = \sum_i < v_i \frac{\partial \chi}{\partial x_i} > = -\sum_i < v_i(\frac{\partial \mathbf{u}}{\partial x_i}, m(\mathbf{v} - \mathbf{u})) >$$

$$= -\sum_{i,j} < v_i \frac{\partial u_j}{\partial x_i} m(v_j - u_j) > = -m \sum_{i,j} \frac{\partial u_j}{\partial x_i} < v_i(v_j - u_j) >$$

$$= -m \sum_{i,j} \frac{\partial u_j}{\partial x_i} < (v_i - u_i)(v_j - u_j) >$$

$$= -\frac{m}{2} \sum_{i,j} (\frac{\partial u_j}{\partial x_i} + \frac{\partial u_i}{\partial x_j}) < (v_i - u_i)(v_j - u_j) >$$

Or

$$n < \mathbf{v}.\nabla\chi > = -\frac{1}{m} \sum_{i,j} \Lambda_{ij} P_{ij}$$

where

$$\Lambda_{ij} = \frac{m}{2}(\frac{\partial u_j}{\partial x_i} + \frac{\partial u_i}{\partial x_j})$$

is the symmetric stress tensor of the gas. The pressure tensor $P_{ij} = \rho < (v_i - u_i)(v_j - u_j) >$ has been introduced earlier. Finally,

$$< \nabla_{\mathbf{v}}.(\chi\mathbf{F}) > = < \mathbf{F}.\nabla_{\mathbf{v}}\chi > = < \mathbf{F}.m(\mathbf{v} - \mathbf{u}) >$$

$$= m < \mathbf{F}.(\mathbf{v} - \mathbf{u}) >$$

For the forces of electromagnetic origin, we can easily evaluate the above average.

$$\mathbf{F}.(\mathbf{v} - \mathbf{u}) = e(\mathbf{E}.(\mathbf{v} - \mathbf{u}) + \mathbf{v} \times \mathbf{B}.(\mathbf{v} - \mathbf{u}))$$

The average of the first term here is zero, i.e.

$$< \mathbf{E}.(\mathbf{v} - \mathbf{u}) > = \mathbf{E}. < \mathbf{v} - \mathbf{u} > = 0$$

The average of the second term can be evaluated:

$$< \mathbf{v} \times \mathbf{B}.(\mathbf{v} - \mathbf{u}) > = < (bfv - \mathbf{u}) \times \mathbf{B}.(\mathbf{v} - \mathbf{u}) > = 0$$

We thus arrive at the heat transfer equation

$$\frac{\partial}{\partial t}(3nkT/2) + \nabla.(\mathbf{q}/m + \frac{3}{2}nkT\mathbf{u}) + \frac{1}{m}\sum_{i,j=1}^{3}\Lambda_{ij}P_{ij} = 0$$

Multiplying both sides by $2m/3$, and using $\theta = kT$ gives us finally the heat transfer equation in the following form:

$$\frac{\partial}{\partial t}(\rho\theta) + \frac{2}{3}\nabla.\mathbf{q} + \nabla.(\rho\theta\mathbf{u}) + \frac{2}{3}\mathbf{\Lambda}.\mathbf{P} = 0$$

Making use of the continuity equation, this reduces to

$$\rho(\frac{\partial\theta}{\partial t} + \mathbf{u}.\nabla\theta) + \frac{2}{3}\nabla.\mathbf{q} + \frac{2}{3}\mathbf{\Lambda}.\mathbf{P} = 0$$

Study project 22

In statistical mechanics, the equilibrium distribution of a system is computed as the Gibbs distribution. If there is just one particle and its energy as a function of its coordinate and velocity is given by $H(x,v)$, then its partition function is given by $Z(\beta) = \int exp(-\beta H(x,v))dxdv$ assuming that the particle is moving in one dimension. The probability of finding the particle in the region $[x, x+dx]$ and with a velocity in the range $[v, v+dv]$ is given by $\frac{exp(-\beta H(x,v))}{Z(\beta)}$ where $\beta = 1/kT$ with T as the temperature. This distribution is used for computing statistical aggregates like energy, entropy, enthalpy at thermodynamic equilibrium. The average energy of the system is for example given by the formula

$$U = \int H(x,v)exp(-\beta H(x,v))dxdv/Z(\beta) = -\frac{1}{Z(\beta)}\frac{dZ(\beta)}{d\beta} = -\frac{d}{d\beta}log(Z(\beta))$$

The evolution of the density of the system when the equilibrium is perturbed is given by the Boltzmann equation which is the fundamental equation of the kinetic theory of gases. This equation can for example be used to describe the dynamics of a plasma which is a gas of ionized atoms. If $f(t, \mathbf{r}, \mathbf{v})$ denotes the density of particles in phase space at time t, and $\mathbf{F}(\mathbf{r}, \mathbf{v})$ is the force on a particle which has mass m as a function of its position and velocity, then the Boltzmann equation is given by

$$\frac{\partial f}{\partial t} + (\mathbf{v}, \nabla_{\mathbf{r}}f) + \frac{1}{m}(\mathbf{F}, \nabla_{\mathbf{v}}f) = 0$$

assuming no collisions occur. This is also called the collisionless Boltzmann or the Vlasov equation. When collisions occur, the right side is replaced by a collision term which depends on the scattering cross section for a collision. Suppose a binary collision occurs between a particle traveling with a velocity $\mathbf{u}_1$ and another particle traveling with a velocity $\mathbf{u}$, causing the original particle to travel with a velocity $\mathbf{u}_1'$ the second to travel with a velocity $\mathbf{u}'$ after the collision, then we choose our reference frame as that in which the second particle traveling with the velocity $\mathbf{u}$ before the collision is originally at rest. The first particle travels with a velocity $\mathbf{u}_1 - \mathbf{u}$ relative to the second prior to the collision. The first particle is scattered into the solid angle $d\Omega$ relative to its original direction of motion relative to the second particle. The incident flux is $|\mathbf{u}_1 - \mathbf{u}|$ and hence if $\sigma(\mathbf{u}_1, \mathbf{u}|\mathbf{u}_1', \mathbf{u}')$ is the scattering cross section computed using the interaction potential, the number of particles scattered into the solid angle $d\Omega$ is given by $|\mathbf{u}_1 - \mathbf{u}|\sigma(\mathbf{u}_1, \mathbf{u}|\mathbf{u}_1', \mathbf{u}')d\Omega$. This expression is used to derive the expression for the rate of increase in the particle number density due to binary collisions. As a result of this collision, the particle moving with velocity $\mathbf{u}$ moves out of its velocity volume element. The rate at which the number density at $(\mathbf{r}, \mathbf{u})$ decreases due to such collisions is given by

$$\int |\mathbf{u}_1 - \mathbf{u}|\sigma(\mathbf{u}_1, \mathbf{u}|\mathbf{u}_1', \mathbf{u}')f(t, \mathbf{r}, \mathbf{u}_1)f(t, \mathbf{r}, \mathbf{u})d^3\mathbf{u}_1 d\Omega$$

Similarly, we can calculate the rate of increase of particles due to collisions. The rate at which the number of particles at the phase point $(\mathbf{r}, \mathbf{u})$ per unit phase volume increase due to collisions is given by

$$\int |\mathbf{u}_1' - \mathbf{u}'|\sigma(\mathbf{u}_1', \mathbf{u}'|\mathbf{u}_1, \mathbf{u})f(t, \mathbf{r}, \mathbf{u}_1')f(t, \mathbf{r}, \mathbf{u}')d^3\mathbf{u}_1 d\Omega$$

Momentum conservation implies $|\mathbf{u}_1' - \mathbf{u}'| = \mathbf{u}_1 - \mathbf{u}|$ and the symmetry of the scattering process implies $\sigma(\mathbf{u}_1', \mathbf{u}'|\mathbf{u}_1, \mathbf{u}) = \sigma(\mathbf{u}_1, \mathbf{u}|\mathbf{u}_1', \mathbf{u}')$. Thus, the total net rate of increase of particles per unit phase volume is obtained as the difference of the above two expressions:

$$(\frac{\partial f(t, \mathbf{r}, \mathbf{u})}{\partial t})_{coll} = \int |\mathbf{u}_1 - \mathbf{u}|\sigma(\mathbf{u}_1, \mathbf{u}|\mathbf{u}_1', \mathbf{u})(f(t, \mathbf{r}, \mathbf{u}')f(t, \mathbf{r}, \mathbf{u}_1'))$$

$$-f(t, \mathbf{r}, \mathbf{u})f(t, \mathbf{r}, \mathbf{u}_1))d\Omega d^3\mathbf{u}_1$$

The final result of these computations is the Boltzmann equation

$$\frac{\partial f}{\partial t} + (\mathbf{v}, \nabla_\mathbf{r} f) + \frac{1}{m}(\mathbf{F}, \nabla_\mathbf{v} f)$$

$$= (\frac{\partial f(t, \mathbf{r}, \mathbf{u})}{\partial t})_{coll}$$

Study project 23

Obtain the general tensor of conductivity for a plasma in a uniform magnetic field. The conductivity tensor is defined by the relation $J_\alpha = \sum_{\beta=1}^{3} \sigma_{\alpha\beta} E_\beta$.

Ans: The equation of motion of the plasma particle is given by

$$m\frac{d\mathbf{v}}{dt} = e(\mathbf{E} + \mathbf{v} \times \mathbf{B}) - \gamma\mathbf{v}$$

where γ is the frictional coefficient arising due to collisions of the plasma particle with the surrounding ions. At equilibrium, we have $\frac{d\mathbf{v}}{dt} = 0$. We are assuming that $\mathbf{B}$ is a constant vector. The condition for equilibrium gives

$$e(\mathbf{E} + \mathbf{v} \times \mathbf{B}) = \gamma \mathbf{v}$$

This can be regarded as a linear equation for $\mathbf{v}$ with $\mathbf{B}$ as a parameter. The current density is given by $\mathbf{J} = ne\mathbf{v}$, where n is the number density of the charged particles. Write $\mathbf{v} = \mathbf{v}_{\parallel} + \mathbf{v}_{\perp}$, where $\mathbf{v}_{\parallel}$ denotes the component of the velocity parallel to $\mathbf{B}$ and $\mathbf{v}_{\perp}$ is the component of the velocity perpendicular to $\mathbf{B}$. Then the above equation assumes the form

$$e(\mathbf{E} + \mathbf{v}_{\perp} \times \mathbf{B}) = \gamma(\mathbf{v}_{\parallel} + \mathbf{v}_{\perp})$$

Taking the cross product on both sides with $\mathbf{B}$ gives

$$e(\mathbf{E} \times \mathbf{B} - B^2 \mathbf{v}_{\perp}) = \gamma \mathbf{v}_{\perp} \times \mathbf{B}$$

From the original equation, we can also deduce that

$$e\mathbf{E}_{\parallel} = \gamma \mathbf{v}_{\parallel}$$

and

$$e(\mathbf{E}_{\perp} + \mathbf{v}_{\perp} \times \mathbf{B}) = \gamma \mathbf{v}_{\perp}$$

Substituting for $\mathbf{v}_{\perp} \times \mathbf{B}$ from the second equation, i.e. the equation obtained after forming the cross product with $\mathbf{B}$ gives

$$e\mathbf{E}_{\perp} + (e^2/\gamma)\mathbf{E} \times \mathbf{B} - (e^2 B^2/\gamma)\mathbf{v}_{\perp} = \gamma \mathbf{v}_{\perp}$$

which can be rearranged as

$$(\gamma + (e^2 B^2/\gamma))\mathbf{v}_{\perp} = e\mathbf{E}_{\perp} + (e^2/\gamma)\mathbf{E} \times \mathbf{B}$$

so that

$$\mathbf{v}_{\perp} = \frac{1}{\gamma + (e^2 B^2/\gamma)}(e\mathbf{E}_{\perp} + (e^2/\gamma)\mathbf{E} \times \mathbf{B})$$

Assume that $\mathbf{B}$ is directed along the z direction. Then,

$$v_z = \frac{e}{\gamma}E_z$$

$$v_x = \frac{1}{\gamma + (e^2 B^2/\gamma)}(eE_x + (e^2/\gamma)E_y B)$$

$$v_y = \frac{1}{\gamma + (e^2 B^2/\gamma)}(eE_y - (e^2/\gamma)E_x B)$$

These equations can be arranged in the following matrix form:

$$\begin{pmatrix} v_x \\ v_y \\ v_z \end{pmatrix}$$

$$= \begin{pmatrix} \dfrac{e}{\gamma+e^2B^2/\gamma} & \dfrac{e^2B}{\gamma^2+e^2B^2} & 0 \\ -\dfrac{e^2B}{\gamma^2+e^2B^2} & \dfrac{e}{\gamma+e^2B^2/\gamma} & 0 \\ 0 & 0 & \dfrac{e}{\gamma} \end{pmatrix} \begin{pmatrix} E_x \\ E_y \\ E_z \end{pmatrix}$$

This equation is expressed as $\mathbf{v} = \mathbf{ME}$, with $\mathbf{M}$ as the matrix occurring in the above equation. $\mathbf{M}$ can be termed as the mobility tensor. The conductivity tensor is given by $\Sigma = ne\mathbf{M}$.

Study project 24

If X, Y are independent random variables uniformly distributed over the interval $[0, a]$, determine the distribution and density of the random variable $Z = |X - Y|$. The random variable Z clearly assumes values in the interval $[0, a]$. For $z \in [0, a]$, we have

$$P(Z \leq z) = P(|X - Y| \leq z) = P(Y - z \leq X \leq Y + z)$$

Since X assumes values in the interval $[0, a]$ and since X and Y are independent, it follows that

$$P(Y - z \leq X \leq Y + z | Y = y) = P(max(y - z, 0) \leq X \leq min(y + z, a))$$

$$= \frac{1}{a}(min(y + z, a) - max(y - z, 0))$$

assuming that $max(y - z, 0) \leq min(y + z, a)$. This latter condition is always satisfied for $z \in [0, a]$ and $y \in [0, a]$. Thus,

$$P(Z \leq z) = \int_{-\infty}^{\infty} P(Y - z \leq X \leq Y + z | Y = y) f_Y(y) dy$$

$$= \frac{1}{a^2} \int_0^a (min(y + z, a) - max(y - z, 0)) dy$$

The evaluation of the two integrals would proceed along the following lines:

$$\int_0^a min(y + z, a) dy = \int_0^{a-y} (y + z) dy + \int_{a-y}^a a\, dy$$

$$\int_0^a max(y - z, 0) dy = \int_0^z 0.dy + \int_z^a (y - z) dy$$

Application to statistical mechanics: Suppose nothing is known about the energy of a system except that its minimum value is zero and maximum value is E_1. Then, one may assume that the energy X of this system is uniformly distributed over the interval $[0, E_1]$. Similarly another system has energy Y uniformly distributed over the interval $[0, E_2]$. The magnitude of the difference $|X - Y|$ between the two energies measures how different the two systems are. Given an $\epsilon > 0$, one can compute the probability that $P(|X - Y| > \epsilon)$. This will give us the probability that the two systems differ from each other by an amount greater than a certain threshold.

Bibliography

[1] Schaum outline series, theory and problems of electromagnetics, Tata McGraw Hill, Second edition, 2003.

[2] David J.Griffiths, Introduction to electrodynamics, Prentice Hall of India, 1984.

[3] A.Papoulis, Probability, random variables and stochastic processes, McGraw Hill, Second Edition, 1984, Reprint, 1989.

[4] John.S.Townsend, A modern approach to quantum mechanics, McGraw Hill Inc., 1992.

[5] Gaonkar, Microprocessor, Architecture, Programming and applications with 8085/8080 A, Wiley Eastern, 1987.

[6] Erwin Kreyszig, Introductory functional analysis with applications, John Wiley and Sons, 1978.

[7] Erwin Kreyszig, Advanced Engineering Mathematics, John Wiley and Sons, Eighth edition, 2004.

[8] Simon Haykin and Barry Van Veen, Signals and Systems, John Wiley and Sons, 2003.

[9] Patrick Billingsley, Probability and Measure, John Wiley and Sons, 1991.

[10] J.Honerkamp, Statistical Physics, An Advanced approach with applications, Second Edition, Springer, 2002.

[11] J.A.Bittencourt, Fundamentals of Plasma Physics, Pergamon press, 1986.

[12] H.C.Corben and Philip Stehle, Classical Mechanics, Second Edition, John Wiley and Sons, 1960.

[13] V.S.Varadarajan, Lie groups, Lie algebras and their representations, Prentice Hall, Inc., 1974.

[14] D.F.Lawden, An introduction to tensor calculus, relativity and cosmology, John Wiley and Sons, 1982.

[15] Ajoy Ghatak, Optics, Tata McGraw Hill, 2005.

INDEX